## Lecture Notes in Earth Sciences 133



For further volumes:
http://www.springer.com/series/772

Stephen G. Evans · Reginald L. Hermanns ·
Alexander Strom · Gabriele Scarascia-Mugnozza
Editors

# Natural and Artificial Rockslide Dams

Springer

*Editors*
Stephen G. Evans
University of Waterloo
Department of Earth and
Environmental Sciences
Landslide Research Programme
University Avenue W. 200
N2L 3G1 Waterloo Ontario
Canada
sgevans@sciborg.uwaterloo.ca

Dr. Reginald L. Hermanns
International Centre for Geohazards
Geological Survey of Norway
Landslide Department
Trondheim
Norway
reginald.hermanns@ngu.no

Alexander Strom
Russian Academy of Sciences
Institute of the Geospheres Dynamics
Leninskiy Avenue 38
119334 Moscow
Bldg. 1
Russia
a.strom@g23.relcom.ru

Dr. Gabriele Scarascia-Mugnozza
University of Rome "La Sapienza"
Department of Earth Sciences
Piazzale Aldo Moro 5
00185 Rome
Italy
gabriele.scarasciamugnozza@uniroma1.it

ISSN 0930-0317
ISBN 978-3-642-04763-3 e-ISBN 978-3-642-04764-0
DOI 10.1007/978-3-642-04764-0
Springer Heidelberg Dordrecht London New York

Library of Congress Control Number: 2011930516



*Cover design*: Integra Software Services Pvt. Ltd., Pondicherry

Printed on acid-free paper

Springer is part of Springer Science+Business Media (www.springer.com)

# Dedication

We wish to dedicate this volume to Dr. John Neville Hutchinson, Professor Emeritus at the Department of Civil and Environmental Engineering, Imperial College, London. He is a leading scientist and a master practitioner in the field of engineering geomorphology and is internationally recognised as one of the foremost landslide researchers of the last five decades.

After graduating in Civil Engineering in 1947, he began his professional career working in construction and structural design. In 1957 he joined the Swedish Geotechnical Institute where he mainly worked on road and airport foundations but also investigated the Gota quick clay landslide with Sven Odenstad. In 1958 he moved to the Norwegian Geotechnical Institute where he initially worked on settlements and friction piles in soft clays as a research engineer under the guidance of L. Bjerrum and O. Eide. In 1959 he led the investigation of two large-scale quick clay landslides in Central Norway at Vibstad and Furre [2]. He published the results of his work on the Furre landslide in *Geotechnique* in 1961 [5]. In doing so he began a 50 year research career in engineering geomorphology with a specific focus on the mechanisms and behaviour of landslides.

In 1961, John returned to England joining the Building Research Station. Concurrently, he studied for a PhD at Cambridge University completing his dissertation on the stability of coastal cliffs in south–east England under the supervision of the noted coastal geomorphologist, Professor J.A. Steers. In 1965 he joined the Soil Mechanics Section at the Department of Civil Engineering, Imperial College London, becoming Professor of Engineering Geomorphology in 1977. At Imperial College he was both a gifted teacher and successful research mentor and he became

Professor Emeritus in 1992. In his early years at Imperial College he made important contributions to the understanding of landslide mechanisms [6] and to the subject of landslide classification culminating in the State-of-the-Art report at the 1969 ISSMFE Conference co-authored with Professor A.W. Skempton [19]. The themes of landslide mechanisms, the role of geology in instability and aspects of landslide classification were further developed in later State-of-The-Art lectures at International Symposia on Landslides in 1988 [10] and 1992 [12], and in a multi-authored paper on landslides of the flow-type published in 2001 [4].

Amongst other achievements in landslide research, we recall six major contributions. His research on undrained loading together with R. K. Bhandari published in 1971 [16] represents a milestone in the investigation of mudslide processes. The undrained loading mechanism, discovered by Hutchinson and Bhandari through detailed field measurements of pore pressure, is not only fundamental in mudslide behavior but is now recognised as an important mechanism in some long run-out landslides. His work on the re-examination of the Folkestone Warren landslides produced a benchmark paper published in 1969 [6] as did his work on the geomorphological evolution of London Clay cliffs published in 1973 [8]. Fourthly, John Hutchinson's work on the stability of Chalk cliffs led to a greater understanding of landslide mechanisms in actively-eroding steep rock slopes [14]; his study of the Joss Bay Chalk fall [7] remains a classic in rock slope failure analysis. In addition, work on the Senise Landslide, Italy, with M. Del Prete, published in 1988 [3] documented an important mechanism of brittleness in landslide initiation. Lastly, in the early 1990s, his studies of a suite of coastal landslides in the Isle of Wight Undercliff showed the importance of careful geomorphological survey and geological interpretation in understanding the mechanisms and development of deep compound retrogressive landslides [11].

John Hutchinson took part in the investigation of the 1966 Aberfan disaster [1], one of the most significant geotechnical events of the late twentieth century, as part of a team led by Professor A.W. Bishop. Partly based on this work, his highly-cited sliding-consolidation model [9] has provided insight into catastrophic flowslides and other flow-type slides both in the subaerial and submarine environment. His work (with E. Kojan) on the 1974 Mayunmarca rock avalanche [13, 17, 18] in the Peruvian Andes provided key data on a giant catastrophic landslide and the behavior of one of the major rockslide dams that formed and failed in the twentieth century. He was also involved in field and laboratory research on other catastrophic rock slope failures, such as the Vaiont rockslide [20].

In 2001 he gave the 4th Glossop Lecture entitled "*Reading the ground: morphology and geology in site appraisal*". His lecture [13] painted a vast canvas and included a strong message on the importance of integrating Quaternary geology and geomorphology into engineering geology to enhance site assessment for engineering works. John was awarded the Varnes Medal of the International Consortium of Landslides in 2004.

In 2002 we invited Professor Hutchinson to present a State-of-the-Art review of landslides from massive rock slope failures at the NATO Advanced Research

Workshop convened in Celano, Italy. There, he identified some of the critical research issues in the understanding of catastrophic rock slope failures and identified new directions for future research activities in this field [15]. On that occasion, as always, he participated with his usual enthusiastic curiosity, thirst for knowledge, and a strong direct desire to discuss new data and fresh views related to catastrophic landslides. Once again he showed the breadth of his research mind, his open and supportive personality, and his modest character.

These qualities make him a true "maestro" to all of us and it is a pleasure to dedicate this volume to Professor J.N. Hutchinson.

Waterloo, ON, Canada — Stephen G. Evans
Trondheim, Norway — Reginald L. Hermanns
Moscow, Russia — Alexander Strom
Rome, Italy — Gabriele Scarascia-Mugnozza

## References

1. Bishop, A.W., Hutchinson, J.N., Penman, A.D.M. and Evans, H.E. (1969) Geotechnical investigation into the causes and circumstances of the disaster of 21st October, 1966: A selection of Technical Reports submitted to the Aberfan Tribunal, London, Welsh Office.
2. Bjerrum, L. and Hutchinson, J.N. (1962): Skredet ved Furre i Namdalen. Norwegian Geotechnical Institute. Publication No. 49, 11 pp.
3. Del Prete, M. and Hutchinson, J.N. (1988) La frana di Senise del 26-7-1986 nel quadro morfologico del versante meridionale della collina Timpone, *Rivista Italiana di Geotecnica* **22**, 7–33.
4. Hungr, O., Evans, S.G., Bovis, M.J. and Hutchinson, J.N. (2001) A review of the classification of landslides of the flow type, *Environmental and Engineering Geoscience* **7**, 221–238.
5. Hutchinson, J.N. (1961) A landslide on a thin layer of quick clay at Furre, Central Norway, *Geotechnique* **11**, 69–94.
6. Hutchinson, J.N. (1969) A reconsideration of the coastal landslides at Folkestone Warren, Kent, *Geotechnique* **19**, 6–38.
7. Hutchinson, J.N. (1972) Field and laboratory studies of a fall in Upper Chalk cliffs at Joss Bay, Isle of Thanet, Roscoe Memorial Symp., G.T. Foulis & Co. Ltd., Cambridge, pp. 692–706.
8. Hutchinson, J.N. (1973) The response of London Clay cliffs to differing rates of toe erosion, *Geologia Applicata e Idrogeologia* **8**, 221–239.
9. Hutchinson, J.N. (1986) A sliding consolidation model for flow slides, *Canadian Geotechnical Journal* **23**, 115–126.
10. Hutchinson, J.N. (1988) General Report: Morphological and geotechnical parameters of landslides in relation to geology and hydrogeology. *Proc. 5th Int. Symp. Landslides*, Lausanne, Vol. 1, pp. 3–35
11. Hutchinson, J.N. (1991) The landslides forming the South Wight Undercliff, *Proc. Int. Conf. Slope Stability Engineering – Developments and Applications*, Thomas Telford, London, pp. 157–168.
12. Hutchinson, J.N. (1995) Keynote paper: Landslide hazard assessment, *Proc. 6th Int. Symp. Landslides*, Christchurch, Vol. 3, pp. 1805–1841
13. Hutchinson, J.N. (2001) Reading the ground: morphology and geology in site appraisal (The Fourth Glossop Lecture), *Quarterly Journal of Engineering Geology and Hydrogeology* **34**, 7–50.

14. Hutchinson, J.N. (2002) Chalk flows from the coastal cliffs of North West Europe, in S.G. Evans and J.V. DeGraff (eds.) *Catastrophic Landslides: Effects, Occurrence, And Mechanisms*, Reviews in Engineering Geology, Geological Society of America, Vol. **XV**, pp. 257–302.
15. Hutchinson, J.N. (2006) Massive rock slope failure: perspectives and retrospectives on State-of-the-Art, in S.G. Evans, G. Scarascia-Mugnozza, A. Strom and R. Hermanns (eds.), *Landslides from Massive Rock Slope Failure*. NATO Science Series IV, Earth and Environmental Sciences Vol. 49, Springer, Dordrecht, pp. 619–662.
16. Hutchinson, J.N. and Bhandari, R.K. (1971) Undrained loading, a fundamental mechanism of mudflows and other mass movements, *Geotechnique* **21**, 353–358.
17. Hutchinson, J.N. and Kojan, E. (1975) The Mayunmarca landslide of 25th April, 1974, Peru, UNESCO Report Serial No. 3124, UNESCO.
18. Kojan, E. and Hutchinson, J.N. (1978) Mayunmarca rockslide and debris flow, Peru, in B. Voight (ed.), *Rockslides and Avalanches 1,* Developments in Geotechnical Engineering **14A**, Elsevier, Amsterdam, pp. 315–361
19. Skempton, A.W. and Hutchinson, J.N. (1969) State-of-the-Art Report: Stability of natural slopes and embankment foundations, *Proc. 7th Int. Conf. Soil Mech. & Foundn Engrg*, Mexico, State-of-the-Art Volume, 291–340.
20. Tika, T. and Hutchinson, J.N. (1999) Ring shear tests on soil from the Vaiont landslide slip surface, *Geotechnique* **49**, 59–74.

# Preface

In the last 100 years, a number of catastrophic events associated with rockslide dam formation and failure have occurred in the mountain regions of the world including the European Alps, the Himalayas, the mountains of Central Asia, the mountainous margin of the Tibetan Plateau in China, and the Andes of South America. These events illustrate the global importance of the process as a natural hazard and highlight the need for scientific and engineering knowledge concerning the characteristics and behaviour of rockslide dams and the hazards that they pose. The global importance of the formation of rockslide dams and their behaviour has been highlighted by the creation of many impoundments in the 2008 Wenchuan Earthquake, the most critical of which were successfully mitigated by Chinese authorities, and the formation of a massive rockslide-dammed lake in the Hunza valley (Northern Pakistan) in 2010. As of July 25, 2010 (200 days after impoundment) the Hunza Lake continues its stable overflow through an excavated spillway as Pakistan authorities consider engineering options to reduce the lake level.

Rockslide dams are a type of natural dam, and are created in bedrock landscapes when landslides resulting from rock slope failure block drainage leading to the formation of rockslide-dammed lakes upstream from the landslide site. As dramatically illustrated in the case of the 2010 Hunza rockslide-dammed lake, rising impounded waters flood areas upstream and form landslide-dammed lakes that vary in volume from <1 $Mm^3$ to >10 $Gm^3$. Lake Sarez, Tajikistan, formed in 1911 by the blockage of the Murgab River by the massive earthquake-triggered Usoi rockslide, has a volume of 17 $km^3$ and is the largest landslide-dammed lake presently in existence. Landslide-dammed lakes can be stable elements of the landscape that may persist for millennia. They can also fail within days, months, or years after their formation leaving remnant debris in the bottom of river valleys. If they breach suddenly the resulting outburst flood may devastate valley floors downstream. Thus outburst floods from rockslide-dammed lakes are a significant element of landslide hazard in mountain terrain and extend the area of potential damage by a rockslide (which we use in this book in a general sense to describe any mass movement involving a significant initial volume of rock) to beyond the debris of the landslide itself.

As the historical record shows, outburst floods from the failure of rockslide dams can cause destruction of linear infrastructure (roads, pipelines, railways, bridges) and communities in populated areas along the flood path causing a high loss of life. In fact, outbursts from rockslide dams have caused some of the most destructive (in terms of life loss) landslide-related natural disasters in recent history. In 1786, for example, an earthquake-triggered rockslide dammed the Dadu River in Sichuan, China. 10 days later, the dam breached releasing a flood of waters downstream; 100,000 people perished in the outburst flood and the eighteenth century Dadu event remains the most destructive single-event landslide disaster in history. Other notable rockslide dam outbursts occurred in 1841 on the Indus River, Pakistan and in 1933 on the Minjiang River, again in Sichuan, China. Together, these outburst floods took the lives of over 6,000 people. It is also noted that upstream flooding by rising waters during lake filling may submerge communities, infrastructure, and agricultural lands adjacent to river channels.

The formation of potentially unstable rockslide-dammed lakes may be an important secondary effect of major earthquakes and are a major component of the hazard associated with earthquake-triggered landslides. Because of this, lakes formed by earthquake-triggered rockslides are sometimes called "Quake Lakes". In 1959, for example, a large rockslide triggered by the M7.1 Hegben Lake earthquake, blocked the Madison River in Montana, USA, to form an extensive rockslide-dammed lake. The lake was stabilised by engineering works and today it is officially called Earthquake Lake. In 2008 the devastating M7.9 Wenchuan Earthquake struck eastern Sichuan Province, China resulting in over 85,000 fatalities. The major secondary effect of the earthquake was the formation of over two hundred "Quake Lakes" which blocked the narrow, deep valleys of rivers draining southeast off the Tibetan Plateau. Fortunately, the effective mitigation of these lakes resulted in controlled drainage or partial drainage of the impoundments and no catastrophic outburst ensued.

Generally, the failure of rockslide dams is initiated by overtopping which leads rapidly to the formation and enlargement of a breach in the debris dam leading to the catastrophic release of impounded lake waters. The engineering mitigation of rockslide-dammed lakes mainly consists of controlled overtopping, usually achieved by the excavation of a spillway across the debris dam. However, case histories show that this is not always successful and may result in triggering a catastrophic breach by initiating uncontrolled erosion of the debris dam. In other examples, however, controlled overflow has been successfully achieved through a constructed spillway, thus reducing the volume of impounded waters. Other engineering solutions include the construction of by-pass tunnels and high-capacity pumping. Where mitigation is not possible, and outburst is considered to be imminent, downstream warning and evacuation measures may be implemented to reduce life loss.

Rockslide debris has similar geotechnical properties to engineered rockfill used to construct conventional dams. As a result, stable rockslide dams have been utilised as foundations for conventional artificial dams in mountainous areas of the world including Canada, USA, New Zealand, at several locations in the European Alps, the Himalayas, and in the Andes. Constructed artificial dams increase the storage capacity of a landslide-dammed lake by increasing the natural dam height. In addition, the emplacement of artificial valley-blocking rockslides by the explosive initiation of massive rock slope failure has been utilised to form stable rockfill dams (so-called blast-fill dams) for water storage and debris flow defence. The technology was developed in the former Soviet Union and its most recent utilisation was in Kyrgyzstan in late 2009.

This book examines the subjects noted above, investigates the characteristics and behaviour of natural and artificial rockslide dams, presents a detailed verified database of major rockslide dams that have formed and/or failed since 1840, reports new data on important rockslide dam case histories (including the 2010 Hunza event), examines mitigation strategies, and reviews the impact of rockslide-damming on the landscape. To begin, Evans et al. present a comprehensive state-of-the-art review of our global understanding of the formation, characteristics, and behaviour of natural and artificial rockslide dams up to July 2010 (including a short review of the rockslide dams emplaced by the 2008 Wenchuan Earthquake, Sichuan, China, and the 2010 Hunza rockslide dam noted above). Evans et al. also examine the mitigation of rockslide-dammed lakes. Davies and McSaveney explore ideas of hazard assessment whilst overviews of approaches to rockslide dam risk mitigation, illustrated with case histories from around the world, are provided in two papers by Schuster and Evans, and Bonnard. The book contains detailed regional studies of rockslide dams in India, Nepal and China (Weidinger), the Upper Indus region of Pakistan (Hewitt), the northwest Himalayas and adjacent Pamirs (Delaney and Evans), Southern Alps of New Zealand (Korup), and the southern Andes of Argentina (Hermanns et al.). Capra reviews the occurrence and behaviour of rockslide dams associated with large-scale instability of volcanoes in the volcanic belts of the world. The formation and behaviour of rockslide-dammed lakes ("Quake Lakes") formed during the 2008 Wenchuan Earthquake (East Sichuan, China) are summarised in an extensive paper by Cui et al. Detailed case histories of well-known historic and prehistoric rockslide dams provide examples of investigations of rockslide dam behaviour, stability, and characteristics; these form chapters on the Scanno, Italy (Bianchi-Fasani et al.), Val Pola, Italy (Crosta et al.), Usoi, Tajikistan (Ischuk), Dadu, China (Lee and Dai), La Josefina, Ecuador (Plaza et al.), Tsao-Ling, Taiwan (Chang et al.) and Flims, Switzerland (Poschinger) rockslide dams. The formation and stability of rockslide dams is examined in analytical papers by Hungr, Eberhardt and Stead, and Dunning and Armitage. Dunning and Armitage also investigate the sedimentology of dam-forming rockslide debris as do Davies and McSaveney. Manville and Hodgson analyse break-out floods from volcanogenic lakes and review hydrological methods of estimating break-out flood magnitude and behaviour from natural dams. Several papers illustrate the use of remote sensing data (including satellite imagery and digital data from the Shuttle Radar Topography Mission (SRTM)) in the characterisation of rockslide-dammed lakes. This is examined with specific reference to the 2000 Yigong Zangbo rockslide dam (Tibet, China) in a paper by Evans and Delaney and a new approach to the classification of rockslide dams is introduced by Hermanns et al. Finally, a unique section of the book summarises Russian and Kyrgyz experience with blast-fill dam construction in two papers by leading authorities on the technology (Adushkin and Korchevskiy et al.).

This book is the first published on the general topic of rockslide dams and is the first book in English that encompasses a treatment of both natural and artificial rockslide dams. The volume contains 26 papers by 56 authors from 17 countries including most of the recognised world authorities on the subject. The volume will be of interest to geologists, geographers, geomorphologists, hydrologists, and engineers involved in the hazard assessment and mitigation of rockslide dams, to emergency preparedness personnel in the management of rockslide dam emergencies, to natural disaster specialists, and to earth scientists in general who require a detailed outline of the occurrence and behaviour of natural and artificial rockslide dams.

Waterloo, ON, Canada
July 25, 2010

Stephen G. Evans
Reginald L. Hermanns
Alexander Strom
Gabriele Scarascia-Mugnozza

# Contents

# Contributors


**K.E. Abdrakhmatov** Institute of Seismology, Bishkek, Kyrgyzstan, kanab53@yandex.ru

**V.V. Adushkin** Institute of Geospheres Dynamics, Russian Academy of Sciences, 119334 Moscow, Russia, adushkin@idg.chph.ras.ru

**P.J. Armitage** Department of Earth and Ocean Science, University of Liverpool, Liverpool, L69 3GP, UK, peter.armitage@liverpool.ac.uk

**M. Barbieri** Dipartimento di Scienze della Terra, Università degli Studi di Roma "La Sapienza", 00185 Rome, Italy, maurizio.barbieri@uniroma1.it

**G. Bianchi-Fasani** Dipartimento di Scienze della Terra, Università degli Studi di Roma "La Sapienza", 00185 Rome, Italy, gianluca.bianchifasani@uniroma1.it

**C. Bonnard** Soil Mechanics Laboratory, Swiss Federal Institute of Technology, Lausanne, Switzerland, christophe.bonnard@epfl.ch

**Éric Cadier** Institut de Recherche pour le Développement, Quito, Ecuador, higeodes@server.epn.edu.ec

**Lucia Capra** Centro de Geociencias, UNAM, Queretaro, México, lcapra@dragon.geociencias.unam.mx

**E. Cardarelli** Dipartimento di Idraulica, Trasporti e Strade, Università degli Studi di Roma "La Sapienza", 00184 Rome, Italy, ettore.cardarelli@uniroma1.it

**M. Cercato** Dipartimento di Idraulica, Trasporti e Strade, Università degli Studi di Roma "La Sapienza", 00184 Rome, Italy, michele.cercato@uniroma1.it

**Yu-Chang Chan** Institute of Earth Sciences, Academia Sinica, Taipei, Taiwan, Republic of China, epidote@ntut.edu.tw

**Kuo-Jen Chang** Department of Civil Engineering, National Taipei University of Technology, Taipei, Taiwan, Republic of China, epidote@ntut.edu.tw

**Dang Chao** Key Laboratory of Geological Hazards on Three Gorges Reservoir Area, China Three Gorges University, Yichang, China, dangchao1982@126.com

**Rou-Fei Chen** Institute of Earth Sciences, Academia Sinica, Taipei, Taiwan, Republic of China; Department of Geology, Chinese Culture University, Taipei, Taiwan, Republic of China, epidote@ntut.edu.tw

**Xiaoqing Chen** The CAS Key Laboratory of Mountain Hazards and Earth Surface Process, Institute of Mountain Hazards and Environment, Chengdu, China, xqchen@imde.ac.cn

**G.B. Crosta** Dipartimento di Scienze Geologiche e Geotecnologie, Università degli Studi di Milano – Bicocca, 20126 Milano, Italy, giovannib.crosta@unimib.it

**Peng Cui** The CAS Key Laboratory of Mountain Hazards and Earth Surface Process, Institute of Mountain Hazards and Environment, Chengdu, China, pengcui@mde.ac.cn

**F.C. Dai** Institute of Geology and Geophysics, Chinese Academy of Sciences, Beijing 100029, China, daifc@mail.iggcas.ac.cn

**T.R. Davies** Department of Geological Sciences, University of Canterbury, Christchurch, New Zealand, tim.davies@canterbury.ac.nz

**Keith B. Delaney** Landslide Research Programme, Department of Earth and Environmental Sciences, University of Waterloo, Waterloo, ON, Canada N2L 3G1, kbdelane@uwaterloo.ca

**G. Di Filippo** Dipartimento di Idraulica, Trasporti e Strade, Università degli Studi di Roma "La Sapienza", 00184 Rome, Italy, gerarda.difilippo@uniroma1.it

**Stuart A. Dunning** Discipline of Geography and Environment, Northumbria University, Newcastle Upon Tyne NE1 8ST, UK; Department of Geography, University of Durham, DH1 3LE, Durham, UK, stuart.dunning@northumbria.ac.uk

**E. Eberhardt** Geological Engineering/Earth and Ocean Sciences, University of British Columbia, Vancouver, BC, Canada V6T 1Z4, eeberhar@eos.ubc.ca

**C. Esposito** Dipartimento di Scienze della Terra, Università degli Studi di Roma "La Sapienza", 00185 Rome, Italy, carlo.esposito@uniroma1.it

**Stephen G. Evans** Landslide Research Programme, Department of Earth and Environmental Sciences, University of Waterloo, Waterloo, ON, Canada N2L 3G1, sgevans@sciborg.uwaterloo.ca

**Luis Fauqué** Servicio Geológico Argentino, Buenos Aires, Argentina, fauqueluis@yahoo.com.ar

**Andres Folguera** Laboratorio de Tectónica Andina, Universidad Buenos Aires, Buenos Aires, Argentina, andresfolguera2@yahoo.com.ar

**P. Frattini** Dipartimento di Scienze Geologiche e Geotecnologie, Università degli Studi di Milano – Bicocca, 20126 Milano, Italy, paolo.frattini@unimib.it

**N. Fusi** Dipartimento di Scienze Geologiche e Geotecnologie, Università degli Studi di Milano – Bicocca, 20126 Milano, Italy, nicoletta.fusi@unimib.it

**Yongshun Han** Hunan University of Science and Technology, Xiangtan, China, yongshunhan@yahoo.com.cn

**Reginald L. Hermanns** International Centre for Geohazards, Geological Survey of Norway, Trondheim, Norway, reginald.hermanns@ngu.no

**Kenneth Hewitt** Department of Geography and Environmental Studies, Cold Regions Research Centre, Wilfrid Laurier University, Waterloo, ON, Canada N2L 3C5, khewitt@wlu.ca

**K.A. Hodgson** Western Heights High School, Rotorua, New Zealand, khodgson@whhs.school.nz

**O. Hungr** Earth and Ocean Sciences Department, University of British Columbia, Vancouver, BC, Canada V6T 1Z4, ohungr@eos.ubc.ca

**A.R. Ischuk** Institute of Earthquake Engineering and Seismology, Academy of Sciences of the Republic of Tajikistan, Dushanbe, Tajikistan, a_ischuk@yahoo.com

**A.V. Kolichko** LLC Hydrospecproject, 10917 Moscow, Russia, avkolichko@rambler.ru

**V.F. Korchevskiy** LLC Hydrospecproject, 10917 Moscow, Russia, korchevskiy@yandex.ru

**O. Korup** Institute of Earth and Environmental Sciences, University of Potsdam, D-14776 Potsdam, Germany, oliver.korup@geo.uni-potsdam.de

**C.F. Lee** Department of Civil Engineering, The University of Hong Kong, Hong Kong, China, leecf@hkucc.hku.hk

**Hou-Yen Lee** Department of Civil Engineering, National Taipei University of Technology, Taipei, Taiwan, Republic of China, epidote@ntut.edu.tw

**V. Manville** Institute of Geological and Nuclear Sciences, Wairakei Research Centre, Taupo, New Zealand, v.r.manville@leeds.ac.uk

**M.J. McSaveney** Institute of Geological and Nuclear Sciences, Lower Hutt, New Zealand, mcsaveney@gns.cri.nz

**Samuel Niedermann** GeoForschungsZentrum Potsdam, D-14473 Potsdam, Germany, nied@gfz-potsdam.de

**Ivanna Penna** Laboratorio de Tectónica Andina, Universidad Buenos Aires, Buenos Aires, Argentina, ivannapenna@hotmail.com

**L.M. Pernik** Institute of Geospheres Dynamics, Moscow, Russia, Pernik@idg.chph.ras.ru

**M. Petitta** Dipartimento di Scienze della Terra, Università degli Studi di Roma "La Sapienza", 00185 Rome, Italy, marco.petitta@uniroma1.it

**Galo Plaza** Escuela Politécnica Nacional, Quito, Ecuador, higeodes@server.epn.edu.ec

**Andreas von Poschinger** Bavarian Environment Agency, D-80696 München, Germany, Andreas.Poschinger@lfu.bayern.de

**Gabriele Scarascia-Mugnozza** Department of Earth Sciences, University of Rome "La Sapienza", Piazzale Aldo Moro 5, 00185 Rome, Italy, gabriele.scarasciamugnozza@uniroma1.it

**Robert L. Schuster** U.S. Geological Survey, Denver, CO 80225, USA, rschuster@usgs.gov

**R. Sosio** Dipartimento di Scienze Geologiche e Geotecnologie, Università degli Studi di Milano – Bicocca, 20126 Milano, Italy, rosanna.sosio@unimib.it

**D. Stead** Department of Earth Sciences, Simon Fraser University, Burnaby, BC, Canada V5A 1S6, dstead@sfu.ca

**A.L. Strom** Institute of Geospheres Dynamics, Russian Academy of Sciences, 119334 Moscow, Russia, a.strom@g23.relcom.ru

**Alfredo Taboada** Laboratoire Géosciences Montpellier, Université Montpellier 2, Montpellier, France, epidote@ntut.edu.tw

**J.T. Weidinger** Department of Geography and Geology, Erkudok© Institute in the K-Hof Museums Gmunden, Salzburg University, Gmunden, Austria, j_weidinger@hotmail.com

**Othon Zevallos** Empresa Metropolitana de Alcantarillado y Agua Potable de Quito, Quito, Ecuador, higeodes@server.epn.edu.ec

# Chapter 1
# The Formation and Behaviour of Natural and Artificial Rockslide Dams; Implications for Engineering Performance and Hazard Management

**Stephen G. Evans, Keith B. Delaney, Reginald L. Hermanns, Alexander Strom, and Gabriele Scarascia-Mugnozza**

## 1 Background

Rockslide dams block surface drainage to form upstream lakes (Fig. 1.1). They may be formed naturally due to landslides in rockslopes or as a result of engineered rock-slope failures. Natural rockslide dams form as a consequence of rockslope instability in mountain terrain when landslide debris blocks river channels (Figs. 1.2 and 1.3). Impoundments develop upstream of the blocking debris inundating valley bottom terrain to form rockslide-dammed lakes (Figs. 1.1, 1.2, and 1.3). As evidenced by the 2010 Hunza River rockslide dam, northern Pakistan (Fig. 1.1; [151], Chap. 7 by Delaney and Evans, this volume), the stability of natural rockslide dams with respect to sudden breaching is a major safety and security issue in mountain areas, yet unbreached rockslide-dammed lakes may persist in the landscape for millennia (e.g. [87, 128, 287], Chap. 5 by Hermanns et al., this volume; Fig. 1.4).

A rockslide dam impounds one of the largest lakes on Earth, that being Lake Sarez in Tajikistan (Fig. 1.3). The lake is dammed by the 2.2 $Gm^3$ Usoi rockslide (Fig. 1.5 [63, 79, 96, 236, 238]; triggered by the 1911 Pamir earthquake [203, 223, 255]; Chap. 7 by Delaney and Evans, this volume). Lake Sarez currently contains about 17 $km^3$ of water behind the impounding rockslide [259].

As discussed in detail below, rockslide dams are commonly emplaced in major earthquakes (Fig. 1.5). Rockslide dams are also commonly associated with major mass movements on volcanoes (e.g. [23, 271, 280], Chap. 9 by Capra, this volume) when large-scale failures block surface drainage.

Rockslide dams, and their effects, may have a major impact on human activity in mountain areas. In historic time, breached natural rockslide dams have caused some of the biggest disasters associated with single event landslides ([57, 76]; Chap. 13 by

S.G. Evans (✉)
Landslide Research Programme, Department of Earth and Environmental Sciences, University of Waterloo, Waterloo, ON, Canada N2L 3G1
e-mail: sgevans@sciborg.uwaterloo.ca

S.G. Evans et al. (eds.), *Natural and Artificial Rockslide Dams*, Lecture Notes in Earth Sciences 133, DOI 10.1007/978-3-642-04764-0_1, © Springer-Verlag Berlin Heidelberg 2011

**Fig. 1.1** EO-1 satellite image of the 2010 Hunza rockslide-dammed lake, northern Pakistan obtained on July 7, 2010. Lake formed as a result of the Attabad rockslide (A) which occurred on January 4, 2010. The rockslide-dammed lake overtopped the debris dam on May 29, 143 days after impoundment, when the volume of the lake was approximately 430 $Mm^3$. Flow in the spillway is visible on the image. The overtopping took place through an excavated spillway across the debris dam (Chap. 7 by Delaney and Evans, this volume). As of July 25, 2010, 200 days after impoundment, the integrity of the rockslide dam was still intact and the extensive rockslide-dammed lake remained in place as stable overtopping continued

Lee et al., this volume). Rockslide dams have also impacted on prehistoric human activity. In northwest North America, for example, rockslide dams have blocked the migration routes of andromonous Pacific Salmon, periodically resulting in the cultural collapse of indigenous people due to the interruption of their food supply

**Fig. 1.2** Rockslide-dammed lake (Earthquake Lake) formed by landslide (est. vol. 20 $Mm^3$) triggered by the M7.3 Hegben Lake earthquake, Montana, USA, August 17, 1959. Maximum volume of Earthquake Lake was 102 $Mm^3$ [269]. Photograph was taken after construction of spillway (visible as a light area in the low centre of the debris) had begun (USGS photograph by John R. Stacy)

[119, 120, 231]. In the Peruvian Andes, Contreras and Keefer [47] suggest a lake impounded by a rockslide dam provided the water supply for the Chavin people at the site of the ancient city of Chavín de Huantar during the first millennium B.C.

Rockslide dams may form favourable foundations for artificial dams constructed for the purposes of water supply, flood control, or hydroelectric power generation [17, 122, 226, 237, 265]. Artificial rockslide dams can be created by the planned explosive-triggering of massive rock slope failure [7, 5, 146]; Chap. 25 by Adushkin, this volume; [250] or, in the past, as a secondary effect of underground nuclear weapons testing [6, 152, 252]. In December 2009, a 3 kt explosion initiated a ca. 3 $Mm^3$ rockslide that blocked the Naryn River, Kyrgyzstan, upstream of the Toktogul Reservoir. The 60 m high artificial rockslide dam was the initial stage of construction of the 360 Mw Kambarata 2 hydroelectric project, the first use of blast-fill technology in Central Asia since the dissolution of the Soviet Union.

Floods from catastrophic rockslide dam failures may be orders of magnitude greater than normal floods (e.g. [29, 170, 199, 200, 201]). From a geomorphic perspective, the interaction between fluvial systems and rockslide dam formation has been the subject of much recent research (e.g. [132, 162, 164–166, 205]) in relation to the evolution of drainage networks [132, 134, 136, 161], the development of relief [165, 166] and as a major process of sediment delivery to river channels [88, 89, 158, 159]. A key step in assessing the role of rockslide dams in landscape development, as well as in mapping for hazard assessment, is the accurate identification of remnant

**Fig. 1.3** Oblique photograph taken from International Space Station in September 2001 of Lake Sarez, Pamir Mountains, Tajikistan. Lake Sarez is a rockslide-dammed lake impounded by the 1911 Usoi rockslide (A) ([223]; Chap. 7 by Delaney and Evans, this volume). Lake Sarez has a length of ca. 60 km and a volume of ca. 17 $Gm^3$ (NASA photograph ISS002-E-7524)

deposits [133]. Types and classification of rockslide dams have been discussed by Swanson et al. [261], Hejun et al. [124], Dunning et al. [65] and Chap. 24 by Hermanns et al., this volume.

In this opening chapter our objectives are sevenfold; (1) we review developments in the understanding of rockslide dam formation and behaviour in the 22 years since the publication of the benchmark Costa and Schuster paper in 1988 [53], (2) we summarize major rockslide dam events since that publication, including the most recent rockslide-dammed lake crisis, the 2010 Hunza rockslide dam in northern Pakistan (3) we present a verified database of rockslide dam events in the period 1840–2010 involving impoundments and/or outburst events of 100 $Mm^3$ of water or more (4) we review the characteristics of nine case histories of rockslide dam formation and/or failure selected from this data base (5) we evaluate the methods in use to estimate discharge and downstream flood effects of the breaching of rockslide dams (6) we review measures to mitigate hazard associated with rockslide-dammed lakes, the utility of natural rockslide dams as foundations for constructed dams, and the formation and behaviour of artificial rockslide dams, and lastly

**Fig. 1.4** Two rockslide-dammed lakes impounded by rockslides that have occurred along the linear trough marking the strike-slip Fergahna Fault, Kyrgyzstan (Google Earth image). The larger lake, Karasu Lake, is impounded by the Karasu rockslide (est. vol. 250 $Mm^3$), the slide from the south side of the trough, as described by Strom [257, 258]

**Fig. 1.5** View to northwest of Usoi rockslide scar and debris, Tajikistan, triggered by the 1911 Pamir Earthquake, taken by Sir Aurel Stein in August 1915 (reproduced from Fig. 1.24 in [255]). Clouds of dust are seen rising as rockfalls occur on the rockslide scarp. Rockslide debris (*bottom* foreground and *bottom left*) blocked the Murgab (Bartang) River and rockslide-dammed Lake Sarez can be seen forming at bottom right. At the time of Stein's traverse over the rockslide, Lake Sarez was about 25 km long and was still "spreading up the valley" (Chap. 7 by Delaney and Evans, this volume)

(7) we summarise recent work on the relationship between fluvial systems, rockslide dams, and landscape development.

## 2 Natural Rockslide Dam Formation and Failure

Summary data on 18 documented rockslide dams that formed and/or failed world-wide in the period 1840–2010 involving impoundment volumes in excess of 100 $Mm^3$ is presented in Table 1.1. The availability of high-resolution digital terrain data (including SRTM-3 data obtained by the Shuttle in February 2000 – [81] and satellite images provided the opportunity to calculate credible volumes of some rockslide-dammed lakes for the first time (e.g., the 1841 Indus event), to verify the published volumes of other impoundments (e.g., the Barrancas rockslide-dammed lake that breached in 1914), and to obtain more accurate volumes than previously published estimates for some other rockslide-dammed lakes (e.g., the 1945 Condor-Sencca and the 1974 Mayunmarca rockslide dammed lakes).

### *2.1 Rates of Rockslide Dam Formation*

Rockslide dams are common throughout the mountain regions of the world (e.g. [54, 103, 164, 235, 258]), since rockslides frequently impound surface drainage (Figs. 1.1–1.6).

Regional studies of rockslides, rockslide dams, and individual case studies of valley-blocking rockslides in the mountainous regions of the world indicate a significant rate of historical rockslide dam formation as well as the widespread distribution of prehistoric rockslide dams (Fig. 1.4). Such studies include work carried out in the Alps [2, 18, 49, 68, 122, 189], the Appennines [22, 26, 194, 216], Sicily [193], Central Asia, including Turkey ([61, 64, 71, 174, 224, 248, 258]; Chap. 7 by Delaney and Evans, this volume), the Himalayas [67, 131, 134, 135]; Chap. 7 by Delaney and Evans (this volume), New Zealand [4, 113, 156, 157, 191, 211, 281], China [30, 124, 172, 173], Japan [147, 179, 204, 245, 261, 262, 264], the North American Cordillera [42, 75, 169, 199, 207, 241, 260] (Figs. 1.2–1.6), Central America [190], the Caribbean [59], and the Andes ([12, 50, 83, 103, 129, 212, 217, 218, 289]; Chap. 5 by Hermanns et al., this volume).

Regional studies of landslide dams form a basis for estimating the rate of dam formation in a given area. Supplemented by additional geomorphometric analysis of landslide dam geometry, impoundment and drainage basin characterisitics, these studies can provide quantitative data for the assessment of rockslide dam formation and for the assessment of rockslide dam stability. Korup [157], for example, discussed an inventory of 232 landslide dams in New Zealand. Rock avalanches formed 27% of the dams in the inventory and a further 14% involved some form of rock slope failure. Some 40 dams (17%) in the inventory were formed co-seismically during the 1929 Murchison Earthquake and 75% of all dated occurrences fall in the twentieth century.

In the Himalayas, the work of Hewitt [132, 134–139], Fort [85, 86] and Fort et al. [87] has documented large numbers of rockslide dams and related

**Table 1.1** 18 documented rockslide dams that formed and/or failed in the period 1840–2010 where impounded and/or outburst volume was 100 $Mm^3$ or more

| Locality | | Location (Lat./Long.) | Date of formation | Date of breaching or partial breaching | Volume of rockslide deposit ($Mm^3$) | Volume of lake ($Mm^3$) | Volume of outburst ($Mm^3$) | Comments | References |
|---|---|---|---|---|---|---|---|---|---|
| 1 | Indus, Pakistan | 35°30′32″N/ 74°37′01″E | 1840 or 1841 (?) | 1841 (June) | 500 | 6,500 | 6,500 | Earthquake-triggered (?) rockslide blocked the Indus in December (?) 1840 forming a ca. 60 km long lake; breach of the dam in June 1841 released a massive flood into the Indus (*The Great Indus Flood* of Mason [183]). Extensive destruction downstream. Indus rose over 25 m at Attock, 425 km downstream | Drew [62], Mason [183], Chap. 7 by Delaney and Evans, this volume; Code and Sirhindi [44], Shroder et al. [247], Shroder [246] |
| 2 | Hunza, Pakistan | 36° 18′04″N/ 74° 46′50″E | 1858 | 1858-08-09 | 500 | 805 | 805 | Rockslide into the Hunza River; breach of the dam after 6 months released a massive flood into the Hunza-Indus (*The Second Great Indus Flood* of Mason [183]). Indus rose 17 m at Attock, 585 km downstream | Becher [15], Henderson [125], Mason [183], Schroder, [246], Chap. 7 by Delaney and Evans, this volume |

**Table 1.1** (continued)

| Locality | | Location (Lat./Long.) | Date of formation | Date of breaching or partial breaching | Volume of rockslide deposit ($Mm^3$) | Volume of lake ($Mm^3$) | Volume of outburst ($Mm^3$) | Comments | References |
|---|---|---|---|---|---|---|---|---|---|
| 3 | Gohna, India | 30°22′32″N/ 79°29′51″E | 1893-09-03 | 1894-08-25 | 150 | 470 | 283 | Two rockslides in close succession blocked the Bireh Ganga River forming Gohna Lake on Sept 3, 1893. Rockslide dam failed on Aug. 25, 1894 releasing 60% of impounded lake volume. Massive flood downstream but no loss of life due to implementation of warning and evacuation plan. Srinigar, 116 km downstream totally destroyed by flood. | Glass [99], Holland [140], Gupta [109], Weidinger [282], Chap. 1 by Evans et al., this volume |
| 4 | Usoi, Tajikistan | 38°16′23″N/ 72°36′14″E | 1911 | n/a | 2,200 | 17,000 | n/a | Massive earthquake-triggered rockslide dammed Murgab River to form 60 km-long Lake Sarez. Rockslide dam has remained stable since. Monitoring and warning system has been implemented in recent years and dam stability subject to intense analysis. | Gaziev [96], Schuster and Alford [238], Droz and Sapsic-Gril [63], Chap. 16 by Ischuk, this volume. |

**Table 1.1** (continued)

| | Locality | Location (Lat./Long.) | Date of formation | Date of breaching or partial breaching | Volume of rockslide deposit (Mm$^3$) | Volume of lake (Mm$^3$) | Volume of outburst (Mm$^3$) | Comments | References |
|---|---|---|---|---|---|---|---|---|---|
| 5 | Rio Barrancos, Argentina | 36°30′56″S/ 70°09′15″W | PREHISTORIC | 1914-12-29 | 1,300 | ? | 1,550 | Prehistoric rockslide dam impounding Lago Carri Lauquen breached on Dec. 29, 1914. Outburst flood devastated Rio Colorado valley in its 900 km-long passage downstream to the Atlantic. Death toll officially estimated at 175. | Groeber [108], Gonzalez Diaz et al. [102], Hermanns et al. [127], Chap. 1 by Evans et al. this volume. |
| 6 | Gros Ventre, Wyoming, USA | 43°37′55″N/ 110° 32′47″W | 1925-06-23 | 1927-05-18 | 40 | 105 | 74 | Rock avalanche dammed Gros Ventre River on June 23, 1925. Rockslide dam breached following overtopping on May 18, 1927 causing downstream flood. The town of Kelly swept away-about 7 deaths. | Emerson [69], Alden [8], Engineering News Record [70], Voight [270]. Hayden [121] |
| 7 | Deixi, China | 32°02′25″N/ 103° 40′29″E | 1933-08-25 | 1933-10-07 | 150 | 400 | ? | Earthquake-triggered rockslide dammed Min River, Sichuan Province. Dam breached on Oct. 7, 1933 after lake having reached its maximum extent. Downstream flood killed at least 2,423 people in its 253 km of passage. | Li et al. [173], Li and Wang [172] |

**Table 1.1** (continued)

| Locality | | Location (Lat./Long.) | Date of formation | Date of breaching or partial breaching | Volume of rockslide deposit ($Mm^3$) | Volume of lake ($Mm^3$) | Volume of outburst ($Mm^3$) | Comments | References |
|---|---|---|---|---|---|---|---|---|---|
| 8 | Tsao-Ling 1, Taiwan | 23°34′35″N/ 120°40′19″E | 1941 | 1951 | 100–150 | 157 | 120 | Earthquake-triggered rockslide dammed Ching-Shui River forming a lake upstream. Rockslide dam augmented by second rockslide (150-200 $Mm^3$) in 1942. Lake contained 157 $Mm^3$. Composite rockslide dam failed in 1951 during spillway excavation. 120 $Mm^3$ of water released. 137 deaths and 1,200 houses destroyed downstream. | Hung et al. [143] |
| 9 | Cerro Condor-Sencca, Peru | 12°48′04″S/ 74°17′11″W | 1945-08-16 | 1945-10-28 | 5.35 | 721 | 664 | Rockslide dammed the Mantaro River to form 21 km long lake that extended into tributary valley. Breach resulted from erosional retrogression of hand-excavated spillway. Extensive damage to infrastructure downstream. Volumes reported here, calculated from SRTM-3 DEM, are more than double previously published values. | Snow [253], Delaney (unpublished data) |

**Table 1.1** (continued)

| Locality | | Location (Lat./Long.) | Date of formation | Date of breaching or partial breaching | Volume of rockslide deposit ($Mm^3$) | Volume of lake ($Mm^3$) | Volume of outburst ($Mm^3$) | Comments | References |
|---|---|---|---|---|---|---|---|---|---|
| 10 | Subansiri River, Assam, India | 27°36 00″N/ 94°12′42″E | 1950-08-15 | 1950-08-19 | ? | 300 | 300 | Rockslide triggered by M8.6 Great Assam earthquake blocked Subansiri River in a gorge section at Sipumukh, stopping the flow of the river. Dam breached 4 days later causing massive outburst flood downstream. Villages swept away. Over 500 deaths. | Mathur [184]; Goswami et al. [104]; J.N. Sarma (Personal Communication, 2010); |
| 11 | Madison Canyon, USA | 44°49′44″N/ 111°25′39″W | 1959-08-17 | n/a | 28-32 | 101 | n/a | Earthquake-triggered rockslide on Aug. 17 blocked the Madison River forming Earthquake Lake. Spillway successfully constructed over rockslide debris resulting in stable overtopping of lake through spillway flow on Sept. 10. Spillway continues to function. | U.S. Corps of Engineers [269], Hadley [111], Barney [14]; |

**Table 1.1** (continued)

| Locality | | Location (Lat./Long.) | Date of formation | Date of breaching or partial breaching | Volume of rockslide deposit ($Mm^3$) | Volume of lake ($Mm^3$) | Volume of outburst ($Mm^3$) | Comments | References |
|---|---|---|---|---|---|---|---|---|---|
| 12 | Tanggudong, China` | 29°24′46″N/ 101°07′06″ | 1967-06-08 | 1967-06-17 | 68 | 680 | 640 | Rockslide dammed Yalong River, Sichuan Province on June 8. Dam breached on June 17 as a result of overtopping and caused large downstream flood (max. discharge 53,000 $m^3/s$) that reached Yibin, a distance of 1,000 km. No deaths due to downstream warning and evacuation; property damage considerable. | Li et al. [173], Chen et al. [37]. |
| 13 | Mayunmarca, Peru (Mantaro River) | 12°36′55″S/ 74°39′25″W | 1974-04-25 | 1974-06-08 | 1,600 | 352 | 350 | Massive rockslide blocked Mantaro River. 31 km long lake formed upstream. Overtopping occurred on June 6 and breaching inititiated on June 8, 44 days after impoundment. Peak discharge ~13,700 $m^3/s$. No fatalities due to downstream warning and evacuation; much damage to infrastructure up to 50 km downstream. Volumes | Lee and Duncan [170], Taype Ramos [263], Kojan and Hutchinson [155], Ponce and Tsivoglou [220]; Delaney (unpublished data) |

**Table 1.1** (continued)

| | Locality | Location (Lat./Long.) | Date of formation | Date of breaching or partial breaching | Volume of rockslide deposit ($Mm^3$) | Volume of lake ($Mm^3$) | Volume of outburst ($Mm^3$) | Comments | References |
|---|---|---|---|---|---|---|---|---|---|
| | | | | | | | | reported here, calculated from SRTM-3 DEM, are much less than previously published values. | |
| 14 | Mt St Helens, USA | 46° 16′07″N/ 122° 09′49″W | 1980-05-18 | n/a | 2,800 | 259 (Spirit Lake) | n/a | Eruption-related flank collapse (2.5 $Gm^3$) resulted in damming of drainage to form Spirit (259 $Mm^3$); mitigation works initiated by pumping and followed by construction of spillways and drainage tunnels stabilized lake levels (preventing overtopping) and still operate. | Meyer et al. [185, 186]; Sager and Chambers [232]; Voight et al. [271] |
| 15 | Rio Paute, Ecuador (La Josefina) | 2°51′08″S/ 78° 50′44″W | 1993-03-29 | 1993-05-01 | 25 | 210 | 175 | Rockslide dammed Rio Paute forming a 10 km long lake. Impounded waters flooded numerous buildings. Breaching resulted from retrogressive erosion of spillway excavated over blocky debris. Peak discharge ~10,000 $m^3$/s. Outburst flood devastated downstream terrain causing | Plaza and Zevallos [218], Canuti et al. [21], Sevilla [243], Zevallos et al. [289], Harden [114], Chap. 14 by Plaza et al. this volume. |

**Table 1.1** (continued)

| Locality | | Location (Lat./Long.) | Date of formation | Date of breaching or partial breaching | Volume of rockslide deposit ($Mm^3$) | Volume of lake ($Mm^3$) | Volume of outburst ($Mm^3$) | Comments | References |
|---|---|---|---|---|---|---|---|---|---|
| | | | | | | | | $192 USD damage. No lives lost because of warning and evacuation. Outburst flood impacted on Paute hydroelectric plant which at the time produced 70% of Ecuador's electricity. | |
| 16 | Yigong, China | 30° 10′58″N/ 94°55′53″E | 2000-04-09 | 2000-06-10 | 100 | 2000 | 2000 | Massive rock avalanche blocked Yigong Zangbo River forming large impoundment upstream and enlarging rockslide-dammed lake formed in 1900. After construction work on a spillway had begun on May 3 catastrophic breach occurred on June 10, 62 days after formation. Massive outburst of 2 $Gm^3$ of impounded water swept down Yigong River and into India reaching the floodplain of the Brahmaputra, ca. 500 km distant. Large scale damage in China and India. | Shang et al. [244]; Zhou et al. [291], Zhu and Li [292]; Chap. 7 by Evans and Delaney, this volume. |

**Table 1.1** (continued)

| Locality | | Location (Lat./Long.) | Date of formation | Date of breaching or partial breaching | Volume of rockslide deposit ($Mm^3$) | Volume of lake ($Mm^3$) | Volume of outburst ($Mm^3$) | Comments | References |
|---|---|---|---|---|---|---|---|---|---|
| 17 | Tangjiashan rockslide dam and lake, Sichuan, China | 31° 50′38″N/ 104° 25′55″E | 2008-05-12 | 2008-06-10 | 24 | 247 | 160 | The largest of many rockslide-dammed lakes emplaced during the M8.0 Wenchuan earthquake; the 23 km long lake threatened many communities downstream. Successfully mitigated by excavation of spillway and controlled, partial-drainage of impoundment that affected magnitude and timing of downstream flood | Liu et al., [176]; Chen et al. [36], Cui et al. [55, 56] Chap. 10 by Cui et al., this volume |
| 18 | Hunza River, Pakistan | 36° 18′24″N/ 74° 49′17″E | 2010-01-04 | n/a | 52 | 430 | n/a | Rockslide dammed the Hunza River completely, immediately forming a rockslide dammed lake. Dam height above river bed approximately 120 m. Lake gradually filled to a total volume of ca. 430 $Mm^3$ and a lake length of 21 km. | Chap. 7 by Delaney and Evans, this volume; Kargel et al. [151] |

**Table 1.1** (continued)

| Locality | Location (Lat./Long.) | Date of formation | Date of breaching or partial breaching | Volume of rockslide deposit ($Mm^3$) | Volume of lake ($Mm^3$) | Volume of outburst ($Mm^3$) | Comments | References |
|---|---|---|---|---|---|---|---|---|
| | | | | | | | Stable overtopping through an excavated spillway occurred on May 29, 143 days after impoundment. 22 km of Karakoram Highway submerged disrupting trade with China and movement of local population. As of July 25, 2010 (200 days after impoundment) rockslide-dammed lake continued stable overtopping and maximum water level maintained. | |

**Fig. 1.6** Rockslide-dammed lake in the Coast Mountains of British Columbia formed where a rock avalanche of unknown age has blocked the Nahatlatch River (49° 53′ 32″ N/ 121° 50′ 19″W)

lacustrine sediments accumulated in rockslide-dammed lakes. Hewitt [132] reported on 115 rockslides of which at least 73 (65%) had dammed rivers of the Upper Indus. Later, updating this inventory, Hewitt [135] presented data on 182 catastrophic rockslides, 174 of which were reconstructed from ancient deposits in the fluvial zone; he later updated the total of rockslides in his study area to 322, many of which blocked major rivers in the region [137, 138] (Chap. 6 by Hewitt, this volume).

In the Andes of Venezuela, Ferrer [83] studied 35 cases of landslide blockages, 63% of these were triggered by earthquakes, while 20% were triggered by torrential rainfalls. However, it is not clear how many of the landslide dams were rockslide dams. In the Argentine Andes, systematic studies were carried out in the Mendoza valley [82], in NW Argentina [103], and northern Patagonia (Chap. 5 by Hermanns et al., this volume). In the Mendoza valley out of 15 temporary dams formed by glacial surges, moraines and rock avalanches, 9 were related to rock avalanches. At all sites several meters to several tens of meters thick lake deposits attest that the dams were stable for a considerable time. However, at none of these sites did the rockslide dam basin fill up completely with sediments to the dam crest indicating some long term instability [82]. In NW Argentina out of 55 rock avalanches only 20 impounded lakes [126, 130]; (Chap. 5 by Hermanns et al., this volume) while in northern Patagonia out of a dataset of 51 rock slope failures, 90% resulted in valley impoundment [126]. This reflects the different geomorphic settings of the two regions; NW Argentina is characterized by tectonic basins several kilometers wide connected partially through narrow gorges whilst northern Patagonia is characterized by glacial valleys.

Massive rock slope failures from volcanoes frequently dam drainage systems to form upstream impoundments (e.g. [23, 24, 78, 185, 242, 271, 280], Chap. 9 by Capra, this volume).

Evans [77] presents an analysis of the global record of large twentieth century non-volcanic rock avalanches (1900–2000) over a threshold volume of $20 \times 10^6$ m$^3$ [77]. There are 38 in the dataset. Of this total, 19 (50%) rock avalanches formed significant landslide dammed lakes, a global rockslide-dam formation rate of 1 per 5.4 years.

With respect to impoundment volumes, Evans [77] suggests, based on the analysis of the historical record, that landslide dams which impound a volume of $10^8$ m$^3$ or greater, form roughly every 30 years, globally. Landslide dams that impound ca. $10^9$ m$^3$ appear to occur every 80 years, while impoundments of ca. $10^{10}$ m$^3$ may occur with a frequency of roughly a century.

However, data on the rate of formation of rockslide dams remains constrained by the completeness of an inventory in a given region, the accurate characterization of the damming landslide, as well as the lack of information on the age of a given impoundment in either prehistoric or historic time.

## 2.2 *Rockslide Dams and Earthquakes*

Rockslide dams are commonly emplaced during strong earthquakes (e.g. [4, 66, 96, 105, 115, 116, 173, 193], Figs. 1.2 and 1.5) and is an important element of hazard associated with earthquake-triggered landslides. Of the 18 major rockslide dams listed in Table 1.1, six (33%) were emplaced during major earthquakes.

The largest recorded disaster (in terms of loss of life) associated with a single event landslide [76] occurred in 1786 in Sichuan, China [57]; (Chap. 13 by Lee and Dai, this volume) as a result of the failure of a rockslide dam emplaced during a major earthquake. The details of this case were reconstructed from historical documents by Dai et al. [57] and Chap. 13 by Lee and Dai (this volume). These documents indicate that a rockslide dam was emplaced during the M7.7 Kangding-Luding earthquake and dammed the Dadu River on June 1, 1786 [57]; (Chap. 13 by Lee and Dai, this volume). A lake filled behind the impounding rockslide dam for a period of 8 days before overtopping on June 9. Triggered by an aftershock on June 10, the rockslide dam suddenly failed releasing a massive volume of water downstream which reached at least as far as Luzhou, ca. 500 km downstream. In the ensuing flood a total of approximately 100,000 lives were lost, including many in Leshan City, 250 km downstream from the rockslide dam. These data are suggestive of a massive outburst flood; an estimate of the volume of the rockslide-dammed lake may be obtained by taking the average discharge of the Dadu River [57] and the length of time the Dadu River was dammed. This gives a volume in excess of 500 Mm$^3$ which is approximately 10 times greater than that estimated by [57] and Chap. 13 by Lee and Dai (this volume).

Many landslides triggered by the 2008 M8.0 Wenchuan earthquake formed landslide dams on rivers draining the Tibet Plateau [38, 41, 56, 177, 285, 290]; (Chap. 10 by Cui et al., this volume) thus creating numerous landslide-dammed lakes, known as "barrier lakes" in the Sichuan context (Fig. 1.7). Approximately 257 landslide dams were formed with 24 lakes over 1 Mm$^3$ in volume [56]; (Chap. 10

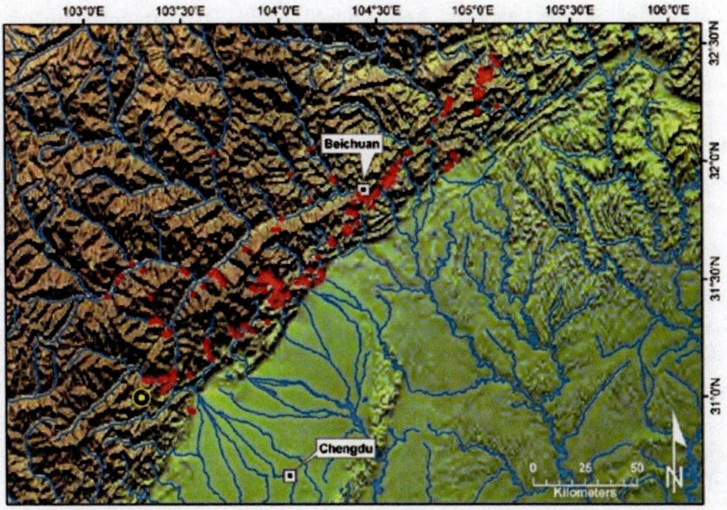


**Fig. 1.7** Map showing location of landslide-dammed lakes (red dots) emplaced during the 2008 Wenchuan Earthquake, Sichuan, China. Data plotted from Table 1.1 in Cui et al. [56]. Epicentre is yellow dot in yellow circle

by Cui et al., this volume); from published tabular data it is evident that five lakes had impounded volumes in excess of 10 $Mm^3$ (e.g. [56, 285]).

The largest barrier lake created was Tangjiashan Lake (Table 1.1) impounded by an earthquake-triggered rockslide, 4 km upstream of Beichuan county town on the Tongkou River, Sichuan Province (Fig. 1.7). As described in detail below, Tangjiashan Lake reached a maximum storage of 247 $Mm^3$ in 29 days of impoundment and posed significant hazard to communities downstream [36, 55, 176]; (Chap. 10 by Cui et al., this volume), including the possible flooding of parts of Mianyang, Sichuan's second largest city located 80 km down river. Controlled partial drainage of the lake was achieved by the excavation of a spillway across the rockslide debris prior to overtopping, thus reducing the magnitude of the outburst flood.

In another major earthquake, many landslide dams were emplaced by the August 15, 1950 Great Assam Earthquake (M8.6) in northeast India, and resulted in total blockage of several tributaries of the Brahmaputra including the Subansiri, Dibang, Lohit, and Tiding rivers [184, 219]. The main channel of the Tsangpo, the main source tributary of the Brahmaputra, was blocked above Pasighat. A major landslide completely blocked the Subansiri River, the largest tributary of the Brahmaputra, in a gorge section of the river, 2 km upstream of the boundary of the Foothills [104], at an epicentral distance of about 260 km. Since the earthquake occurred during the monsoon season (May to September), the Subansiri River was swollen with monsoon run-off. The river was blocked for a total of 4 days. If we assume a reasonable monsoon discharge of the Subansiri River, say 15,000 $m^3/s$ (e.g. [149]), the blockage would have created an impoundment of approximately 300 $Mm^3$ (see Table 1.1). The rockslide dam breached on the night of August 19, releasing the impounded volume of water into the lower reaches of the Subansiri River inundating an area of 775 $km^2$ [219]. Some villages, tea gardens and other cultivated areas were destroyed and about 500 people lost their lives. The outburst flood from the landslide dam caused dramatic changes in the channel of the Subansiri River immediately downstream of the blockage [104]. India's largest hydroelectric project (Lower Subansiri) involving a 116 m high concrete gravity dam and a 2,000 MW powerhouse is currently (2010) under construction in the same gorge section of the Subansiri river as affected by the 1950 rockslide dam blockage.

Other important rockslide dams emplaced during earthquakes that impounded significant volumes of water (i.e., greater than 100 $Mm^3$) include (Table 1.1) the 1841 blockage of the Indus (Pakistan) (Chap. 7 by Delaney and Evans, this volume), the 1911 Usoi rockslide (Tajikistan) noted above, the 1933 Deixi landslide (China), the 1941 Tsao-Ling rockslide (Taiwan), the 1959 Madison Canyon rockslide (USA) discussed in detail below, and the Tangjiashan rockslide (China), one of the many rockslide dams emplaced in the 2008 Wenchuan Earthquake (China), as noted above.

## 2.3 *Rockslide Dam Failure*

Landslide dams fail by a number of processes including (1) overtopping (e.g., [37, 98, 170]), which may lead to progressive upstream erosion and lateral and vertical enlargement of the overflow channel (e.g. [113]) (2) piping (e.g. [100, 186]), and (3) the sliding collapse of the downstream face in association with high water and/or seepage pressures (e.g. [67, 99]. Overtopping may be initiated by static overflow, in which the water level in the lake simply exceeds the level of the lowest point of the debris dam (e.g., Chap. 7 by Delaney and Evans, this volume) or by dynamic overflow in which a catastrophic overtopping by a landslide-generated wave within the impounded lake (e.g., [128]; [279]) occurs. Earthquake shaking [57], Chap. 13 by Dai and Lee, this volume)) may also lead to instability in the upstream or downstream faces of the debris dam or to piping through internal deformation of the debris. We also note that overtopping need not necessarily result in rockslide dam failure since stable overflow may persist for a period after overtopping (Chap. 7 by Delaney and Evans, this volume) and may be lead to a permanent condition in landscape terms (see discussion in Sect. 2.5 below).

The effect of processes 1–3 may be amplified or accelerated by an engineering attempt to excavate a spillway over the debris dam (e.g. [154]; Chap. 2 by Schuster and Evans, this volume; Chap. 14 by Plaza et al., this volume; Chap. 7 by Delaney and Evans, this volume). In the process of overtopping the overflow channel develops and enlarges initially by vertical downcutting which then evolves into lateral widening [46]. Widening is accelerated by the collapse of the spillway channel walls as they are undercut and the debris is transported downstream. Continued erosion maintains a feedback process occurs in which the channel cross-section increases thus allowing greater discharge and more erosive power in the overflow [46] until some stable lake level is reached.

Again using the twentieth-century rockslide data set, Evans (unpublished data) found that of the 19 landslide dams formed by the rock avalanches/rockslides over the threshold volume of $20 \times 10^6$ m$^3$, 9 (47%) failed (or partially failed) catastrophically, a rate of 1 per 11.3 years globally. In the twentieth-century data set, time to failure varied from 7 days (Diexi, China) to 3,435 days (Tsao-Ling 2, Taiwan) and released impoundment volumes ranged from 40 Mm$^3$ to 2 Gm$^3$.

Regional studies also provide data on the failure rates of landslide dams. Canuti et al. [22] report that 21% of the documented historical landslide dams in the northern Appenines failed within 1 day of their formation, 30% failed within a month, 58% within 1 year and only 3% of the lakes lasted for more than a century. However only 28% of the cases studied were the result of rockslides. In contrast, an analysis of a landslide dam inventory in New Zealand by Korup [157] found that only 37% of 232 landslide dams in the dataset had failed, the oldest landslide dam in the data set dating to the Pleistocene.

In NW Argentina only at one site out of 20 prehistoric rockslide dams is a lake presently preserved; 2 breached catastrophically, 3 rockslide dam basins filled completely with deposits and 14 cases were eroded, with the type and rate of erosion unknown. In northern Patagonia the ratio of eroded and stable prehistoric rockslide dams is markedly different. 41 rockslide dams exist in the Patagonia dataset, 18 dams impound lakes at present (44%) with no sign of any partial breach, while 5 breached partially and 18 (44%) were eroded without any indications of the form of erosion preserved (Chap. 5 by Hermanns et al., this volume).

These data suggest that landslide dam failure rates are between one third and one half the landslide dam formation rate.

Studies of prehistoric catastrophic rockslide dam failure have been based on an analysis of downstream sediments deposited during the breach event and have yielded important data on breaching mechanisms and downstream flood processes. Examples include the work of O'Connor [198], Wayne [279], Hermanns et al. [127]; (Chap. 5 by Hermanns et al., this volume), Wassmer et al. [277], Schneider et al. [234], Capra and Macias [24], Waythomas [280], and O'Connor and Beebee [199].

## *2.4 Outburst Floods from Rockslide-Dammed Lakes*

As noted above, the discharge of floods from catastrophic rockslide dam failures may be orders of magnitude greater than that of normal river floods [29, 53, 153, 170, 199, 200, 201] and have produced some of the largest floods on Earth (cf. [200]). Outburst volumes of some prehistoric rockslide-dammed lakes in the Himalayas, for example, may have exceeded 20 Gm$^3$ (Chap. 7 by Delaney and Evans, this volume) which would produce a peak discharge, estimated from Evans [74], in excess of 200,000 m$^3$/s.

Historical data on outburst volumes from landslide-dammed lakes (Table 1.2), indicate the largest outburst is that of the *Great Indus Flood* of 1841 (est. vol. 6.5 Gm$^3$ [1, 80, 183, 284]; Chap. 7 by Delaney and Evans, this volume). Peak discharge during the breach may have approached 115,000 m$^3$/s (Chap. 7 by

**Table 1.2** Major outbursts from rockslide-dammed lakes 1841–2010 (Data from Table 1.1; Chap. 7 by Delaney and Evans, this volume; Chap. 22 by Evans and Delaney, this volume)

| Rank | Location | Date of outburst | Impounded volume ($Mm^3$) | Outburst volume ($Mm^3$) |
|---|---|---|---|---|
| 1 | Indus River, Pakistan | 1841 | 6,500 | 6,500 |
| 2 | Yigong Zangbo, Tibet | 2000 | 2,015 | 2,015 |
| 3 | Rio Barrancas, Argentina | 1914 | 1,550 | 1,550 |
| 4 | Hunza River, Pakistan | 1858 | 805 | 805 |
| 5 | Tanggudong, China | 1967 | 680 | 640 |
| 6 | Condor-Sencca, Peru | 1945 | 721 | 664 |
| 7 | Mayunmarca, Peru | 1974 | 352 | 350 |

Delaney and Evans, this volume). The Yigong Zangbo outburst volume (Chap. 22 by Evans and Delaney, this volume; Table 1.2) exceeds that of the 1914 outburst from Lago Carri Lauquen, in the Rio Barrancas, Argentina Andes (1.55 G $m^3$; [127]). These three volumes exceed that of the greatest recorded outburst from an artificial dam [77].

## 2.5 Rockslide Dam Longevity

As suggested by the data in Section 2.3, not all landslide dams are unstable or have failure potential. Indeed the majority of rockslide dams that form do not fail. Some landslide dams impound lakes that persist in the landscape for millennia and become permanent features of the landscape (e.g. [182, 194, 283, 287]), whilst other rockslide dam sites have been utilized as foundations of dams constructed for water supply or hydroelectric power generation (e.g. [48, 122, 237]).

Rockslide dam longevity was examined by Schuster and Costa [239] and Costa and Schuster [53]. In the latter work they summarized the longevity of seventy-three landslide-dam failures (some of which were rockslide dams) resulting in their well-known, often-reproduced, and sometimes misinterpreted Fig. 1.6 ([53]; p. 1059). As Costa and Schuster [53] emphasise, their Fig. 1.6 summarises the longevity of landslide dams that actually failed.

In Fig. 1.8 we have plotted the longevity of 15 rockslide dams that have failed in the period 1841–2010 (Evans, unpublished data). The plot is suggestive of a robust power law with a negative exponent. 50% of the rockslide dams that failed did so within 75 days of their formation. Less than 20% of the rockslide dams that failed did so after 1 year and only 6% of the dams that failed did so after a period of 2 years after they formed.

Taking these observations together with our estimate of the rate of rockslide dam failure compared to the rate of rockslide dam formation, we can roughly state that between 17 and 25% of the rockslide dams that form globally (roughly 1 in 5) fail within 75 days of their formation.

Lacustrine deposits formed in rockslide-dammed lakes have been reported from many mountainous regions of the world (Figs. 1.9 and 1.10). Such deposits have been described in the Alps [11, 16], the Andes [114, 208, 267], the North American

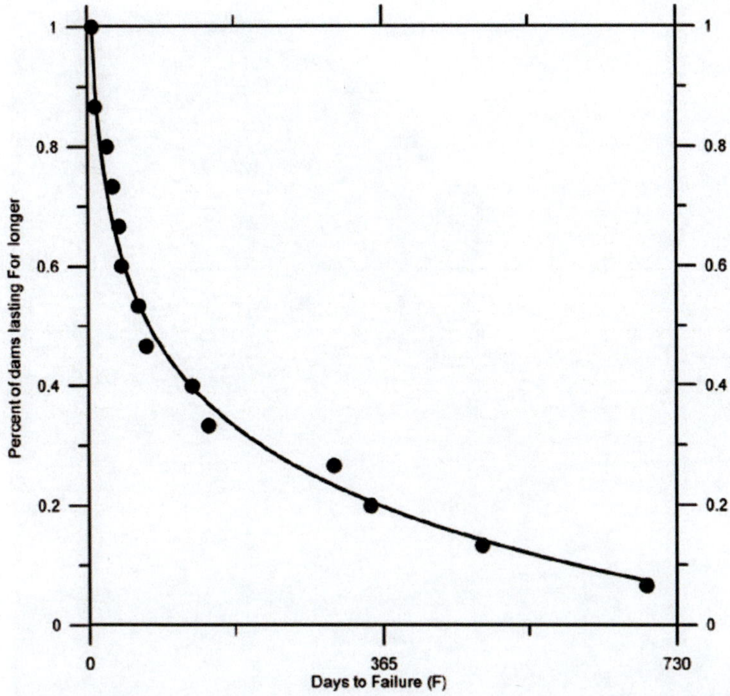


**Fig. 1.8** The longevity of 15 rockslide dams for which the dates of formation and failure are known in the period 1841–2010 (Evans, unpublished data)

**Fig. 1.9** Lacustrine sediments deposited in a lake formed when a flank-collapse of Mount Cayley volcano blocked the Squamish River, southwestern British Columbia. The impoundment was dated to ca. 4,800 y BP [19, 78]

**Fig. 1.10** Lacustrine sediments deposited in Saidmarreh Lake which was formed when the gigantic Saidmarreh rock avalanche blocked the Saidmarreh River, Zagros Mountains, Iran [118]. See Fig. 1.33. Photograph by N.J. Roberts, 2006

Cordillera [19, 78, 227, 230], the Zagros Mountains of Iran (Fig. 1.10; [118]), the Himalayas of Pakistan and India (e.g. [20, 85, 86, 87, 134, 135, 206, 214, 222]). These studies indicate that rockslide-dammed lakes can persist in the landscape to the extent that they become important sediment storage sites (e.g. [182]), and, as discussed below, impact on the geomorphic evolution of mountain landscapes (cf. [165, 166]).

The flat surface of sediments accumulated in former rockslide-dammed lakes form attractive sites for agriculture and settlement in rugged mountain terrain (e.g., [118]). The surface may be exposed when the lake finally fills with sediment (e.g. [109]), or when the lake is drained either by natural processes [16] or constructed drainage works after a long period of sedimentation [95].

Rockslide dams and rockslide-dammed lakes can become integrated into both the geomorphic and socio-economic development of the landscape even though the rockslide damming event occurred relatively recently. Alleghe Lake (Fig. 1.11) was formed by a rock avalanche (est. vol. 20 $Mm^3$) which blocked the Cordevole River

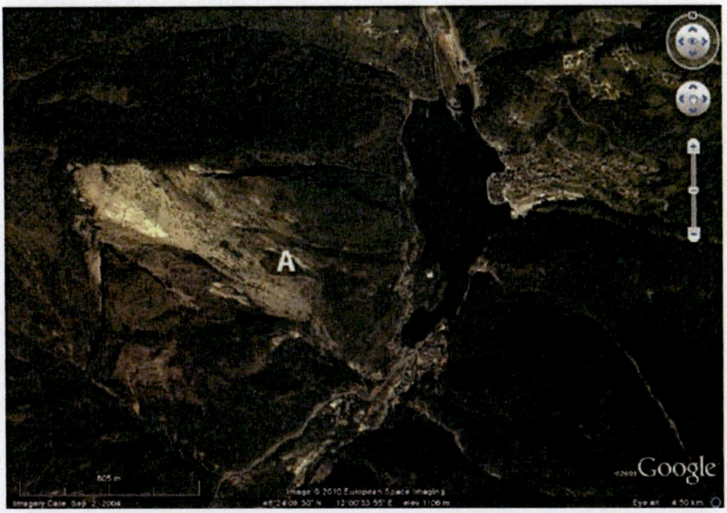


**Fig. 1.11** Google Earth image of Alleghe Lake in the Italian Dolomites (46° 23′ 56″N; 12° 00′ 47″E). Lake formed when rockslide (A) occurred on January 11, 1771 and blocked Cordevole River [68, 182]. Note location of community of Alleghe on rockslide debris

in the Italian Dolomites on January 11, 1771. A rockslide-dammed lake formed upstream flooding farmland and villages and still exists today (Fig. 1.11). Since 1771 the lake has decreased in areal extent, maximum depth and volume due to sedimentation [182]. Sediment deposition in the period 1933–2001 averaged 37,000 $m^3$/yr [182]. In 1938 the lake was developed for hydro power and a spillway was constructed to regulate lake level. The lake has also developed as a tourist resource. To preserve the aesthetc and engineering potential of the rockslide-dammed lake local authorities have resorted to removal of sediment by dredging [182].

# 3 Selected Case Histories of Rockslide Dam Formation and/or Failure 1894–2010

Nine case histories of rockslide dam formation and/or failure have been selected to illustrate a number of important characteristics of the formation and behaviour of rockslide dams.

## *3.1 Gohna Rockslide Dam Failure, India, 1894*

On September 6th, 1893, near the village of Gohna, two large rockslides occurred in close succession from the same source and blocked the Bireh Ganga River, forming part of the headwaters of the Ganges in the Garhwal Himalayas, northern India

[140, 256]. The damming occurred at 30°22′32″N; 79°29′51″E. The formation of the Gohna rockslide dam, the filling of Gohna Lake and its subsequent partial failure, as well as the meticulously planned warning and evacuation measures devised and instituted by government authorities, is described in some detail by Holland [140] and Glass [99] as well as in regular contemporary reports in 1894 in the journal *Nature* and *The Times* newspaper. These reports constitute the first detailed documentation of the formation and failure of a rockslide dam in the scientific literature and are worthy of review.

The rockslide took place in two main stages and involved the movement of a total volume of about 150 $Mm^3$ of limestone [123] from a steep high mountainside [282]. The crest of the rockslide dam was 305 m above the valley floor and the debris dam extended about 3 km along the thalweg. A lake formed upstream and reached an impounded volume of 470 $Mm^3$ before the breach of the rockslide dam took place in the night of August 25, 1894 [99, 175]. During the breach, 283 $Mm^3$ of water was released from the lake, leaving a residual water volume of 187 $Mm^3$. The drawdown of the lake was 119 m [99] and took place over a period of four and a half hours, equivalent to an average breach discharge of 17,479 $m^3/s$ (calculated from [99]). Heavy seepage had been noted at the toe of the rockslide dam before the end of July. The breach was apparently accelerated by a slip in the downstream face of the rockslide dam on August 9 [99, 175].

Based on detailed measurements of the rate of filling, officers of the Indian Geological Survey had predicted that overtopping and a possible breach would occur in August 1894 [140]. This allowed the planning and implementation of a detailed emergency response plan as outlined by Glass [99]. The downstream effects were substantial and included the destruction of buildings and property. However, there was no loss of life with the exception of a fakir and his family who were overwhelmed at the site of the rockslide dam itself after refusing an evacuation order.

The impact was mitigated by the warning system and response plan put in place by the Royal Engineers. The first component of the warning system was the marking out of the anticipated flood level by masonry pillars, at half-mile intervals along the valley approximately 60 m above ordinary flood level, above which was considered to be safe ground [99]. In addition, a number of bridges were dismantled in anticipation of the outburst flood. As part of the mitigation strategy, the lake level and the state of the rockslide dam were closely monitored.

The first signs of the breach were heralded by the sound of a bell. "*The warning was communicated through the whole of the threatened territory almost instantaneously by means of telegraphic messages, bonfires, rockets, the beating of drums . . . and the people immediately fled, with all their cattle . . . to higher ground*" (extract from Note to *Nature, v. 50, September 20, 1894, p. 501–502)*.

The passage of the flood is summarized in Table 1.3. It reached Srinagar, 116 km downstream from the rockslide dam in about 4 h, an average velocity of 8 m/s. Srinagar, with a population of about 2,000 people was totally destroyed by the flood [175]. Buildings destroyed included the Raja's palace, the police station and the Dak Bungalow (traveller's lodge). However, because of the warning and

**Table 1.3** Passage of outburst flood wave resulting from the partial breach of the Gohna rockslide dam, India, at about 23:30 h on August 25, 1894. Time denotes first arrival of flood. Elapsed time is time since breach. Rise is height above "ordinary flood level". Distance (D) is distance from the Gohna dam. Velocity (V) is D/Elapsed Time. Data derived from Glass [99]

| | Location | (D) Miles | (D) Klm | Local Time | Elapsed Time (h) | V (km/h) | V (m/s) | Rise (ft) | Rise (m) |
|---|---|---|---|---|---|---|---|---|---|
| 1 | Chamoli | 13 | 21 | 23:55:00 | 0.42 | 50.21 | 13.95 | 160 | 49 |
| 2 | Nandpryag | 20 | 32 | 00:17:00 | 0.78 | 41.09 | 11.41 | 113 | 34 |
| 3 | Karnpryag | 30 | 48 | 00:50:00 | 1.67 | 28.97 | 8.05 | 130 | 40 |
| 4 | Rudrapryag | 51 | 82 | 02:00:00 | 2.50 | 32.83 | 9.12 | 140 | 43 |
| 5 | Srinagar | 72 | 116 | 03:25:00 | 3.92 | 29.58 | 8.22 | 42 | 13 |
| 6 | Hardwar | 150 | 241 | 08:45:00 | 9.75 | 24.76 | 6.88 | 11 | 3 |

evacuation procedures which had been implemented no human casualties were incurred.

A second destructive outburst from Gohna Lake occurred in 1924 [109]. By the early 1970s the lake had completely silted up and had disappeared [282], largely due to the massive Alaknanda flood of 1970 that filled the lake with sediment and debris.

## *3.2 Rio Barrancas Rockslide Dam Failure and Flood in the Rio Colorado, Argentina, 1914*

On December 29, 1914, the failure of a prehistoric rockslide dam (est. rockslide volume 1.3 $Gm^3$; [127]; Fig. 1.12) took place on the Rio Barrancas in the Andes of Argentina at 36°30′56″S; 70°09′15″W [50, 102, 108, 127]. The event is of interest because it involved (1) the breaching of a rockslide dam that was formed at least 425 years earlier [127][1] (2) the outburst volume was in excess of 1 $Gm^3$, and (3) because the released waters of Lago Carri Lauquen formed a massive outburst flood which traveled about 900 km[2] down the Rio Colorado to the Atlantic Ocean [102, 108].

The 1914 Barrancas rockslide dam failure was initially described in a scientific report by Pablo Groeber published in 1916 [108]. Groeber had visited the landslide dam prior to, and few weeks after, the catastrophic failure and provided detailed drawings (Figs. 1.13 and 1.14) of the landslide dam and the landslide dammed lake in his publication. He correctly identified the dam to be of a landslide origin, gave a very rough first estimate of the water volume (2 $Gm^3$) released by the breach of the rockslide dam, and described the morphology of the dam breach in some detail.

[1]Cosmogenic surface age dating of the Barrancas rockslide indicated an age of 2.1 ka [98].
[2]This distance is incorrectly given by [102] as 1250 km.

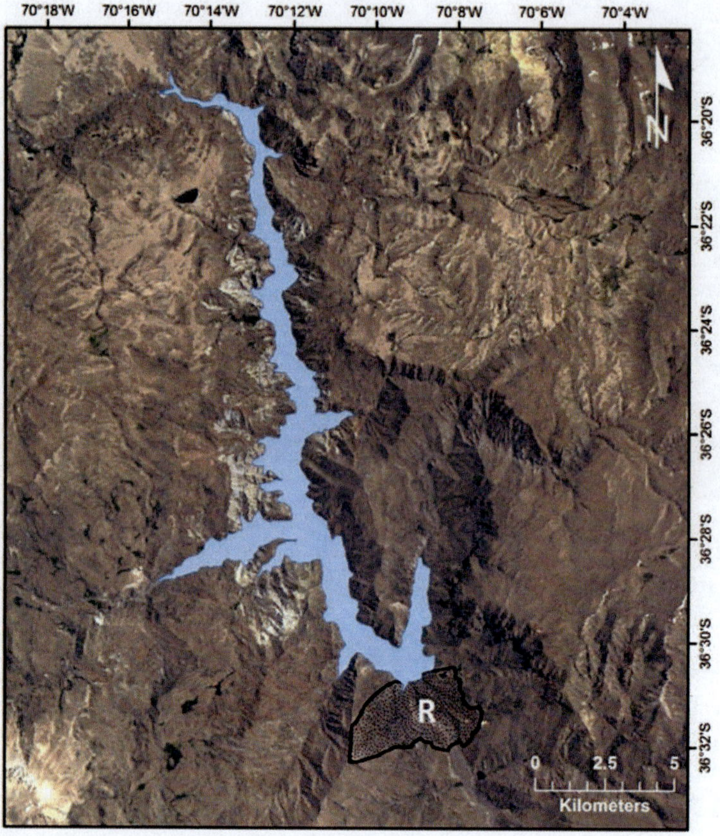


**Fig. 1.12** LANDSAT-7 image obtained in 2001 of the Barrancas rockslide (R), Argentina, showing the outline of rockslide dammed lake (Laguna de Carri Lauquen) that breached in 1914. The pool elevation of the reservoir was taken from Gonzalez Diaz et al. ([102], p. 468) as 1,500 m a.s.l. Elevation based on SRTM-3 data. The rockslide dam impounded 1.55 $Gm^3$ of water

Groeber [108] observed the loss of agricultural land downstream of the breach in areas "*where there was once fertile field there is today a rock-strewn wasteland*" (Fig. 1.10). He also predicted that the aggradation in the valley bottom below the rockslide dam would impede any agriculture in the area for decades afterwards, which unfortunately has turned out to be the case.

The volume released during the 1914 breach was recalculated in later work by (Coleman [45] – 1.8 $Gm^3$) and Hermanns et al. ([127] – 1.55 $Gm^3$). The drawdown following the breach (Fig. 1.9) was estimated at about 95 m [108] and measured in the field as 85 m by Hermanns et al. [127].

González Díaz, et al. [102] summarized all published reports on the downstream effects of the outburst flood within the Barrancas and Colorado valleys giving some insight into the effects of the large magnitude outburst and the behaviour of the flood

**Fig. 1.13** Drawing of Lago Carri Lauquén dammed by the deposits of a rock avalanche before breaching. This view is downstream towards SE with the dam crest in the mid-foreground and the break-away scarp of the rock avalanche in the centre-right background. S marks the locality of "salida" ("outlet") of the Barrancas river (reproduced from [108])

**Fig. 1.14** Drawing of a relict Carri Lauquén lake after breach of the landslide dam reproduced from Groeber [108]. In the drawing a paleoshoreline is clearly visible as a horizontal white line that indicates the water level of the rockslide-dammed lake prior to failure. Water level dropped 85 m with respect to this paloeshoreline during the outburst [127]

wave over a distance of 900 km to the Atlantic (Fig. 1.14). In addition, interviews with the last surviving witnesses of the event gave important information on the failure mechanism of the rockslide dam [102].

From Groeber's report [108] it is not evident whether the landslide dam was overtopped by the Barrancas river (a) for an extended time prior to failure, (b) sporadically or episodically overtopped or (c) overtopped only shortly before catastrophic failure. In one of his drawings (Fig. 1.13) Groeber marked the location of "water escape" on the dam crest ("S" in Fig. 1.13) indicating that the water escape occurred at that place. However, whether it occurred by overflow or seepage remains unclear in his report.

Better information came from the last surviving witness of the disaster from the area close to the dam, a Ms. Avelina Canale (age 93 years). In an interview with Gonzalez-Diaz et al. [102] she reported that the lake level had risen substantially and that overtopping started to take place prior to failure. She also stated that the 21.5 km long lake had always maintained its level by seepage through the rockslide dam.

However, immediately prior to the catastrophic failure there was strong concern amongst the local people in the valley about the stability of the dam to the extent that a special watchman was employed to give a warning when a certain lake level was reached [102]. On the day of the dam failure no warning was given because the watchman was incapacitated with drink. The breach reportedly started at 4 p.m. and the first narrow reach of the valley 11.6 km downriver from the dam filled up with water up to a height of 32 m above normal river level [102]. Striations from boulder transport during the flood are still observable on the valley wall at this location where the valley has a width of ~300 m.

Four hours after initiation of failure, the 17 m high outburst flood wave arrived at the village of Barrancas where the valley width is 780 m. The distance between the landslide dam and the village of Barrancas is 49.6 km (Fig. 1.15). Considering

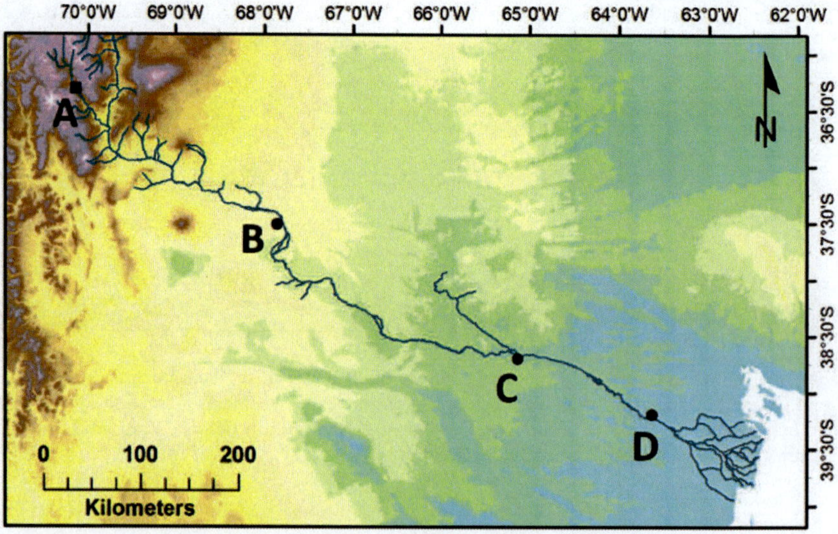


**Fig. 1.15** The Colorado River valley showing the path of the outburst flood from the Rio Barrancas rockslide dam (**A**) to the Atlantic Ocean (*bottom right*), a distance of about 900 km. Breach of Barrancas rockslide dam occurred on December 29, 1914 at about 16:00 h. Flood wave arrived at Colonia 25 de Mayo (**B**) on December 30, 1914 at 14:00 h (7 m high), at Pichi Mahuida (**C**) on January 3, 1915 (3.5 m high). At Buena Parada (**D**) the river rose about 7 m between January 6–8, 1915 (summarized after [102])

a time lapse for breach development between the initiation of breach and formation of the large wave this gives a minimum flow velocity of 3.4 m/s and estimated peak discharges of ~33,000 $m^3$/s at km 11.6 down river and of ~45,600 $m^3$/s at Barrancas village. This discharges compare to the peak discharge during breaching of 48,721 $m^3$/s calculated using the emprical formula of Evans [74].

14.6 km below Barrancas village (64.2 km downstream from the rockslide dam) the confluence of Barrancas River with Rio Grande valley occurs forming the Colorado River. Hermanns et al. [127] mapped terraces deposited by the outburst flood and indicated average terrace thicknesses between landslide dam and this confluence. The volume deposited in these first 64.2 river km is estimated to be 160 $Mm^3$. A smaller volume of debris (120 $Mm^3$) was eroded from the rockslide dam during the breach event [127]. Taking into account that further downstream several m thick terraces of sediment were also deposited, it becomes clear that the mass transport and subsequent deposition of sediment during outburst flood events, such as the 1914 Barrancas outburst, impact strongly on the long term sediment yield of a river.

The passage of the flood and its effects downstream in the remaining distance to the Atlantic Ocean is summarized in detail by [102]. The flood wave arrived at Colonia 25 de Mayo (Fig. 1.15), ~300 km downriver of the confluence of Barrancas and Grande river the next day (December 30) at 14 h as a 10 m high wave, an average velocity of 3.8 m/s from the breach. The flood arrived at Pichi Mahuida (ca. 650 km from the breach) as a flood wave 3.5 m high on January 3, 1915, an average velocity of ca. 3 m/s from the breached Barrancas rockslide dam. Further downstream flooding was more prolonged in time and several villages remained flooded for several days. For example, the village of Buena Parada (D in Fig. 1.15) had its peak flood with the river 7 m above normal on January 6 to 8. Even closer to the Atlantic Ocean the flooding continued to January 25 [102].

Along the entire Barrancas and Colorado river 175 persons were found dead and 118 disappeared [102]; however the informal number might be higher. 100 km of railroad, a large number of houses and a large area of fertile agricultural land were destroyed in the outburst flood [102]. The total economic costs of the disaster were never calculated.

Rockslide dam formation is interpreted to be related to seismic activity. A concentration of large landslides exists in this area along active faults [50].

Most reports associate the increase in lake level which lead to dam breach with a very wet winter preceding the austral summer 1914/1915 [45, 108, 127]. Climatic data from 5 Argentine weather stations surrounding that site indicate that the winter indeed was wetter than average (Fig. 1.16). However such precipitation accumulations have a return interval of 5–17 years in that area. With respect to temperature, data from the same weather stations indicate that the austral summer 1913/1914 was the hottest summer (Fig. 1.16) in the 145 year long record (1858–2003) suggesting that increased inflow of glacial melt water in the preceding summer in combination with higher than average precipitation in the austral winter 1914 caused very high water levels and the rockslide-dammed lake to overtop.

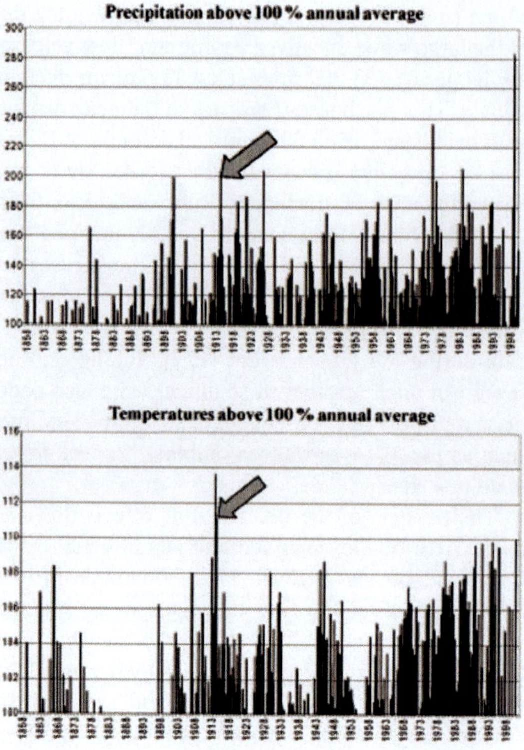


**Fig. 1.16** The 1914 Barrancas rockslide dam outburst flood – precipitation and temperature above 100% annual average from 5 weather stations from Central to Patagonian Argentina (1854–1998) showing that the austral summer 1913/1914 was the warmest in the Argentine record (*arrow*) while the winter 1914 was indeed wetter than normal (*arrow*) but the anomaly is within a recurrence interval of 5–17 years for these 5 stations. (Year measured as from July to June to fit to austral summer. Data provided by Climatic Research Unit, University of East Anglia, Norwich, UK)

## *3.3 Cerro Condor-Sencca Rockslide Dam, Mantaro River, Peru, 1945*

In the Central Andes of Peru, the Amazon-draining Mantaro River was dammed [253] by a rockslide (12°48′04″S; 74°17′11″W) on August 16, 1945, approximately 60 km downstream of the 1974 blockage described below. The volume of the rockslide was 5.35 $Mm^3$ of which 3.5 $Mm^3$ formed the rockslide dam in the Mantaro valley. Snow (253) reports that the volume of water impounded behind the rockslide dam reached 300 $Mm^3$ before the dam was overtopped and catastrophic breaching occurred 73 days after impoundment [253]. Snow [253] reports that the ensuing flood was 50 times greater than the annual flood of the Mantaro River.

A re-analysis of the extent and volume of the 1945 Mantaro rockslide-dammed lake (Delaney, unpublished data) suggests that the data published by Snow [253] seriously underestimates the magnitude of the Condor-Sencca rockslide dammed lake. We note that the rockslide occurred about 9 km downstream of the confluence of the Mantaro and the Huarpa, a major tributary (Fig. 1.17). The rockslide-dammed lake had a maximum pool elevation of 2,216 m a.s.l. (data derived from Snow [253]). At this elevation the lake extended not only up the main Mantaro valley but also up

**Fig. 1.17** LANDSAT-7 image showing location of 1945 Condor-Sencca rockslide dam (C at top right) and extent of rockslide-dammed lake at pool elevation of 2,216 m a.s.l. (data derived from [253]). Note the lake extends southward into the Huarpa valley. Volume of the lake at this maximum extent is calculated to be 721 $Mm^3$, more than twice the volume estimated by Snow [253]. Elevation data derived from SRTM-3 DEM

into the tributary valley of the Huarpa (Fig. 1.17) covering a total area of 19.5 $km^2$. The SRTM-3 DEM indicates that the volume of the rockslide-dammed lake at this elevation had a volume of 721 $Mm^3$. The drawdown created by the October 28 outburst of the lake was about 56 m, creating a residual lake of 57 $Mm^3$. Thus the outburst volume was in the order of 664 $Mm^3$ (Delaney, unpublished data). Use of Evans' empirical formula [74] for estimating peak discharge during landslide dam breaching indicates a peak discharge of about 34,000 $m^3/s$. This compares favourably with the estimate of peak discharge during breaching of 35,300 $m^3/s$ reported by Snow [253].

At Luichos, 90 km downstream in the Mantaro Valley, the passage of the outburst flood resulted in a river level 26 m above the normal stage [253].

## *3.4 Mayunmarca Rockslide Dam, Mantaro River, Peru, 1974*

On April 25, 1974 one of the largest landslides (est. vol. 1.6 $Gm^3$) of the twentieth century occurred in the Peruvian Andes at 12°36′55″S; 74°39′25″W [145, 155, 170, 263]; (Chap. 3 by Bonnard, this volume). The rockslide caused the death of

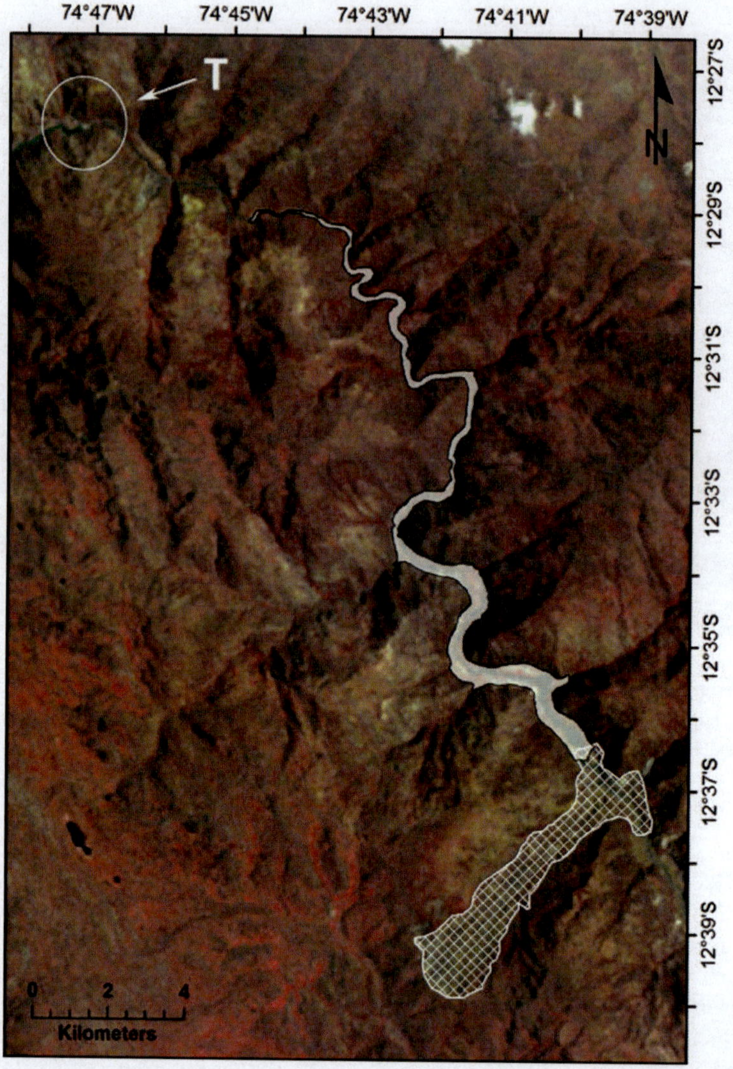


**Fig. 1.18** LANDSAT-5 image obtained in 1980 showing outline of 1974 Mayunmarca rockslide and the rockslide-dammed lake at its full pool elevation. Extent of the lake is based on SRTM-3 digital terrain data using a maximum pool elevation of 2,625 m a.s.l [145]. The volume of the rockslide-dammed lake at this pool elevation is calculated to be 352 $Mm^3$ (Delaney, unpublished data). T is location of the Tablachacca hydroelectric dam mentioned in the text

680 persons [84] and blocked the Amazon-draining Mantaro River, forming a lake upstream (Fig. 1.18). The 1974 rockslide dam was located 60 km upstream of the 1945 blockage described above. The sinuous impoundment (Fig. 1.18) extended for 31 km and reached a reported estimated volume of 670 $Mm^3$ [145, 155, 170, 263] before overtopping after 42 days on June 6, 1974. The rockslide-dammed lake extended to within 4 km of the Tablachaca hydroelectric dam which had been

constructed in 1967–1971. In addition, the community of Pururo was submerged by the impounded waters.

When the rockslide dam was overtopped on June 6, significant erosion of the debris did not immediately occur. However, at about noon on June 8 (44 days after impoundment), catastrophic breaching was initiated and accelerated during that afternoon. A peak discharge of 13,700 $m^3/s$ was reached [220]. This discharge is more than 8 times the discharge of the peak recorded natural flood in the Rio Mantaro [170]. 170 $Mm^3$ of water reportedly remained in a substantial residual lake. Thus the total reported outburst volume was 500 $Mm^3$ which drained in about 24 h. This is equivalent to a mean breach discharge of approximately 11,500 $m^3/s$.

Timely evacuation of downstream communities resulted in no loss of life in the outburst flood. However, much damage to infrastructure and communities occurred up to 50 km downstream [170].

However, the published volume of the 1974 rockslide-dammed lake appears to be in substantial error. Using SRTM-3 digital terrain data we re-analysed the dimensions and volume of the 1974 Mayunmarca rockslide-dammed lake (Fig. 1.18 – Delaney, unpublished data). Using a maximum pool elevation of 2,625 m a.s.l [145] in ArcGIS, we calculate that the lake had an area of 7.32 $km^2$ and a volume of 352 $Mm^3$, almost half the previously reported volume, at its maximum extent (Fig. 1.18). We note that the rockslide dammed lake was filled mainly by discharge from the Tablachacca Dam which was estimated at 74 $m^3/s$ during the period of lake filling [145]. This suggests a minimum volume of impounded water of 281 $Mm^3$.

During the outburst, the drawdown of the rockslide dammed lake was approximately 90 m [145] indicating a pool elevation of the post-outburst residual lake at 2,535 m a.s.l. Again using a DEM generated with SRTM-3 data this indicates an outburst volume of about 350 $Mm^3$ and a residual lake volume of ca. 2 $Mm^3$. It is this data that is entered in Table 1.1.

## *3.5 La Josefina Rockslide Dam, Ecuador, 1993*

On March 29, 1993 a rockslide (est. vol. 25 $Mm^3$) occurred on the Amazon-draining Rio Paute at 2°51′08″S, 78° 50′44″W in the Andes of Ecuador, damming its flow [21, 114, 218, 243, 289]; (Chap. 14 by Plaza et al. this volume). The crest of the rockslide dam was 120 m above the river channel and the dam had a length of 1,100 m along the thalweg of the Rio Paute.

A lake formed upstream and at its maximum extent was about 10 km long and contained, at its maximum pool elevation, 210 $Mm^3$ of water. The impoundment was partially breached 33 days later (May 1st) following engineering activities on the surface of the rockslide dam. These activities were designed to reduce further upstream flooding by arresting the rise of the lake and prevent a possible catastrophic breach by draining the lake through a spillway channel. In 14 days, 160,000 $m^3$ of debris were removed, lowering the rockslide dam crest. The resulting channel was 6 m wide at its base, 18 m deep, and 407 m in length but the channel did not perform well due to instability of its steep sides, the blocky nature of the debris and was frequently blocked. Retrogressive erosion was eventually initiated at the downstream end of the overflow channel and catastrophic breaching resulted. During the breach about 175 $Mm^3$ of water was released into the Rio Paute in about 8 h, with a peak discharge of approximately 10,000 $m^3/s$. About 25 $Mm^3$ of water remained in a residual lake after a drawdown of 40 m [243].

The catastrophic breach devastated the Paute valley floor downstream from the rockslide dam. Damage included the partial destruction of the town of Paute, the destruction of 4 concrete road bridges, 10 km of roads along the Paute valley and 700 buildings. Total losses were estimated to be in the order of $192 M USD [243]. No life loss ensued because of the evacuation of the population from valley floor sites.

Upstream of the dam, numerous buildings (including a thermal generating plant) were flooded in the impoundment [289]. The rockslide dam formed 77 km upstream from the Amaluza Dam and the 1,200 MW Paute hydroelectric plant, which at the time produced about 70% of Ecuador's electrical power [243]. The Amaluza reservoir was drawn down by 31 m to accommodate 51 $Mm^3$ of water in the expected outburst flood. However the reservoir could not contain all the outburst volume and the excess was released through the spillway with a maximum discharge of 4,000 $m^3/s$. The flood wave propagated further downstream causing severe erosion and damage to its entry into the Nomangoza River, more than 120 km downstream from the rockslide dam [21]. In the Paute hydroelectric plant, turbines were damaged by excessive concentration of suspended solids which resulted in nationwide power cuts in Ecuador. Further, sedimentation in the Amaluza reservoir dramatically decreased its storage capacity [114].

The Rio Paute-La Josefina rockslide dam case is notable because (1) the attempted engineered draining of the impoundment, (2) the extensive damage due to both upstream and downstream flooding in the Rio Paute valley, and (3) the impact of the outburst flood on the operation and capacity of the Amaluza Reservoir and Paute hydroelectric plant which in 2006 produced 60% of Ecuador's electrical power.

## *3.6 Tsao-Ling Rockslide Dam, Taiwan, 1999*

On September 21, 1999 a rockslide triggered by the Chi-Chi earthquake ($M = 7.3$) dammed the Chingshuichi River in central Taiwan (23°34′35″N; 120°40′19″E) (Chap. 23 by Chang et al., this volume). The rockslide was the fifth to have occurred on the Tsao-Ling dipslope since 1862 (Table 1.4). All of these rockslides dammed the Chingshui River [32, 35, 40, 141, 143]; (Chap. 23 by Chang et al., this volume). Breaches of the rockslide dams occurred in 1898, 1951, 1979, and controlled overflow took place following renewed impoundment in 1999 [141]; (Table 1.4).

The 1999 Tsao-Ling rockslide had a volume of 125 $Mm^3$ [34, 40] and the debris dam ranged in height between 50 and 150 m [171]. An extensive lake formed behind the dam in a basin with a maximum storage capacity of 46 $Mm^3$ [171]; the lake filled to a maximum volume of 42 $Mm^3$ during Typhoon Bilis (August 2000). Controlled overflow was established by December 1999 through a constructed armoured spillway which was excavated over the debris [171]. Check dams were

**Table 1.4** Rockslide dams on the Chingshui River formed by rockslides on the Tsao-Ling dip slope, Central Taiwan 1862–1999. The dip slope dips at 14° and consists of Pliocene sandstones and shales (data from [35, 40, 141, 143, 171])

| Date of rockslide | Trigger | Volume and dam characteristics | Volume of rockslide-dammed lake ($Mm^3$) | Fate of rockslide dammed lake | Remarks |
|---|---|---|---|---|---|
| 1862-06-06 | Earthquake (M = 6–7) | Unknown | Unknown | Breached in 1898 | |
| 1941-12-17 | Earthquake (M = 7.1) | 150 $Mm^3$; dam 140 m in height | | Breached by overtopping in 1951 during spillway construction following heavy rains; 120 $Mm^3$ of water released | |
| 1942-08-10 | Rainfall | 200 $Mm^3$ – enlarged 1941 blockage increasing dam height to 217 m. Lake contained 157 $Mm^3$ | 120 | Breached by overtopping in 1951 during spillway construction following heavy rains; 120 $Mm^3$ of water released | 137 deaths amongst spillway workers and 1,200 houses destroyed downstream |
| 1979-08-15 | Rainfall | 5 $Mm^3$ – dam 90 m in height. Lake contained 40 $Mm^3$ | 40 | Breached in 1979 | Downstream villages evacuated |
| 1999-09-21 | Earthquake (M = 7.6) | 125 $Mm^3$ | 42 | Lake of 42 $Mm^3$ formed; controlled overflow established in December 1999. Lake filled in after Typhoon Mindulle August 2004. | Controlled overflow through excavated armoured spillway in December 1999; volume of lake reduced by sedimentation and erosion-induced outlet lowering. |

constructed downstream to reduce movement of sediment originating in the erosion of the rockslide dam.

The volume of the 1999 rockslide-dammed lake gradually decreased as a result of sedimentation in the lake resulting from a succession of typhoons from 2000 to 2004. Also during the typhoons, overflow discharge increased causing incision and widening of the spillway [141], lowering the outlet elevation of the lake. The combined effect of sedimentation and outlet lowering successively reduced the volume of the lake. Following Typhoon Toraji (July 2001) the lake volume was reduced to 20 $Mm^3$; following Typhoon Nari (September 2001) the lake volume had reduced to 3 $Mm^3$. By August 2004, following Typhoon Mindulle (July 2004), the 1999 Tsao-Ling rockslide-dammed lake had almost completely filled with sediments from upstream sources in the Chingshui River [141].

## 3.7 *Yigong Zangbo River, Tibet, China, 2000*

On April 9, 2000 a large-scale rock avalanche occurred in the mountains of southeastern Tibet at 30°10′58″N, 94°55′53″E (Fig. 1.19). The rock avalanche, which had entrained material in its path increasing the initial volume to an estimated

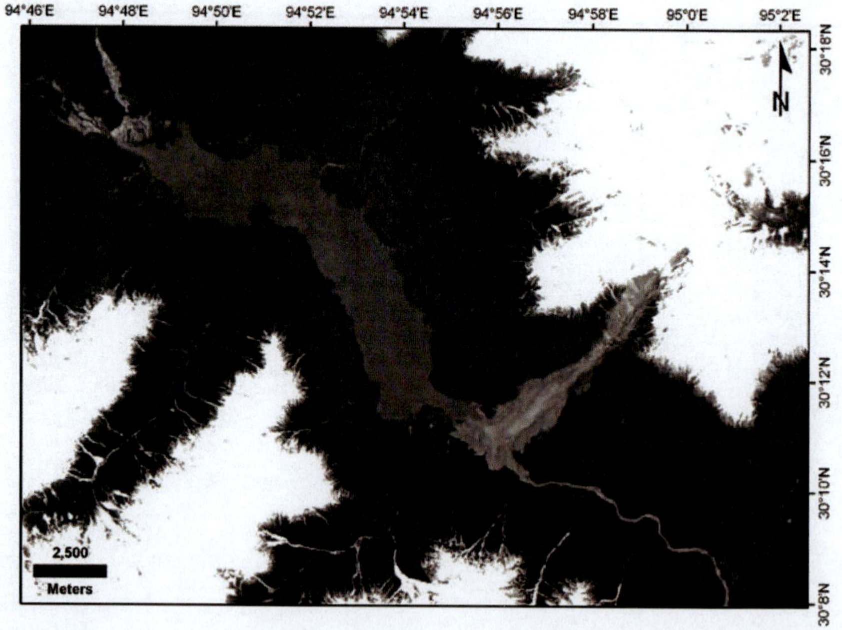


**Fig. 1.19** LANDSAT-7 image obtained on May 4, 2000 (25 days after impoundment) showing valley blocking Yigong landslide and landslide-dammed Yigong Lake forming upstream. Volume of lake when image obtained is approximately 313 $Mm^3$, compared to its final volume of ca. 2 $Gm^3$ when breaching occurred on June 10, 2000 (Chap. 22 by Evans and Delaney, this volume)

100 $Mm^3$ of debris (Chap. 22 by Evans and Delaney, this volume), blocked the Yigong Zangbo River [112, 244, 291, 293] and the outlet to Yigong Lake which had been formed by a previous rock avalanche dam in 1900. The debris spread out in the Yigong Valley and formed an irregular dam, 60–100 m in height, which led to a rapid increase in volume of Yigong Lake. The upstream flooding threatened several villages, a school and the famous Yigong Tea Garden with a total of potentially affected people of 4,000. The landslide debris consisted of large blocks of rock set in matrix that included gravel size particles as well as finer material.

On May 3 construction work on a channelised overflow channel was begun [244]. The channel was designed to be 1,000 m long, 24 m deep, and 150 m wide at the top and 20 m wide at its base. On June 3 the channel was completed after the removal of 1.36 $Mm^3$ of material. On June 10 water broke through the channel initiating a catastrophic breach of the rockslide dam, 62 days after formation. By June 12 an estimated 2 $Gm^3$ of water had discharged from the rockslide-dammed lake (Chap. 7 by Evans and Delaney, this volume).

The breached volume entered the Yigong Zangbo River and created a massive flood. At 2:50 am on June 11, the maximum instantaneous discharge at the Tongmai Bridge, 17 km downstream was 120,000 $m^3/s$ and the flood water level was 32 m above the bridge deck [244]. The Tongmai Bridge, five other bridges, and a 4.5 km section of the Qinghai-Tibetan were destroyed by the flooding [244, 291]. The flood wave ran about 216 km to the China-India border and then a further 284 km through Arunachal Pradesh to the flood plain of the Brahmaputra in Assam, India, a total distance of about 500 km. In India 20 bridges, including a large steel suspension bridge over the Siang River were destroyed and 50,000 people were rendered homeless. At Pasighat in Arunacal Pradesh, India, 454 km downstream from the rockslide dam, the flashflood involved a 5 m rise in the river stage at the Siang River gauging station [266]. Damage in India alone was estimated at $23 M [292].

In the historical record (Table 1.2) the outburst volume of the 2000 Yigong Zangbo rockslide dammed lake is only superseded by that of the 1841 outburst of the rockslide-dammed lake on the Indus River, Pakistan (Chap. 7 by Delaney and Evans, this volume).

## *3.8 Tangjiashan Lake, Sichuan, China, 2008*

As noted above, many landslides triggered by the May 12, 2008, M8.0 Wenchuan earthquake formed landslide dams [38, 41, 56, 177, 276, 285, 290]; (Chap. 10 by Cui et al., this volume) thus creating numerous landslide-dammed lakes (known as "barrier lakes" in the Sichuan context) (Fig. 1.7) in the narrow deep valleys that drain the Tibetan Plateau (cf. [205]).

The largest barrier lake created was Tangjiashan Lake [36, 55] impounded by an earthquake-triggered rockslide at 31° 50′38″N, 104° 25′55″E, 4 km upstream of Beichuan county town on the Tongkou River, Sichuan Province (Fig. 1.20). The damming rockslide was ca. 24 $Mm^3$ in volume and buried the village of Yuanheba

**Fig. 1.20** Aerial view of Tangjiashan rockslide dam and lake forming upstream on May 28, 2010 (Day 16 of impoundment). Lake level was approximately 730 m a.s.l. and the lake volume was approaching 200 $Mm^3$, approximately 81% of the maximum volume reached before controlled breaching on June 10, 2008 (photograph by Xinhua News Agency, China)

resulting in 84 deaths [177]. The height of the dam was approximately 100 m [142, 176] and the maximum possible reservoir storage capacity, determined by the elevation of the crest of the rockslide dam at 752.2 m a.s.l., was 320 $Mm^3$ [176]. The filling of Tangjiashan Lake is summarised in Fig. 1.21. Tangjiashan Lake posed a significant hazard to communities downstream [176], including the possible flooding of parts of Mianyang, Sichuan's second largest city located 80 km down river. The maximum potential dam break flood was calculated by N. Lui et al. [176] as 46,000 $m^3/s$.

Beginning on May 25 (Day 13 of the impoundment), mitigation works were initiated to excavate a spillway across the debris both to reduce the maximum possible volume in the lake basin and to initiate a controlled drainage, which was in effect a controlled partial breaching, of the rockslide-dammed lake [176]. The 13 m deep spillway had a trapezoid cross section with channel side slopes of 1:1.5 and was 8 m wide at the base [176]. The bottom of the 475 m long discharge channel was set at 740.7 m a.s.l. at its upstream limit. Construction techniques combined excavation and blasting and involved over 1,000 workers and 26 pieces of heavy machinery [176]. The work, which took 7 days (completed on June 1-Day 20) removed 135,000 $m^3$ of debris from the rockslide dam (Chap. 10 by Cui et al., this volume).

The water level in the barrier lake exceeded the elevation of the entrance of the excavated spillway at 0700 h on June 8 (Day 27) and began to overflow. In the first 24 h approximately 5.38 $Mm^3$ of water was released. However, in the first period of overflow, lake level continued to increase because inflow exceed outflow. The peak pool elevation (743.1 m a.s.l.) was reached on June 10 (Day 29) equivalent to a maximum lake volume of 247.4 $Mm^3$, 73 $Mm^3$ (23%) less than the maximum

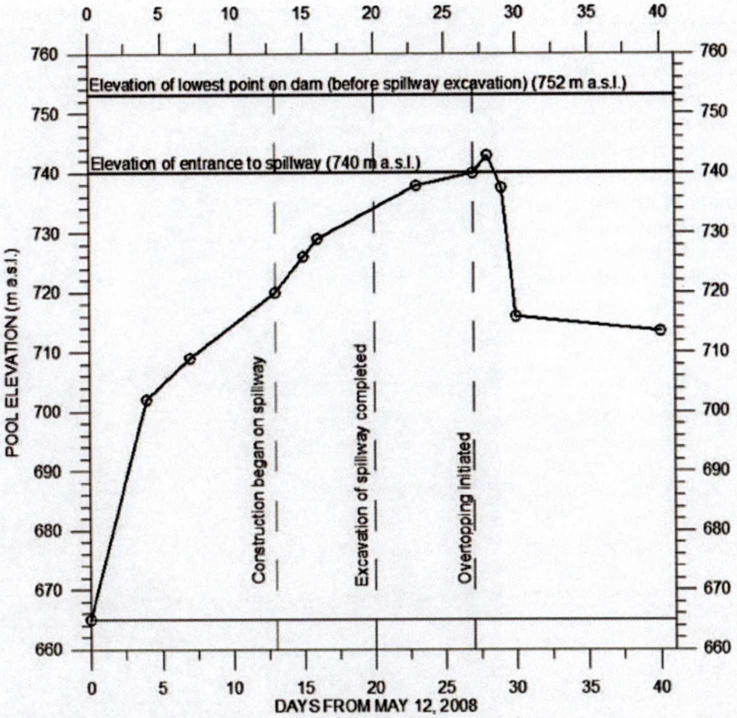


**Fig. 1.21** Filling curve for Tangjiashan rockslide-dammed lake showing rise in pool level after damming on May 12, 2008. Excavation of spillway lowered maximum lake level by 12 m, equivalent to 73 $Mm^3$ of storage (based on data in [177] and [176])

possible volume of the lake before the excavation of the spillway. Breaching flow was initiated early on June 10 (Fig. 1.22) when significant erosion of the spillway began. Lake level began to drop during the day as flow increased in the spillway (Fig. 1.21). A peak discharge of 6,500 $m^3/s$ was reached during the drainage event on June 10 and a total of 160 $Mm^3$ of water (65% of the maximum lake volume) was released from the lake by June 11 leaving a residual lake of ca. 90 $Mm^3$ (Fig. 1.21). During the breach the spillway was considerably enlarged by widening and deepening (Fig. 1.22). The post-drainage spillway channel was 100 m wide in places and scour depth was between 40 and 60 m [176]. We note that the post-partial breaching lake level was 713 m a.s.l. on June 21, 27 m lower than the finished spillway elevation.

## *3.9 Hunza Rockslide-Dammed Lake, Pakistan 2010*

A rockslide (the Attabad rockslide) blocked the Hunza River, northern Pakistan, on January 4, 2010 at a location (34° 18′ 24″ N/74° 49′ 17″E) only 2.5 km upstream of a large-scale rockslide which occurred in 1858 and blocked the Hunza River

**Fig. 1.22** Aerial view of flow in Tangjiashan spillway at about 9 a.m. June 10, 2008, during controlled partial drainage/breaching of rockslide dam. Discharge is approximately 2,000 $m^3/s$. Note effects of spillway erosion and construction machinery for scale (photograph by Xinhua News Agency, China)

[151, 183, 246]; (Chap. 7 by Delaney and Evans, this volume). The 1858 rockslide occurred on the south side of the Hunza valley and impounded a lake ca. 800 $Mm^3$ in volume (Chap. 7 by Delaney and Evans, this volume). In August 1858 the rockslide dam failed and released the impounded waters into the Hunza-Gilgit-Indus river system giving rise to the *Second Great Indus Flood* of Mason [183] (Chap. 7 by Delaney and Evans, this volume).

The 2010 rockslide originated on the north side of the Hunza valley and the volume of the debris is estimated to be in the order of 55 $Mm^3$ [151]. The rockslide mobilised valley fill sediments from the valley floor of the Hunza that liquefied and travelled as a rapid mudflow over the valley floor up to a distance of 2.7 km downstream. Part of the mudflow debris swept up the valley side overwhelming part of a village downstream from the rockslide, killing about 20 people (Figs. 1.1, 1.23).

The debris immediately blocked the flow of the Hunza River and a rockslide dammed lake began to form upstream (Fig. 1.23). The impoundment continued to fill until May 29, a period of 143 days, until, as discussed below, overtopping of the debris occurred (Chap. 7 by Delaney and Evans, this volume). As the waters of the lake rose upstream they flooded numerous villages, large tracts of agricultural land

**Fig. 1.23** EO-1 satellite image obtained on March 16, 2010 of rockslide-dammed lake impounded by Attabad rockslide that occurred on January 4, 2010, Hunza Valley, northern Pakistan. On this day, lake level was ca. 2,392 m a.s.l. and the estimated volume of the lake was ca. 135 $Mm^3$

adjacent to the Hunza, and 22 km of the Karakoram Highway linking Pakistan and China disrupting road travel and commercial traffic between the two countries, as well as disrupting the lives of the inhabitants of the Hunza Valley and region. To achieve an eventual controlled overflow, Pakistan authorities began the excavation of a spillway over the debris on January 29 (Day 25) completing it around May 15 (Day 128) (Fig. 1.24). The excavation of the spillway lowered the effective crest of the dam by around 15 m.

**Fig. 1.24** Excavation of spillway in the debris of the Attabad rockslide dam, Hunza valley, northern Pakistan. The Hunza rockslide dammed lake is visible in the background. Photograph taken on April 18 (Day 102 of impoundment) when the lake level was at about 2,405 m a.s.l., 30 m below the eventual overflow elevation (*Pamir Times* photograph)

The elevation of the Hunza River at the blockage site is estimated at 2,325 m a.s.l. (Chap. 7 by Delaney and Evans, this volume). The minimum height of the dam is estimated as ca. 120 m for an overflow crest elevation of about 2,445 m.a.s.l. (Chap. 7 by Delaney and Evans, this volume). A filling curve was approximated for the Hunza rockslide dammed lake (Fig. 1.25; Chap. 7 by Delaney and Evans, this volume).

The excavation activity resulted in a spillway elevation of 2,433 m a.s.l. At this elevation, we estimate from our SRTM-3 digital terrain model that the rockslide-dammed lake had a volume of ca. 430 $Mm^3$, an area of 9.74 $km^2$, and a length of about 22 km (Chap. 7 by Delaney and Evans, this volume). The volume of the 2010 rockslide-dammed lake was thus approximately half that of the 1858 lake.

The Hunza rockslide-dammed lake overtopped on May 29 (Day 143) [151] by overflow through the excavated spillway. Despite this the lake level continued to rise a further 3 m (an equivalent additional volume of ca. 30 $Mm^3$) in the 5 days up to June 3, since lake inflow (estimated from filling data and the SRTM-3 digital terrain model as ca. 70 $m^3/s$) exceeded lake outflow. By June 5, however, the lake level had stabilised (Fig. 1.25) as discharge in the spillway increased to about 95 $m^3/s$ and seepage through the dam itself had continued at 10 $m^3/s$ for a total outflow of ca. 105 $m^3/s$. On the day of June 5, the total outflow from the lake was just over 9 $Mm^3$. Photographs published by the *Pamir Times* show the enlargement of the spillway, first by deepening by headward erosion from the lowest part of the spillway and

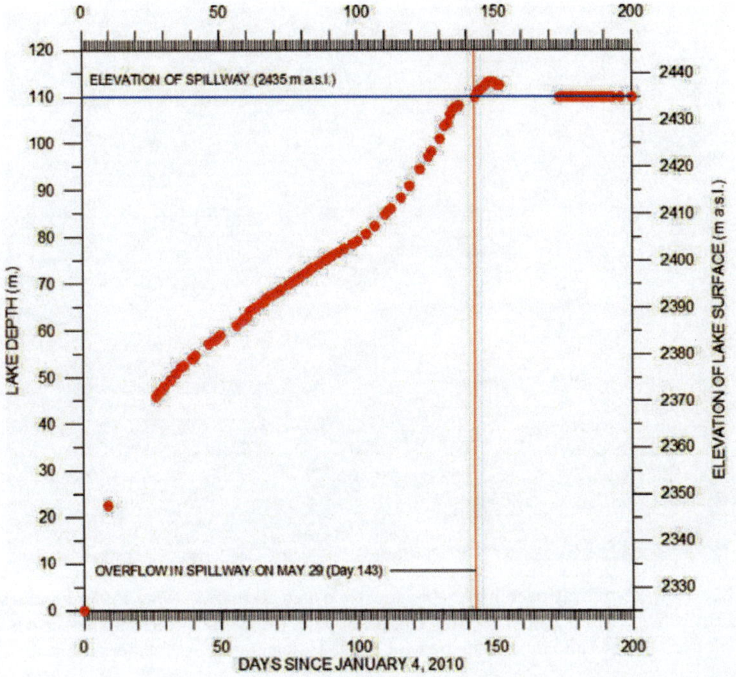


**Fig. 1.25** Filling curve of 2010 Hunza rockslide-dammed lake, northern Pakistan. Lake began to form on January 4, 2010 as a result of the river-blocking Attabad rockslide and overtopped through an excavated spillway on May 29 (Day 143). As of July 20, 2010, stable overflow continues as Pakistan authorities consider other drainage/partial drainage strategies. (Data derived from the National Disaster Management Authority (NDMA) Pakistan and Chap. 7 by Delaney and Evans, this volume)

then by widening by lateral undercutting and seepage erosion of the spillway walls (Fig. 1.26).

As of July 25, 2010 (Day 200) stable overflow continues and the lake level was within a metre of its maximum elevation at overflow on May 29, as Pakistan authorities consider other drainage/partial drainage strategies to reduce the lake level.

## 4 Rockslide Dam Hazard Assessment

Rockslide dam hazard assessment involves a consideration of (1) the extent of the flooded area upstream within the impoundment created by the dam, (2) the stability of the dam with respect to catastrophic breaching resulting in an outburst flood, (3) the extent of the area downstream that would be subject to inundation following a catastrophic outburst, the inundation being the result of a water flood, debris flood or in some cases a debris flow [235] depending on the solids concentration of the flow.

**Fig. 1.26** View of spillway over the Attabad rockslide dam on June 14 (Day 159 of impoundment) as stable overflow continued after the May 29 overtopping. Hunza rockslide-dammed lake is visible in the left background (Photograph courtesy of Focus Humanitarian Assistance, Pakistan)

A number of methods exist for the hazard assessment of these aspects of rockslide dam formation and failure. These form seven broad groups as follows; (1) indeces of stability of the rockslide dam itself based on geomorphometric analysis, (2) estimates of potential upstream inundation, (3) geotechnical and hydrogeological analysis of rockslide dam stability, (4) three dimensional gemorphological assessment of dam form in relation to valley profile, (5) empirical estimates of estimating peak discharge during rockslide dam breach, (6) physical-based hydraulic analysis of rockslide-dam breaching mechanism, and (7) simulation of the downstream propagation of a potential outburst flood (or debris flow) resulting from a rockslide dam failure.

## *4.1 Geomorphometric Analysis of Rockslide Dams*

The use of geomorphometric analysis to identify rockslide dam failure potential was initiated by work on landslide dams in Japan by Swanson et al. [261] and further developed by work on landslide dams in the Appennines and New Zealand [22, 26, 72, 157, 158, 216]. In this approach to predicting landslide dam behavior, geomorphic parameters such as drainage basin area, impoundment volume, landslide dam height, and landslide dam volume are plotted against each other and fields occupied by failed and unfailed dams are delineated. Early attempts focused on the ratio of two geomorphic parameters, the landslide dam volume and the upstream watershed area, which were considered major factors in determining dam stability.

This was further developed by the Dimensionless Blockage Index (DBI) [72] and is defined as DBI = log ($A_bH_d/V_d$) where $A_b$ is the upstream catchment area ($km^2$), $H_d$ is the dam height (m), $V_d$ is the landslide dam volume ($m^3$). An analysis of 84 landslide dams from around the world [72] indicated that three domains could be distinguished; DBI < 2.75 – stability domain, 2.75 < DBI < 3.08 uncertain domain, DBI > 3.08 instability domain.

However, some important anomalies in rockslide dam behavior occur outside these domains. Comparative geomorphic indeces have been developed by Korup [157] with mixed results. Other geomorphic parameters may be integrated into a GIS-based hazard assessment methodology which is found to give crude, order-of-magnitude information critical for estimates of rockslide dam stability and outburst flood potential [158].

Related work reporting the discriminant analysis of the geomorphic characteristics of landslide dams in a Japanese dataset of 43 landslide dams has achieved promising results [60].

## *4.2 Potential Upstream Inundation*

Attempts have been made to predict the form and geometry of the river-blocking landslide and thus the maximum pool level of the landslide-dammed lake [43]; Chap. 18 by Hungr, this volume; [181]. Using GIS, this data may be used to map out inundation levels within the landslide-dammed lake basin and to identify facilities and infrastructure that may be flooded. This information may in turn be used as input into decision-making about possible mitigation strategies such as upstream evacuation or lake level control.

## *4.3 Geotechnical and Hydrogeological Analysis*

The stability of rockslide dams may also be investigated by means of conventional geotechnical site investigation and analytical techniques. Probably the best examples of this approach to hazard assessment are the investigations of the rockslide dams formed in the North Fork Toutle River Valley, Washington, as a result of the eruption-related flank collapse of Mt St Helens in 1980 [101, 187, 288]. Detailed sub-surface investigations and groundwater monitoring were carried out together with geotechnical characterization of materials. This data formed the basis for the slope stability analysis of upstream and downstream faces (e.g. [101], as well as the evaluation of piping and heave potential [186]).

Several workers have attempted to characterize the complete grain size distribution of rockslide dams to assist in geotechnical and hydraulic analysis [27]; (Chap. 17 by Davies and McSaveney, this volume; Chap. 19 by Dunning and

Armitage, this volume) and outline the challenges of recording the volumetric contribution of the wide range of grain sizes present in a typical rockslide dam deposit.

### 4.4 Three Dimensional Geomorphological Assessment of Dam Form in Relation to Cross-Valley Profile

The three dimensional relation between the rockslide dam and the valley profile is decisive in cases where a breach may form due to overtopping and hence the potential of enlarging the breach and catastrophic water release exists. Rockslide dams for example, which push the lake outflow over bedrock are rather stable because only slow downcutting is possible. In contrast, a symmetry between rockslide deposit and cross-valley profile give potential for complete downcutting of the dam during overtopping which could result in the entire volume of water being released in a single event. This influence is discussed in more detail in the context of a new rockslide dam classification proposed in this volume (Chap. 24 by Hermanns et al., this volume).

### 4.5 Empirical Methods of Estimating Breach Discharge

Empirical approaches to the estimate of peak discharge during the breaching of constructed embankment and natural rockslide dams have been discussed by Evans [74], Costa [51], Costa and Schuster [53], Froehlich [93], Walder and O'Connor [274], O'Connor and Beebee [199], and Pierce et al. [215]. The approach is useful since it gives a first-approximation to the peak discharge during the breach process, an estimate that can be used to calibrate physical-based hydraulic analyses of the breach (see below) and as an initial discharge in downstream flood propagation analysis.

### 4.6 Physically-Based Hydraulic Models for Breach Analysis

Early attempts to develop mathematical models based on the principles of hydraulics, sediment transport, and soil mechanics indicated strong potential for hazard assessment of rockslide dams (e.g. [220]). This led to the development of simulation software [272] for the analysis of the breaching of both conventional and landslide dams (e.g., HEC-1 [268], DAMBRK, BREACH [90]). The software has been applied with some success to the analysis of the breaching failure of rockslide dams [28, 58, 90, 113, 180]. In the model, sediment transport in evolving flow conditions under lowering head determines the rate of erosion of the breach and hence its size and geometry through time. The model also predicts the breach-outflow hydrograph, an important part of the simulation of the downstream propagation of the subsequent outburst flood. The results are very sensitive to input parameters for the geotechnical properties of the dam material, such as grain size distribution and shear strength of the debris [90].

Analytical techniques have also been developed for the behavior of rockslide dams when overtopped by a landslide-generated wave (e.g. [229]). In addition, breach mechanisms and processes have been investigated in laboratory experiments involving the failure of scaled embankments in flumes (e.g. [46, 107, 188]), and also in the retrospective analysis of case histories of dam breaches [31, 178, 180, 220, 272, 273]. Both these activities have provided valuable calibration data for numerical methods [94, 272].

Breach modeling is a major field of civil and hydraulic engineering and the reader is referred to the extensive literature in that field (e.g. [209, 249, 272]).

### *4.7 Simulation of the Downstream Propogation of Potential Outburst Flood or Debris Flow*

A number of numerical methods for the simulation of the downstream propogation of an outburst flood, debris flood or debris flow, resulting from the breaching of a rockslide dam and the catastrophic release of its impounded waters have been successfully applied to actual or potential landslide dam breaches (e.g. [33, 229]). Available software, including FLO-2D [196, 197], FLDWAV (e.g. [91, 92, 150]) and CASTOR [210] are used in conjuction with GIS. Using reasonable assumptions, these methods delineate inundation areas, estimate depth and velocity of flow, and give downstream travel time data for the flood wave or debris flow, results that may be used as a framework for developing land-use control in threatened areas and warning system strategies. LAHARZ [148] is an automated GIS-based statistical model utilizing digital elevation models which can approximate inundation levels and extent of deposits.

## 5 Measures to Mitigate Hazard Associated with Rockslide-Dammed Lakes

Measures to mitigate the hazard associated with the stability of rockslide dams and the potential of catastrophic outburst floods from rockslide-dammed lakes include monitoring and early warning-based systems, pumping of water from the impoundment to reduce water levels, direct measures to control breaching involving the construction of a spillway, and the construction of by-pass tunnels to control impoundment levels in newly formed rockslide dams or to prevent impoundment of surface waters by potential rockslide dams. These measures may be applied individually or in combination [52, 286]; (Chap. 2 by Schuster and Evans, this volume).

## 5.1 Monitoring and Early Warning of Rockslide Dam Failure

The monitoring and early warning system developed and successfully implemented in the case of the Gohna rockslide dam in the late nineteenth century has been described in detail above. Similar principles govern the design of present-day telemetric satellite-based systems. As detailed by Droz and Spasic-Gril [63], an advanced monitoring and early warning system was installed at the Usoi rockslide dam, Tajikistan, in 2004, funded by the World Bank. Monitored parameters include lake level and abnormal wave occurrence (measured by pressure cells), outflow discharge (measured by radar sensors and flow meter), downstream flood occurrence (flood sensors), and the movement of debris in body of rockslide dam (GPS manual measurements). These are linked to alert activation criteria which, in the event of a developing outburst would activate sirens at downstream locations to initiate evacuation of threatened areas [63]. These monitoring measures are in addition to strong motion accelerographs, basic meteorological measurements, and the continual visual monitoring of the dam by local people.

An important additional component of current monitoring systems is the use of high-resolution remotely-sensed data obtained by satellites [151]. Access to real-time satellite imagery also not only allows the initial detection of a rockslide-dammed lake, but also provides a means for the accurate tracking of the filling of a rockslide-dammed impoundment [151, 291]; (Chap. 22 by Evans and Delaney, this volume; Chap. 7 by Delaney and Evans, this volume).

## 5.2 High-Capacity Pumping of Rockslide-Dammed Lake

High capacity pumping has been attempted at number of rockslide-dammed lakes in an attempt to retard the rate of filling or to assist in impoundment drainage [52]; (Chap. 2 by Schuster and Evans, this volume). In the case of Spirit Lake, formed by the 1980 flank collapse of Mt St Helens, USA, a temporary 20-pump system with a maximum capacity of 5 $m^3/s$ was installed to pump water from the lake through a pipeline across the dam crest. The pumping was an interim measure prior to the construction of a 2,500 m drainage tunnel as a long-term measure to control lake level [232]; (Chap. 2 by Schuster and Evans, this volume).

## 5.3 Construction of Spillways Over Rockslide Debris

Direct measures to initiate, assist or control the stable draining of rockslide-dammed lakes, without catastrophic breaching occurring, have been implemented since ancient times. These measures include a range of construction activities from the hand-excavation of a spillway over the debris to the mechanical excavation of a spillway by heavy machinery. In other cases explosives have been used to aid in the excavation of the spillway and/or to initiate a controlled breach of the rockslide-dammed impoundment. These attempts have not always proved successful in the construction of a stable spillway or avoiding the catastrophic breaching of the rockslide dam in question.

In the case of the 2010 Hunza rockslide-dammed lake, overflow through an excavated spillway resulted in stable overtopping. As of July 25, 2010 (200 days after impoundment) the maximum pool elevation had not changed dramatically and large areas of agricultural land, several villages, and a 22 km segment of the Karakoram Highway remained submerged by lake waters (Chap. 7 by Delaney and Evans, this volume).

The best documented case of the construction of a spillway over a rockslide dam is the mitigation of the hazard of Earthquake Lake formed by the Madison Canyon rockslide (44°49′44″N; 111°25′39″W) triggered by the M7.1 Hegben Lake Earthquake on August 17, 1959 (Figs. 1.2, 1.27; [14, 111, 117, 269]) at an epicentral distance of 32 km. The source rock mass had an in-situ volume of between 21.4 $Mm^3$ [111] and 29.8 $Mm^3$ [269] bulking to between 28.3 $Mm^3$ [111] and 32.9 $Mm^3$ [269] respectively during emplacement. This corresponds to a bulking factor of between and 10 and 32%.

The rockslide dammed the Madison River and a lake began to rapidly fill behind the rockslide barrier, natural river flow being augmented by post-earthquake safety-related releases from the Hegben Dam, 10 km upstream (Fig. 1.27). The rockslide dam had a maximum height of around 120 m with the lowest point in the debris being 67 m above the original streambed (Fig. 1.28; [269]). Construction started on a spillway (Fig. 1.28) on August 25 with the objective of ensuring stable overflow of the rockslide debris [14]. The spillway had a capacity of 283 $m^3/s$ and had a

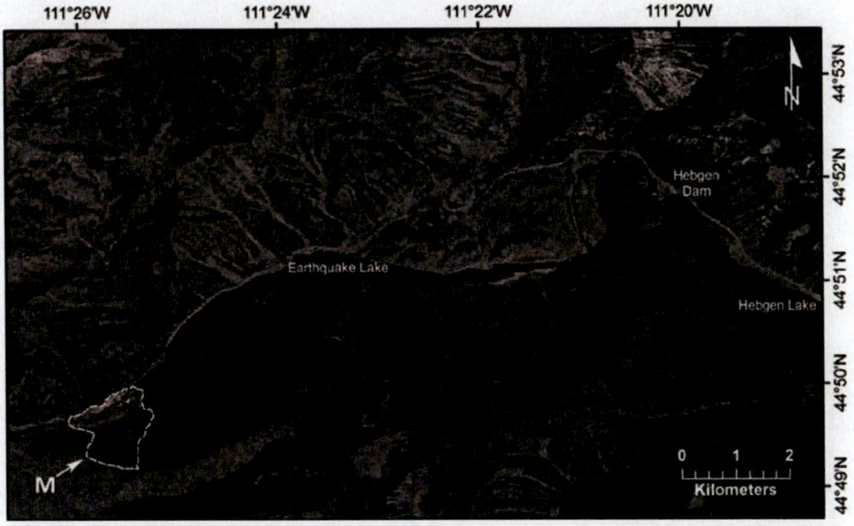


**Fig. 1.27** LANDSAT-7 image of 1959 Madison Canyon rockslide (M), Earthquake Lake impounded behind it (impounded volume ~101 $Mm^3$), the site of the Hebgen Dam constructed in 1914, and Hegben Lake impounded behind it (impounded volume is 475 $Mm^3$)

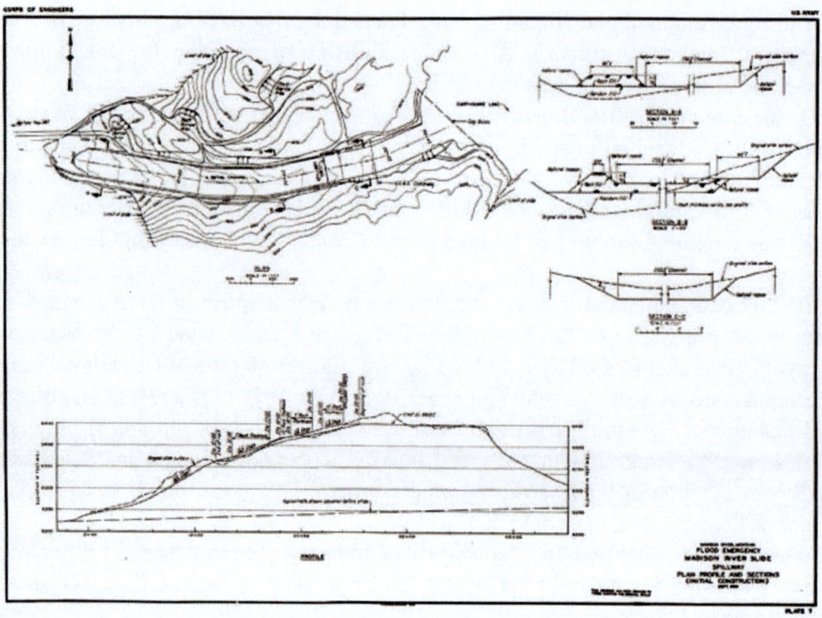

**Fig. 1.28** Construction drawings of the spillway excavated over the 1959 Madison Canyon rockslide dam by the US Army Corps of Engineers. The earthquake-triggered rockslide impounded Earthquake Lake which had a maximum volume of 101.8 $Mm^3$ before overtopping into the constructed spillway (reproduced from [269])

width of 76 m to spread the flow and keep velocities as low as possible. Lake level reached the elevation of the spillway (1,966 m a.s.l.) on September 10 after which the construction effort was directed to retain the integrity of the spillway, preventing gullying and erosion. The maximum volume of the lake (101.8 $Mm^3$) occurred on September 20 [269]. The spillway was subsequently armoured with quartzite and limestone boulders from the debris and further modified by rock weirs. Construction activity ceased on October 30, 1959 and the spillway continues to function in 2010.

## 5.4 In-Debris and Valley Wall By-Pass Tunnels

In-debris tunnels have been constructed at a number of locations worldwide to assist in draining or controlling the level of an existing rockslide dammed lake [232]; (Chap. 2 by Schuster and Evans, this volume). As noted above a 2,500 m tunnel was constructed through the Spirit Lake blockage to control the level of the rockslide dammed lake.

Valley-wall by-pass tunnels have also been constructed to route a river through locations where a future rockslide could have the potential of blocking a river. At the Randa rockslides, Switzerland [106], a 3,770 m long tunnel has been constructed [233] to prevent the formation of another lake that in 1991 flooded the village of

Randa, should another damming rockslide occur. Also in the Alps, similar works have been carried out at the 1987 Val Pola rock avalanche; two 3.5 km long tunnels, with a combined capacity of 500 $m^3$/s, were constructed through rock in the valley wall to carry the Adda River around the rock debris dam [52].

### *5.5 The Example of the 2002 Baipaza Rockslide Dam, Tajikistan*

Triggered by an earthquake in March 2002, the Baipaza rockslide (est. vol. 5-10 $Mm^3$) partly blocked the Vakhsh River, approximately 5 km downstream of the 600 MW Baipaza Hydroelectric power plant, part of the Nurek-Vakhsh River hydropower cascade in western Tajikistan. The water level in the Vakhsh rose 7 m, threatening to flood the Baipaza power plant. Generation of power and supply of irrigation water was disrupted [13]. Downstream communities would be threatened in the event of an outburst flood. Tajikistan authorities took the immediate measure of blasting the debris from the Vakhsh channel to avoid the flooding of the Baipaza plant.

A mitigation program designed to prevent a complete blockage of the Vaksh and to restore the Vaksh flow to its pre-March 2002 levels was undertaken beginning in September 2002. It consisted of stabilization of the rockslide by removing material from the head of the landslide forming a series of benches in the headscarp area and the installation of open-trench drains in the upper part of the landslide. It also involved the removal of landslide debris from the river channel to restore channel capacity. A radio-controlled disaster mitigation system was set up (including monitoring) and a long-term solution, which involves the possible construction of a 1,180 m by-pass tunnel through the steep left bank of the Vakhsh, was designed. Total cost of the mitigation was $5.47 M and was completed in June 2005 [13]. The Baipaza hydroelectric dam is a 63 m high blast-fill structure (see below).

## 6 Natural Rockslide Dams As Foundations for Constructed Dams

As first noted by Heim [122], rockslide dams may, under favourable circumstances, prove to be effective foundations for conventional constructed dams. In these circumstances, the construction of an artificial dam on top of the rockslide debris increases the capacity of the natural impoundment for water storage or hydroelectric generation. Several rockslide dams have been utilised for hydroelectric power in the Alps [2, 122]. The 30 m high Rhodannenberg Dam, for example, is built on the Klontalersee rockslide dam (Glarnisch-Gleiter rockslide of Heim [122]) in Switzerland. Completed in 1910, the plant has a generating capacity of 60 MW and the dam impounds a reservoir of 56.4 $Mm^3$. In Serbia, the rockslide dam that formed in 1963 on the Visočica River forms the foundation of the 80 m high Zavoj (Pirot) Dam which impounds Lake Zavoj. Lake Zavoj has a volume of 170 $Mm^3$ compared

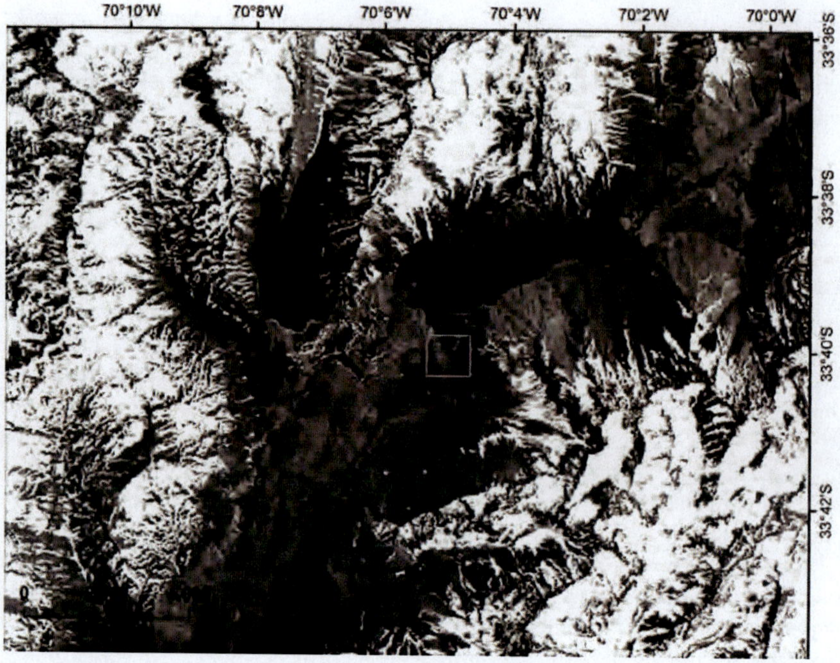


**Fig. 1.29** El Yeso dam (*white rectangle*) built on rockslide debris to increase storage and regulate water level in the Yeso reservoir (*right*), one of two rockslide-dammed lakes formed by the massive (est. vol. 4.5 $km^3$) Meson Alto-Yeso rockslide in the Chilean Andes (33° 39′58″S; 70° 05′ 05″W). Surface-age dating indicates the age of the rockslide is about 4,500 ka [10]. The rockslide is described in detail by Abele [3] and Antinao and Gosse [10]

to the original volume of the rockslide-dammed lake of 30 $Mm^3$. The powerplant began operation in 1990.

Other examples of dams being constructed on rockslide dams include the Bonneville Dam (USA) [17, 198, 237, 240], the Daisy Lake Dam (Canada) [265], the El Yeso Dam, Chile (Fig. 1.29) and the Waikaremoana Dam (New Zealand) [9, 25, 225–226, 228]. As in the cases of Tafjord, Norway, or Laguna Aricota, Peru, there are likely several places in the world where mountain lakes formed by prehistoric rockslide dams were used for hydro power generation without design engineers ever considering the mode of formation of the lake. A recent publication by R.L. Schuster [237] has provided an extensive survey of the interaction of dams and landslides.

## 7 Artificial Rockslide Dams

Artificial rockslide dams are created by massive rock slope failure induced by large-scale explosion [5–6, 7, 250, Chap. 25 by Adushkin, this volume]. These may occur as an incidental or intended consequence of explosion. An example of a rockslide dam forming as an incidental consequence of explosion occurred on Novaya Zemlya (73°20′13″N; 54°56′21″E) in September 1973 when a massive rock slope failure (est. vol. 80 $Mm^3$), triggered by a 4.2 Mt underground nuclear explosion, dammed a glacial stream forming a 2 km long lake [5, 152, 252].

Planned massive rock slope failure by directed large-scale explosion is achieved by setting explosives in a system of galleries driven into the rock mass which it is intended to displace [5, 73]. This construction technique has been used in the former Soviet Union, India, and China [146]. Rockslide dams created by this method are also termed "blast-fill" or "explosive-fill" dams. The objective of the blasting is not only to deliver large volumes of rockfill to a valley bottom design site but to produce a fill that has strength, deformability, density, permeability, particle size, and seismic response characteristics appropriate for a stable and operational dam structure [213, 250].

As part of the development of blast-fill technology in the Soviet Union an experimental blast-fill dam was emplaced in 1975 on the Burlykiya River, 8.5 km northeast of Torkent, Kyrgyzstan (Chap. 25 by Adushkin, this volume). The planned explosions created a 50 m-high rockslide dam with a volume of 300,000 $m^3$ [167]. The properties of the debris (e.g., particle size, debris density) were intensively investigated in a series of shafts and galleries (Chap. 26 by Korchevskiy et al., this volume), and a full-scale analysis of seepage through the rockslide debris was conducted when the impoundment filled [192]. The results of these investigations give valuable insight into the behaviour of both natural and artificial rockslide dams (Chap. 26 by Korchevsky et al. this volume).

## *7.1 Medeo Debris Flow Retention Dam – A Partial Blast-Fill Structure in Kazakhstan*

The best known and largest blast-fill dam yet constructed, is the Medeo debris retention dam constructed in 1966–1967 in a canyon section of the Malaya Almaatinka river, Kazakhstan (43° 09' 03''N; 77° 03' 44''E). The structure was built to protect the city of Alma Ata, ca. 8 km downstream, from debris flows originating in the Zailiysky Alatau Mountains of the northern Tien Shan [195]. Debris flows had impacted the city in 1841, 1921, 1950, 1956 resulting in many hundreds of deaths and major property damage [195].

The present dam (Fig. 1.30) is the result of a three-phase construction in which only the initial phase involved blast-fill construction techniques. The elevation of the thalweg of the Malaya Almaatinka at the dam location is about 1,750 m a.s.l. Phase 1 involved explosions in chambers and adits excavated in the granitoid rock slopes on both sides of the canyon at the planned dam location and the formation of an artificial rockslide dam in the 1960s [5, 251, 254]; (Chap. 25 by Adushkin, this volume). The first blast took place on the right bank in 1966 resulting in the placement of 1.7 $Mm^3$ of debris in the valley bottom and an initial 62 m high embankment [254]. The second blast took place on the left bank in 1967 creating 0.7 $Mm^3$ of blast

**Fig. 1.30** View upstream of the Medeo debris retention dam constructed in multiple phases between 1966 and 1975 on the Malaya Almaatinka River, Zailiysky Alatau Mountains, Kazakhstan. The present debris flow retention capacity of the structure is 12.6 $Mm^3$. Note the bypass tunnel and spillway at the right of the photograph designed to drain water content of debris flows through the dam to the downstream channel (reproduced from *High Dams of the USSR*, Technopromexport, 1989)

rockfill within the dam design area, increasing the height of the 1966 embankment by 10–20 m.

Detailed studies were undertaken of the geotechnical properties of the blast-fill debris, including the sinking of shafts into the surface of the debris [251]. The debris consisted of rock fragments up to 5 m in diameter (average of 60%) set in a finer <2 mm matrix (average 15%). Numerous gaping cavities, up to 1 m in diameter, were observed in rubble horizons. The average porosity of the debris is estimated at 20% which is equivalent to the bulking of the in-situ bedrock due to blasting and emplacement. This value of bulking is comparable to values of bulking (19%) reported from a detailed study of a natural rockslide dam, the 1999 Tsao-Ling rockslide, by Chen et al. [34]. The angle of internal friction of the debris has an average of 37° [251].

The initial blastfill structure was finished in Phase 2, which involved the mechanical placement of an outer shell (Fig. 1.31) that consisted of high-quality rockfill (est. volume 1.5 $Mm^3$) obtained from the debris created by the earlier blasting [254]. This phase of construction ended in 1972 [97]. The final volume of the pre-1973 rockslide dam was ca. 3.9 $Mm^3$, its crest elevation was 1,860 m a.s.l., 110 m above the thalweg of the river (Fig. 1.31). The debris flow retention capacity (solid and water components) of the structure was 6.5 $Mm^3$ [97].

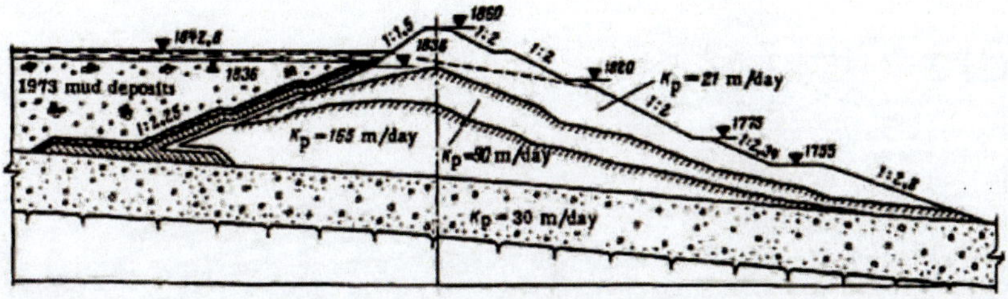


**Fig. 1.31** Cross-section of Medeo rockfill debris flow retention dam, as completed in Phase 2 of construction in 1972, after retaining the 1973 debris flow in the Malaya Almaatinka River, Kazakhstan (reproduced after Fig. 1.1 in [39]). Crest elevation is 1,860 m a.s.l

On July 15, 1973 the Medeo Dam blocked a massive debris flow with a solids volume at the dam of ca. 4 $Mm^3$ [39, 195, 221]. The debris flow was initiated by the outburst of a moraine-dammed lake 10 km upstream as described in detail by Niyazov and Degovets [195]. The debris filled the retention basin up to 1,836 m a.s.l., 86 m above the thalweg, and within 14 m of the dam crest (Fig. 1.31). The rockfill dam withstood the dynamic loading of the debris flow impact and the static loading of the deposited debris despite the fact that the maximum discharge of the debris flow at the dam was 5,100 $m^3/s$ and its velocity was 10–12 m/s [195].

The debris blocked multi-level dewatering intakes built into the structure and a bypass tunnel through the left abutment leading to a situation where rising water behind the dam threatened to overtop the structure. About 1 $Mm^3$ of water accumulated due to river inflow as the pool rose to within 6 m of the lowest point of the dam crest [168]. High capacity pumps were placed on pontoons in the rising waters to draw down the lake forming upstream and this was successfully achieved by July 24. When the pumping started, the Medeo retention basin contained ca. 5.5 $Mm^3$ of water, mud, and boulders [195], 85% of its retention capacity.

In the aftermath of the debris flow, dramatic seepage developed in the full extent of the downstream face of the rockfill dam. Despite heavy and pervasive seepage through the downstream face (seepage rates were estimated at 4.5 $m^3/s$) no piping failure ensued [39, 97]. Observation of seepage rate and location gave valuable insight into permeability contrasts within the debris retention structure. Rubble zones, originating in the initial blastfill phase (containing up to 60% coarse fragments) were characterized by permeability coefficients of 165 m/day whereas the mechanically placed outer fill had a permeability coefficient of 22 m/day. This contrast led to seepage through the downstream face.

Following the 1973 debris flow event, Phase 3 of the construction of the Medeo Dam took place in 1973–1975 to raise the crest of the original structure thus restoring its debris retention capacity. This phase involved the mechanical emplacement of ca. 1.25 $Mm^3$ screened rockfill and 1973 debris flow deposits to raise the crest (Fig. 1.32). The add-on structure, founded partly on the 1973 debris retained behind the dam, increased the height of the dam by 40 m to a design crest elevation of 1,900 m a.s.l., 150 m above the thalweg and 68 m above the surface of the 1973 deposits.

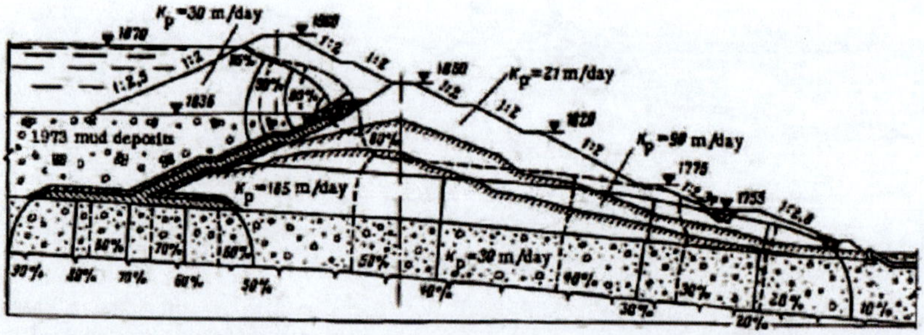


**Fig. 1.32** Cross-section of Medeo rockfill debris flow retention dam, as completed in Phase 3 of construction in 1975. Note add-on crest structure is founded partially on consolidated 1973 debris flow deposits (reproduced after Fig. 1.3 in [39])

The final volume of the Medeo debris retention dam is roughly 5 $Mm^3$ and the resulting debris flow storage capacity is 12.6 $Mm^3$ [294]; (Fig. 1.32).

## 8 Fluvial Systems, Rockslide Dams, and Landscape Development

Valley filling by large volumes of landslide debris has significant impact on the geometry and development of fluvial systems [179, 202, 245]. Some recent geomorphic research on rockslide dams has focused on the impact of rockslide damming on fluvial systems draining active mountain landscapes (e.g [49, 86, 88, 132, 134, 136, 160, 164–165, 166, 205]). Damming, or obstruction, of river channels by large rockslides creates a local base level for upstream elements of a drainage network and a high-energy knick point for downstream elements [179]. This commonly leads to aggradation upstream of the rockslide dam resulting in a local decrease in channel gradient [144, 161, 245]; the knick point creates conditions for upstream incision into the rockslide mass and its possible breaching [275]. Rockslide dams thus locally distort energy relationships in fluvial systems and establish spatial and temporal variability in incision rates, even when uplift is steady and uniform [205]. Obstructing rock avalanches divert or displace river channels and also directly affect river channel morphology (e.g. [159–160, 179]). In addition, rockslide deposits which are higher than the lowest water divide in the upstream catchment may cause a reorganization of the entire drainage system should the level of the rockslide-dammed lake exceed the elevation of the divide (Chap. 24 by Hermanns et al., this volume).

Rockslide damming has a dramatic effect on the sediment budget of a drainage basin [49, 163, 245]. The occurrence of the damming rockslide itself represents the injection of a sediment pulse into the basin. Upstream of rockslide dams, where extensive lakes previously existed, massive volumes of fine-grained lacustrine sediments (e.g. [20, 87, 134, 135, 206, 208, 214]), are significant sediment sources

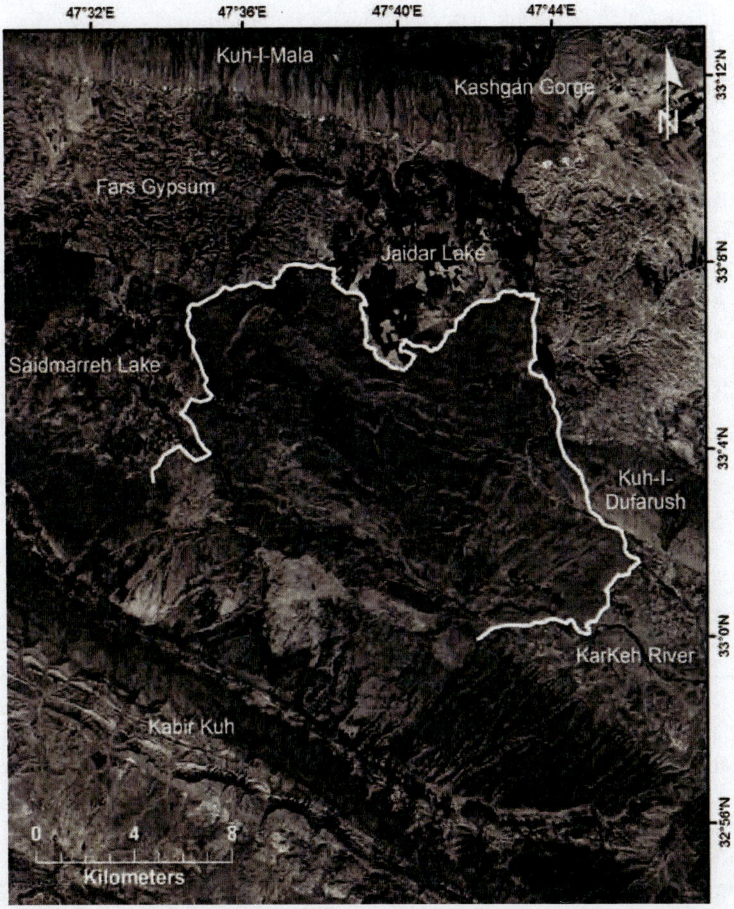


**Fig. 1.33** SPOT image of the gigantic Seymareh (Saidmarreh) rock avalanche, Zagros Mountains, Iran. The limit of the debris is outlined with the white line and the regional drainage runs downstream to the mid-lower *right*. The rock avalanche took place roughly 10,000 years ago [278] and resulted in the damming of local drainage forming Jaidar Lake and the much larger Saidmarreh Lake [118]. These lakes subsequently drained leaving extensive fine-grained lacustrine deposits which have been used for agriculture for millennia

for the fluvial system in question. Downstream of rockslide dams outburst floods, that are significantly greater than maximum normal river floods, may mobilize large volumes of sediment and thus have a long term impact on sediment flux as noted in the Barrancas case outlined above. Sediments deposited in landslide-dammed lakes record rates of aggradation in mountain watersheds, and where incision into a new channel has taken place, provide data for bedrock incision rates [222].

The key factor in controlling the effects of rockslide dams on landscape evolution is their longevity or persistence, which is largely a function of volume [110].

Gigantic rockslide dams and associated lacustrine sedimentation may affect landscape development at the intra-drainage basin scale in mountain landscapes at time scales up to $10^4$ years (Fig. 1.33; [118, 278]).

## 9 Conclusions

Rockslide dams, and associated impoundments, are found to be common throughout the mountain regions of the world. In addition to the security issue associated with their formation and possible failure, they are increasingly being recognized as important geomorphic elements of mountain landscapes. This is evidenced by the 2010 Hunza rockslide dam emergency in the Karakoram Mountains of northern Pakistan which at the time of writing (July 25, 2010) remains to be resolved. The emplacement of landslide dams is an important component of the hazard associated with earthquake-triggered landslides. They are also an important element of hazard associated with the instability of volcanoes. Outburst floods resulting from the breach of rockslide dams have caused some of the most devastating landslide disasters in history. The process of rockslide failure is complex. Overtopping need not necessarily result in immediate breaching and stable overflow may result.

Our data suggests that the rate of rockslide dam failure is between one third and one half that of the rate of rockslide dam formation. Roughly 80% of the rockslide dams that do fail, do so within 1 year of formation, and about 60% of this group fail after 6 months or less. However, there are important examples of historical rockslide dam failures that have occurred hundreds of years after dam formation. The majority of rockslide dams remain stable and may impound lakes that, on a human scale, become permanent features of the landscape.

Geomorphometric measures have been recently proposed in an attempt to identify the potential for rockslide dam failure. Methods of estimating breach discharge are largely empirical but techniques are available for the analysis of breach mechanics and the simulation of the downstream propagation of possible outburst floods. Used together these four methods can be useful for the first order characterization of rockslide dam failure hazard. Measures to mitigate the hazard include the active measures of constructing a spillway across the rockslide debris (with the objective of achieving a controlled draining of the impoundment) and the construction of bypass tunnels at the rockslide site (or potential rockslide site) to prevent or limit the formation of a rockslide-dammed lake. Passive measures include monitoring and warning systems designed to give enough time for evacuation from threatened areas. We note that these measures have enjoyed mixed success in the historical period. Natural rockslide dams have proved to be desirable and successful foundations for constructed dams in many parts of the world.

Artificial rockslide dams are created by massive rock slope failure induced by large scale explosion. These include massive rockslide dams formed by underground nuclear explosions in the Russian Arctic during the Soviet era and the largest blast fill dam in the world, the Medeo debris flow retention dam built in Kazakhstan during the 1960s. The Medeo Dam remains in operation today and has performed successfully in protecting Alma Ata from debris flows.

Finally, the focus of much recent research has been the legacy of rockslide dams in the landscape. Rockslide-dammed lakes create fragmented fluvial systems and act as major sediment traps. Rockslide dams themselves become major sediment sources. Together they form significant storage sites in the routing of the products of primary denudation in mountain terrain and directly affect the evolution of river drainage systems and mountain landscapes.

## References

1. Abbott, J. (1848) Inundation of the Indus, taken from the lips of an eye-witness A.D. 1842, *Journal of the Asiatic Society of Bengal* **17**, 230–232.
2. Abele, G. (1974) Bergstürze in den Alpen, ihre Verbeitung, Morphologie und Folgeerscheinungen, *Wissenschaftliche Alpenverein* **25**, 1–230.
3. Abele, G. (1984) Derrumbes de montana y morrenas en los Andes chilenos, *Revista de Geografía Norte Grande* **11**, 17–30.
4. Adams, J. (1981) Earthquake-dammed lakes in New Zealand, *Geology* **9**, 215–219.
5. Adushkin, V.V. (2000) Explosive initiation of creative processes in nature, *Combustion, Explosion, and Shock Waves* **36**, 695–703.
6. Adushkin, V.V. (2006) Mobility of rock avalanches triggered by underground nuclear explosions, in S.G. Evans, G., Scarascia-Mugnozza, A. Strom and R. Hermanns (eds.), *Landslides from Massive Rock Slope Failure*, Vol. 49 NATO Science Series IV, Earth and Environmental Sciences. Springer, Dordrecht, pp. 267–284.
7. Adushkin, V.V., Pernik, L.M. and Zykov, Y.N. (1993) Modeling of explosion-triggered rock slides for construction of the Kambarata 1 hydroelectric power dam, Republic of Kyrgyzstan, *Landslide News* **7**, 7–9.
8. Alden, W.C. (1928) Landslide and flood at Gros Ventre, Wyoming, *Transactions American Institue of Mining and Metallurgical Engineers* **76**, 347–360.
9. Anderson, G.P. (1948) Waikaremoana; the problem of lake control, *Proceedings, New Zealand Institution of Engineers* **34**, 508–526.
10. Antinao, J.L. and Gosse, J. (2009) Large rockslides in the southern central Andes of Chile (32–34.5°S): Tectonic control and significance for Quaternary landscape evolution, *Geomorphology* **104**, 117–133.
11. Antognini, M. and Volpers, R. (2002) A Late Pleistocene age for the Chironico rockslide (Central Alps, Ticino, Switzerland), *Bulletin of Applied Geology* **7**, 113–125.
12. Asanza, M., Plaza-Nieto, G., Yepes, H., Shuster, R.L. and Ribadeneira, S. (1992) Landslide blockage of the Pisque River, northern Ecuador. *Proc. 6th International Symp. on Landslides*, Christchurch, New Zealand, pp. 1229–1234.
13. Asian Development Bank (2006) Tajikistan: Emergency Baipaza Landslide Stabilization Project. Completion Report Project 36233, 25 p.
14. Barney, K.R. (1960) Madison canyon slide, *Civil Engineering*, August 1960, 72–75.
15. Becher, J. (1859) The flooding of the Indus. Letter addressed to R.H. Davies, Secretary to the Government of Punjab and its dependencies, *Journal of the Asiatic Society of Bengal* **28**, 219–228.
16. Becker, A., Davenport, C.A., Haeberli, W., Burga, C., Perret, R., Flisch, A. and Keller, W.A. (2000) The Fulnau landslide and former Lake Seewen in the northern Swiss Jura Mountains, *Eclogae Geologicae Helvetiae* **93**, 91–305.
17. Berkey, C.P. (1935) Foundation conditions for Grand Coulee and Bonneville projects, *Civil Engineering* **5**, 67–71.

18. Bromhead, E.N., Coppola, L. and Rendell, H.M. (1996). Valley-blocking landslide dams in the eastern Italian Alps, *Proc. 7th Int. Symp. on Landslides*, Trondheim, Vol. 2, pp. 655–660.
19. Brooks, G.R. and Hickin, E.J. (1991) Debris avalanche impoundments of Squamish River, Mount Cayley area, southwestern British Columbia, *Canadian Journal of Earth Sciences* **28**, 1375–1385.
20. Burgisser, H.M., Gansser, A. and Pika, J. (1982) Late Glacial lake sediments of the Indus valley area, northwestern Himalayas, *Eclogae Geologicae Helvetiae* **75**, 51–63.
21. Canuti, P., Frassoni, A. and Natale, L. (1994) The 1993 La Josefina rockslide and Rio Paute landslide dam, Ecuador. 2) Failure of the Rio Paute landslide dam, *Landslide News* **8**, 6–7.
22. Canuti, P., Casagli, N. and Ermini, L. (1998) Inventory and analysis of landslide dams in the northern Appenine as a model for induced flood hazard forecasting, in K. Andah (ed.), *Managing Hydro-Geological Disasters in a Vulnerable Environment*. CNR-GNDCI Publication 1900, pp. 189–202.
23. Capra, L. (2007) Volcanic natural dams: Identification, stability, and secondary effects, *Natural Hazards* **43**, 45–61.
24. Capra, L. and Macías, J.L. (2002) The cohesive Naranjo debris-flow deposit (10 $km^3$): A dam breakout flow derived from the Pleistocene debris-avalanche deposit of Nevado de Colima volcano (Mexico), *Journal of Volcanology and Geothermal Research* **117**, 213–235.
25. Carter, N.R. (1948) Construction work at Waikaremoana upper power development, *Proceedings, New Zealand Institution of Engineers* **34**, 172–230.
26. Casagli, N. and Ermini., L. (1999) Geomorphic analysis of landslide dams in the northern Appennines, *Transaction Japanese Geomorphological Union* **20**, 219–249.
27. Casagli, N., Ermini, L. and Rosati, G. (2003) Determining grain size distribution of the material composing landslide dams in the Northern Appennines: Sampling and processing methods, *Engineering Geology* **69**, 83–97.
28. Cencetti, C., Fredduzzi, A., Marchesini, I., Nacccini, M. and Tacconi, P. (2006) Some considerations about the simulation of breach channel erosion on landslide dams, *Computational Geosciences* **10**, 201–219.
29. Cenderelli, D.A. (2000) Floods from natural and artificial dam failures, in E.E. Wohl (ed.), *Inland Flood Hazards; Human, Riparian, and Aquatic Communities*. Cambridge University Press, New York, NY, pp. 73–103.
30. Chai, H.J., Liu, H.C., Zhang, Z.Y. and Wu, Z.W. (2000) The distribution, causes and effects of damming landslides in China, *Journal of the Chengdu Institute of Technology* **27**, 1–19.
31. Chang, D.S. and Zhang, L.M. (2010) Simulation of the erosion processes of landslide dams due to overtopping considering variations in soil erodibility along depth, *Natural Hazards and Earth System Sciences* **10**, 933–946.
32. Chang, S.C. (1984) Tsao-Ling landslide and its effect on a reservoir project, *Proc. 4th International Symposium on Landlsides*, Toronto, ON, 1: pp. 469–473.
33. Chen, C.-Y., Chen, T.-C., Yu, F.-C. and Hung, F.-Y. (2004) A landslide dam breach induced debris flow – a case study on downstream hazard areas delineation, *Environmental Geology* **47**, 91–101.
34. Chen, R.-F., Chang, K.-J., Angelier, F., Chan, Y.-C., Deffontaine, B., Lee, C.-T. and Lin, M.-L. (2006) Topographical changes revealed by high-resolution airborne LIDAR data: The 1999 Tsaoling landslide induced by the Chi–Chi Earthquake, *Engineering Geology* **88**, 160–172.
35. Chen, T.-C., Lin, M.-L. and Hung, J.-J. (2003) Pseudostatic analysis of Tsao-Ling rockslide caused by Chi–Chi earthquake, *Engineering Geology* **71**, 31–47.
36. Chen, X-Q., Cui, P., Li, Y. and Zhao, W-Y. (2010) Emergency response to the Tangjiashan landslide-dammed lake resulting from the 2008 Wenchuan Earthquake, China, *Landslides*, doi10:1007/s10346-010-0236-6.
37. Chen, Y.-J., Zhou, F., Feng, Y. and Xia, Y.C. (1992) Breach of a naturally embanked dam on Yalong River, *Canadian Journal of Civil Engineering* **19**, 811–818.

38. Cheng, G., Wang, X., He, X., Fan, J. and Fan, J. (2008) Outburst risk of barrier lakes in Sichuan, China, *Journal of Mountain Science* **5**, 189–193.
39. Chichasov, V.Y. and Shapovalov, G.I. (1977) Seepage regime in the Medeo Dam following accumulation of a mudflow, *Hydrotechnical Construction* **11**, 1216–1219.
40. Chigira, M., Wang, W.-N., Furuya, T. and Kamai, T. (2003) Geological causes and geomorphological precursors of the Tsaoling landslide triggered by the 1999 Chi–Chi earthquake, Taiwan, *Engineering Geology* **68**, 259–273.
41. Chigira, M., Wu, X., Inokuchi, T. and Wang, G. (2010) Landslides induced by the 2008 Wenchuan earthquake, Sichuan, China, *Geomorphology* **118**, 225–238.
42. Clague, J.J. and Evans, S.G. (1994) Formation and failure of natural dams in the Canadian Cordillera, *Geological Survey of Canada Bulletin* **464**, 35 pp.
43. Clerici, A. and Perego, S. (2000) Simulation of the Parma River blockage by the Corniglio landslide (Northern Italy), *Geomorphology* **33**, 1–23.
44. Code, J.A. and Sirhindi, S. (1986) Engineering implications of impoundment of the Upper Indus river, Pakistan, by an earthquake-induced landslide, in R.L. Schuster (ed.), *Landslide Dams: Processes, Risk, and Mitigation*, Geotechnical Special Publication No. 3. American Society of Civil Engineers, New York, NY, pp. 97–109.
45. Coleman, A. (1949) Al vaciarse un lago cordillerano, provoca una gran inundación, Capitulo III, *Mi vida de ferroviario inglés en la Argentina: 1887–1948*, Bahia Blanca, 1949, pp. 241–250.
46. Coleman, S.E., Andrews, D.P. and Webby, M.G. (2002) Overtopping breaching of noncohesive homogeneous embankments, *Journal of Hydraulic Engineering* **128**, 829–838.
47. Contreras, D.A. and Keefer, D.K. (2009) Implications of the fluvial history of the Wacheqsa River for hydrologic engineering and water use at Chavín de Huántar, Peru, *Geoarchaeology* **24**, 589–618.
48. Coppola, L. and Bromhead, E.N. (2008) Fossil landslide dams and their exploitation for hydropower in the Italian Dolomites, *Bollettino della Societa Geologica Italiana* **127**, 163–171.
49. Cossart, E. and Fort, M. (2008) Consequences of landslide dams on alpine river valleys: Examples and typology from the French Southern Alps, *Norsk Geografisk Tidsskrift* **62**, 75–88.
50. Costa, C.H. and Gonzalez Diaz, E.F. (2007) Age constraints and paleoseismic implication of rock avalanches in the northern Patagonian Andes, Argentina, *Journal South American Earth Sciences* **24**, 48–57.
51. Costa, J.E. (1988) Floods from dam failures, in V.R. Baker, R.C. Kochel, and P.C. Patton (eds.), *Flood Geomorphology*. John Wiley and Sons, New York, NY, pp. 439–463.
52. Costa, J.E. (1991) Nature, mechanics, and mitigation of the Val Pola landslide, Valtellina, Italy, 1987–1988, *Zeitschrift für Geomorphologie*, N.F. **35**, 15–38.
53. Costa, J.E. and Schuster, R.L. (1988) The formation and failure of natural dams, *GSA Bulletin* **100**, 1054–1068.
54. Costa, J.E. and Schuster, R.L. (1991) Documented historical landslide dams from around the world. *United States Geological Survey Open-File Report* 91–239, 486 p.
55. Cui, P., Dang,C., Zhuang, J-Q., You, Y., Chen, X-Q. and Scott, K.M. (2010) Landslide-dammed lake at Tangjiashan, Sichuan province, China (triggered by the Wenchuan Earthquake, May 12, 2008): risk assessment, mitigation strategy, and lessons learned, *Environmental Earth Science*, doi 10.1007/s12665-010-0749-2.
56. Cui, P., Zhu, Y.-Y., Han, Y.-S., Chen, X.-Q. and Zhuang, J.-Q. (2009) The 12 May Wenchuan earthquake-induced landslide lakes: Distribution and preliminary risk evaluation, *Landslides* **6**, 209–223.
57. Dai, F.C., Lee, C.F., Deng, J.H. and Tham, L.G. (2005) The 1786 earthquake-triggered landslide dam and subsequent dam-break flood on the Dadu River, southwestern China, *Geomorphology* **65**, 205–221.

58. Davies, T.R., Manville, V., Kunz, M. and Donadini, L. (2007) Modeling landslide dambreak flood magnitudes:case study, *Journal of Hydraulic Engineering* **133**, 713–720.
59. DeGraff, J.V., James, A. and Breheny, P. (2010) The formation and persistence of the Matthieu Landslide-Dam lake, Dominica, W.I., *Environmental and Engineering Geoscience* **16**, 73–89.
60. Dong, J.-J., Tung, Y.-H., Chen, C.-C., Liao, J.-J. and Pan, Y.-W. (2009) Discriminant analysis of the geomorphic characteristics and stability of landslide dams, *Geomorphology* **110**, 162–171.
61. Doyle, B.C., Levine, N.S., Newhard, J.M.L. and Kyer, J.A. (2010) The Mahras Dag complex landslide, Upper Goksu River Valley, Turkey, *Environmental and Engineering Geoscience* **16**, 91–105.
62. Drew, F. (1875) *The Jummoo and Kashmir Territories; A Geographical Account*. Edward Stanford, London, 568 p.
63. Droz, P. and Spasic-Gril, L. (2006) Lake Sarez mitigation project: A global risk analysis, *Proceedings 22nd Congress on Large Dams, Q.36-R75, Barcelona, Spain.*
64. Duman, T.Y. (2009) The largest landslide dam in Turkey: Tortum landslide, *Engineering Geology* **104**, 66–79.
65. Dunning, S.A., Petley, D.N. and Strom, A.L. (2005) The morphologies and sedimentology of valley confined rock-avalanche deposits and their effect on potential dam hazard, in O. Hungr, R. Couture, E. Eberhardt, and R. Fell (eds.), *Landslide Risk Management*. Balkema, Amsterdam, pp. 691–704.
66. Dunning, S.A., Mitchell, W.A., Rosser, N.J. and Petley, D.N. (2007) The Hattian Bala rock avalanche and associated landslides triggered by the Kashmir Earthquake of 8 October 2005, *Engineering Geology* **93**, 130–144.
67. Dunning, S.A., Rosser, N.J., Petley, D.N. and Massey, C.R. (2006) Formation and failure of the Tsatichhu landslide dam, Bhutan, *Landslides* **3**, 107–113.
68. Eisbacher, G.H. and Clague, J.J. (1984) Destructive mass movements in high mountains: Hazard and management, Geological Survey of Canada, *Paper 84–16*, 230 p.
69. Emerson, F.B. (1925) 180-ft. dam formed by landslide in Gros Ventre Canyon, *Engineering News Record* **95**, 467–468.
70. Engineering News Record (1927) Further data on failure of Gros Ventre "Dam", *Engineering News Record* **99**, 600–601.
71. Engineering News Record (1964) Russians blast through landslide dam, *Engineering News Record* **172**, p. 24.
72. Ermini, L. and Casagli, N. (2003) Prediction of the behaviour of landslide dams using a geomorphological dimensionless index, *Earth Surface Processes and Landforms* **28**, 31–47.
73. Etkin, M.B. and Azarkovich, A.E. (2001) Determination of optimal blast parameters for construction of blast-created rock-fill dams, *Hydrotechnical Construction* **35**, 421–427.
74. Evans, S.G. (1986) The maximum discharge of outburst floods caused by the breaching of man-made and natural dams, *Canadian Geotechnical Journal* **23**, 385–387.
75. Evans, S.G. (1986) Landslide damming in the Cordillera of western Canada, in R.L. Schuster (ed.), *Landslide Dams: Processes, Risk and Mitigation*. American Society of Civil Engineers Geotechnical Special Publication No. 3, pp. 111–130., ASCE, New York, N.Y.
76. Evans, S.G. (2006) Single-event landslides resulting from massive rock slope failure: Characterizing their frequency and impact on society, in S.G. Evans, G. Scarascia-Mugnozza, A. Strom and R. Hermanns (eds.), *Landslides from Massive Rock Slope Failure*, Vol. 49 NATO Science Series IV, Earth and Environmental Sciences. Springer, Dordrecht, pp. 53–73.
77. Evans, S.G. (2006) The formation and failure of landslide dams: an approach to risk assessment, *Italian Journal of Engineering Geology and Environment*, **Special Issue 1**, 15–19.
78. Evans, S.G. and Brooks, G.R. (1991) Prehistoric debris avalanches from Mount Cayley volcano, British Columbia, *Canadian Journal of Earth Sciences* **28**, 1365–1374.

79. Evans, S.G., Scarascia-Mugnozza, G., Strom, A. and Hermanns, R.L. (2006) Landslides from massive rock slope failure and associated phenomena, in S.G. Evans, G. Scarascia-Mugnozza, A.L. Strom and R.L. Hermanns (eds.), *Landslides from Massive Rock Slope Failure*, Vol. 49 NATO Science Series IV, Earth and Environmental Sciences. Springer, Dordrecht, pp. 3–52.
80. Falconer, H. (1841) Letter to the Secretary of the Asiatic Society, on the recent cataclysm of the Indus, *Journal of the Asiatic Society of Bengal* **10**, 615–620.
81. Farr, T.G., Rosen, P.A., Caro, E., Crippen, R., Duren, R., Hensley, S., Kobrick, M., Paller, M., Rodriguez, E., Roth, L., Seal, D., Shaffer, S., Shimada, J., Umland, J., Werner, M., Oskin, M., Burbank, D. and Alsdorf, D. (2007) The Shuttle Radar Topography Mission, *Reviews of Geophysics* **45**, RG2004, doi: 10.1029/2005RG000183.
82. Fauqué, L.E., Baumann, V., Rosas, M., González, M.A., Coppolecchia, M., Di Tommasso, I., Wilson, C.G.J. and Hermanns, R.L. (2005) Natural dams in the Mendoza River Basin, Mendoza, Province, Argentina, *Proc. Int. Conf. on Landslide Risk Management*, Canada, Vancouver, Balkema, Rotterdam, CD.
83. Ferrer, C. (1999) Represamientos y rupturas de embalses naturales (lagunas de obstrución) como efectos cosísmicos: Algunos ejemplos en los Andes venezolanos, *Revista Geográfica Venezolana* **40**, 109–121.
84. Fidel, L. (2007) Deslizamiento traslacional – flujo de detritos de Mayunmarca, Perú, in *Movimientos en Masa en la región Andina: Una guía para la evaluación de amenazas*, Proyecto Multinacional Andino: Geosciencias para las Comunidades Andinas, *Servicio Nacional de Geología y Minería, Publicación Geológica Multinacional* **4**, 213–217.
85. Fort, M. (1987) Sporadic morphogenesis in a continental subduction setting: An example from the Annapurna Range, Nepal Himalaya, *Zeitschrift fur Geomorphologie* Suppl.-Bd. **63**, 9–36.
86. Fort, M. (2000) Glaciers and mass wasting processes: Their influence on the shaping of the Kali Gandaki valley (higher Himalaya of Nepal), *Quaternary International* **65/66**, 101–119.
87. Fort, M., Burbank, D.W. and Freytet, P. (1989) Lacustrine sedimentation in a semiarid alpine setting: An example from Ladakh, northwest Himalaya, *Quaternary Research* **31**, 332–350.
88. Fort, M., Cossart, E. and Arnaud-Fassetta, G. (2010) Catastrophic landslides and sedimentary budgets, in I., Alcantara-Ayala, and A.S., Goudie (eds.), *Geomorphological Hazards and Disaster Prevention*. Cambridge University Press, Cambridge, United Kingdom, pp. 75–85.
89. Fort, M., Cossart, E. and Arnaud-Fassetta, G. (2010) Hillslope-channel coupling in the Nepal Himalayas and threat to man-made structures: the middle Kali Gandaki valley, *Geomorphology* **124**, 178–199.
90. Fread, D.L. (1988) *BREACH; an Erosion Model for Earthen Dam Failures, US National Weather Service Report*. NOAA, Silver Spring, MD, 31 p.
91. Fread, D.L. (1993): NWS FLDWAV Model: The replacement of DAMBRK for dam-break flood prediction, *Proceedings: 10th Annual Conference of the Association of State Dam Safety Officials*, Kansas City, Missouri, pp. 177–184.
92. Fread, D.L. and Lewis, J.M. (1988) FLDWAV: A generalized flood routing model, *Hydraulic Engineering*, Proceedings of 1988 Conference, HY Div, ASCE, Colorado Springs, CO, pp. 668–673.
93. Froehlich, D.C. (1995) Peak outflow from breached embankment dam, *Journal of Water Resources Planning and Management* **121**, 90–97.
94. Froehlich, D.C. (2008) Embankment dam breach parameters and their uncertainties, *Journal of Hydraulic Engineering* **134**, 1708–1721.
95. Galadini, F., Colini, L., Giaccio, B., Messina, P., Salvi, S. and Sposato, A. (2003) Persisting effects of the Colfiorito (central Italy) Pleistocene paleo-landslide in the planning of land use: Upper Palaeolithic and proto-historical coexistence and Antique-Modern modifications, *Env. Geol.* **43**, 621–634.

96. Gaziev, E. (1984) Study of the Usoy Landslide in Pamir. *Proc. 4th Int. Symp. on Landslides*, Toronto, ON, Vol. 1, pp. 511–515.
97. Gerasimov, G.B., Zinevich, Y.N. and Shapovalov, G.I. (1979) The Medeo mudflow-protection dam, *Hydrotechnical Construction* **13**, 914–921.
98. Gerasimov, V.A. (1963) The Issyk catastrophe of 1963 and its impact on the geomorphology of the Issyk River valley, *Izv. Vses. Geograf. Kratkiye Soobs* **97**, 541–547.
99. Glass, J.H. (1896) The great landslip at Gohna, in Garhwal, and the measures adopted to prevent serious loss of life, *Journal. of Society of Arts* **44**, 431–445.
100. Glazyrin, G.Y. and Reyzvikh, V.N. (1968) Computation of the flow hydrograph for the breach of landslide lakes, *Soviet Hydrology* **5**, 492–496.
101. Glicken, H., Meyer, W. and Sabol, M. (1989) Geology and ground-water hydrology of Spirit Lake Blockage, Mount St Helens, Washington, with implications for lake retention: *United States Geological Survey Bulletin* 1789, 33 p.
102. González Diaz, E.F., Giaccardi, A.D. and Costa, C.H. (2001) La avalancha de rocas del rio Barrancas (Cerro Pelan), norte del Neuquen; su relacion con la catastrofe del Rio Colorado (29/12/1914), *Revista de la Asociación Geologica Argentina* **56**, 466–480.
103. González Díaz, E.F., Folguera, A., Costa, C.H., Wright, E. and Ellisondo, M. (2006) Los grandes deslizamientos de la región Septentrional Neuguina entre los 36°–38°S: Una propuesta de inducción sísmica, *Bulletin de la Asociación Geologica Argentina* **61**, 197–217.
104. Goswami, U., Sarma, J.N. and Patgiri, A.D. (1999) River channel changes of the Subansiri in Assam, India, *Geomorphology* **30**, 227–244.
105. Goto, A.S., Muramatsu, T. and Teraoka, Y. (2010) Timing of the landslide-dammed lake triggered by earthquake, at Toyama River, Central Japan, *Radiocarbon* **52**, 1090–1097.
106. Gotz, A. and Zimmermann, M. (1993) The 1991 rock slides in Randa: Causes and consequences, *Landslide News* **7**, 22–25.
107. Gregoretti, C., Maltauro, A. and Lanzoni, S. (2010) Laboratory experiments on the failure of coarse homogeneous sediment natural dams on a sloping bed, *Journal of Hydraulic Engineering* **136**, 868–879.
108. Groeber, P. (1916) *Informe sobre las causas que han producido las crecientes del Rio Colorado (Territorios del Neuquen y La Pampa) en 1914*, Direccion General de Minas, Geologia e Hidrologia, 11 (Serie B, Geologia), Buenos Aires, 29 p.
109. Gupta, A.C. (1974) Lakes of sorrow, *The Journal of Civil Engineering (India)* **58**, 6–11.
110. Guthrie, R.H. and Evans, S.G. (2007) Work, persistence, and formative events: The geomorphic impact of landslides, *Geomorphology* **88**, 266–275.
111. Hadley, J.B. (1964) Landslides and related phenomena accompanying the Hegben Lake Earthquake of August 17, 1959. *United States Geological Survey Professional Paper* 435-K, pp. 107–138.
112. Han, Z.S. (2003) Large-scale landslide-debris avalanche in Tibet, China (1) April-June 2000 Yigong Landslide, Tibet China, *Landslide News* **14–15**, 22–23.
113. Hancox, G.T., McSaveney, M.J., Manville, V.R. and Davies, T.R. (2005) The October 1999 Mt. Adams rock avalanche and subsequent landslide dam-break flood and effects in Poera River, Westland, New Zealand, *New Zealand Journal of Geology and Geophysics* **48**, 683–705.
114. Harden, C.P. (2001) Sediment movement and catastrophic events: The 1993 rockslide at La Josefina, Ecuador, *Physical Geography* **22**, 305–320.
115. Harp, E.L. and Crone, A.J. (2006) Landslides triggered by the October 8, 2005, Pakistan Earthquake and associated landslide-dammed reservoirs, *United States Geological Survey Open File Report 2006–1052*, 10 p.
116. Harp, E.L., Wilson, R. and Wieczorek, G. (1981) Landslides from the February 4, 1976, Guatemala Earthquake, *United States Geological Survey Professional Paper 1203-A*, 35 p.
117. Harrison, A. (1974) Madison Canyon slide modification by the U.S. Army Corps of Engineers, in B. Voight and M.A. Voight (eds.), *Rock Mechanics and the American Northwest*. pp. 138–141.

118. Harrison, J.V. and Falcon, N.L. (1938) An ancient landslip at Saidmarreh in southwestern Iran, *Journal of Geology* **46**, 296–309.
119. Hayden, B. and Ryder, J.M. (1991) Prehistoric cultural collapse in the Lillooet Area, *American Antiquity* **56**, 50–65.
120. Hayden, B. and Ryder, J.M. (2003) Cultural collapses in the Northwest: A reply to Ian Kuijt, *American Antiquity* **68**, 157–160.
121. Hayden, E.W. (1956) The Gros Ventre Slide (1925) and the Kelly Flood (1927), 11th Annual Field Conference, Wyoming Geological Association, pp. 20–23.
122. Heim, A. (1932) *Bergsturz und Menschenleben*. Fretz and Wasmuth Verlag, Zurich, 218 p.
123. Heim, A. and Gansser, A. (1939) *Central Himalaya: Geological Observations of the Swiss Expedition 1936, Denk. Der Schweiz. Natur. Gesell.*, Vol. 73. p. 245.
124. Hejun, C., Hanchao, L. and Zhuoyuan, A. (1998) Study on the categories of landslide-damming of rivers and their characteristics, *Journal of the Chengdu Institute of Technology* **25**, 411–416.
125. Henderson, W. (1859) Memorandum on the nature and effects of the flooding of the Indus, 10th August 1858, as ascertained at Attock, *Journal of the Asiatic Society of Bengal* **28**, 199–228.
126. Hermanns, R.L., González Díaz, E.F., Folguera, A. and Mardones, M. (2003) Large massive rock slope failures, landslide dams, related valley evolution, and their association with the tectonic setting in the Argentine and Chilean Andes between 36 and 38°S, *10th Congreso Geologico Chileno, Concepción*, CD, 5 p.
127. Hermanns, R.L., Naumann, R., Folguera, A. and Pagenkopf, A. (2004) Sedimentologic analyses of deposits of a historic landslide dam failure in Barrancas valley causing the catastrophic 1914 Rio Colorado flood, northern Patagonia, Argentina, *Proc. 9th Int. Symp. on Landslides*, Rio de Janeiro, Balkema, Rotterdam, Vol. 2, pp. 1439–1445.
128. Hermanns, R.L., Niedermann, S., Ivy-Ochs, S. and Kubik, P.W. (2004) Rock avalanching into a landslide-dammed lake causing multiple dam failure in Las Conchas valley (NW Argentina) – evidence from surface exposure dating and stratigraphic analyses, *Landslides* **1**, 113–122.
129. Hermanns, R.L., Niedermann, S., Villanueva Garcia, A. and Schellenberger, A. (2006) Rock avalanching in the NW Argentine Andes as a result of complex interactions of lithologic, structural and topographic boundary conditions, climate change and active tectonics, in S.G. Evans, G. Scarascia-Mugnozza, A. Strom and R. Hermanns (eds.), *Landslides from Massive Rock Slope Failure*, v. 49, NATO Science Series IV, Earth and Environmental Sciences. Springer, Dordrecht, pp. 497–520.
130. Hermanns, R.L. and Schellenberger, A. (2008) Quaternary tephrochronology helps define conditioning factors and trigger mechanisms of rock avalanches in NW Argentina, *Quaternary International* **178**, 261–275.
131. Hewitt, K. (1968) Records of natural damming and related events in the Upper Indus basin, *Indus: Journal of Water Power Devevelopment Authority* **10**, 11–19.
132. Hewitt, K. (1998) Catastrophic landslides and their effects on the Upper Indus streams, Karakoram Himalaya, northern Pakistan, *Geomorphology* **26**, 47–80.
133. Hewitt, K. (1999) Quaternary moraines vs catastrophic rock avalanches in the Karakoram Himalya, northern Pakistan, *Quaternary Research* **51**, 220–237.
134. Hewitt, K. (2002) Postglacial landform and sediment associations in a landslide-fragmented river ystem: The TransHimalayan Indus streams, Central Asia, in K. Hewitt, M.L. Byrne, M. English, and G. Young (eds.), *Landscapes of Transition*. Kluwer Academic, Dordrecht, pp. 63–91.
135. Hewitt, K. (2002) Styles of rock avalanche depositional complexes conditioned by very rugged terrain, Karakoram Himalaya, Pakistan, in S.G. Evans and J.V. DeGraff (eds.), *Catastrophic Landslides: Effects, Occurrence, and Mechanisms*, Vol. XV. Reviews in Engineering Geology, Geological Society of America, Boulder, CO, pp. 345–377.

136. Hewitt, K. (2006) Disturbance regime landscapes: Mountain drainage systems interrupted by large rockslides, *Progress in Physical Geography* **30**, 365–393.
137. Hewitt, K. (2009) Catastrophic rock slope failures and late Quaternary developments in the Nanga Parbat-Haramosh massif, Upper Indus basin, northern Pakistan, *Quaternary Science Reviews* **28**, 1055–1069.
138. Hewitt, K. (2009) Glacially conditioned rock-slope failures and disturbance-regime landscapes, Upper Indus Basin, northern Pakistan. in J. Knight and S. Harrison (eds.), *Periglacial and Paraglacial Processes and Environments*, Geological Society London, Special Publication **320**, 235–255.
139. Hewitt, K. (2010) Gifts and perils of landslides, *American Scientist* **98**, 410–419.
140. Holland, T.H. (1894) Report on the Gohna Landslip, Garhwal, *Records of Geological Survey of India* **27**, 55–65.
141. Hsu, Y.-S. and Hsu, Y.-H. (2009) Impact of earthquake-induced dammed lakes on channel evolution and bed mobility: Case study of the Tsaoling landslide dammed lake, *Journal of Hydrology* **374**, 43–55.
142. Hu, X., Huang, R., Shi, Y., Lu, X., Zhu, H. and Wang, X. (2009) Analysis of blocking river mechanism of Tangjiashan Landslide and dam-breaking mode of its barrier dam, *Chinese Journal of Rock Mechanics and Engineering* **28**, 181–189.
143. Hung, J.-J., Lee, C.-T. and Lin, M.-L. (2002) Tsao-Ling rockslides Taiwan, in S.G. Evans and J.V. DeGraff (eds.), *Catastrophic Landslides: Effects, Occurrence, and Mechanisms*, Vol. XV. Reviews in Engineering Geology, Geological Society of America, Boulder, Co. pp. 91–115.
144. Hutchinson, J.N. (2001) Reading the ground: Morphology and geology in site appraisal (The Fourth Glossop Lecture), *Quarterly Journal of Engineering Geology and Hydrogeology* **34**, 7–50.
145. Hutchinson, J.N. and Kojan, E. (1975) *The Mayunmarca landslide of 25th April, 1974, Peru*. UNESCO Report Serial No. 3124 Paris, UNESCO.
146. ICOLD (2000) Blast fill dams. ICOLD Bulletin E01, 74 p.
147. Inoue, K., Mori, T., Itou, T. and Kabeyama, Y. (2005) Outburst and disasters of Takaisoyama and Hose landslide dams (1892) in East Shikoku (in Japanese with English abstract), *Journal of Japan Society of Erosion Control Engineering* **58**, 3–12.
148. Iverson, R.L., Scilling, S.P. and Vallance, J.W. (1998) Objective delineation of lahar-inundation hazard zones, *GSA Bulletin* **110**, 972–984.
149. Jain, S.K., Agarwal, P.K. and Singh, V.P. (2007) *Hydrology and Water Resources of India*. Springer, The Netherlands, 1258 p.
150. Jin, M. and Fread, D.L. (1997) One-dimensional routing of mud/debris flows using NWS FLDWAV model, *Proc. 1st International Conference on Debris Flow Hazards Mitigation: Mechanics, Prediction, and Assessment*, San Francisco, CA, pp. 687–696.
151. Kargel, J.S., Leonard, G., Crippen, R.E., Delaney, K.B., Evans, S.G. and Schneider, J. (2010) Satellite monitoring of Pakistan's rockslide-dammed Lake Gojal, *EOS* **91**, 394–395.
152. Khalturin, V.I., Rautian, T.G., Richards, P.G. and Leith, W.S. (2005) A review of nuclear testing by the Soviet Union at Novaya Zemlya, 1955–1990, *Science and Global Security* **13**, 1–42.
153. Khan, M. (1969) Influence of Upper Indus basin on the elements of the flood hydrograph at Tarbela-Attock, in *Floods and their Computation*, Vol. 2. IAHS Publication 85, pp. 918–925.
154. King, J., Loveday, I. and Schuster, R.L. (1989) The 1985 Bairaman landslide dam and resulting debris flow, Papua New Guinea, *Quarterly Journal of Engineering Geology* **22**, 257–270.
155. Kojan, E. and Hutchinson, J.N. (1978) Mayunmarca rockslide and debris flow, Peru, in B. Voight (ed.), *Rockslides and Avalanches, 1, Natural Phenomena*. Elsevier, Amsterdam, NY, pp. 315–361.
156. Korup, O. (2002) Recent research on landslide dams – a literature review with special attention to New Zealand, *Progress in Physical Geography* **26**, 206–235.

157. Korup, O. (2004) Geomorphometric characteristics of New Zealand landslide dams, *Engineering Geology* **73**, 13–35.
158. Korup, O. (2005) Geomorphic hazard assessment of landslide dams in South Westland, New Zealand: Fundamental problems and approaches, *Geomorphology* **66**, 167–188.
159. Korup, O. (2005) Large landslides and their effect on sediment flux in South Westland, New Zealand, *Earth Surface Processes and Landforms* **30**, 305–323.
160. Korup, O. (2005) Geomorphic imprint of landslides on alpine river systems, southwest New Zealand, *Earth Surface Processes and Landforms* **30**, 783–800.
161. Korup, O. (2006) Rock-slope failure and the river long profile, *Geology* **34**, 45–48.
162. Korup, O. and Tweed, F. (2007) Ice, moraine, and landslide dams in mountain terrain, *Quaternary Science Reviews* **26**, 3406–3422.
163. Korup, O., McSaveney, M.J. and Davies, T.R.H. (2004) Sediment generation and delivery from large historic landslides in the Southern Alps, New Zealand, *Geomorphology* **61**, 189–207.
164. Korup, O., Strom, A.L. and Weidinger, J.T. (2006) Fluvial response to large rock-slope failures: Examples from the Himalayas, the Tien Shan, and the Southern Alps in New Zealand, *Geomorphology* **78**, 3–21.
165. Korup, O., Montgomery, D.R. and Hewitt, K. (2010) Glacier and landslide feedbacks to topographic relief in the Himalayan syntaxes, *Proceedings National Academy of Sciences*, www.pnas.org/cgi/doi/10.1073/pnas.0907531107.
166. Korup, O., Densmore, A.L. and Schlunegger, F. (2010) The role of landslides in mountain river evolution, *Geomorphology* **120**, 77–90.
167. Kuperman, V.L., Kornakov, G.I. and Korchevskii, V.F. (1977) Construction of an experimental dam on Burlykiya River, *Hydrotechnical Construction* **11**, 465–469.
168. Kuzmin, K.K. (1974) The catastrophic flash flood of 1973 and the Medeo Dam, *Hydrotechnical Construction* **8**, 203–206.
169. Lawrence, D.B. and Lawrence, E.G. (1958) Bridge of the Gods legend; its origin, history and dating, *Mazama* **40**, 33–41.
170. Lee, K.L. and Duncan, J.M. (1975) *Landslide of April 25, 1974 on the Mantaro River, Peru.* National Academy of Sciences, Washington, DC, 72 p.
171. Li, M.-H., Hsu, M.-H., Hsieh, L.-S. and Teng, W.-H. (2002) Inundation potentials analysis for Tsao-Ling landslide lake formed by Chi–Chi earthquake in Taiwan, *Natural Hazards* **25**, 289–303.
172. Li, T. and Wang, S. (1992) *Landslide Hazards and their Mitigation in China*. Science Press, Beijing, 84 p.
173. Li, T., Schuster, R.L. and Wu, J. (1986) Landslide dams in south-central China, in R.L. Schuster (ed.), *Landslide Dams: Processes, Risk and Mitigation.* American Society of Civil Engineers Geotechnical Special Publication No. 3, pp. 146–162., ASCE, New York, N.Y.
174. Litovchenko, A.F. (1964) The catastrophic mudflow on the Issyk River, *Meteorologiya i Gidrologiya* **1**, 39–42 (in Russian).
175. Lubbock, G. (1894) The Gohna Lake, *The Geographical Journal* **4**, 457.
176. Lui, N., Zhang, J.-X., Lin, W., Cheng, W.-Y. and Chen, Z.-Y. (2009) Draining Tangjiashan Barrier Lake after Wenchuan Earthquake and the flood propagation after the dam break, *Science in China Series E; Technological Sciences* **52**, 801–809.
177. Lui, L.Y., Wu, Y.H., Zuo, Z.L., Chen, Z.C., Wang, X.X. and Zhang, W.J. (2009) Monitoring and assessment of barrier lakes formed after the Wenchuan earthquake based on multitemporal remote sensing data, *Journal of Applied Remote Sensing* **3**, pp. 031665-031665-12.
178. MacDonald, T.C. and Langridge-Monopolis, J. (1984) Breaching characteristics of dam failures, *Journal Hydraulic Engineering* **110**, 567–586.
179. Machida, H. (1966) Rapid erosional development of mountain slopes and valleys caused by large landslides in Japan, *Geographical Reports of Tokyo Metropolitan University* **1**, 55–78.

180. Manville, V. (2001) *Techniques for evaluating the size of potential dam-break floods from natural dams*, Institute of Geological and Nuclear Sciences Science Report 2001/28, 72 p.
181. Mandrone, G., Clerici, A. and Tellini, C. (2007) Evolution of a landslide creating a temporary lake: Successful prediction, *Quaternary International* **171–172**, 72–79.
182. Marchi, L., Michieli, F. and Zuppi, G.M. (2006) The Alleghe Lake (Dolomites, Italy): Environmental role and sediment management, in J., Krecek, and M., Haigh (eds.), *Environmental Role of Wetlands in Headwaters*, Vol. 63, NATO Science Series IV Earth and Environmental Sciences. Springer, Dordrecht, p. 161–172.
183. Mason, K. (1929) Indus Floods and Shyok Glaciers, *Himalayan Journal* **1**, 10–29.
184. Mathur, L.P. (1953) The Assam Earthquake of 15th August, 1950; a short note on factual observations, in M.B. Ramachandra Rao *A Compilation of papers on the Assam Earthquake of August 15, 1950*. The Central Board of Geophysics (India), compiler, Publication No. 1, pp. 56–60.
185. Meyer, W., Sabol, M.A. and Schuster, R.L. (1986) Landslide dammed lakes at Mount Saint Helens, Washington, in R.L. Schuster (ed.), *Landslide Dams: Processes, Risk and Mitigation*. American Society of Civil Engineers, New York, NY, Geotechnical Special Publication No. 3, pp. 21–41.
186. Meyer, W., Schuster, R.L. and Sabol, M.A. (1994) Potential for seepage erosion of landslide dam, *Journal of Geotechnical Engineering* **120**, 1211–1229.
187. Meyer, W., Sabol, M.A., Glicken, H.X. and Voight, B. (1985) The effects of groundwater, slope stability, and seismic hazard on the stability of the South Fork Castle Creek blockage in the Mt St Helens area, Washington. *USGS Professional Paper* 1345, 42 p.
188. Mizuyama, T., Satohuka, Y., Ogawa, K. and Mori, T. (2006) Estimating the outflow discharge from landslide dam outbursts, Proceedings INTERPRAEVENT Symposium *Disaster Mitigation of Debris Flows, Slope Failures and Landslides*, Niigata, Japan, Vol. 1, pp. 365–377.
189. Montadon, F. (1933) Chronologie des grands éboulements alpins du début de l'ère chrétienne à nos jours, *Matériaux pour l'étude des calamités* **32**, 271–340.
190. Mora, S., Madrigal, C., Estrada, J. and Schuster, R.L. (1993) The 1992 Rio Toro landslide dam, Costa Rica, *Landslide News* **7**, 19–22.
191. Nash, T., Bell, D., Davies, T. and Nathan, S. (2008) Analysis of the formation and failure of Ram Creek landslide dam, South Island, New Zealand, *New Zealand Journal of Geology and Geophysics* **51**, 187–193.
192. Nedriga, V.P., Pokrovskii, G.I., Korchevskii, V.F. and Petrov, G.N. (1978) Full-scale investigations of seepage in an experimental blast-fill dam, *Hydrotechnical Construction* **12**, 679–683.
193. Nicoletti, P.G. and Parise, M. (2002) Seven landslide dams of old seismic origin in southeastern Sicily (Italy), *Geomorphology* **46**, 203–222.
194. Nicoletti, P.G., Parise, M. and Miccadei, E. (1993) The Scanno rock avalanche (Abruzzi, south-central Italy), *Boll. Soc. Geol., It.* **112**, 523–535.
195. Niyazov, B.S. and Degovets, A.S. (1975) Estimation of the parameters of catastrophic mudflows in the basins of the Lesser and Greater Almatinka Rivers, *Soviet Hydrology* **2**, 75–80.
196. O'Brien, J.S. (2003) Reasonable assumptions in routing a dam break mudflow, *Proc. 3rd International Conference on Debris Flow Hazards Mitigation: Mechanics, Prediction, and Assessment*, Davos, **1**, 683–693.
197. O'Brien, J.S., Julien, P.Y. and Fullerton, W.T. (1993) Two-dimensional water flood and mudflow simulation, *Journal of Hydraulic Engineering*, ASCE, **119**, 244–261.
198. O'Connor, J.E. (2004) The evolving landscape of the Columbia River Gorge: Lewis and Clark and cataclysms on the Columbia, *Oregon Historical Quarterly* **105**, 390–421.
199. O'Connor, J.E. and Beebee, R.A. (2009) Floods from natural rock-material dams, in D.M. Burr, P.A. Carling and V.R. Baker (eds.), *Megafloods on Earth and Mars*. Cambridge University Press, Cambridge, United Kingdom, p. 128–171.

200. O'Connor, J.E. and Costa, J.E. (2004) The world's largest floods, past and present: Their causes and magnitudes, *United States Geological Survey Circular* 1254, 13 p.
201. O'Connor, J.E., Grant, G.E. and Costa, J.E. (2002) The geology and geography of floods, in P. Kyle House, R.H. Webb, V.R. Baker, D.R. Levish (eds.), *Ancient Floods, Modern Hazards: Principles and Applications of Paleoflood Hydrology* Water Science and Application Volume 5. American Geophysical Union, Washington, DC, pp. 359–385.
202. Ohmori, H. (1992) Dynamics and erosion rate of the river running on a thick deposit supplied by a large landslide, *Zeitschrift fur Geomorphologie* **36**, 129–140.
203. Oldham, R.D. (1923) The Pamir Earthquake of 18th February, 1911, *Quarterly Journal of the Geological Society* **79**, 237–245.
204. Ouchi, S. and Mizuyama, T. (1989) Volume and movement of Tombi landslide in 1858, Japan, *Transaction Japan Geomorphological Union* **10**, 27–51.
205. Ouimet, W.B., Whipple, K.X., Royden, L.H., Sun, Z. and Chen, Z. (2007) The influence of large landslides on river incision in a transient landscape: Eastern margin of the Tibetan Pateau (Sichuan, China), *GSA Bulletin* **119**, 1462–1476.
206. Owen, L.A. (1996) Quaternary lacustrine deposits in a high-energy semi-arid mountain environment, Karakoram Mountains, northern Pakistan, *Journal of Quaternary Science* **11**, 461–483.
207. Palmer, L. (1977) Large landslides of the Columbia River Gorge, Oregon and Washington, *Geological Society of America, Reviews in Engineering Geology* **III**, 69–83.
208. Page, W.D. and Mattsson, L. (1981) Landslide lakes near Santa Fe de Antioquia, *Revista CIAF* **6**, 469–478.
209. Paquier, A. (2005) Dam break modelling, in D.W. Knight and A.Y. Shamseldin (eds.), *River Basin Modelling for Flood Risk Mitigation*. Taylor Francis, pp. 463–472.
210. Paquier, A. (2006) Case studies in dam break modelling, in D.W. Knight and A.Y. Shamseldin (eds.), *River Basin Modelling for Flood Risk Mitigation*. Taylor Francis, London, pp. 473–480.
211. Perrin, N.D. and Hancox, G.T. (1992) Landslide dammed lakes in New Zealand-preliminary studies on their distribution, causes and effects, *Proc. 6th International Symposium on Landslides*, Christchurch, New Zealand, 1457–1466.
212. Perucca, L.P. and Angillieri, M.Y.E. (2009) Evolution of a debris-rock slide causing a natural dam: The flash flood of Rio Santa Cruz, Province of San Juan – November 12, 2005, *Natural Hazards* **50**, 305–320.
213. Petrov, G.N., Reifman, L.S. and Khusankhodzhaev, F.Z. (1975) Dam construction by blasting, *Hydrotechnical Construction* **9**, 938–944.
214. Phartiyal, B., Sharma, A., Srivastava, P. and Ray, Y. (2009) Chronology of relict lake deposits in the Spiti River, NW Trans Himalaya:Implications to Late Pleistocene-Holocene climate-tectonic perturbations, *Geomorphology* **108**, 264–272.
215. Pierce, M.W., Thornton, C.I. and Abt, S.R. (2010) Predicting peak outflow from breached embankment dams, *Journal of Hydrologic Engineering, ASCE* **15**, 338–349.
216. Pirocchi, A. (1992) Laghi di sbarramento per frana nelle Alpi: Tipologia ed evoluzione, *Atti I Convegno Nazionale Giovani ricercatori in Geologia Applicata*, Gargnano. CUEM, Ric. Sci. Educaz. Perm., Suppl., 93; 127–136.
217. Plaza-Nieto, G., Yepes, H. and Schuster, R.L. (1990) Landslide dam on the Pisque River, northern Ecuador, *Landslide News* **4**, 2–4.
218. Plaza-Nieto, G. and Zevallos, O. (1994) The 1993 La Josefina rockslide and Rio Paute landslide dam, Ecuador. 1) The La Josefina rockslide, *Landslide News* **8**, 4–6.
219. Poddar, M.C. (1953) A short note on the Assam Earthquake of 15th August, 1950, In M.B. Ramachandra Rao *A Compilation of papers on the Assam Earthquake of August 15, 1950.* The Central Board of Geophysics (India), compiler, Publication No. 1, pp. 38–42.
220. Ponce, V.M. and Tsivoglou, A.J. (1981) Modeling gradual dam breaches, *Journal of the Hydraulics Division, American Society of Civil Engineers* **107(HY7)**, 829–838.

221. Popov, N. (1990) Debris flows and their control in Alma-Ata, Kazakh SSR, USSR, *Landslide News* **4**, 25–27.
222. Pratt-Sitaula, B., Garde, M., Burbank, D.W., Oskin, M., Heimsath, A. and Gabet, E. (2007) Bedload-to-suspended load ratio and rapid bedrock incision from Himalayan landslide-dam lake record, *Quaternary Research* **68**, 111–120.
223. Preobrajensky, J. (1920) The Usoi Landslide, *Geol. Comm., Papers on Applied Geol., 14*, 21 p. (In Russian).
224. Pushkarenko, V.P. and Nikitin, A.M. (1988) Experience in the regional investigation of the state of mountain lake dams in Central Asia and the character of breach mudflow formation, In E.A. Koslovskii (ed.), *Landslides and Mudflows*, Vol. 2. UNEP/UNESCO, Moscow, USSR, pp. 10–19.
225. Read, S.A.L. (1979) Lake Waikaremoana outlet; engineering geological studies of factors related to leakage through the natural dam. *Engineering Geology Report EG-336, New Zealand Geological Survey*, Lower Hutt. 33 p.
226. Read, S.A.L., Beetham, R.D. and Riley, P.B. (1992) Lake Waikaremoana barrier – a large landslide dam in New Zealand, *Proc. 6th Intern. Symposium on Landslides*, Christchurch, New Zealand, Vol. 2, pp. 1481-1487, Balkema, Rotterdam.
227. Reneau, S.L. and Dethier, D.P. (1996) Late-Pleistocene landslide-dammed lakes along the Rio Grande, White Rock Canyon, New Mexico, *GSA Bulletin* **108**, 1492–1507.
228. Riley, P.B. and Read, S.A.L. (1992) Lake Waikaremoana present day stability of landslide barrier. *Proc. 6th Intern. Symposium on Landslides*, Christchurch, New Zealand, Vol. 2, pp. 1249–1256, Balkema, Rotterdam.
229. Risley, J.C., Walder, J.S. and Denlinger, R.P. (2006) Usoi dam wave overtopping and flood routing in the Bartang and Panj rivers, Tajikistan, *Natural Hazards* **38**, 375–390.
230. Ryder, J.M. and Church, M. (1986) The Lillooet terraces of Fraser River: A palaeoenvironmental enquiry, *Canadian Journal of Earth Sciences* **23**, 869–884.
231. Ryder, J.M., Bovis, M.J. and Church, M. (1990) Rock avalanches at Texas Creek, British Columbia, *Canadian Journal of Earth Sciences* **27**, 1316–1329.
232. Sager, J.W. and Chambers, D.R. (1986) Design and construction of the Spirit Lake outlet tunnel Mt St Helens Washington, in R.L. Schuster (ed.), *Landslide Dams: Processes Risk and Mitigation.* American Society of Civil Engineers Geotechnical Special Publication **3**, New York, NY, pp. 42–58.
233. Sartori, M., Baillifard, F., Jaboyedoff, M. and Rouiller, J.-D. (2003) Kinematics of the 1991 Randa rockslides (Valais, Switzerland), *Natural Hazards and Earth System Sciences* **3**, 423–433.
234. Schneider, J.-L., Pollet, N., Chapron, E., Wessels, M. and Wassmer, P. (2004) Signature of Rhine Valley sturzstrom dam failures in Holocene sediments of Lake Constance, Germany, *Sedimentary Geology* **169**, 75–91.
235. Schuster, R.L. (2000) A worldwide perspective on landslide dams, in D. Alford and R.L. Schuster (eds.), *Usoi Landslide Dam and Lake Sarez*, **1**, ISDR Prevention Series. United Nations, Tajikistan, pp. 19–22.
236. Schuster, R.L. (2002) Usoi landslide dam, southeastern Tajikistan. *Proc. Int. Symp. on Landslide Risk Mitigation and Protection of Cultural and Natural Heritage, Kyoto, Japan*, pp. 489–505.
237. Schuster, R.L. (2006) Interaction of dams and landslides-case studies and mitigation. *United States Geological Survey Professional Paper 1723*, 107 p.
238. Schuster, R.L. and Alford, D. (2004) Usoi landslide dam and Lake Sarez, Pamir Mountains, Tajikistan, *Environmental and Engineering Geoscience Quarterly* **10**, 151–168.
239. Schuster, R.L. and Costa, J.E. (1986) A perspective on landslide dams, in R.L. Schuster (ed.), *Landslide Dams: Processes, Risk and Mitigation*. American Society of Civil Engineers, New York, NY, Geotechnical Special Publication No. 3, pp. 1–20.

240. Schuster, R.L. and Pringle, P.T. (2002) Engineering history and impacts of the Bonneville landslide, In J., Rybar, J., Stemberk, and P., Wagner (eds.), *Landslides: Proceedings, 1st European Conference on Landslides*. A. A. Balkema, Lisse. Prague, pp. 59–78.
241. Schuster, R.L., Logan, R.L. and Pringle, P.T. (1992) Prehistoric rock avalanches in the Olympic Mountains, Washington, *Science* **258**, 1620–1621.
242. Sekiya, S. and Kikuchi, Y. (1889) The eruption of Bandai-san: Tokyo Imperial University, College, *Science Journal* **3**, 91–172.
243. Sevilla, J.H. (1994) The Josefina landslide and its implications in the electrical service for the Republic of Ecuador. *Proc. 7th IAEG Cong*., Lisbon, Vol.3, Balkema, Rotterdam, pp. 1801–1810.
244. Shang, Y., Yang, Z., Li, L., Liu, D., Liao, Q. and Wang, Y. (2003) A super-large landslide in Tibet in 2000: Background, occurrence, disaster, and origin, *Geomorphology* **54**, 225–243.
245. Shimazu, H. and Oguchi, T. (1996) River processes after rapid valley-filling due to large landslides, *GeoJournal* **38**, 339–344.
246. Shroder, J.F. (1998) Slope failure and denudation in the western Himalaya, *Geomorphology* **26**, 81–105.
247. Shroder, J.F., Cornwell, K. and Khan, M.S. (1991) Catastrophic break-out floods in the western Himalaya, Pakistan. Geological Society of America Annual Meeting, Program with Abstracts, Vol. 23, p. A87.
248. Shroder, J.F. and Weihs, B.J. (2010) Geomorphology of the Lake Shewa landslide dam, Badakhshan, Afghanistan, using remote sensing data, *Geografiska Annaler* **92A**, 469–483.
249. Singh, V.P. (1996) *Dam Breach Modeling Technology*. Kluwer, Dordrecht.
250. Sinyakov, V.K. (1971) Construction of a rockfill dam for the Baipaz hydraulic system by mass blasting, *Hydrotechnical Construction* **5**, 411–414.
251. Skorobogatov, A.V. (1978) Character of the texture of the material placed in the Medeo Rock Fill Dam, *Hydrotechnical Construction* **12**, 354–355.
252. Skorve, J. (1997) The environment of the nuclear test sites on Novaya Zemlya, *The Science of the Total Environment* **202**, 167–172.
253. Snow, D.T. (1964) Landslide of Cerro Condor-Sencca, Department of Ayacucho, Peru, in G.A. Kiersch (ed.), *Geological Society of America Reviews in Engineering Geology*, Vol. 5, Geological Society of America, Boulder, Co., pp. 1–6
254. Sokolov, Y.F. and Shapovalov, G.I. (1968) Building a rockfill dam by directional blasting, *Hydrotechnical Construction* **2**, 385–389.
255. Stein, A. (1916) A third journey of exploration in Central Asia, 1913–1916 (continued), *The Geographical Journal* **48**, 193–225.
256. Strachey, R. (1894) The landslip at Gohna, in British Garwhal, *The Geographical Journal* **4**, 162–170.
257. Strom, A.L. (2006) Morphology and internal structure of rockslides and rock avalanches; grounds and constraints for their modelling, in S.G. Evans, G. Scarascia-Mugnozza, A.L. Strom and R.L. Hermanns (eds.), *Landslides from Massive Rock Slope Failure* NATO Science Series IV, Earth and Environmental Sciences, Vol. 49. Springer, Dordrecht, pp. 305–326.
258. Strom, A. (2010) Landslide dams in Central Asia Region, *Journal of the Japan Landslide Society*, **47**, 309–324.
259. Stucky Consulting Engineers (2001) Lake Sarez Mitigation Project; Design Report, 149 p.
260. Suter, M. (2004) A neotectonic-geomorphologic investigation of the prehistoric rock avalanche damming Laguna de Metztitlan (Hidalgo State, east-central Mexico), *Revista Mexicana de Ciencias Geológicas* **21**, 397–411.
261. Swanson, F.J., Oyagi, N. and Tominaga, M. (1986) Landslide dams in Japan, in R.L. Schuster (ed.), *Landslide Dams: Processes, Risk, and Mitigation*, Geotechnical Special Publication No. 3. American Society of Civil Engineers, New York, NY, pp. 131–145.

262. Tabata, S., Mizuyama, T., Inoue, K. and Sugiyama, M. (2000) Sediment movement caused from the Tombi landslide, by the 1858 Hietu Earthquake (M 7.0 – 7.1) in Central Japan, *Journal of Japan Society of Erosion Control Engineering* **53**, 59–70.
263. Taype Ramos, V. (1976) Mecanica del deslizamiento de Ccochay en el Rio Mantaro, *Bol. de la Soc. Geol. del Peru* **52**, 73–90.
264. Terado, T. (1970) Landslide of Mt. Takaiso, Tokushima Prefecture, and prevention of reservoir disaster (in Japanese with English abstract), *Geographical Sciences* **14**, 22–28.
265. Terzaghi, K. (1960) Storage dam founded on landslide debris, *Journal of the Boston Society of Civil Engineers* **47**, 64–94.
266. Tewari, P. (2004) A study of soil erosion in Pasighat town (Arunachal Pradesh) India, *Natural Hazards* **32**, 257–275.
267. Trauth, M.H., Alonso, R.A., Haselton, K.R., Hermanns, R.L. and Strecker, M.R. (2000) Climate change and mass movements in the NW Argentine Andes, *Earth and Planetary Science Letters* **179**, 243–256.
268. U.S. Army Corps of Engineers (1995) *Hydrologic engineering requirements for flood damage reduction studies*, Engineer Manual EM-1110-2-1419.
269. U.S. Army Corps of Engineers (Omaha District) (1960) *Madison River, Montana – Report on Flood Emergency Madison River Slide*, Vol. 2. Omaha, NE.
270. Voight, B. (1978) The Lower Gros Ventre slide, Wyoming, USA, in B. Voight (ed.), *Rockslides and avalanches, 1, Natural Phenomena*. Elsevier, Amsterdam, NY, pp. 113–166.
271. Voight, B., Janda, R.J., Glicken, H. and Douglass, P.M. (1983) Nature and mechanics of the Mount St. Helens rockslide-avalanche of 18 May, 1980, *Géotechnique* **33**, 243–273.
272. Wahl, T.L. (1998) *Prediction of embankment dam breach parameters; a literature review and needs assessment*, U.S. Bureau of Reclamation Dam Safety Office Dam Safety Research Report DSO-98-004, 60 p.
273. Wahl, T.L. (2004) Uncertainty of predictions of embankment dam breach parameters, *Journal of Hydraulic Engineering* **130**, 389–397.
274. Walder, J.S. and O'Connor, J.E. (1997) Methods for predicting peak discharge of floods caused by the failure of natural and constructed earthen dams, *Water Resources Research* **33**, 2337–2348.
275. Wang, Z., Cui, P., Yu, G-A. and Zhang, K. (2010) Study of landslide dams and development of knick points, *Environmental Earth Science*, doi 10.1007/s12665-010-0863-1.
276. Wang, G.-Q., Liu, F., Fu, X.-D. and Li, T.-J. (2008) Simulation of dam breach development for emergency treatment of the Tangjiashan Quake Lake in China, *Science in China Series E; Technological Sciences* **51**, 82–94.
277. Wassmer, P., Schneider, J.L., Pollet, N. and Schmitter-Voirin, C. (2004) Effects of the internal structure of a rock-avalanche dam on the drainage mechanisms of its impoundment, Flims sturzstrom and Ilanz paleo-lake, Swiss Alps, *Geomorphology* **61**, 3–17.
278. Watson, R.A. and Wright, H.E., Jr. (1969) The Saidmarreh landslide, Iran, *Geological Society of America, Special Paper* **123**, 115–139.
279. Wayne, W.J. (1999) The Alemania rockfall dam: A record of a Mid-Holocene earthquake and catastrophic flood in northwestern Argentina, *Geomorphology* **27**, 295–306.
280. Waythomas, C.F. (2001) Formation and failure of volcanic debris dams in the Chakachatna River valley associated with eruptions of the Spurr volcanic complex, Alaska, *Geomorphology* **39**, 111–129.
281. Webby, M.G. and Jennings, D.N. (1994) Analysis of dam break flood caused by failure of Tunawaea landslide dam, *Proceedings of International Conference on Hydraulics in Civil Engineering 1994, University of Brisbane, Australia*, 163–168.
282. Weidinger, J.T. (1998) Case history and hazard analysis of two lake-damming landslides in the Himalayas, *Journal of Asian Earth Sciences* **16**, 323–331.
283. Weidinger, J.T., Wang, J. and Ma, N. (2002) The earthquake-triggered rock avalanche of Cui Hua, Qin Ling Mountains, P.R. of China – the benefits of a lake-damming prehistoric natural disaster, *Quaternary international* **93–94**, 207–214.

284. Weil, B. (2006) The rivers come: Colonial flood control and knowledge systems in the Indus Basin, 1840s–1930s, *Environment and History* **12**, 3–29.
285. Xu, Q., Fan, X.-M., Huang, R.-Q. and van Westen, C. (2009) Landslide dams triggered by the Wenchuan Earthquake, Sichuan Province, south west China, *Bulletin of Engineering Geology and the Environment* **68**, 373–386.
286. Yang, X., Yang, Z., Cao, S., Gao, X. and Li, S. (2010) Key techniques for the emergency disposal of quake lakes, *Natural Hazards* **52**, 43–56.
287. Yi, C., Zhu, L., Seong, Y.B., Owen, L.A. and Finkel, R.C. (2006) A late glacial rock avalanche event, Tianchi Lake, Tien Shan, Xinjiang, *Quaternary International* **154–155**, 26–31.
288. Youd, T.L., Wilson, R.C. and Schuster, R.L. (1981) Stability of blockage in North Fork Toutle River, in P.W. Lipman and D.R. Mullineaux (eds.), *The 1980 eruptions of Mt St Helens, Washington*. United States Geological Survey Professional Paper 1250, pp. 821–829.
289. Zevallos, O., Augusta Fernandez, M., Plaza Nieto, G., and Klinkicht Sojos, S. (Eds.) (1996) *Sin Plazo Para la Esperanza; Reporte Sobre el Desastre de la Josefina-Ecuador, 1993*. Escuela Politecnica Nacional, Quito, Ecuador, 348 p.
290. Zhang, F., Zhang, Z., Han, Z., Sun, J. and Xiao, G. (2009) Risk assessment and treatment countermeasures for the barrier lakes of Wenchuan Earthquake on May 12th, 2008, *Acta Geologica Sinica* **83**, 826–833.
291. Zhou, C.H., Yue, Z.Q., Lee, C.F., Zhu, B.Q. and Wang, Z.H. (2001) Satellite image analysis of a huge landslide at Yi Gong, Tibet, China, *Quarterly Journal of Engineering Geology and Hydrogeology* **34**, 325–332.
292. Zhu, P.Y. and Li, T. (2001) Flash flooding caused by landslide dam failure. ICIMOD Newsletter No.38.
293. Zhu, P.Y., Wang, C.H. and Wang, Y.C. (2003) Large-scale landslide-debris avalanche in Tibet, China. (2) Formation of an exceptionally serious outburst flood from a landslide dam in Tibet, *Landslide News* **14–15**, 23–25.
294. Zinevich, Y.N. (1981) Mudflow detention dam at Mynzhilki, *Hydrotechnical Construction* **15**, 762–767.

# Chapter 2
# Engineering Measures for the Hazard Reduction of Landslide Dams

Robert L. Schuster and Stephen G. Evans

## 1 Modes of Flooding Resulting from Landslide Dams

Landslide dams cause two types of floods: (1) upstream (backwater) flooding as the impoundment fills (Fig. 2.1), and (2) downstream flooding resulting from failure of the dam. Although less common than upstream flooding, downstream flooding is usually more serious and may be catastrophic. The world's worst landslide-dam disaster occurred when the 1786 Kangding-Louding earthquake in Sichuan Province, China, triggered a huge landslide that dammed the Dadu River [26]. After 10 days, the landslide dam breached, creating a flood that extended 1,400 km downstream and drowned 100,000 people. A similar disaster resulted from the 1513 failure of a rock-avalanche dam on a tributary of the Ticino River in southern Switzerland. The resulting flood engulfed the city of Biasca with a surge of debris and water that continued down the Ticino valley for 35 km, drowning about 600 people [31]. These are examples of extreme events that probably could not have been prevented even with modern technology. However, mitigative measures can prevent the failure of many landslide dams, or at least can reduce the severity of flooding by reducing the volume of water in the impoundment. These mitigative measures often are combined with hazard warning to downstream communities.

## 2 Characteristics of Failure of Landslide Dams

A landslide dam and its impounded lake may last for a few hours or for thousands of years [37]. An analysis of failed landslide dams indicate that 83% failed within 6 months of formation (Fig. 2.2). Longevity of landslide dams and the landslide-dammed lake they impound depends on;

- Rate of inflow to the lake, which is based on the size of the upstream drainage basin and on the amount and rate of precipitation into the basin.

R.L. Schuster (✉)
U.S. Geological Survey, Denver, CO 80225, USA
e-mail: rschuster@usgs.gov

S.G. Evans et al. (eds.), *Natural and Artificial Rockslide Dams*, Lecture Notes in Earth Sciences 133, DOI 10.1007/978-3-642-04764-0_2, © Springer-Verlag Berlin Heidelberg 2011

**Fig. 2.1** Flooding in the Hunza Valley, northern Pakistan due to the filling of the rockslide-dammed lake impounded behind the Attabad rockslide which blocked the Hunza River on January 4, 2010 [23]. Lake waters submerged several villages, vast areas of agricultural lands adjacent to the Hunza, and 22 km of the Karakoram Highway. Photograph taken on May 28, 2010, a day before overtopping of rockslide dam (*Pakistan Tribune* photograph)

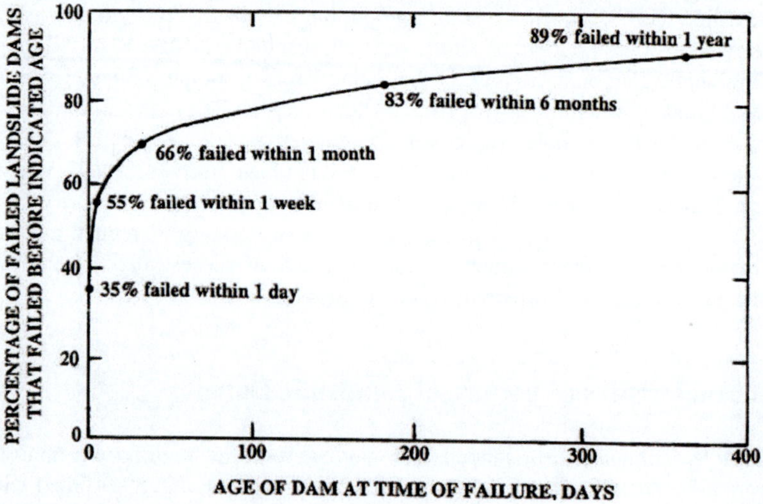


**Fig. 2.2** Length of time that failed landslide dams exist, based on 187 historic cases of failed dams (Modified from Schuster and Costa [37])

- Size and shape of the dam. Obviously, a high dam will take longer to fill than a low one, and a wide dam will be more resistant to failure upon overtopping than will a narrow one. Note that landslide dams are almost always wider (dimension parallel to the stream valley) in respect to height than are constructed earthfill dams, a factor that aids their stability.

- Behaviour during overtopping (if overtopping occurs)
- Physical character of the materials that comprise the dam. In general, the coarser the material (more rock fragments and boulders) comprising the dam, the more resistant it will be to erosion during overflow.
- Rate of seepage through the dam. Although only a few landslide dams have been known to have failed due to internal erosion, piping constitutes a possible cause of failure.

Longevity of landslide dams whose impoundments have filled depends mainly on resistance to erosion, either at the dam surface from surface overflow or in the interior of the dam from internal erosion (piping). Landslide dams consisting of large rock fragments are more resistant to erosion than dams comprised of high percentages of unconsolidated, easily eroded geologic materials. Occasionally, dams with considerable amounts of rock fragments and/or boulders will partially fail due to removal of fines, but the downward cutting of the breaching mechanism will be stopped by "self armoring" by the remaining coarse material. An outstanding example of this was the 1928 partial failure of the Gros Ventre landslide dam, Wyoming, USA, in which breaching was stopped by boulders after about one half of the dam had failed [1].

Some landslide dams do not fail quickly, often requiring months, or even years, to fail. An example is the 1985 200 $Mm^3$ Bairaman River landslide in Papua New Guinea, which formed a 200-m-high dam that required 16 months to fill. This dam breached within 1 day after overtopping [25]. Other dams never fail; instead, their impoundments slowly fill with lacustrine sediments. Examples of well-known, very long-lived landslide-dammed lakes that were formed by the following landslides are: Saidmarreh, Iran (10,000–11,000 yrs B.P; lake completely filled with sediment) [39]; Waikaremoana, New Zealand (2,200 yrs B.P.) [33]; Slumgullion, Colorado, USA (800–900 yrs B.P.) [35]; and Lake Crescent, State of Washington, USA (> 650 yrs B.P.) [27].

## 3 Mitigative Measures to Prevent Floods from Landslide Dams

When the possible failure of a landslide dam threatens people or property either upstream or downstream, immediate measures must be taken either to stabilize the blockage or to prevent lake levels from reaching dangerous elevations. These mitigative measures can be accomplished by one or more of the following:

- *Diversion of inflow water upstream from the lake* – This can be accomplished by diverting water from the stream into upstream reservoirs or irrigation systems. Although usually only a temporary measure, diversion may slow the filling of the lake enough to allow implementation of longer-term measures.
- *Temporary drainage from the lake by pumps or siphons* – Rising lake levels can be controlled temporarily by means of pumps or siphons that raise water over

the low point of the dam crest. This is usually a short-term (less than 1–2 yrs) measure that provides time for more extensive, long-term solutions.

- *Construction of an erosion-resistant spillway* – The most common method of stabilizing a landslide dam is to construct an erosion-resistant open-channel spillway across the dam or an adjacent abutment. When overtopping occurs, flow is controlled by the spillway, as is the case for emergency spillways on engineered dams. In addition, an excavated spillway lowers the elevation of the dam crest, thus reducing the extent of upstream flooding and the volume of impounded water. However, spillways are not always successful in preventing breaching and downstream flooding; they sometimes fail due to retrogressive erosion (i.e., headward erosion from the spillway outlet to its intake) caused by high-velocity overflow. To prevent erosion by minimizing flow velocity, the spillway should be wide and shallow. If possible, it should be lined with erosion-resistant materials (commonly rip rap), especially at the outlet. Often "check dams" or sills (structures that slow stream velocity in the channel) are installed along steeper grades of the spillway to prevent erosion. Spillways that fail due to erosion may have been partially successful because they have limited the total volume of the impoundment, thus reducing total discharge. Open-channel spillways across the landslide dam commonly are excavated by bulldozers; however, draglines, backhoes, explosives, and hand labor all have been used.
- *Drainage tunnel through an abutment* – A long-term method of preventing overtopping and breaching of a landslide dam is construction of a diversion tunnel through an adjacent dam abutment. Because large landslide dams commonly occur in mountain canyons, they usually have bedrock abutments; thus, rock-tunneling methods commonly are used. Usually, tunnel construction begins at the downstream portal and progresses upstream. It may be necessary to prevent the rising lake from encroaching on construction of the upstream portal of the tunnel by pumping and/or siphoning or by construction of a cofferdam. In some cases, it may be necessary for the tunnel to enter the lake below the level of the lake surface, thus requiring construction of a lake tap.
- *Drainage conduit through the dam* – Occasionally, a drainage conduit (commonly a culvert, pipe, or lined tunnel) is installed through the landslide dam to limit the elevation to which the lake can rise. To allow time for "dry" installation of the conduit, pumps and/or siphons can be used to slow the rate of filling of the lake.

## 4 Case Histories Utilizing the Above Mitigative Measures

### 4.1 *Cordèvolle Torrent, Veneto Province, Italy, 1771 (Monte Forca Rock Slide-Rock Avalanche)*

On 11 January 1771, a 20 $Mm^3$ rock slide–rock avalanche in carbonate rocks dammed the Cordèvolle River in northern Italy, impounding 4-km-long Lake Alleghe [13, p. 148]. In February 1771, an engineering mission contemplated the possibility of draining the lake, but soon realized that this was not possible. Later, stone masonry and concrete linings were installed to stabilize the overflow channel across the rocky debris that formed the dam crest. The Lake of Alleghe still exists as discussed by Marchi et al. [30], and (Chap. 1 by Evans et al., this volume).

## 4.2 Passer River, Alto Adige Province, Italy, 1774 (Ganderberg Rock Slide)

In 1404, the large Ganderberg rock slide blocked the flow of the Passer River, impounding the Passer Wildsee ("wild lake") ([13], p. 93). In 1419, a major part of the debris barrier failed; the ensuing outburst flood killed at least 400 people. On 17 September 1772, a regional rainstorm and plugging of the outlet by a log jam caused the lake to rise dramatically, releasing another flood. It was realized that the only lasting remedy against future debris floods would be to permanently lower the level of the lake. As a result, in 1773 a reinforced outlet channel and flood gates were constructed at the narrowest part of the outlet in order to control runoff during critical periods of sudden rainstorms or snowmelt. However, on 22 October 1774 the supervisor of the operation opened the sluice gates too far and the waters of the rain-swollen lake poured into the outlet channel. As a result, the undercut embankments collapsed and the lake drained in 12 h. The lake has not reappeared, but the Ganderberg slope is still moving. Over the years a great variety of training dikes, masonry linings, and check dams have been added to prevent similar disasters; their maintenance continues to be costly.

## 4.3 Madison River, Montana, USA, 1959 (Madison Canyon Rock Slide)

In 1959, the earthquake-induced, 21 $Mm^3$ Madison Canyon rock slide formed a 70-m-high blockage (Fig. 2.3) of the Madison River, southwestern Montana [17]. Fearing an overtopping failure of the natural dam, the US Army Corps of Engineers quickly constructed an open-channel spillway across the crest of the rock-slide dam. This 75-m-wide spillway was designed for a discharge of 280 $m^3/s$ and velocities that would only slowly erode the rock-fragment size that comprised the landslide dam [20; Chap. 1 by Evans et al., this volume]. In the 50 yrs since construction of the spillway, erosion has degraded the lake outlet only about 6–7 m.

## 4.4 Visocica River, Serbia, 1963 (Zavoj Landslide)

In February 1963, the 4 $Mm^3$ Zavoj landslide blocked the Visocica River, in eastern Serbia, forming a 35-m-high natural dam that impounded a 30 $Mm^3$ lake [3].

**Fig. 2.3** The landslide dam formed by the 1959 earthquake-induced Madison Canyon rock slide, Montana, USA, and Earthquake Lake, which it impounded (Photo by J.R. Stacey, US Geological Survey)

Inflow to the lake was estimated at 26 $m^3/s$, a rate that necessitated construction of a temporary open-channel spillway because there was not time to complete a permanent diversion tunnel before overtopping of the dam would occur. Long-term mitigation was accomplished by means of a 600-m-long, 2.5-m-diameter tunnel excavated through the left sandstone/marl abutment, starting at the down-stream end. The below-lake-level intake for the tunnel was accomplished by a lake tap, which was constructed by blasting through the last 10 m of rock. Today, the Zavoj rockslide dam forms the foundations of the Zavoj hydroelectric dam, constructed in 1988–1990, that impounds the enlarged Lake Zavoj (storage ~170 $Mm^3$).

## *4.5 Vaiont River, Friulia-Venezia-Giulia Region, Northern Italy, 1961 (Vaiont Landslide)*

In 1961, a 2.3-km-long diversion tunnel was constructed along the north bank of Vaiont Reservoir during early stages of its filling (Fig. 2.4). The tunnel was constructed as a precautionary measure because of concern that a landslide might block flow through the reservoir, a possibility posed by cracks along the south valley wall during early stages of filling [24]. In October 1963, as the lake neared capacity, a 270 $Mm^3$ rock slide from the south valley wall blocked the reservoir, thus damming the Vaiont River. The diversion tunnel was ideally located in that neither portal was buried by the slide.

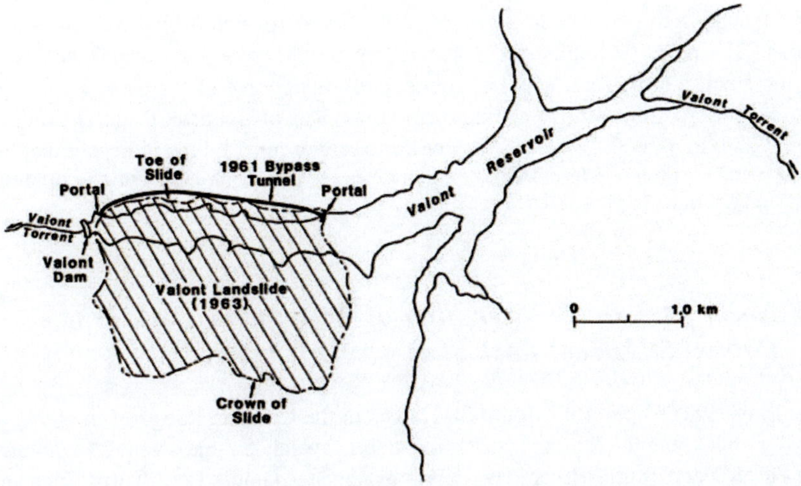


**Fig. 2.4** Map of the 1963 Vaiont rock slide indicating location of precautionary bypass tunnel that was constructed in 1961 (after [24])

## *4.6 Zeravshan River, Tajikistan, 1964 (Zeravshan Landslide)*

In 1964, large-scale blasting and bulldozing were used to excavate an open-channel spillway across a 15 $Mm^3$ landslide that dammed the Zeravshan River upstream from the ancient city of Samarkand [14]. The landslide dam was 220 m high, 400 m long (normal to the river), and 1,800 m wide. Two blasts, utilizing 250 tons of conventional explosives, followed by bulldozing, excavated a 40- to 50-m-deep open- channel spillway across the blockage, allowing the Zeravshan River to resume pre-landslide flow.

## *4.7 Mantaro River, South-Central Peru, 1974 (Mayunmarca Landslide)*

On 25 April 1974, the massive Mayunmarca landslide occurred in the valley of the Mantaro River in the Andes Mountains of south-central Peru, killing about 450 people. This landslide, with a volume of about 1.6 $km^3$, dammed the Mantaro River, forming a lake that reached a depth of about 170 m and a length of about 31 km before overtopping on 6–8 June 1974 [21]. Before overtopping, a trapezoidal spillway channel 3 m deep and 250 m long was excavated by bulldozers. The channel floor and sides were protected by steel netting, anchored by four rock-filled gabions. On the downstream face of the dam, three "energy dissipation" structures, consisting of long rock-filled cages of netting, were constructed.

The first trickle of water over the crest of the dam occurred on 6 June. On 8 June, the dam breached catastrophically by erosion of the channel, 44 days after the Mantaro River had been blocked by the slide. As noted by Hutchinson and Kojan [21], p. 7: "In retrospect, it seems that the measures that were taken were on too small a scale to effect [sic] significantly the course of the overtopping and subsequent erosion of the landslide dam." Estimates of the peak discharge ranged from 7,000 to 15,000 $m^3$/sec. An unconfirmed report on 11 June indicated that the flood wave by then had traveled 800 km downstream at an average velocity of about 11 km/hr (3 m/s) [21].

## 4.8 North Fork Toutle River, State of Washington, USA, 1980 (Mount St. Helens Rock Slide–Debris Avalanche)

The May 1980 eruption of Mount St. Helens in the Cascade Range of the State of Washington created a 2.8-$km^3$ rock slide–debris avalanche that swept 24 km down the North Fork Toutle River. The debris avalanche dammed the North Fork and tributaries, impounding four landslide-dammed lakes, the largest being Spirit Lake (Fig. 2.5). A government panel concluded that the 70-m-high Spirit Lake dam could not hold water above elevation 1,060 m without breaching, and estimated that the lake would reach this level by March 1983. By August 1982, 330 $Mm^3$ of water had been impounded and lake level had reached 1,055 m. Using a mathematical dam-break model, Swift and Kresch [38] predicted a flood of catastrophic proportions if the dam were to fail. The estimated peak discharge of 75,000 $m^3$/s would have

**Fig. 2.5** View of Mount St. Helens and Spirit Lake a few days after the 1980 eruption. The crest of the debris-avalanche dam that impounds Spirit Lake (foreground) is indicated by the *small arrows*. The location of the intake portal for the diversion tunnel (Fig. 2.6) through the volcanic-rock right abutment of the dam is indicated by the single *large arrow* (Photo by R.M. Krimmel, US Geological Survey)

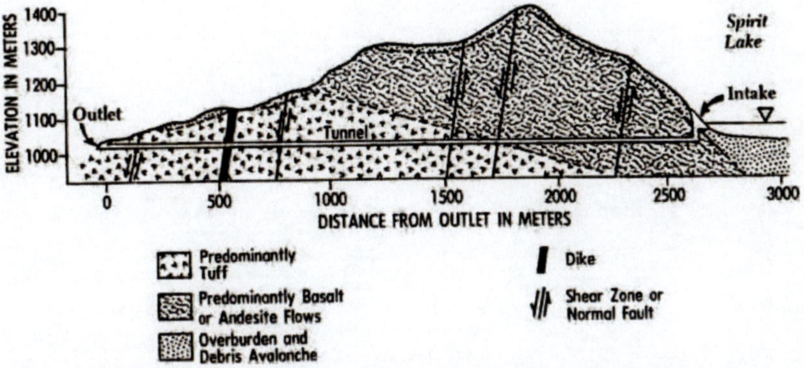


**Fig. 2.6** Geologic cross section along the Spirit Lake diversion tunnel (after [34])

inundated towns downstream along the Toutle and Cowlitz Rivers to depths of as much as 20 m.

As Spirit Lake slowly rose, in November 1982 the US Army Corps of Engineers installed a temporary pumping system to prevent overtopping. The 20-pump facility, with a maximum capacity of 5 $m^3$/s, pumped water through a 1,112-m-long by 1.5-m-diameter buried pipe placed across the dam crest [34]. Four possible alternatives were considered for a permanent solution to the Spirit Lake hazard: (1) a permanent pumping facility (2) an open-channel spillway (3) a buried conduit, and (4) a diversion tunnel through the bedrock right abutment. Each alternative was evaluated in regard to (1) stability in relation to the debris avalanche (2) stability in case of volcanic or seismic activity (3) efficacy of construction (4) environmental disturbance of the site, and (5) cost. Based on these considerations, it was decided to maintain permanent lake level at elevation 1,048 m by means of a 2.59-km-long, 3.4-m-diameter tunnel (Fig. 2.6) through the volcanic-ridge right abutment immediately west of Spirit Lake.

Excavation of the tunnel was accomplished by the tunnel-boring machine (TBM) method from the downstream portal to the intake. Because the intake was below lake level, it was constructed using staged excavation (Fig. 2.7). A bedrock "cofferdam" (i.e., intake plug) left in the intake portal was removed by explosives.

## *4.9 Coldwater Creek, Tributary of the North Fork Toutle River, Washington, USA, 1980 (Mount St. Helens Rock Slide–Debris Avalanche)*

Coldwater Creek, a tributary of the North Fork Toutle River, was dammed at its outlet by the 1980 Mount St. Helens debris avalanche (Fig. 2.8) [29]. If it had been allowed to overtop naturally, Coldwater Lake would have developed an estimated

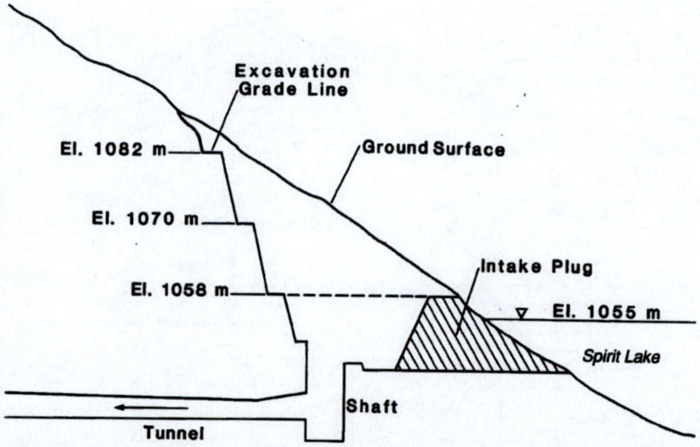


**Fig. 2.7** Diagrammatic cross section through the intake for the Spirit Lake diversion tunnel after [34]

**Fig. 2.8** Coldwater Lake and the debris-avalanche blockage of Coldwater Creek, near Mount St. Helens. Location of the excavated channel spillway across the right volcanic-bedrock abutment is indicated by the *white arrows*

maximum volume of 128 Mm$^3$ by late 1981 [34]. However, to prevent catastrophic breaching of the debris dam, the US Army Corps of Engineers constructed a permanent spillway for the lake across the erosion-resistant volcanic-bedrock right abutment. This spillway, which continues to function satisfactorily, has stabilized the volume of Coldwater Lake at 83 Mm$^3$.

## 4.10 Spanish Fork River, Utah, USA, 1983 (Thistle Debris Slide)

The April 1983 Thistle debris slide (volume: 22 $Mm^3$) in central Utah was a reactivation of a prehistoric landslide in sedimentary rocks caused by above-normal spring snowmelt. The debris slide blocked the Spanish Fork River, forming 63-m-high Thistle landslide dam (Fig. 2.9), which impounded a lake with a volume of 74 $Mm^3$ [19]. Upstream flooding submerged the village of Thistle, destroying 15 businesses, 10 homes, and railroad switching yards.

Soon after the landslide occurred, officials of the State of Utah decided to drain the lake to prevent overtopping and probable failure of the natural dam. A temporary 145-m-long, 2.4×3.0-m-diameter, drainage tunnel through the dam at the right abutment was completed on 4 May 1983 to prevent overtopping [5, 19]. Shortly thereafter, it was decided to drain the lake by means of a permanent diversion tunnel through the sandstone right abutment. The plan included (1) a 670-m-long, 4.25-m-diameter, horseshoe-shaped tunnel excavated through the right abutment at stream level, and (2) a 5- to 6-m-diameter vertical shaft from lake level to the upstream terminus of the tunnel. A lake tap was installed by excavating a channel between the lake and the shaft. This channel was deepened sequentially as lake level was lowered (Fig. 2.10).

## 4.11 Adda River, Lombardia Region, Northern Italy, 1987 (Mount Zandila Rock-Slide–Rock Avalanche)

On 28 July 1987, the 35 $Mm^3$ Mount Zandila (Val Pola) rock slide–rock avalanche (Fig. 2.11) formed a natural dam 30–60 m high, 1,200 m long (normal to the river),

**Fig. 2.9** The 1983 Thistle debris slide and its temporary impoundment, Thistle Lake. The lake was drained by a tunnel constructed through the mountain at the *lower left*

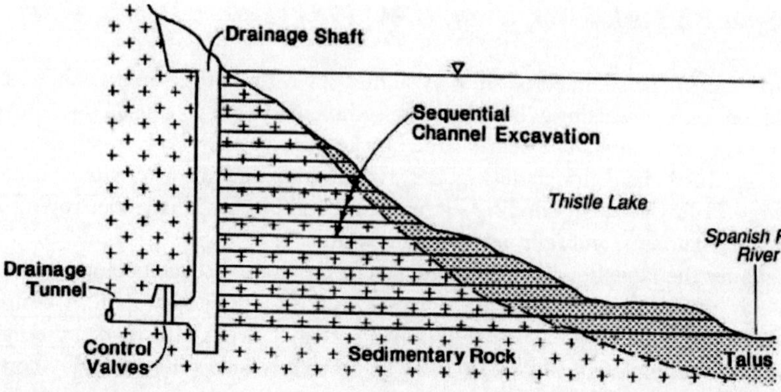


**Fig. 2.10** Cross section showing construction sequence for intake of the Thistle landslide-dam diversion tunnel through the right bedrock abutment (after [5])

**Fig. 2.11** Looking downstream across the 1987 Mount Zandila rock slide–rock avalanche and the Val Pola landslide dam. An open-channel spillway is under construction across the landslide dam. A twin-bore diversion tunnel (Fig. 2.13) passes through the granite bedrock valley wall on the lower left

and 2,500 m wide on the Adda River in northern Italy [7, 16]. Because of the danger of breaching of the landslide dam upon overtopping, 80,000 people living downstream were evacuated during August, a 2-km-long open-channel spillway was dug across the crest of the dam, and check dams were constructed on the steeper parts of this channel (Fig. 2.12). As a temporary measure, pumps and siphons were installed to drain Lake Val Pola, a function that was completed by the end of 1987.

**Fig. 2.12** Open-channel spillway, concrete sills, and a check dam under construction on the downstream slope of the Val Pola landslide dam

**Fig. 2.13** Exit portals for the twin-bore diversion tunnel that carries the Adda River around the Val Pola landslide dam. The tunnel on the right was operational at the time of the photo, and the left-hand tunnel was near completion

Soon after the landslide occurred, construction by drilling and blasting began on two diversion tunnels through the granite left abutment. These tunnels (Fig. 2.13), with diameters of 5.0 and 4.2 m, are 3.5 km long and have a total discharge capacity of 540 $m^3/s$ [7]. Because Lake Val Pola had been drained by pumping and siphoning before tunnel construction began, no lake tap was needed. In addition, timing of construction of the tunnels was not as critical as at most other landslide dams because there was little or no danger of the tunneling operation being inundated during construction. The tunnels became operational in 1988.

## *4.12 Pisque River, Northern Ecuador, 1990 (Pisque Landslide)*

On 2 January 1990, a 3.6 $Mm^3$ landslide, triggered by irrigation wastewater, dammed the Pisque River 30 km northeast of Quito, creating a 58-m-high natural dam [4]. The dam was composed of easily erodible, uniform silty sand derived from volcanic tuff; thus, it was considered to be very susceptible to erosional failure upon overtopping.

Consideration was given to pumping water over the dam to prevent overtopping by the lake, but this action was not feasible because pumps large enough for such an operation were not available in Ecuador. Instead, it was decided to bulldoze a deep open-channel spillway across the dam, thus lowering the dam crest. By 6 January, construction of an access road began over rough terrain to allow heavy equipment to reach the site. On 16 January, the access road had been completed and two bulldozers reached the site. It required 7 days to complete the channel under dangerous working conditions. The completed, nearly horizontal channel was 11 m wide, 9 m deep, and 100 m long. It was not lined because of lack of time to procure and place erosion-resistant rip rap. Because the landslide materials offered little resistance to erosion, it was recognized that this channel spillway would not prevent erosional failure of the dam. Instead, its function was to reduce the severity of the expected flood by limiting the depth of the lake.

One day after lake level rose to the channel invert, spillway erosion had caused a breach at least 50 m wide and 30 m deep, draining more than 60% of the lake at a peak discharge of 700 $m^3/s$. Had the spillway not been constructed, a larger flood undoubtedly would have occurred.

## *4.13 Vispa River, Switzerland, 1991 (Randa Rockfall)*

The Randa rockfall (volume: approximately 30 $Mm^3$), which occurred on 18 April and 9 May 1991, formed a 100-m-high dam of the Vispa River 10 km downstream from the tourist resort of Zermatt, Switzerland [6]. Four factors contributed to limit the rising of the lake impounded by this blockage:

- During the spring of 1991, the weather was fairly cold and dry so that snowmelt did not provide a major contribution to the flow of the Vispa River.
- A large part of the drainage area of the Vispa River is included in the catchment area of the Grand Dixence hydroelectric facility; thus, it was possible to divert part of the river's flow thanks to the Z'mutt pumping station.

- A road was quickly constructed to the site, which allowed the Swiss Army and several contractors to install 36 pumps that pumped water across the dam.
- An open-channel spillway was constructed across the dam.

These measures have been successful; the lake has not overtopped the dam thus far.

## *4.14 Paute River, Southern Ecuador, 1993 (La Josefina Rock Slide)*

On 29 March 1993, the 30 $Mm^3$ La Josefina rock slide formed a 70-m-high blockage of the Paute River, 25 km downstream from the city of Cuenca in southern Ecuador [9, Chap. 14 by Plaza et al., this volume]. The impoundment soon inundated upstream homes, farmland, and roads. It was not possible to obtain pumps in Ecuador that could handle the 50-150 $m^3/s$ inflow to the lake; thus, a decision was made to excavate a 23-m-deep open-channel spillway across the dam crest. By 8 April, bulldozers were working 24 h a day to reach design depth of the spillway before the lake rose to the working level. However, because of heavy rains, the lake rose faster than anticipated, and the excavated slope was in danger of failing. On 13 April, excavation efforts were abandoned at a depth of 18 m because of risk to bulldozer operators and equipment. Figure 2.14 shows the proposed and as-constructed cross sections of the excavation; the final excavated volume was 150,000 $m^3$. At this time, the opinion was that, after overtopping, erosion would be fairly rapid for the first few meters, but later would slow due to "self armoring" by material too coarse to be eroded from the spillway. Thus, partial breaching with a maximum flow of <3,000 $m^3/s$, not a catastrophic flood, was predicted.

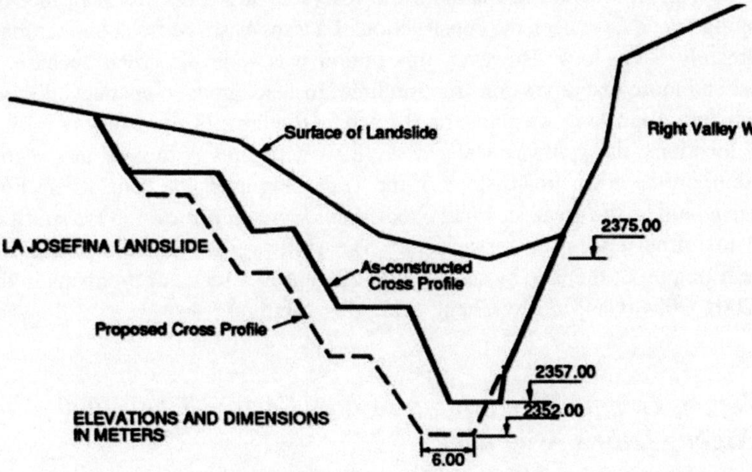


**Fig. 2.14** Proposed and as-constructed cross profiles for the open-channel spillway across the La Josefina (Paute River) landslide dam after [9]

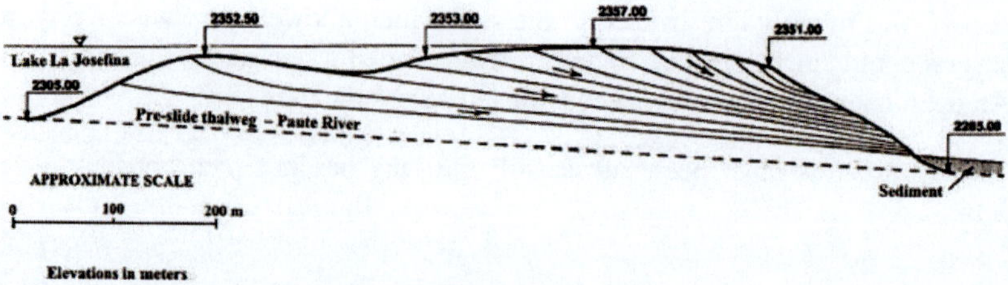


**Fig. 2.15** Cross section showing headward progress of erosion along the open-channel spillway across the La Josefina landslide dam after [9]

On 25 April, the lake overflowed the spillway at elevation 2,357.0 m; flow in the morning was about 1 $m^3$/s. On 30 April, flow increased to 35–40 $m^3$/s, and spillway erosion began working headward along the channel. Early on 1 May, retrogressive erosion (Fig. 2.15) reached the channel entrance, and the spillway failed, leading to a catastrophic flood estimated at 8,250 $m^3$/s [8]. The flood inundated homes, farmland, and transportation facilities before entering Amaluza Reservoir of the Paute Hydroelectric Project, 50 km downstream; an estimated 4 $Mm^3$ of sediment was deposited in the reservoir.

## *4.15 Chin-Shui River, Yin-Lin County, West-Central Taiwan, 1999 (Tsao-Ling Landslide)*

Large landslides dammed the Chin-Shui River at Tsao-Ling in west-central Taiwan in 1862, 1941, 1942, and 1979 [10]. Most of these landslide dams failed catastrophically, causing major damage downstream. On 21 September 1999, a M7.3 earthquake in central Taiwan caused another large (volume: $\sim$125 $Mm^3$) landslide [22]. A permanent solution to maintain the reservoir at a level less than that for overtopping would have been the construction of a trans-basin bedrock outlet tunnel approximately 5 km long. However, this option was soon discarded because of high cost and more than 2 yrs construction time. Instead, an open-channel spillway approximately 4 km long was constructed across the crest of the dam. At critical channel locations, the spillway walls were lined with rip rap to prevent erosion. In addition, on the downstream slope of the Tsao-Ling landslide dam, where flow velocities would be the greatest, small check dams were constructed in the spillway channel to further ensure against erosion. The spillway performed satisfactorily until the filling up of the lake by sediment in 2004 after a series of typhoons in the period 2001–2004 (Chap. 23 by Chang et al., this volume).

## *4.16 Yigong Zangpo River, Bomi County, Eastern Tibet, 2000 (Yigong Debris Avalanche)*

On 9 April 2000, a 100 $Mm^3$ debris avalanche dammed the Yigong River in eastern Tibet [18]. The natural dam, which was 60–100 m high, 2.5 km long, and 2.5 km wide, impounded Yigong Lake [42]. Following the advice of geologists and engineers, a spillway channel was dug across the dam by Chinese Army troops. By 7 June, rapidly rising Yigong Lake attained a volume of 2 $km^3$ (Chap. 22 by Evans and Delaney, this volume). On 10 June, the dam overtopped and the flood rapidly eroded the newly dug spillway channel, resulting in catastrophic failure of the dam. The 45-m-deep outburst flood lasted 6 h, damaging a highway downstream in Tibet and destroying many bridges. The flood rushed downstream into the valley of the upper Brahmaputra River in northern India where 94 people were killed and 2.5 million were left homeless [42].

## 5 Possible Mitigation of a Long-Lived Landslide Dam: Murgab River, Gorno-Badakhshan Province, Southeastern Tajikistan (Usoi Landslide)

In 1911, the 2 $km^3$ earthquake-triggered Usoi rock slide blocked the Murgab River, southeastern Tajikistan, forming a still-existing 600-m-high natural dam (Figs. 2.16 and 2.17) – the highest dam, natural or constructed, in the world [2, 36, Chap. 16 by Ischuk, this volume]. Lake Sarez, impounded by this blockage, is 60 km long, with a maximum depth of 550 m and a volume of approximately 17 $km^3$. This lake, which has never overtopped the dam, exits the downstream face as a series of large springs that regroup as the Murgab River. Freeboard between the lake surface and the lowest

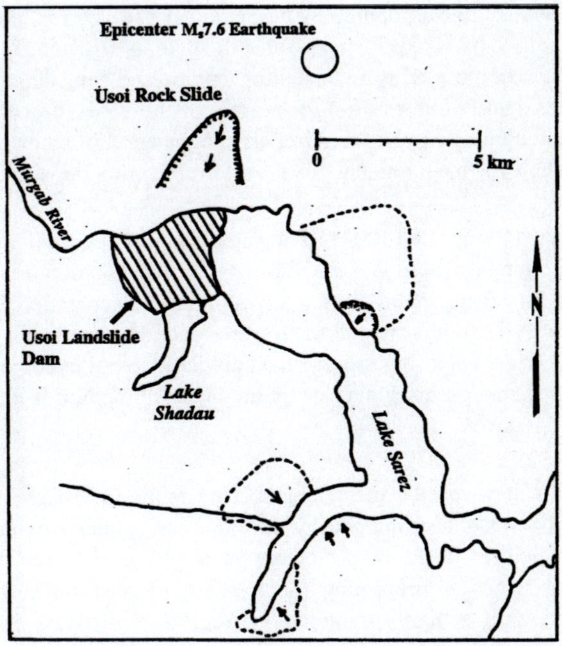


**Fig. 2.16** Map of the Lake Sarez, Tajikistan, area, showing the location of the 1911 $M_S$7.6 earthquake epicenter, source area of the Usoi rock slide, and rough outline of the Usoi landslide dam, after [15]

**Fig. 2.17** View of the Usoi rock-slide dam (*arrows*) and the downstream (west) end of Lake Sarez from the right (north) valley wall of the lake (1999 photo by J. Hanisch)

point on the dam crest currently is approximately 50 m and the lake is rising at an average rate of 18.5 cm/yr. Although the possibility of dam failure is slight, it potentially could result from (1) seismic shaking (2) catastrophic overtopping caused by a landslide entering the lake at high velocity (3) surface erosion caused by natural overtopping by the rising lake (4) internal erosion (piping) (5) instability caused by lake pressure against the dam, or (6) slope instability of the dam faces.

The possibility of installing engineered remedial measures to reduce the risk of dam failure resulting from any of the above processes was given considerable thought by Soviet scientists and engineers. Zolotarev et al. [44], p. 156 noted that the two "most rational" methods of protecting the dam were:

- "[C]ontrolled 100–150 m water level drawdown in the lake to eliminate overtopping by [a] high wave, through construction of a tunnel spillway on the left bank for irrigation in dry years and for power generation."
- "[R]aise the crest of the lowered section [i.e., lowest part of the dam crest] at the right bank [by] moving the boulder material over the construction using the construction machinery or by the blast fill method from the exposed scarps located above."

However, the major engineering required to lessen the hazard posed by Lake Sarez has been judged by development agencies as far too expensive to be realistic. As noted by Zolotarev et al. [44], p. 156: "The engineering protection activities at preventing the Ussoi [sic] obstruction from break-through with a view of a highly mountainous region, remoteness and complicated engineering geology are difficult to realize and are rather costly." The Usoi landslide dam is located approximately 200 km by road and track from the town of Khoroq, the capital of Gorno-Badakhshan Province [36]. Immediate access to the dam and lake lies over approximately 130 km of an undeveloped, one-lane track and 20 km by footpath through a very rugged canyon. It has been estimated by Periotto [32] that rehabilitation of the 130 km of one-lane road to permit transit of heavy construction equipment would cost from US$300,000 to US$600,000 per kilometer. Periotto noted that the feasibility of constructing a road over the 20 km of footpath through the Usoi River canyon to the dam has not yet been established; so the cost of such a road cannot even be estimated. Thus, it is clear that the cost of road construction through this extreme, mountainous terrain currently renders major remedial modifications to the dam and lake financially impractical [36].

# 6 Hazard Reduction Measures at Recent Rockslide Dams 2008–2010

## *6.1 Mitigation of Rockslide Dams Triggered by 2008 Wenchuan Earthquake*

The May 12, 2008 M8.0 Wenchuan Earthquake triggered thousands of landslides in the rugged topography of the southeastern Tibetan Plateau in Sichuan Province China [11, 12]. Many of these landslides dammed rivers draining to the southeast forming hundreds of "quake lakes" [12, 40]. 104 of these landslide dams impounded more than 100,000 $m^3$ of water [41] and their stability became a subject of great concern to Chinese authorities [12, 43] in the aftermath of the devastating earthquake. Critical lakes were identified [12] and mitigation works at a number of sites were begun as the lakes filled.

The measure most widely adopted by Chinese authorities to reduce both outburst risk and the magnitude of the hazard was the excavation of spillways over the surface of the damming debris. In the case of Tangjiashan Lake, the largest of the rockslide-dammed lakes formed in the aftermath of the earthquake, a major civil engineering operation was involved in the excavation of a spillway. Construction lasted 7 days during which time 135,000 $m^3$ of material were excavated to form the 425 m long spillway. On June 8, the lake reached the elevation of the spillway and overflow began [28]. The volume of the lake at overflow was 247.4 $Mm^3$ compared to a maximum lake basin volume (without spillway) of 320 $Mm^3$, a reduction in maximum lake volume of 23%. Flow in the spillway increased and a controlled partial breach occurred on June 10 involving a breached volume of 160 $Mm^3$ and a maximum discharge of 6,500 $m^3/s$ (Fig. 2.18). During the controlled breaching, lateral and vertical erosion of the spillway occurred (Fig. 2.18) but complete catastrophic failure of the dam did not take place. The controlled partial breaching left a significant remnant lake with an estimated volume of 87 $Mm^3$ [28].

**Fig. 2.18** Controlled breaching of Tangjiashan Lake, Sichuan, China taking place on June 10, 2008

The success of the Tangjiashan mitigation operation was repeated at several other rockslide dams in the earthquake affected area and no catastrophic outbursts occurred [Chap. 10 by Cui et al., this volume].

## *6.2 Measures Taken to Reduce Risk at 2010 Hunza Rockslide-Dammed Lake*

A major rockslide-dammed lake formed in the Hunza River valley of northern Pakistan in the first half of 2010 [23]. It was impounded by the Attabad rockslide (est. vol. 55 $Mm^3$) which blocked the Hunza River on January 4, 2010. As the waters of the lake rose they flooded several villages, large tracts of agricultural land adjacent to the Hunza, and 22 km of the Karakoram Highway linking Pakistan and China (Fig. 2.1). To achieve an eventual controlled overflow, Pakistan authorities began the excavation of a spillway over the debris on 29 January (Day 25 of impoundment) completing it around May 15 (Day 129). The excavation of the spillway lowered the effective crest of the dam by around 15 m.

Overtopping of the debris through flow in the spillway occurred on May 29 (Day 143 of impoundment) at an elevation of 2,435 m a.s.l. At this maximum pool elevation the lake area was 9.74 $km^2$ and its volume was 430 $Mm^3$ [23, Chap. 7 by Delaney and Evans, this volume]. However, the lake level continued to rise a further 3 m in the 5 days up to June 3, since lake inflow exceeded lake outflow. By June 5, however, the lake level had stabilised (Fig. 2.19) as discharge in the spillway increased. Photographs published by the *Pamir Times* after overtopping show the enlargement of the spillway, first by deepening through headward erosion from the

**Fig. 2.19** Spillway in Attabad rockslide debris, shortly after overtopping began on May 28, 2010. Note vertical erosion of spillway and retrogression of "knickpoint". As of July 25, 2010 (200 days after impoundment) stable overtopping continues with lake level more or less unchanged [Photograph courtesy of Focus Humanitarian, Pakistan]

lowest part of the spillway and then by widening through lateral undercutting and seepage erosion of the spillway walls. As of July 25, 2010 stable overflow through the spillway is continuing and lake level remains about the same as it was at the time of overflow. Whilst lake waters still submerge vast tracts of land in the Hunza Valley the stable overflow has thus far avoided a catastrophic draining/partial draining of Hunza Lake.

Pakistan authorities are actively considering engineering solutions which can result in at least a partial lowering of the lake surface. They have also been active in mitigating the hazard at the rockslide dam and the risk to downstream communities. In addition to excavating the spillway over the debris to effect a controlled overtopping as described above, they have also been diligent in monitoring the lake, delimiting possible flood zones downstream, evacuating threatened areas, and installing a siren-based warning system in the event of a catastrophic breach being initiated.

## 7 Conclusions

The cases presented illustrate the difficulties in mitigation of hazards from landslide dams. Open-channel spillways are often excavated across the dam to provide controlled overtopping. In some cases, such open-channel spillways have not been successful because of rapid retrogressive erosion. However, even for failure of a dam with a constructed open-channel spillway, these surface drains have reduced both upstream and downstream flooding by limiting the amount of water impounded. Spillways constructed across bedrock abutments usually provide long-term mitigation.

Bedrock diversion tunnels constructed through an abutment have almost always been successful as long-term remedial measures for landslide dams. However, these tunnels are expensive and usually require considerable construction time. The extra time needed to construct such tunnels can be gained by utilizing upstream diversion, pumps and/or siphons, temporary conduit drains (through the dam), or open-channel spillways (across the dam). In some cases, the inlets for these tunnels are constructed below lake level, requiring construction of lake taps.

## References


1. Alden, W.C. (1928) Landslide and flood at Gros Ventre, Wyoming, *Institute of Mining and Metallurgical Engineers Transactions* **76**, 347–360.
2. Alford, D. and Schuster. R.L. (eds.) (2000) *Usoi Landslide dam and Lake Sarez – An Assessment of Hazard and Risk in the Pamir Mountains, Tajikistan*. United Nations International Strategy for Disaster Reduction (ISDR), Geneva, ISDR Prevention Series No. 1, 115 pp.
3. Anagnosti, P., Cavounides, S. and Petrides, G. (1989) Landslides in southeastern Europe: Extent and economic significance, in E.E. Brabb and B.L. Harrod (eds.), *Landslides: Extent and Economic Significance*, Proceedings, 28th International Geological Congress, Symposium on Landslides, Washington, D.C., pp. 381–385.
4. Asanza, M., Plaza-Nieto, G., Schuster, R.L., Yepes, H. and Ribadeneira, S. (1992) Landslide blockage of the Pisque River, northern Ecuador, in D.H. Bell (ed.), *Landslides*, Vol. 2, Proceedings, 6th International Symposium on Landslides, Christchurch, 10–14 February, pp. 1229–1234.
5. Bischoff, J.A., Essex, R.J. and Freeman, S.T. (1984) Lake Thistle – evaluation of lake tap alternatives, *Tunneling Technology Newsletter*, National Research Council, Washington, DC **45**, 1–7.
6. Bonnard, C. (2004) Technical and human aspects of historic rockslide dammed lakes and landslide dam breaches, in K. Abdrakhmatov, S.G. Evans, R. Hermanns, G. Scarascia-Mugnozza and A.L. Strom (eds.), *Security of Natural and Artificial Rockslide Dams*, Extended Abstracts Volume, NATO Advanceed Research Workshop, Bishkek, Kyrgyzstan, 8–13 June, pp. 13–19.
7. Cambiaghi, A. and Schuster, R.L. (1989) Landslide damming and environmental protection – a case study from northern Italy, in *Proceedings, 2nd International Symposium on Environmental Geotechnology*, Shanghai, Vol. 1, pp. 381–385.
8. Canuti, P., Frassoni, A. and Natale, L. (1999) The 1993 La Josefina rockslide and Rio Paute landslide dam, Ecuador – 2) Failure of the Rio Paute landslide dam, in K. Sassa (ed.), *Landslides of the World*. Kyoto University Press, Kyoto, pp. 359–361.
9. Chamot, P. (1993) *El Deslizamiento La Josefina en el Valle del Rio Paute, Cuenca, Ecuador*. Report to the United Nations and Swiss Red Cross, Zurich.
10. Chang, S.C. (1984) Tsao-Ling landslide and its effect on a reservoir project, in *Proceedings, 4th International Symposium on Landslides,* Toronto, Vol. 1, pp. 469–473.
11. Chigira, M., Wu, X., Inokuchi, T. and Wang, G. (2010) Landslides induced by the 2008 Wenchuan earthquake, Sichuan, China, *Geomorphology* **118**, 225–238.

12. Cui, P., Zhu, Y.-Y., Han, Y.-S., Chen, X.-Q. and Zhuang, J.-Q. (2009) The 12 May Wenchuan earthquake-induced landslide lakes: Distribution and preliminary risk evaluation, *Landslides* **6**, 209–223.
13. Eisbacher, G.H. and Clague, J.J. (1984) *Destructive Mass Movements in High Mountains: Hazard and Management*. Geological Survey of Canada Paper 84–16, Canada, 230 pp.
14. *Engineering News-Record* (1964) *Russians Blast Through Landslide Dam*. The McGraw-Hill. Companies, Inc., America, 7 May, p. 24.
15. Gaziev, E. (1984) Study of the Usoy landslide in Pamir, in *Proceedings, 4th International Symposium on Landslides*, Toronto, 16–21 September, **1**, 511–515.
16. Govi, M. (1999) The 1987 landslide on Mount Zandila in the Valtellina, northern Italy, in K. Sassa (ed.), *Landslides of the World*. Kyoto University Press, Kyoto, pp. 47–50.
17. Hadley, J.B. (1964) Landslides and related phenomena accompanying the Hebgen Lake earthquake of August 17, 1959, in *The Hebgen Lake, Montana, Earthquake of August 17, 1959*, U.S. Geological. Survey Professional Paper 435, pp. 107–138.
18. Han, Z.S. (2003) Large-scale landslide-debris avalanche in Tibet, China (1) April-June 2000 Yigong landslide, Tibet, China, *Landslide News* **14/15**, 22–23.
19. Hansen, D.C. and Morgan, R.L. (1986) The 1983 landslide dam at Thistle, Utah, in R.L. Schuster (ed.), *Landslide Dams: Processes, Risk, and Mitigation*. American Society of Civil Engineers Geotechnical Special Publication 3, New York, NY, pp. 59–83.
20. Harrison, A. (1974) Madison Canyon slide mass modification by the U.S. Army Corps of Engineers, in B. Voight (ed.), *Rock Mechanics – the American Northwest*, 3rd International Congress on Rock Mechanics, International Society Rock Mechanics, Pennsylvania State University, College of Earth and Mineral Sciences Experiment Station Special Publication, pp. 138–143.
21. Hutchinson, J.N. and Kojan, E. (1975) *The Mayunmarca Landslide of 25 April 1974*. UNESCO, Paris, February, Serial No. 3124/RMO.RD/SCE, 23 pp. plus figures.
22. Kamai, T., Wang, W.N. and Shuzui, H. (2000) The landslide disaster induced by the Taiwan Chi–Chi earthquake of 21 September 1999, *Landslide News* **13**, 8–12.
23. Kargel, J.S., Leonard, G., Crippen, R.E., Delaney, K.B., Evans, S.G. and Schneider, J. (2010) Satellite monitoring of Pakistan's rockslide-dammed Lake Gojal, *EOS* **91**, 394–395.
24. Kiersch, G.A. (1964) Vajont reservoir disaster, geologic causes of tremendous landslide accompanied by destructive tidal wave, *Civil Engineering, American Society of Civil Engineers* **34**(3), 32–39.
25. King, J., Loveday, J. and Schuster, R.L. (1989) The 1985 Bairaman River landslide dam and resulting debris flow, Papua New Guinea, *Quarterly Journal of Engineering Geology* **22**, 257–270.
26. Li, T. (1989) Landslides: Extent and economic significance in China, in E.E. Brabb and B.L. Harrod (eds.), *Landslides: Extent and Economic Significance*, Proceedings, 28th International Geological Congress Symposium on Landslides, Washington, DC, pp. 271–287.
27. Logan, R.L. and Schuster, R.L. (1991) Lakes divided: The origin of Lake Sutherland and Lake Crescent, Clallam County, Washington, *Washington Geology*, Washington Department of Natural Resources, Division of Geology and Earth Resources **19**(1), 38–42.
28. Lui, N., Zhang, J.-X., Lin, W., Cheng, W.-Y. and Chen, Z.-Y. (2009) Draining Tangjiashan Barrier Lake after Wenchuan Earthquake and the flood propagation after the dam break, *Science in China Series E; Technological Sciences* **52**, 801–809.
29. Meyer, W., Sabol, M.A. and Schuster, R.L. (1986) Landslide-dammed lakes at Mount St. Helens, Washington, in R.L. Schuster (ed.), *Landslide Dams: Processes, Risk, and Mitigation*. American Society of Civil Engineers Geotechnical Special Publication 3, 21–41.
30. Marchi, L., Michieli, F. and Zuppi, G.M. (2006) The Alleghe Lake (Dolomites, Italy): Environmental role and sediment management. In J. Krecek, and M. Haigh (eds.), *Environmental Role of Wetlands in Headwaters*, Vol. 63, NATO Science Series IV Earth and Environmental Sciences. Springer, Dordrecht, pp. 161–172.

31. Montandon, F. (1933) Chronologie des grands eboulements alpins, du debut de l'ere chretienne a nos jours, *Materiaux pour l'Etude des Calamites*, Societe de Geographie Geneve **32**, 271–340.
32. Periotto, B. (2000) Accessibility of the Bartang River valley and the Usoi Dam/Lake Sarez, in D. Alford and R.L. Schuster (eds.), *Usoi Landslide Dam and Lake Sarez – An Assessment of Hazard and Risk in the Pamir Mountains, Tajikistan.* United Nations Secretariat for International Strategy for Disaster Reduction, Geneva, pp. 73–82.
33. Read, S.A.L., Beetham, R.D. and Riley, P.B. (1992) Lake Waikaremoana barrier – a large landslide dam in New Zealand, in D.H. Bell (ed.), *Landslides*, Proceedings, 6th International Symposium on Landslides, Christchurch, 10–14 February, 2, pp. 1481–1487.
34. Sager, J.W. and Chambers, D.R. (1986) Design and construction of the Spirit Lake outlet tunnel, Mount St. Helens, Washington, in R.L. Schuster (ed.), *Landslide Dams: Processes, Risk, and Mitigation.* American Society of Civil Engineers Geotechnical Special Publication 3, pp. 42–58.
35. Schuster, R.L. (1996) Slumgullion landslide dam and its effects on the Lake Fork, in D.J. Varnes and W.Z. Savage (eds.), *The Slumgullion Earth Flow: A Large-Scale Natural Laboratory*. U.S. Geological Survey Bulletin 2130, Washington, pp. 35–41.
36. Schuster, R.L. and Alford, D. (2004) Usoi landslide dam and Lake Sarez, Pamir Mountains, Tajikistan, *Environmental and Engineering Geoscience* **10**, 151–168.
37. Schuster, R.L. and Costa, J.E. (1986) A perspective on landslide dams, in R.L. Schuster (ed.), *Landslide Dams: Processes, Risk, and Mitigation.* American Society of Civil Engineers Geotechnical Special Publication No. 3, New York, NY, pp. 1–20.
38. Swift, C.H., III and Kresch, D.L. (1983) *Mudflow Hazards Along the Toutle and Cowlitz Rivers from a Hypothetical Failure of Spirit Lake Blockage*. U.S. Geological Survey Water-Resources Investigation Report 82–4125, Tacoma, WA, 10 pp.
39. Watson, R.A. and Wright, H.E. (1969) The Saidmarreh landslide, Iran, *Geological Society of America Special Paper* **123**, 115–139.
40. Xu, Q., Fan, X.-M., Huang, R.-Q. and van Westen, C. (2009) Landslide dams triggered by the Wenchuan Earthquake, Sichuan Province, south west China, *Bulletin of Engineering Geology and Environment* **68**, 373–386.
41. Yang, X., Yang, Z., Cao, S., Gao, X. and Li, S. (2010) Key techniques for the emergency disposal of quake lakes, *Natural Hazards* **52**, 43–56.
42. Zhu, P.Y., Wang, C.H. and Wang, Y.C. (2003) Large-scale landslide-debris avalanche in Tibet, China (2) formation of an exceptionally serious outburst flood from a landslide dam in Tibet, *Landslide News* **14/15**, 23–25.
43. Zhang, F., Zhang, Z., Han, Z., Sun, J. and Xiao, G. (2009) Risk assessment and treatment countermeasures for the barrier lakes of Wenchuan Earthquake on May 12th, 2008, *Acta Geologica Sinica* **83**, 826–833.
44. Zolotarev, G.S., Fedorenko, V.S. and Vinnichenko, S.M. (1986) Ancient seismogenic and recent slides and slumps in Tajikistan (Central Asia), in V. Cotecchia (ed.), *Geologia Applicata e Idrogeologia*, Proceedings of the International Symposium on Engineering Geology Problems in Seismic Areas, **21**, pp. 149–157.

# Chapter 3
# Technical and Human Aspects of Historic Rockslide-Dammed Lakes and Landslide Dam Breaches

**C. Bonnard**

## 1 Introduction

Landslide dam lakes are fairly frequent phenomena in mountainous regions, but are usually unknown or not recorded because of their often temporary character, following their rapid destruction or filling with sediments, especially if the reservoir is of a limited volume. Indeed, detailed data on landslide dams are quite scarce in the case of rapid dam breaches. However, the major phenomena have left significant traces in history following the destructive consequences induced by dam breaches which are liable to transform an initially localised event into a regional, if not international, disaster, as could happen with the largest landslide dam in the world, Lake Sarez in Tadjikistan [10, 13].

Indeed, in most of the cases for which information is available, the landslide itself caused only minor damage to persons and property, but the formation of the lake upstream and the flooding downstream following the dam breaches have provoked momentous or permanent destruction that is reported in archive documents or in historic books relating past disasters [1, 7, 12, 16], or are known through expert reports.

A general description of landslide dam processes and characteristics is given in (Schuster (ed.) [22]), but most of the cases presented in this book refer to America and Asia, so that it was deemed interesting to present a dozen major historical European cases which have ocurred over the last 2,000 years. The paper will then give more detailed data on three fairly recent landslide dams: in Switzerland (Randa, 1991), Peru (Mayunmarca, 1974 – thanks to a personal contribution of Prof. Hutchinson, 2004, personal communication) and Ecuador (La Josefina, 1995). In these cases, preventive human action was essential to the mitigation of the consequences of the potential or effective dam breaches. The paper will then conclude with some perspectives related to the management of potential landslide dams, illustrated by a significant case in France (Séchilienne).

C. Bonnard (✉)
Soil Mechanics Laboratory, Swiss Federal Institute of Technology, Lausanne, Switzerland
e-mail: christophe.bonnard@epfl.ch

S.G. Evans et al. (eds.), *Natural and Artificial Rockslide Dams*, Lecture Notes in Earth Sciences 133, DOI 10.1007/978-3-642-04764-0_3, © Springer-Verlag Berlin Heidelberg 2011

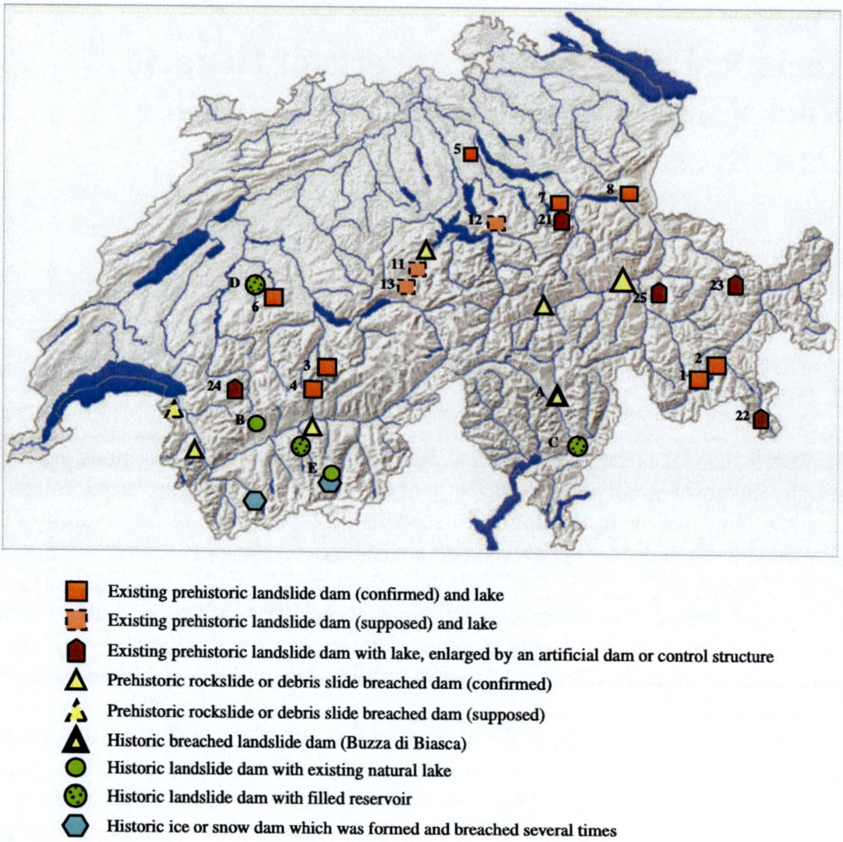


**Fig. 3.1** Map of landslide dams in Switzerland (the cases presented in the following text are marked by letters A–E; see the names of numbered cases in Table 3.1)

Landslide dams are quite ancient phenomena and their existence since the glacial periods is still evidenced by several natural lakes of large size in Switzerland (e.g., Lake Sils, Lake Silvaplana, Lake Poschiavo, Lake Schwarzsee, Lake Klöntal, Lake Davos, Lake Arnon; for the last three, the reservoir has even been increased for hydro-electric purposes by the construction of a dam on the landslide mass itself) (Fig. 3.1). The most ancient reported events date from the beginning of the Christian Era (the Turedunum Landslide, in the Rhone Valley, reported by Bishop Marius of Lausanne in the fifth century [21]; the Platé Massif Landslide, in the Arve Valley, as commented by Mougin [17] – see below).

Some of these ancient events are even better recorded as they occurred several times at the same site; either they were caused by repetitive massive debris flows or the first main scarp induced regressive failures and consecutive rockfalls. Many reported cases put forward several typical characteristics, either with respect to their mechanism or with respect to the reaction of the affected population or authorities

**Table 3.1** List of major confirmed or supposed landslide dam lakes in Switzerland, still existing

| No. | Name | Canton | Area ($km^2$) | Max. El. (m a.s.l.) | Max. depth (m) | Volume of lake or reservoir ($M m^3$) | Type | Observations |
|---|---|---|---|---|---|---|---|---|
| *Confirmed existing prehistoric landslide dam and lake* | | | | | | | | |
| 1 | Lake Sils | Graubünden | 4.11 | 1,797 | 71 | 137 | A | Max. level regulated |
| 2 | Lake Silvaplana | Graubünden | 3.16 | 1,791 | 77 | 132 | A | Max. level regulated |
| 3 | Lake Oeschinen | Bern | 1.15 | 1,578 | 56 | | B | |
| 4 | Lake Dauben | Wallis | 0.69 | 2,205 | | | A | No overflow is observed |
| 5 | Lake Türler | Zürich | 0.49 | 643 | 22 | | A | |
| 6 | Lake Schwarzsee | Freiburg | 0.47 | 1,046 | 10 | | A1 | |
| 7 | Obersee | Glarus | 0.24 | 989 | 10 | | A | |
| 8 | Lake Voralp | St. Gallen | 0.15 | 1,123 | | | | |
| *Supposed existing prehistoric landslide dam and lake* | | | | | | | | |
| 11 | Lake Sarnen | Obwald | 7.38 | 469 | 52 | 244 | A | According to A. Heim |
| 12 | Lake Lauerz | Schwyz | 2.99 | 447 | 14 | | A | |
| 13 | Lake Lungern | Obwald | 2.01 | 689 | 68 | 65 | A | |
| *Existing prehistoric landslide dam with lake enlarged by an artificial dam or control structure* | | | | | | | | |
| 21 | Lake Klöntal | Glarus | 3.29 | 847 (832)[a] | 47 | 56 | A | Dam built in 1908 |
| 22 | Lake Poschiavo | Graubünden | 1.98 | 962 | 84 | 111 | A | Control structure built in 1908 |
| 23 | Lake Davos | Graubünden | 0.59 | 1,559 | 51 | 15 | A | Dam built in 1920 |
| 24 | Lake Arnen | Bern | 0.45 | 1,543 | 50 | 12 | A1 | Dam built in 1942 |
| 25 | Heidsee | Graubünden | 1.0 | 1,484 | 5 | 0.5 | A | Lake on the landslide mass |

Notes: Among the 35 natural lakes in Switzerland (including those enlarged by dams) which cover (or covered) an area above approx. 0.5 $km^2$, 12 can reliably be considered as landslide dam lakes.

[a]Original level before construction of the dam

Explanation of types according to Heim [12] (modified):

A A main valley is dammed by a landslide originating in an adjacent slope or a lateral valley

A1 A-type landslide dam formed by two opposed landslides

B A lateral valley is dammed by a landslide extending in the main valley and the technical actions undertaken in order to reduce the harmful consequences of these sudden dramatic phenomena.

## 2 Historical Survey of Major Landslide Dams in the Alps

This survey, which does not aim at being exhaustive, will be made from a chronological perspective (always considering the earliest event reported at a specific site), without consideration of the respective countries (which have in any event changed in the Alpine region over the last 2,000 years), so as to describe the evolution of the respective impacts and human reactions. For each of the three periods considered in the following paragraphs, four cases will be presented and generally illustrated.

### *2.1 First to Fifteenth Centuries (Roman Era and Middle Ages)*

One of the most ancient reported cases originated below the limestone Massif of Platé, near Mont-Blanc, some 1,500 m above the Arve River (Fig. 3.2). During the early Christian Era a large slide mass blocked the gorge of the Arve, an affluent of the Rhone River, impounding a lake which overflowed, not into the original gorge, but into a parallel gorge of Le Châtelard, some 500 m to the south. The lake seems to have persisted until the thirteenth century, that is more or less 1,000 years, which can be explained by the fact that the initial overflow did not occur on the slide mass itself. The failure of the landslide dam finally reestablished the old course of the river.

In February 1471 a part of the slope failed again, raising the water level some 150 m higher than the river bed, due to the narrowness of the gorge. Before the nearby village of Servoz was affected, its inhabitants excavated a ditch on the slide

**Fig. 3.2** Perspective view on the slides affecting the slopes of the Massif of Platé. The village of Servoz is shown by a star

mass, thus draining the lake without further damage. This last action proves that the best engineering technique to control landslide dams was already practiced more than 500 years ago.

In August 1191, some 150 km to the south of the first-mentioned site, a large rock avalanche and debris flow elevated a narrow part of the Romanche Valley which meets the Isère River near Grenoble, and rose the level of the already existing shallow lake near the town of Bourg d'Oisans by 10–15 m, apparently without causing fatalities. 28 years later, on September 24, 1219, the downstream face of the debris dam was eroded, leading to a collapse. The wave of water and debris destroyed the settlements all along the Romanche Valley down to the city of Grenoble, located 30 km downstream, which was partially flooded, as it was built on a flat plain. The channels of the Romanche and Isère Rivers were displaced and the flood was noted down to the Mediterranean Sea. Thousands of casualties were reported.

The debris flow dam phenomenon repeated itself in 1449 (when a new lake appeared), 1465, 1617 and 1666, but without downstream flooding, as a channel was quickly excavated after the last two episodes, limiting the flooding upstream and the surge downstream. Over the last three centuries, the lake bed has been progressively filled with sediments and has become a swampy area along a channelized river course. In the nineteenth century, drainage ditches transformed the plain into fertile agricultural land and later permitted the town to expand.

This case is useful to remember because, some 10 km downstream, the Séchilienne Landslide presently threatens the Romanche Valley (see Sect. 4 below).

On January 25, 1348 a significant earthquake shook the Dobratsch massif (el. 2,000 m a.s.l.) and caused large rockslides, one of which, approx. 30 $M\,m^3$ in volume, along a fault surface dipping 45° to the south, blocked the 2.5 km wide Gail Valley (el. 500 m a.s.l.) (Fig. 3.3). This valley, which corresponds to a major Alpine fault (Periadriatic Lineament) had already been affected by several rockfalls at the end of the last glaciation following the erosion of the toe of the north side of the valley. Several settlements were buried by the rockfall and flooded upstream, and a short time later the dammed lake burst across the dam and devastated the valley below. At present the hydraulic head created by the remaining debris lobe is used in a hydroelectric scheme and villages are developing in the valley despite risks related to the presence of large sinkholes and fissures behind the highest Dobratsch cliff face.

Finally, some 250 km to the west of the Dobratsch massif, in the narrow Passer Valley upstream of the town of Merano, the large Ganderberg rockslide blocked the valley in 1404 and created a lake 50 m deep and 1 km long, called Wildsee (or Wild Lake) (Fig. 3.4). This lake lasted for almost 400 years, despite several serious overflow episodes: in September 1419, a section of the debris dam failed and the resulting flood killed 400 people, down to the town of Merano, 25 km south of the lake; in September 1503, the walls of the town of Merano were completely demolished; in September 1512, the city tower of Merano collapsed and in May 1572, the reconstructed walls were breached again. This series of events is probably due to the constantly sagging movement affecting the Ganderberg slope. After a quiet period of nearly 150 years, a section of the dam failed again in June 1721 and

**Fig. 3.3** View of the top scarp of the Dobratsch rockslide. Large fissures can be seen in the forest on the top

a flood dashed downstream; at that time plans were prepared to design a protection structure, but the works were not carried out.

In September 1772, after a severe regional rainstorm which brought many logs to the lake, the outlet was plugged and the level of the lake rose, until 3 M $m^3$ of water and debris swept down the valley. The following year, a drainage channel equipped with gates was built to control the runoff, but in October 1774, the gates were opened too far and the flood flows undercut the embankments which then collapsed and provoked complete drainage of the lake in 12 h. Several buildings in Merano were demolished. The lake did not reappear, but over the last century several dykes, check dams and linings have been built and require permanent maintenance.

These first four ancient cases, whose details have been obtained from Eisbacher and Clague [7], demonstrate that the problems posed by landslide dammed lakes are always the same, whatever their age may be:

- dams may last very short or very long periods, therefore, the threat usually does not decrease with time;
- they cause dramatic consequences upon failure at a large distance downstream;
- possibilities of limiting the damage do exist provided that well and quickly built structures are carried out.

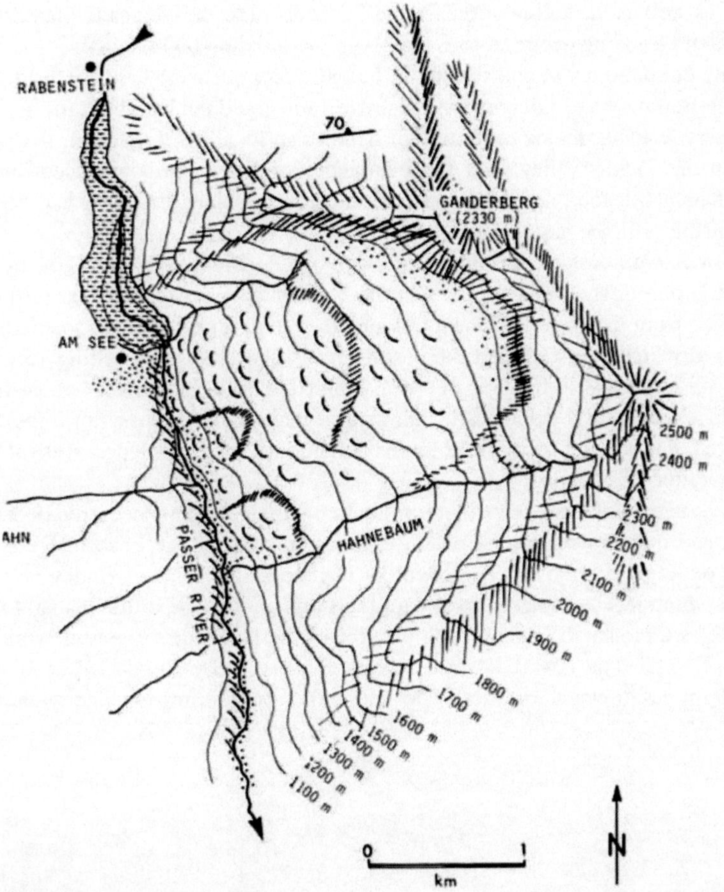


**Fig. 3.4** Sketch of the Ganderberg rockslide [7]

## *2.2 Sixteenth to Seventeenth Centuries*

On September 30, 1513, a large rockslide 10–20 M $m^3$ collapsed from the north-western steep slope of Pizzo Magno (el. 2,329 m a.s.l.) (Case A in Fig. 3.1) and blocked the Blenio Valley at el. 400 m, just upstream of the town of Biasca where the Ticino and Brenno Rivers (flowing in Val Blenio) merge. This landslide dam, called Buzza di Biasca [3], first destroyed several buildings down to the outskirts of Biasca, then created a lake extending 5 km upstream, which reached a volume of 100 M $m^3$ and drowned the communities of Malvaglia and Semione. During nearly 20 months, this lake, first well observed, seemed to have become a permanent feature of the valley. But on May 20, 1515, a few months before the battle of Marignan, the spring thaw overflow eroded the dam, causing its collapse. The city of Biasca was swept away, as well as the valley of the Ticino River down to Bellinzona (a main fortified town), and flooding occurred to Lago Maggiore, causing 600 fatalities.

This dramatic event had two main human consequences: first, the bridge connecting Bellinzona to Locarno was destroyed and could not be rebuilt for more than a century, causing major economic difficulties in local trade. Second, the population of the Ticino Valley took the inhabitants of Biasca to court, accusing them of witchcraft. It took several years until the court decided that they had no direct responsibility in the disaster.

To the south-east of the mountain ridge of Les Diablerets (el. 3,209 m a.s.l.), which is one of the highest peaks in the western Swiss Alps, several million $m^3$ detached from the slope in June 1714 and after that, repeated small rockfalls were observed which overwhelmed the pasture lands of Derborence, killing 15 persons (Fig. 3.5 and Case B in Fig. 3.1); one man trapped in his chalet, where he was making cheese, survived and after digging through the slide mass for 3 months he emerged, but was not recognized by his neighbours, as recounted in a novel by the Swiss writer C.F. Ramuz.

In the same area, a renewed larger failure of some 30 $M m^3$ occurred on July 23, 1749, and moved over a total distance of 5 km, creating the Lake of Derborence at the outlet of a nearby creek called the Derbonne, the Lake of Godey in another creek, which has been raised by a dam built just on the side of the landslide mass, as well as three small lakes just on the side of the landslide mass itself which still exist [9]. The total landslide mass has been assessed by Heim [12] at 50 $M m^3$. Smaller rockfall events occurred later but without major impact. The main natural

**Fig. 3.5** Lower part of the Derborence rockslide as seen from Les Diablerets mountain. The major lake is visible on the right side of the photo

lake, still existing today, is now nearly filled with sediment and its depth does not exceed 3 m, so that it no longer represents a threat (Fig. 3.5).

Some years later, in 1771, not very far from the recent Vaiont disaster at the limit of the Southern Alps, a dip slope rockslide of some 20 M $m^3$ blocked the Cordèvole Valley, climbing more than 50 m up the opposite slope, destroying three remote villages and killing their 49 inhabitants. With the formation of the Lake of Alleghe, 5 remote villages were drowned upstream as the lake extended over some 4 km. On May 1, 1771, a part of the main scarp cliff fell and plunged into the lake from a height of 1,000 m, causing a wave that surged over the opposite slope and up a lateral valley formed by the Zunaina torrent. In the nearby village of Alleghe, it damaged the church and killed three people. Happily, the rockslide dam was not affected by the wave; it was later reinforced with a stone masonry and concrete lining. As in the preceding case, the sediments of Cordèvole torrent are progressively filling the lake, the depth of which has decreased from 50 to 15 m. Despite residual risks of rockfall, the area of the dam has been developed as a vacation resort (Fig. 3.6).

Some 40 km south of the Dobratsch rockslide, at the toe of the southern Alps near the town of Udine, a large dip slope rockslide of some 30 M $m^3$ fell from the slope of Monte Auda, at an elevation of 1,500 m a.s.l., down to the Tagliamento Valley at an elevation of 500 m a.s.l., and climbed 100 m up the opposite side of the valley. This rockslide, which destroyed the village of Borta and killed 53 inhabitants in August 1692, is represented on one of the most ancient sketches of landslide disasters, produced by Pascolo Pascoli 1 month after the event [12], quoting [5] (Fig. 3.7). It created a lake almost 7 km long which reached a maximum depth of 80 m in 2 months; it then experienced two overflows on October 4 and 20, 1692,

**Fig. 3.6** Lake of Alleghe with a developing vacation resort on the landslide dam

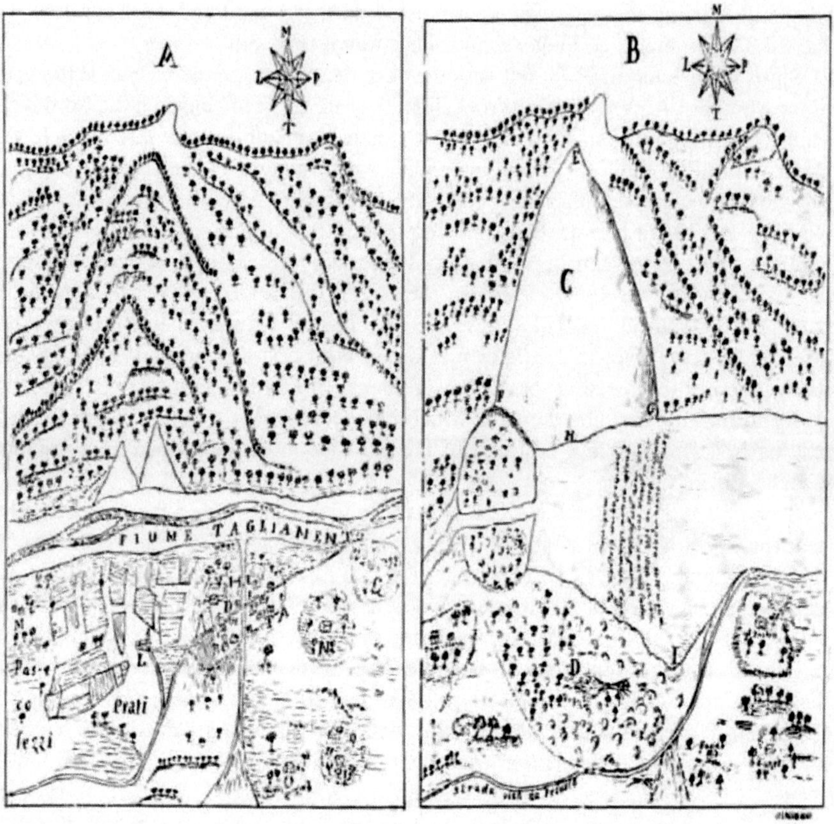


**Fig. 3.7** Sketch of the Tagliamento Valley rockslide produced by Pascolo Pascoli 1 month after the event, i.e. in 1692 [7]

which lowered the lake level. The lake persisted for some time, then filled with sediments over a depth of 30 m which were eroded to form a deep channel across the river deposits and rockslide mass. Other events affected the Tagliamento Valley over the last two centuries, mainly due to seismic activity, but they did not modify the stability conditions of the slope that had destroyed the village of Borta.

These last four cases, whose details were mainly obtained from Eisbacher and Clague [7] and Heim [12], confirm some well-known major long-term characteristics of landslide dam lakes:

- When the drainage area of the dammed valley is large, such as in the case of the Biasca and Tagliamento Valleys, the reservoir usually does not last very long.
- The presence of lakes in touristic areas constitutes an attraction even though residual risk related to the event which formed the lake often still exists.

- In three out of the four cases presented, warning signs of abormal activity of the slope had been perceived before the disaster, but the potential impact of the main landslide had been completely underestimated; the falling mass can thus extend high up on the slope facing the landslide.

## 2.3 Twentieth Century Cases

It is remarkable that very few major landslide dams are reported in the last century in the Alpine range. Therefore, only four particular cases bringing a new perspective to some problems related to landslide dams will be briefly mentioned.

The first one refers to the Motto d'Arbino rockfall, located some 20 km south of the Buzza di Biasca (see above), which occurred in October 1928, when 30–40 M $m^3$ of rock blocked the valley of Arbedo resulting in a small lake 1.5 km long (Fig. 3.8 and Case C in Fig. 3.1). Despite several debris flows which reached the village of Arbedo downstream, the gorge formed by the Traversagna torrent was not eroded. Between 1930 and 1950, a 30 m high debris retention dam was built so that the lake is now totally filled with sediments and the former debris course has been occupied by residential buildings. The most interesting aspect of this case is

**Fig. 3.8** View of the Motto d'Arbino rockfall dam in the Arbedo Valley, taken just after the event, in 1928

that, since 1925, a routine triangulation survey on the Motto d'Arbino monument at the top of the mountain showed a clear preparatory movement, which was monitored in detail in 1927 and 1928, supplying the first scientific and duly measured evidence of the acceleration phase before the rockfall [8], allowing the evacuation of the 16 houses that were demolished by the rockfall.

In the second case, the La Clapière rockslide in the south of the French Alps near the Italian border, the slope movement caused by a large toppling mechanism was first reported in 1976, but despite several critical periods of acceleration, in particular in 1987 and 2001 [Durville, 2002, personal communication, 19], the landslide mass displaced the river bed of the Tinée River towards the west, but never truly dammed it (Fig. 3.9). However, due to the high risks involved for the village of St. Etienne de Tinée immediately upstream and for the highly developed valley downstream, reaching the Mediterranean Sea in the outskirts of Nice, a diversion tunnel was built. The human aspects justified many preventive actions (sophisticated monitoring systems, significant modification of the access road, construction of a long diversion tunnel).

The third case, in the Valtellina Valley in the north of Italy, deserves mention for two reasons. First, it is a dramatic example of a two-phase landslide: an initial debris flow originating in the adjacent Val Pola, following the extraordinary rainfall of July 18 and 19, 1987, dammed the Adda Valley and increased the height of an existing small lake. A week later, as a long crack appeared in the slope of Monte Zandila and frequent rock blocks fell below this crack, four villages upstream were evacuated.

**Fig. 3.9** Recent photo of the La Clapière rockslide which permanently threatens the Tinée Valley, north of Nice (photograph supplied by J.-L. Durville)

But on July 28, at 7:23 a.m., a huge rockslide (approx. 40 M $m^3$) occurred, fell in the valley mainly downstream of the lake, the level of which had increased, and went up the opposite slope some 300 m above the valley floor; the upstream part of the mass fell directly in the lake, causing a 95 m high wave of water and debris which moved more than 2 km upstream, provoking the destruction of two villages which had not been evacuated initially and, therefore, the death of 27 persons [11] (Chap. 12 by Crosta et al., this volume).

This deserves mention here as a series of technical measures were taken to control the risks downstream induced by the 90 m high natural dam. First, a channel was excavated in the landslide mass to control the overflow. Pumping stations were installed to reduce the lake water level. Two 3 km long diversion tunnels were built within a year and a large spillway was built on the mass to allow the floods to pass without destroying the dam (see Chap. 12 by Crosta et al., this volume).

The last recent case worth reporting occurred in August 1994, due to the Chlöwena Landslide [23] which dammed the Höllbach torrent, not far from the city of Freiburg, after a progressive acceleration phase which lasted at least 4 months (Fig. 3.10 and Case D in Fig. 3.1). Even though the lake was of very limited volume, the risk was high as the dam height reached nearly 40 m, as the torrent was known to experience violent floods (hence his name: Höllbach, i.e. Hell torrent) and as there were several industrial plants along the river downstream in the town of Marly.

Due to the progressive movement of the slide (a maximum displacement of 250 m was recorded) and the creation of the dam, time was available for the computation of the formation of the dam with a viscous mechanical model and for the execution of hydraulic model tests to check the overflow conditions. Finally, the flow found its way along the edge of the landslide mass without causing any damage up to the present.

**Fig. 3.10** Lower part of the Chlöwena landslide and the destroyed village of Falli Hölli. The lower part of the slide evolved in a mudslide to form a dam in the nearby river

The last four cases allow the formulation of the following conclusions:

- Modern surveying and modelling techniques enable engineers to develop fairly reliable predictions related to the formation and behaviour of landslide dams, if time is available.
- Several engineering works can be designed and carried out to limit the consequences of landslide dams, but they require sophisticated design and appropriate building technologies.
- A fast preventive action is required as soon as the risks are assessed, because the value of the exposed objects, in particular downstream, tends to increase exponentially with time.

# 3 Recent Cases in Which Human Preventive Action Was Essential to Save Lives and Property

Three cases, namely one in Switzerland and two in South America, have been selected for discussion as their management proves that efficient measures may be taken, even in difficult situations, to decrease the risks induced by landslide dams.

## *3.1 Randa Rockfall, Switzerland (1991)*

This large rockfall (approx. 30 M $m^3$ in two main stages) which occurred on April 18 and May 9, 1991, formed a large dam in the Vispa Valley, some 10 km downstream of the famous tourist resort of Zermatt, interrupting railway and road access [18, 20].

The dam had a minimum height of 100 m above the river bed in the central part of the rockfall mass, whereas the lowest point of the "crest" of the dam, at el. 1,421 m a.s.l., on the opposite slope formed by the alluvial fan of the Dorfbach torrent, was 44 m higher than the upstream toe of the dam (see Fig. 3.11 and Case E in Fig. 3.1). Despite the very coarse material observed at the surface, the first excavations for the construction of a spillway channel revealed a compact material, which was later confirmed by the high natural compaction of the mass (swell factor of 13%), so that the seepage through the huge fallen mass was negligible.

Fortunately three factors contributed significantly to limit the rising of the lake:

1. The weather was fairly cold and not very wet during the spring of 1991 so that the snowmelt did not represent a major contribution to the flow of the Vispa River.
2. A large part of the drainage area of this river is included in the general catchment area of the Grande Dixence hydroelectric scheme, so that it was possible to divert a large part of its flow out of the Vispa River through the continuous operation of the Z'mutt pumping station.
3. The Swiss army, as well as several contractors, quickly intervened on site, once a provisional road was reestablished, to install up to 36 pumps in the lake which was beginning to rise and to evacuate the pumped waters downstream of the rockfall mass.

**Fig. 3.11** General view of the Randa rockfall causing the damming of the Vispa River. The mitigation works have been already carried out

These comprehensive mitigation actions, combined with the construction of a provisional channel, limited the flooding of the lower part of the village of Randa: only on two occasions, after intense storms, was the village partially under water; in particular, on the second occasion, the main cause was the plugging of the diversion channel by a debris flow coming from the Dorfbach, facing the Randa rockfall (Fig. 3.12).

Later, a 3 km long diversion tunnel was built in the left bank to avoid flooding, so that no catastrophic dam breach occurred. However, the sewage plant used for Randa and the upstream villages, located along the river, was completely flooded during the year 1991, which caused additional environmental impact (see lower part of Fig. 3.12).

**Fig. 3.12** Due to a debris flow which blocked the provisional excavation channel, the plain below the village of Randa was flooded for a short while. The road was maintained by a floating bridge set up by the Swiss army (photograph supplied by CREALP)

## *3.2 Mayunmarca Rockslide, Peru (1974)*

In April 1974, a gigantic and fast-moving rockslide with a volume estimated at 1 to 1.3 $\mathrm{M\,m^3}$ dammed the Mantaro River in the Peruvian Andes, causing more than 450 casualties and forming a 130–150 m high dam [14, 15].

The lake was rapidly filled to el. 2,625 m a.s.l. reaching a volume of 670 $\mathrm{M\,m^3}$ and extending about 30 km upstream from the natural dam. On June 8, 1974, 44 days after the rockslide, it breached (Fig. 3.13); the flow had passed first, during the previous day, through a shallow channel up to 3 m deep, which had been excavated by bulldozer over a length of 250 m. The maximum discharge is estimated between 7,000 and 15,000 $\mathrm{m^3/s}$, which eroded a deep gorge whose longitudinal profile had an inclination of not more then 1.5° (Fig. 3.14).

Thanks to well-coordinated action from the Civil Defence, all the potentially exposed persons downstream, as well as along the lake shores where slides could occur (on a zone 3 km wide), were evacuated. No lives were lost during the dam

**Fig. 3.13** Aerial view of the Mayunmarca landslide dam (from the Servicio Aerofotográfico Nacional del Perú)

breach, despite the very high wave, which reached 20 m at an approximate distance of 100 km downstream from the slide dam, at the site of the Mantaro Power Station (the corresponding Tablachacca Dam was only 1–2 km upstream of the farthest point impounded by the landslide lake).

This example, corresponding to one of the largest landslide dams, again shows the brittleness of landslide dams, especially as the zone of origin was already affected by deep rotational slides which decreased the shear strength of these materials. This led to a short duration lake which was quickly emptied as soon as overtopping began. But good management of the situation downstream nevertheless allowed the protection of all persons threatened by the dam breach. In this case, however, the low population density of the affected zone reduced the effective damage.

## 3.3 *La Josefina Landslide, Ecuador (1993)*

After a long period of intense rain, a large rockslide with a volume of 20–30 M $m^3$ occurred during the night of March 29, 1993 at La Josefina, downstream from the city of Cuenca, Ecuador, damming the Paute River as well as a small affluent called

**Fig. 3.14** Erosion gorge in the Mayunmarca Landslide dam after the outflow, more than 100 m deep

the Jadán River, causing 80 deaths [2]. It can be proved that it was a slide because, in a house located at the top of the landslide mass, a baby was left behind when his parents fled, but was found safe the next morning some 300 m lower.

Despite very rapid action from the army which excavated an 18 m deep channel with a volume of 150,000 $m^3$ in 2 weeks, the lake reached a volume of 185 M $m^3$ in 1 month, flooding important infrastructures, including the access road to the lower Amazonian Valleys, the Panamerican highway and a power plant upstream.

On May 1, 1993, the flow in the channel rapidly increased the erosion so that nearly 90% of the reservoir was emptied in 10 h with a maximum discharge of 8,000–10,000 $m^3/s$, causing major damage to the communities of Gualaceo and Paute downstream, but no fatalities (Fig. 3.15). The most important hydro-electric plant of Ecuador, located downstream and which had been partially emptied during the month of April, was rapidly filled with water and sediments (equivalent to 1 year of normal siltation) and the spillway could safely evacuate some 4,500 $m^3/s$ at the maximum, saving the dam. The only major damage was the destruction of the platform in front of the entrance to the underground power plant, so that it was made inaccessible by vehicles.

This example once more proves the efficiency of the rapid construction of a channel in the slide mass to reduce the damage downstream and the possibility to save all lives even over a large area if the action of the Civil Defence is well

**Fig. 3.15** View of the La Josefina landslide dam, some monthes after breaching. A deeper artificial channel was excavated to decrease the level of the residual lake

organized, but provided that the overflow occurs a few weeks after the slide. If the dam fails after a much longer period, it is more difficult to keep the entire threatened population away from the endangered zone. The second lesson from this case is that good coordination among all the public administrations involved may reduce the damage and help solve many problems created by the landslide lake. However, the action of these administrations was much more difficult after the disaster, to manage

**Fig. 3.16** Excavation of a new channel in the landslide mass after the May 1st, 1993 overflow. The unstable left slope that required a stabilizing scheme can be observed

the reconstruction and provide all the affected persons with partial help depending on the resources gathered through national and international assistance. One of the mitigation actions included the excavation of a new channel in order to decrease the level of the upstream residual lake and the stabilization of the river bed downstream to limit river erosion (Fig. 3.16). It was only after these preventive measures that the road in the valley could be rebuilt.

## 4 Conclusions and Perspectives

The consideration of historic landslide dam events and of the processes through which they occur leads to the conclusion that an active intervention to stop the landslide would have been impossible in most of the cases. After occurrence, a passive acceptance of the situation is unacceptable when the potential dam breach represents a high risk for the population and properties downstream. It is thus essential to take immediate and even preventive measures, when possible, to protect exposed population and infrastructure.

**Fig. 3.17** Zone of impact of the Séchilienne rockfall

A good example of these preventive actions is given by the Séchilienne Landslide in the valley of the Romanche River as already mentioned above in Section 2.1 (Fig. 3.17). The evidence of increasing movements in the slope from 1985 perceived by the fall of individual blocks and the evolution of the morphology led the authorities to develop extensive monitoring systems, to change the layout of the exposed national road leading to important tourist resorts and to create a new river bed protected by a dam, to retain a limited volume of rockfall material [6]. As the monitoring results after some years led to the conclusion that the potential rockfall volume could be larger than expected, a tunnel was built to allow a partial diversion of the river flow in case of the formation of a landslide dam. Finally, detailed modelling of the formation of the dam, studied by two different dynamic approaches, allowed the determination of the dam height for several scenarios and thus the analysis of the flow conditions downstream. Other preventive measures are still being studied in order to limit the impact of the eventual flood wave on the city of Grenoble and its surroundings, where the presence of important industrial plants imply very high indirect risks.

It is clear that, for potential large landslide dams, the early identification of eventual scenarios based on a complete analysis of their triggering and run-out conditions is essential to organize appropriate preventive actions in a timely manner [4]. But a detailed knowledge of past events, such as those mentioned above, will always be useful to help determine their probability of occurrence, considering that a large number of examples are lacking. It is, therefore, interesting to continue analyzing the major landslide dams which occurred in the past.

**Acknowledgements** Sincere thanks are expressed to my collegues Jordi Moreno who contributed to the preparation of the figures and to Karine Barone who mastered the typing and editing tasks of this paper.

## References

1. Abele, G. (1974) Bergstürze in den Alpen, *Wissenschaftliche Alpenvereinshefte, München* **H25**, 230 S.
2. Basabe, P. et al. (1996) Prevención de desastres naturales en la cuenca del río Paute. *Libro "Sin plazo para la esperanza"*. Escuela politécnica nacional, Quito-Ecuador, 271–287.
3. Bonnard, Ch. (2004) *La Buzza di Biasca*. Contribution au Dict. Historique de la Suisse.
4. Bonnard, Ch., Forlati, F. and Scavia, C. (eds.) (2004) *Identification and Mitigation of Large Landslide Risks in Europe: Advances in Risk Assessment*. Balkema, Leiden, 317 p.
5. Cavallin, A. and Martinis, B. (1974) Studio geologico della grande frana di Borta (Ampezzo), *Cronaca del Soc. Alp. Friuliana* **58**, 297–319.
6. Durville, J.-L., Effendiantz, L., Pothérat, P. and Marchesini, P. (2004) The Séchilienne Landslide, in C. Bonnard, F. Forlati, and C. Scavia (eds.) *Identification and Mitigation of Large Landslide Risks in Europe: Advances in Risk Assessment*. Balkema, Leiden, pp. 253–269.
7. Eisbacher, G.H. and Clague, J.J. (1984) *Destructive Mass Movements in High Mountains: Hazard and Management*. Geol. Survey of Canada, Canada, 230 p.
8. Engel, Th. (1986) *Nouvelles méthodes de mesure et d'analyse pour l'étude des mouvements du sol en terrains instables*. Thèse n°601, Ecole Polythechnique Fédérale de Lausanne.

9. Gabus, J.H. (1990) Atlas géologique de la Suisse, feuille No 1285, Les Diablerets, Office fédéral des eaux et de la géologie, Bienne.
10. Gaziev, E. (1984) Study of the Usoi Landslide in Pamir, *Proc. 4th Int. Symp. on Landslides, Toronto* **1**, 511–514.
11. Govi, M. (1990) Mouvements de masse récents et anciens dans les Alpes italiennes, *Proc. 5th Int. Symp. on Landslides, Lausanne* **3**, 1509–1514.
12. Heim, A. (1932) *Bergsturz und Menschenleben*. Fretz & Wasmuth Verlag, Zürich, 227 p.
13. ISDR (2000) *Usoi Landslide Dam and Lake Sarez*, UN-ISDR Prevention Series N° 1. 115 p.
14. Kojean, E. and Hutchinson, J.N. (1978) Mayunmarca rockslide and debris flow, Peru, in B. Voight (ed.), *Rockslides and Avalanches*, Vol. 1. Elsevier, Amsterdam, pp. 315–361.
15. Lee, K.L. and Duncan, J.M. (1975) *Landslide of April 25, 1974 on the Mantaro River, Peru*. Committee on Natural Disasters, National Research Council. 72 p. Washington: National Academy of Sciences.
16. Montandon, F. (1933) Chronologie des grands éboulements alpins, du début de l'ère chrétienne à nos jours. Soc. de Géographie, Genève, *Matériaux pour l'Etude des Calamités* **N° 32**, 271–340.
17. Mougin, P. (1914) *Les Torrents de la Savoie*. Société d'Histoire naturelle de Savoie, Grenoble, 1251 p.
18. Noverraz, F. and Bonnard, Ch. (1992) L'écroulement rocheux de Randa, près de Zermatt, *Proc. 6th Int. Symp. on Landslides, Christchurch* **1**, 167–170.
19. Pilot, G. and Durville, J.-L. (1990) Les mouvements de terrain dans les Alpes françaises, *Proc. 5th Int. Symp. on Landslides, Lausanne* **3**, 1515–1537.
20. Schindler, C. and Eisenlohr, T. (1992) *Bergsturz Grossgufer bei Randa*. Geologisch-geotechnische Expertise. ETH-Zürich, 83 p. + Annexes.
21. Schöneich, Ph.(2000) Impact of Rockfalls on soft water saturated floodplain sediments. Three examples from the Swiss Rhone Valley. *Proc. 8th Int. Symp. on Landslides, Cardiff*, Vol. 3, pp. 1327–1332. London, Thomas Telford.
22. Schuster, R.L. (ed) (1986) *Landslide Dams: Processes, Risk and Mitigation*, Geot. Spec. Publ. No 3. American Society of Civil Engineers, New York, NY, 164 p.
23. Vulliet, L. and Bonnard, Ch. (1996) The Chlöwena landslide: Prediction with a viscous model, *Proc. 7th Int. Symp. on Landslides, Trondheim* **3**, 1525–1530.

# Chapter 4
# Rockslide and Rock Avalanche Dams in the Southern Alps, New Zealand

**O. Korup**

## 1 Introduction

Studying the blockage of mountain rivers by bedrock landslides offers both fascinating and challenging prospects. It allows us to investigate not only some of the largest processes of mass movement on the Earth surface, but also the fluvial response to the emplacement of many millions of tons of rock and debris. The decay process of river-blocking debris links geomorphology to hazard and risk assessment, given the potential for catastrophic outburst floods and debris flows from rockslide-dammed lakes [26].

There is a steadily growing number of studies on landslide dams in high mountain belts throughout the world [4, 16]. In New Zealand, the occurrence of landslide dams is regionally clustered and largely controlled by relief, lithology, and landslide trigger mechanisms [24]. Landslide dams were mainly mapped in the East Cape, Taranaki, and Wanganui regions of North Island, and the Southern Alps and Fiordland mountains of South Island [17]. This pattern is partly an artefact of the regional landslide-triggering effect of historic earthquakes.

This study is intended to give a state-of-the-art overview on natural dams formed by large rockslides and rock avalanches in the Southern Alps of New Zealand. The term "large" refers here to an arbitrarily defined threshold of a mobilised volume $>10^6$ $m^3$. This contribution augments and updates previous seminal studies [24, 30]. Geomorphic evidence from various sources is presented and analysed with respect to blockage type, geomorphometric characteristics, distribution, post-failure morphodynamics, and fluvial adjustment. The discussion focuses on constraints on blockage types, regional and global comparison, trigger mechanisms, dam longevity, geomorphic long-range and long-term effects, as well as some hazard and risk implications. Other recent research issues on landslide dams in New Zealand were reviewed earlier [16].

O. Korup (✉)
Institute of Earth and Environmental Sciences, University of Potsdam,
D-14776 Potsdam, Germany
e-mail: oliver.korup@geo.uni-potsdam.de

S.G. Evans et al. (eds.), *Natural and Artificial Rockslide Dams*, Lecture Notes in Earth Sciences 133, DOI 10.1007/978-3-642-04764-0_4, © Springer-Verlag Berlin Heidelberg 2011 

## 2 The Southern Alps of New Zealand

The geomorphic setting of the Southern Alps results from the close coupling of tectonic uplift and erosion, creating a mountain range asymmetric in cross-section [15, 31]. Dextral transpression along the Alpine Fault, which marks the Australian-Pacific plate boundary through the South Island of New Zealand, rapidly exhumes partially metamorphic mid-late Triassic greywacke of the Torlesse Terrane along a ramp-like structure [20]. The resulting mountain range attains elevations up to nearly 3,700 m a.s.l. with deeply dissected valleys. A major backthrust structure is evident in the Main Divide Fault Zone along the topographic backbone of the Southern Alps, and major alpine valleys adapted to the course of numerous oblique reverse, normal, and strike-slip faults [21]. Uplift and erosion rates vary throughout the Southern Alps and are highest at <11 mm/year near the Alpine Fault, and lowest in the east at ~1 mm/year [23, 27]. Metamorphic overprint on greywacke is exposed in NW-ward increasing grade, attaining amphibolite schist facies at the Alpine Fault, whereas subgreenschist and unaltered greywacke/argillite facies dominate east of the divide. The southwestern fringe of the Southern Alps is characterised by accreted and fault-bended terranes, including rocks of ultramafic lithology [23].

The mountain belt acts as a barrier to moisture-bearing winds from the Tasman Sea, causing orographically-enhanced precipitation of up to 14 m/year west of the main divide [13]. During the Quaternary, the Southern Alps were subject to numerous glaciations, and the terminal moraines of valley glaciers on the western fringe were drowned by postglacial sea-level rise. Based on physiographic characteristics, the Southern Alps can be divided into a Western (WSA), Axial (ASA), and Eastern (ESA) part [31]. Valleys in the WSA are generally steep, V-shaped, closely spaced, and rejuvenated by coseismic base-level changes along the Alpine Fault. The ASA have the highest elevations and are dominated by glacial landforms. In the ESA, basin areas are generally larger, and tributaries feed into broad moraine-fringed valley trains, which retain extensive amounts of glacigenic deposits and large ice-scoured intramontane lake basins (Fig. 4.1).

Fluvial sediment yields vary between $10^4$ and $10^2$ t/km$^2$/year, west and east of the main divide, respectively [9]. The contemporary regional sediment yield from the WSA makes up 30% of the total export from the New Zealand land mass [14].

Important predisposing factors for rock-slope instability in the study area include gravitational stress, deterioration of rock-mass strength through repeated earthquake shaking, hanging-wall shattering, and incision-driven loss of lateral support. The term "rockslide" is used here in a general sense, and comprises slope failures in bedrock, involving an arbitrary minimum volume of $10^6$ m$^3$ of translational/rotational sliding, toppling, or other types of deep-seated slope deformation. The term "rock avalanche" is used in its strict sense, and refers to large volumes (>$10^6$ m$^3$) of fragmenting rock particles in extremely rapid flow movement [6].

Trigger mechanisms include seismic ground shaking, high-intensity rainstorms, and fluvial undercutting. Several historic rock avalanches have occurred without

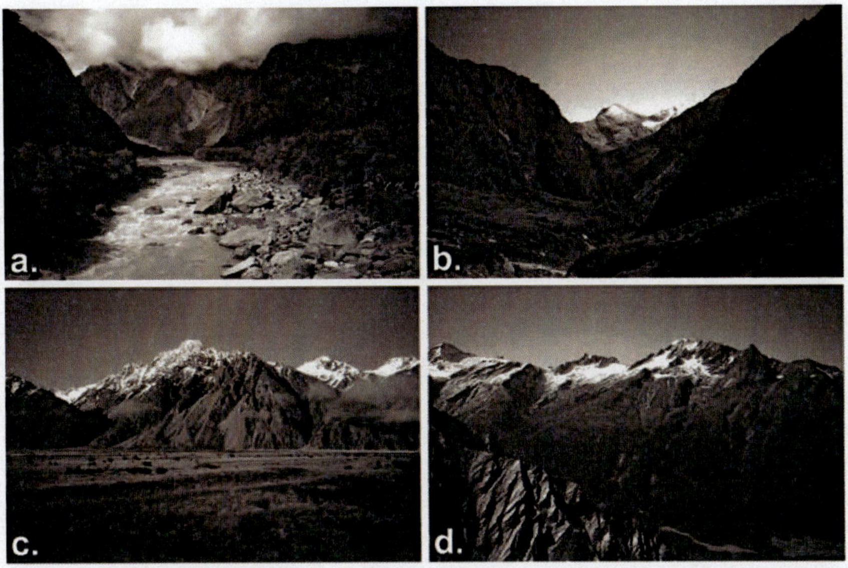


**Fig. 4.1** Typical valley geometries of the Southern Alps, New Zealand (**a**) Glacial trough valley, lower Fox River, WSA; (**b**) Fluvially-incised V-shaped valley near the main divide, upper Copland River, WSA; (**c**) Extensive valley train fringed by debris-mantled slopes, Tasman River, Mt Cook, ASA; (**d**) Alpine valley walls developed in Otago Schist, West Matukituki River, ESA

any observed triggers and may have resulted from exceedance of intrinsic stress thresholds, possibly representing the catastrophic culmination of rock creep [22]. Very large rock avalanches can be caused by $M>6.5$ earthquakes, on slopes >25°, and >100 m high, and especially on strongly shaken high narrow ridges [12].

Significant ($M \geq 5.5$) historic earthquakes in the study area occurred near Arthur's Pass, ASA, in 1929 and 1995. Large ($M$~8) prehistoric (i.e., pre-1850 in New Zealand) earthquakes along the Alpine Fault were inferred from several palaeoseismological studies to occur every 250–300 year on average [2, 29].

Regional studies on landslides and landslide dams in the Southern Alps [1] and New Zealand [24], augmented with new case studies and results from geomorphometric analyses [17], were major sources for this compilation. There are currently 52 documented rock avalanches (>$10^6$ $m^3$) in the Southern Alps, several of which also had dammed ephemeral lakes [22, 30]. Few further studies addressed the consequences of rockslide and rock avalanche dams in the region. Recently, landslide dam-break failures were addressed in quantitative hazard and risk assessments for Franz Josef Glacier Township, an increasingly popular tourist destination at the fringe of the WSA [5, 7]. Finally, sediment budgets indicate extreme debris discharge from historic rockslides and described geomorphic off-site consequences for downstream reaches [19].

# 3 Rockslide and Rock Avalanche Dams in the Southern Alps

## 3.1 Identification and Compilation

The study area covers some 15,000 km$^2$ of mountainous terrain between 168 and 172°E, and 42° 30′–44° 30′S. Its northern and southern boundaries roughly coincide with the Alpine Fault trace in the Taramakau valley and a line from the upper Cascade River to Lake Wanaka, respectively. The western and eastern boundaries are defined by the Westland piedmont and Lakes Pukaki and Tekapo, respectively (Fig. 4.2). The locations and geomorphic characteristics of 43 rockslide and rock avalanche dams are part of a national inventory of over 240 landslide dams compiled from various sources [1, 12, 17, 24]. More than half of all recorded landslide dams in New Zealand were formed by large rock-slope failures, with most of them occurring in the Southern Alps and adjacent Fiordland. Almost 30% of all recorded landslide dams in New Zealand were formed by rock avalanches [17].

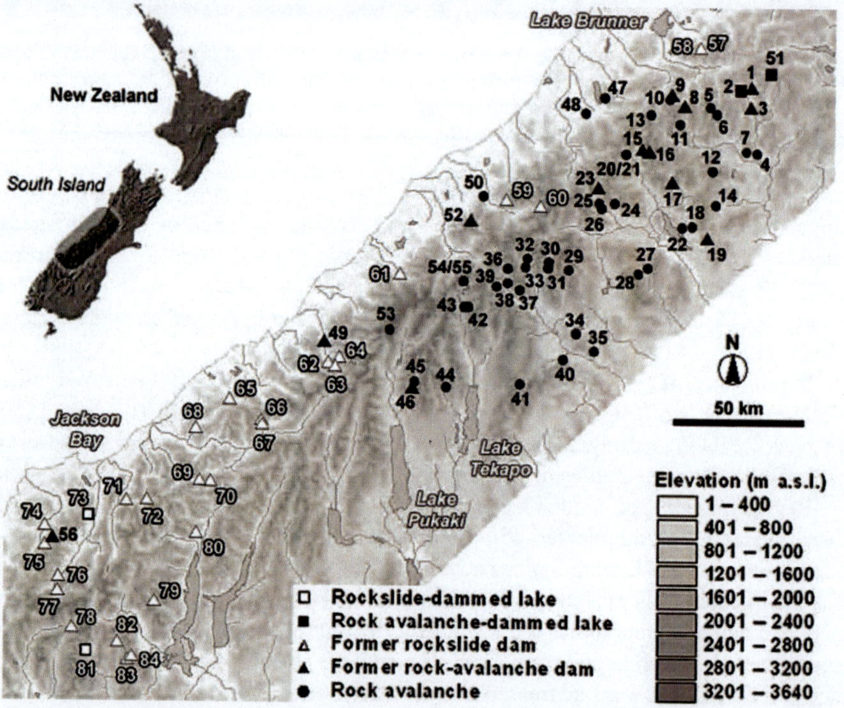


**Fig. 4.2** Map showing the distribution of 43 rockslide (△, □) and rock avalanche dams (▲, ■) in the Southern Alps, New Zealand. Numbers refer to rock avalanche (*black*) and rockslide (*white*) locations, including all rock avalanches >10$^6$ m$^3$ (•) currently reported. Selected sites are described in Table 4.1. Note that previously reported rock avalanches [30] retain their original numbering. Mean density is 0.003 dams/km$^2$

**Table 4.1** Characteristics of selected rockslide and rock avalanche dams in the Southern Alps, New Zealand. Numbers (#) refer to locations in Fig. 4.2. $V_L$ = deposit volume; $H_D$ = dam height; $L_D$ = dam length; $W_D$ = dam width; $A_C$ = contributing catchment area

| # | Location | Type[a] | $V_L$ ($10^6$ m$^3$) | $H_D$ (m) | $L_D$ (km) | $W_D$ (km) | $A_C$ (km$^2$) | Remarks |
|---|---|---|---|---|---|---|---|---|
| 23 | Boundary Ck | Ra (II) E | 280 | 80 | 1.3 | 2.6 | 55 | $H_D$ max. 110 m |
| 16 | Burnett Str | Ra (II) E | 7 | 30 | 0.3 | 0.5 | 12 | $H_D$ max. 40 m |
| 3 | Casey Hut | Ra (III) E | 23 | 10 | – | – | 161 | |
| 78 | Dart Hut | Rs (II) E | >22 | – | – | 0.4 | 49 | episodic failures |
| 51 | Ellis Str | Ra (II) E | 5 | 5 | – | – | 2 | |
| 83 | Harris (Pt 1845) | Rs (II) E | 8 | 30 | 0.2 | 0.8 | 3 | |
| 84 | Harris (Pt 1870) | Rs (II) E | – | – | – | 1.1 | 7 | |
| 46 | Jollie | Ra (III) E | 77 | 50 | 1.0 | 1.1 | 86 | $H_D$ max. 60 m |
| 81 | Lochnagar | Rs (II) L | 690 | 340 | 2.1 | 2.0 | 21 | $H_D$ max. 400 m |
| 79 | Minaret Burn | Rs (II) E | 25 | | 0.3 | 1.2 | 30 | |
| 2 | Minchin | Ra (II) L | 27 | 150 | 0.6 | 0.8 | 25 | |
| 82 | Polnoon Burn | Rs (II) E | 140 | 120 | 1.2 | 1.5 | 32 | $H_D$ max. 140 m |
| 1 | Thompson Str | Ra (II) E | 18 | 40 | 0.6 | 0.7 | 37 | |
| 80 | Tiel Ck | Crs (II) E | 36 | – | 0.3 | 1.2 | 24 | |
| 17 | Triangle Ck | Ra (II) E | 60 | 50 | 1.1 | 1.6 | 105 | $H_D$ max. 110 m |
| 15 | Weka Str | Ra (II) E | 19 | – | – | – | 4 | |
| 59 | Carls Ridge | Crs (II) E | 25 | 20 | 0.1 | 0.4 | 3 | debris-flow dam |
| 72 | Crevice Gorge | Rs (II) E | – | 90 | 0.4 | 0.7 | 25 | |
| 57 | Crooked A | Crs (VI) E | 170 | 20 | 0.3 | 1.2 | 38 | |
| 58 | Crooked B | Rs (VI) E | 40 | 20 | – | 1.0 | 38 | |
| 62 | Dirty Ck | Rs (II) E | – | – | 0.5 | 1.8 | 43 | exact size unknown |
| 64 | Douglas | Wrs (II) E | 14 | 50 | 0.8 | 0.9 | 60 | 120-m deep gorge |
| 75 | Falls Ck | Rs (III) E | 130 | 50 | – | 1.5 | 22 | |
| 74 | Hope Blue R. | Rs (III) E | >130 | 200 | 1.8 | 2.6 | 177 | 200-m deep gorge |
| 70 | Howe Ck | Rs (II) E | 5 | – | 0.2 | 0.6 | 24 | |
| 9 | Hunts Ck | Ra (III) E | 35 | 40 | – | – | 11 | |
| 63 | Lame Duck Flat | Rs (II) E | 62 | – | – | 1.0 | 38 | |
| 77 | McArthurs Flat | Rs (II) E | 45 | – | 0.3 | 1.0 | 179 | |

**Table 4.1** (continued)

| # | Location | Type[a] | $V_L$ ($10^6$ m$^3$) | $H_D$ (m) | $L_D$ (km) | $W_D$ (km) | $A_C$ (km$^2$) | Remarks |
|---|---|---|---|---|---|---|---|---|
| 56 | McKay Ck | Ra (III) E | – | 100 | – | 0.8 | 10 | |
| 52 | Poerua | Ra (III) E | 10–15 | 120 | 0.5 | 0.7 | 50 | |
| 69 | Princes Ck | Rs (II) E | 50 | – | – | 1.6 | 3 | |
| 71 | Te Naihi | Rs (II) E | – | 30 | 0.5 | 1.0 | 76 | rapids at breach site |
| 60 | Whirlwind Spur | Rs (II) E | – | – | 0.2 | 1.4 | 12 | |
| 66 | Zeilian Ck A | Rs (II) E | 8 | – | – | 1.0 | 13 | |
| 67 | Zeilian Ck B | Rs (VI) E | – | – | – | 1.2 | 23 | |
| 8 | Zig–Zag | Ra (III) E | 43 | 100 | – | 1.4 | 15 | $H_D$ max. 130 m |

[a]Dam type [2] in brackets, E = partly eroded, L = presently retains lake; Ra = rock avalanche; Rs = rockslide; Crs = complex rock slide; Wrs = wedge failure/rock slide.

Many of the Southern Alps headwaters show ample evidence of channel diversion and short-lived blockage in response to lateral debris input from landslides and steep tributary streams. Key geomorphic indications of rockslide and rock avalanche dams chiefly comprise amphitheatre-shaped, often scree-mantled, detachment scars with associated extensive hummocky, furrowed, and blocky deposits covering the valley floor [30].

Conspicuous changes in channel pattern and morphology limited to rockslide and rock avalanche deposits or their adjacent reaches are further diagnostic landforms. Non-catastrophic rock-slope deformation was identified by tensional scarp depressions, bulging (toe) slopes, or conspicuous lack of fluvial dissection. Often, place names such as "Rotten Ridge" (Roaring Swine Ck, Haast River), "Trembling Spur", or "Seismic Rift" (Palmer Ck, Waiatoto River) also indicate potential sites of rock-slope instability.

Several large rockslides occur at tributary junctions. There, enhanced downcutting following base-level adjustments and debuttressing along two oblique erosional fronts may favour dilatational slope movement.

Size estimates of rockslides and rock avalanches are expressed as volume of displaced material, $V_L$, or total affected area, $A_L$ (scarp and deposit), allowing for an average error of $\pm 30\%$. Rockslide and rock avalanche dam height, $H_D$, was estimated from interpolating river longitudinal profiles along the blockage site, and the elevation difference between the deposit top and the present river bed. A measurement error of $\pm 15\%$ was assumed, due to often quite substantial differences between maximum and minimum elevations on top of rockslide and rock avalanche dams relative to the (breach-) channel bed.

The maximum width and length of rockslide and rock avalanche dams, $W_D$ and $L_D$, respectively, were estimated with an accuracy of $\pm 100$ m from geomorphic sketches, air photos, topographic maps, and digital elevation models. Dense vegetation cover – particularly in the WSA – as well as unquantified erosion and deposition following dam formation made the accurate reconstruction of some dam geometries difficult. In many cases, the size of dams and formerly dammed lakes requires further validation and reconstruction from field evidence. Thus, certain characteristics of rockslide and rock avalanche dams are not equally available, so that mostly sub-sets of the database can be compared (Table 4.1).

Data presented here include do not include minor (<0.02 km$^2$) rockslide ponds perched on top of the Thomson Range rock-mass collapse, Arawhata River (#73, Fig. 4.2), or the 1929 Falling Mountain rock avalanche, Otehake River (#5, Fig. 4.2). Lake Lyes, interpreted as a crown-scarp or "neck" pond in a large fault-controlled schist rockslide in the Whitcombe catchment [24], is also not considered here. Similarly, the Acheron rock avalanche (#19, Fig. 4.2) [30] is excluded from the dataset since it caused blockage of minor streams only. At three further omitted locations of former blockage on the Callery, Karangarua (McTaggart Ck), Moeraki (Horseshoe Flat), and Thomas (Haast) Rivers (#61, #49, #65, and #68, respectively, Fig. 4.2), rockslide-dam size could not be determined reliably.

### 3.2 Characteristics

The range of river-blocking rockslide and rock avalanche volumes lies between an arbitrarily set minimum of $10^6$ $m^3$ and nearly $10^9$ $m^3$. About half of the examined river-damming failures measure >1 km from scarp to toe and >2 km in runout, their average thickness being 30–40 m. At least 28 of them formed type II dams [4], and there are ten type III dams typical of rock avalanches. However, evidence of such additional tributary blockage was often more difficult to establish. Reliable estimates of dam height, $H_D$, and length, $L_D$, are available for about half of the sites, while dam width, $W_D$, could be determined for nearly 75% of the data (Table 4.1).

$H_D$ ranges between 10 and ~400 m; with more than half of the dams being >50 m, and seven dams >100 m high. Irregular dam surfaces can create significant differences in $H_D$ (Table 4.1). Mean $L_D$ is 0.7 km, and mean $W_D$ is 1.2 km. Only at two sites, i.e. Lochnagar and Hope Blue River Range, do $L_D$ and $W_D$ exceed 1.5 km and 2 km, respectively (Table 4.1). On average, rockslide and rock avalanche dams are more than 20 times as wide as they are high, thus forming very broad and often asymmetric cross-sections. This $W_D/H_D$ ratio does not differ significantly with regard to failure type, i.e. rock sliding or avalanching, where one might expect higher ratios for catastrophic long-runout deposits. The total river length affected by a given dam, including backwater and downstream aggradation, is generally much longer, and may in few demonstrable cases reach up to 5 km [19]. Some of the dams were formed by large secondary debris flows fed by rockslide and rock avalanche debris, such as at Carls Ridge on Hendes Ck, Wanganui River (#59, Fig. 4.2).

Seven out of eleven available estimates on the volume of former rockslide-dammed lakes – mainly based on extrapolations of lake strandlines or lacustrine deposits [24] – exceed $10^7$ $m^3$. Geomorphic evidence of smaller or ephemeral lakes is rarely reported. Absolute dates are available for 16 blockage sites. Six of these dams formed in historic times, with four during the 1929 Arthur's Pass earthquake alone [12]. Most of the remaining occurred during the late Holocene [30]. Trigger mechanisms for prehistoric dam formation remain speculative, although the spatial clustering of several rock-slope failures and river blockage strongly indicates regional triggering events [2, 24].

Only two of the rockslide and rock avalanche dams still retain lakes, of which Lochnagar in the Shotover catchment is the largest, with a lake volume of ~$10^8$ $m^3$[24]. Being alternatively interpreted as a rock avalanche [30] or rock-block slide [24], it is one of the largest river-blocking rock-slope failures in the study area ($A_L$ >3.4 $km^2$; $V_L = 690 \times 10^6$ $m^3$), forming the highest intact dam ($H_D$ ~340–400 m), although its exact size is disputed [24, 30] (#81, Fig. 4.2).

The only other rockslide-dammed lakes of comparable size in the South Island are Lake Christabel on Blue Grey River ($A_{Lake} = 2.6$ $km^2$), just north of the study area, Lake Constance ($A_{Lake} = 0.9$ $km^2$) in the North West Nelson Alps, and numerous others in the Fiordland mountains. The fact that these rockslide and rock avalanche dams remained so well preserved, may add bias to the geomorphometric database, since erosion and deposition associated with failed dams may affect estimates of their original size.

Lake Minchin in the Poulter catchment [30] is the second-largest, presently existing lake dammed by a rock avalanche in the Southern Alps, covering an area of only 0.18 $km^2$ (#2, Fig. 4.2). There are no existing rockslide-dammed lakes west of the main divide. There, lakes are mostly ephemeral due to extremely high sediment yields which cause rapid infilling of both landslide- and glacially-dammed reservoirs in trunk valleys. Many of the rock avalanches in the ESA appear to have formed temporary dams, although geomorphic evidence of lake sedimentation is only found at 13% of the sites [30].

Several bivariate indices were used to empirically characterise "stability domains" for regional landslide-dam inventories [3, 17]. The low number of existing lakes in the Southern Alps precludes further testing of this approach. The majority of the rockslide and rock avalanche dams discussed here have not remained intact, and show evidence of breaching and fluvial erosion. Similarly to observed trends for New Zealand-wide data, the minimum deposit volume to retain a lake over several years or decades is estimated at $100 \times 10^6$ $m^3$ and $10^9$ $m^3$, for 10 and 100-$km^2$ upstream catchment areas, respectively [17].

## 3.3 Distribution

Rockslide and rock avalanche dams occur on both sides of the main divide in similar numbers (Fig. 4.2). Although rock avalanches appear to be less frequent in the WSA [30], rivers there are blocked by large complex rockslides. Scatter plots show no significant differences in the relationships between deposit volume and dam dimensions on either side of the divide. Rockslide or rock avalanche volume $V_L$ explains almost 60% of the (low) variance in dam width $W_D$. Dam height and length, $H_D$ and $L_D$, respectively, may vary by over one order of magnitude for a given $V_L$ (Fig. 4.3).

The majority of rockslide and rock avalanche dams occur in the upper reaches of alpine drainage basins with upstream catchment areas $A_C$ <50 $km^2$, and only three blocked major channels ($A_C$ >100 $km^2$). Similarly to dam geometry, there is no recognisable difference in this measure of relative catchment position between the western and eastern catchments of the Southern Alps. However, given generally smaller basin areas in the WSA, there, a river-blocking rockslide or rock avalanche of a given area $A_L$ would be expected to affect a relatively larger portion of the catchment. This is reflected in the ratio $A_L/A_C$, which is on average 5% in the ESA, and 9% in the WSA.

## 3.4 Post-Failure Morphodynamics and Fluvial Adjustment

Most of the dams studied were breached. In many cases, breach channels evolved to steep meandering gorges cut through rockslide and rock avalanche debris, a

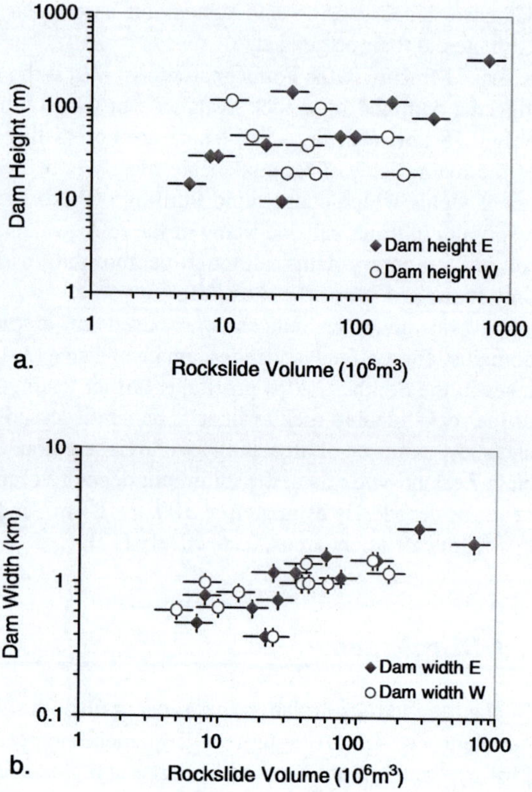


**Fig. 4.3** Rockslide- (and rock avalanche) dam height $H_D$, and width $W_D$, as a function of rockslide volume $V_L$, for selected cases in the eastern (*black dots*) and western (*white dots*) Southern Alps. Bars allow for errors of ±30% and ±15% on *x*- and *y*-axes, respectively

characteristic which can aid visual reconnaissance from air photos. Impressive examples include the gorges at Falls Ck (160 m deep), below the Hope Blue River Range (200 m), and the Douglas River (120 m, Table 4.1). The "Zig–Zag" rock avalanche created a 55-m (at its highest point near the crest) to perhaps 130 m-high dam on upper Otira River, WSA, some 2,000 year ago (#8, Figs. 4.2 and 4.4a).

The dam impounds Pegleg Flat, forming a distinctive horizontal step ca. 0.8 km long and extending over 0.07 $km^2$. The river has cut a headward-eroding and 50–60-m deep gorge with an average gradient $S = 0.19$ through the debris (Fig. 4.5a). Geotechnical drilling conducted during the construction for a state highway viaduct through the rock avalanche deposition zone revealed that the bridge pylons required grounding in rock avalanche debris >30 m thick [25]. The Otira River thus has not adjusted to its former gradient since blockage occurred. Net fluvial downcutting along Pegleg Flat was as high as 0.1 m/year historically [18], and may be related to headward erosion at the dam crest.

This and other examples of breached dams allow insights into the state of fluvial recovery to blockage, assuming that the original channel gradient would have not had any major steps at the blockage site. For instance, the Polnoon Burn rockslide

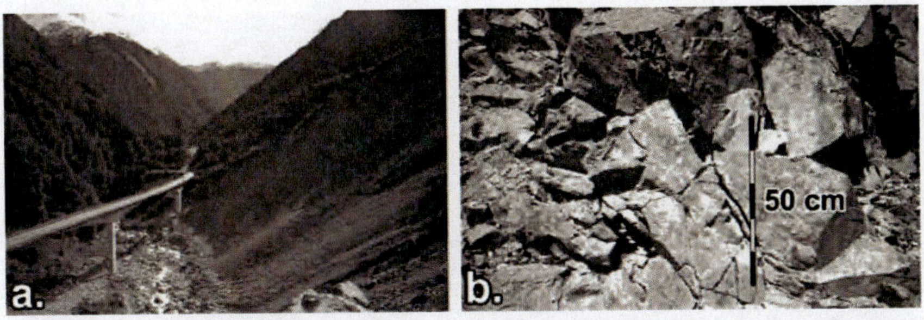


**Fig. 4.4** Zig–Zag rock avalanche, Otira River, WSA. **(a)** Downstream view from the former rock avalanche dam crest at Death's Corner of the State Highway 6 viaduct located within the breach-channel gorge. Scree slope on the right is >100 m high and lower part of continuosly eroding rock avalanche debris. Note that bridge pylons are grounded in rock avalanche debris for >30 m. **(b)** Details of former roadcut in fragmented angular debris

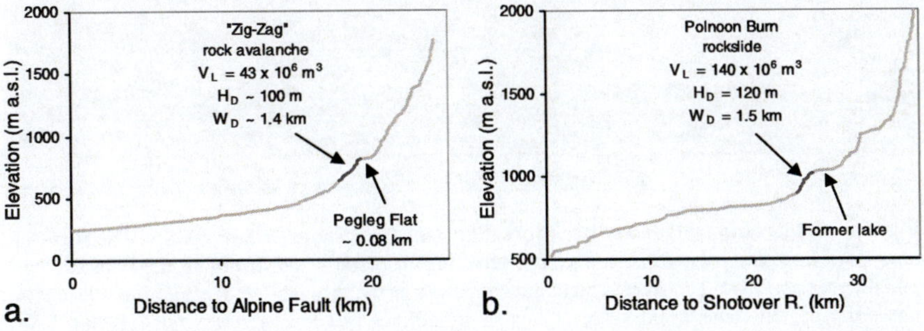


**Fig. 4.5** Longitudinal profiles of **(a)** Otira River (WSA), and **(b)** Polnoon Burn (ESA). The ~2,000-year old "Zig–Zag" rock avalanche dam [30] continues to form a prominent step, with Pegleg Flat marking the extent of aggradation behind the dam. A similar step is evident in the upper Polnoon Burn

($A_L$ >1.9 km$^2$, $V_L = 140 \times 10^6$ m$^3$) in the Shotover catchment, ESA, formed a dam between 120 and 140 m high, judging from the depth of the present 1.7-km long river gorge dissecting the deposit (#82, Figs. 4.2 and 4.6).

The 4,000-year old dam [24] is ~1.2 km long and >1.5 km wide, forming a conspicuous >100 m high and slightly convex step in the river long profile. In the downstream section, this step is occupied by the breach channel ($S = 0.12$), while the flat upstream section results from backwater aggradation of ~3 km, representing the infilling of a former lake that lasted ~2,000 year [24] (Figs. 4.5b and 4.6). This knickpoint indicates that fluvial recovery at this site has also remained incomplete. The amount of rockslide debris eroded by the river is estimated between 20 and 40 × 10$^6$ m$^3$. With an upstream catchment area of 32 km$^2$, this would be equal to an average specific sediment yield between 2.8 and 5.6 × 10$^2$ t/km$^2$/year, while estimating bulk density at 1.8 t/m$^3$.

**Fig. 4.6** Orthophoto of Polnoon Burn rockslide dam (rd), Shotover River, ESA. The rockslide originated from the western ridgeline (A), affecting >1.9 km$^2$, and damming the Polnoon Burn ~4,000 year ago [24]. The river has cut a meandering gorge through the deposit after a lacustrine phase that lasted some 2,000 year, evident in strandlines and level terraces (sl) (cf. Fig. 4.5b). Note that another 1.3-km$^2$ landslide (B) on the opposite upstream bank did not cause any obvious blockage due to low displacement. Image courtesy of Land Information New Zealand, Crown Copyright reserved (F40 Wanaka 2002/03, www.linz.govt.nz)

These figures may well be twice as high, if assuming that most of the dam erosion took place in the last 2,000 year, following the draining of the lake. Large bedrock landslides are prolific in the area. Another large ($A_L = 1.3$ km$^2$) complex landslide on the opposite bank just upstream did not cause any obvious blockage, possibly due to its low degree of displacement (Fig. 4.6).

In other cases, large cross-valley deposits appear to have been completely dissected, with the formerly dammed river having re-attained a profile gradient similar to that of its adjacent up- and downstream reaches. Some former breach channels thus have equilibrated to form broad gravel-bed rivers. This is for instance the case at the 2.6-km wide Boundary Ck rock avalanche dam ($V_L = 260 \times 10^6$ m$^3$; #23, Fig. 4.2), Mathias River, ESA, where the flat breach-channel floor is now ~200 m wide.

Similarly, the Cascade River, WSA, has cut a 3-km long and 200-m deep gorge through a catastrophic rockslide deposit below the Hope Blue River Range (#74, Fig. 4.2). The rockslide(s) affected ~9 km$^2$ and the debris is truncated by the Alpine

Fault trace, indicating a failure age >280 year, assuming that the fault was subject to surface rupture in this area in 1717 AD [29]. The eroded volume from the deposit is estimated at $130 \times 10^6$ m$^3$, theoretically resulting in a maximum sediment discharge rate of $4.6 \times 10^5$ m$^3$/year, or a specific sediment yield of 4,700 t/km$^2$/year.

Often, the results of river blockage are not as well-defined. High rates of erosion and deposition in the WSA make the identification of remnants of former rockslide and rock avalanche dams difficult. Also, several large rock-slope failures are associated with only minor impact on alpine rivers. This is exemplified in the Crooked River basin, where two large rockslides form a 20-m high type VI dam [4] (#57, #58, Figs. 4.2 and 4.7). The larger of the two ("Crooked A") is a complex failure, mobilising ~$170 \times 10^6$ m$^3$, and extending over 3.7 km$^2$. A 300-m wide extensional scarp depression along the ridge crest exposes NE-trending schist foliation, and indicates failure along a schistosity plane with an estimated angle of $\varphi = 27° \pm 9°$. Compared to the total length of the failure (2.1 km), this scarp also shows that it has not moved very far. Steep ravines dissect and divide the failure complex into at least three individual slope portions. The maximum thickness of the rockslide exceeds 100 m, which is typical of many similar large-scale valley-side collapses in the WSA [30]. The rockslide is in direct contact with 1.8 km of Crooked River and 1.2 km of an unnamed right-hand tributary. Secondary slope failure(s) involving ~$16 \times 10^6$ m$^3$ formed a debris fan, which has partly encroached onto the tributary junction ("df" in Fig. 4.7).

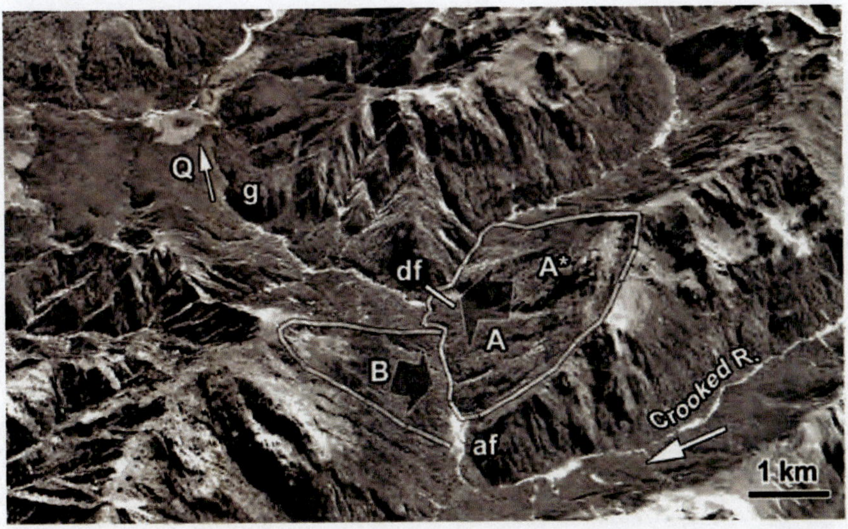


**Fig. 4.7** Large rockslides (A: 170, and B: $40 \times 10^6$ m$^3$) forming a 20-m high type VI dam [4], and a small alluvial flat (af) on Crooked River, WSA. Note secondary failure scar (A*) and associated, partly trimmed, debris fan (df). Q = Quaternary glacigenic, fluvial, and debris-fan gravels cut by the Crooked R. gorge (g). View is to the north, scale approximate only. Image courtesy of Land Information New Zealand, Crown Copyright reserved (K32 Lake Brunner 1995/96, www.linz.govt.nz)

The "Crooked B" landslide is slightly juxtaposed and smaller ($V_L = 40$ to $50 \times 10^6$ m$^3$, with some $12 \times 10^6$ m$^3$ remaining as deposit on the lower slopes). Its motion may well have been triggered, or at least enhanced, by fluvial undercutting, following a possible initial channel diversion in response to the larger rockslide. Fluvial aggradation behind both landslides extends for up to ~1.2 km, forming a 0.3-km$^2$ intramontane alluvial flat fringed by forested terraces. The river longitudinal profile does not show any obvious step. The presence of knickpoints here is ambiguous, since they could also be the result of local channel adjustment at this major tributary junction.

Similar geomorphic impacts on river channels are common in the WSA, where large deep-seated rockslides have diverted river channels around their toes without creating persistent or well-defined blockage. At several such occlusion sites, fluvial trimming has oversteepened landslide toes, and triggered nested secondary rotational failures. Blockage of rivers by these secondary failures is interpreted to be mainly ephemeral and in close feedback with fluvial erosion, where no major obstructing deposits remain (Fig. 4.8).

## 3.5 Sediment Discharge

Once failed, rockslide and rock avalanche dams are subject to fluvial and mass-wasting erosion. Sediment discharge rates from eroding rock avalanches in the

**Fig. 4.8** Downstream view of remnant of deep-seated rockslide ($V_L$ ~$22 \times 10^6$ m$^3$) on Dart River near Dart Hut (#78, Fig. 4.2), ESA (February 2001). Repeated rock-slope failure at this site may have caused short-lived river blockage several times

central Southern Alps lie between $10^2$ and $1.5 \times 10^5$ $m^3$/year [30]. Averaged discharge from the breached 930-year old rock avalanche-dam on Jollie River amounts to $2.8 \times 10^4$ $m^3$/year or $7 \times 10^2$ t/$km^2$/year [30] (#46, Fig. 4.2), surpassing historically gauged catchment sediment yield by > 60%. Annual to decadal sediment discharge from failed rock avalanche dams may reach peak values of up to $1.8 \times 10^6$ $m^3$/year. Such specific yields are extreme ($8 \times 10^4$ t/$km^2$/year), and were estimated in the months and years immediately following the failure of the 1999 Poerua rock avalanche dam below Mt Adams, WSA [19] (#52, Fig. 4.2). Such major sediment pulses can cause infilling of steep bedrock gorges and extend for several km downstream of the blockage site.

If extrapolated over longer periods (>1,000 year), the apparent average sediment discharge from eroding rockslide and rock avalanche dams is biased by formation age, though still ranges from $5 \times 10^2$ to $8 \times 10^5$ $m^3$/year. The highest specific yield from a prehistoric rock avalanche is $2.9 \times 10^3$ t/$km^2$/year (Boundary Ck, Mathias R.; #23, Fig. 4.2). These figures highlight the importance of rockslide and rock avalanche dams as major sediment sources in the Southern Alps, especially when considering, that over half of the studied failures affected only ~5% of their respective upstream catchment areas.

Predicting the sediment flux from breached rockslide dams would be useful for assessing the amount of consequential downstream aggradation on floodplains or alluvial fans. Although rockslide-dam height, assumed to be a proxy for hydraulic head, statistically explains 45% of the variation in mean specific sediment yield ($n = 10$), it is insensitive to the timing of delivery, which is indeed the most important aspect for hazard and risk assessments and river management. The breach of the Poerua rock avalanche dam demonstrated that sediment delivery was highest in the first months and years after the failure, and substantially declined afterward [19].

# 4 Discussion

## 4.1 Types and Occurrence of River Blockage

Review and compilation of geomorphic data from various sources sheds some new light on the distribution and characteristics of 43 rockslide and rock avalanche dams currently reported in the Southern Alps of New Zealand. A major finding is that evidence of river blockage occurs on both sides of the main divide in comparable numbers. Although this notion is open to further refinement due to the likely detection of additional former dam sites, it modifies earlier notions on the distribution and geomorphic impacts of rock avalanches in the WSA [30]. The 1999 Mt Adams rock avalanche, Poerua River, has shown the rapidity of dam erosion, and thus, removal of geomorphic evidence [19].

The geomorphic impact of rockslides and rock avalanches on alpine rivers is characterised by two distinctive types: (a) Rock-slope failures that produced clearly (and often catastrophically) detached deposits, which formed type II and

III dams [4]; and (b) Deep-seated, low-displacement, suspended or chronic, often sackung-type failures causing river diversion and occlusion around their toe zones. This type of landslide-induced river occlusion not always forms cross-valley barriers, though may involve ephemeral blockage through nested secondary rockslides or large surficial debris flows (Fig. 4.8).

There are several sites, where large rockslides or rock avalanches in direct contact with major channels did not produce any obvious effects of blockage. For instance, the Falling Mountain rock avalanche ($V_L$ ~55 × $10^6$ $m^3$), Otehake River (#5, Fig. 4.2), did not cause any major blockage, because it completely buried an alpine headwater branch.

Furthermore, several historic large rock avalanches also ran out on large glaciers, not causing any impact on the drainage system [22]. Several deep-seated complex rock-slope failures in schist lithology with volumes $\geq 10^8$ $m^3$ were not found to be associated with any blockage despite being located at the confluence of major rivers. The degree of displacement, velocity, and type of emplacement, are thus additional important controls on river blockage rather than the mobilised volume alone.

## 4.2 Regional Comparison

Assuming comparable numbers of occurrence on both sides of the divide, the geometry of rockslide and rock avalanche dams may be expected to reflect one or combined differences in external controls such as lithology, climate, valley geometry, or erosion rates. For instance, the runout of large rockslides and rock avalanches may be assumed to be less constrained by valley geometry in the trunk valleys east of the divide. The notion, that a given deposit volume, $V_L$, would thus produce a lower dam height, $H_D$, on average in the ESA, is not confirmed by the data (Fig. 4.3a).

Moreover, most of the mean geomorphometric characteristics of rockslide and rock avalanche dams are not strikingly different either side of the main divide. Dams have a slightly higher mean length, $L_D$, east of the main divide, possibly reflecting the influence of wider valley-floors on average. Differences in the mean or median width of dams, $W_D$, between the WSA and ESA, are however minute (i.e., <10%). This is interpreted as a result of fewer topographic constraints on dam formation in along-valley direction. Also, there is no detectable difference with regard to respective catchment location, as indicated by the contributing catchment area, $A_C$.

Extreme geomorphic process rates in the WSA constrain the longevity of rockslide-dammed lakes, and rapid dam-breaching or infilling is assumed. There, small contributing catchment areas and steep channels prevent the formation of large lakes, and high sediment yields shorten natural reservoir lifetimes to <100 year [8]. Process rates are one order of magnitude lower in the ESA, yet very few lakes are presently existing there. Judging from few published dates, these and other preserved large lakes north of the study area, may be $\geq$1,000-year old on average.

Comparison with the characteristics of numerous rockslide dams preserved in the Fiordland mountains highlights the controls of bedrock geology, seismicity, and sediment yield. There, large deep-seated rockslides have retained nearly 40 lakes [24]. Generally lower sediment yields [14] may explain the longer reservoir lifetimes. The higher rock-mass strength of the Fiordland crystalline basement implies, that deep-seated rock-slope failures require very high earthquake magnitudes to be triggered [10].

Thus an additional explanation for the better preservation of rockslide-dammed lakes in this region may be that of a fairly recent (nineteenth century) earthquake in the region [24]. The prominent glacial U-shape of many Fiordland valleys may also explain the better preservation of rockslide- and rock avalanche dammed lakes on broad and flat valley floors, although some of the valley fills may have also formed as a direct consequence of blockage.

## *4.3 Comparison with Other High Mountain Belts*

A currently compiled inventory of 160 rockslide and rock avalanche dams in mountain belts throughout the world helps to put the geomorphic characteristics examined here into a global perspective. Scatter plots suggest that, for example, the relationship between dam height, $H_D$, and deposit volume, $V_D$, in the New Zealand Southern Alps does not differ much from those in other high mountain regions. The cases studied here occupy a "mid-field" position within the broad scatter of data collected from the European Alps, Central and High Asia, and the Andes, while any significant clustering is not evident (Fig. 4.9). Extremely large river-damming

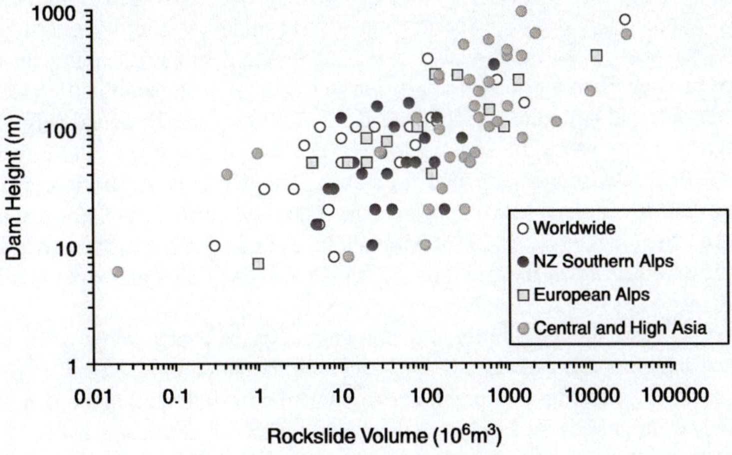


**Fig. 4.9** Comparison of relationship between dam height, $H_D$, and deposit volume, $V_L$, for selected data from the New Zealand Southern Alps, the European Alps, Central and High Asia, and others. Data from the New Zealand Southern Alps take a "mid-field" position within the general scatter

rockslides ($V_L > 10^9$ m$^3$) or such that caused major drainage reversals are not documented in the study area. Site-specific conditions of blockage, such as e.g. valley geometry, lithology, climate, or deposit sedimentology subsumed in this scatter plot would need to be compared more directly in order to characterise regionally distinct types of rockslide and rock avalanche dams.

## *4.4 Implications for Trigger Mechanisms and Longevity*

Judging from the number and area distribution of mapped landslides in the WSA, there may be a few hundred rockslides with volumes $>10^6$ m$^3$ in the Southern Alps, and it is likely that more rockslide and rock avalanche dams will be detected. More important however is the observation that many large rock-slope failures in the WSA do not show any geomorphic evidence of river blockage, either due to eradication of this evidence or close interaction between slope movement and fluvial erosion. Understanding better the dynamics of large-scale slope deformation is an important issue for future research, especially with regard to deep-seated sackung-type failures that are potentially subject to episodic or catastrophic reactivation by fluvial undercutting or earthquake shaking.

The next $M$ ~8 earthquake on the Alpine Fault is likely to cause landsliding over some 10,000 km$^2$ of mountainous South Island terrain [12, 29]. Studies of historic New Zealand earthquakes illustrate the coseismic and synchronous formation of several rockslide dams. Thus the present situation of very few existing rockslide-dammed lakes may not be representative, given the short time it would take to obliterate evidence of such lakes, particularly in the WSA.

Assuming an age of ~1,000 year for the Lochnagar rockslide dam implies that the dam remained intact despite experiencing several major Alpine Fault earthquakes [29]. It remains speculative, whether repeated strong ground shaking has contributed to strengthening of the integrity of the dam. Unfortunately, little is known about the internal structure of rockslide dams until they are breached (Fig. 4.4b), although in many cases the sedimentologic characteristics of the dam govern its stability [28]. Intensely fragmented rock avalanche debris is much more prone to erosion than a coherent rock-block deposit, once the surficial boulder armour is breached. The longevity and size of dammed lakes on major rivers in the WSA may however be constrained more tightly by high sediment yields and river gradients, respectively.

There is also the possibility of spontaneous rock avalanching without any observed trigger [22], and deep-seated gravitational slope deformation (e.g. rock flow or sackung) acting as a precursor to catastrophic failure. The contemporary frequency of large rock avalanches in the ASA and ESA is estimated at 1 per 20–30 year [22], and this episodic river-damming potential should be regarded in addition to that from coseismic failures.

## 4.5 Long-Term and Long-Range Effects

The estimates of sediment discharge from rockslide dams in the Southern Alps confirm earlier observations [30], that large rockslides and rock avalanches are important (both prehistoric and contemporary) sources of sediment for alpine rivers. Reworking and temporary storage of eroded dam debris downstream can lead to adverse off-site effects of floodplain and fan-head aggradation, and channel avulsion, where agricultural land, infrastructure, or settlements are affected [18, 19]. Given a short transport distance and insufficient attenuation space, catastrophic sediment pulses pose lasting problems to land use and infrastructure, even several kilometres away from the actual rockslide or rock avalanche dam (Fig. 4.10).

Such long-term and long-range geomorphic effects necessitate the recognition of, and dealing with, a broader range of hazards from rockslides and rock avalanches. Experiences made with the Poerua rock avalanche dam give an indication of what

**Fig. 4.10** Catastrophic aggradation is one long-term and long-range effect following the failure of rock avalanche dams, e.g. the >100-m high blockage (rd) below Mt Adams, Poerua River, WSA, in 1999 [11]. Sc = Scarp area; ds = secondary debris slide, induced by aggradation. Coarse sedimentation on Poerua River alluvial fan has buried pristine floodplain forests up to 6 m, with numerous dead trees remaining in growth position (photo taken by J. Thomson in March 2004, i.e. four-and-a-half years after dam failure)

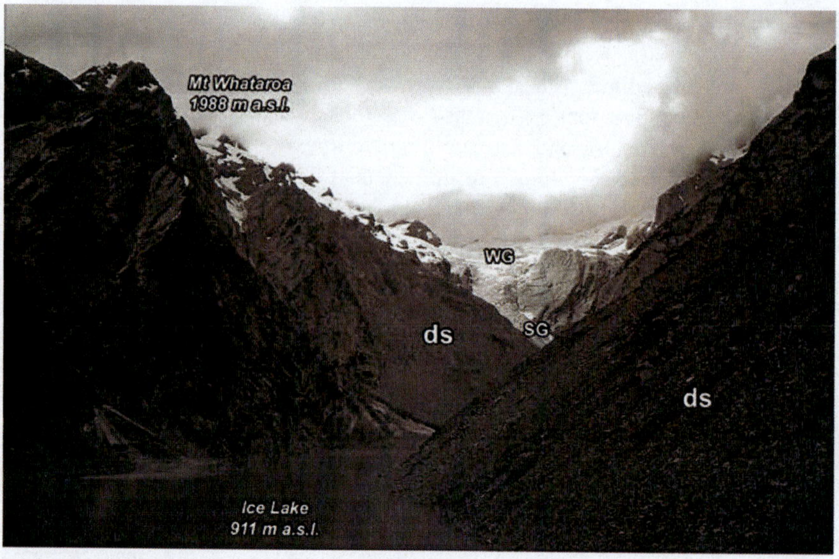


**Fig. 4.11** Ice Lake on upper Butler/Whataroa River, near the main divide, WSA (March 2003). WG = Whataroa Glacier; SG = Shackleton Glacier. Instability on steep rock- and debris slopes (ds) may trigger landslide-induced displacement waves in this proglacial lake, potentially causing downstream flooding

to possibly expect at several locations, following the next major earthquake in the Southern Alps [11].

Similarly, rockslide and rock avalanche-induced displacement waves have not yet been addressed in detail in landslide (hazard) research in the Southern Alps. Rock avalanches from Mt Fletcher (1992), CSA, and a rockslide near Mt Aspiring (1988), have displaced millions of $m^3$ of water in proglacial lakes, causing downstream flooding [22]. Rockslide-dammed and proglacial lakes such as Ice Lake (Butler, Whataroa, WSA; Fig. 4.11), or several sites in the Mt Cook area (ASA, CSA) could give rise to displacement waves. However most of the lakes in question are situated in headwater areas at sufficient distances away from infrastructure or settlements, so that flood waves can attenuate, and a low destruction potential be assumed.

## 5 Conclusions

This study reviewed and compiled geomorphic evidence from 43 large (>$10^6$ $m^3$) river-damming rockslides and rock avalanches in the Southern Alps. A major finding is their occurrence in comparable numbers to the west and east of the main divide. Regional variations in lithology, precipitation, and valley geometry, are not particularly well reflected in geomorphometric characteristics of the dams. Catastrophic rock avalanches represent the dominant type of river-blocking slope failure in the greywacke terrain of the Southern Alps. In the schist lithology of the western Southern Alps, evidence of river-damming rock avalanches is rarely observable. Instead, deep-seated complex failures with low displacement cause ephemeral blockage or, more prominently, channel occlusions and diversions. High rates of sediment yield and erosional processes constrain the longevity of rockslide- and rock avalanche-dammed lakes, of which very few presently exist. Given the currently high probability for a $M$ ~8 earthquake [29], the present picture of rockslide- and rock avalanche impact on rivers in the Southern Alps may typify the final stage of alpine landscape recovery from episodic high-magnitude seismicity.

Post-failure morphodynamic processes of rockslide- and rock avalanche-dam decay can substantially extend the impact range of river blockage. Despite affecting small portions of a given catchment, the amount of debris supplied from breached dams can be massive, and even exceed catchment background yields for prolonged periods. Geomorphic consequences of catastrophic aggradation and channel instability pose persisting problems to downstream land-use, infrastructure, or settlements on floodplains and alluvial fans at the fringes of the Southern Alps.

The low population density in the New Zealand Southern Alps however keeps risk levels from direct rockslide or rock avalanche impact low. The piedmont west of the WSA is particularly vulnerable to outburst floods and post-failure aggradation from river-blocking rockslides and rock avalanches, where the short longevity of dammed lakes requires appropriate and rapid response immediately upon their formation. The broad and less steep valley trains of the ESA tend to reduce this potential by offering sufficient attenuation space.

**Acknowledgements** I thank the organizers of the 2004 Bishkek NATO ARW for their invitation to a very inspiring meeting. Thanks are also extended to R.L. Schuster, M. Crozier, N. Perrin, G. Dellow, T.R. Davies, M. McSaveney, and S.G. Evans for many encouraging discussions.

## References

1. Adams, J.E. (1980) Contemporary uplift and erosion of the Southern Alps, New Zealand. Part 2, *Geological Society of America Bulletin* **91**, 1–114.
2. Bull, W.B. and Brandon, M.T. (1998) Lichen dating of earthquake-generated regional rockfall events, Southern Alps, New Zealand, *Geological Society of America Bulletin* **110**, 60–84.
3. Casagli, N. and Ermini, L. (1999) Geomorphic analysis of landslide dams in the Northern Apennine, *Transactions Japanese Geomorphological Union* **20**, 219–249.
4. Costa, J.E. and Schuster, R.L. (1988) The formation and failure of natural dams, *Geological Society of America Bulletin* **100**, 1054–1068.
5. Davies, T.R.H. (2002) Landslide-dambreak floods at Franz Josef Glacier township, Westland, New Zealand: A risk assessment, *Journal of Hydrology (New Zealand)* **41**, 1–17.
6. Davies, T.R. and McSaveney, M.J. (2002) Dynamic simulation of the motion of fragmenting rock avalanches, *Canadian Geotechnical Journal* **39**, 789–798.
7. Davies, T.R.H. and Scott, B.K. (1997) Dambreak flood hazard from the Callery River, Westland, New Zealand, *Journal of Hydrology (New Zealand)* **36**, 1–13.
8. Einsele, G. and Hinderer, M. (1997) Terrestrial sediment yield and the lifetimes of reservoirs, lakes, and larger basins, *Geologische Rundschau* **86**, 288–310.
9. Griffiths, G.A. (1979) High (suspended) sediment yields from major rivers of the Western Southern Alps, New Zealand, *Nature* **282**, 61–63.

10. Hancox, G.T., Cox, S.C., Turnbull, I.M. and Crozier, M.J. (2003) Reconnaissance studies of landslides and other ground damage caused by the $M_w$ 7.2 Fiordland earthquake of 22 August 2003, *Institute of Geological & Nuclear Sciences Science Report* **2003/30**, 39 pp.
11. Hancox, G.T., McSaveney, M.J., Davies, T.R. and Hodgson, K. (1999) Mt Adams rock avalanche of 6 October 1999 and subsequent formation and breaching of a large landslide dam in Poerua River, Westland, New Zealand, *Institute of Geological and Nuclear Sciences Report* **99/19**, Lower Hutt, 22 pp.
12. Hancox, G.T., Perrin, N.D. and Dellow, G.D. (1997) Earthquake-induced landsliding in New Zealand and implications for MM intensity and seismic hazard assessment, *Institute of Geological and Nuclear Sciences Client Report* **43601B**, prepared for Earthquake Commission Research Foundation. Lower Hutt, 85 pp.
13. Henderson, R.D. and Thompson, S.M. (1999) Extreme rainfalls in the Southern Alps of New Zealand, *Journal of Hydrology (New Zealand)* **38**, 309–330.
14. Hicks, M., Shankar, U. and Mckerchar, A. (2003) Sediment yield estimates: A GIS tool, *NIWA Water & Atmosphere* **11**, 26–27.
15. Koons, P.O. (1989) The topographic evolution of mountain belts: A numerical look at the Southern Alps, New Zealand, *American Journal of Science* **289**, 1041–1069.
16. Korup, O. (2002) Recent research on landslide dams – a literature review with special attention to New Zealand, *Progress in Physical Geography* **26**, 206–235.
17. Korup, O. (2004) Geomorphometric characteristics of New Zealand landslide dams, *Engineering Geology* **73**, 13–35.
18. Korup, O. (2004) Landslide-induced river channel avulsions in mountain catchments of southwest New Zealand, *Geomorphology* **63**, 57–80.
19. Korup, O., McSaveney, M.J. and Davies, T.R.H. (2004) Sediment generation and delivery from large historic landslides in the Southern Alps, New Zealand, *Geomorphology* **61**, 189–207.
20. Little, T.A., Holcombe, R.J. and Ilg, B.R. (2002) Kinematics of oblique collision and ramping inferred from microstructures and strain in middle crustal rocks, central Southern Alps, New Zealand, *Journal of Structural Geology* **24**, 219–239.
21. Long, D.T., Cox, S.C., Bannister, S., Gerstenberger, M.C. and Okaya, D. (2003) Upper crustal structure beneath the eastern Southern Alps and the Mackenzie Basin, New Zealand, derived from seismic reflection data, *New Zealand Journal of Geology and Geophysics* **46**, 21–39.
22. McSaveney, M.J. (2002) Recent rockfalls and rock avalanches in Mount Cook National Park, New Zealand, in S.G. Evans and J.V. DeGraff (eds.), *Catastrophic Landslides: Effects, Occurrence, and Mechanisms*, Reviews in Engineering Geology **XV**, Geological Society of America, Boulder, CO, pp. 35–70.
23. Norris, R.J. and Cooper, A.F. (2000) Late Quaternary slip rates and slip partitioning on the Alpine fault, New Zealand, *Journal of Structural Geology* **23**, 507–520.
24. Perrin, N.D. and Hancox, G.T. (1992) Landslide-dammed lakes in New Zealand preliminary studies on their distribution, causes and effects, in D.H. Bell (ed.), *Landslides. Glissements de terrain*, Balkema, Rotterdam, pp. 1457–1466.
25. Ramsay, G. (2000) Otira Gorge rock avalanche, in E. Bromhead, N. Dixon and M.L. Ibsen (eds.), *Landslides in Research, Theory and Practice*, Thomas Telford, London, pp. 1274–1280.
26. Schuster, R.L. (2000) Outburst debris flows from failure of natural dams, in G.F. Wieczorek and N.D. Naeser (eds.), *2nd International Conference on Debris-Flow Hazard Mitigation*, Balkema, Rotterdam, pp. 29–42.
27. Tippett, J.M. and Kamp, P.J.J. (1995) Quantitative relationships between uplift and relief parameters for the Southern Alps, New Zealand, as determined by fission track analysis, *Earth Surface Processes and Landforms* **20**, 153–175.
28. Weidinger, J.T., Wang, J. and Ma, N. (2002) The earthquake-triggered rock avalanche of Cui Ha, Qin Ling Mountains, P.R. of China – the benefits of a lake-damming prehistoric natural disaster, *Quaternary International* **93–94**, 107–214.

29. Wells, A., Duncan, R.P., Stewart, G.H. and Yetton, M.D. (1999) Prehistoric dates of the most recent Alpine fault earthquakes, New Zealand, *Geology* **27**, 995–998.
30. Whitehouse, I.E. (1983) Distribution of large rock avalanche deposits in the central Southern Alps, New Zealand, *New Zealand Journal of Geology and Geophysics* **26**, 272–279.
31. Whitehouse, I.E. (1988) Geomorphology of the central Southern Alps, New Zealand: The interaction of plate collision and atmospheric circulation, *Zeitschrift für Geomorphologie NF* **69**, 105–116.

# Chapter 5
# Landslide Dams in the Central Andes of Argentina (Northern Patagonia and the Argentine Northwest)

**Reginald L. Hermanns, Andres Folguera, Ivanna Penna, Luis Fauqué, and Samuel Niedermann**

## 1 Introduction

The Andean Cordillera of Argentina is more than 3,500 km long, 50–550 km wide and characterized by steep relief throughout. Except for a few valleys, most of this region is sparsely populated. Hence geologic mapping was mainly carried out on a small scale (1:200,000 or 1:250,000) which is often too small to map out various Quaternary units and therefore to highlight relations of landslide deposits to lakes or lake deposits. Relatively few studies have focused on large landslide events [e.g. 6, 14, 22, 34]. However, during the last decade declared by the United Nations as the decade to reduce the impact of natural disasters, more landslide studies were carried out in this area. Most of them focused on large landslides of the rock avalanche type which are frequently related to formation of natural dams forming large lake basins in the narrow valleys of the Argentine Andes, often with surface areas of several km$^2$ to several 100 km$^2$ in area [e.g., 8, 20, 21, 23, 28, 47, 48]. Argentina's largest landslide disaster was related to the catastrophic failure of a rockslide dam on December 29, 1914 in Barrancas valley that causing downstream destruction and at least 155 casualities in the Barrancas and Colorado river valleys over a distance of ~1,250 km [21]. Despite this, little attention has been given to landslide dams and their stability. Only in recent years studies on landslide dam stability have been carried out [9, 16, 20, 26, 27, 29].

In this paper we will focus on two regions: one in the Argentine NW (24–27.5°S) and the other in northern Patagonia (36–38°S) (Fig. 5.1). Our aim is to summarize studies carried out in these regions previous, and to provide new, previously unpublished case studies and morphological data. In both regions, the authors have carried out systematic studies related to landslides with volumes in excess of 10$^6$ m$^3$. These landslides formed at least 20 and 41 landslide dams in the two regions, respectively. In the following details of some cases will be described, which will help to better

R.L. Hermanns (✉)
International Centre for Geohazards, Geological Survey of Norway, Trondheim, Norway
e-mail: reginald.hermanns@ngu.no

S.G. Evans et al. (eds.), *Natural and Artificial Rockslide Dams*, Lecture Notes in Earth Sciences 133, DOI 10.1007/978-3-642-04764-0_5, © Springer-Verlag Berlin Heidelberg 2011

**Fig. 5.1** Locations of study areas in Argentina. Boxes represent extents of study areas shown in Figs. 5.2 and 5.3, NW Argentina and N Patagonia, respectively

understand large landslide dams and their stability in the Argentinian Andes and other regions.

# 2 Geological Setting

## *2.1 NW-Argentine Andes Between 25 and 27.5°S*

The study area in NW Argentina lies at the transition between the Puna Plateau to the west, and the northern Sierras Pampeanas, and the Cordillera Oriental to the south and east (Fig. 5.2). The Puna Plateau and the northwestern Sierras Pampeanas are characterized by mountains formed as basement-blocks with elevations between 4,000 and >6,000 m and intervening basins at elevations between 1,500 and 3,500 m (Fig. 5.2). The basement-blocks are bounded by reverse-faults and are composed of late Precambrian-Palaeozoic low- to high-grade metamorphic rocks and Palaeozoic granites [e.g. 43]. The intervening basins contain Tertiary to Quaternary basin-fill deposits [e.g. 45]. The Cordillera Oriental (Fig. 5.2) is a fold-and-thrust belt of Precambrian basement and overlying unmetamorphosed Cambrian to Tertiary

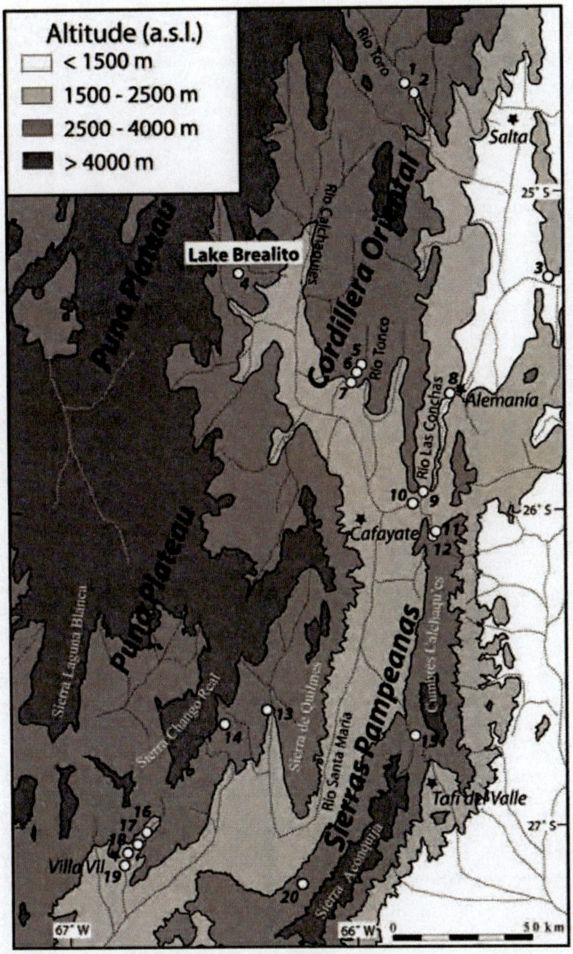


**Fig. 5.2** Topographic map of study area in NW Argentina with distribution of mountain ranges, intramontane basins, narrow valleys, and palaeo-landslide dams (*circles*) as well as Lake Brealito dammed by a rock avalanche. Names and geomorphic parameters of each dam are given in Table 5.1

sediments [e.g. 44]. Common to these valleys and basins are large rock slope failures of the rock-avalanche type comprising > 0.01 km$^3$ rock [24]. These occur only in three lithologies: granites, low grade metamorphic rocks, and coarse clastic sedimentary rocks. Their deposits dammed lakes in narrow valley environments in prehistoric times (Fig. 5.2) [24, 25, 46], which extended up to ~600 km$^2$ in area and depths >200 m [1, 24, 37].

## 2.2 Northern-Patagonian Argentine Andes Between 36 and 38°S

The study area in N Patagonia lies at the southernmost part of the Southern Central Andes (33–39°S). It is comprised of continental volcaniclastic and volcanic rocks of Permian to Neogene age [e.g. 12, 39–40, 50]. These units are associated with Upper Pliocene to Holocene volcanoes [e.g. 12, 39] throughout the region.

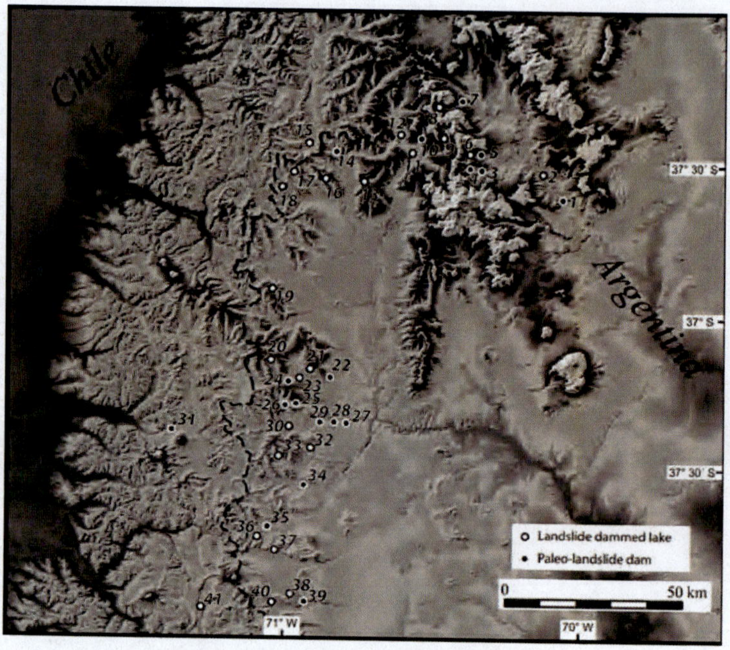


**Fig. 5.3** Digital elevation model of N Patagonia showing distribution of mountain ranges, narrow valleys, as well as landslide dams and palaeo-landslide dams. Names and geomorphic parameters of each dam are given in Table 5.2

Mountains are 2,500–4,000 m high and valleys lie at altitudes between 1,000 and 3,000 m. Large deposits of mountain front failures involving more than 0.1 $km^3$ are predominantly of the rock-avalanche and rotational slide type. However, other failure types such as toppling failures and deep gravitational deformations also exist. All of these landslides were prehistoric and they originated predominantly in basaltic flow deposits with horizontal attitude [26]. In addition, the deposits line up along active Quaternary faults [3, 12, 20] and caused frequent valley impoundments [26].

## 3 Methods

The formation of all lakes in the study area were studied. Landslide-dammed lakes were recognized by the hummocky, lobate, or step-like surface of the landslide deposits using Landsat TM imagery, depending on whether landslide barriers were formed by rock avalanches, rock-block slides or rotational slides, respectively. Landslide barriers that were eroded and, consequently the impounded lake had drained in the geological past, were recognized by mapping the distribution of its remnants debris within the valley. In the same way, landslide lake basins that have

**Table 5.1** Geomorphometric data of landslide dams in NW Argentina

| Number | Lake, river or site name | Catchment area (km$^2$) | Dam volume (km$^3$) | Dam height (m) | Status | Dam type[a] |
|---|---|---|---|---|---|---|
| 1 | Río Toro | 2,530 | ND | 50 | ND | ND |
| 2 | Río Toro | 2,542 | ND | 35 | ND | ND |
| 3 | Río Pasaje | 31,900 | ND | ND | ND | II |
| 4 | Lake Brealito | 26 | 0.03 | 120 | Lake | II |
| 5 | Río Tonco | 220 | 0.07 | 35 | ND, IN | III |
| 6 | Río Tonco | 275 | ND | 10 | ND | II |
| 7 | Río Tonco | 280 | ND | ND | ND | ND |
| 8 | Río Las Conchas | 19,820 | 0.06 | 75 | BR | III |
| 9 | Río Las Conchas | 19,760 | 0.16 | 220 | BR | III |
| 10 | Río Las Conchas | 19,730 | 0.44 | ~190 | ND | III |
| 11 | Río La Yesera | 51 | ND | ND | ND | IV |
| 12 | Río La Yesera | 16 | 0.03 | ND | IN | II |
| 13 | Río Toroyacu | 975 | ND | ND | IN | III |
| 14 | El Rincón | 2 | ND | ND | ND | pond |
| 15 | Río Amaicha | 112 | ND | ND | ND | II |
| 16 | Río El Bolson | 222 | 0.18 | ND | ND | II |
| 17 | Río El Bolson | 287 | 0.37 | 100 | IN | New |
| 18 | Río Villa Vil | 290 | 0.24 | >12 | ND | II |
| 19 | Río Villa Vil | 347 | 0.03 | ND | ND | II |
| 20 | Loma de la Aspereza | 4 | ND | ND | ND | pond |

(ND = not defineable, IN = infilled, BR = breached).
[a]Dam type after [2], new = does not fit with classification (see below), pond = water body entirely on top of landslide deposit.

filled up with sediments and are now dry basins, were found by searching for landslides lying at the end of flat segments of the thalweg of the valley. In these latter cases spectral contrasts of lake deposits and basin fill with the surrounding bedrock helped to locate such palaeo-landslide dams. In a second step, the inferred landslide origin of the dams was verified on stereo pairs of aerial photographs. Finally, all of the landslide dam sites were visited in the field in NW Argentina and 45% in N Patagonia.

To calculate morphologic parameters such as dam height, dam volume and drainage area upriver of the landslide dam, Shuttle Radar Topographic Mission (SRTM) data with a horizontal resolution of 90 m and a vertical accuracy of 10 m were used as well as topographic maps of 1:100,000 and 1:250,000 with 50 and 100 m contours in Patagonia and NW Argentina, respectively. The dam height was constructed by plotting the valley profile and interpolating the knick in the thalweg between the upper end of the lake and the foot of the landslide dam. The dam height was measured as the difference in altitude between the valley profile and the interpolated line. In two cases, we obtained actual measurements of the depth of the lake. In both cases these were within 10% accuracy of the constructed lake depth. The altitude of eroded dams was estimated by either measuring the highest lake deposits

**Table 5.2** Geomorphometric data of landslide dams in N Patagonia

| Number | Lake, river or site name | Catchment area (km$^2$) | Dam volume (km$^3$) | Dam height (m) | Status | Dam type[a] |
|---|---|---|---|---|---|---|
| 1 | Río Barrancas | 21,173 | 0.03 | 30 | ND | II |
| 2 | L. Cari Lauquen | 1,923 | 1.30 | 160[b] | BR, lake | III, pond |
| 3 | A. Los Díaz | 45 | 0.21 | 58 | ND | II |
| 4 | A. Los Díaz | 34 | 0.30 | 66 | ND | II |
| 5 | A. Blanco | 54 | ND | ND | ND | II |
| 6 | L. Cochico | 25 | 0.10 | 48 | Lake | II |
| 7 | A. los Nevados | 66 | 0.06 | 80 | ND | II |
| 8 | L. los Nevados | 8 | 0.20 | ND | Lake | pond |
| 9 | L. Cachaico | 30 | 0.57 | 94 | lake | V |
| 10 | A. La Crianza | 18 | 0.01 | 49 | ND | II |
| 11 | L. Varvarco Tapia | 380 | 1.45 | 85[b] | BR, lake | III, pond |
| 12 | L. Varvarco C. | 300 | 3.1 | 188 | lake | new, pond |
| 13 | L. La Leche | 26 | 0.19 | 91 | lake | II |
| 14 | A. Catrinao | 31 | 0.05 | 69 | ND | II |
| 15 | L. Dial | 91 | 0.50 | 164 | BR, lake | II |
| 16 | L. Frutillar | 5.5 | 0.06 | 79 | lake | II |
| 17 | L. Molinos | 3 | 0.03 | ND | lake | II |
| 18 | L. Navarrete | 26 | 0.13 | 80[b] | BR, lake | III |
| 19 | L. Los Pajaritos | 16 | 0.03 | 164 | lake | II |
| 20 | L. La Tregua | 12.2 | 0.19 | 52 | lake | II |
| 21 | L. Los Rojos | 11 | ND | ND | lake | pond |
| 22 | Río Lileo | 296 | 0.40 | 44 | ND | III |
| 23 | L. N.N. | 222 | 0.37 | ND | ND | II, pond |
| 24 | A. Buta Mallín | 142 | ND | ND | ND | III |
| 25 | Río Guañaco | 42 | 0.21 | 39 | ND | II |
| 26 | Río Guañaco | 31 | ND | ND | ND | III |
| 27 | Río Reñileuvú | 527 | 0.07 | 36 | ND | II |
| 28 | Río Reñileuvú | 516 | 0.05 | 30 | ND | II |
| 29 | Río Reñileuvú | 482 | 0.81 | 58 | ND | III |
| 30 | L. Pelahuenco | 335 | 0.94 | 65 | lake | III |
| 31 | Río Laja | 106 | 1.00 | 800 | BR | III |
| 32 | L Lauquen Mallín | 31 | 0.16 | 67 | lake | II |
| 33 | L. Negra | 24 | 0.06 | 47 | lake | III |
| 34 | A. Picunleo | 255 | 0.50 | 75 | ND | III |
| 35 | A. Pillum Chaya | 138 | ND | ND | ND | III |
| 36 | L. N.N. | 23 | 0.02 | 18 | lake | II |
| 37 | L. Trolope | 81 | 0.01 | 9 | lake | I |
| 38 | L. Agrio | 94 | 0.48 | 88 | lake | new |
| 39 | A. Hualcupén | 157 | 0.05 | 10 | ND | III |
| 40 | L. Hualcupén | 31 | 0.32 | 61 | lake | III |
| 41 | L. El Barco | 56 | 0.17 | 101 | lake | VI |

L. = Lake, A. = Arollo, N.N. = no name, ND = not defineable, BR = breached.
[a]Dam type after [2], new = does not fit with classification (see below), pond = water body entirely on top of landslide deposit.
[b]Dam height before failure.

or palaeoshorelines upriver of the dam, or by interpolating between the altitudes of the remnants of the landslide deposits and the present river bed. Field measurements were carried out with an altimeter with 1 m relative precision. For those landslides not visited in the field, dam height calculations were based on interpolations using the topographic data. Except in two cases where a palaeoshoreline was found, we suspect that the uncertainty of our palaeoshorline estimates is twice to three times higher than the uncertainty of the constructed dam heights.

Landslide dam volumes were estimated using the product of surface area of the landslide, obtained from satellite image analysis, and the dam height, which was then calibrated for average landslide morphology using a standard formula [49]:

$$\mathrm{VOL} = \pi \mathrm{DWL}/6$$

where VOL is the volume of the landslide, D the depth of the displaced mass, W the width of the displaced mass, and L the length of the displaced mass. As all landslides have been prehistoric and no topographic information previous to the landslide exists, there is no economically feasible method to control these estimates; however we expect that the uncertainty of our volume estimates is within ~50%.

# 4 Status, Types of Landslides and Landslide Dams and Geomorphometric Parameters

## *4.1 Landslides Dams in NW-Argentinean Andes*

All 20 landslide dams in our inventory from NW Argentina formed in prehistoric time. Their ages are between >> 32 and < 3.7 ka; however some of them are poorly defined in their age [25, 31, 46]. Some lasted for several 1,000 years [31, 47]. Only one of the dams is impounding a lake at present (Brealito Lake in Fig. 5.2), all the other were eroded in prehistoric time. They are either partially eroded, totally eroded, or have lake basins filled up with sediments (Fig. 5.4, Table 5.1). Among the eroded dams, lake sediments were found behind 12 palaeo-dams; the other dams could be reconstructed from the distribution of the landslide deposits. The mode of dam failure could be defined for two cases only ([29], see below).

Except for landslide dams 16–19 (Table 5.1), which were formed by rock-block slides with distal flow components similar to rock avalanches, and landslide 13 (Fig. 5.2), which is a compound block failure with translational and rotational components, all landslide dams were formed by rock avalanches [8, 24]. Due to the variable degree of erosion not all dams could be classified. Using the classification system proposed by Costa and Schuster [2], type II and III dams are the most frequent one. However, ponds on top of landslide deposits were also found, as well as one other landslide dam type which cannot be adequately described by the Costa-Schuster classification system (Fig. 5.5, and below).

Due to the strong erosion of most dams, landslide dam geomorphometric parameters are often difficult to reconstruct (Table 5.1).

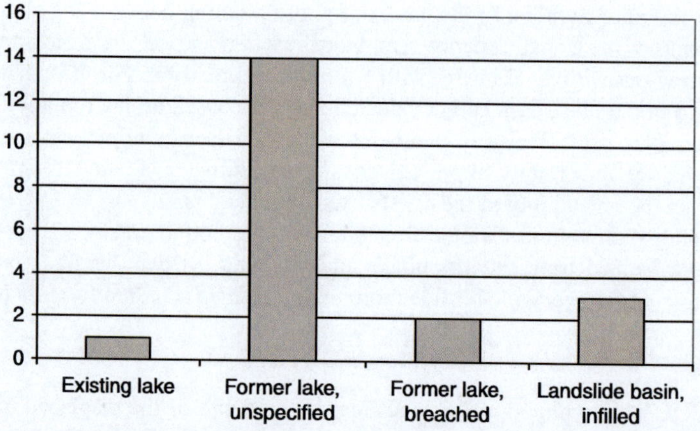


**Fig. 5.4** Present status of $n = 20$ landslide dams in NW Argentina (see Table 5.1)

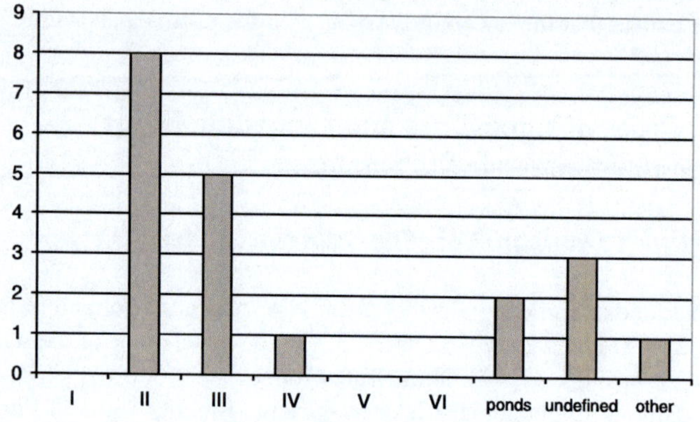


**Fig. 5.5** Landslide dam types in NW Argentina following [2]

## 4.2 Landslides Dams in the N-Patagonian Andes

From the 41 landslide dams in the N-Patagonian Andes 38 lie within Argentina, 2 in Chile and one landslide originated on the Argentine-Chilean border but formed a lake within Chile. Again, all landslide dams are believed to be prehistoric. Their ages are poorly constrained but some of the impounding lakes several tens of km$^2$ in area formed 30 to 60 ka ago [3]. Twenty-two landslide dammed lakes exist in this region at present, 18 dams were eroded in the geological past, 5 dams have clear indications of partial failure (Fig. 5.6). None of the landslide dam basins are filled up with sediments. Among the partial failures is the failure of December 29, 1914 that produced the Río Barrancas/Río Colorado flood down to the Atlantic Ocean [16, 21].

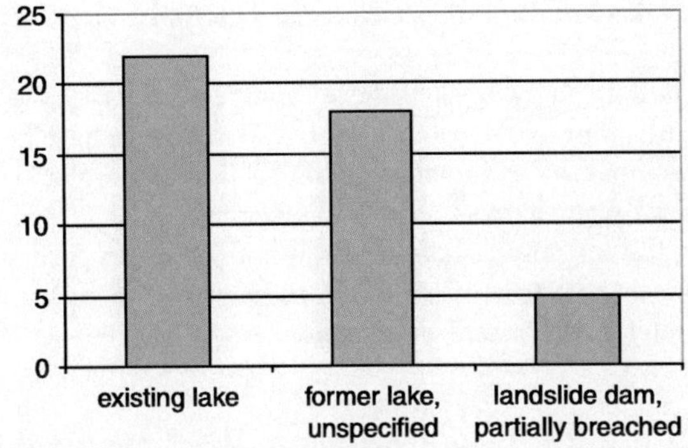


**Fig. 5.6** Present status of $n = 41$ landslide dams in N Patagonia

Landslide dams were formed by rock avalanches, rotational slides (32, 41 in Fig. 5.3) and deep gravitational spreads (sagging, also "Talzuschub" in German, 36 in Fig. 5.3) with volumes between 0.01 and 3.1 $km^3$ (Table 5.2). There have been more landslides in this range of volumes in the region [21, 26]. However they did not cause the damming of rivers, or their deposits are so strongly eroded that it is impossible to reconstruct any valley impoundments.

Most of the preserved landslide dams could be classified (Fig. 5.7); however, some of them (Table 5.2) did not fit with the Costa-Schuster landslide dam classification in use [2]. New types thus had to be defined (see below). At several sites there have been larger water bodies impounded on top of the landslide deposit (Fig. 5.7). Due to the good preservation of most of these dams a larger number of geomorphometric dam parameters could be reconstructed as compared to NW Argentina (Table 5.2).

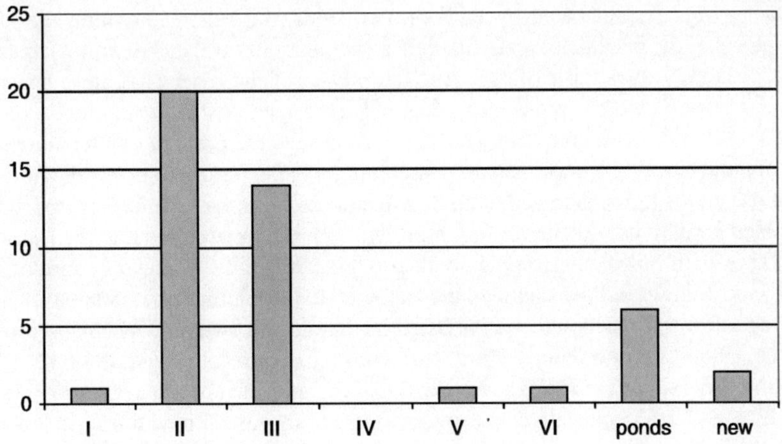


**Fig. 5.7** Landslide dam types in Northern Patagonia following [2]

# 5 Examples of Rockslide Dams in the Argentine Andes

In both study areas of the Argentine Andes rockslide dams were formed in prehistoric times. While the dams in NW Argentina have been eroded to a large extent, and geomorphic parameters are difficult to reconstruct (Table 5.1), these data are more easily generated in the N-Patagonian Andes (Table 5.2). In the following we will illustrate some examples from both areas. In particular we will focus on landslide dams which have breached, as catastrophic rockslide dam failure is of particular interest for hazard assessment.

## 5.1 *Rock Avalanching into Rock Avalanche-Dammed Lakes in Las Conchas Valley (NW Argentina)*

At the upper end of Las Conchas valley, the mountain Cerro Zorrito is located. It is composed of conglomeratic rock and has collapsed repeatedly in prehistoric time, forming rock avalanches at its E, S, and W sides [24]. Las Conchas valley was dammed by the deposits of these rock avalanches at two sites, El Paso and Casa de los Loros (Fig. 5.8). The El Paso dam had a volume of 0.21 $km^3$ and formed 15.3±2.0 ka ago [29]. The height of the dam at that time could not be reconstructed. Casa de los Loros dam had a volume of 0.16 $km^3$ and formed 13.6±0.9 ka ago, hence both dams might have formed simultaneously at ca. 13.8±0.8 ka ago [29]. The Casa de los Loros dam impounded the valley up to the contour of 1,700 m a.s.l. as indicated by lake deposits upriver. Therefore the dam must have been 220 m high, causing a lake with a surface area of ~600 $km^2$ which was subdivided by the El Paso dam into a large lake basin upriver and a lake basin ~8 $km^2$ in size between both dams (Fig. 5.8).

Although the area was affected twice by strong earthquakes with estimated magnitudes ~M 7, as indicated by fault offsets, folds and seismites within the lake sediments [30], both dams existed until a second rock avalanche with a volume of 0.23 $km^3$ occurred at El Paso. This fell on top of the older dam and impacted into La Yesera lake basin 4.8±0.5 ka ago. This impact pushed water, in the order of $10^7$–$10^8$ $m^3$, out of the lake, causing a tsunami wave probably high enough to overtop the Casa de los Loros dam. Based upon coarse terrace deposits downriver, the Casa de los Loros dam and flood deposits of the same age in a further landslide-dammed lake 30 km downriver (no. 8 in Fig. 5.2), it is interpreted that the Casa de los Loros dam failed catastrophically due to this tsunami [29]. The dam formed by both landslides at El Paso dammed the valley up to the contour of 1,680 m a.s.l. and eroded a few 100 years later as indicated by 10-m-high swamp deposits on top of the landslide [29] and redeposited tephra with a calibrated $^{14}C$ age of 3,632 ± 70 from within redeposited lake sediments [25]. Because lake sediments terminate at an altitude of 1,635 m a.s.l., the basin obviously never filled up with lake sediments and the El Paso dam was most likely eroded by overtopping. This can be temporally related to climatic change in this part of the Andes towards more humid climate conditions ([31, 47] and references therein).

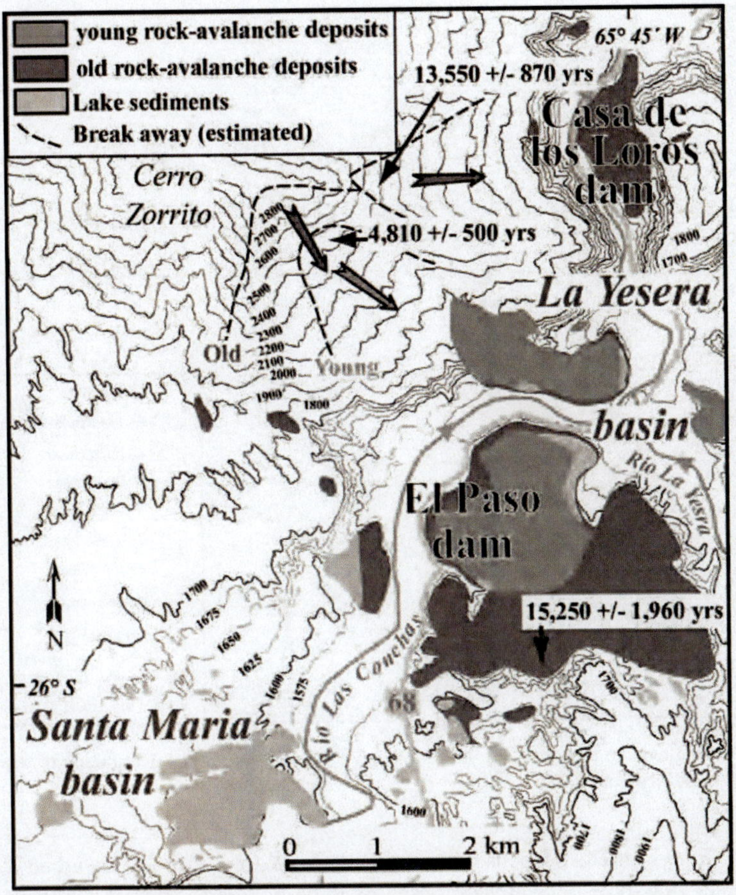


**Fig. 5.8** Topographic map of Cerro Zorrito showing the distribution of rock avalanche deposits and related lake sediments as well as surface exposure ages (SED) of break away scars and landslide deposits

The geological history of this area can be deduced from the name "Cafayate" of the largest village, which lies within the 600 km$^2$ area of the Holocene lake. It is a name in the local Indian Kakan language in which "Cafa" means "water" and "Yate" means "land in elevation".

## *5.2 Landslide Dams in the El Bolson and Villa Vil Valleys (NW Argentina)*

Six landslide deposits, which originated as translational slides in 24° northwest-dipping Tertiary volcanic breccias [8, 24], occur in the Villa Vil and El Bolson

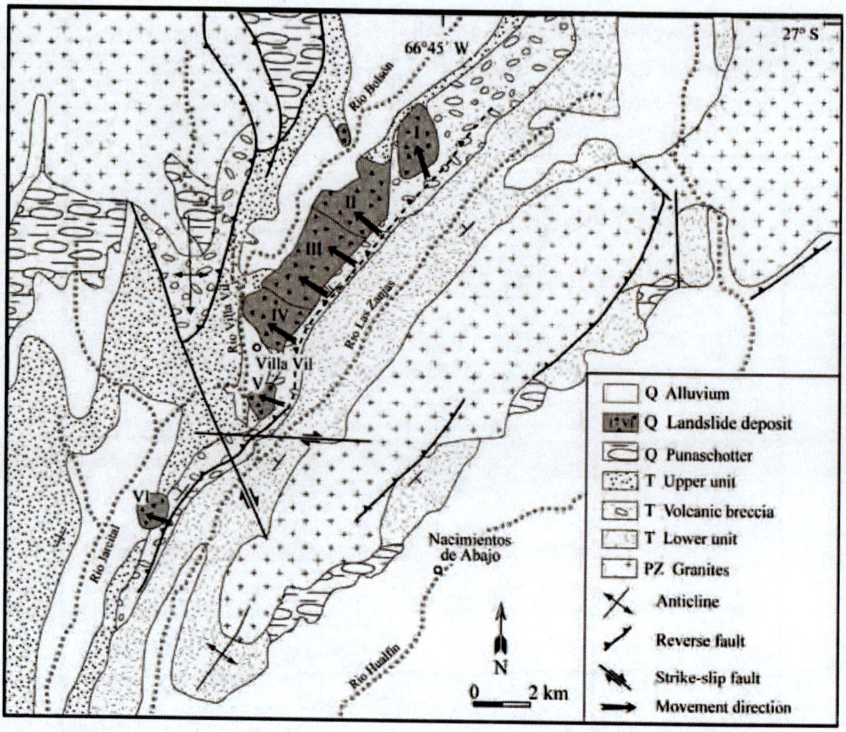


**Fig. 5.9** Geologic map of the El Bolson – Villa Vil valley showing the distribution of landslide deposits (modified after [23])

valleys (Fig. 5.2). The deposits are composed of proximal strongly disturbed blocks and distal flow-type deposits composed of large boulders. Four of the landslides have impounded the valley as indicated by preserved basin fill and lake sediments (Fig. 5.9). Landslide deposit II is found on both sides of the valley; however its central part is not exposed because it is submerged by deposits accumulated upriver the dam formed by landslide III. Landslide III spans the entire valley and formed a dam. This landslide dam is higher than the opposite slope of the valley which is composed of alternating Tertiary sandstones, conglomeratic rocks and siltstone. The river did not cut into this rock and the lake basin behind the landslide entirely filled up with sediments. This dam is estimated to have been ~100 m high [8].

Despite the large volume (0.24 km$^3$) of landslide IV, it formed only a small basin a few 100 m long. The landslide deposit is topped by a volcanic ash with a calibrated $^{14}$C age of 3,632 ± 70 a [25]. The 12-m-thick lake sediments within the landslide-dammed basin contain charcoal with a calibrated $^{14}$C age of 1,310 ± 130 a [8], indicating therefore a lake lifetime of more than 2 ka and hence a low deposition rate, that is clearly due to the landslide barrier upriver. The lake sediment thickness gives a minimum estimate of the dam height.

Likewise the distribution of the landslide deposit V (Fig. 5.9) indicates that the Villa Vil river was most likely impounded by that slide; however, no lake deposits upriver this dam are preserved.

## 5.3 *Landslide Dam Failure in Barrancas Valley (N Patagonia)*

The failure of the rockslide dam on December 29, 1914 in the Barrancas valley is the only failure of a large landslide dam that has occurred in Argentina in the last century. It was described in detail by Groeber [21]. The outburst flood destroyed low lying areas in the Barrancas and Colorado river valleys down to the Atlantic Ocean 1,250 km away [16]. The landslide dam was formed by a rock avalanche with a volume of 1.3 $km^3$ in layered volcanic rocks of basaltic to andesitic composition, dipping 30° towards the valley [27] (Fig. 5.10). This landslide impounded the Barrancas and tributary valleys and formed the ~25 km long Carri Lauquén lake. The dam existed for $\geq$ 427 years as indicated by the AMS $^{14}C$ calibrated date of 1408–1487 A.D. obtained on organic material sampled from delta deposits [27], which propagated into the ancient lake. This age is supported by a cosmogenic nuclide age of 2.2 $\pm$ 06 ka of the rock avalanche deposit [3]. During the 1914 outburst a steep-walled breach was eroded 85 m deep into the landslide dam, removing 0.12 $km^3$. This material was deposited downriver of the dam in an outburst flood deposit inundating the entire valley floor and decreasing from > 21 m thick, 1 km below the dam, to 3.5 m thick, 60 km below the dam. The outburst flood deposit impounded tributary valleys forming a chain of secondary lakes. The Quili Malal lake (300 m long) 1 km downriver of the failed dam is still impounded by that deposit. The amount of water released by the outburst was estimated to be 2 $km^3$ [21]; Hermanns et al. [27] calculated a volume of 1.55 $km^3$. At present only a 2-km-long relict of Cari Lauquén lake is preserved.

For almost 100 years it was believed that the abnormally wet austral winter of 1914 resulted in excessive snow melt causing the lake to rise and the dam to fail by overtopping [16, 21]. However, the analysis of climatic data suggests that this is not likely the only reason. Winters with similar amounts of precipitation occurred in the past 120 years with return intervals of 5–17 years, and therefore likely also occurred frequently in the previous centuries. Instead, temperature records from most of Argentina indicate that the austral summer 1913/1914 was the warmest on record, suggesting that the combination of glacial retreat in the upper catchments and melting of snow accumulation, in the previous summer, in combination with a relatively wet winter, caused the surface of Cari Lauquén lake to rise to a critical level [27].

## 5.4 *Rock Avalanche Deposits Impounding Lake Varvarco Campos and Varvarco Tapia (N Patagonia)*

Among the largest landslide-dammed lakes at present in N Patagonia are the Varvarco Campos and Varvarco Tapia lakes. Both are formed by rock-avalanche

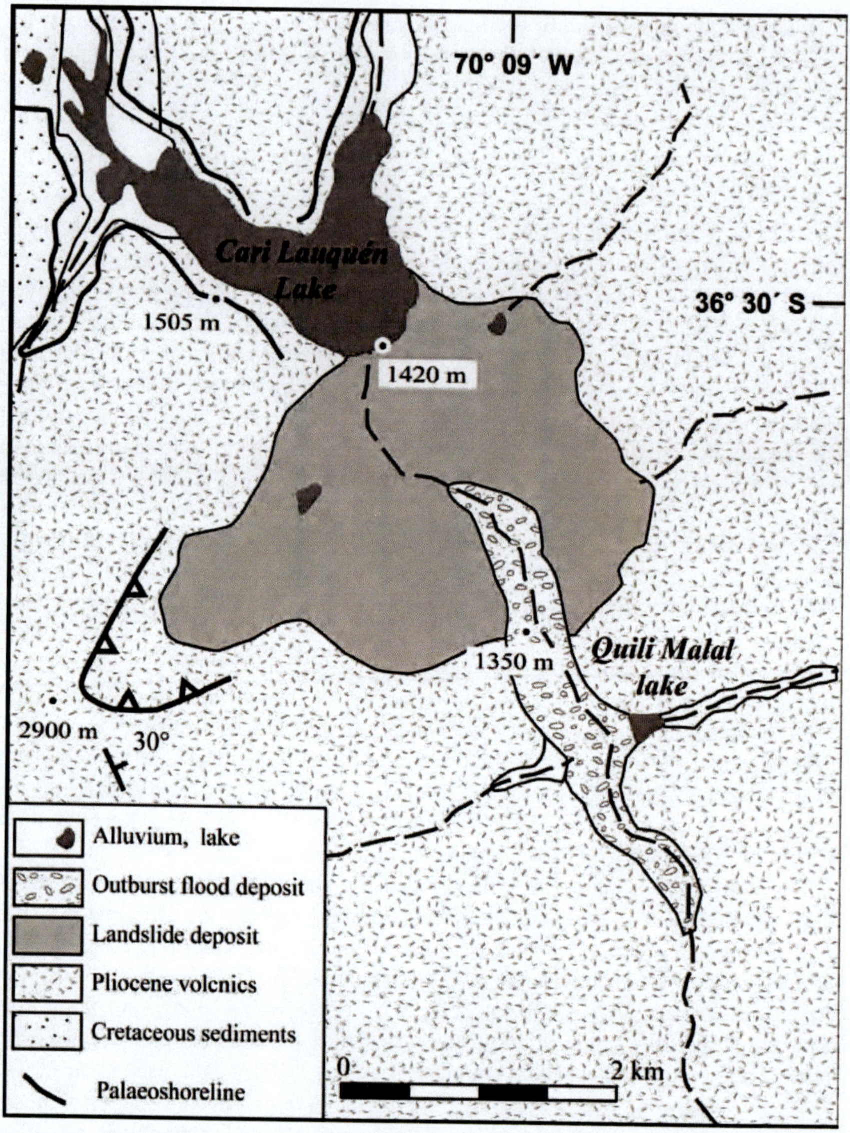


**Fig. 5.10** Simplified geologic map of the Barrancas valley showing the distribution of landslide and outburst flood deposits as well as related lakes

deposits which originated in inter-layered Late Pliocene basaltic volcanic rocks and dacitic tephra deposits with horizontal attitude [50] (Fig. 5.11). These lakes are 11 and 3.5 km long and landslide dams are 188 and 85 m high, respectively, following our reconstruction. The depth of the Varvarco Campos lake has been measured in detail by González Díaz et al. [15] and its volume was reconstructed based upon these measurements. The lake is 155 m deep (Fig. 5.11) at its upper

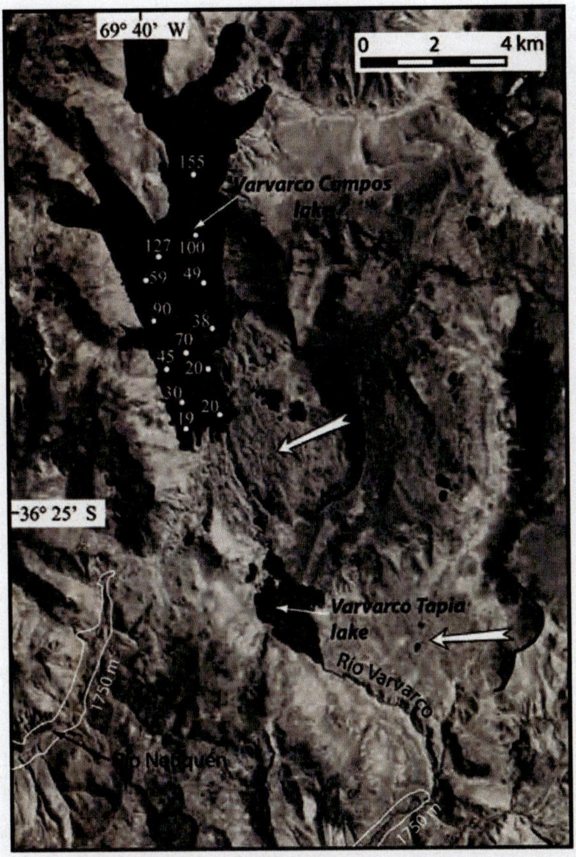


**Fig. 5.11** LANDSAT TM image of Varvarco Campos and Varvarco Tapia lakes and impounding rock-avalanche deposits. Bathimetric data of Laguna Varvarco Campo from [15], 1,750 m contour from 1:100,000 topographic map "Pichi Neuquén" (3772–11 + 12). The rock avalanche impounding Varvarco Campos lake deposited in the drainage divide between the Río Varvarco and the Río Neuquén valleys and most likely caused a drainage capture

end and has a volume of 3.51 $km^3$. Taking the valley slope into account this fits within our given precision levels in our reconstruction method. Due to the large surface area, both landslide deposits have large ponds on their surfaces (Fig. 5.11). On top of the Varvarco Campos dam the pond size reaches up to 1 km length (Fig. 5.11).

The ages of both rock avalanches were determined by cosmogenic nuclide dating yielding ages for the Varvarco Campos rock avalanche of 29–37 ka and of the Varvarco Tapia rock avalanche of 63–79 ka [3]. This fits well with the different degree of degradation of their surfaces (Fig. 5.11). As the Varvarco Campos rock avalanche deposit also lies within the Varvarco Tapia lake, this indicates that this younger rock avalanche impacted into the lower landslide dammed lake and likely caused a tsunami wave. This interpretation is further supported by the erosion of

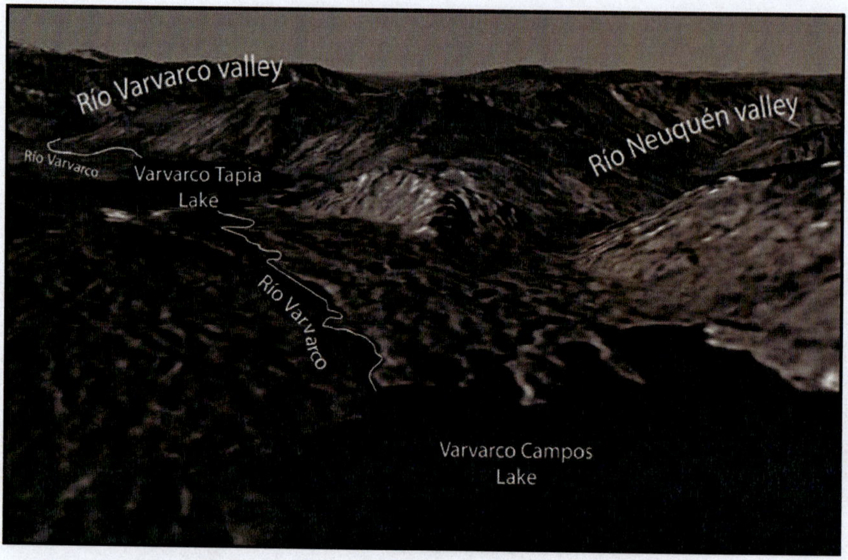


**Fig. 5.12** Oblique view towards the south of the Río Neuquén and Río Varvarco valleys, and over the rock-avalanche deposit which impounds Varvarco Campos lake. This rock avalanche occurred in the drainage divide between both valleys. The oblique view suggests that drainage from this catchment before landsliding was towards the Río Neuquén valley, in contrast to the present drainage direction of the Río Varvarco

the Varvarco Tapia dam down to the underlying rock and important terrace deposits downriver from the breach of the dam [15, 32].

Unique to the Varvarco Campos rock avalanche deposit is its location within the drainage divide between the Río Varvarco and the Río Neuquén valleys. At present Río Varvarco drains from the Varvarco Campos lake into the Varvarco Tapia lake and then into the Río Varvarco valley (Figs. 5.11 and 5.12). However, this was not necessarily the case before landsliding. Actually, the same altitude (1,750 m a.s.l. contour) of the valley floor in the Río Neuquén valley is closer to the Varvarco Campos lake than in the Río Varvarco valley (Fig. 5.11). Furthermore, the catchment area upriver of that altitude in the Río Neuquén valley is at present only 12 km$^2$ in size, while in the Río Varvarco valley it is 625 km$^2$. Both the altitude of the valley floor and the size of the present day catchments in combination suggest that the Río Neuquén fluvial basin was larger before landsliding and that the drainage upriver of the Varvarco Campos dam was towards the Río Neuquén valley, as also visually suggested by the oblique view in Fig. 5.12.

## 5.5 *Partial Breach of Lake Navarrete Rock Avalanche Dam (N Patagonia)*

Lake Navarrete is impounded by a rock avalanche deposit with a volume of ~0.13 km$^3$. This rock avalanche formed in prehistoric time in inter-layered Late

Miocene volcanic rocks and tephra deposits of basaltic to dacitic composition dipping with an angle of ~5° towards the valley [31, 41, 42]. The landslide dam is partially eroded and drainage of the lake occurs through a 40-m-deep steep-walled canyon. Below that canyon there is a flood surface spanning the entire width of the valley and covering also the down-valley part of the hummocky rock-avalanche deposit (Fig. 5.13). The inundation surface is dissected itself by a steep walled gorge. Lake Navarrete is surrounded by a palaeoshoreline 40 m above the present lake level indicating that the lake level was stable for a prolonged time at

**Fig. 5.13** Air photo with interpretation of Navarrete Lake and partially eroded rock avalanche dam (*dashed line*) after [31, 41, 42]. It is interpreted that the dam breached in a catastrophic event as indicated by the steep walled canyon eroded into the crown of the rock avalanche deposit (*large black arrows*) and the inundation surface spanning the entire width of the valley surrounding highest hummocks of the rock-avalanche deposit downriver the canyon, which in turn is cut by a steep walled gorge (*small arrows*). A palaeoshoreline surrounding the lake at a relative elevation of 40 m (*white triangles*) indicates that the lake level was stable for a long time before dam failure. A possible reason for failure is landsliding into the lake by a massive debris flow from the E-side of the lake (*large black arrow*), generating a displacement wave overtopping the dam crown and causing intensive erosion of the dam. This debris flow deposit appears to be young and does not show the palaeoshoreline; hence it was deposited simultaneously with or subsequently to dam failure

**Fig. 5.14** Photo of large debris flow alluvial fan in front of a small tributary which is not eroded by the palaeo-shoreline indicating that it was deposited simultaneously or subsequently to the partial dam failure of lake Navarrete dam. A related displacement wave is a possible reason for the dam failure (photo by L. Oliva)

a higher elevation. Based upon these observations we interpret that the dam partially breached in a catastrophic event. The exact reason of dam failure is hard to define as the breach apparently occurred in prehistoric time. However, within the lake there is an alluvial fan deposit composed of debris flow deposits (Fig. 5.14). These deposits do not contain the palaeoshoreline noted above, indicating that the debris flow occurred subsequently to or contemporaneously with the dam failure. The catchment area behind the dam with a size of 0.86 $km^2$, is relatively small in comparison to the size of the alluvial fan (0.16 $km^2$). Moreover there are fresh landslide scarps in the upper part of the catchment indicating that a landslide may have occurred in this valley, producing a large debris flow which impacted into the lake causing a displacement wave (tsunami wave) that overtopped the crown of the Lake Navarrete dam, causing its catastrophic partial failure [41, 42].

## *5.6 Lake Dial Rock Avalanche Dam (N Patagonia)*

Lake Dial landslide dam is one of the most difficult dams to study in N Patagonia because of its remote location (36°25′22″ S/70°55′32″ W). It was formed by a rock avalanche that originated at the present-day border between Argentina and Chile and moved into Chile forming the dam. Because of the location on the border, there

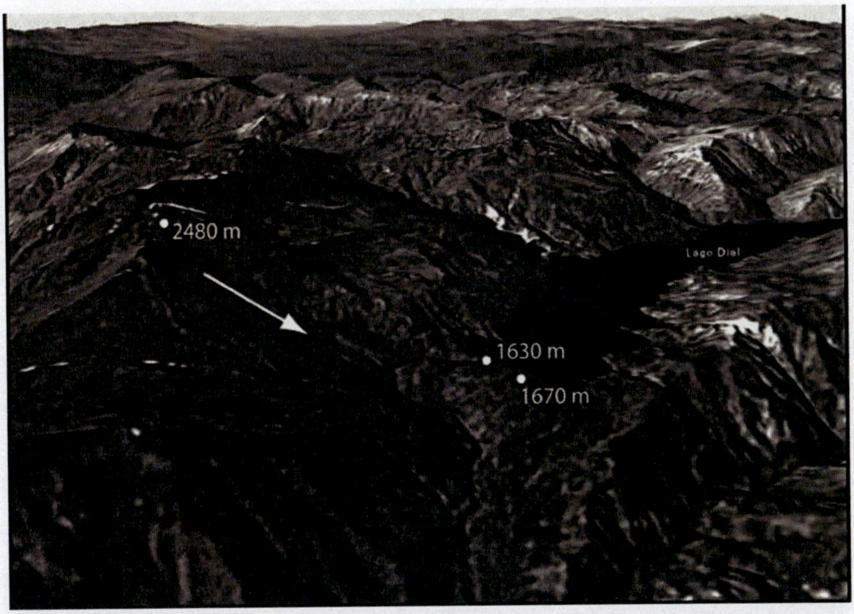


**Fig. 5.15** Oblique view towards the south (*GoogleEarth*) of the rock-avalanche deposit impounding Lake Dial. Note the obvious asymmetric distribution of the landslide deposit within the valley profile with a larger volume in the distal part of the path. Because of this asymmetry dam erosion occurs on the proximal slope where erosion rapidly cuts down to the underlying bedrock

are no air photos covering the entire landslide dam and lake, and access is only possible on a several day trip using a 4×4 vehicle, horse back riding, and a canoe trip on the 8.5-km-long lake. Because of the enhanced weathering of the dam and the erosion of a channel into the landslide dam, it is suspected that this lake formed in prehistoric time. Here, we describe the dam because of the asymmetric distribution of landslide deposits within the valley. Due to the steep valley slopes and the relative drop height of the rock avalanche of >1,000 m (relative difference of landslide crown to lake level of 850 m plus dam height of 164 m), the velocity of the landslide was so high that most of the material was deposited on the slope opposite the landslide scar (Fig. 5.15). Hence dam erosion started at the proximal part of the landslide deposit almost at the contact between bedrock and the deposit. The dam has breached partially as suggested by deposits inundating the entire valley width downriver of the dam, which in turn are eroded by a channel (Fig. 5.15).

## 5.7 *Los Rojos Lake, a Large Pond on Top of a Landslide Deposit (N Patagonia)*

The largest surface area of continuous landslide deposits (48 $km^2$) in the study area exists at Los Cardos volcano [19]. The failure of that volcano is complex, involving deep gravitational deformation, rock avalanching and sliding. It has not been assessed if failure of this volcanic centre occurred in one or in a number of events. The failure occurred in layered volcanic rocks of basaltic composition which had been deformed by young tectonic activity (1.7–1.2 Ma) with evidence of present deformation [11].

The landslide deposit is dotted with several ponds of various sizes; Laguna Los Rojos with a surface area of 0.57 $km^2$ being the largest one (Fig. 5.16). This lake lies on top of a tilted block composed of volcanic rock and overlying moraine deposits. The lake is mainly drained by a creek flowing over the volcanic rocks; there is also some infiltration through the moraine deposits. The crown of the dam is only 20 m wide and terminates in the secondary scarp of the mass movement (Fig. 5.16). A further retreat of the secondary scarp in a future landslide could cause the lake to drain. Due to the steep slope below the lake and a vertical difference of 540 m to the valley floor of the Río Lileo valley, this would cause a high energy flood accompanied by intense scouring on the slope, which is likely to affect the trans-Andean gas pipeline located within the valley below.

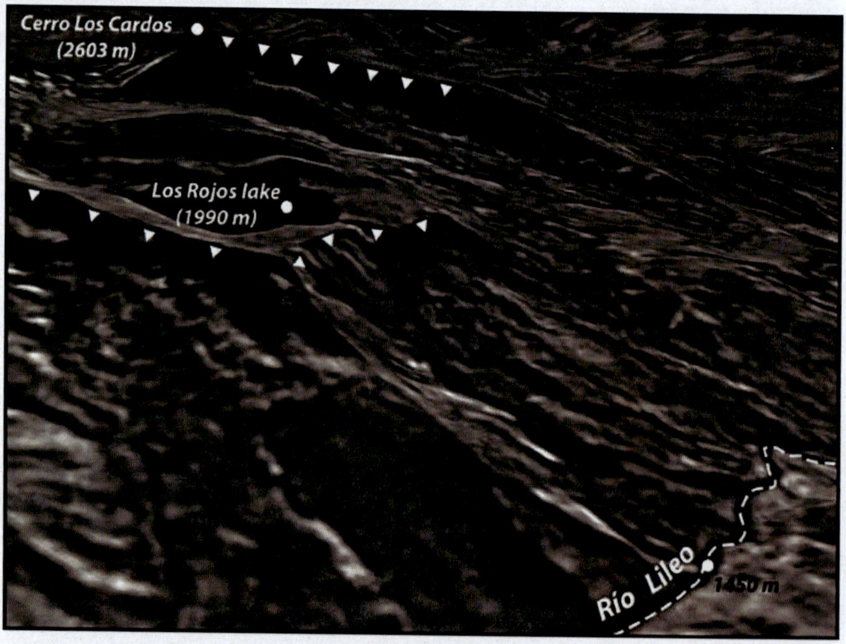


**Fig. 5.16** Oblique view towards ENE (*GoogleEarth*) showing the Los Rojos lake, the summit of Cerro Los Cardos and the valley of Río Lileo. The Cerro Los Cardos is a 3.2–3.4 Ma volcanic edifice that has collapsed in a highly complex mode involving deep gravitational deformation, rotational sliding and rock avalanching. The Los Rojos lake is a pond with a surface area of 0.57 $km^2$ on top of a tilted block composed of volcanic rocks and overlying moraines. In the Río Lileo valley one of the trans-Andean gas pipelines connecting Argentina and Chile passes through. *White triangles* indicate main and secondary scarps

## 5.8 Lake Lauquén Mallín and Toppling into a Landslide Dammed Lake (N Patagonia)

Massive rock slope failures formed Lauquén Mallín lake in the Ñireco valley in prehistoric time in nearly horizontally lying interlayered volcanic rocks and tephra with basaltic to dacitic composition [18]. However, at this site the landslides are not of the rock-avalanche type as for the other landslide dams described from northern Patagonia. The landslide that formed the dam originated in the S slope of the valley and was a rotational slide with a main scarp at the crown, and secondary sliding scarps within the landslide mass (Fig. 5.17). The deposit has a width of 2 km and forms a massive dam within the narrow valley. Landslide velocity was apparently relatively slow in comparison to the rock avalanches previously described, as there is no run-up at the opposite slope and stratification of the slided block is well preserved. On the north shore of the lake there is a 1.2-km-long deposit of a toppling failure. This tilted block is poorly fractured and basaltic layers, which previous to toppling had a horizontal attitude now dip ~35° towards the lake.

The breakaway of the toppling failure is apparently less weathered and eroded than the breakaway of the rotational slide which indicates that the toppling failure is younger, suggesting that the landslide dropped into the pre-existing lake. The velocity of this landslide is unknown but was most likely fast considering the strong tilt of the block. We hypothesize that this landslide caused displacement waves within the

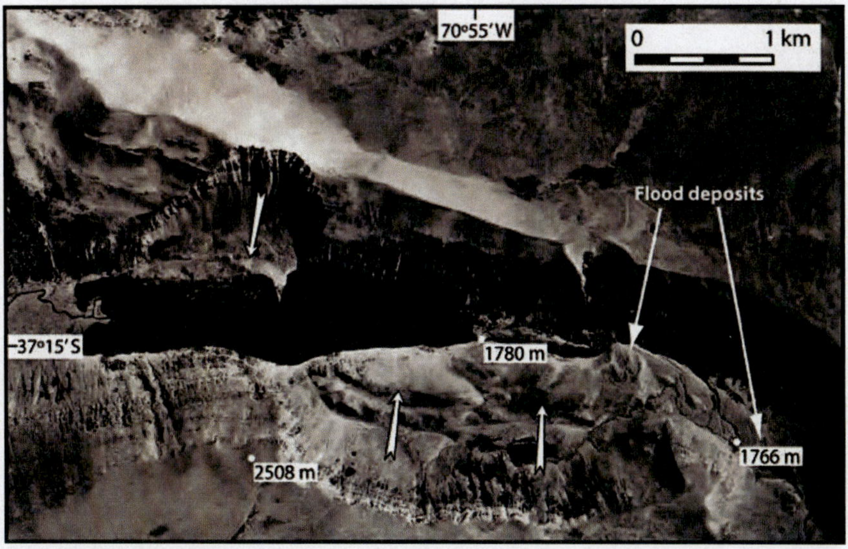


**Fig. 5.17** Air photo of Ñireco lake showing rotational slide deposit (south side) damming the valley and the deposit of a large toppling failure within the lake (north side). Restricted flood deposits downriver the landslide dam suggest that the toppling failure occurred posterior to landslide damming and that a related tsunami wave washed over the dam but did not cause its catastrophic failure

lake. Displacement waves overtopping the dam are also suggested by limited flood deposits (only 1.2 km in extent) downriver from the dam. Even though a channel is eroded into the dam crown, it is not deep and not as steep-walled as the canyons eroded into the breached rock avalanche dams described above.

## 5.9 Lake Agrio in the Agrio Caldera (N Patagonia)

The Agrio caldera is controlled by a transtensional basin formed by normal faults surrounding the SE and N sides of the depression [12] (Fig. 5.18). Along the steep sided walls of the caldera and in the neighbouring valleys rock avalanche deposits are common (Fig. 5.18); several of them dam lakes or dammed lakes in the geological past [17]. Here we focus on the rock-avalanche deposit emplaced in the SW corner of the caldera (Fig. 5.18). This rock avalanche has formed in volcanic rocks of basaltic composition with horizontal attitude. The rock-avalanche deposit lies within

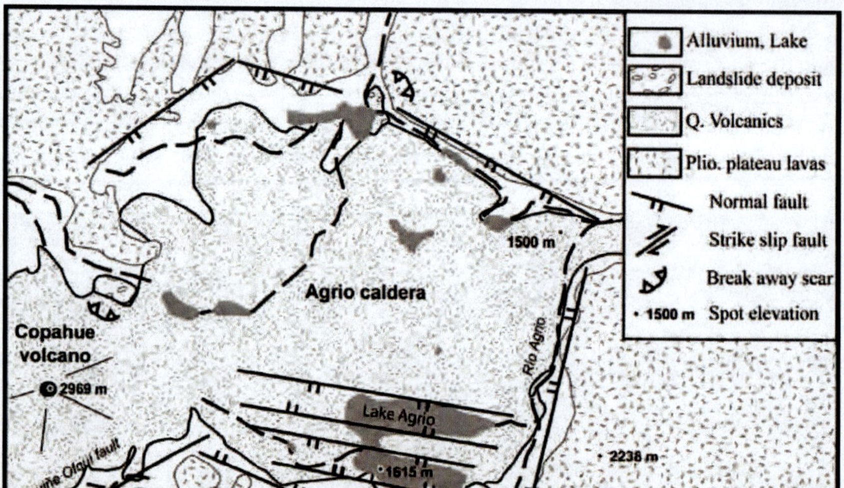


Fig. 5.18 Generalized geologic map of the Agrio Caldera showing the distribution of faults and landslide deposits after [12, 17]. Note the landslide deposit in the W entrance of the Cajon Chico valley, which is obstructed by a large landslide deposit damming Lake Agrio and causing the diversion of Río Agrio into a northern valley

the entrance to the Cajon Chico valley, which is eroded 600 m deep into the basaltic plateau; however it is nearly dry. Only a small creek runs in the centre of the valley. This is related to the landslide deposit forming Agrio Lake that causes the diversion of the Agrio river into the adjacent drainage towards the north. The maximum depth of Agrio Lake was measured to be 49 m (Scharf, personal communication, 2003).

# 6 Discussion and Conclusion

Our study shows that although other investigations based on data sets of landslide dams from all over the world do not contain many examples from the Andes [3, 5], landslide damming is a frequent phenomenon in this mountain system. In our two study areas, even though representing only a small portion of the Andes, we found 1 landslide-dammed lake and remnants of 19 landslide dams in NW Argentina and 22 landslide-dammed lakes and an additional 19 remnants of landslide dams in northern Patagonia. Similar numbers of landslide dams have also been reported from other places of the Andes [8, 10]. Landslide dams in the Andes show some interesting particularities not previously described from other regions.

## *6.1 New Types of Landslide Dams*

As in other parts of the world, type II and III dams of the classification by Costa and Schuster [2] are the most frequent types of landslide dams; other types are relatively rare. More importantly, our data show that not all Argentine landslide dams fit into this well-established classification (types I–VI, Fig. 5.19) and new types of landslide dams must be defined (types VIIa–IX, Fig. 5.19) to adequately describe them, understand their stability, and/or understand their impact on valley evolution.

For example, the rock avalanche dam impounding Lake Agrio did not occur in the lowest part of the valley profile, causing a shift of the drainage away from the original streambed and into areas where a new drainage system develops a drainage diversion into the neighbouring valley (type VIIa, Fig. 5.19). Such a drainage diversion due to landslide damming has previously been found in other regions of the Argentine Andes [7]. In a similar way, a new drainage establishes over bedrock in the case that the dam is higher than the bedrock slopes as in the El Bolson river in NW Argentina, that resulted in the development of a new drainage cutting into bedrock (type VIIb, Fig. 5.19). In both cases the landslide dam is very stable; a failure of the landslide dam is unlikely because no dam overtopping can take place. Overtopping is the most frequent process of catastrophic dam erosion [3, 5]. At the case study site in the El Bolson valley, this type of dam has persisted for several tens of thousands of years as indicated by the history of lake basin fill [8]. Likewise, the dam of Agrio Lake does not show any signs of erosion.

A further landslide dam type (type VIII, Fig. 5.19) which needs to be defined is a dam forming exactly at the drainage divide. In this case catastrophic dam failure

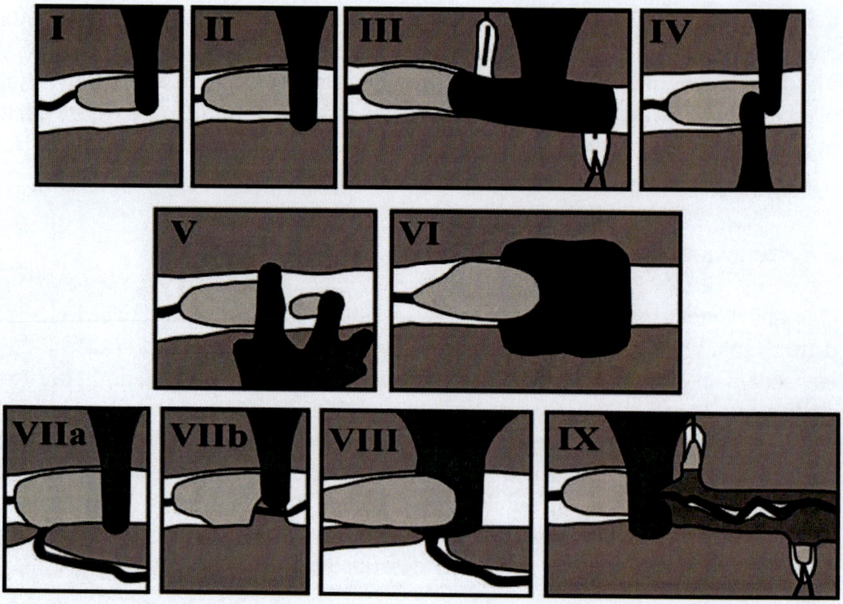


**Fig. 5.19** Landslide dam types I–VI after Costa and Schuster [2] and new proposed dam types VIIa–IX

might occur into both valleys, in the one where superficial drainage occurs and by seepage in the other one. In the only known example from the Argentine Andes (the Varvarco Campos lake) the dam composed of $3.5 \times 10^9$ $m^3$ rock avalanche debris, existed for 30 ka with a powerful river discharge over the dam [3, 15]. However, in this valley, dam stability is further controlled by the fact that river discharge over the dam is directly into another large landslide-dammed lake, preventing toe erosion of the landslide dam. These landslides affecting drainage systems (types VIIa and VIII) are in contrast to observations from Papua New Guinea [33]. In our example from the Varvarco Campos lake in northern Patagonia massive landsliding caused redistribution of the drainage system and river capture.

A further type of landslide dam (type IX, Fig. 5.19) or chains of landslide dams must be defined based upon the case history of the catastrophic failure of a rock-avalanche dam in Barrancas valley in 1914. During this event $1.2 \times 10^8$ $m^3$ of debris were eroded from the landslide dam forming a giant debris flow, which deposited terraces decreasing in thickness with increasing distance from the dam toe from 20 to 3.5 m at 60 km downriver of the dam. These terraces impounded tributary valleys of the Barrancas valley. Most of the impoundments were eroded shortly after formation; however the tributary valley closest to the failed dam is still impounded up to a depth of $\geq 20$ m forming a lake with a surface area >1 $km^2$. Similar chains of lakes have been reported on the opposite slope of the Andes in Chile, related to the giant outburst flood generated by the failure of the dam formed by the Antuco volcano collapse (landslide dam no. 34 in Fig. 5.3, Table 5.2) [38].

In addition to these newly defined cases, small lakes frequently occur as ponds on the hummocky surface of landslide deposits in the Andes. Their size is mainly related to the size of the landslide deposit. They can be present in all types of landslide dams and are in general very stable; however they might represent a local hazard when eroding rivers get close to their limits, e.g. Los Rojos lake in northern Patagonia. Considering a medium lake depth of only 2 m, a dam collapse would release a volume of ~$10^6$ $m^3$ at this lake. Much smaller jökulhlaups in the Canadian Rockies have caused considerable damage [35].

The Lake Dial landslide dam clearly indicates that the horizontal distribution of landslide deposits within the valley is not the only factor that affects the stability of a landslide dam. At this place the dam thickness is asymmetric, with a large amount of debris on the opposite slope and only a thin deposit at the proximal slope. In this situation landslide dam erosion starts close to the valley wall. Scouring of the landslide mass will cut rapidly down to the underlying bedrock, where erosion will slow down significantly making these dams more stable. Similar landslide dam behaviour has been observed in other mountains [4].

In total, these cases show that (a) the classification system [2] used in the past 20 years has to be extended in plan with additional plan view cases (Fig. 5.19), and (b) that the three dimensional relation of rockslide mass and valley morphology has to be considered to adequately assess the long term stability of rockslide dams (The classification of rockslide dams 24 by Hermanns et al., this volume).

## 6.2 *Landslide Dam Stability in the Argentine Andes*

In the NW Argentine focus region most of the dams have been eroded in the geological past. There is only one dam still impounding a lake [31], while four lake basins have been filled up with sediments. In the N Patagonia focus region ~50% of the landslide dams, all of them formed in prehistoric times, still impound the main valley. In addition there are several localities where water bodies occur on top of landslide deposits. Therefore, a large group of our landslide dams have longevity of several hundreds, thousands or even tens of 1,000 years. In most cases of eroded dams it could not be established if they breached catastrophically or were slowly eroded (Fig. 5.20).

When compared to landslide dam stability concepts as suggested by Ermini and Casagli [5], which based on geomorphic parameters are expressed in a Dimensionless Blockage Index (DBI), the stability of most of the Argentine landslide dams behaved as indicated by stability/instability boundaries obtained from worldwide data sets (Fig. 5.20)

$$\mathrm{DBI} = \log\,(V \times \mathrm{A} / \mathrm{H})$$

with A being the area of the drainage upriver the dam, H the dam height and V the dam volume (Fig. 5.20). However there are many exceptions, where landslide dams existed for several hundreds or thousands of years but finally failed (named

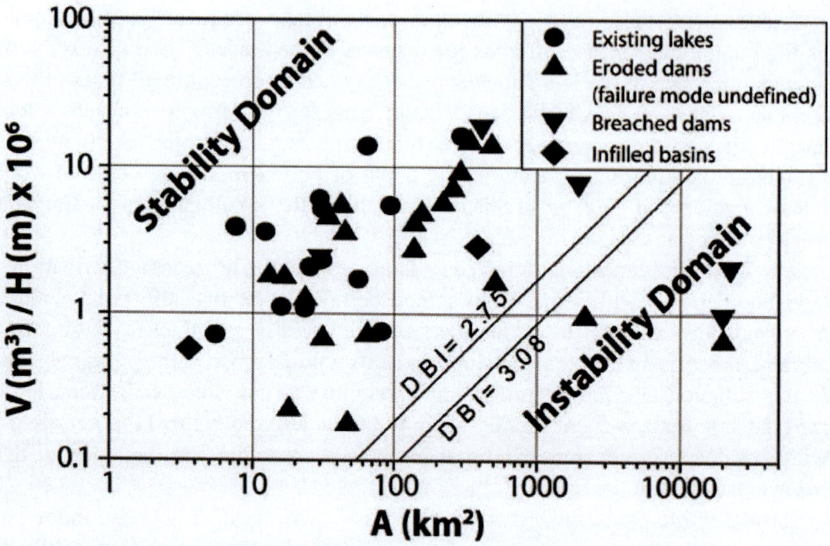


**Fig. 5.20** Geomorphic parameters of Argentine landslide dams in comparison with stability/instability boundaries of the Dimensional Blockage Index (DBI) of worldwide data from Ermini and Casagli [5]. Although most of our data plot in the correct domains, there are a few exceptions of breached dams that plot in the stability domain

"quasi-stable dams" [29]) or dams failed although their geomorphic parameters plot within the stability domain in Fig. 5.20. Hence a simple division between stable and unstable dams based upon whether a dam has breached in the past is questionable. For example, Casa de los Loros dam in the Las Conchas valley in NW Argentina with a DBI of 4.3 would have been exceptionally unstable according to these concepts; however it existed for 7 ka. Our examples rather suggest that landslide dam longevity is mainly controlled by the climatic situation of the region and the probability of landsliding into the existing landslide-dammed lake.

A good example of the influence of climatic variability is the catastrophic dam failure in Barrancas valley in 1914 where, as noted above, exceptional climatic conditions occurred the year before dam failure; the dam had existed for more than 427 years [27] and had a DBI of 2.39 (stability domain in Fig. 5.20). In a contrasting way, the Casa de los Loros dam in Las Conchas valley existed for ~7 ka although it clearly plots within the instability domain. This dam lies within a region of arid climate with yearly precipitation of $\leq$ 200 mm. With the present day climate situation, the lake would not have filled up to the landslide dam crest but would have remained 70 m below that level [1]. The El Paso dam located in the same valley, only 2 km upriver, is only 30 m lower; however it existed for only a few 100 years after the breach of the Casa de los Loros dam [29]. The type of erosion is unknown, but the landslide dam failed when climate changed towards more humid conditions in this part of the Andes. Other impoundment indices have been proposed [36], however they require additional geomorphometric data which are more difficult to obtain.

All breached dams, except for the catastrophic failure in the Barrancas valley in 1914, are related to landsliding into the landslide-dammed lake. The Casa de los Loros dam in Las Conchas valley failed by the impact of a rock avalanche into the lake only 1 km upriver of the dam, and the related outburst flood filled the basin behind the Alemanía landslide dam only 30 km downriver, causing, in turn, its catastrophic failure [29]. In northern Patagonia, the Varvarco Tapia lake dam and the Lake Navarete dam breached catastrophically. In both cases there are voluminous landslide deposits within the lake that are younger than the landslide dam. It is very likely that displacement waves generated by the impact of the landslide into the lake overtopped the dam crest causing catastrophic failure. These interpretations can only be tested at the Navarete dam by dating the landslide deposit and at the Varvarco Tapia dam by sedimentologically characterising the voluminous terrace deposits downriver the dam. Dam failure due to landslide impact into a landslide-dammed lake causing displacement waves has previously been described in western Nevada [13].

The landslide dam forming the Ñireco dam apparently experienced overtopping by tsunami waves as well; however this dam is not deeply eroded and does not seem to have breached. More likely the dam was overtopped by a displacement wave that caused restricted flooding downriver but did not cause dam failure. This is most likely related to the type of landslide that formed the dam. At this site the dam was formed by a rotational slide, which indicates only minor internal deformation of the basaltic masses constituting the landslide. In comparison with rock avalanche deposits forming the other dams that failed by overtopping by a tsunami wave, these landslide deposits have significantly less fragmentation, making the dam more stable.

**Acknowledgement** This project was sponsored by the GeoForschungsZentrum Potsdam and by the Deutsche Forschungsgemeinschaft as part of the Collaborative Research Center 267 and Graduate College 450 grant to Hermanns. We especially emphasize that Mario A. Deza inspired R.L.H. to start landslide studies in N-Patagonia. Mario died during our first campaign in November 2002 crossing Lileo river. This paper is dedicated to Mario who conducted landslide hazard research in Neuquén Province as a private effort in his spare time. Although we did not know Mario well, we were impressed by his professional dedication, by his open and always friendly way in interacting with people, and the way he loved the Neuquén Andes. We also thank A. Villanueva Garcia and R. Alonso for fruitful discussions, field assistance and logistcal support in NW Argentina and M.R. Strecker and M. Trauth for discussions in Potsdam. A. Pagenkopf helped to assemble data on landslide dams in N Patagonia. R.L.H. acknowledges B. Merz, G. Borm, J. Erzinger, K. Czurda, and F. Gehbauer for their encouragement and help, and C. Hickson for her support.

## References

1. Bookhagen, B., Haselton, K. and Trauth, M.H. (2001) Hydrological modelling of a Pleistocene landslide-dammed lake in the Santa Maria Basin, NW Argentina, *Palaeogeography Palaeoclimatology Palaeoecology* **169**, 113–127.
2. Costa, J.E. and Schuster, R.L. (1988) The formation and failure of natural dams, *Geological Society of America Bulletin* **100**, 1054–1068.

3. Costa, C.H. and González Díaz, E.F. (2007) Age contraints and paleoseismic implication of rock avalanches in northern Patagonian Andes, Argentina, *Journal of South American Earth Sciences* **24**, 48–57.
4. Dunning, S., Petley, D., Rosser, N. and Strom, A. (2005) The morphology and sedimentology of valley confined rock-avalanche deposits and their effect on potential dam hazard, in O. Hungr, R. Fell, R. Couture and E. Eberhardt (Eds.), *Landslide Risk Management*. Taylor and Francis Group, London, pp. 691–701.
5. Ermini, L. and Casagli, N. (2003) Prediction of the behaviour of landslide dams using a geomorphological dimensionless index, *Earth Surface Processes and Landforms* **28**, 31–47.
6. Fauqué, L. and Strecker, M.R. (1988) Large rock avalanche deposits (Sturzstrome, sturzstroms) at Sierra Aconquija, northern Sierras Pampeanas, Argentina, *Eclogae Geologicae Helvetiae* **81**, 579–592.
7. Fauqué, L., Cortés, J., Folguera, A. and Etcheverría, M. (2000) Avalanchas de roca asociadas a neotectónica en el valle del río Mendoza, al sur de Uspallata, *Revista de la Asociación Geológica Argentina* **55**, 419–423.
8. Fauqué, L. and Tchilinguirian, P. (2002) Villavil rockslides, Catamarca Province, Argentina, in S.G. Evans and J.V. DeGraff (eds.), *Catastrophic landslides: Effects, Occurrence, and Mechanism*, Reviews in Engineering Geology XV, Geological Society of America, Boulder, CO, pp. 303–324.
9. Fauqué, L.E., Baumann, V., Di Tomasso, I., Rosas, M., Hermanns, R.L., González, M., Coppolecchia, M. and Wilson, C.G.J. (2005) Evidencia de paleoendicamientos en la cuenca del río Mendoza, Argentina, *XVI Congreso Geológico Argentino, La Plata* **3**, 507–514.
10. Ferrer, C. (1999) Represamientos y rupturas de embalses naturales (lagunas de obstrución) como efectos cosísmicos: Algunos ejemplos en los Andes venezolanos, *Revista Geográfica Venezolana* **40**, 109–121.
11. Folguera, A., Ramos, V.A., González Díaz, E. and Hermanns, R. (2006) Miocene to quaternary evolution of the Guañacos fold-and-thrust belt in the Neuquen Andes between 37 and 37°30′S, in S. Kay and V.A. Ramos (eds.), *Evolution of the Andean Margin: A Tectonic and Magmatic View from the Andes to the Neuquen Basin (35–39° S lat)*, Vol. 407. GSC, Boulder, CO, Special Publication, pp. 247–266.
12. Folguera, A., Ramos, V.A., Hermanns, R.L. and Naranjo, J.A. (2004) Neotectonics in the foothills of the southernmost central Andes (37–38°S): Evidence of strike-slip displacement alng the Atiñir-Copahue fault zone, *Tectonics* **23**, doi:10.1029/2003TC001533.
13. Glancy, P.A. and Bell, J.W. (2000) Landslide-induced flooding at Ophir Creek, Washoe County, western Nevada, May 30, 1983, *U.S. Geological Survey Professional Paper* **1617**, 1–94.
14. González Díaz, E.F. and Fauqué, L. (1987) Provenencia del material componental del torrente de barro de "El Volcan", Quebrada de Huamahuaca (Jujuy), República Argentina, *X Congreso Geologico Argentino* **1**, 309–312.
15. González Díaz, E.F., Fauqué, L., Giaccardi, A. and Costa, C. (2000) Las lagunas de Varvar Co Campos y Varvar Co Tapia (N del Neuquén, Argentina): Su relación con avalanchas de rocas, *Revista de la Asociacion Geológica Argentina* **55**, 147–164.
16. González Díaz, E.F., Giaccardi, A. and Costa, C. (2001) La avalancha de rocas del río Barrancas (Cerro Pelán), norte del Neuquén: Su relación con la catástrofe del río Colorado (29/12/1914), *Revista de la Asociacion Geológica Argentina* **56**, 466–480.
17. González Díaz, E.F. (2003) El englazamiento en la región de la caldera de Caviahue-Copahue (Provincia del Neuquén): Su reinterpretación, *Revista de la Asociación Geológica Argentina* **58**, 356–366.
18. González Díaz, E.F. and Folguera, A. (2005) Reconocimiento y descripción de avalanchas de rocas prehistóricas en el área neuquina delimitada por los paralelos 37°15′ y 37°30′S y los meridianos 70°55′ y 71°05′S, *Revista de la Asociación Geológica Argentina* **60**, 446–460.
19. González Díaz, E.F., Folguera, A. and Hermanns, R.L. (2005) La avalancha de rocas del cerro Los Cardos (37°10′S, 70°53′O) en la región norte de la provincia del Neuquen, *Revista de la Asociacion Geologica Argentina* **60**, 207–220.

20. González Díaz, E.F., Folguera, A., Costa, C.H., Wright, E. and Ellisondo, M. (2006) Los grandes deslizamientos de la región septentrional neuquina entre 36–38°S: Una propuesta de inducción sísmica, *Revista de la Asociacion Geologica Argentina* **61**, 197–217.
21. Groeber, P. (1916) Informe sobre las causas que han producido las crecientes del río Colorado (Territorios del Neuquén y La Pampa) en 1914, Dirección General de Minas, Geología e Hidrogeología, 11, Serie B, Geología, 29 p.
22. Harrington, H.J. (1946) Las corrientes de barro (mud flows) de "El Volcan", Quebrada de Huamhuaca, Jujuy, *Revista de la Asociación Geológica Argentina* **1–2**, 149–165.
23. Hermanns, R.L. (1999) Spatial-temporal distribution of mountain-front collapse and formation of giant landslides in the arid Andes of northwestern Argeintina (24–28° S, 65–68° W) [Ph.D. thesis]: Universität Potsdam, 123 p.
24. Hermanns, R.L. and Strecker, M.R. (1999) Structural and lithological controls on large Quaternary rock avalanches (sturzstroms) in arid northwestern Argentina, *Geological Society of America Bulletin* **111**, 934–948.
25. Hermanns, R.L., Trauth, M.H., Niedermann, S., McWilliams, M. and Strecker, M.R. (2000) Tephrochronologic constraints on temporal distribution of large landslides in northwest Argentina, *Journal of Geology* **108**, 35–52.
26. Hermanns, R.L., González Díaz, E.F., Folguera, A. and Mardones, M. (2003) Large massive rock slope failures, landslide dams, related valley evolution, and their association with the tectonic setting in the Argentine and Chilean Andes between 36 and 38° S, *10th Congreso Geologico Chileno, Concepción*, CD, 5 p.
27. Hermanns, R.L., Naumann, R., Folguera, A. and Pagenkopf, A. (2004) Sedimentologic analyses of deposits of a historic landslide dam failure in Barancas valley causing the 1914 Rio Colorado flood, northern Patagonia, Argentina, in W.A. Lacerda, M. Ehrlich, S.A.B. Fontoura and A.S.F. Sayão (eds.), *Landslides, Evaluation and Stabilization*. Balkema, Leiden, pp. 1439–1445.
28. Hermanns, R.L., Niedermann, S., González Díaz, E.F., Fauque, L., Folguera, A., Ivy, O.S. and Kubik, P.W. (2004) Landslide dams in the Argentine Andes, *NATO Advanced Research Workshop: Security of natural and artificial rockslide dams, Bishkek, Kyrgyzstan*, Abstract Volume, pp. 79–85.
29. Hermanns, R.L., Niedermann, S., Ivy-Ochs, S. and Kubik, P.W. (2004) Rock avalanching into a landslide-dammed lake causing multiple dam failure in Las Conchas valley (NW Argentina) – evidence from surface exposure dating and stratigraphic analyses, *Landslides* **1**, 113–122.
30. Hermanns, R.L. and Villanueva Garcia, A. (2005) Did large earthquakes cause massive landsliding and near-surface deformation in the Calchaquíes valleys, NW Argentina? *XVI Congreso Geológico Argentino, La Plata*, CD, 8 p.
31. Hermanns, R.L., Niedermann, S., Villanueva Garcia, A. and Schellenberger, A. (2006) Rock avalanching in the NW Argentine Andes as a result of complex interactions of lithologic, structural and topographic boundary conditions, climate change and active tectonics, in S.G. Evans, G. Scarascia Mugnozza, A.L. Strom and R.L. Hermanns (eds.), *Landslides from massive rock slope failure*, NATO Science Series IV, v. **49**, Springer, Dordrecht, 539–569.
32. Hermanns, R.L., Blikra, L.H. and Longva, O. (2009) Relation between rockslide dam and valley morphology and its impact on rockslide dam longevity and control on potential breach development based on examples from Norway and the Andes, 2nd International Congress: Long term behavior of dams, Graz, pp. 789–794.
33. Hovius, N. (1998) Landslide-driven drainage network evolution in a pre-steady-state mountain belt: Finisterre Mountains, Papua New Guinea, *Geology* **26**, 1071–1074.
34. Igarzabal, A.P. (1978) "Los flujos densos de la Quebrada de Escoipe", *VII Congreso Geologico Argentino, Neuquen* **Actas**, 109–117.
35. Jackson, L., Hungr, O., Gardner, J. and Mackay, C. (1989) Cathedral Mountain debris flows, Canada, *Bulletin of the International Association of Engineering Geology* **40**, 35–54.
36. Korup, O. (2004) Geomorphometric characteristics of New Zealand landslide dams, *Engineering Geology* **73**, 13–35.

37. Malamud, B.D., Jordan, T.E., Alonso, R.A., Gallardo, E.F., Gonzalez, R.E. and Kelley, S.A. (1996) Pleistocene lake Lerma, Salta Province, NW Argentina, *XIII Congreso Geologico Argentino y III Congreso de Exploracion de Hidrocarburos, Buenos Aires* **4**, 103–114.
38. Mardones Flores, M. (2002) Evolución morfogenética de la hoya del río Laja y su incidencia en la geomorfología de la región del Biobío, Chile, *Revista Geográfica de Chile Terra Australis* **47**, 97–127.
39. Narciso, V., Santamaría, G. and Zanettini, J.C. (1999) Hoja Geológica preliminar 3769-I, Barrancas, Servicio Geológico Minero Argentino, scale 1:250.000.
40. Niemeyer, H. and Muñoz, J. (1988) The Quaternary volcanic belt of the southern continental margin of South America: Transverse structural and petrochemical variations across the segment between 38 and 39°S, *Journal of South American Earth Sciences* **1**, 147–161.
41. Penna, I., Hermanns, R.L. and Folguera, A. (2007) Determinación del área inmediata afectada por el desagote catastrófico de la laguna Navarete en relación a la actividad tectónica del norte Neoquíno, *Revista de la Asociación Geológica Argentina* **62**, 460–466.
42. Penna, I., Hermanns, R.L. and Folguera, A. (2008) Remoción en masa y collapso catastrófico de diques naturales generados en el frente orogénico andino (36–38°S9): Los casos Navarrete y río Barrancas, *Revista de la Asociación Geológica Argentina* **63**, 172–180.
43. Rapela, C.W. (1976) El basamento metamórfico de la region de Cafayate, Provincia de Salta, *Aspectos petrológicos y geoquímicos, Revista de la Asociación Geológica Argentina* **21**, 203–222.
44. Salfity, J.A. and Marquillas, R.A. (1994) Tectonic and sedimentary evolution of the Cretaceous-Eocene Salta Group Basin, Argentina, in J.A. Salfity (ed.), *Cretaceous Tectonics of the Andes*. Friedrich Vieweg und Sohn, Berlin, pp. 266–315.
45. Strecker, M.R., Cerveny, P., Bloom, A.L. and Malizzia, D. (1989) Late cenozoic tectonism and landscape development in the foreland of the Andes: Northern Sierras Pampeanas (26–28°S), Argentina, *Tectonics* **8**, 517–534.
46. Trauth, M.H. and Strecker, M.R. (1999) Formation of landslide-dammed lakes during a wet period between 40,000 and 25,000 yr BP in northwestern Argentina, *Palaeogeography Palaeoclimatology Palaeoecology* **153**, 277–287.
47. Trauth, M.H., Alonso, R.A., Haselton, K.R., Hermanns, R.L. and Strecker, M.R. (2000) Climate change and mass movements in the NW Argentine Andes, *Earth and Planetary Science Letters* **179**, 243–256.
48. Wayne, W.J. (1999) Alemanía rockfall dam: A record of a mid-Holocene earthquake and catastrophic flood in northwestern Argentina, *Geomorphology* **27**, 295–306.
49. WP/WLI (1990) A suggested method for reporting a landslide, *Bulletin of the International Association of Engineering Geology* **3**, 101–110.
50. Zanettini, J.C. (2000) Carta Geologica preliminar: Las Ovejas, SEGEMAR, scale 1:250.000.

# Chapter 6
# Rock Avalanche Dams on the Trans Himalayan Upper Indus Streams: A Survey of Late Quaternary Events and Hazard-Related Characteristics

**Kenneth Hewitt**

## 1 Introduction

Rivers of the Trans Himalayan Indus basin are subject to damming by glaciers and a variety of mass movements. There are many catastrophic outburst floods on record. Most involved ice dams, but the two largest were due to early failure of landslide dams, those of 1841 and 1858 [23, 26, 58]. The more frequent impoundments are by debris flows and snow avalanches, and by ice and moraine along glacier margins. Several damaging cases occur in most years [12, 37, 38, 41, 43]. However, these tend to be relatively small, the inundations or outburst floods of fairly local significance. The larger natural dams in the late Quaternary have been due to rock slide-rock avalanches. At this time, 161 have been identified, probably a modest fraction of all those that occurred in the Holocene. Only a few small lakes exist today, but many former lakes exceeded 20 km in length and some more than 100 km long. A few impounded more than one third of the basin (Fig. 6.1). Two or more may have coincided at certain times to impound almost all of its roughly 80,000 km$^2$. Immediately south of the region examined here, several catastrophic landslides from the west flank of the Nanga Parbat Range impounded all of the Upper Indus in the Karakoram behind barriers several hundred metres high. Those at Shatial, Habdas and Raikot Bridge lasted several centuries, at least [33, 34]. Sediment yields here are among the highest in the world, suggesting the impacts on fluvial processes and landforms have been profound [29]. The chapter will focus on the nature of the rock avalanche dams and hazards they involve.

It is noted that in January 2010, the Hunza River was blocked by the Attabad rockslide that resulted in the subsequent formation of rockslide-dammed Lake Gojal upstream of the dam [35, 42; Chap. 7 by Delaney and Evans, this volume].

K. Hewitt (✉)
Department of Geography and Environmental Studies, Cold Regions Research Centre,
Wilfrid Laurier University, Waterloo, ON, Canada N2L 3C5
e-mail: khewitt@wlu.ca

S.G. Evans et al. (eds.), *Natural and Artificial Rockslide Dams*, Lecture Notes in Earth Sciences 133, DOI 10.1007/978-3-642-04764-0_6, © Springer-Verlag Berlin Heidelberg 2011

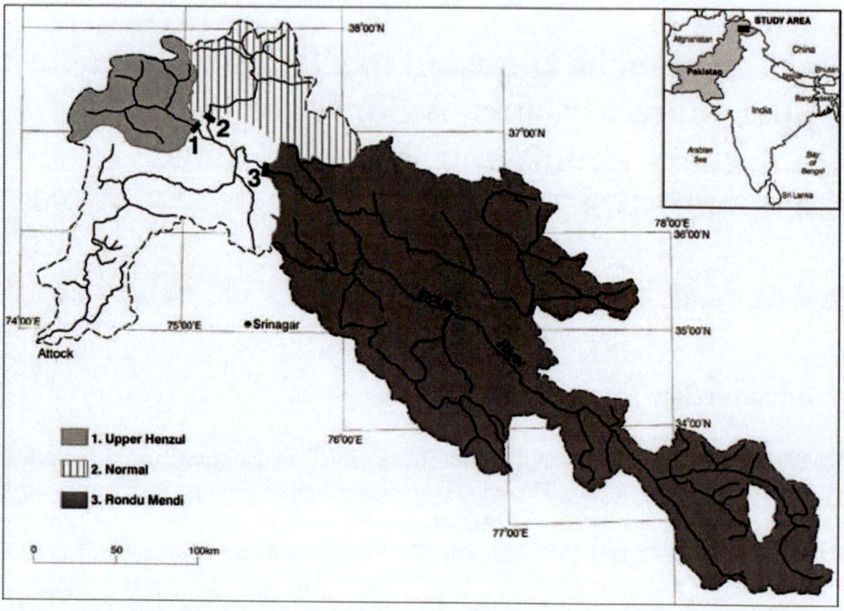


**Fig. 6.1** The Trans Himalayan Upper Indus Basin, showing its location and basin areas impounded by selected rock avalanche dams (after [29]). **1**. Upper Henzul is one of several that impounded the Gilgit Basin, **2**. The giant Nomal Complex impounded most of the tributary Hunza Basin, and **3**. Rondu-Mendi dammed the (Middle) Indus (Dam dimensions are in Table 6.2)

## *1.1 Incidence and Dimensions of Rock Avalanche Dams*

In surveys between 1993 and 2009, in the Karakoram and Hindu Raj Mountains, a total of 322 [31, 33] rock avalanches were identified [31]; this total has been increased to more than 370 in recent work. Most are prehistoric events reconstructed from their deposits. They occur throughout the basin, in all elevation zones, and a wide variety of lithologies and geological terranes. As noted, at least 161 formed cross-valley barriers impounding rivers have been investigated in the field (Table 6.1).

Almost 90% of the rock avalanches identified exceeded $10 \times 10^6$ m$^3$ in volume. Those forming the larger dams exceed $10^7$ m$^3$ and at least two over $10^9$ m$^3$ (Table 6.2). Vertical displacements, from the head of the detachment zone to the farthest run out of debris, were generally over 1,000 m, in some more than 2,000 m. Maximum horizontal displacements exceeded 5 km, in some cases more than 10 km [25, 26, 27, 28]. The table identifies the cross-valley and down-valley lengths of rock avalanche deposits forming barriers. It needs to be emphasized that deposit dimensions and morphologies are strongly modified by run out in rugged terrain or over deformable substrates.

Most dams had an initial effective height – the lowest part of the cross-valley barrier – greater than 10 m. More than half exceeded 25 m, about 20% exceeded

**Table 6.1** Cross-valley rock avalanche barriers in the Upper Indus Basin

| River | Cross-valley RA deposit | Existing lakes | Existing sediment barrier | Lake deposits | Epigenetic gorge |
|---|---|---|---|---|---|
| *Chitral basin* | | | | | |
| Upper Yarkhun[a] | 7 | | 7 | 7 | 7 |
| Laspur | 3 | | 3 | 3 | 1 |
| *Gilgit basin* | | | | | |
| Ghizar | 4 | | 3 | 4 | 2 |
| Gilgit | 17 | | 13 | 14 | 3 |
| Yasin | 5 | | 5 | 4 | 1 |
| Ishkoman-Karambar | 10 | 1 | 7 | 7 | 2 |
| Hirali | 1 | | 1 | 1 | – |
| Bagrot | 6 | | 5 | 6 | 1 |
| Batkor | 1[b] | | 1 | 1 | – |
| Kargah | 3 | | 2 | 3 | 1 |
| Upper Hunza[c] | 12 | | 7 | 9 | 3 |
| Chapursan | 6 | | 6 | 6 | 3 |
| Hispar | 1 | | 1 | 1 | 1 |
| Lower Hunza | 6 | | 4 | 6 | 4 |
| Chalt | 2 | | 1 | 2 | 2 |
| Naltar | 5[d] | 1 | 4 | 4 | 2 |
| *Indus basin* | | | | | |
| Middle Indus[e] | 21 | | 8 | 11 | 8 |
| Right bank tributaries: | | | | | |
| Stak | 2 | | 1 | | |
| Tormik | 2 | | 2 | 2 | 1 |
| Shigar | 6 | | 1 | 6 | – |
| Baumaharel | 2 | | 2 | 2 | 1 |
| Basha | 2 | | 1 | 2 | – |
| Braldu | 7 | | 5 | 7 | 2 |
| Dumordo | 2 | | 2 | 2 | – |
| Shyok | 4 | | 4 | 4 | 1 |
| Thalle | 7 | | 6 | 5 | – |
| Hushe | 9 | | 7 | 8 | 2 |
| Saltoro (lower) | 2 | | 2 | 2 | 1 |
| Kondus | 1 | | 1 | 1 | 1 |
| Left bank tributaries: | | | | | |
| Shigarthang | 3 | | 3 | 3 | 1 |
| Satpara | 3 | 1 | 2 | 2 | – |
| Totals | 161 | 3 | 117 | 115 | 45 |

[a] Parwak to Karambar Pass; [b] Also dammed Gilgit R.; [c] Baltit to Sost; [d] Includes Nomal barrier, which dammed mouth at Hunza R.; [e] From the junction with Gilgit R. to junction with Shyok R.

**Table 6.2** Estimated dimensions of some major rock avalanches that have dammed the Upper Indus streams

| Rock Av. | River[a] | Vol. of deposits ($\times 10^6$ m$^3$) | Area of deposits (km$^2$) | Height: original (m) | Height: breached (m) | Length: cross valley (m) | Width: down valley (m) | Lake length: Original (km) |
|---|---|---|---|---|---|---|---|---|
| *Shyok Basin (Baltistan):* | | | | | | | | |
| Litak | Hushe | 80 | 4+ | 120 | 120 | 2,000 | 2,000+ | 7 |
| Haldi | Saltoro | 600 | 25 | 260 | 240 | 2,000 | 5,000 | 20+ |
| Yugo-Kurphak | Shyok | 100+ | 3+ | 60+ | 10 | 1,500 | 2,000+ | 32 |
| Kunis-Gwa | Shyok | 1,000+ | 25+ | 200+ | 150 | 2,000 | 11,000 | 50+ |
| *Middle Indus Basin (Baltistan):* | | | | | | | | |
| Gol-Ghone 'A' | Indus | 600+ | 18+ | 600+ | 550 | 2,500 | 8,000+ | 80+ |
| Gol-Ghone 'B' | Indus | 300+ | 15 | 550 | 500 | 2,500 | 6,000+ | 70+ |
| Tsok | Dumordo | 150 | 8 | 110 | 95 | 1,800 | 4,500 | 6 |
| Satpara-Skardu | Satpara | 400 | 22 | 120 | 60 | 2,000 | 5,000+ | 7 |
| Katzarah | Indus | 200+ | 20 | 150+ | 55 | 4,500 | 9,000 | 55 |
| Rondu-Mendi | Indus | 1,500+ | 49+ | 1,100 | 950 | 4,400 | 14,000 | 100+ |
| *Gilgit-Hunza Basin:* | | | | | | | | |
| Dulung Bar | Yasin | 450 | 20 | 150 | 120 | 1,500 | 2,500 | 6 |
| Gupis | Ghizar | 160 | 5+ | 30 | 30 | 2,000 | 2,000 | 15 |
| Bhurt | Karam-bar | 110 | 8 | 25 | 20 | 2,500 | 3,000 | 10 |
| Upper Henzul | Gilgit | 150+ | 16 | 70+ | 50 | 2,500 | 3,500+ | 15+ |
| Batkor | Gilgit | 900 | 17 | 110 | 110 | 2,200 | 3,000 | 90 |
| Nomal | Hunza | 1,500+ | 40+ | 1,000+ | 600 | 5,000 | 6,000 | 150 |
| Naltar Lakes | Naltar | 300 | 14 | 20 | 20 | 2,500 | 6,000 | 40 |
| Baltit-Sumayar | Hunza | 400 | 8 | 150 | 150 | 3,000 | 3,000 | 20 |
| Yashbandan-Barut | Hunza | 250 | 21 | 50 | 50 | 6,000 | 3,000 | ? |
| *Chitral Basin:* | | | | | | | | |
| Miragram-Parwak | Yarkhun | 600 | 18 | 120 | 100 | 3,000 | 7,000 | 35 |

[a]see Fig. 6.1

**Table 6.3** Estimated initial and effective heights[a] of landslide dams on the Trans Himalayan streams

| Height class | Main dam | Secondary[b] | Sum | % of all |
|---|---|---|---|---|
| Partial dam (Type I)[c] | 13 | 7 | 20 | 9 |
| Cross-valley dams | | | | |
| < 10 m | 5 | 6 | 11 | 5 |
| 10–50 m | 82 | 46 | 128 | 56 |
| 50–100 m | 28 | 6 | 34 | 15 |
| 100–250 m | 7 | – | 7 | 12 |
| 250–500 m | 2 | 1 | 3 | 1 |
| 500–1,000 m | 2 | 1 | 3 | 1 |
| > 1,000 m | 2 | – | 2 | 1 |
| Totals | 161 | 67 | 228 | 100 |

[a]The lowest part of the cross-valley barrier; [b]Partial inventory only (see text); [c]See Table 6.2.

100 m, and two were over 1,000 m (Table 6.3). Dam height is positively correlated with rock avalanche size and width measured parallel to the stream valley – except in a few cases where the bulk of the landslide material moved along or sub parallel to valley direction. In general there is considerable variability related mainly to terrain.

On average, one cross-valley rock avalanche barrier was found for every 15 km of river course surveyed. Actual spacing is highly variable. There is a firm date for only one prehistoric event, a $^{14}C$ age of 7,150 BP for Ghoro Choh I [27]. However, it is clear most were emplaced on ice-free valley floors; that is, since the last major glaciation when the areas surveyed were ice-covered. This suggests all are late Quaternary events. A majority could be Holocene. Meanwhile, even the degraded dams continue to play a major role in fluvial developments, flood routing, sediment delivery and intermontane sedimentation [25, 29].

Surveys to date cover barely 20% of the roughly 80,000 km$^2$ ice-free area of the Trans Himalayan Upper Indus Basin in the Karakoram. Since each extension of the survey has identified similar densities of events, it should be fairly representative, suggesting as many as a thousand late Quaternary events await discovery in these valleys. The Karakoram tributaries of the Yarkand River, draining northwards to China's Xinjiang Province, has similar features and rock avalanches should also be found there [44]. The same applies to the high mountain basins in the Hindu Kush [57]. Many more have been discovered in the Pamir, Alai and Tien Shan Ranges to the west and north of the Karakoram [2, 52].

## 2 The Varieties of Impoundment and Landslide Dam Classification

The plan forms and geometry of the Upper Indus dams, their composition and the topography of landslide deposits, exhibit great diversity. All those examined here originated with catastrophic rockslides and rock avalanches. The resulting barriers reflect critical modifications by rugged terrain, especially, and in some cases topoclimatic and substrate conditions. These modifications have been a major influence on the character and fate of the dams. Indeed, their role in this and other high mountain regions suggests a need to revise and extend existing classifications of landslide dams (cf. Chap. 24 by Hermanns et al., this volume).

## 2.1 *The Role of Complex Emplacement in Barrier Morphology and Composition*

Interaction of the rock avalanche with rugged terrain can critically affect barrier height, geometry and compaction, resistance to and patterns of erosion. Topographic interference involves both longitudinal and transverse confinement, the effects of sudden changes in slope, valley geometry, junctions and bifurcations. These can block or stall the rock avalanche, and channel debris streams or cause them to split into two or more distributaries. As a result, there is also great diversity of constructional landforms, internal heterogeneity of the landslide material, and topographic blocking that can cause greater compaction of the barriers emplaced. These features have been described in detail elsewhere, and will be identified here as they affect dam stability and histories [30, 31].

The highest barriers tend to be where relatively large volumes of rock avalanche debris have stalled as cross-valley ramps against an opposing slope. However, in most cases secondary debris streams also develop spreading up and down the valley. These emplace extensive supporting masses above and below the dam crest. Relatively low cross-profiles are created even in barriers of great height, an important element of dam stability [39, p. 20].

Multiple major and minor lobes are also found that record splitting around obstacles. Movement of material into or out of two or more valleys is a common occurrence. Responses to rugged terrain can also create longitudinal and transverse pressure ridges tens of meters higher or thicker than adjacent rock avalanche debris. One or more of the ridges may form the initial highest level of the dam. Where the barrier itself is drowned, complicated lake morphologies and sedimentation arise in the lower several kilometers 95 the water inundates the rock avalanche debris [29]. Depressions between ridges become sites of ponds and small lakes, or develop into spillways carrying initial overflow.

The shape of the cross-valley profile, which determines effective dam height and affects breaching patterns, exhibits wide diversity. Where large volumes of debris travel across relatively narrow gorges, most of the material tends to be deposited against the opposing slope – Heim's [21] "main accumulation" or *Hauptanlage* [26, 30]. In such cases the lowest part of the cross-valley barrier often occurs in the proximal region, below the source slope (Fig. 6.2). According to Heim [21] this is typical of examples in the European Alps [5, 49]. In the Karakoram it occurs mainly in rock avalanches derived from the wall of the valley impounded, or a hanging valley high above it. In those issuing from a tributary to block a main valley, the lowest part of the cross-valley ramp is more commonly in the central or distal part.

**Fig. 6.2** Topographic interference and barrier form. A view looking downstream to the Ziarat rock avalanche barrier in Chapursan Valley, Hunza. The rockslide descended from the valley wall to the left of the photograph. Run up of some 150 m on the right, opposing flank, emplaced the bulk of the mass there (H = *Hauptanglage*). The flats around the white building, a shrine to a holy man, record the former lake bed. Overflow and breaching were in the lowest, proximal part of the barrier (*arrow*), below the source slope [29]

This can reflect sudden slowing and spreading over a sediment fan at the valley junction, after confinement in a narrow canyon, or a limited amount of material reaching the main valley. The Dulung Bar barrier in Darkot valley shows the effect (Fig. 6.7c).

## 2.2 The Case of Composite or "Hybrid" Dams

In 27 of the Upper Indus blockages, the barriers were found to be partly rock avalanche and partly other materials, sometimes in distinct sections, sometimes with complex mixing. Schneider [56] reports similar cases in the Southern Pamir, which he refers to as "composite" dams. Weidinger (Chap. 8, this volume), describes others in Nepal. Such dams raise special problems for stability analyses and reconstruction of dam histories.

Three main classes of hybrid dam were identified in the region (Table 6.4):

(i) Where a rock avalanche over-ran glacier ice and glacial deposits.
(ii) Where rock avalanches disturbed and incorporated large masses of valley fill materials, or created thickened, "tectonised" areas in them.
(iii) Rock avalanches with lobes transformed to debris avalanches or debris flows by entrained moisture and wet sediment.

**Table 6.4** Hybrid dams identified on the Upper Indus streams. In each case parts of the barrier consists of materials of different provenance, one of which is rock avalanche. Three types are recognized

1. Rock avalanche, glacier ice and/or moraine (8)

| River | Glacier | Rock avalanche | Initial height |
|---|---|---|---|
| Shimshal | Malangutti | Malangutti | 80 m+ |
| Hunza | Batura | (3) "Bilgim Platt" | 20–30 m |
| Hunza | Pasu (?Ghulkin) | Yashbandan-Bharut | 50 m+ |
| Karambar | Chillinji | Chillinji Charagah | ? |
| Karambar | Chhateboi | Qalandar Uwin | 100 m+ |
| Karambar | unnamed | Karambar Lake E. | 20–25 m |
| Yarkhun | unnamed | Karambar Lake W/ | 15–20 m |
| Hushe | Charakusa | Ghorsit | 50+ |

2. Rock avalanche, disturbed (i.e. "tectonised") and/or transported masses of valley fill sediments (7)

| River | Rock avalanche | Initial height |
|---|---|---|
| Naltar | Naltar Lakes | 20 m+ |
| Shigar | Ghoro Choh I | 10 m+ |
| Shigar | Ghoro Tsar | 15 m+ |
| Shigar | Yarbah Tsoh | 5–10 |
| Indus | Goma Thurgon W. | ? |
| Indus | Ghomba Thurgon | ? |
| Dumordo | Tsok-Panmah | 80 m |

3. Rock avalanche with areas or lobes saturated by entrained moisture and/or wet sediment to become debris avalanches or debris flows (10)

| River | Rock avalanche | Debris Av./Debris flow | Initial height |
|---|---|---|---|
| Ishkoman | Asambar | DA | ? |
| Gilgit | Gakutch | DA | ? |
| Bagrot | Sinakar-Hope Complex | DA and DF | ? |
| Chapursan | Rishijerab | DA | ? |
| Chapursan | Shikarjerab | DA and DF | ? |
| Naltar | Naltar Lakes | DA and DF | 20 m+ |
| Shyok | Surmo-Khapalu | DA and DF | ? |
| Braldu | Ste–Ste | DA | ? |
| Braldu | Gomboro Complex II | DA and DF | ? |
| Braldu | Ghoro Tsar | DF | 15 m+ |

### 2.2.1 Mixed Rock Avalanche and Glacial Barriers

Eight rock avalanches were identified that spread over glacier termini at or near their present-day positions. The impoundments consist partly of ice and/or moraines. Like the most dangerous ice dams in the Karakoram, all involve ice of a tributary glacier spreading into a main stream valley [23]. In all except two cases, glacial deposits or ice formed the larger part of the cross-valley barrier. The rock avalanches were, however, decisive in sealing the dam and creating relatively long-lasting impoundments. This applies to the best-known cases at the termini of the Batura and Malangutti Glaciers in Hunza, although they have always been attributed solely to the glaciers. In each case the rock avalanches, descending from rock walls undercut by the ice tongue or river forced against them by the glacier and ice-margin deposition, sealed the narrows at the upstream flank of the terminus (Fig. 6.3).

Since, in these and most other cases, the lakes or lacustrine deposits recording them were formerly attributed to glacier damming alone, the discoveries reported here radically change their interpretation and implications. The rock avalanche material was missed or mistaken for moraine [4, 18, 40, 47, 51]. The same glaciers

**Fig. 6.3** A hybrid rock avalanche-glacier dam in Shimshal Valley. The view looks along the lateral moraine complex and ice of the Malangutti Glacier which extends across the main Shimshal River valley. The view is taken from remnants of the rock avalanche (RA), which descended from the rock slope behind to seal the dam. It spread over the ice, and moraines to the left of the picture. Remnants of extensive, deep lacustrine sediments occur upstream. In the foreground the Shimshal river is superimposed in bedrock within a complex pattern of overflow and breaching

have, or may have, formed ice dams too but known examples have rarely survived as much as 1 year without draining [23]. However, deposits involving glaciers and recording long-lived lakes are all associated specifically with the hybrid rock-avalanche and glacial barriers. This raises new issues for hazard assessment as well as glacier change in the late Quaternary.

### 2.2.2 Barriers with "Tectonised" and Transported Substrate Materials

Rock avalanches moving over deformable substrates can transfer strain to them, generating complex folding and faulting [1]. Brittle, dry units may be dislodged and transported. Wet, fine-grained sediment may undergo liquefaction and extrusion.

In the wide, sediment-filled intermontane basins of Skardu and Shigar, Baltistan, ten rock avalanches are known that traveled into shallow lakes, over former lake beds, or extensive, braided river reaches [27]. Some descended over large, low-angle sediment fans and fan deltas. In fact, the visible valley fill was impounded behind other, older rock avalanches; in particular the Katzarah event blocking the exit from Skardu Basin. In each case there were complex interactions with, and transport of, deformable substrates [27, 29, 30]. As a result, thickened zones and transported substrate masses, were important elements of the cross-valley barriers (Fig. 6.4).

**Fig. 6.4** The Yarbah Tsoh hybrid dam, Shigar valley, Baltistan. The view looks over the surface of the dam from the distal rim towards the detachment zone of the landslide on the distant Marshakala Peak (D). In the foreground is rock avalanche debris that skidded over or ploughed into (*arrow*) the fine-grained alluvium of the valley floor. Unseen parts of the rock avalanche sliced under the alluvium creating complex folding, thickening and transporting it in the direction of movement

**Fig. 6.5** Disturbed ("tectonised") and transported valley fill alluvium (A), mixed with rock avalanche material (R). This is part of the former, Gomba Thurgon landslide barrier on the main Indus River in Skardu Basin [27]

In some cases, melange-like regions developed with rock avalanche and substrate materials chaotically interspersed and mixed (Fig. 6.5). The geometry, stability and fate of these barriers differ from those of rock avalanche materials alone (see below).

### 2.2.3 Rock Avalanches Transformed to Debris Avalanches or Debris Flows

In other regions there are some well-known examples of catastrophic rock slope failures that were partly or wholly transformed to debris avalanches or debris flows through incorporation of large amounts of substrate materials, or moisture. They include the Cerro Rabicano 1987 event in Chile [19], the Yungay 1970 and Mayunmarca 1974 events in Peru [59], a number that moved over glaciers [14, 16, 46], and rock avalanches from dissected volcanoes [9, 15]. Most of the examples are in humid climates, or triggered by heavy rainfall, snow and ice melt. In the Upper Indus Basin, where ten barriers were found to have sections consisting of debris avalanche and/or debris flow materials, the main factor was movement over valley fill of the main river valley or descent over glaciers (Table 6.4). The Naltar Lakes event is in an exceptionally humid valley for the region.

While significant where they occur, developments incorporating deformable substrate or large amounts of moisture, apply to a relatively small proportion of the examples. This reflects, in part, the severely rain-shadowed, arid conditions in the Upper Indus stream valleys. The climate of the fluvial zone, where the major dams have formed, is presently arid and semi-arid. Most of the moisture here derives from higher, more humid zones. Rock avalanches in dry climates are thought to be a distinctive class [62], and aridity reduces the chances of moisture or wet sediment being entrained.

Also, it may be important that Karakoram rock avalanches are virtually all derived from of hard, massive bedrock. Mud rocks are almost unknown except across the northern fringe. In the landslides rock units behave as rigid units, and are broken and crushed by brittle fracture. Moderate amounts of moisture will be vaporized and expelled. The behavior and final compaction of the mass depend upon internal grain collisions and brittle fragmentation.

## 2.3 *The Varieties of Dam: Suggestions Toward a Revised Classification*

The number and diversity of events with complex emplacement or those identified as hybrid dams are sufficient to indicate their significance both for the regional history of landside dams and the implied risks. In order to compare them to dams reported elsewhere and in the literature it is useful to classify them [54].

Various classifications have been developed. That of Costa and Schuster [11] emphasises barrier morphology in relation to valley topography. Since these are key aspects of Karakoram examples, their typology is adopted and extended here, but modified to take account of complex emplacement (Table 6.5). Their types I, II, and III are adopted unchanged when applied to single impoundments, but making an added distinction between landslides that originate in the valley that is dammed and those from a tributary valley that dam the main valley. The proportions for these types are roughly similar to Costa and Schuster's [11] world-wide survey, although less than a third of their examples involved rock avalanches (Table 6.6). Meanwhile, their Types IV and V seem rare or absent in the Karakoram, and are reassigned as new classes associated with complex emplacement and multiple impoundments (Figs. 14 and 24 in [29]). These comprise 34% of the inventory and, more importantly, most of the largest dams including 82% of those over 100 m in height, and all barriers over 250 m high.

The classification does not directly take into account differences in dam size or basin parameters and hydrology, nor specifics of barrier geometry and composition. Two dams in the same morphological class can differ in each or all of these respects, and in ways that can be even more significant for the assessment of dam stability

**Table 6.5** Rock Avalanche Dams: a morphological classification taking account of conditions in the Upper Indus Basin (after [11, 29]). Types Ia, IIa, IIIa, IIIa, and VI, are the same as Types I, II, III and VI respectively of Costa and Schuster [11]. Examples (italics) are from the region

**A. Single Rock Avalanche lobe/single impoundment**

**Type I** partial or groyne-like
a) main valley source (i.e. wall of valley dammed) – *Gomba Thurgon, Skardu Basin, Baltistan*
b) tributary valley source – *Kepchun, Skardu Basin, Baltistan*
**Type II** full cross-valley barrier
a) main valley source – *Dhak Chauki, Gilgit Basin*
b) tributary valley source – *Litak. Hushe Valley, Baltistan*

**B. Double lobe or "deformed T-type"/single impoundment**

**Type III** full barrier spreading up and down valley floor
a) main valley source – *Upper Henzul, Gilgit Basin*
b) tributary valley source – *Nuh, Darkot valley, Gilgit Basin*

**C. Complex emplacement dams: multiple lobes/ two or more impoundments**

**Type IV** enters and dams both main valley and tributary
a) main valley source – *Gol-Gone "A", Indus valley, Baltistan*
b) tributary valley source – *Katzarah, Skardu Basin, Baltistan*
**Type V** enters and dams three or more valleys
a) main valley source – *Haldi, Saltoro valley, Baltstan*
b) tributary valley source – *Dulung Bar, Darkot valley, Gilgit Basin*

**Type VI** combines deep-seated rockslide and rock avalanche(s) – *Nomal Complex, Hunza valley, Gilgit basin*

**Table 6.6** Morphological classification of rock avalanches that dammed upper Indus streams

| Category | (a) Valley wall source | (b) Tributary valley source | Sum | % |
|---|---|---|---|---|
| Type I | 11 | 8 | 19 | *12* |
| Type II | 35 | 14 | 49 | *30* |
| Type III | 29 | 8 | 37 | *24* |
| Type IV | 19 | 18 | 37 | *24* |
| Type V | 8 | 10 | 18 | *10* |
| Type VI | 1 | – | 1 | – |
| | 103 (*78%*) | 58 (*32%*) | 161 | 100 |

and history than the parameters used in the classification. Thus, while morphological typologies are useful for field surveys and regional comparisons, they are only a beginning. However, since the results here are already cumbersome it may be preferable, instead, to work with a list of parameters and generate measures of their relative significance, perhaps identifying classes based on clustering into distinct analytical categories.

# 3 Elements of Dam Stability

The best known and largest historical flood from the Upper Indus, in 1841, involved early and catastrophic failure of a rock avalanche dam [22, 58, Chap. 7 by Delaney and Evans, this volume]. The second largest, in 1858, also involved early collapse of a landslide dam [20, 50]. These supported an emphasis on catastrophic failure and outburst floods [10, 23]. However, most of the newly-discovered rock avalanches have multi-year lacustrine deposits and extensive valley fill sediments against and far upstream, indicative of relatively long-lived impoundments [29, 30]. While most are drained and degraded today, the evidence is of gradual breaching or failure in several stages. At the same time three-fourths (116) are not fully cut through and continue to influence stream hydrology and sediment delivery, showing that their fluvial system roles are not confined to the time of full impoundment or sudden collapses. What are the conditions that have given rise to these longer-lived dams?

## *3.1 Composition and Compaction*

In fact, rock avalanches comprise a large share of all long-lived landslide dams around the world. This reflects their sudden emplacement, composition, compaction, and morphology. They are mass movements derived from the failure of largely intact bedrock slopes, those in the Karakoram being hard and massive rocks of plutonic, metamorphic and metasedimentary types. Collapse and run out processes are significant in barrier composition as they serve to fracture, crush and pulverize the rock material, and through the way emplacement of the mass occurs.

The surface of an undisturbed rock avalanche deposit usually consists of open-work rubble, often with boulders exceeding 10 m in diameter. This horizon is porous, and the surface boulders usually sit in an upper "carapace" of crushed materials which may not be fully compacted, allowing water infiltration. However, if and when the impoundment is filled and water overflows the barrier, the coarse debris resists erosion. The situation below the surface carapace is very different. The main body of a rock avalanche deposit comprises a densely compacted, impermeable mass, with larger clasts embedded in matrix material.

Since the fragments, from coarsest to finest, are angular or very angular, they might seem to preclude tight packing. However, if compacting forces are great enough, such angular material can be packed with far less void space than more rounded and well-sorted particles. In its final moments there is usually an abrupt deceleration of a rock avalanche causing sudden, shock-like compaction. This is likely to be intensified when the landslide impacts an opposing valley wall. It also tends to create a tight, impermeable seal over pre-existing terrain.

### 3.1.1 Internal Structure and Composition of Complex Emplacement Events

Interaction with rugged terrain, especially blocking and emplacement in relatively thick masses, can also involve a range of distinctive modifications of the barrier materials. In any rock avalanche deposit exceeding about 10 m in thickness, changes in structure and grain size distribution are observed at depth, especially in those that stall while several tens or hundreds of meters thick (Fig. 6.6). This may, but need not, involve lithological variations. Even where lithology remains the same, variations with depth and horizontally arise from modification of bedrock material by forces operating within the mass movement. Where run-out is blocked by an opposing slope the material at depth may be thoroughly crushed, with great distortion, stretching, folding or faulting of the rock mass, but no mixing of lithologies [29]. The features probably represent intermediate stages of break-up or cataclasis – internal

**Fig. 6.6** Vertical profiles through the Dhak Chauki rock avalanche, Gilgit valley [28]. **(a)** Upper part of rock avalanche showing change from surface rubble down to more highly crushed and consolidated material at depth. **(b)** Longitudinal section cut by the river in the same area showing, at depths of 10–25 m, folded and flame structures in crushed quartz (*arrows*), indicative of cataclasis and deformation patterns along the line of rock avalanche movement

crushing and grinding, that all rock slides – rock avalanches go through, but preserved by topographic stalling and diversion of movement. Parts of these deposits may consist of largely intact bedrock masses seemingly carried along within surrounding, thoroughly crushed material like translational slides or block glides. Such units are often fractured, but with minimal relative displacement, forming crackle-, or jigsaw-brecciated masses [45, Fig. 6.5b in 29]. If two or more source lithologies are present, which is common in larger events, they can introduce further sources of heterogeneity if the rock types have differing mechanical properties. In Karakoram examples, quartz veins are often crushed more fully than surrounding rock, even powdered amid sand-sized and coarser materials but, again, without mixing of different lithological units [30]. From the perspective of dam formation and stability such properties, or the distinctive facies they give rise to, affect how different parts of cross-valley barriers respond to erosional forces and introduce the problem of the "weakest link" among the complex of differing constituents [60].

### 3.1.2 Hybrid Barriers

In all Karakoram examples of large, combined landslide and glacier dams examined, the rock avalanche barrier is decisive in the stability and survival of the dam as well as its sealing. Yet, ice or moraine often comprises the largest part of the total barrier. All but a small part of the cross-valley ramps against which large lakes formed at Malagutti and Batura glaciers consist of lateral moraine complexes that formed on the upvalley flank of the ice. While the impoundments were intact, melt water from the glaciers continued to drain downstream of the barrier. The lateral moraine complexes are relatively strong and impermeable, buttressed along their downvalley flank by active ice. Overflow and breaching occurred against the valley wall opposite the tributary glacier entrance and across the rock avalanche although some overflow had occurred over low parts of the moraines. The extensive lacustrine deposits behind the "Bilgim Plat"-Batura example suggest it may have lasted decades, and the Malangutti-Shimshal case a century or two [28].

Dams involving entrained or tectonised substrate materials are rarely as high as those where there is topographic blocking, or a cross-valley glacier component. Deformed and transported valley fill sediments are not as resistant to erosion as rock avalanche material. Nevertheless, events such as Yarba Tsoh and Ghoro Tsar in the Shigar Valley, Baltistan, formed impermeable impoundments for some years. Although breached, they continue to constrain the Shigar River and up valley sedimentation over tens of kilometers. In these cases, disturbed valley fill forms a substantial part of full height and surface of the barriers. A different situation is illustrated by the Dhak Chauki rock avalanche which dammed the combined waters of the Gilgit and Hunza Basins just below their junction (Fig. 6.6). The landslide material travelled across the valley to reach and climb some tens of metres up the far side. It passed over deep valley fill sediments, leaving them buried under 5–25 m of rock avalanche material over an area of some 9 $km^2$. Valley fill material is disturbed, eroded, folded and mixed into the base of the rock avalanche. It seems likely that the cause of the failure of the dam was due to seepage, explaining the full-height breach following the mid-line of the valley. However, lacustrine beds against and over the rock avalanche indicate that the dam lasted some decades.

### *3.2 Speed of Emplacement*

Since rock avalanches only last two or three minutes, rapid emplacement of debris is important for the compaction and sealing of a dam. It can also offset the tendency of large rivers, or movement over snow and ice, to saturate and transform the mass movement to a wet flow, which tends to produce thinner and less stable barriers. They seem less frequent than in more humid and heavily forested mountain environments.

Timing and provenance are also important. In summer, stream flows are huge at the sites of most of the Indus dams, due to melting of heavy seasonal snowfall and extensive glacier covers at elevations above 4,000 m a.s.l. [37]. In effect, there is an annual window of 8–10 months when, if earthquake-, or weather-triggered landslides occur, streams are at low flow and sub-zero temperatures prevail over the glaciers. Snow cover is generally small or negligible in river valleys. A rock avalanche at this time will likely emplace a dry, stable, well-sealed dam. In summer, however, if a rock avalanche enters a major river valley, or descends over a glacier valley, despite rapid movement the potential for conversion to a debris flow is quite high.

## 4 Impoundment Histories

Impoundments have been found in every conceivable stage of infilling and degradation. Once recognized, this helps in reconstructing the phases an episode of damming may go through [32].

### *4.1 Landslide Lakes*

Only a few small lakes exist today, but many former lakes were over 20 km long, some over 90 km with depths attaining hundreds of meters (Table 6.1). Lake morphologies have been as diverse as the rock avalanche barriers impounding them (Fig. 6.7). Together, the morphology of the landslide barrier and river reach determine the form of the lake. Its survival and history depend on the hydrology of the contributing basin and sediment delivery.

### *4.2 Infilling Water and Sediment Delivery*

Since most of the dams are breached by overtopping, inflow volumes, or flow rates once the dam is full, play critical roles. Today, as noted, the basins that were or may be impounded have extensive humid areas and glacier covers at higher elevations.

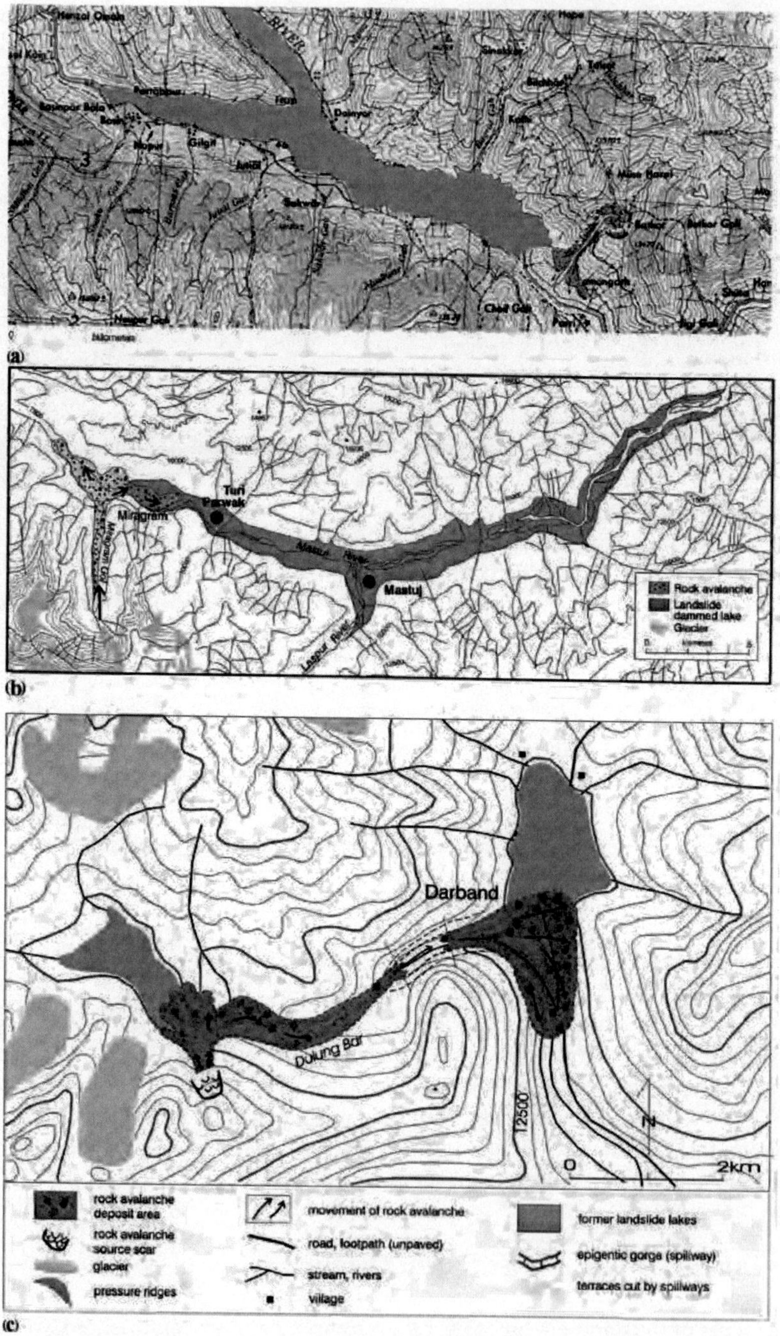
(a)
(b)
Turi
Parwak
Miragram
Mastuj
Rock avalanche
Landslide dammed lake
Glacier
(c)
Darband
Dulung Bar
12500
N
0
2km
rock avalanche deposit area
rock avalanche source scar
glacier
pressure ridges
movement of rock avalanche
road, footpath (unpaved)
stream, rivers
village
former landslide lakes
epigentic gorge (spillway)
terraces cut by spillways

The rainfall contribution to runoff is negligible. River regimes are glacio-nival, some with a greater glacier, others a greater snowmelt, component. Over 90% of inflow to any impounded river reach occurs in less than 3 months, and may be concentrated in a few weeks of summer where glacier basins dominate yields [36].

The Upper Indus has one of the highest sediment yields among basins of its size. Sediment transport is even more concentrated in time than stream flow. In many years, as much as half the annual yield is carried in as little as 10 days of the highest flows [22]. It is not clear how far sediment yields represent present-day climatic conditions and how far they depend upon the storage and release of intermontane sediments, especially related to past and existing natural dams [17, 30]. Even more than the effects of deglaciation and climate change in the Holocene, the history of dams requires caution in using present-day sediment transport rates to estimate inflows and sedimentation for prehistoric dams, as well as contemporary rates of denudation.

Fine grained lacustrine sediment (silt and clay grades mainly derived from contemporary or reworked glacial outwash) covers the upstream slope of the barriers. It may be important in reducing permeability, especially in the upper 10–20 m of the rock avalanche deposit, which can be more permeable. Sediment input also affects the longevity of dams through rates of infilling. Spillway erosion may be slow or absent until infilling is complete, since the lake water lacks a sediment load to abrade the resistent surface boulders. Once fully silted up, however, abundant, coarser sediment of these mountain rivers is carried through and can attack the resistant rubble of the rock avalanche surface – a critical threshold recently reached, for example, at the main Naltar Lakes dam on the Naltar River, and Jut dam on the Kargah, both tributaries of the Gilgit River [28].

The balance between sediment delivery and deposition by the axial stream, and from tributaries or valley walls, can also be critical. In several cases localised fan development from tributary valleys divided up the body of water to create, in effect, secondary impoundments and complicate the history of breaching [29]. The role of the Parwak fan in the Miragram-Parwak dam on the Yarkun River, Chitral, is an example (Fig. 6.7b). The fan eventually built across the valley and severed the main lake and contributing basin from the rock avalanche barrier. The breaching of this sediment fan ultimately controlled the draining of the lake.

## *4.3 Overflow Paths and Breaching Styles*

In most of these dams overtopping preceded, and was the main source of, the first phase of breaching. This has involved impounded water flowing along one of

**Fig. 6.7** Reconstruction of lakes behind Upper Indus rock avalanche dams as indicated by initial overflow patterns, and lacustrine deposits. Topography based on the 1/250,000 AMS 2nd Edition topographical maps and [28]. **(a)** Batkor event, damming the Gilgit and Hunza Rivers just below their junction. **(b)** Miragram-Parwak event on the Yarkun River, near Mastuj, Chitral, **(c)** Dulung Bar rock avalanche on the Yasin River near Darkot four possible overflow paths – proximal, medial or distal parts of the cross-valley barriers, or diagonally.

At first, overflow channels reflect dam morphology, generally where the lowest part of the cross-valley barrier lies, hence the importance of rock avalanche emplacement (Fig. 6.8). A number of the Upper Indus dams preserve evidence of initial drainage and breaching sequences; for example, the Litak, Gol-Ghone "B" and Haldi events in Baltistan, the Nomal and Gakutch events on Gilgit River, and Miragram-Parwak and Barok-Laspur in Chitral [25, 26, 29, 30]. Equally, however, the preservation of the spillways itself shows how, in the later stages of breaching, barrier surface morphology can become irrelevant. In each of the cases just cited, the main breach developed elsewhere, usually along one or other valley flank between the rock wall and rock avalanche debris. This applies quite widely and probably relates to the fifth possible path, seepage and piping through or beneath the barrier. Breaching, percolation or piping along one or other valley wall, or sapping and erosion of the down valley flank, which seem rare in early stages can become decisive later. They replace the initial, topography-related overflow channels, by a cut along either the proximal (e.g. Satpara Lake, and Gol-Ghone "B") or distal rim (Dulung Bar-Darkot, Rondu Mendi). Water seeping along the bedrock-rock avalanche interface may reflect a less firm seal there. In several of these examples, erosion has

**Fig. 6.8** Well-preserved initial overflow channel (*arrows*) of the Gakutch rock avalanche dam on the Gilgit River. Thick deposits of lacustrine sediments on the upvalley flank indicate the spillway operated for some decades at least. It was replaced by a breach against the proximal valley wall to the right of the photograph. The outflow is now in a rock gorge superimposed on a bedrock spur (see text Sect. 4.4.)

exposed substrate plastered against steep valley walls in irregular patches that may form lines of weakness and seepage.

In cases like Tsok-Panmah in Baltistan (Fig. 6.9), initial overflow was across the lowest part but the main breach seems quite indifferent to barrier topography. It cuts through the crest of the cross-valley ramp and follows neither valley wall. A likely explanation is that seepage developed through wedges of former valley fill sediments incorporated within the body of the rock avalanche at depth. The Dhak Chauki barrier near Gilgit, mentioned above, and the Yarbah Tsoh barrier near Shigar (Fig. 6.4), record initial overflows controlled by rock avalanche topography, but the main and present day breaches are at mid-valley locations with no obvious relation to barrier morphology. These channels are incised in valley fill that lay beneath the barriers.

The examples also suggest that determining long-term stability of large rock avalanche dams cannot settle for a single cause or breach pattern. The sources of instability may change after years or decades, in these cases through slower processes of infiltration patterns and seepage, or sapping and erosion of the downstream flank. These are concerns at Usoi [52]. Alternatively, it has been suggested that the stability of such dams may be suddenly increased by seismic activity, common in high mountains, that can cause settling and increased compaction.

Where breaching occurs gradually or with less than full-height breaches, there is a progressive upstream extension of stream incision and migration of the effective

**Fig. 6.9** View looking downstream along Durmordo River to where it cuts through the Tsok rock avalanche dam (RA). The initial overflow was in the lowest, medial part of the barrier (*arrow*). Sediments at left were deposited in the former lake

barrier location. In many Upper Indus dams only partly cut through, the controlling barrier height is now found several kilometers up-valley of the original dam crest.

Further complications arise in the the many Upper Indus examples where a single rock avalanche impounded two or more valleys, since the history of each dam can be, and usually is, quite distinct. The dam on the main or largest stream will usually have a decisive role, but there are examples, such as the Naltar Lakes and Dulug Bar events, where it has been breached and removed while the smaller barriers on tributaries remain intact (Fig. 6.7c).

### 4.4 Bedrock Controlled Outlets

Another significant and not uncommon development involves spillways superimposed upon bedrock flanks or spurs of the pre-existing valley. The size and morphology of rock avalanche dams make it unlikely an overflow channel will exactly match the buried river's course. As a result, spillways can be let down on bedrock spurs to create superimposed or "epigenetic" gorges [26]. If this happens, rather than the landslide barrier, the bedrock gorge comes to control lake drainage, the trenching and movement of upstream sediments (Fig. 6.10). Some 45 of these epigentic gorges have been identified with rock avalanche dams (Table 6.1).

**Fig. 6.10** Epigentic gorge where the Naltar River was deflected and superimposed on bedrock of its left flank by the Nomal rockslide complex (*arrows*). The river makes a sharp turn to the left creating an anomalous junction, entering the main Hunza River facing upstream. Its original channel is buried under the landslide [28]

## 5 Contemporary Risks and Assessment Problems

Nearly all the rock avalanche deposits have settlements on or near them. The famous fortified palace and other facilities of Baltit, Karimabad and Sumaiyar settlements in Hunza lie on and adjacent to a rock avalanche. So does the Ziarat shrine in Chapursan (see Fig. 6.2), Bhorit Jheel (lake) near Pasu, "Shangri La" Hotel at Katzarah and tourist facilities at Satpara Lake, both near Skardu in Baltistan. These are all major tourist destinations. Skardu, capital of Baltistan, and Khapalu, capital of Ganchen District are built over rock avalanches [26, 28]. Meanwhile, hundreds of villages and most of the region's agricultural land lie on river flats, former lake beds, terraces and sediment fans built up behind the landslide [35]. The same applies to most modern infrastructure, including roads, airports, hydro-electric dams, hotels and other tourist facilities.

Almost all permanent habitation is in the fluvial zone, and today there are more people, wealth and infrastructure at risk there than ever before. The prospect of further catastrophic landlides, inundations above a dam or catastrophic outburst floods, appears unusually threatening. Moreover, this presents a dramatically new and different risk picture from that described prior to the last decade, when most of the rock avalanches had been classified and mapped as moraines or debris flows [27, 48, Chaps. 11 and 12 in 55, 51]. The related lacustrine sediments and impoundments they record, had been attributed to former glacier dams [7]. As such, the hazards they represent within the inhabited areas of the Upper Indus Basin were assumed to have disappeared with the glaciers. Instead, it now appears that most of the late Quaternary and Recent impoundments recorded by sediments along the Indus streams occurred after the last major glaciation, and many are Holocene events. We are faced with a need to reassess danger based on formerly unrecognized hazards from catastrophic landslides and large impoundments.

Major significance has been accorded to the hazards of slope instability and mass movements along highways [41]. However, past hazard assessments had also missed the role the degraded rock avalanches and related features as sources of these problems, notably unstable material contributing to debris flows (Fig. 6.11).

Meanwhile, landforms and sedimentary features associated with rock avalanche barriers actually dominate geomorphic developments in much of today's fluvial zone, and hence the scope of other geohazards. Long after being breached and drained the barriers continue to constrain stream flow and sediment movement.

Risk assessment is hampered by enormous difficulties of slope stability analysis, even for single sites. The high relief, steep rock walls and otherwise inhospitable terrain, create difficulties, as do large changes in slope process environments with elevation, geological structures and lithologies. Meanwhile, catastrophic landslides are classes of rare, extreme events demanding risk profiles based on millennial time scales, at least. It is likely that the major landslides causing dams on the Upper Indus – landslides with volumes in excess of $100 \times 10^6$ m$^3$ – have recurrence intervals of several centuries for the whole region, and some millennia in any given sub-basin. Now that we have a large and fairly representative set of known events, covering most environments in the region, there is the opportunity for a comparative

**Fig. 6.11** Mountain road cut through a rock avalanche, the Bilchar event, in Bagrot valley, Gilgit District. The vertical cuts testify to the strength and compaction of the rock avalanche material, but rainfall and snowmelt can generate sudden collapses or debris flows at the surface. The landslide once formed a dam on the Bagrot River [28]

analysis of risk-related features. However, lack of information about what triggered the landslides and, especially, their actual incidence or recurrence intervals in time, severely limit what can be done.

A number of models to interpret the magnitude, temporal and spatial frequencies of catastrophic rock slides may apply and offer a first approach to the problem. Those directed mainly at triggering mechanisms tend to represent them as rare, essentially random, responses to extreme seismic or weather events. An alternative, relevant for post-glacial conditions on a Holocene time scale or longer, proposes some sort of temporal relaxation or "exhaustion" process. The numbers and scale of events will decline as unstable sites (created or exposed by, say, glaciation or rapid incision) are used up [13]. The temporal shape is similar to Church's [8] model for paraglacial sedimentation. It might seem to apply to the main fluvial zone of the Upper Indus Basin, mostly confined to elevations below about 4,000 m a.s.l., where deglaciation occurred in the late Pleistocene and climate desiccation through the Holocene.

However, the strong vertical organization of these high mountain geomorphic environments, and tendency of environmental changes to involve vertical shifts, also means that different magnitude/frequency distributions can apply simultaneously in different elevation bands [24]. Meanwhile, landslide incidence is expected to approximate a steady–state if uplift and denudation are more or less balanced.

Rapid rates of uplift, and the maintaining of steep rock walls, appear broadly true for the Karakoram and adjacent ranges [6]. Recall that two thirds of the surface area of the Upper Indus basin lies above 3,800 m a.s.l. in elevation, and an even larger fraction of the steep rock wall terrain [22].

A more explanatory approach suggests that landslides occur at sites "pre-designed" by tectonics, lithology and terrain development [53, 61]. Yet, this too can result in a range of distributions over time. Evidently, any development that slows or accelerates controlling variables, including glaciation and deglaciation, extreme weather or seismicity, can maintain, increase, decrease or set a new probability distribution for landslide incidence. Moreover, these landslide processes involve "polycausal phenomena" rather than single explanations, and not only where there is human intervention [3].

An outstanding problem is the absence of data to confirm or refute these models, or test what they imply about ongoing risk. We do not known whether conditions that applied when the known landslides occurred have changed substantially or no longer apply, and have only indirect evidence to suggest possibilities. However, now that there is a large inventory of recognized rock avalanche events, there are more promising ways to assess contemporary landslide risk. If dates can be established for a sizeable sample of known events, they should reveal the probability distribution(s) applying in the Holocene and likely shape of the continuing threat.

## References

1. Abbot, P.L., Kerr., D.R., Steven, E.B., Washburn, J.L. and Rightmer, D.A. (2002) Neogene sturzstrom deposits, Split Mountain area, Anza-Borrego Desert State Park, California, in S.G. Evans and J.V. DeGraff (eds.), *Catastrophic Landslides: Effects, Occurrence, and Mechanisms*, Vol. 15. Geological Society of America Reviews in Engineering Geology, Boulder, Colorado, pp. 379–400.
2. Abdrakhmatov, K., Evans, S.G., Hermanns, R., Scarascia-Mugnozza, G. and Strom, A.L. (eds.) (2004) *Security of Natural and Artificial Rockslide Dams: Extended Abstracts Volume*. NATO Advanced Research Workshop, Bishkek, Kyrgyzstan, June 8–13.
3. Alexander, D.E. (1993) Landslides as polycausal phenomena, in J. Nemec (ed.), *Prediction and Perception of Natural Hazards*. Kluwer Academic, Dordrecht, pp. 93–99.
4. Batura Glacier Investigation Group (1979) The Batura Glacier in the Karakoram Mountains and its Variations, *Scientia Sinica* **22**(8), 958–975.
5. Bonnard, C. (ed.) (1988) *Landslides: Proceedings, 5th International Symposium on Landslides, 11–15 July, Lausanne, Switzerland*. Balkema, Rotterdam.
6. Burbank, D.W., Leland, J., Fielding, E., Anderson, R.S., Brozovic, N., Reid, M.R. and Duncan, C. (1996) Bedrock incision, rock uplift and threshold hillslopes in the northwestern Himalayas, *Nature* **379**, 505–510.
7. Bürgisser, H.M., Gansser, A. and Pika, J. (1982) Late Glacial lake sediments of the Indus valley area, northwestern Himalayas, *Eclogae geologische Helvetica* **75**, 51–63.
8. Church, M. (2002) Fluvial sediment transfer in cold regions, in K. Hewitt, M.L. Byrne, M. English and G. Young (eds.), *Landscapes of Transition: Landform Assemblages and Transformations in Cold Regions*. Kluwer, Dordrecht, pp. 93–117.
9. Clague, J.J. and Souther, J.G. (1982) Dusty Creek landslide on Mount Cayley, British Columbia, *Canadian Journal of Earth Science* **19**, 524–539.

10. Code, J.A. and Sirhindi, S. (1986) Engineering implications of impoundment of the Indus by an earthqauke-induced landslide, in R.L. Schuster (ed.), *Landslide Dams: Processes, Risk anmd Mitigation*, Vol. 3.American Society of Engineers, Geotechnical Special Publication, New York, NY, pp. 97–110.
11. Costa, J.E. and Schuster, R.L. (1987) *The Formation and Failure of Natural Dams*. U.S. Geological Survey, Open File Report 87–392.
12. Craddock-Henry, N.A. (2001) *Risk, Vulnerability and Environmental Hazards in the Village of Darkot, Northern Pakistan*, [unpublished Masters dissertation]. Wilfrid Laurier University, Waterloo, ON.
13. Cruden, D.M. and Hu, X.-Q. (1993) Exhaustion and steady-state models for predicting landslide hazard in the Canadian Rockies, *Geomorphology* **21**, 18–30.
14. Evans, S.G. and Clague, J.J. (1988) Catastrophic rock avalanches in glacial environments, in C. Bonnard (ed.), *Landslides: Proceedings, 5th International Symposium on Landslides, 11–15 July, Lausanne, Switzerland*. Balkema, Rotterdam, pp. 1153–1158.
15. Evans, S.G., Hungr, O. and Clague, J.J. (2001) Dynamics of the 1984 rock avalanche and associated distal debris flow on Mount Cayley, British Columbia, Canada: Implications for the landslide hazard assessment on dissected volcanoes, *Engineering Geology* **61**, 29–51.
16. Fahnestock, R.K. (1978) Little Tahoma Peak rockfalls and avalanches, Mount Rainier, Washington, U.S.A, in B. Voight (ed.), *Rockslides and Avalanches, 1 Natural Phenomena*. Elsevier, New York, NY, pp. 181–196.
17. Ferguson, R.J. (1984) Sediment load of the Hunza River, in K.J. Miller (ed.), *The International Karakoram Project*, Vol. 2. Cambridge University Press, Cambridge, pp. 374–382.
18. Goudie, A.S., Jones, D.K.C. and Brunsden, D. (1984) Recent fluctuations in some glaciers of the western Karakoram Mountains, Pakistan, in K.J. Miller (ed.), *The International Karakoram Project*, Vol. 2. Cambridge University Press, Cambridge, pp. 411–455.
19. Hauser, A. (2002) Rock avalanche and resulting debris flow in Estero Parraguire and Rio Colorado, Region Metropolitana, Chile), in S.G. Evans J.V. Degraff (eds.), *Catastrophic Landslides: Effects, Occurrence, and Mechanisms*, Vol. 15.Geological Society of America Reviews in Engineering Geology, Boulder, Colorado, pp. 135–148.
20. Henderson, W. (1859) Memorandum on the nature and effects of the flooding of the Indus on 10th August 1858 as ascertained at attock and its neighbourhood, *Journal of the Asiatic Society of Bengal* **28**, 199–219.
21. Heim, A. (1932) *Bergsturz und Menschenleben*. Fretz and Wasmuth, Zürich.
22. Hewitt, K. (1968) Records of natural damming and related events, *Indus: Journal of the Water and Power development Authority, Pakistan* **10**, 2–10.
23. Hewitt, K. (1982) Natural dams and outburst floods of the Karakoram Himalaya, in J. Glen (ed.), *Hydrological Aspects of Alpine and High Mountain Areas*, Vol. 138.IAHS/AISH Publication, pp. 259–269.
24. Hewitt, K. (1993) The altitudinal organization of Karakoram geomorphic processes and depositional environments, in J.F. Shroder (ed.), *Himalaya to the Sea: Geology, Geomorphology and the Quaternary*. Routledge, New York, NY, pp. 159–183.
25. Hewitt, K. (1998) Himalayan Indus streams in the Holocene: Glacier-, and landslide-"interrupted" fluvial systems, in I. Stellrecht (ed.), *Karakoram-Hindukush-Himalaya: Dynamics of Change*, Vol. 4/1. Rüdgers Köppe Verlag, Köln, pp. 3–28
26. Hewitt, K. (1998) Catastrophic landslides and the Upper Indus streams, Karakoram Himalaya, northern Pakistan, *Geomorphology* **26**, 47–80.
27. Hewitt, K. (1999) Quaternary moraines vs catastrophic rock avalanches in the Karakoram Himalaya, Northern Pakistan, *Quaternary Research* **51**, 220–237.
28. Hewitt, K. (2001) Catastrophic rock slides and the geomorphology of the Hunza and Gilgit Basins, Karakoram Himalaya, *Erdkunde* **55**, 72–93.
29. Hewitt, K. (2002) Styles of rock avalanche depositional complex in very rugged terrain, Karakoram Himalaya, Pakistan, in S.G. Evans and J.V. DeGraff (eds.), *Catastrophic Landslides: Effects, Occurrence, and Mechanisms*, Vol. 15. Geological Society of America Reviews in Engineering Geology, Boulder, Colorado, pp. 345–378.

30. Hewitt, K. (2002) Postglacial landform and sediment associations in a landslide-fragmented river system: The transHimalayan Indus streams, in K. Hewitt, M.L. Byrne, M. English and G. Young (eds.), *Landscapes of Transition: Landform Assemblages and Transformations in Cold Regions*. Kluwer, Dordrecht, pp. 63–91.
31. Hewitt, K. (2004) Rock avalanche dams on the transHimalayan Upper Indus streams: A survey of hazard-related characteristics, in K. Abdrakhmatov, S.G. Evans, R.L. Hermanns, G. Scarascia-Mugnozza, and A.L. Strom (eds.), *Security of Natural and Artificial Rockslide Dams: Extended Abstracts Volume*. NATO Advanced Research Workshop, Bishkek, Kyrgyzstan, pp. 86–90.
32. Hewitt, K. (2006) Disturbance regime landscapes: Mountain drainage systems interrupted by large rockslides, *Progress in Physical Geography* **30**, 365–393.
33. Hewitt, K. (2009) Catastrophic rock slope failures and late Quaternary developments in the Nanga Parbat-Haramosh massif, Upper Indus basin, northern Pakistan, *Quaternary Science Reviews* **28**, 1055–1069.
34. Hewitt, K. (2009) Glacially conditioned rock-slope failures and disturbance-regime landscapes, Upper Indus Basin, northern Pakistan, in Knight, J. and Harrison, S.(Eds.), *Periglacial and Paraglacial Processes and Environments*, Vol.320. Geological Society, London, Special Publication, pp. 235–255.
35. Hewitt, K. (2010) Gifts and perils of landslides, *American Scientist* **98**, 410–419.
36. Hewitt, K. and Young, G.J. (1993) Glaciohydrological features of the Karakoram Himalaya; measurement possibilities and constraints, *International Association Scientific Hydrology Publication* **218**, 273–283.
37. Hewitt, K., Wake, C.P., Young, G.J. and David, C. (1989) Hydrological investigations at Biafo Glacier, Karakoram Himalaya, an important source of water for the Indus River, *Annals of Glaciology* **13**, 103–108.
38. Hughes, R. and Nash, D. (1986) The Gupis debris flow and natural dam, July 1980, *Disasters* **10**, 8–14.
39. International Strategy for Disaster Reduction (2000) *Usoi Landslide Dam and Lake Sarez: An Assessment of Hazard and Risk in the Pamir Mountains, Tajikistan*. United Nations, Geneva.
40. Iturrizaga, L. (1998) Preliminary results of field observations on the typology of Post-glacial debris accumulations in the Karakoram and Himalaya Mountains, in I. Stellrecht (ed.), *Karakoram-Hindukush-Himalaya: Dynamics of Change*, Vol. 4/1. Rüdgers Köppe Verlag, Köln, 71–98.
41. Jones, D.K.C., Brunsden., D. and Goudie, A.S. (1983) A preliminary geomorphological assessment of part of the Karakoram Highway, *Quarterly Journal of Engineering Geology* **16**, 331–355.
42. Kargel, J.S., Leonard, G., Crippen, R.E., Delaney, K.B., Evans, S.G. and Schneider, J. (2010) Satellite monitoring of Pakistan's rockslide-dammed Lake Gojal, *EOS* **91**, 394–395.
43. Kreutzmann, H. (1994) Habitat conditions and settlement processes in the Hindukush-Karakoram, *Petermanns Geographische Mitteilungen* **138**, 337–356.
44. Kuhle, M. (2004) The Pleistocene glaciation in the Karakoram Mountains: Reconstruction of past glacier extensions and ice thicknesses, *Journal of Mountain Science* **1**, 3–17.
45. Laznicka, P. (1988) *Breccias and Coarse Fragmentites: Petrology, Environments, Associations, Ores. Developments in Economic Geology*, Vol. 25. Elsevier, New York, NY.
46. McSaveney, M.J. (2002) Recent rockfalls and rock avalanches in Mount Cook National Park, New Zealand, in S.G. Evans and J.V. DeGraff (eds.), *Catastrophic Landslides: Effects, Occurrence, and Mechanisms*, Vol. 15.Geological Society of America Reviews in Engineering Geology, Boulder, Colorado, pp. 35–70.
47. Meiners, S. (1998) Preliminary results concerning historic to Post-glacial stages in the NW-Karakorum (Hispar Muztagh, Batura Muztagh, Rakaposhi Range), in I. Stellrecht (ed.), *Karakoram-Hindukush-Himalaya: Dynamics of Change*, Vol. 4/1. Rüdgers Köppe Verlag, Köln, pp. 49–70.

48. Miller, K.J. (ed.) (1984) *The International Karakoram Project*, Vol. 2. Cambridge University Press, Cambridge.
49. Mollard, J.D. (1977) Regional landslide types in Canada,, in D.R.Coates (ed.), *Landslides*, Vol. 3. Geological Society of America Reviews in Engineering Geology, Boulder, Colorado, pp. 29–56.
50. Montgomerie, T.G. (1860) Memorandum on the great flood of the River Indus which reached Attock on 10th August, 1858, *Journal, Asiatic Society of Bengal* **29**, 128–135.
51. Owen, L. (1991) Mass movement deposits in the Karakoram Mountains: Their sedimentary characteristics, recognition and role in Karakoram landform evolution, *Zeitschrift fur Geomorphologie N. F.* **35**, 401–424.
52. Raetzo, H. (2004) Hazard assessment of Lake Sarez rockslides and Usoy dam in Pamir Mountains (Tajikistan), in K. Abdrakhmatov, S.G. Evans, R.L. Hermanns, G. Scarascia-Mugnozza, and A.L. Strom (eds.)., *Security of Natural and Artificial Rockslide Dams: Extended Abstracts volume*, NATO Advanced Research Workshop. Bishkek, Kyrgyzstan, pp. 145–149.
53. Scheidegger, E. (1998) Structural predesign of mass movements, *Geomorphology* **6**, 82–93.
54. Schuster, R.L. (ed.) (1986) *Landslide Dams: Processes, Risk and Mitigation*, Geotechnical Special Publication 3. American Society of Civil Engineers, New York, NY, 164 p.
55. Searle, M.P. (1991) *Geology and Tectonics of the Karakoram Mountains*. Wiley, New York, NY.
56. Schneider, J.F. (2004) Risk assessment of remote geohazards in central and southern Pamir/Tajikistan, in K. Abdrakhmatov, S.G. Evans, R.L. Hermanns, G. Scarascia-Mugnozza, and A.L. Strom (eds.)., *Security of Natural and Artificial Rockslide Dams: Extended Abstracts Volume*. NATO Advanced Research Workshop, Bishkek, Kyrgyzstan, pp. 164–169.
57. Shroder, J.F. (1989) Hazards of the Himalaya, *American Scientist* **77**, 564–573.
58. Shroder, J.F. (ed.) (1993) *Himalaya to the Sea: Geology, Geomorphology and the Quaternary*. Routledge, London, 429 p.
59. Voight., B. (1978) *Rockslides and Avalanches, 1 Natural Phenomena*. Elsevier, New York, NY.
60. Wassmer, P., Schneider, J.L., Pollet, N. and Schmitter-Voirin, C. (2004) Effects of the internal structure of a rock-avalanche dam on the drainage mechanism of its impoundment, Flims Sturzstrom and Ilanz paleo-lake, Swiss Alps, *Geomorphology* **61**, 3–17.
61. Weidinger, J.T. (2004) Case histories of Pleistocene to Recent rockslide dams in the Himalayas (India, Nepal) and Qin Ling Mountains (China), in K. Abdrakhmatov, S.G. Evans, R.L. Hermanns, G. Scarascia-Mugnozza, and A.L. Strom (eds.)., *Security of Natural and Artificial Rockslide Dams: Extended Abstracts Volume*. NATO Advanced Research Workshop, Bishkek, Kyrgyzstan, pp. 189–201.
62. Yarnold, J.C. and Lombard, J.P. (1989) A facies model for large rock-avalanche deposits formed in dry climates, in I.P. Colburn, P.L. Abbott and J. Minch (eds.), *Conglomerates in Basin Analysis: A Symposium dedicated to A.O. Woodford*, Vol. 62. Pacific Section S.E.P.M., pp. 9–31.

# Chapter 7
# Rockslide Dams in the Northwest Himalayas (Pakistan, India) and the Adjacent Pamir Mountains (Afghanistan, Tajikistan), Central Asia

**Keith B. Delaney and Stephen G. Evans**

## 1 Introduction

The remains of rockslide dams are widespread in the northwest Himalayas ([33, 34]; Chap. 6 by Hewitt, this volume) and the adjacent Pamir Mountains [44] of Afghanistan and Tajikistan, Central Asia. The region contains in excess of three hundred known rockslide deposits of, as yet, unknown age that have interrupted surface drainage and previously dammed major rivers in the region in prehistoric time. Most of these dams have been breached and only remnant debris, and in some cases lacustrine deposits from formerly impounded lakes, are visible today in the deep valleys of the region.

In addition, the region contains (1) the highest rockslide dam (Usoi rockslide) in the world that dams the largest rockslide-dammed lake (Lake Sarez) on Earth, (2) the largest documented outburst flood associated with a historical rockslide dam outburst (the 1841 Indus Flood), and (3) the world's most recent (2010) rockslide-dammed lake emergency, the Attabad rockslide dam on the Hunza River [47], in the Upper Indus basin of northern Pakistan (Fig. 7.1). The outburst flood hazard presented by Lake Sarez has attracted global scientific attention [78].

The spatial distribution of rockslide dam deposits, the rate and frequency of rockslide dam formation, the subsequent behaviour of the rockslide dam and outburst flood potential are important elements in landslide hazard assessment in the region. To some extent the accuracy of the hazard assessment is conditioned by the interpretation of rubbly valley fill deposits as representing rockslide debris as opposed to glacial deposits, a debate that continues to simmer today (e.g. [53]), and the conceptualising and analysis of various breaching scenarios. In addition, a consideration of rockslide dam hazard is important for the assessment of hydroelectric potential in the region and in some cases, as noted below, the suitability of natural dam debris for foundations of constructed dams.

K.B. Delaney (✉)
Landslide Research Programme, Department of Earth and Environmental Sciences, University of Waterloo, Waterloo, ON, Canada N2L 3G1
e-mail: kbdelane@uwaterloo.ca

S.G. Evans et al. (eds.), *Natural and Artificial Rockslide Dams*, Lecture Notes in Earth Sciences 133, DOI 10.1007/978-3-642-04764-0_7, 

**Fig. 7.1** Lake Gojal [47], Hunza Valley, northern Pakistan, the rockslide-dammed lake impounded by the Attabad rockslide that occurred on January 4, 2010, Hunza Valley, northern Pakistan. Extent shown on EO-1 satellite image obtained on October 03, 2010 (Day 270 of impoundment). White rectangle is enlarged in inset, which shows Attabad rockslide and heavy flow in spillway over rockslide dam debris (for further details, see note added in proof at the end of the Chapter)

In this paper we (1) review previous work carried out on the presence of rock-slide debris that formed prehistoric rockslide dams in the region, (2) report new data on previously unknown rockslide dams, and (3) using remote sensing imagery and SRTM-3 digital terrain data [23], we determine and analyse the geometry, geomorphology and behaviour of four rockslide dams that have formed in historical time, i.e., in 1841 (Indus), 1858 (Hunza), 1911 (Murgab/Bartang), and 2010 (Hunza).

## 2 Methodology

We carried out a review of the literature on rockslides and rock avalanches in the northwest Himalyas and the adjacent Pamirs (Fig. 7.2). This has included a review of a number of original contemporary accounts of rockslide dam events including those on the Indus (e.g. [22]), Hunza (e.g. [31]), and the Murgab (Bartang) Rivers (e.g. [62]).

For topographic data we utilised the SRTM-3 (3 arc second) data set obtained during the Shuttle Radar Topography Mission in February 2000 [23] to develop

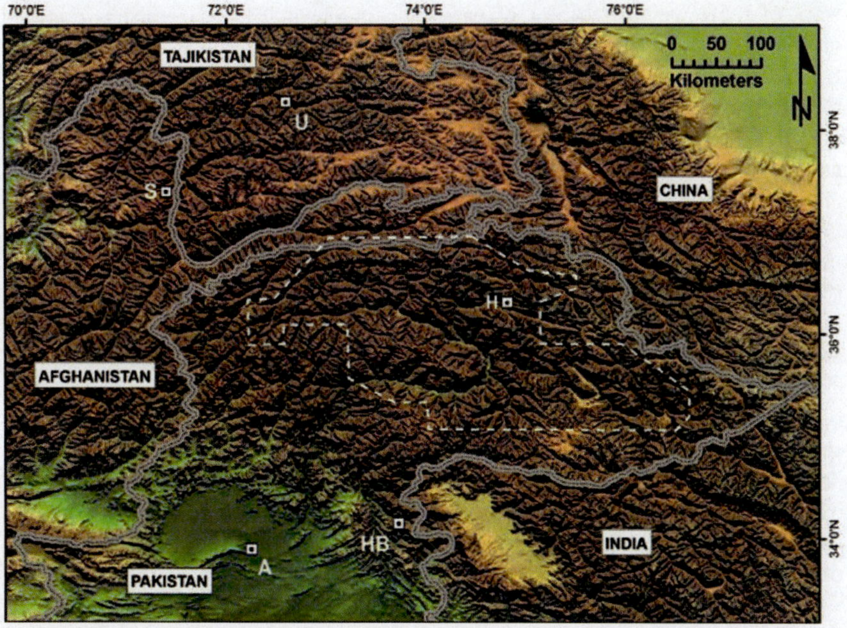


**Fig. 7.2** Map of northwest Himalayas (India and Pakistan) and adjacent Pamir mountains (Tajikistan and Afghanistan). Key to locations discussed in text is as follows; A-Attock on the Lower Indus River, H – 2010 Hunza River rockslide dam, HB – 2005 Haittan Bala rockslide dam, *S* – Shiwa Lake rockslide dam, and U – 1911 Usoi rockslide dam and Lake Sarez. *Dashed white line* outlines area mapped by K. Hewitt in a series of papers since 1998 (see text for references and Fig. 7.3 for further details)

digital elevation models (DEMs) for the rockslide-dammed lakes discussed in both the northwest Himalayas and the Pamirs. To assess the accuracy of the SRTM data in the region of interest, we compared the elevation of three runways at Gilgit and Skardu obtained from the SRTM digital terrain models to the official air navigation elevations of both ends of the three runways (cf. [7]). The mean error of 6 SRTM elevations was less than 1 m suggestive of high vertical accuracy.

SRTM-3 data has been used in conjunction with Google Earth and LANDSAT-7 satellite imagery to illustrate geomorphic features of rockslide dam deposits and to determine the extent and volume of historic and prehistoric rockslide-dammed lakes.

## 3 Rockslide Dams in the Northwest Himalayas (Pakistan and India)

### *3.1 Previous Work*

Our knowledge of rockslide dams in the northwest Himalayas is largely due to the work of K. Hewitt. In a series of papers beginning in 1998, Hewitt has reported on the occurrence of rockslide and rock avalanche deposits in the Karakoram Himalaya, Pakistan. His work has (1) successively mapped the distribution of rockslide and rock avalanche deposits (Fig. 7.3), (2) developed diagnostic criteria for the identification of rockslide dam deposits thus correcting the apparent misinterpretation of the genesis of bouldery debris in Karakoram valleys that had previously been mapped as moraines, (3) described the field geomorphology and inferred landslide behaviour of selected rockslides and rock avalanches, (4) outlined the impact of rockslides on river systems, specifically river damming, and (5) developed a conceptual model of fragmented/interrupted river systems in which landslide damming impacts on the geomorphic evolution of mountain river systems and mountain relief.

We note that with respect to (1) no formal inventory, involving a complete listing of rockslide sites, has been published by Hewitt. Each successive published map shows an incremental increase in the number of identified rockslide/rock avalanche deposits; 104 events in 1998 and 1999 [33, 34], and 143 in 2002 and 2006 [36, 38–39], and 90 in 2009 [41, 42]. Over 80% of these rock avalanches/rockslides formed valley-blocking rockslide dams [40] that impounded the Indus, its tributaries, or other major rivers in the region.

We collated the Hewitt maps and imported the location of rockslides shown on them into a GIS referenced framework (Fig. 7.3) in an attempt to quantify his mapping in a regional and global context. By our count, there are total of 233 rock avalanches identified on Hewitt's maps (This should be considered a minimum since Chap. 6 by Hewitt, this volume, reports a new total of 322 within his study area). With the addition of the 2010 Hunza rockslide, this makes a total of 234; three of these events have taken place in the period of 1841–2010, a frequency of occurrence

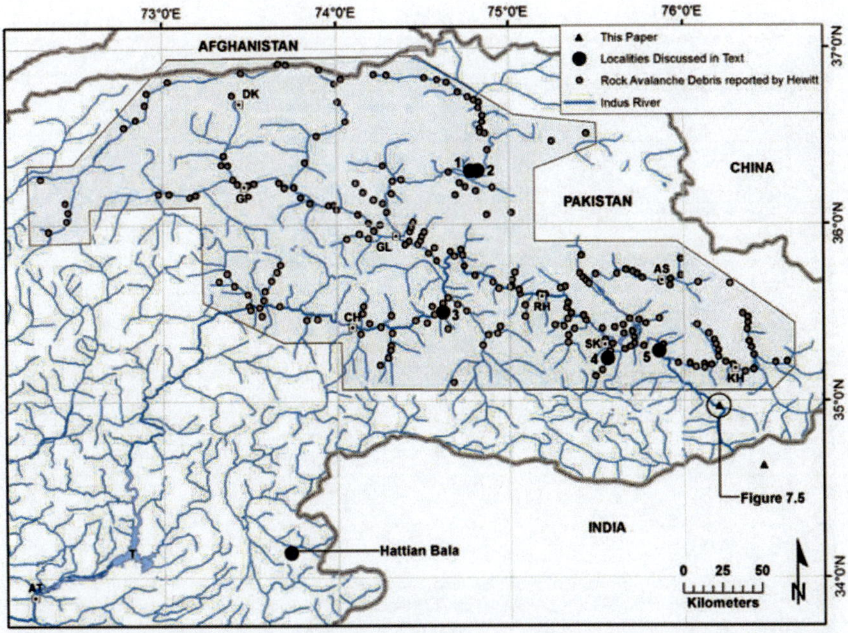


**Fig. 7.3** Map showing distribution of 233 rockslide deposits (*grey dots*) extracted from maps published by K. Hewitt between 1998 and 2009 (see text for references) and imported into GIS. *Shaded area* approximates area mapped by Hewitt (see Fig. 7.2 for location). Also shown are two additional large scale rockslide deposits in the upper Indus (*black triangles*) mapped in this study. Key to towns and villages; AS-Askole, AT-Attock, CH-Chilas, DK-Darkot, GL-Gilgit, GP-Gupis, KH-Khapalu, RH-Rhondu, SK-Skardu. The Tarbela Reservoir is labelled T. Location of historic landslide dams and prehistoric rockslide deposits discussed in text (*black dots*) are *1*: Hunza 1858, *2*-Hunza-2010, *3*-Indus 1841, *4*-Satpara Lake, and *5*-the Gol Ghone rockslides. The location of the earthquake-triggered 2005 Hattian Bala rock avalanche and rockslide-dammed lake (Fig. 7.4) is also shown

of 1 in 56 years. For the area mapped by Hewitt, estimated in Fig. 7.3 as 49,783 km$^2$, this is equivalent to roughly 50 major rockslides (interpreted from their deposits) per 10,000 km$^2$. In Fig. 7.3, two previously undocumented rock avalanches that once blocked the Upper Indus are shown as is the location of the 2005 Hattian Bala rockslide dam and the 2010 blockage of the Hunza River.

Other work has been carried out on the geomorphology of rockslides of the northwest Himalaya by Shroder [70, 71], Shroder et al. [73], Owen [58], and more recently by Seong et al. [69] and Dortch et al. [14]. In addition, investigations of lacustrine deposits formed in rockslide-dammed lakes have been reported by Burgisser et al. [10] in the Skardu area of Pakistan, and Fort et al. [25] and Phartiyal et al. [61] in the Ladakh Himalayas of northwest India.

We also note the occurrence of the Hattian Bala rock avalanche (34° 08′ 24″N/ 73° 43′ 46″E), located in Fig. 7.3 and illustrated in Fig. 7.4. The landslide (est. volume 85 Mm$^3$ [19]) was triggered by the M7.6 October 2005 Pakistan Earthquake

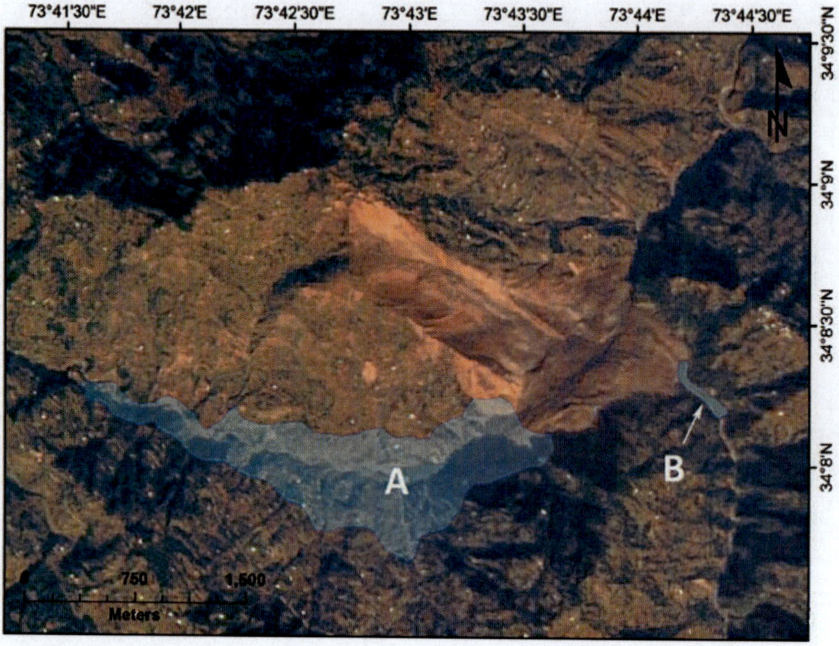


**Fig. 7.4** The Hattian Bala rock avalanche (est. vol. 85 $Mm^3$) triggered by the 2005 M7.6 Kashmir earthquake at 34° 08′ 24″N/73° 43′ 46″E. The rock avalanche debris blocked two streams forming two rockslide-dammed lakes, A (Karli Lake) and B (Tang Lake). The maximum volume of Karli Lake was estimated at 65 $Mm^3$. EO-1 image obtained on November 4, 2005. Lake outlines based on SRTM-3 DEM. Karli Lake drained, after heavy rains, in February 2010

[19, 59, 67] and dammed surface drainage to form significant impoundments in a tributary watershed to the Jhelum River (Fig. 7.3). The larger lake (Karli Lake) reached a volume of about 65 $Mm^3$ [19, 67] before overtopping through a 425 m-long excavated spillway in April 2007 [19]. Following overtopping the lake level was more or less stable as overflow passed through the spillway. The smaller lake (Tang Lake) reached a volume of about 3.5 $Mm^3$ before overtopping through a second excavated spillway, 130 m in length, over the debris [67].

## 3.2 *Prehistoric Rockslide Dams and Deposits: Two Examples in the Skardu Region (Upper Indus)*

The deep valleys of the northwestern Himalayas and the adjacent Pamirs frequently contain complex valley fills characterised by chaotic rubbly debris (Fig. 7.5) as well as stratified silts (e.g. [10]). The interpretation of these deposits as remnant rock-slide debris and lacustrine sediments deposited in rockslide-dammed lakes only became established after work by M. Fort and co-workers [24, 25] and K. Hewitt

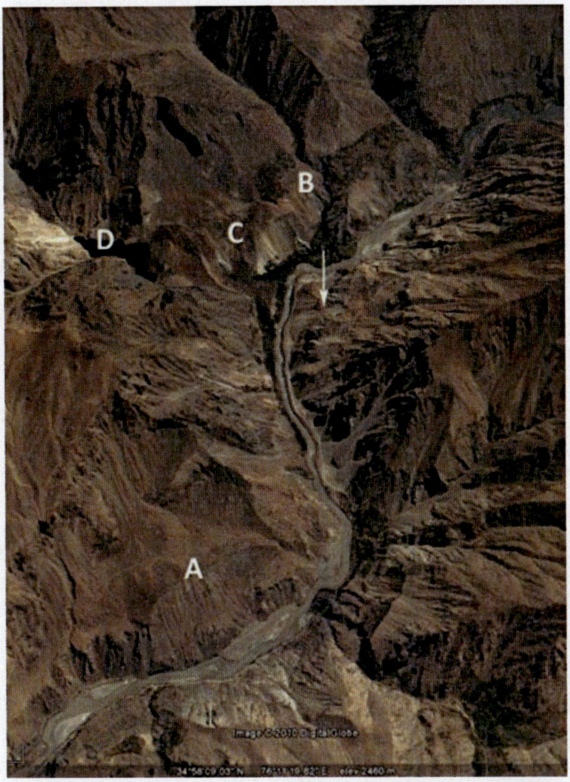


**Fig. 7.5** Typical example of complex valley fill deposits in the Upper Indus basin, northern Pakistan. Previously unmapped deposits include (1) older rockslide deposits at A and B that both dammed the Indus River, (2) younger rockslide deposts (C) that have dammed a tributary forming a rockslide-dammed lake at D and ran up the opposite valley slide to a limit indicated by *vertical arrow*. This rockslide also dammed the Indus. Length of Indus (flowing from top *right* to bottom *left*) seen in image is 13 km. *Google Earth* image at 34° 58′ 09″N/76° 11′ 19″E, 65 km SW of Skardu. Locality is indicated in Fig. 7.3

between 1989 and 1999 [33, 34]. Previously, beginning with the earliest scientific observations (e.g. [15]), they had generally been interpreted as moraine deposits and lake sediments deposited in glacial lakes, respectively (see full discussion in [34]). A notable exception to this was a discussion contributed to Drew's paper (Drew [15], p. 471) by W.T. Blanford who, in disagreeing with Drew, argued that the rubbly deposits in the Upper Indus were rockslide deposits and that the origin of the laminated fine-grained sediments was in rockslide-dammed lakes.

The re-interpretation of the origin of these deposits was significant in four ways; (1) the occurrence of major rockslides was established as a relatively common phenomenon over a large area of mountainous terrain and thus a critical geomorphic process, (2) that in the past, the Indus and other major rivers in the region had been dammed by large volumes of rockslide debris, and, in some cases had formed very large, persistent rockslide-dammed lakes, (3) the geometry of some rubbly deposits indicated that they had formed massive rockslide dams that impounded gigantic rockslide-dammed lakes, and (4) the rubbly deposits (and associated lacustrine sediments) were Holocene rather than Pleistocene (Late Glacial) in age, i.e., were much younger than previously thought. Together these implications have led to a new interpretation of the mountain landscape in the region in which rockslides and their effects are both critical geomorphic processes in landscape evolution (e.g. [50]), and also pose significant hazards to communities and infrastructure with a spatial and temporal frequency that had not been recognised hitherto [43]. The presence of remnant rockslide deposits is also a major factor in the high sediment yield in river basins of the region (e.g. [4, 5]).

These implications will be examined with reference to two examples from the Skardu region (Fig. 7.3).

### 3.2.1 Gol Ghone Rockslides

The Gol Ghone rockslides (35° 17′ 12″N/75° 51′ 34″ E) were first identified by Hewitt [33]. They consist of two massive rock avalanches that occurred on the west side of the Indus River, 28 km upstream of Skardu, and 9 km downstream from the confluence of the Indus and Shyok rivers (Fig. 7.3); their deposits overlap and the older larger deposit is designated Gol Ghone A, the younger (southern) deposit is designated Gol Ghone B [33]. The deposits of the two rock avalanches, which as Hewitt ([33], p. 63) notes were previously interpreted as moraines, filled the Indus River for a distance of 5 km along the thalweg.

The geomorphology of the deposits were examined on satellite imagery and in the SRTM-3 DEM. This examination indicates that the younger Gol Ghone rockslide dam had a minimum dam height of 2,565 m a.s.l., 300 m above the Indus at this point. This is taken as a minimum pool level for the impoundment formed by the rockslide damming of the Indus. At this pool elevation, a massive rockslide-dammed lake would have formed in the Indus and Shyok Rivers (Fig. 7.6) that was 219 $km^2$ in area and 25.7 $Gm^3$ in volume.

The Gol Ghone example exemplifies the massive impoundment volumes implied by the presence of very large remnant rockslide dams in the Indus and other valleys of the Karakoram Himalaya (cf. Chap. 6 by Hewitt, this volume).

### 3.2.2 Satpara Rockslide – Dammed Lake and Satpara Dam

Satpara Lake (35° 14′ 17″N/75° 37′ 51″E), located 6 km south of Skardu (Fig. 7.3), is impounded by a natural debris dam (Fig. 7.7). The natural dam is utilised as the foundation for the Satpara Dam which (as of 2010) is under construction; the 40 m high earthcore rockfill dam will augment the storage capacity of Satpara Lake and will have a gross capacity of 115 $Mm^3$.

When completed the Satpara Project will generate 17.36 MW of hydropower, irrigate 6,287 ha of lands and supply 14 M litres daily of drinking water to Skardu [53].

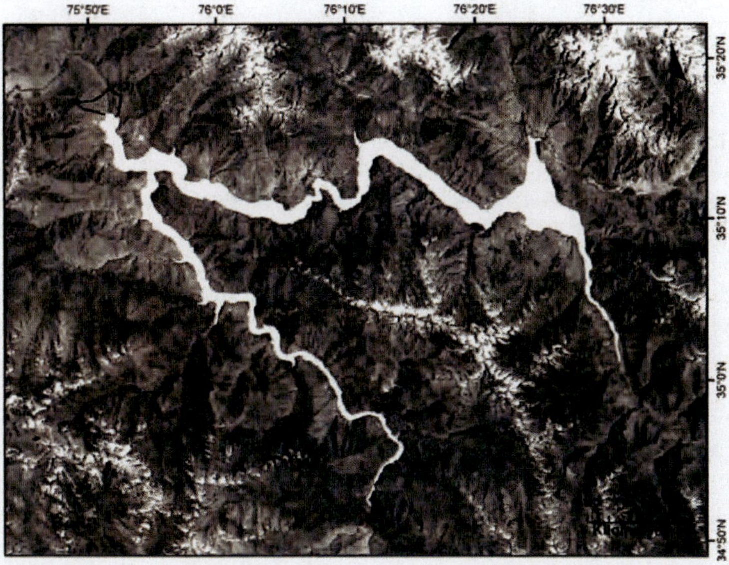


**Fig. 7.6** LANDSAT-7 satellite image showing extent of rockslide-dammed lake (*white fill*) that would have formed in the Indus and Shyok Valleys as a result of damming by the Gol Ghone B rockslide (*black stippled* pattern) at a spillway elevation of 2,565 m a.s.l. Extent of lake based on SRTM-3 DEM. The Gol Ghone rockslides [33] are located in Fig. 7.3

**Fig. 7.7** *Google Earth* image of the site of Satpara Dam (*S*), northern Pakistan, currently under construction (2010). The structure will utilise a prehistoric rockslide dam deposit that impounded Satpara Lake. Source of rockslide is at A. Satpara Lake is located in Fig. 7.3. See text for further discussion

Satpara Lake was identified as a rockslide-dammed lake by Hewitt [33] in interpreting the debris forming the dam as a rockslide deposit. He also identified a clear source area for the debris on the east side of Satpara Valley (Fig. 7 in Hewitt [33]; Fig. 7.7). As reviewed by Hewitt [34], a succession of workers dating back to 1906 had previously identified the Satpara debris dam as being morainic in origin. The lines of evidence that support the interpretation of the Satpara debris as rock avalanche in origin include: (a) a clear source scar on the mountain slope above the debris, (b) run-up of debris on the opposite valley side, (c) the rim-like boundary of the upper limit of debris run-up, (d) the nature of the debris as described from surface exposures by Hewitt [34]. These features strongly suggest a rock avalanche origin for the Satpara debris.

Despite this evidence, other workers have more recently at least partially disputed this interpretation suggesting that the identification of major rockslide deposits in the valleys of the northwest Himalayas is still not a routine geomorphological exercise. The dispute is discussed by Malik et al. [53]. These authors prefer the interpretation of Kuhle [51] who suggests that the debris dam is hybrid in origin, consisting of a mixture of late-glacial moraine deposits and rock avalanche debris.

The characteristics of the debris reported by Malik et al. [53], based on a detailed sub-surface investigation of the natural dam, indicate: (1) that the materials are a mixture of large angular boulders in a finer matrix that consist of sand, silt and clay, (2) that bedrock is over 100 m deep in the middle of the valley, (3) the presence of high permeability zones along the sides and the base of the rock avalanche deposit, (4) that within the debris dam, permeability generally increases towards the surface where it approaches $10^{-3}$ m/s, and (5) that during permeability testing some zones of excessive water losses were encountered indicating the presence of open-work zones within the debris.

These characteristics appear consistent with a geomorphological setting in which rock avalanche debris overlies a valley fill consisting of morainic and alluvial materials which were probably partially incorporated into the landslide debris during emplacement by entrainment.

## *3.3 Historical Rockslide Dam Blockages of the Indus and the Hunza Rivers in 1840, 1858, and 2010*

### 3.3.1 Historical Rockslide Dams I – 1840–1841 Indus

The cataclysm on the Indus resulting from the breach of a rockslide dam in 1841 was first reported by Falconer [22]. Early reports, including Falconer's, describing the circumstances and effects of the formation and failure of the 1840–1841 Indus rockslide dam were summarised in the benchmark paper on great floods of the Indus by Mason [54].

In describing what Mason termed "*The Great Indus Flood*" the following facts emerge: (1) In December 1840 or 1841 a rockslide dammed and blocked the Indus completely, (2) the rockslide, apparently triggered by an earthquake, originated in the Lechar Spur of Nanga Parbat, (3) a rockslide-dammed lake formed and filled; in April 1841 local leaders issued warnings of an impending breach to downstream locations, (4) most probably at the beginning of June 1841, the 300 m high dam was "liberated" and the impounded waters, nearly 65 km in length, were catastrophically released in a period of about 24 h, (5) the outburst flood devastated the Indus valley downstream of the blockage, sweeping away many villages, with "thousands of human beings and cattle"; at Chach, near Attock, a Sikh Army was overwhelmed by the waters of the Indus that suddenly rose 24 m, (6) during the blockage the Indus at Attock was easily fordable; in spite of this, and the warnings received, no notice was taken, and (7) the height of the flood wave at Attock is believed to have been approximately 13 m above the "normal high flood of summer".

We reviewed the original sources used by Mason [54], and other earlier reports, to obtain further details as follows;

(a) *The timing of the blockage*; some uncertainty surrounds the precise timing of the blockage. However, Abbott [1], Montgomerie [55] and Drew [16] remark upon the fact the Indus had been observed from December 1840 to be unusually low at Attock. Drew [16] also reports an interview with a local man that put the date of the blockage in the late autumn of 1840.
(b) *The location and geometry of the 1840–1841 rockslide dam*; as Mason [54] notes, there was much contemporary discussion on the location of the rockslide blockage (e.g. [22, 31]). Based on reports by Becher [8], Montgomerie [55] and Drew [16], Mason [54] concluded that there is "*not the slightest doubt that the dam was directly caused by a great fall of rock from the western spurs of Nanga Parbat*" (p. 17).
(c) *The rockslide-dammed lake*; Mason [54] reports that at its maximum extent the lake was 65 km in length. Drew [16] reports that the lake extended far up valley and rose up to the level of Bawanji (Bunji) Fort, ca. 100 m above the level of adjacent Indus river level. This indicates a pool elevation of 1,335 m a.s.l. Drew [16] also notes that the lake extended up the Gilgit valley to "*where the Hunza stream falls in*". The elevation of this confluence is approximately 1,415 m a.s.l. We suggest that these historical observations bracket the actual maximum pool elevation of the 1841 rockslide dammed lake.
(d) *Timing of the breach*; Drew [16] reports the breach as occurring in the beginning of June, "as near as I can make out". We take this estimate as being the most credible estimate of the timing of the breach suggesting that the rockslide dam may have existed for roughly 170 days.
(e) *Characteristics and effects of outburst flood*; the massive outburst flood devastated communities in the Indus valley [8]. The flood suddenly appeared at Attock, 406 km downstream as an "*overwhelming irrestible wall (a "bore") of discoloured waters*" [8] and swept away at least 500 soldiers of a Sikh army encamped on the flood plain [1]. Witnesses describe a distant sound, similar to that of cannons or thunder before the river appeared as an "absolute wall of

mud" ([1] p. 231). Vast agricultural areas of the Indus Plain were covered with sand deposited by the flood.

(f) *Extent of flood downstream of Attock*; Henderson [31] reports that the 1841 flood was observable at least as far as Kalabagh (157 km downstream of Attock), a settlement on the Indus about 563 km downstream of the 1841 rockslide dam.

With respect to point (b) above, even given this knowledge from earlier work, there has been disagreement in more recent literature on the precise location of the rockslide dam (cf [11, 12, 57, 70, 73]). However, a consensus appears to have emerged (e.g., [41]) on the location of 1840–1841 rockslide dam, as being that initially suggested by Code and Sirhindi [13] and Butler et al. [12]. Shroder [71] describes the rockslide location in detail and we locate it at 35° 30′ 36″N/74° 37′ 22″E on the left bank of the Indus (Fig. 7.8).

As suggested above, historical observations bracket the maximum credible pool elevation as being between 1,335 and 1,415 m a.s.l. We also note that Code and Sirhindi [13] in describing the effects of the Bunji slide give the approximate elevation of the town of Bunji as 1,387 m a.s.l.; this suggests an alternative credible minimum elevation and a range of 1,387–1,415 m a.s.l.

We examined satellite imagery of the site in conjunction with a SRTM-3 DEM, and together with the results of the review of the literature summarised above, we estimate the maximum pool elevation of the 1840–1841 Indus landslide-dammed lake to be a minimum of 1,390 m a.s.l. and a maximum of 1,400 m a.s.l. (cf. Code and Sirhindi [13]). For a first approximation in this paper we take the most probable maximum pool elevation of the 1841 rockslide-dammed lake as 1,390 m a.s.l.

From the SRTM-3 DEM we calculate that an impoundment to the minimum pool elevation (1,390 m a.s.l.) formed a lake (Fig. 7.8) a maximum of 57 km in length, 88 $km^2$ in area with a volume of 6.5 $Gm^3$. Our volume estimate is compared to that of Code and Sirhindi [13] who obtained a volume of 4.6 $Gm^3$ for a maximum pool elevation of 1,387 m a.s.l. and to that of 1.2 $Gm^3$ estimated by Shroder et al. [72].

If our estimates of lake volume and duration are correct, the mean discharge of the Indus River into the lake over the 170 days of impoundment was 442 $m^3/s$ for the most probable maximum pool elevation estimated above. This discharge is far below the mean streamflow estimates for the Indus for the months inferred for the blockage in the vicinity of the rockslide dam reported by Code and Sirhindi ([13], p. 101). In addition, using the monthly mean instantaneous discharge data from Code and Sirhindi [13] we calculate that for the 170 days of impoundment suggested by historical accounts, assuming that the Indus was totally blocked, that a lake of 9.7 $Gm^3$ in volume would have been formed. This clearly is not the case given the geometry of the blocking landslide, the topography of the Indus valley in the area where it is known that the lake existed, and the fact that it is certain that the rockslide-dammed lake did not reach Gilgit and therefore did not exceed 1,415 m a.s.l. in water surface elevation. Both these discrepancies suggest a shorter duration for the existence of the 1841 rockslide dammed lake.

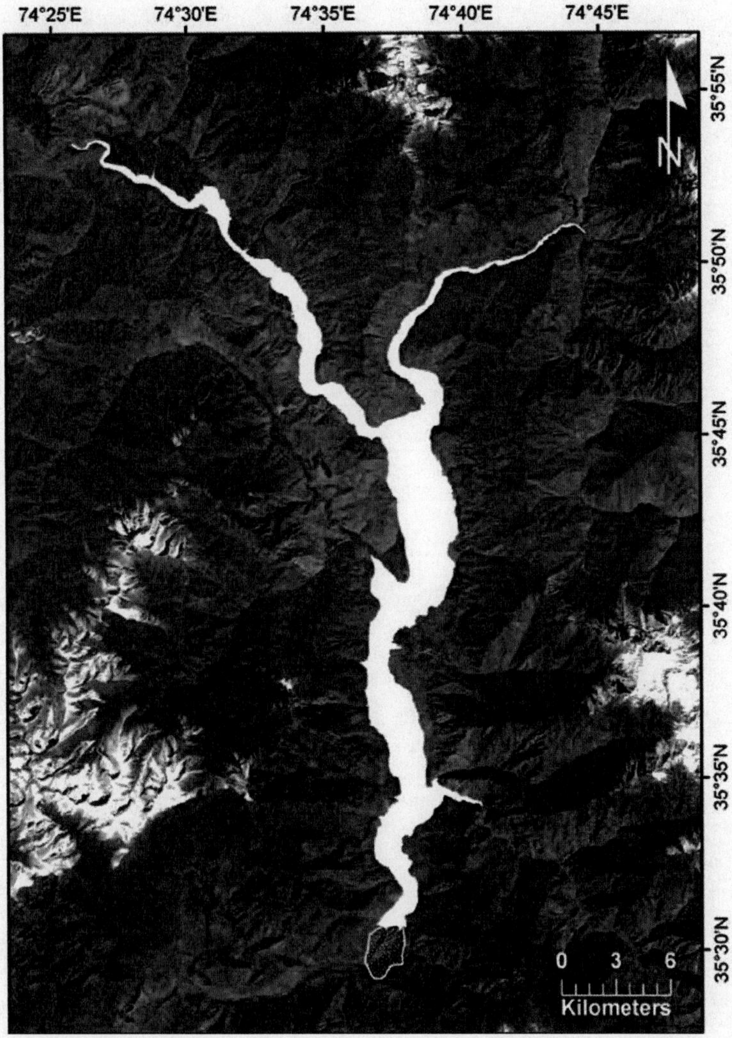


**Fig. 7.8** LANDSAT-7 image showing extent of rockslide-dammed lake (*white fill*) that formed in 1841 in the Indus River, northern Pakistan as a result of a rockslide (*white stippled* pattern), 9 km upstream of Gor. The extent of the rockslide dammed lake (88 $km^2$ in area) is based on a SRTM-3 DEM and assumes a pool elevation of 1,390 m a.s.l. At this elevation the rockslide dammed lake has a calculated volume of 6.5 $Gm^3$. The dam breached in June 1841 and the full impounded volume was catastrophically released causing a massive flood downstream (*The First Great Indus Flood* of Mason [54]). The 1841 event was the largest documented outburst from a rockslide-dammed lake in historical time

If we assume that Drew's account [16] of the date[1] of the outburst flood at Attock is more or less correct, say June 5, then working back, and again using the mean streamflow data from Code and Sirhindi [13] for the preceding months, we obtain a lifetime of about 65 days for the lake and the date of the damming event as around the end of March under conditions of comparatively low flow in the Indus. A hypothetical filling curve for this scenario is illustrated in Fig. 7.9 suggesting that failure

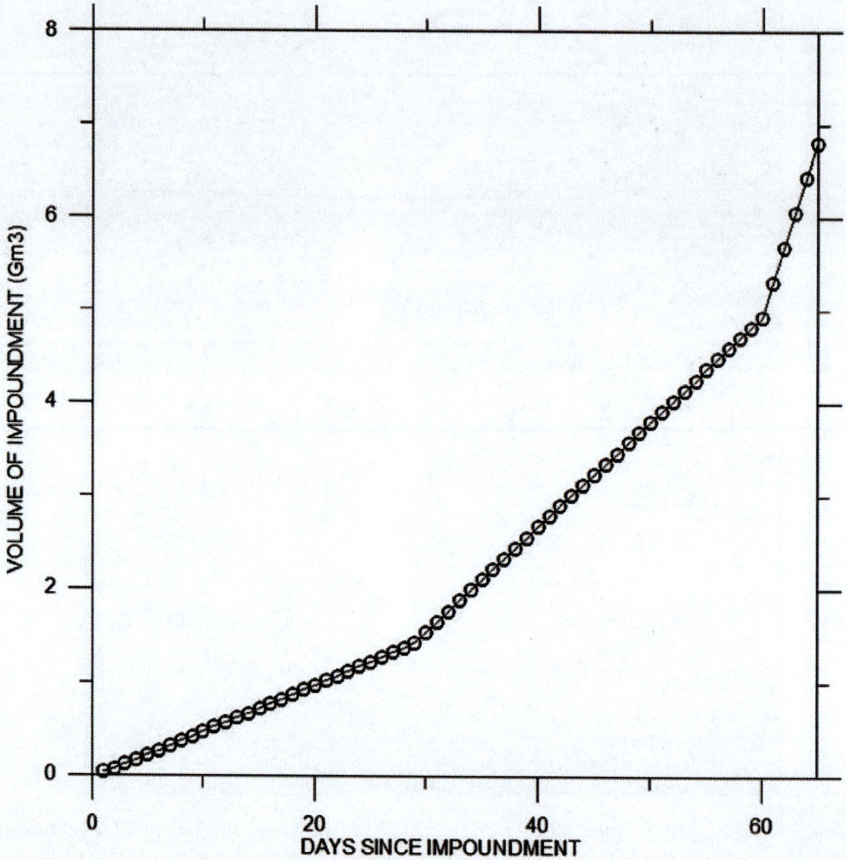


**Fig. 7.9** Hypothetical filling curve for 1841 rockslide-dammed lake on the Indus River, northern Pakistan. The rockslide dam is assumed to have breached on June 5 (based on [16]) with a volume of 6.5 $Gm^3$ at a maximum pool elevation of 1,390 m a.s.l. The filling curve was derived by working backwards from that date using the discharge data of the Indus for a reach just downstream from the rockslide blockage (see Table 1 in Code and Sirhindi [13]). It is assumed that the Indus was totally blocked by the rockslide

[1] We would expect the rockslide dam to have failed as stream flow increased due to snow and ice melting.

took place after a sharp rise in lake level as a result of a summer-related increase in discharge. We note that the first warnings were sent downstream in April of 1841 by local leaders who had observed the filling landslide-dammed lake (e.g. [16, 54]) and that Falconer's original letter, dated 6 July, stated that the rockslide dammed lake had existed for 4 months prior to its writing, suggesting a lake being formed in the previous March. For a lake volume of 6.5 $Gm^3$, and a lifetime of about 65 days, the average discharge into the lake would be about 1,150 $m^3/s$.

Using the relationship developed by Evans [20] a peak discharge of 114,334 $m^3/s$ is calculated for an outburst volume of 6.5 $km^3$. Further, if we assume that the entire impoundment drained off in 24 h, an average breach discharge of 75,230 $m^3/s$ is obtained.

Hewitt [32] has estimated the discharge of the 1841 outburst flood at Attock as 56,640 $m^3/s$. The Indus River has been gauged at Attock since 1868 [48]. Prior to the construction of the Tarbela Dam, the highest rainfall floods measured at Attock were in 1882 (22,087 $m^3/s$) and 1929 (24,352 $m^3/s$) whilst the highest snowmelt floods measured were in 1878 (21,634 $m^3/s$) and 1924 (19,538 $m^3/s$). These discharges compare to those of glacial outburst floods, originating in the Shyok River in 1926 and 1929 (19,482 $m^3/s$). The volume of the 1929 flood obtained from the Attock hydrograph was 1.2 $Gm^3$, corresponding to a maximum increase in stage of 8.7 m [48]. If we assume that the base June discharge at Attock before the Tarbela Dam was built was around 6,000 $m^3/s$, from the descriptions of the flood stage in 1841 we estimate the discharge of the flood at Attock to be roughly 38,000 $m^3/s$.

### 3.3.2 Historical Rockslide Dams II – 1858 Upper Hunza

The documentation of the outburst flood, that was observed at Attock on August 10, 1858 as a result of the breaching of a rockslide dam on the Hunza River is considerable, largely because the experience of the 1841 flood described above was very much on the mind of local people and officers of the British colonial forces present in Pakistan at the time. These early reports are summarised in the paper by Mason [54] noted above.

In describing what he termed the "*Second Great Indus Flood*", the following facts emerge from Mason's review, largely based on Henderson [31] but augmented with important detail on the location of the blockage supplied by a Mr. H. Todd, the Political Agent at Gilgit; (1) the flood resulted from the secondary effect of a rockslide that dammed the Hunza River, a tributary of the Gilgit River, which in turn runs into the Indus at 450 km above Attock, (2) the rockslide (reported as the Ghammesar landslide) originated on the left (south) bank of the Hunza, 11 km above Altit and 1.6 km below Attabad (scene of the 2010 rockslide dam described below). The landslide reportedly resulted from the action of snow and rain in the winter of 1858, (3) a rockslide-dammed lake formed and filled and stretched back as far as Pasu, roughly 30 km upstream and where, according to the local Mir, a line of silt observed on the valley side marked the maximum lake level; local leaders issued warnings of an impending breach to downstream locations, (4) in August, the rockslide dam was breached and the impounded waters were catastrophically released, (5) the outburst flood caused damage in the Hunza and Indus, including the destruction of Nomal Fort, located about 23 km upstream of the Hunza/Gilgit confluence, (6) during the blockage no significant variation of the Indus stage at Attock was observed, (7) the rising of the flood wave at Attock was witnessed by Henderson [31], who was in a rowboat on the river during the period of rising waters; the river began to rise at 6 a.m. on August 10, 1858 and eventually rose to a maximum level at 1:30 p.m. that day, (8) as in 1841 the rising Indus waters flowed into the Kabul River reversing its flow and flooding areas up to 48 km upstream from the confluence.

We reviewed the original sources used by Mason [54], other earlier reports, and more recent literature to clarify details related to these points;

(a) *The timing of the blockage*; as in 1841, some uncertainty surrounds the precise timing of the blockage. However, Henderson [31] relates that the first news of the blockage reached Attock about the middle of July 1858. He refers to a July report that indicates the Hunza had been closed for 3 months, inferring that the blockage took place at the beginning of April, 1858. Dismissing this estimate, Henderson [31] argues for the blockage to have taken place in the middle of May. Becher [8] on the other hand suggests that the rockslide "*dammed up the narrow bed of the river for 6 months*", indicating a rockslide damming event in early February.

(b) *The location and geometry of the 1858 rockslide dam*; as in 1841 there was much contemporary discussion on the location of the rockslide blockage (e.g. [16, 31, 55]). The report by Mr. Todd quoted in Mason [54] above gives almost a precise location, which we estimate as 36° 18′ 16″N, 74° 46′ 54″E (see also [71]). At this location the bed of the Hunza is approximately 2,250 m a.s.l.

(c) *The rockslide-dammed lake*; Reports that the lake reached Pasu are noted by Mason [54]. Thus the rockslide dam would have had a spillway elevation, and a maximum pool elevation, of at least 2,450 m a.s.l., 200 m above the bed of the Hunza at the site of damming rockslide. At this pool elevation the lake would extend roughly 30 km upstream. Based on the observations of Shroder [71], an examination of satellite imagery and the SRTM-3 DEM, we estimate the maximum pool elevation of the 1858 Hunza landslide-dammed lake to be 2,450 m a.s.l. From the SRTM-3 DEM we calculate that this pool elevation formed a lake (Fig. 7.10), 28 km long, 15.6 $km^2$ in area with a volume of approximately 805 $Mm^3$ (Fig. 7.10).

(d) *Timing of the breach and characteristics and effects of outburst flood*; the outburst flood reached Attock at around 6 a.m. on August 10, 1858. Montgomerie [55] suggests that the flood passed Bunji, 118 km downstream of the rockslide dam, on August 9, at 9 or 10 a.m. This indicates an average velocity to Attock, 426 km downstream of roughly 5–6 m/s and the time of the breach as being early on August 9, 1858. The 1858 outburst flood destroyed many forts and villages and extensive areas of agricultural lands in the Hunza and Indus valley [8] but no major loss of life occurred. This was due to warnings given and the fact that the events of 1841 were still fresh in the memory of the local people [8, 31].

**Fig. 7.10** LANDSAT-7 image showing extent of 1858 Hunza rockslide-dammed lake (*white fill*) impounded behind rockslide (*white stipple*). Lake extent is based on a maximum pool elevation of 2,450 m a.s.l. which was established in a SRTM-3 DEM. Volume of lake at this maximum pool elevation is calculated to be 805 $Mm^3$

(e) *Observations at Attock*; several writers published observations on the 1858 flood at Attock in the Journal of the Asiatic Society of Bengal. The behavior of the rising water was quite different than in 1841; instead of the arriving flood waters approximating a bore or a wall of mud and water (as in 1841) the 1858 flood was more like a rising wave that rose, peaked and passed. When the flood wave arrived the river was lower than normal; Henderson [31] notes that "*at first it came welling up quite rapidly, not less than 1 ft/min*". Beginning at 6 am it rose to its peak of 17 m above the ordinary July-August flood level [31]. Henderson [31] mentions substantial deposition of sediment in the bed of the Indus as reported by boatmen.
(f) *Damage downstream of Attock*; Henderson [31] reports 100 houses were demolished by the flood at Mokkad (Makhad), about 130 km downstream of Attock, where the river reached 3.5 m above annual flood level. Damage was also recorded at Marree (Mari) (154 km downstream; river level 2.5 m above the annual flood level) and at Kalabagh (157 km downstream of Attock), a settlement about 700 km downstream from the 1858 rockslide dam. According to Henderson [31] Kalabagh was also the limit of the 1841 flood, ca. 563 km downstream of the 1841 rockslide dam.

As noted above, we estimate the volume of the 1858 lake to be in the order of 805 $Mm^3$. If we assume the Hunza was blocked completely during the impoundment, for blockages of 6, 4, and 3 months this impounded volume would be created by average discharges in the Hunza of 52, 78, and 106 $m^3/s$ respectively. These discharges are significantly less than the values of the mean monthly instantaneous discharge of the Hunza for the possible months of the blockage reported by Archer [6] at the Dainyor Bridge, 98 km downstream from the 1858 rockslide dam.

Using the relationship developed by Evans [20] a peak breach discharge of 37,792 $m^3/s$ is calculated for an outburst volume of 805 $Mm^3$ in the 1858 Hunza event.

### 3.3.3 Historical Rockslide Dams III – 2010 Upper Hunza at Attabad

A second major rockslide (the Attabad rockslide) blocked the Hunza River [47] at about 13:00 h on January 4, 2010 at a location (34°18′ 24″N/74° 49′ 17″E) (Figs. 7.1 and 7.11) only 2.5 km upstream of the 1858 blockage described above. In 2010 the rockslide originated on the north side of the Hunza valley and the volume of the debris is estimated to be in the order of 55 $Mm^3$. The rockslide mobilised valley fill sediments from the valley floor of the Hunza that liquefied and travelled as a rapid mudflow over the valley floor up to a distance of 2.7 km downstream. Part of the mudflow debris swept up the valley side overwhelming part of a village downstream of Attabad, killing about 20 people (Fig. 7.1). The rockslide blocked the Hunza River and a rockslide-dammed lake (Lake Gojal) immediately began to form upstream. The elevation of the Hunza River at the blockage site is estimated from our SRTM-3 DEM at 2,325 m a.s.l. The minimum height of the dam is estimated as ca. 125 m for an overflow crest elevation of about 2,450 m.a.s.l.

**Fig. 7.11** Aerial view of Attabad rockslide (est. vol. 55 $Mm^3$) that blocked the Hunza River, northern Pakistan on January 4, 2010. Rockslide slid from right. Rockslide-dammed lake is filling behind the debris on January 17 (Day 13 of impoundment). View is downstream. [*Pamir Times* photograph]

As the waters of the lake rose they flooded several villages, large tracts of agricultural land adjacent to the Hunza, and 22 km of the Karakoram Highway linking Pakistan and China, disrupting road travel and commercial transport between the two countries (Fig. 7.12).

To achieve an eventual controlled overflow, Pakistan authorities began the excavation of a spillway over the debris on 29 January (Day 25 of impoundment) completing it around May 15 (Day 128). The excavation of the spillway lowered the effective crest of the dam by around 15 m, equivalent to a maximum possible pool height of 2,435 m a.s.l.

We remotely tracked and quantified the filling of the 2010 Hunza rockslide-dammed lake by utilising (1) SRTM-3 digital terrain data, (2) multi-temporal satellite imagery, and (3) data published on the websites of the *Pamir Times* (http://pamirtimes.net/) and the National Disaster Management Authority (NDMA) of Pakistan. As noted above, from the SRTM data we took the elevation of the valley floor at the site of the rockslide to be 2,325 m a.s.l. Taking this elevation as the local

**Fig. 7.12** Aerial view of 2010 Hunza rockslide-dammed lake in northern Pakistan. At the time of the photograph (May 13, 2010; Day 127 of impoundment) lake waters are still rising (pool el. 2,423 m a.s.l.), filling the Hunza Valley. View is upstream. Lake would rise another 12 m before overtopping on May 29, 2010 (*Pamir Times* photograph)

datum, i.e., a lake depth of 0 m, a filling curve for Lake Gojal was plotted (Fig. 7.13) based on the depth of the lake reported by the NDMA on their website from January 31 (Day 27). To obtain the pool elevation in metres above sea level (Fig. 7.13) we added the reported depth of the lake to the elevation of the local datum as given above. We also tracked the development of the lake through multi-temporal satellite imagery obtained on 16 March (EO-1; Fig. 7.1), 5, 22, and 25 May and 1 June (All ASTER) (Fig. 7.14) as summarised in [47].

No large scale topographic maps were available to us. In addition to the calibration of the SRTM-3 DEM based on runway elevations mentioned in Section 2 above, we further checked the accuracy of the DEM by referring to news reports that indicated that the waters of the landslide dammed lake reached the piers of the Karakoram Highway Bridge across the Hunza River located at Gulmit, 10.5 km upstream of the rockslide dam on February 10 (Day 37). The elevation of the bridge piers is estimated at 2,375 m a.s.l. which is within 1 m of the estimate of the lake level on that day from the NDMA data; the SRTM-3 DEM indicates that the lake volume on Day 37 was 66.55 $Mm^3$ at the pool elevation of 2,375 m a.s.l.

During filling, seepage began to appear in the downstream face on approximately March 5 (Day 60) at a pool elevation of ca. 2,387 m a.s.l. Initial estimates of seepage were made on March 9 (Day 64) and approximated 18,348 $m^3$/day. This seepage had increased to 40,367 $m^3$/day by March 17 (Day 72). Seepage then appeared at various locations in the downstream face and discharge through the debris had

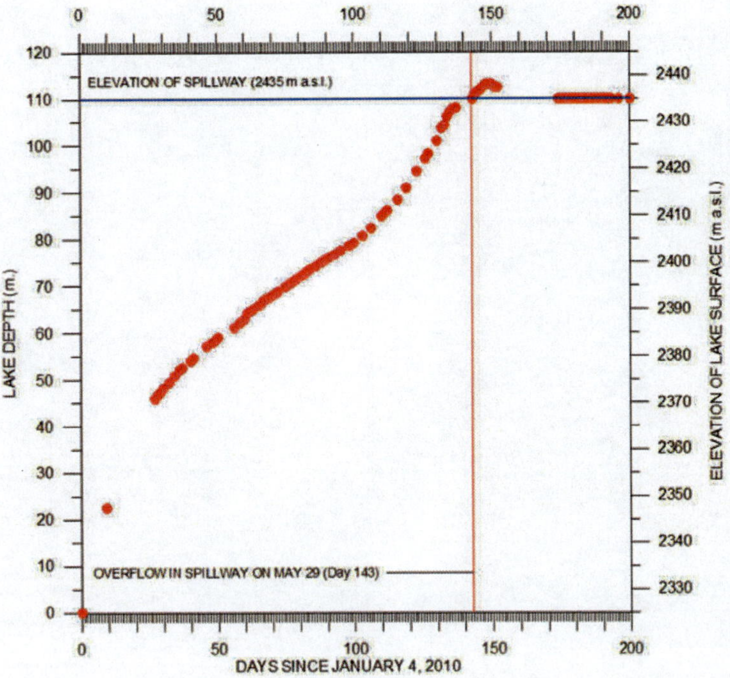


**Fig. 7.13** Filling curve of 2010 Hunza rockslide-dammed lake, northern Pakistan. Lake began to form on January 4, 2010 as a result of the river blocking Attabad rockslide and overtopped through an excavated spillway on May 29 (Day 143). As of July 20, 2010, stable overflow continues as Pakistan authorties consider other drainage/partial drainage strategies (Data derived from the National Disaster Management Authority (NDMA) Pakistan)

notably increased. This resulted in the occurrence of internal erosion within the rockslide debris and the formation of sink holes on the surface of the debris. Despite this evidence of substantial seepage forces the integrity of the rockslide remained intact and the lake level continued to rise until overtopping on May 29, through the excavated spillway.

The elevation at overtopping on 29 May (Day 143) was estimated from the SRTM DEM to be 2,435 m a.s.l. At this maximum pool elevation the lake area was 9.74 $km^2$ and its volume was 430 $Mm^3$. To verify our maximum pool elevation we superimposed the 2,435 m contour from the SRTM DEM on the ASTER June 1 image; the contour and the lake outline corresponded almost exactly.

After overtopping the lake level continued to rise a further 3 m (an equivalent additional volume of ca. 30 $Mm^3$) in the five days up to June 3, since lake inflow exceeded lake outflow (Fig. 7.13). By 5 June, however, the lake level had stabilised (Fig. 7.13) as discharge in the spillway increased, and stable overtopping continued (Fig. 7.15). Photographs published by the *Pamir Times* after overtopping show the enlargement of the spillway, first by deepening through headward erosion from the lowest part of the spillway and then by widening through lateral undercutting and seepage erosion of the spillway walls.

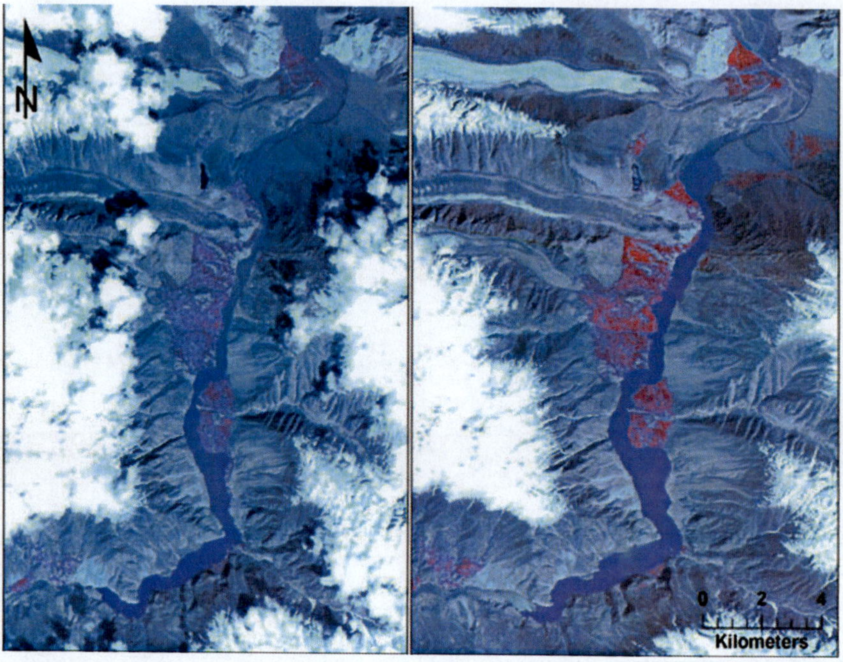


**Fig. 7.14** ASTER satellite images showing filling of 2010 Hunza rockslide-dammed lake between May 5 (*left*) and May 25 (*right*). Lake level on May 5 is 2,416 m a.s.l. and May 25 is 2,433 m a.s.l. The lake overtopped the rockslide dam on May 29 at a spillway elevation of 2,435 m a.s.l

Pakistan authorities have been active in mitigating the hazard at the rockslide dam and the risk to downstream communities. In addition to excavating the spillway over the debris to effect a controlled overtopping and to reduce the maximum volume of the lake, they have also been diligent in monitoring the lake, delimiting possible flood zones downstream, evacuating threatened areas, and installing a siren-based warning system in the event of a catastrophic breach being initiated.

As of July 25, 2010 (Day 200 of impoundment) the Hunza rockslide-dammed lake continued its stable overflow of the spillway over the debris. Lake level has been more or less maintained at around 2,435 m a.s.l. as Pakistan authorities consider engineering strategies to lower the level of the lake thus alleviating the upstream flooding of large tracts of the Hunza Valley and the Karakoram Highway (see note added in proof at the end of the Chapter for update to February 11, 2011).

## 4 Rockslide Dams in the Pamir Mountains (Tajikistan and Afghanistan)

Rockslide-dammed lakes are widespread throughout the Pamirs of Tajikistan and Afghanistan (e.g. [44]). They include Yashilkul Lake, Lake Sarez, and Shiwa Lake. In addition the remnants of large rockslide debris masses that previously blocked

**Fig. 7.15** View of overflow of 2010 Attabad rockslide dam and the Hunza rockslide-dammed lake in the background. Date of photograph is June 10, 2010 12 days after overtopping on May 29 (143 days after impoundment). Volume of lake is ca. 430 $Mm^3$. Note limited erosion of spillway. Photograph by Focus Humanitarian Assistance, Pakistan

surface drainage, comparable in geomorphic character and scale to those examples noted in the Himalayas of India and Pakistan, are evident on satellite imagery at many locations in the Pamir Mountains of Tajikistan and Afghanistan. These include the gigantic Pasor rock avalanche, Tajikistan (38° 23′ 50″N/73° 34′ 33″E) that has previously blocked the upper Bartang River (Fig. 7.16), only 12 km north of the 1911 Usoi rockslide described below. However, no detailed field survey has been performed in the region.

## *4.1 Rockslide Dams in the Gunt River Basin, Tajikistan*

The Gunt River Basin drains the south west portion of Tajikistan and joins the Panj River at Khorog, just east of the Afghanistan border. A number of rockslide dams are noted in the Gunt River Basin including the giant Bedurth rockslide (37° 31′ 27″N/71° 36′ 00″E) that blocked the Gunt River in prehistoric time before being breached. The deposits of the rockslide are visible on satellite imagery, 7 km upstream from the city of Khorog. In addition several present-day lakes are dammed by substantial rockslide deposits within the Gunt River basin, for example Lake Rivakul (37° 36′ 59″N/72° 03′48″E).

**Fig. 7.16** *Google Earth* image of the gigantic Pasor rockslide (*A*) that has blocked the Bartang River (flowing from *right* to *left*) in the Pamirs of Tajikistan at 38° 23′ 03″N/72° 34′ 49″E. The rockslide debris extends 2.72 km across the valley floor and has a volume in excess of 500 $Mm^3$

The largest rockslide-dammed lake in the Gunt River basin is Lake Yashilkul, located at 37° 47′ 07″N/72° 44′ 47″E in southwestern Tajikistan; the lake was formed when a rock avalanche from the steep southern slope of the Gunt River valley blocked surface drainage to form the 20 km-long rockslide dammed lake (Fig. 7.17).

The rock avalanche has been studied by Zech et al. [81] who used cosmogenic surface age dating to determine that the landslide occurred ca. 4,400 yr. In recent years an engineered retention structure and spillway have been constructed over the rockslide debris to regulate the lake level and winter outflow of Yashilkul Lake to, in turn, regulate flow in the Gunt River for downstream power generation at the 28 MW run-of-the-river Pamir 1 plant, at Khorog (ca. 130 km downstream), during periods of high winter electricity demand.

## *4.2 Shiwa Lake (Afghanistan)*

In this study we identified Shiwa Lake, located in the Pamirs of northeast Afghanistan (37° 23′ 37″N/71° 22′ 42″E), 20 km southwest of Khorog, as a rockslide-dammed lake (Fig. 7.18). The lake has a maximum length of 9.75 km and is 14 $km^2$ in area. The damming rockslide has an estimated volume of 800 $Mm^3$ and is of unknown age. Shiwa Lake existed as early as 1921 since its presence was recorded by maps produced by the Surveyor-General of India. The rockslide dam is

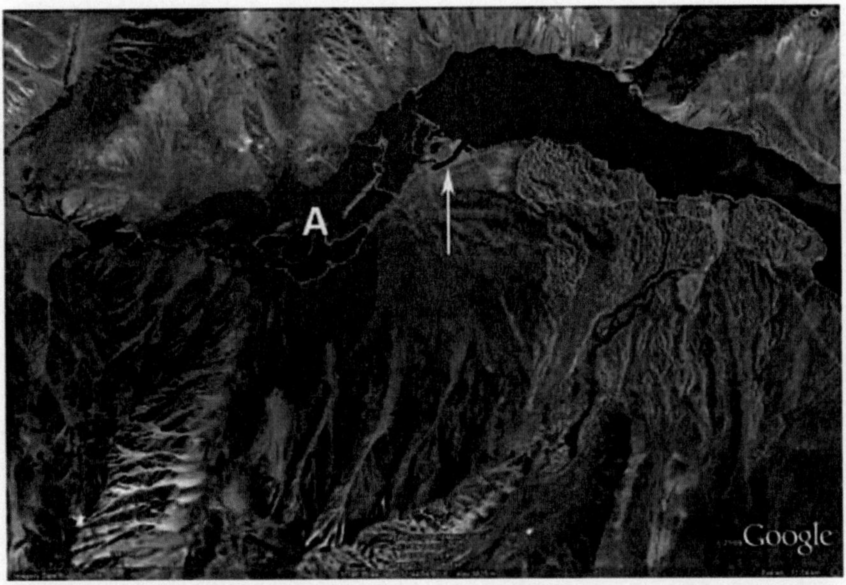


**Fig. 7.17** *Google Earth* image of Lake Yahilkul in the Pamir Mountains of Tajikistan (37° 47′ 37″N/72°45′06″). The lake is formed by rock avalanche debris (A) which has been dated to ca. 4,400 yr. [81]. Note spillway (*white arrow*) that forms part of civil engineering works to regulate lake level and outflow in Lake Yashilkul

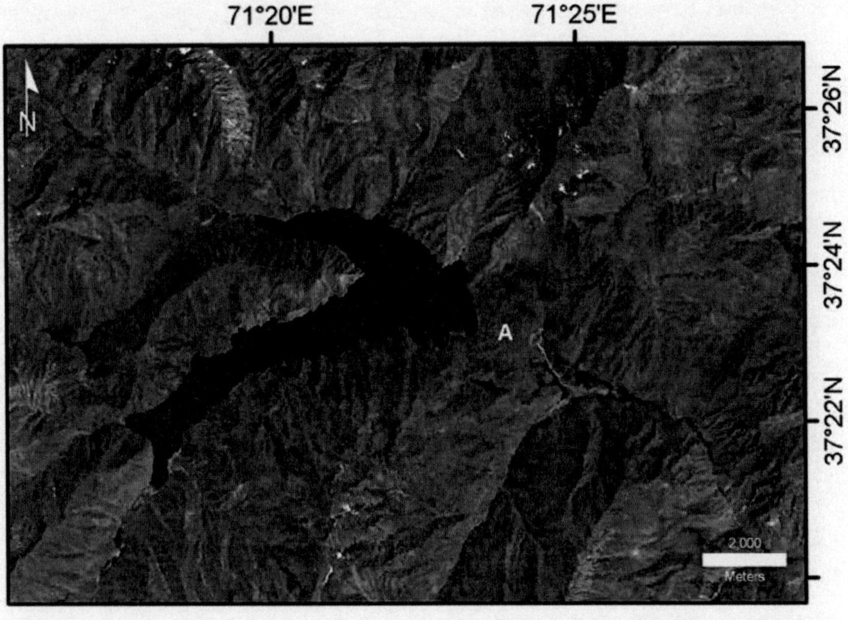


**Fig. 7.18** LANDSAT-7 image of Lake Shiwa, western Pamir Mountains, northeastern Afghanistan (37° 23′ 37″N/71° 22′ 42″E). The valley blocking rockslide (A), of unknown age, is approximately 800 $Mm^3$ in volume and the impounded Lake Shiwa has a surface area of 14 $km^2$

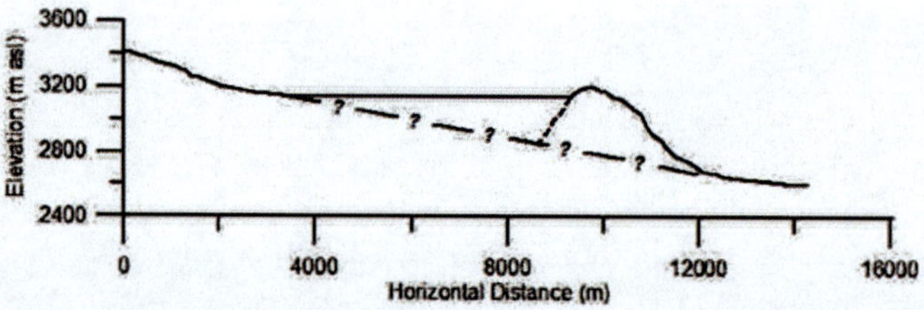


**Fig. 7.19** Longitudinal profile (vertical exaggeration = 4H) of rockslide-dammed Shiwa Lake, Afghanistan, based on SRTM-3 topographic data. Pool elevation of lake is 3,142 m a.s.l. Profile is through the southern arm of the lake (see Fig. 7.18)

400 m high at its mid point (Fig. 7.19) and has a freeboard of about 80 m above the current pool height (ca. 3,142 m a.s.l.). Lake Shiwa (Shewa) was recently described by Shroder and Weihs [74].

## 4.3 1911 Usoi Rockslide and Lake Sarez (Tajikistan)

Lake Sarez was formed when the Bartang (Murgab) River was blocked by the gigantic Usoi rockslide (est. vol. 2 $Gm^3$) triggered by the 18 February 1911 Pamir Earthquake [62, 75–77]. Although much has been published on the rockslide and rockslide-dammed Lake Sarez in the Russian language (see Stucky [79]), only summary reports have been published in English (e.g. [27, 68]). The stability of the Usoi Dam and the potential outburst hazard that it represents has become of global scientific interest [78].

### 4.3.1 18 February 1911 Pamir Earthquake

Despite the fact that the Pamir Earthquake was recorded by seismographs around the world, some uncertainty still surrounds the location and magnitude of the earthquake. As discussed by Oldham [56], Richter [65] and Weichert et al. [80] some initial reports tried to demonstrate that the enormous rockslide had generated the Pamir earthquake [26, 45, 49]. An analysis of seismograms by Galitzin [26] and Klotz [49] showed the epicentre of the 1911 earthquake to be in close proximity to the landslide and it was thus argued that the gigantic landslide had generated the Pamir earthquake. In this argument, the epicentre of the earthquake was considered to be coincident with the location of the rockslide given by Shpilko [75] as 38° 16′N/72° 34′E. The idea that the Pamir earthquake had in fact been caused by the Usoi rockslide was first challenged by Oldham [56] followed by Macelwane [52]. By 1937, Jeffreys [46] had reversed his opinion on the landslide origin of the February 1911 Pamir Earthquake.

In a summary of Soviet earthquake catalogues, Rautian and Leith [64] give the epicentre of the 1911 Pamir earthquake as 38.2°N/72.8°E, at a distance of 19 km southeast from the rockslide dam (Fig. 7.20), in the vicinity of the former village of

**Fig. 7.20** LANDSAT-7 ETM+ image of Lake Sarez (*grey* tone) formed behind the blocking Usoi rockslide (diagonal pattern) emplaced during the 1911 Pamir earthquake. Epicentre (taken from Rautian and Leith [64]), of the earthquake shown as black dot

Sarez [56]. The focal depth is given as 26±5.2 km. Also in the Soviet catalogues [64], the magnitude of the Pamir Earthquake ($M_S$) is given as 7.0±0.3; this compares to recalculated values of $M_s$ 7.3 by Abe and Noguchi [2] and 7.2 by Pacheco and Sykes [60]. We note that the epicentre reported in Gutenberg and Richter [30], Abe and Noguchi [2] and Pacheco and Sykes [60] is incorrect since the epicentre reported in these publications (40.0° N, 73.0° E) is 195 km north of Usoi in Kyrgyzstan.

The 1911 earthquake shook a large area of the Pamirs and other landslides occurred in the epicentral region [56, 76, 77].

### 4.3.2 Usoi Rockslide

The geology of the source area is complex involving NNW steeply-dipping structurally-disturbed Paleozoic (Sarez Formation) and Mesozoic sediments cut by two major geological structures, the Usoi Thrust that steeply dips to the SE, and a second shear zone steeply dipping to the NW ([21, 44], Chap. 16 by Ischuk, this volume). Although the relationship between the geological structure of the source rock mass and the movement mechanism involved in the Usoi rockslide has not been ascertained in detail, the direction and morphology of movement is suggestive of a large scale-wedge failure. The wedge-failure generated a giant rockslide (Fig. 7.21) in which the debris underwent a limited vertical displacement (750 m – [62]) and in which large masses of the source rock mass remained intact during movement [21].

The topography of the rockslide was instrumentally surveyed in August 1915 by Preobrajensky [62] who also reconstructed the pre-rockslide topography of the site with the help of local Tajiks familiar with the pre-rockslide terrain. Using this topographic data he estimated the source volume of the rockslide as 2 $Gm^3$ and the

**Fig. 7.21** Views of 1911 Usoi rockslide dam. A – View of Usoi rockslide debris filling the Murgab (Bartang) valley damming the flow of the river. Note the waters of Lake Sarez filling the impoundment at lower right of photograph, and Lake Shadau filling at lower left. The photograph is reproduced from Preobrajensky [62] and was taken in August 1915 when lake levels were ca. 3,100 m a.s.l. B – Photograph by A. Ischuk taken from a similar location in 2002 when lake levels were ca. 3,260 m a.s.l. In both photographs the source scar of the Usoi rockslide is seen at right

debris volume as 2.2 $Gm^3$, which as discussed by Preobrajensky [62], is equivalent to a 20% bulking factor due to fragmentation.

The rockslide debris fills the Murgab (Bartang) valley, covers an area of ca. 16 $km^2$ and forms the Usoi rockslide dam (Figs. 7.20, 7.21). The dam has a length along the thalweg of 7 km [9], a maximum height of 740 m (3,490 m a.s.l.) above the Bartang river bed and a lowest crest elevation of 3,310 m a.s.l. [63, 79], 560 m above the base of the dam (2,750 m a.s.l.) at its upstream limit.

### 4.3.3 The Filling of Lake Sarez

Preobrajensky [62] summarises the filling of Lake Sarez up to August 1915 when he carried out his survey of the rockslide debris (Fig. 7.21A). He reports the relative

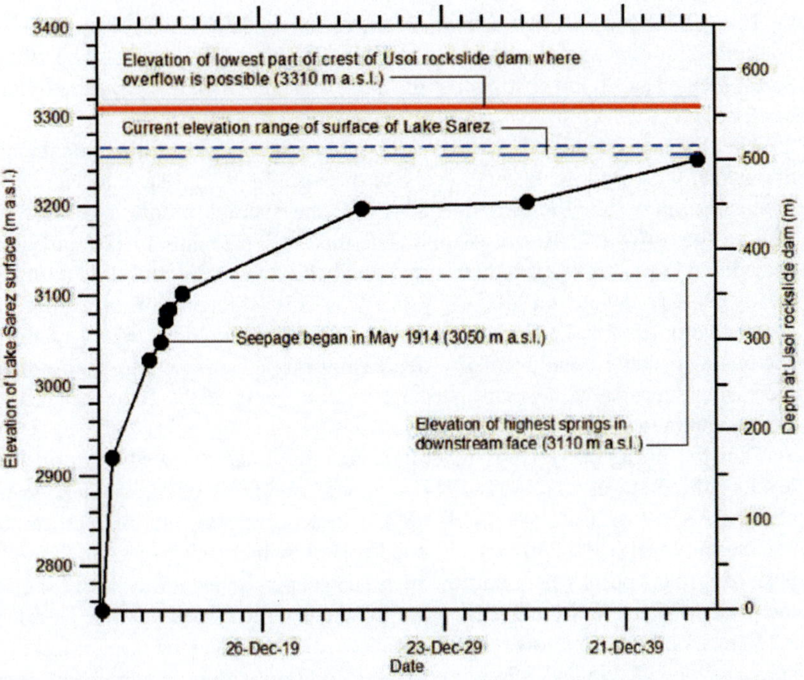


**Fig. 7.22** Filling curve for Lake Sarez from February 1911 to early 1944. Current elevation range of the lake surface is ca. 3,255 to 3,265 m.a.s.l. resulting in 55–45 m freeboard at the Usoi rockslide dam. Data summarised from Preobrajensky [62] and Stucky [79]

lake level elevations in sazhens, an old Russian unit of length equivalent to 7 ft or 2.1336 m [29]. We converted his measurements to metres to generate the initial part of the filling curve in Fig. 7.22 based on a datum elevation of 2,750 m a.s.l. The datum elevation is an estimate of the intersection of the rockslide debris surface with the pre-slide thalweg of the Bartang River based on a comparison of Preobranjenky's maps,[2] information contained in Stucky [79] and the Tajikistan 1:50,000 scale topographic map of Usoi/Lake Sarez.

The filling curve (Fig. 7.22) indicates rapid initial filling up to ca. 3,102 m a.s.l. in August 1915 [28]. According to Shpilko [75] and Preobrajensky [62] it appears that the Usoi rockslide initially blocked the Bartang River completely. Seepage began to appear in the downstream face of the rockslide dam only in the Spring of 1914 [62] when the lake level reached el. 3,050 m a.s.l. (Fig. 7.22), taken as the maximum lake level with zero seepage (cf. [28]). In the autumn of 1914 water appeared in

[2] Preobrajenky's maps had an unknown error in the absolute elevations used in his survey. This was due to a malfunction of barometers at the site and at the control location at Pamir Post. By comparison of his maps with current topography in GIS, we estimated this error in absolute elevation to be – 25 m (or about 10 sazhens). Thus it is necessary to add 25 m to the absolute elevations reported by Preobrajenksy.

the Murgab River again, downstream of the rockslide dam (Chap. 16 by Ischuk, this volume). This corresponded to a lake elevation of about 3,080 m a.s.l. The village of Sarez, located ca. 20 km upstream of the dam, was inundated in the summer-October of 1912.

After the lake reached el. 3,102 m a.s.l. in August 1915, the filling rate decreased dramatically as outflow, through seepage through the dam, began to increase and approach inflow. For the period 1940–1991, the average inflow into Lake Sarez was ca. 45 $m^3/s$ [9]; if we assume that this was the case in the early filling period, 6.55 $Gm^3$ of water had accumulated behind the Usoi rockslide dam before steady seepage developed after August 1915. This is about 38% of the estimated current volume of Lake Sarez. The remainder of the volume (62%) of the lake accumulated under conditions of a decreasing rate of surface rise conditioned by increasing seepage as the more permeable upper parts of the Usoi rockslide dam were encountered.

Thus the filling curve (Fig. 7.22) indicates that Lake Sarez filled to its present level in three stages (1) 1911–1914 – under conditions of total blockage up to about el. 3,050 m a.s.l., (2) 1914–1944 – in a decreasing rate of increase to its pre-present level (ca. 3,250 m a.s.l.) conditioned by increasing seepage through the Usoi dam to the point where outflow by seepage approached inflow, and (3) 1944 to the present where very gradual increase in lake level to a mean surface elevation of 3,260 m a.s.l. Today outflow roughly balances inflow [9] and the level of Lake Sarez varies between 3,255 and 3,265 m a.s.l. (Fig. 7.22), resulting in a freeboard ranging between 45 and 55 m. The highest ever lake level (3,270 m a.s.l.) was reached in 1994 with a minimum freeboard of 40 m. According to Stone [78] this was superseded in 2008 when the freeboard reached 38 m due to unprecedented high lake levels (3,272 m a.s.l.). The area of Lake Sarez as determined from Fig. 7.20 is 87 $km^2$ at a lake level of 3,264 m a.s.l.

The long term filling and the present change in the level of Lake Sarez is a function of the water balance of the lake, broadly inflow and outflow. Since there is no surface outflow over the debris dam all outflow occurs through the debris in subsurface pathways as seepage. Seepage is dependent on the permeability of the rockslide debris which may vary (1) with time (e.g., consolidation of debris, filling of voids through sedimentation, enlarging of voids by seepage erosion) and (2) in the vertical direction from the base of the debris to its surface reflecting the structure of the rockslide debris. With respect to (1) it has been suggested that consolidation of the Usoi debris has resulted in a decrease in permeability, and thus a decrease in seepage, over time (e.g. [28]). With respect to (2) the filling of the lake to the elevation without observed seepage (3,050 m a.s.l.) is an indication of an impermeable zone within the rockslide dam up to that point. Calculations by Gladkov et al. [28], using the observed filling rate of Lake Sarez, indicate a marked increase in permeability above el. 3,140 m a.s.l. to a maximum in the surface carapace of the rockslide dam. It is noted that the highest spring in the downstream face of the Usoi dam is at el. 3,110 m a.s.l. (Fig. 7.22).

Data indicates that between 1949 and 1999 the level of Lake Sarez rose at about 18.5 cm/year, a rate which continues today. The small difference between outflow and inflow in which 98.7% of the inflow is converted into outflow apparently contributes to the increase in lake level [79]. The role of a possible long term increase in inflow (e.g., by melting glaciers in the Sarez watershed) or a long term decrease in the permeability of the Usoi barrier (e.g., consolidation or so-called "mud-clogging" of voids in the debris due to redistribution of finer material) remains uncertain [79].

### 4.3.4 Lake Sarez Hazard Assessment

As summarised in [79] there has been substantial and lengthy investigation of the possible outburst hazard at Lake Sarez and the effects of an outburst flood in the Murgab River downstream. This dates back to the original Russian scientific expeditions of Shpilko in 1913 and Preobrajensky in 1915. Concern has centred on the possibility of some form of catastrophoic breaching resulting in the release of all or part of the impounded lake waters.

The hazard assessment has examined the possible modes of failure of the Usoi rockslide dam in five scenarios ([68, 79]; Chap. 16 by Ischuk, this volume). These consist of (1) slope instability in the downstream and upstream faces under static and/or dynamic conditions, (2) the potential of rockslide dam failure by piping (internal erosion), (3) breach induced by retrogression of present surface erosion in the downstream face into the body of the dam, (4) breaching caused by overtopping involving retrogressive erosion of an overflow channel when the lake level exceeds 3,310 m a.s.l., (5) breaching initiated by dynamic overtopping by a displacement wave generated by a landside into Lake Sarez from its shoreline with the lake at its approximate current level.

With respect to (1) a number of studies have shown that the stability of the upstream and downstream faces is high, due in part to their gentle slopes and the high shear strength of the debris. In (2) the analysis of piping potential indicates that the high hydraulic gradients necessary for piping do not exist in the Usoi Dam. In (3), the retrogression of surface erosion at the head of the so-called "canyon" controlled by springs at about 3,110 m a.s.l. in the downstream face is also considered unlikely since the headward progression of the canyon would have to penetrate the blocky carapace of the debris on a low geomorphic gradient [79]. With respect to (4) it is considered to be very unlikely that Lake Sarez will overflow at el., 3,310 m a.s.l. since this would require an additional 3.2 $Gm^3$ of water to be added to the lake without outflow taking place [79]. In the fifth scenario, the possible breaching initiated by an overtopping displacement wave generated by a large rapid rockslide from the shoreline of Lake Sarez has been investigated in detail. In particular, the stability of the Right Bank Landslide has been analysed and the slopes have been monitored [17, 18, 66]. At the site (circle in Fig. 7.23) rocks of the Sarez Formation dip into the steep slope and evidence of slope movement observed is probably due to toppling with an unknown catastrophic potential.

The hazard assessment has led to the establishment of a monitoring and early warning system for Lake Sarez which became operational in 2004 [17, 18]. The early warning system would provide leadtime for evacuation of downstream

**Fig. 7.23** Usoi rockslide dam (A), Lake Sarez (B), and Lake Shadau (C) in an oblique photograph taken from the International Space Station in September 2001. The Right Bank landslide discussed in text is circled (NASA Photograph ISS002-E-7771)

communities in the event of a flood wave being produced by a breach of the Usoi rockslide dam.

Finally, recent reports indicate a developing focus on the possibility of reducing the hazard at Lake Sarez by drawing down the lake level [78] by the excavation of a drainage tunnel in the left abutment of the Usoi rockslide dam. This drawdown could be up to 100 m [79], which would have the effect of increasing the freeboard of the rockslide dam to as much as 140 m.

## 5 Conclusion

The remains of rockslide dams are widespread in the river valleys of the northwest Himalayas (Pakistan and India) and the adjacent Pamir Mountains of Afghanistan and Tajikistan, Central Asia. The region contains in excess of two hundred known rockslide deposits of, as yet, unknown age that have interrupted surface drainage and previously dammed major rivers in the region in prehistoric time. Most of these dams have been breached and only remnant debris, and in some cases lacustrine deposits from formerly impounded lakes, are visible today in the deep valleys of the region. In addition, the region contains (1) the highest rockslide dam (the 1911 Usoi rockslide, Tajikistan) in the world that dams the largest rockslide-dammed lake (Lake Sarez) on Earth (est. vol. 17 $Gm^3$), (2) the largest documented outburst flood (6.5 $Gm^3$) associated with a historical rockslide dam outburst (the 1841 Indus

Flood, Pakistan), and (3) the world's most recent (2010) rockslide-dammed lake emergency, the Attabad rockslide dam on the Hunza River, in the Upper Indus basin of northern Pakistan. In the Karakoram Himalaya mapping of remnant rockslide deposits by K. Hewitt indicates a density of rockslides of about 50/10,000 $km^2$. This density is approximately 5 times that of the Alps (data in [3]).

A critical step forward in the understanding of these mountain landscapes was the re-interpretation of rubbly debris and lacustrine deposits as, respectively, remnant rock avalanche debris and sediments deposited in rockslide-dammed lakes by M. Fort and coworkers and K. Hewitt between 1989 and 1999. However, recent discussion of the genesis of the Satpara rockslide deposit indicates that the interpretation of rubbly valley fill deposits remains far from routine. We show that some prehistoric rockslide dams in the northwest Himalayas impounded massive lakes with volumes in excess of 20 $Gm^3$, significantly larger than present-day Lake Sarez. Data on rockslide dams in the Pamir Mountains of Afghanistan and Tajikistan is also summarised from primary sources but further detailed mapping is necessary to define the regional landslide dam signature for the Pamirs.

We used SRTM-3 digital terrain data and satellite imagery to analyse four major historical rockslide damming events as follows; (1) our analysis of the 1841 Indus rockslide-damming event indicates that the volume of the impoundment and subsequent outburst was a minimum of 6.5 $Gm^3$, the largest outburst from a rockslide-dammed lake in historical time, (2) the 1858 Hunza Valley rockslide dam impounded about 805 $Mm^3$ before catastrophic outburst in August 1858, (3) the development of the 2010 Hunza rockslide-dammed lake is described in detail. It reached a maximum volume of 430 $Mm^3$ before stable overflow of the rockslide debris began on May 29, 2010. This remains the situation as of July 25, 2010 (200 days after impoundment) as Pakistan authorities consider alternative mitigation strategies to lower the waters of the lake, (4) the filling curve of Lake Sarez was developed from primary sources; it was conditioned by excessive outflow seepage through the debris to the extent that a freeboard of ca. 50 m is still naturally maintained without engineering intervention.

The emplacement of rockslide dams and the behaviour of their impounded lakes are critical hazards to communities and the development of infrastructure, including hydroelectric facilities, in the northwest Himalyas (India and Pakistan) and adjacent Pamir Mountains (Afghanistan and Tajikistan).

**Note added in proof:**

## Update on 2010 Attabad Rockslide Dam, Hunza River, Pakistan, as of February 11, 2011

As of February 11, 2011 (Day 401), the rockslide-dammed lake (Lake Gojal) at Attabad, Hunza Valley, is still in existence. As noted above, in the 143 days from the initial blocking rockslide on January 4$^{th}$ 2010, Lake Gojal filled to overtopping at 2,435 m a.s.l. on May 29, 2010. Stable overtopping continued through June, July, and August without catastrophic failure occurring. Work continued on widening the spillway in an attempt to accelerate erosion of the dam to a point where controlled partial drainage and lowering of lake level would occur.

In conjunction with digital SRTM-3 topography, we estimated the lake level from an EO-1 satellite image (Fig. 7.1 above) obtained on October 03, 2010 (Day 270) as ca. 2425 m a.s.l. This is ca. 10 m lower than the overflow level on May 29, 2010 and resulted from gradual spillway erosion. Our SRTM-3 DEM indicates that the drop in lake level from May 29 to October 3 resulted in the total volume of the lake being reduced to ca. 334 M $m^3$, a reduction of about 96 M $m^3$ (22%) from the maximum volume at overtopping in May.

Landsat-7 imagery of Lake Gojal obtained on December 12, 2010 (Day 340) showed very little change in the outline of the rockslide-dammed lake and minimal change in lake elevation. On this date stable overflow continued, and Pakistani authorities had continued to widen the spillway to effect a controlled lowering the lake level with the twin objectives of restoring travel on the flooded sections of the Karakoram Highway and reducing the area of agricultural land underwater in the Hunza valley upstream of the rockslide barrier. As of February 11, 2011 (Day 401 of impoundment) accelerated erosion of the spillway leading to a significant drop in lake level had not occurred.

## References


1. Abbott, J. (1848) Inundation of the Indus, taken from the lips of an eye-witness A.D. 1842, *Journal of the Asiatic Society of Bengal* **17**, 230–232.
2. Abe, K. and Noguchi, S. (1983) Revision of magnitudes of large shallow earthquakes, 1897–1912, *Physics of the Earth and Planetary Interiors* **33**, 1–11.
3. Abele, G. (1974) Bergstürze in den Alpen, ihre Verbeitung, Morphologie und Folgeerscheinungen; Wissenschafteliche Alpenvereinsheft 25, München, 230 p.
4. Ali, K.F. and de Boer, D.H. (2007) Spatial patterns and variation of suspended sediment yield in the upper Indus river basin, northern Pakistan, *Journal of Hydrology* **334**, 368–387.
5. Ali, K.F. and de Boer, D.H. (2008) Factors controlling specific sediment yield in the upper Indus river basin, northern Pakistan, *Hydrological Processes* **22**, 3102–3114.
6. Archer, D. (2003) Contrasting hydrological regimes in the upper Indus Basin, *Journal of Hydrology* **274**, 198–210.
7. Becek, K. (2008) Investigating error structure of shuttle radar topography mission elevation data product, *Geophysical Research Letters* **35**, L15403, doi: 10.1029/2008GL034592, 2008.
8. Becher, J. (1859) The flooding of the Indus. Letter addressed to R.H. Davies, Secretary to the Government of Punjab and its dependencies, *Journal of the Asiatic Society of Bengal* **28**, 219–228.
9. Biedermann, P. and Attewill, L. (2002) Evolution of seepage through landslide dams: The case of Usoy Dam in Tajikistan. Paper presented at IAHR Syposium, St. Petersburg, 8 p.
10. Burgisser, H.M., Gansser, A. and Pika, J. (1982) Late Glacial lake sediments of the Indus valley area, northwestern Himalayas, *Eclog. Geol. Helveticae* **75**, 51–63.
11. Butler, R.W.H. and Prior, D.J. (1988) Tectonic controls on the uplift of the Nanga Parbat Massif, Pakistan Himalayas, *Nature* **333**, 247–250.
12. Butler, R.W.H., Owen, L. and Prior, D.J. (1988) Flashfloods, earthquakes, and uplift in the Pakistan Himalayas, *Geology Today* **4**, 197–201.
13. Code, J.A. and Sirnhindi, S. (1986) Engineering implications of impoundment of the upper Indus river, Pakistan, by an earthquake-induced landslide, in R.L. Schuster (ed.), *Landslide*

*Dams: Processes, Risk, and Mitigation*, Geotechnical Special Publication No. 3. American Society of Civil Engineers, New York, NY, pp. 97–109.
14. Dortch, J.M., Owen, L.A., Haneberg, W.C., Caffee, M.W., Dietsch, C. and Kamp, U. (2009) Nature and timing of large landslides in the Himalaya and Transhimalaya of northern India, *Quaternary Science Reviews* **28**, 1037–1054.
15. Drew, F. (1873) Alluvial and lacustrine deposits and glacial records of the upper-Indus Basin (with discussion), *Quarterly Journal of the Geological Society* **29**, 441–471 (includes discussion).
16. Drew, F. (1875) *The Jummoo and Kashmir Territories, a Geographical Account*. Edwin Stanford, London, 568 p.
17. Droz, P., Fumagalli, A., Novali, F. and Young, B. (2008) GPS and INSAR technologies: A joint approach for the safety of Lake Sarez. *4th Canadian Conference on Geohazards*, Quebec, QC, Conference CD, 8 p.
18. Droz, P. and Spasic-Gril, L. (2006) Lake Sarez mitigation project: A global risk analysis, *Proceedings 22nd Congress on Large Dams, Q.36-R75, Barcelona, Spain.*
19. Dunning, S.A., Mitchell, W.A., Rosser, N.J. and Petley, D.N. (2007) The Hattian Bala rock avalanche and associated landslides triggered by the Kashmir Earthquake of 8 October 2005, *Engineering Geology* **93**, 130–144.
20. Evans, S.G. (1986) The maximum discharge of outburst floods caused by the breaching of man-made and natural dams, *Can. Geot. Jo.* **23**, 385–387.
21. Evans, S.G., Scarascia-Mugnozza, G., Strom, A. and Hermanns, R.L. (2006) Landslides from massive rock slope failure and associated phenomena, in S.G. Evans, G., Scarascia Mugnozza (eds.), *Landslides from Massive Rock Slope Failure*, NATO Science Series IV, Earth and Environmental Sciences, Vol. 49. Springer, Dordrecht, pp. 3–52.
22. Falconer, H. (1841) Letter to the Secretary of the Asiatic Society, on the recent cataclysm of the Indus, *Journal of the Asiatic Society of Bengal* **10**, 615–620.
23. Farr, T.G., Rosen, P.A., Caro, E., Crippen, R., Duren, R., Hensley, S., Kobrick, M., Paller, M., Rodriguez, E., Roth, L., Seal, D., Shaffer, S., Shimada, J., Umland, J., Werner, M., Oskin, M., Burbank, D. and Alsdorf, D. (2007) The shuttle radar topography mission, *Reviews of Geophysics* **45** RG2004, doi: 10.1029/2005RG000183.
24. Fort, M. and Peulvast, J.-P. (1995) Catastrophic mass-movements and morphogenesis in the Peri-Tibetan ranges: Examples from west Kunlun, east Pamir, and Ladakh, in O. Slaymaker (ed.), *Steepland Geomorphology*. Wiley, New York, NY, pp. 171–198.
25. Fort, M., Burbank, D.W. and Freytet, P. (1989) Lacustrine sedimentation in a semiarid alpine setting: An example from Ladakh, northwest Himalaya, *Quaternary Research* **31**, 332–350.
26. Galitzin, M.B. (1915) Sur le tremblement de terre du 18 février 1911, *Comptes Rendus des Seances de l'Académie des Sciences* **160**, 810–814.
27. Gaziev, E. (1984) Study of the usoy landslide in Pamir. *Proceeding 4th International Symposium on Landslides*, Toronto, Vol. 1, pp. 511–515.
28. Gladkov, E.G., Eletskii, V.S. and Zhabin, V.F. (1990) Prediction of the change in the water level of Lake Sarez and characteristics of seepage through the Usoi Barrier, *Hydrotechnical Construction* **24**, 25–28.
29. Gordin, M.D. (2003) Measure of all the Russias: Metrology and governance in the Russian Empire, *Kritika* **4**, 783–815.
30. Gutenberg, B. and Richter, C.F. (1954) Second Edition, *Seismology of the Earth and Associated Phenomena*. Princeton University Press, Princeton, NJ, 310 p.
31. Henderson, W. (1859) Memorandum on the nature and effects of the flooding of the Indus, 10th August 1858, as ascertained at Attock, *Journal of the Asiatic Society of Bengal* **28**, 199–228.
32. Hewitt, K. (1968) Records of natural damming and related events in the upper Indus basin, *Indus Journal of Water Power Development Authority* **10**, 11–19.
33. Hewitt, K. (1998) Catastrophic landslides and their effects on the upper Indus streams, Karakoram Himalaya, northern Pakistan, *Geomorphology* **26**, 47–80.

34. Hewitt, K. (1999) Quaternary moraines vs catastrophic rock avalanches in the Karakoram Himalaya, northern Pakistan, *Quaternary Research* **51**, 220–237.
35. Hewitt, K. (2001) Catastrophic rockslides and the geomorphology of the Hunza and Gilgit river valleys, Karakoram Himalaya, *Erdkunde* **55**, 72–93.
36. Hewitt, K. (2002) Styles of rock avalanche depositional complexes conditioned by very rugged terrain, Karakoram Himalaya, Pakistan, in S.G. Evans and J.V. DeGraff (eds.), *Catastrophic Landslides: Effects, Occurrence, and Mechanisms*, Reviews in Engineering Geology, Vol. XV, Geological Society of America, Boulder, CO, pp. 345–377.
37. Hewitt, K. (2002) Postglacial landform and sediment associations in a landslide-fragmented river system: The TransHimalayan Indus streams, central Asia, in K. Hewitt, M.L. Byrne, M. English, and G. Young(eds.), *Landscapes of Transition*. Kluwer Academic, Dordrecht, pp. 63–91.
38. Hewitt, K. (2006) Disturbance regime landscapes: Mountain drainage systems interrupted by large rockslides, *Progress in Physical Geography* **30**, 365–393.
39. Hewitt, K. (2006) Rock avalanches with complex run out and emplacement, Karakoram Himalaya, inner Asia, in S.G. Evans, G., Scarascia-Mugnozza (eds.), *Landslides from Massive Rock Slope Failure*, NATO Science Series IV, Earth and Environmental Sciences Vol. 49. Springer, Dordrecht, pp. 521–550.
40. Hewitt, K. (2006) Rock avalanche dams on the Transhimalayan Upper Indus streams: A survey and assessment of hazard-related characteristics, *Italian Journal of Engineering Geology and Environment* Special Issue 1, 61–65.
41. Hewitt, K. (2009) Catastrophic rock slope failures and late Quaternary developments in the Nanga Parbat-Haramosh massif, Upper Indus basin, northern Pakistan, *Quaternary Science Reviews* **28**, 1055–1069.
42. Hewitt, K. (2009) Glacially conditioned rock-slope failures and disturbance-regime landscapes, Upper Indus Basin, northern Pakistan, in Knight, J. and Harrison, S. (eds.), *Periglacial and paraglacial processes and environments*, Special Publication 320. Geological Society, London, 235–255.
43. Hewitt, K. (2010) Gifts and perils of landslides, *American Scientist* **98**, 410–419.
44. Ischuk, A. (2006) Usoy natural dam: Problem of security (Lake Sarez, Pamir Mountains, Tadjikistan), *Italian Journal of Engineering Geology and Environment* Special Issue 1, 189–192.
45. Jeffreys, H. (1923) The Pamir earthquake of 1911 February 18, in relation to the depths of earthquake foci, *Monthly Notices of the Royal Astronomical Society Geophysics Supplement* **1**, 22–31.
46. Jeffreys, H. (1937) On the materials and density of the earth's crust, *Monthly Notices of the Royal Astronomical Society Geophysics Supplement* **4**, 50–61.
47. Kargel, J.S., Leonard, G., Crippen, R.E., Delaney, K.B., Evans, S.G. and Schneider, J. (2010) Satellite monitoring of Pakistan's rockslide-dammed Lake Gojal, *EOS* **91**, 394–395.
48. Khan, M. (1969) Influence of upper Indus basin on the elements of the flood hydrograph at Tarbela-Attock, in *Floods and their computation*, IAHS Publication, 85, Vol. 2. pp. 918–925.
49. Klotz, O. (1915) Earthquake of February 18, 1911, *Bulletin of the Seismological Society of America* **5**, 206–213.
50. Korup, O., Densmore, A.L. and Schlunegger, F. (2010) The role of landslides in mountain river evolution, *Geomorphology* **120**, 77–90.
51. Kuhle, M. (2004) The Pleistocene glaciation in the Karakoram-Mountains: Reconstruction of past glacier extensions and ice thicknesses, *Journal of Mountain Science* **1**, 3–17.
52. Macelwane, J.B. (1926) Are important earthquakes ever caused by impact? *Bulletin Seismological Society of America* **16**, 15–18.
53. Malik, Z.M., Tariq, A. and Anwer, J. (2008) Seepage control for Satpara dam, Pakistan, *Proceedings of the Institution of Civil Engineers Geotechnical Engineering GE5* **161**, 235–246.

54. Mason, K. (1929) Indus floods and Shyok Glaciers, *Himalayan Journal* **1**, 10–29.
55. Montgomerie, T.G. (1860) Memorandum on the great flood of the river Indus which reached Attock on the 10th August 1858, *Journal of the Asiatic Society of Bengal* **29**, 128–135.
56. Oldham, R.D. (1923) The Pamir Earthquake of 18th February, 1911, *Quarterly Journal of the Geological Society* **79**, 237–245.
57. Owen, L.A. (1989) Neotectonics and glacial deformation in the Karakoram Mountains and Nanga Parbat Himalaya, *Tectonophysics* **163**, 227–265.
58. Owen, L.A. (1996) Quaternary lacustrine deposits in a high-energy semi-arid mountain environment, Karakoram Mountains, northern Pakistan, *Journal of Quaternary Science* **11**, 461–483.
59. Owen, L.A., Kamp, U., Khattack, G.A., Harp, E.L., Keefer, D.K. and Bauer, M. (2008) Landslides triggered by the 8 October 2005 Kashmir earthquake, *Geomorphology* **94**, 1–9.
60. Pacheco, J.F. and Sykes, L.R. (1992) Seismic moment catalog of large shallow earthquakes, 1900 to 1989, *Bulletin of the Seismological Society of America* **82**, 1306–1349.
61. Phartiyal, B., Sharma, A., Srivastava, P. and Ray, Y. (2009) Chronology of relict lake deposits in the Spiti River, NW trans Himalaya: Implications to Late Pleistocene-Holocene climate-tectonic perturbations, *Geomorphology* **108**, 264–272.
62. Preobrajensky, J. (1920) The Usoi Landslide, *Geological Committee Papers on Applied Geology* **14**, 21 p. (In Russian).
63. Raetzo, H. (2006) Hazard assessment of Lake Sarez rockslides and Usoy Dam in Pamir Mountains (Tadjikistan), *Italian Journal of Engineering Geology and Environment* Special Issue 1, 193–196.
64. Rautian, T. and Leith, W. (2002) Composite regional catalogs of earthquakes from the Former Soviet Union, United States Geological Survey Open File Report 02–500
65. Richter, C.F. (1958) *Elementary Seismology*. Freeman, San Francisco, 768 p.
66. Risley, J.C., Walder, J.S. and Denlinger, R.P. (2006) Usoi dam wave overtopping and flood routing in the Bartang and Panj rivers, Tajikistan, *Natural Hazards* **38**, 375–390.
67. Schneider, J.F. (2009) Seismically reactivated Hattian slide in Kashmir, Northern Pakistan, *Jo. Seismology* **13**, 387–398.
68. Schuster, R.L. and Alford, D. (2004) Usoi landslide dam and Lake Sarez, Pamir Mountains, Tajikistan, *Environmental and Engineering Geoscience.* **10**, 151–168.
69. Seong, Y.B., Bishop, M.P., Bush, A., Clendon, P., Copland, L., Finkel, R.C., Kamp, U., Owen, L.A. and Shroder, J.F. (2009) Landforms and landscape evolution in the Skardu, Shigar and Braldu Valleys, central Karakoram, *Geomorphology* **103**, 251–267.
70. Shroder, J.F. (1989) Slope failure: Extent and economic significance in Afghanistan and Pakistan, in Brabb, E.E., Harrod, B.L. (eds.), *Landslides: Extent and Econmic Significance in the World.* Balkema, Rotterdam, pp. 325–341.
71. Shroder, J.F. (1998) Slope failure and denudation in the western Himalaya, *Geomorphology* **26**, 81–105.
72. Shroder, J.F., Cornwell, K. and Khan, M.S. (1991) Catastrophic break-out floods in the western Himalaya, Pakistan, *Geological Society of America Annual Meeting Program with Abstracts* **23**, A87.
73. Shroder, J.F., Khan, M.S., Lawrence, R.D., Madin, I.P. and Higgins, S.M. (1989) Quaternary glacial chronology and neotectonics in the Himalaya of northern Pakistan, in L. Malinconico and R.J. Little (eds.), *Tectonics of the western Himalaya*, Geological Society of America Special Paper 232, pp. 275–294.
74. Shroder, J.F. and Weihs, B.J. (2010) Geomorphology of the Lake Shewa landslide dam, Badakhshan, Afghanistan, using remote sensing data, *Geografiska Annaler* **92A**, 469–483.
75. Shpilko, G. (1914) The Pamir Earthquake of 1911 and its consequences: Chronological reference and report of the works of the Military Detachment of the Pamir, *Bulletin Imperial Russian Geographic Society* **50**, 68–94.

76. Stein, A. (1916) A third journey of exploration in central Asia, 1913–16 (continued), *The Geographical Journal* **48**, 193–225.
77. Stein, A. (1928) *Innermost Asia-Detailed Report of Explorations in Central Asia, Kan-Su and Eastern Iran.* Clarendon Press, Oxford, 1159 p.
78. Stone, R. (2009) Peril in the Pamirs, *Science* **326**, 1614–1617.
79. Stucky Consulting Engineers (2001) Lake Sarez Mitigation Project, Design Report, 149 p.
80. Weichert, D.H., Horner, R.B. and Evans, S.G. (1994) Seismic signatures of landslides: The 1990 Brenda Mine collapse and the 1965 Hope rockslides, *Bulletin of the Seismological Society of America* **84**, 1523–1532.
81. Zech, R., Abramowski, U., Glaser, B., Sosin, P., Kubik, P.W. and Zech, W. (2005) Late Quaternary glacial and climatic history of the Pamir mountains derived from cosmogenic $^{10}$Be exposure ages, *Quaternary Research* **64**, 212–220.

# Chapter 8
# Stability and Life Span of Landslide Dams in the Himalayas (India, Nepal) and the Qin Ling Mountains (China)

J.T. Weidinger

## 1 Introduction

To elaborate on the results known so far from lake-damming natural rock blockages, a number of rockslide-, rock avalanche- and landslide dams were investigated. Dating of the events was achieved mostly by means of morpho-stratigraphic analysis of the Quaternary sequences or by historic investigations (Fig. 8.1, Tables 8.1, 8.2, 8.3, 8.4, 8.5, and 8.6). These case studies are analysed here in detail.

Creating a helpful tool for estimating potential hazard after the occurrence of a lake-damming landslide was the desired aim. The resulting "block size stability diagram" is one of the first attempts for a regional comparison of such kind of phenomena.

### *1.1 The Himalayas of India and Nepal*

The Himalayas, the highest mountain range in the world, has been the product of a collision between the Indian Subcontinent and the Eurasian Plate [40], which started 65 Ma ago (Fig. 8.1). The Tibetan zone (Proterocoic – Eocene) is exposed in the north, partly forming the summits of the highest peaks on earth with its basal strata (for example on top of Mt. Everest; 8,848 m). Further south crustal thrust planes formed within the Indian Plate 25–20 Ma ago. The most important of these is the Main Central Thrust south of the High Himalayan Crystalline, the root zone of the Precambrian metasediments and crystalline nappes of the Lesser Himalayas. The ongoing convergence led to the Main Boundary Thrust, which separates the Lesser Himalayas from the Sub-Himalayas (Miocene – Pleistocene), the molasse-basin, mainly composed of the Siwalik Formation. The actual convergence between the continents is documented by the high uplift rates of the Tibetan Plateau, which started 5 Ma ago.

J.T. Weidinger (✉)
Department of Geography and Geology, Erkudok© Institute in the K-Hof Museums Gmunden, Salzburg University, Gmunden, Austria
e-mail: j_weidinger@hotmail.com

S.G. Evans et al. (eds.), *Natural and Artificial Rockslide Dams*, Lecture Notes in Earth Sciences 133, DOI 10.1007/978-3-642-04764-0_8, © Springer-Verlag Berlin Heidelberg 2011

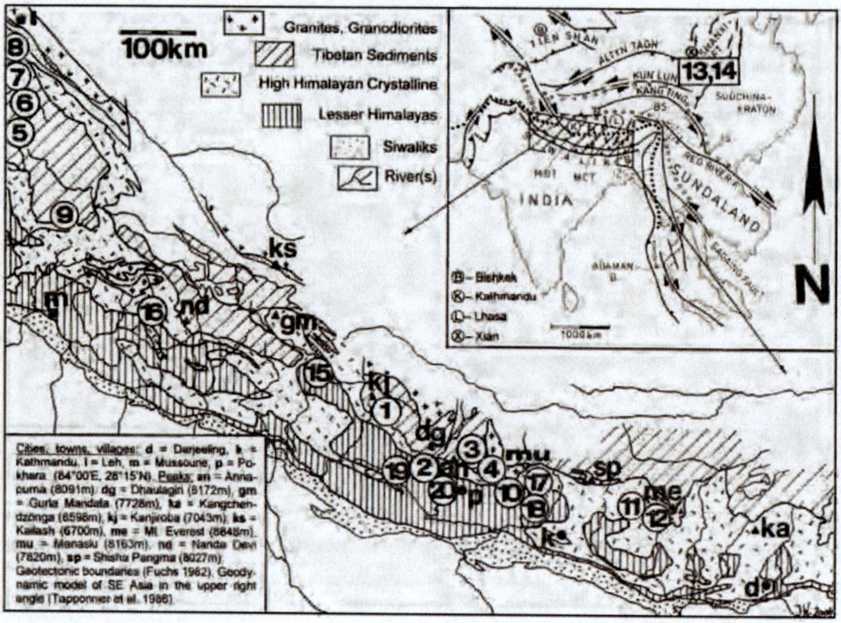


**Fig. 8.1** Locations of rockslide-, rock avalanche- and landslide areas investigated (for names of locations 1–20, see Tables 8.1, 8.2, 8.3, 8.4, 8.5, and 8.6); Geotectonics after [40]

## 2 Lake Damming Giant Rockslides in the Nepal Himalayas

The rare phenomenon of giant rockslides has been found not only in the Indian Himalayas [2], but also in the Nepal Himalayas; e. g. at Ringmo (Suli Gad Valley, Dolpa), Kalopani (Kali Gandhaki Valley, Dhaulagiri Himal), Dukur Pokhari and Latamrang (Marsyandi Valley, Annapurna Himal).

### *2.1 General Description of the Giant Rockslide Areas*

In the upper course of Suli Gad River, 30 km north of Dunai town (Dolpo, Western Nepal), one of the largest rockslide deposits in the Himalayas [44] formed a massive barrier, which has dammed Phoksundo Lake (alt. 3,700 m), the second largest lake in Nepal (Figs. 8.2 and 8.3; Table 8.1). This rockslide deposit around Ringmo village involved Dhaulagiri limestones of the Tibetan sediments [11]. Its detritus is covered by loess, in which intercalations of reddish soil layers were dated at 30–40 ka, the landslide's age and the time of origin of Phoksundo Lake [55].

In the central part of Kali Gandhaki Valley (Fig. 8.1, Table 8.1), which crosses three tectonic units of the Himalayas, geoscientists have investigated a giant

**Table 8.1** Characteristics and stability conditions of giant rockslide dams and life span of dammed lakes in the Nepal Himalayas

| Landslide/parameters | 1. Ringmo | 2. Kalopani | 3. Dukur Pokhari | 4. Latamrang |
|---|---|---|---|---|
| Preparatory causal factors | Folding, fabric, glacial erosion and stress release | Sheeted units, fabric, glacial erosion and stress release | Fabric, glacial erosion, stress release | Sheeted units, fabric, glacial erosion and stress release |
| Age | 30,000–40,000a [55] | > 56.36 +/– 6.95 and 55.84 +/– 5.9 ka [1] | Post-glacial | 5.4 ka [33] |
| Movement type | In-situ collapse | Sledge-like sliding | In-situ collapse, rolling, bouncing | Sledge-like sliding |
| Geographic position | Suli Gad Valley/dolpa, Nepal | Kali Gandhaki Valley/Mustang, Nepal | Marsyandi Valley/Manang, Nepal | Marsyandi Valley/Manang, Nepal |
| Geotectonic position | Tibetan zone of Himalayas | High Himalayan crystalline | Tibetan zone of Himalayas | High Himalayan crystalline, tibetan zone of Himalayas |
| Lithology | Limestone | Augen-, biotite- gneiss | Limestone, shists | Gneiss, quartzite |
| Material | Disintegrated, re-cemented fragments and blocks | Brecciated, cataclastic and re-cemented blocks | Brecciated, cataclastic and re-cemented blocks and boulders boulder size: $\leq$2 m | Brecciated, cataclastic and re-cemented blocks |
| Length, wide, thickness of dam | 2.5/1.5/0.5–0.7 km | <4/<6/<1 km | <2/<3/<0.5 km | <2.5/<3.5/<0.7 km |
| Volume of deposit | 1.5 $Bm^3$ | 3 $Bm^3$ | 1 $Bm^3$ | Rem. 4.5 $Bm^3$ from $\geq$ 5.5 $Bm^3$ |
| Length/width/depth of lake(s) | 5/0.8/ 0.2 km | $\leq$30/1/ 0.1 km | $\leq$3/$\leq$1/$\leq$0.05 km | estimated: 3/0.5/0.1 km |
| Fahrböschung | Ca. 35° | Ca. 17° | Ca. 20° | Ca. 22° resp. 12° |
| Volume of lake | 350–400 $Mm^3$ | 1.5–3 $Bm^3$ | 75 $Mm^3$ | $\leq$8.5 $Mm^3$ |
| Stability of dam | Stable due to re-cementation | Stable due to recementation | Stable due to ce-mentation with sediments | Long time stability due to recementation |
| Life span of lake | Still existing after 30,000–40,000a | Disappeared within late to postglacial times | Filled up with alluvions within short period | Overtopping and piping after <500a [33]. |

**Table 8.2** Characteristics and stability conditions of rock avalanche dams and life spans of lakes in the Tibetan Zone of the Himalayas in India

| Landslide/parameters | 5. Pateo | 6. Sarai Kenlung | 7. Chumik Marpo | 8. Tso Tok Phu |
|---|---|---|---|---|
| Preparatory Causal Factors | Fabric, glacial erosion, stress release | Tectonics, fabric, faulting, glacial erosion | Tectonics, fabric, faulting | Tectonics, fabric, faulting, glacial erosion |
| Age | 8.3+/– 1.5 ka and 7.6+/– 0.7 ka BP [8] | 5.9+/– 0.5 ka BP [31] and 6.3+/– 0.3 ka BP resp [8]. | Post-glacial | Post-glacial to subrecent |
| Type of movement | Rolling, bouncing | Rolling, bouncing | Rolling, bouncing | Rolling, bouncing |
| Geographic Position | Bhaga Valley/Himachal Pradesh, India | Yünan Chu Valley/ Zanskar, India | Lingti Chu Valley/ Zanskar, India | Tributary of Tsarap Chu, 13 km NNE of Phuktal, Zanskar |
| Geotectonic Position | Tibetan Zone of Himalayas | Tibetan Zone of Himalayas | Tibetan Zone of Himalayas | Tibetan Zone of Himalayas |
| Lithology | Limestone, dolomite | Conglomerate, quartzite, shists, dolomites | Quartzite, basalts, sandstone, limestone, dolomite, shales | Limestone, dolomite, marls |
| Composition of Material | Boulders/rock fragments, size: ≤0.5 m | Boulders/rock fragments, size: ≤10 m | Boulders | Boulders |
| Length. wide. thickness of dam | 2/1.5/0.05–0.1 km | 3/1/0.2 km | 0.8/0.8/0.05 km | 0.7/0.7/0.05 km |
| Fahrböschung | Ca. 18° | Ca. 24° | No data | No data |
| Volume of deposit | Ca. 150 $Mm^3$ | Ca. 350 $Mm^3$ | Est. 60 $Mm^3$ (before erosion) | Est. 60 $Mm^3$ |
| Length/width/depth of lake(s) | 1st: 6/≤1/≤0.1; 2nd: 2.5/≤0.5/≤0.1 km | 2.5/≤1/≤0.1 km | 2.5/0.2–0.9/?km | 1.5/0.4/?km |
| Water volume of lake (s) | 1st: 300 $Mm^3$ 2nd: 62.5 $Mm^3$ | 125 $Mm^3$ | est. 110 $Mm^3$ | est. 30 $Mm^3$ |
| Stability of dam | Deep gully erosion through the deposit | Stable due to cementation with sediments | Unstable due to erosion along spillway | Stable due to cementation with sediments |
| Life span of lake | Both disappeared within short periods | ≤3,000–5,000a | Disappeared due to erosion | Still exists |

**Table 8.3** Characteristics and stability conditions of rock avalanche dams and life span of their lakes in the High Himalayan Crystalline

| Landslide/parameters | 9. Bawa pass | 10. Tal | 11. Lamabagar [41] | 12. Gath-Chaunrikharka [41] |
|---|---|---|---|---|
| Preparatory causal factors | Fabric, glacial erosion, stress release | Glacial erosion, stress release | Fabric, glacial erosion, stress release | Fabric, glacial erosion, stress release |
| Age | Post-glacial | ≤1,000a | 3 generations | Post-glacial |
| Type of movement | Rolling, bouncing | Rolling, bouncing | Rolling, bouncing | Rolling, bouncing |
| Geographic position | Garak Chu Valley, Himachal, Inda_ | Marsyandi Valley/Nepal | Tama Koshi Valley/Dolakha, Nepal | Dudh Koshi Valley/ Solukhumbu, Nepal |
| Geotectonic position | High Himalayan crystalline | High Himalayan crystalline | High Himalayan crystalline | High Himalayan crystalline |
| Lithology | Gneiss, granite | Sillimanite gneiss | Gneiss, migmatite | Gneiss |
| Composition of material | Boulders; size: 2–5 m | Boulders; size 0.2–10 m | Boulders/rock fragments | Shattered, mainly cohesive rock |
| Length, wide, thickness of dam | 1/0.5/0.2 km | 1/0.5/0.1 km | 2/1/0.3 km | 3.4/1/0.1 km |
| Volume of deposit | 16 $Mm^3$ | 4.5 $Mm^3$ | ca. 30 $Mm^3$ | 100 $Mm^3$ |
| Fahrböschung | No data | No data | No data | No data |
| Length/width/depth of lake | 1/ca.0.3/ 0.2 km | 1/0.3/ 0.1 km | 3/max.0.1/max. 0.3 km | 5/0.8/0.09–0.1 km |
| Water volume of lake(s) | Ca. 30 $Mm^3$ | 10–15 $Mm^3$ | 18 $Mm^3$ | 90 $Mm^3$ |
| Stability of dam | Stable due to cementation with sediments | Stable due to cementation with sediments | Still stable | Failed by overtopping and erosion |
| Life span of lake | Filled up with alluvions within short periods | Filled up with alluvions within ≤200a | Silted up within short period | Silted up following a first outbreak after 120 years |

**Table 8.4** Characteristics and stability conditions of rock avalanche dams and a lake in the Qin Ling Mountains (China)

| Landslide/parameters | 13. Liu Dshe Tse | 14. Cui Hua |
|---|---|---|
| Preparatory causal factors | Tectonics, erosion, seismic activity | Tectonics, erosion, seismic activity |
| Age | Sub-recent (2,784a) | Sub-recent (2,784a) |
| Type of movement | Rolling, bouncing | Rolling, bouncing |
| Geographic position | Dong Cha Valley/Shaanxi, China | Dong Cha Valley/Shaanxi, China |
| Geotectonic position | Crystalline of the Qin Ling mountains | Crystalline of the Qin Ling Mountains |
| Lithology | Gneiss, granite | Gneiss, granite |
| Composition of material | Boulders diameters: $\leq$20 m | Boulders diameters: $\leq$70 m |
| Length, wide, thickness of dam | 1/1/0.5 km | 1/1/0.3 km |
| Volume of deposit | 1 $Bm^3$ | 350 $Mm^3$ |
| Fahrböschung | No data | No data |
| Length/width/depth of lake(s) | No lake dammed | 1/$\leq$0.3/$\leq$0.02 km |
| Water volume of lake | No stored water | $\leq$3 $Mm^3$ |
| Stability of dam | Stable | Still stable (cementation, human impact) |
| Life span of lake | Due to permeability (tectonic structure) of bottom, no water | Still existing after 2,784a |

rockslide (Fig. 8.4) that covers an area of 10 $km^2$ [5, 10, 15, 52]. The rockslide masses originated SW of the deposition area near Kalopani village and formed a barrier that dammed the Kali Gandhaki River. From Kalopani towards the north the Kali Gandhaki Valley widens as far as the village of Kagbeni to a more than 30 km long former dammed reservoir of the river. Today this is filled with alluvial, mass wasting and lake sediments (Fig. 8.5).

Another giant rockslide in the Himalayas was investigated at Dukur Pokhari (Figs. 8.1 and 8.6, Table 8.1) near the eastern end of the Upper Marsyandi Valley (Annapurna Himal, Nepal) [52]. Its source lies on the southeastern ridge of Naur Himal at elevations of 4,700–4,900 m. The folded and steeply SW to W dipping, well-bedded formations of the Tibetan Sediments served as preparatory causal factors for the failure of this mountain crest in a direction from NE to SW. The dislocated masses dammed the main river (see Fig. 8.7).

Further down the valley, in the middle course of the Marsyandi Valley, one of the biggest rockslide areas in the Himalayas (Fig. 8.7, Table 8.1), which once dammed the main river to form a large lake, was identified [44, 52]. Its deposit stretches over 4 km along the Marsyandi River, from the village of Thanchauk (2,480 m) passing Latamrang (2,400 m), with the most accessible and extensive

**Table 8.5** Characteristics and stability conditions of post-glacial and recent landslide dams and temporary dammed lakes in the Lesser Himalayas

| Landslide/parameters | 15. Ghatta Khola | 16. Gohna Tal | 17. Jagat [28] |
|---|---|---|---|
| Preparatory causal factors | Stress release | Climatic conditions, stress release | Fabric, erosion, stress release, climate |
| Age | Post-glacial | 1893 | Between 1962 and 1979 |
| Type of movement | Rolling, bouncing | Debris stream | Debris stream |
| Geographic position | Ghatta Khola Valley/Western Nepal | Birahi Ganga Valley/Garhwal, India | Buri Gandhaki Valley/Nepal |
| Geotectonic position | Lesser Himalayas | Lesser Himalayas | Lesser Himalayas |
| Lithology | Dolomite, shists | Dolomites, limestone, marls, shists | No data |
| Composition of material | Boulders boulder size: $\leq$5 m | Shattered, pulverized, cataclastic rock | Pulverized and cataclastic rock |
| Length, wide, thickness of dam | 0.3/0.3/0.2 km | 1.1/1/ 0.3 km | No data |
| Volume of deposit | 4.8 $Mm^3$ | 150–200 $Mm^3$ | No data |
| Side effects | | Dammed one lake, its outbreak in 1894 claimed one victim | Dammed one lake |
| Fahrböschung | Ca. 21° | Ca. 35° | No data |
| Length/width/depth of lake(s) | 0.3/0.1/0.1 km | 4/0.35–max.1/ $\leq$0.3 km | Est. 5/0.4/0.1/km |
| Water volume of lake | $\leq$1 $Mm^3$ | 250 $Mm^3$ (both before 1st break) | No data |
| Stability of dam | Stable due to cementation with sediments | Outbreaks and erosion through deposit | Outbreaks and erosion through deposit |
| Life span of lake | Filled up with alluvions within $\leq$100a | Outbreaks/filled with alluvions within 78a | Outbreaks, filled with alluvions <17a |

outcrops of the rockslide, as far as Danaque (2,180 m). This area is part of translationally transported and shattered blocks. The possible scarp is known as the eastern part of the Lamjung Himal (6,988 m). Main directions of movement of the rockslide from SW to NE as well as from SE to NW (or SSE to NNW) can be reconstructed. Fragments of wood within the preserved lake sediments (Fig. 8.8) on the steep flanks of the former reservoir have been dated by Pratt-Sitaula et al. [33] at 5.4 ka.

**Table 8.6** Characteristics and stability conditions of landslide dams and temporary dammed lakes in the Lesser Himalayas

| Landslide/parameters | 18. Labubesi [28] | 19. Dharbang | 20. Tatopani [42] |
|---|---|---|---|
| Preparatory causal factors | Fabric, erosion, stress release, climate | Stress release, human impact, climate | Fabric, erosion, stress release, climate |
| Age | 1968 | 1926, 1988 | 1998 |
| Type of movement | Debris stream | Debris stream | Debris stream |
| Geographic position | Buri Gandhaki Valley/Nepal | Myagdi Valley/Nepal | Kali Gandhaki Valley/Nepal |
| Geotectonic position | Lesser Himalayas | Lesser Himalayas | Lesser Himalayas |
| Lithology | No data | Shists, slates, phyllite, quartzite | Phyllite, quartzite |
| Composition of material | Pulverized and cataclastic rock | Pulverized and cataclastic rock | Pulverized and cataclastic rock |
| Length, wide, thickness of dam | No data | 1.5/0.3/0.1 km (1988) | No data |
| Volume of deposit | No data | 5 $Mm^3$ (1988) | 400,000 $m^3$ |
| Fahrböschung [16] | No data | Ca. 23° | No data |
| Side effects | Dammed one lake | Dammed one lake (claimed 600 victims in both events) | Dammed a lake, which endangered the health resort Tatopani |
| Length/width/depth of lake(s) | Est. 5/0.4/0.2/km | 0.7/0.1/0.05 km | $\leq$1/$\leq$0.1/0.02 km |
| Water volume of lake | 0.05 $km^3$ [28] | $\leq$1.75 $Mm^3$ | $\leq$1 $Mm^3$ |
| Stability of dam | Outbreaks and erosion through deposit | Failed due to overtopping and erosion | Failed due to overtopping and erosion |
| Life span of lake | Outbreaks, filled with alluvions <17a | <6 h | <72 h |

## *2.2 Stability Conditions of Giant Rockslide Dams and Life Span of Their Dammed Lakes*

At Ringmo the failure mechanism has played the major role for the stabilization of the rockslide dam and the life span of Phoksundo Lake. The scarp of the rockslide, an almost vertical wall (Fig. 8.9), indicates an in-situ collapse of the former mountain crest with high angle and a relatively short distance of dislocation. That is why huge, not disintegrated but interlocking blocks of several dozens of meters in diameter dominate the deposit. Together with intercalated, finely grained material that was secondarily compacted by mineralization of mountain waters, they have proven a positive influence on the dam's stability.

The spillway through the deposit is just slightly eroded – only an extraordinary event such as a mountain flank collapse into the lake could cause severe damage [34]. Although the catchment area extends some 150 $km^2$, Phoksundo Lake is still

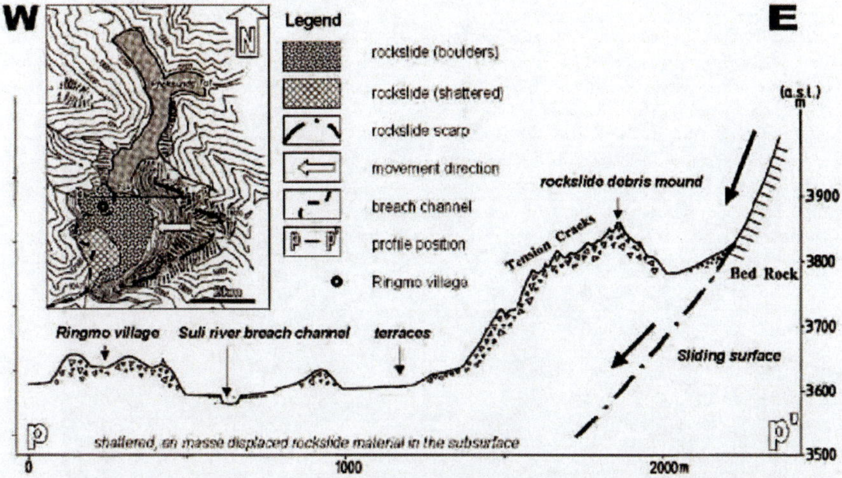


**Fig. 8.2** Geologic sketch map and cross section of Ringmo rockslide (location: 82°56'E, 29°09'N); after [55]

**Fig. 8.3** View of Ringmo rockslide deposit (ld) towards NE; Phoksundo Lake (pl)

not filled up with alluvial and/or lake sediments. This is a matter of the local climatic conditions within the rain shadow behind the Higher Himalayas.

It is assumed that the bulk of Kalopani rockslide material was dislocated en masse, having been shattered and brecciated with decreasing intensity from the bottom to the top and overridden by typical angular boulders. Similar observations in other high mountain localities have been made [19, 20, 36, 39, 45–47, 49, 50]. This kind of natural dam remained stable for a long time. That is why the former reservoir of the dam in the northern part of Kali Gandhaki Valley bears a complex sequence of moraines [26, 27], glacio-fluvial and alluvial sediments, mass wasting deposits

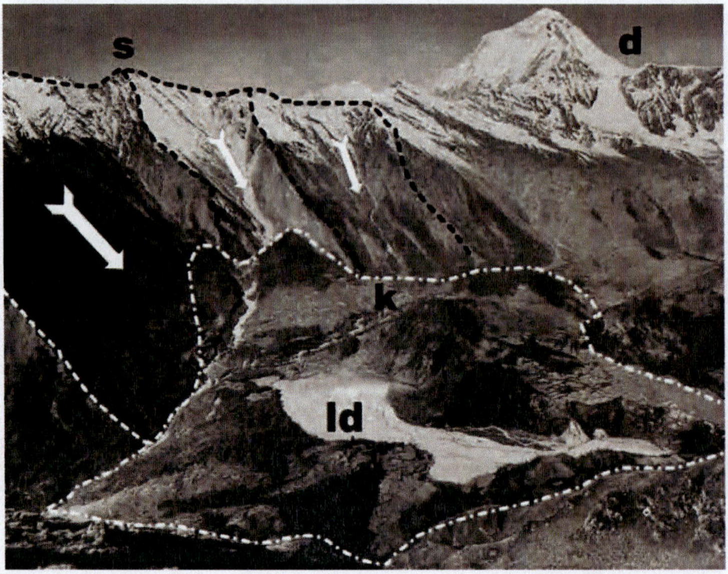


**Fig. 8.4** SE-NW view from Nimek Peak (3,900 m) over the Kali Gandhaki Valley towards Kalopani rockslide area (ld); d = Dhaulagiri (8,172 m), s = Sarpang Dhuri, → = direction of movement, k = Kalopani. Photograph by K. Hormann 1974

from the Nilgiris, periglacial phenomena and the 200 m thick infillings of a huge lake, the Marpha Formation [1]. Hanisch [15] connects the origin of these lake sediments exclusively with Kalopani rockslide, suggesting a former crest of the dam at 3,200 m altitude, which is much higher than the present one, while Weidinger [52] assumes different aged lake sediments generated by Kalopani rockslide dam and ice-debris avalanches from the Nilgiris; at least a former dammed lake at 2,800 m altitude is evident for him (Fig. 8.5).

The angular character of the blocks at Dukur Pokhari rockslide is a consequence of the bedded rocks of the source. The basal zone of this rockslide shows a much higher disintegration (Figs. 8.6 and 8.10), indicating high and destructive mechanical energy during the event. Thus the mechanism of sliding of this composite rockslide occurred in a translational manner at the bottom and in the form of rolling on the top. Despite the formation of a spillway at the distal part of the masses at the border to the hard rock, the dam kept stable. Due to sufficient availability of glacio-fluvial sediments, the basin behind was silted and filled up within a short period.

Due to the extraordinarily high volume of Latamrang rockslide, the natural dam existed for relatively long time, although waters of the dammed lake could seep through the landslide deposit and tunnels were formed by piping, finally leading to the collapse of the dam. In this way a volume of ca. 260 $Mm^3$ has been eroded from the bottom of the valley to the height of the former water level of the lake

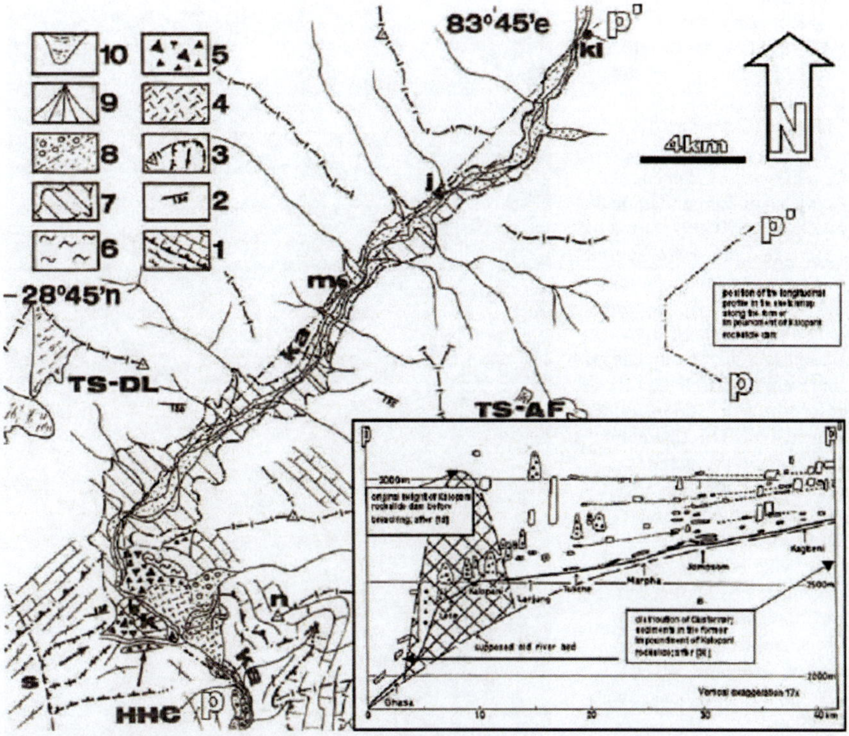


**Fig. 8.5** Geologic-geomorphologic sketch map of the upper Kali Gandhaki valley featuring Kalopani rockslide area and its former dammed lake. Geotectonics after [5]. Legend: *HHC* = high Himalayan crystalline, *TS* = Tibetan sediments (*AF* = Annapurna formation, *DL* = Dhaulagiri limestones), *1*= hard rock and thrust plane, *2* = dip, *3* = peak, crest, scarp and movement direction, *4* = rockslide deposit (shattered), *5* = rock avalanche deposit, *6* = lake sediments, *7* = extension of former dammed lake at 2,800 m, *8* = glacio-fluvial and alluvial gravel, *9* = debris cone, *10* = recent glaciers; *j* = Jomosom, *k* = Kalopani, *ki* = Kagbeni, *kg* = Kali Gandhaki, *m* = Marpha, *n* = Nimek (3,904 m), *ni* = Nilgiris (7,061 m), *s* = Sarpang Dhuri. Insert: longitudinal profile along the recent Kali Gandhaki River and the former impoundment according to [15, 26]

(Fig. 8.11). Terraces of alluvial sedimentation on top of the rockslide deposit at an altitude of 2,600 m serve as good indicators of the former elevation of the water in the reservoir and for the level of the spillway of the lake.

The fluvial response to large rockslide dams was examined along the Kali Gandhaki and the multiple blocked Marsyandi rivers [29]. Both valleys have developed knickslopes and higher steepness indices $k_s$ [21] along breach channels of former blockages (Figs. 8.10 and 8.12).

Long-lived rockslide-dammed lakes may result from the favourable combination of several factors such as very large dam size and (semi-) arid climate in headwaters (Fig. 8.13).

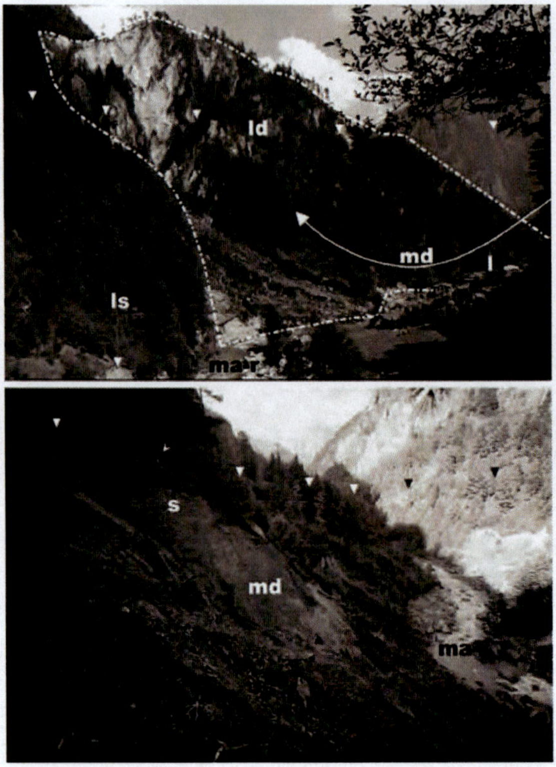


**Fig. 8.6** *Upper Photograph*: W-E view towards the eroded dam of Latamrang rockslide (*ld*) on the left side of Marsyandi River (ma-*r*); *md* = movement direction, *ls* = lacustrine sediments, ↓ = level of former dammed lake (alt. 2,650 m); observation point: 400 m west of Latamrang village (*l*), alt. 2,490 m. *Lower Photograph*: Outcrop of rhythmically sedimented lake sediments (*s*) that were formed in the reservoir (↓ = level of former dammed lake) of Latamrang rockslide. In the summer monsoon of 2003 huge areas of the trekking route (↓ Trekkers) along the Marsyandi River (*ma-r*) were endangered by sliding and mudflow (*md* = direction) processes. Location: right side of Marsyandi River, 1 km W of Thanchouk, alt. 2,500 m

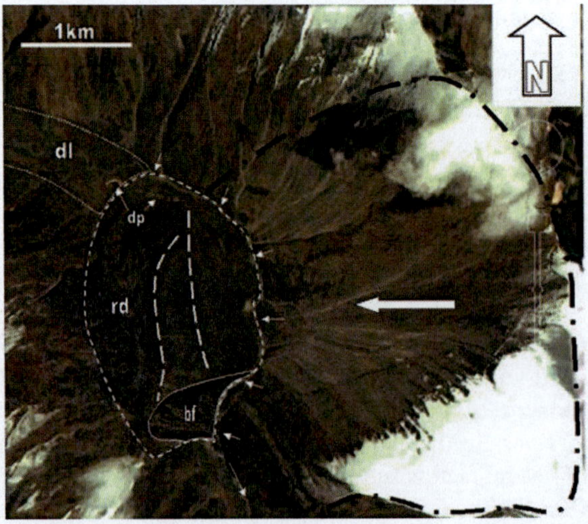


**Fig. 8.7** Satellite image (source: Google Earth) of Dukur Pokhari rockslide dam (*rd*), position: 84°11'E, 28°35'N; (-.-.-) = scarp (←) = direction of movement (—) = 2 ridges of "brandung", *bf* = shattered base facies of rockslide (as seen in Fig. 8.10), *dl* = former dammed lake; white arrows mark proximal spillway along Marsyandy River, *dp* = position of Dukur Pokhari (= small pond, *left* side) and village (*right* side)

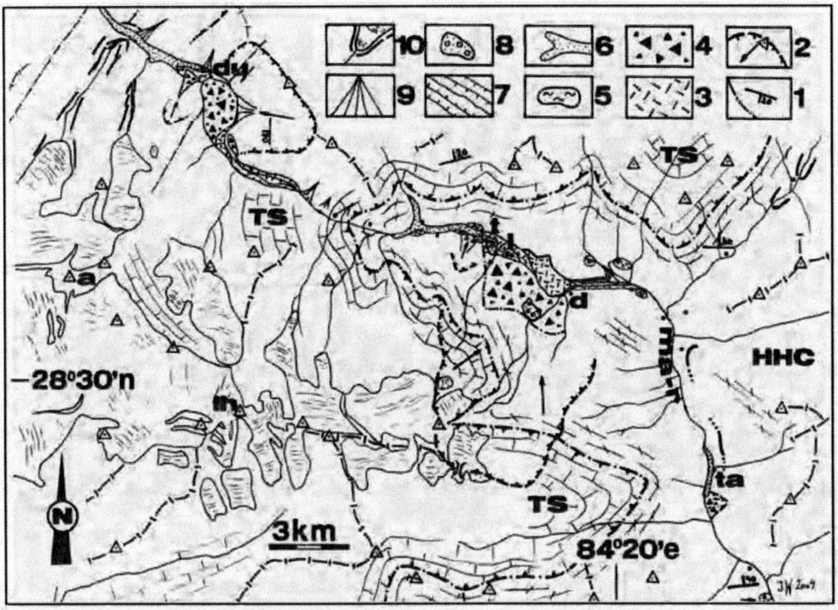


**Fig. 8.8** Geologic-geomorphologic sketch map of the lower Marsyandi River (*ma-r*) valley; rockslides of Dukur Pokhari (*du*) and Latamrang (*l*) and the rock avalanche of Tal (ta); Geotectonics after [5]. Legend: *HHC* = High Himalayan Crystalline, *TS* = Tibetan sediments; *1* = hard rock, fault, dip, *2* = peak, scarp and movement direction, *3* = rockslide deposit (shattered), *4* = rock avalanche deposit, *5* = lake sediments, *6* = extension of former lakes, *7* = terraced lake outflows, *8* = alluvial gravel of former spillway, *9* = debris cones, *10* = recent glaciers and moraines; *a* = Annapurna II (7,937 m), *lh* = Lamjung Himal (6,938 m), *d* = Danaque, *t* = Thanchouk

**Fig. 8.9** S-N view towards the Ringmo rockslide area (*ld*); *s* = scarp, ↓ = direction of movement, *pl* = position of Phoksundo Lake. Photograph by H. J. Ibetsberger 1995

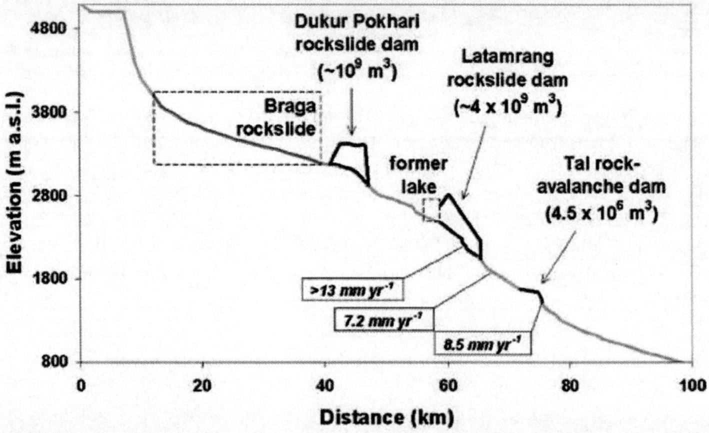


**Fig. 8.10** Upper image: SE-NW view towards Dukur Pokhari rockslide (*ld*). Location: 1.5 km west of Bhratang, alt. 2,900 m (Lower image). Long profile of the Marsyandi River formerly blocked by the large rockslides of Dukur Pokhari and Latamrang as well as by the rock avalanche of Tal. Note the formation of knickpoints and the contrast between high rates of fluvial incision and the preservation of the rockslide deposits (after [29]; $^{14}$C and $^{10}$Be/$^{26}$Al bedrock incision rates after [33])

## 2.3 Results, Conclusions and General Discussion on Giant Rockslide Dams in the Nepal Himalayas

Besides the huge volume of deposited material, the mechanism of sliding has played a major role in the stability of giant rockslide dams. In-situ collapses of rock masses dominate, with relatively short distances of dislocation and/or rapid translational movements causing huge blocks of material shattered by mechanical stress during

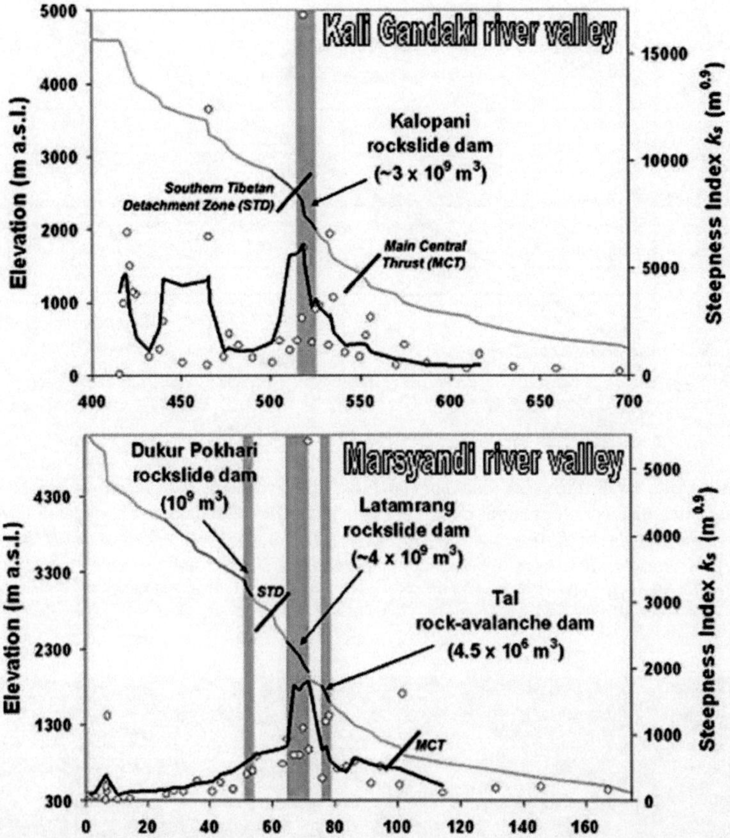


**Fig. 8.11** River long profiles (*grey lines*) with variation of the Steepness Index $k_S$ (*white dots*) and its 5-point running mean (*black lines*) for fixed Concavity Index (=0.45) show spatial coincidence between high values of $k_S$ and deposits of large rockslides (*grey shaded boxes*) (after [29])

sliding. These shattered but not disintegrated masses have been the basis for secondary stabilization of the landslide dams and the life span of the lakes dammed. They underwent recompactation and secondary cementation by mineralized springs and/or mountain waters. Nevertheless, very few areas still have an existing dammed lake in their impoundments. Other geological processes, such as sedimentation into the reservoirs and/or erosion of the dam over thousands of years cause the lakes to disappear.

## 3 Lake Damming Rock Avalanches in the Indian and Nepal Himalayas and in the Qin Ling Mountains of China

Generally, rock avalanches are much more common than giant rockslides, especially in areas with active tectonic history, different lithologies, stress release due to deglaciation, and a high morphologic overprint causing steep valley flanks and deep

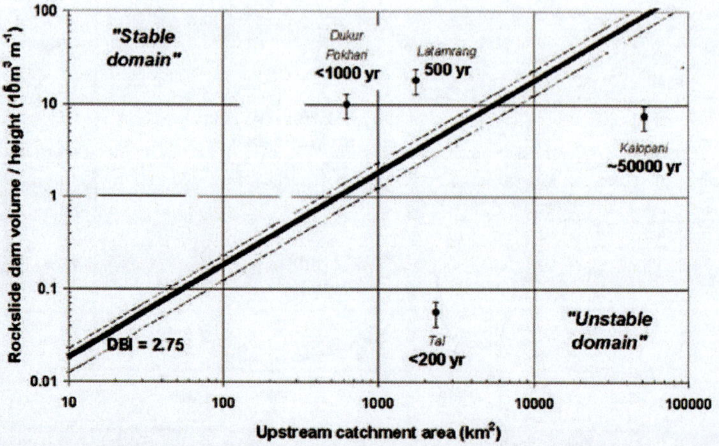


**Fig. 8.12** Volumes of three rockslide dams and one rock avalanche dam scaled by height, versus upstream catchment area for natural river blockages in the Kali Gandaki and Marsyandi River valleys of Nepal Himalayas (*bars* indicate +/–30% error). Dimensionless blockage index ($DBI = 2.75$) after [9] for world-wide data (*diagonal line*), separating "stable" from "unstable" landslide dams (empirically derived); (after [29])

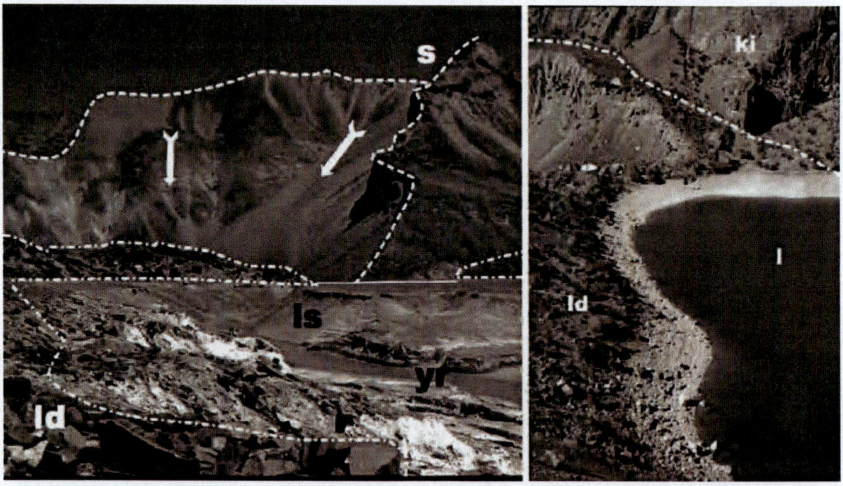


**Fig. 8.13** *Left* side: view of Sarai Kenlung rock avalanche; $s$ = scarp, ↓ = direction of movement, $ld$ = landslide deposit, $ls$ = lake sediments, $yr$ = Yünan Chu River. *Right* side: view of Tso Tok Phu rock avalanche; $ld$ = landslide deposit, $l$ = Tso Tok Phu Lake, $ki$ = Kioto Limestones. Photograph by G. Fuchs, 1985

gorges. This section of the paper presents examples from the Tibetan Zone of the Himalayas and the High Himalayan Crystalline Zone. Data on two rock avalanches from the Qin Ling Mountains of China in a similar litho-tectonic position have been added.

### *3.1 General Description of Lake-Damming Rock Avalanches in the Tibetan Himalayas of India*

Frequently-occurring earthquakes often serve as triggers for rock avalanches in the Tibetan Zone of the Himalayas, e.g., in Southern Zanskar, India. One of the reasons might be the changing lithological conditions of the sedimentary sequence of the Zanskar Synklinorium (Fig. 8.14).

North of Kyelong town (district Lahul) the road from Manali (Himachal Pradesh) to Leh (Ladakh) crosses two deposits of rock avalanches (Fig. 8.14) [30, 31, 45]. At Pateo, where the N-S striking Bhaga River Valley has its confluence with an eastern tributary, a rock avalanche deposit is located between altitudes of 3,850–3,925 m (Fig. 8.15). It covers an area of 2 km$^2$ and once dammed 2 lakes. 5.5 km NNE of

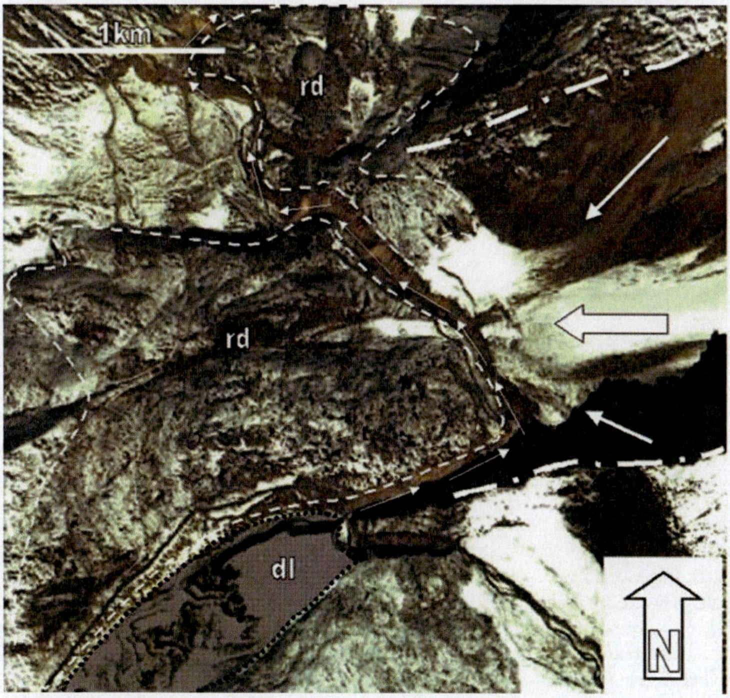


**Fig. 8.14** Satellite image (source: Google Earth) of Sarai Kenlung rock avalanche (77° 27'E, 32° 48'N); ↓ = direction of movement, *rd* = rock avalanche deposits, *dl* = former dammed lake (→) = proximal spillway breach channel along Yünan Chu River

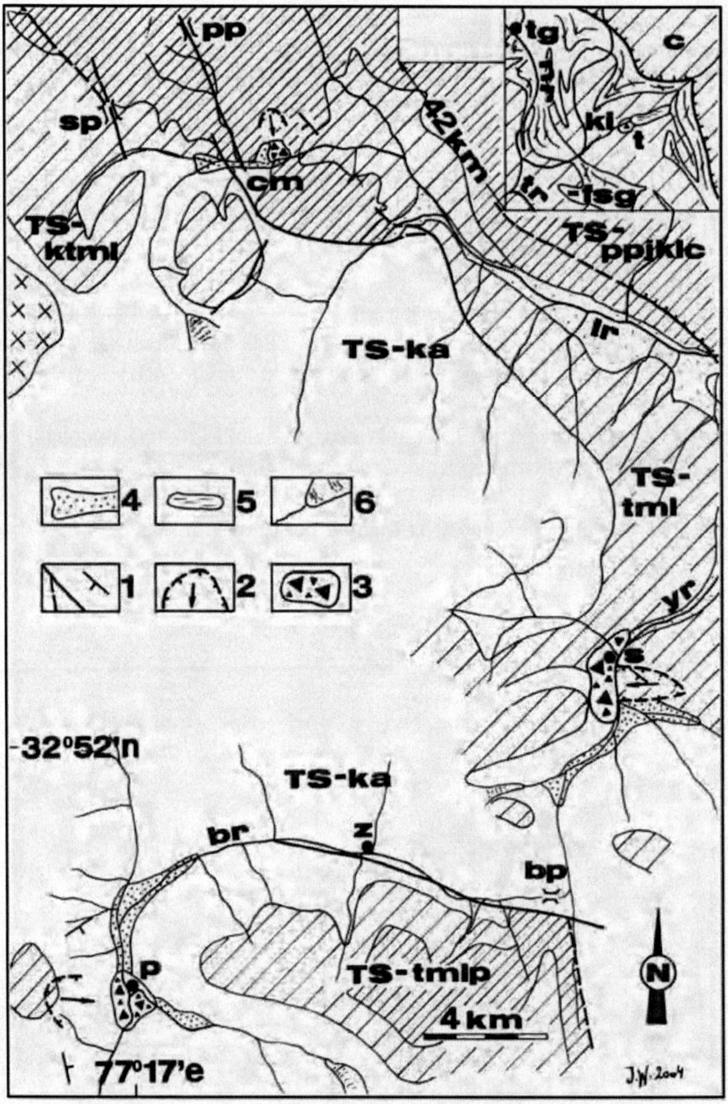


**Fig. 8.15** Geologic sketch map of SE Zanskar [13] with the locations of the lake-damming rock avalanches of Pateo (*p*, 3,740 m), Sarai Kenlung (*s*, 4,540 m), Chimik Marpo (*cm*, 4,650 m) and Tso Tok Phu (*t*); Legend: *TS* = Tibetan sediments (*ka* = Karsha, *k* = Kurgiagh, *t* = Thaple, *m* = Muth, *l* = Lipak, *p* = Po, *pj* = Panjal, *kl* = Kuling, *c* = Triassic-Jurassic carbonate, *f* = Ferruginous, *s* = Spiti, *g* = Giumal formations; *1* – Hardrock, faults, dip, 2 – Scarp, direction of movement, *3* – Rock avalanche deposit, *4* – Former dammed lake, *5* – Recent dammed lake, *6* – Recent glaciers and rivers; *bp* = Bara Lacha Pass (4,810 m), *br* = Bhaga River, *lr* = Ligti Chu River, *nr* = Niri Chu River, *pp* = Phirtse Pass (5,400 m), *sp* = Sirichun Pass, *tg* = Tanak Monastery, *tr* = Tsarap Chu River, *yr* = Yünan Chu River, *z* = Zingzingbar (3,900 m)

the Bara Lacha Pass (4,810 m) another barrier once dammed the main N-S striking valley of Yünan Chu, forming a lake at its confluence with an eastern tributary near today's Sarai Kenlung (Fig. 8.13). Near Chumik Marpo (4,650 m), 6 km ESE of Sirichun Pass and 5 km SE of Phirtse Pass (5,400 m), an area of 0.5 km$^2$ is covered by a blocky stream of boulders in the upper course of the Lingti Chu River. It once dammed a lake valley due W. 42 km further NNW another rock blockage dammed the still existing lake Tso Tok Phu (5 km SE of Tanak monastery, Fig. 8.13, right image) in a northern tributary valley of Tsarap Chu River, affecting an area of about 0.4 km$^2$ (Fig. 8.15).

## 3.2 *Stability Conditions of Rock Avalanche Dams in the Tibetan Himalayas and Life Span of Their Dammed Lakes*

The material of Pateo rock avalanche disintegrates to coarse-grained and irregular boulders due to weathering along the bedding. That is why the dam is composed of a material with high content of fine as well as large angular blocks and boulders. No lacustrine sediments were recognized in the impoundment(s), but several alluvial terraces give evidence of former landslide dammed lakes behind the barrier of the rock avalanche – one to the north and one to the east. Their spillways have formed deep gullies into the deposit (Fig. 8.16), so that the dam still acts as a hazard in the case when landslide material generated by undercutting of the steep flanks blocks

**Fig. 8.16** Deeply eroded gully of Bhaga River (*br*) through the Pateo rock avalanche deposit (*ld*), seen from Pateo (alt. 3,950 m); (↓) = potential masses for collapsing into and damming the river temporarily

the river again. Then new lakes could form, endangering villages down valley by overtopping the new dam.

As seen on the southern rim of the scarp of Sarai Kenlung rock avalanche (Figs. 8.13 and 8.14), the displacement of its material of conglomerates, quartzites, (carbonate) sandstones, slates and thin lenses of dolomite, seems to have started along a fault structure; the sliding surface was created along the bedding of the bedrock, another reason for creating a deposit like a huge blocky stream. Today the former dammed lake behind the rock avalanche is filled with lacustrine and alluvial sediments. The most interesting part of the former lake close to the barrier is composed of eroded terraces of lake sediments with different levels. Additionally, the spillway and the proximal part of the landslide give evidence of several outbreaks of the lake.

Exposures in the scarp area of Chumik Marpo rock avalanche, NNW of the deposit, show a mixture of very hard, thick-bedded rocks and ductile ones, a composite material with big boulders in a finer matrix. This feature is similar to that in Pateo and might be the reason for the erosion of the dam after a short life span of the full lake. The rest of the lake basin was silted up and filled up with alluvial sediments.

The blocky materials of the Kioto Limestone (Fig. 8.13), which mainly composed the deposit of the Tso Tok Phu rock avalanche, obtained its stabilizing input from fine sediments of the seeping waters. Due to dry weather conditions, low rate of sedimentation into the reservoir, and low rates of erosion, the lake still exists.

### *3.3 General Description of Rock Avalanches in the High Himalayan Crystalline Zone of India and Nepal*

South of Bawa Pass (4,800 m, Himachal Pradesh, India), a rock avalanche dam (alt. 3,380–3,600 m) originated from the left (SE) side of the valley, where it changes its trend from E-W to N-S. The blocky material has dammed Garak Chu River close to its confluence with Lanak Chu River. The impact of this main event triggered another, smaller rock avalanche on the opposing valley wall.

One of the most important sites of rock avalanches of the lower Marsyandi Valley is the deposit near the village of Tal (alt. 1,600 m, Fig. 8.8). 7 km south of the confluence of the Dudh Khola River with the main river, this rock avalanche, which occurred perpendicularly to the N-S striking valley, dammed the river to form a lake. Today this former basin is silted up with lake and alluvial sediments (Fig. 8.17).

Similar rock avalanches can be found in eastern Nepal, about 10 km north of the Main Central Thrust, at Lamabagar (86°13'E, 27°55'N; crown elevation of the dam 1,970 m) and Chaunrikharka (86°43'E, 27°42'N; crown elevation of the dam 2,550 m) villages, the latter located close to the eroded Gath-Phakting sediments, deposited in the formerly dammed lake [18, 41].

**Fig. 8.17** View towards N to the former lake (*l*) dammed by the rock avalanche (*ld*) of Tal (*t*); *ma-r* = Marsyandi River. Observation point 1 km S of Tal, alt. 1,650 m

## *3.4 Stability Conditions of Rock Avalanche Dams in the High Himalayan Crystalline Zone and Life Span of Their Dammed Lakes*

The area of the impoundment behind the rock avalanche of Bawa La indicates the damming of a lake at an altitude of 3,560 m the Garak Chu Valley upstream as far as the confluence with the tributary of Puiti Chu River. Due to the small distance of landslide travel, the material of the rock avalanche dam is course and blocky. Flowing and seeping waters did not damage the dam, but stabilized it by depositing sediment between the boulders. Therefore the dam increased its stability over time. Later, a spillway was formed at the distal side of the deposit, between the rock avalanche and the hard rock of the opposing valley wall. The river has cut through this barrier; at least two alluvial terraces along the lake basin document two stages of lake levels, before it disappeared completely.

One of the reasons why the impoundment of Tal (Fig. 8.17) was silted up before dam breaching was the mechanism of the rock avalanche. The dislocated mass contains mostly boulders lacking fine grained material. Additionally, secondary cementation and compactation of the boulders with fine sediments deposited by flowing seeping waters, provided the most important influence on the stability of the dam after emplacement [38]. The dam remained stable and the lake disappeared due to sedimentation into it. The spillway at the distal part of the deposit was successively cut down by erosion. This occurred as the village of Tal (which means lake) was founded on the shore of a still existing lake, at most, 500 years ago.

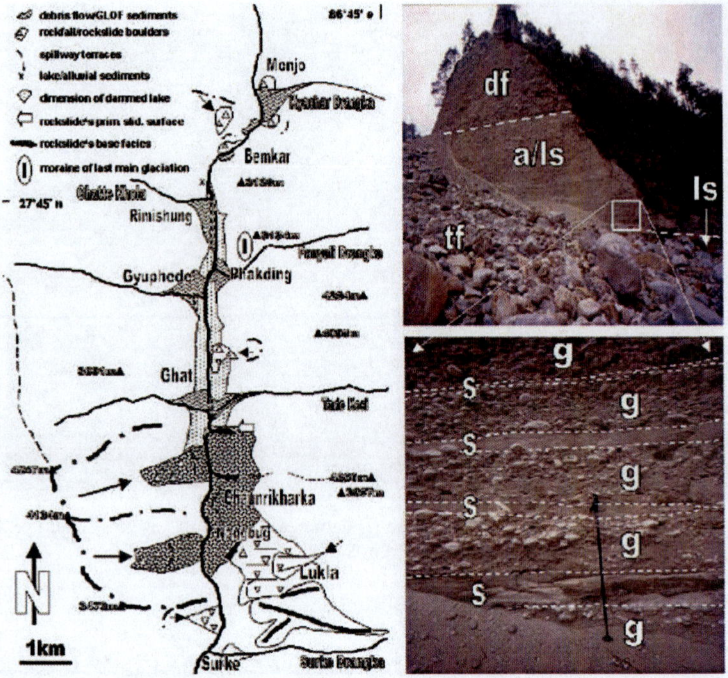


**Fig. 8.18** Geologic sketch map of the Chaunrikharka rockslide (*left* side), and exposed backwater sediment infill of the Dudh Kosi River at Gyuphede village (2,560 m a.s.l). Description of sequence from bottom to top in upper right: (*ls*) = silty sands, interpreted as lake sediments; (*a/ls*) = alternating layers (See enlargement in lower right; with walking stick for scale. Fine sands (*s*) and well-sorted gravel (*g*), possibly generated during changes between stagnating and rapidly flows (thickness 20 m)) (*df*) = several generations of debris-flow sediments (total thickness 50–100 m) (*tf*) = torrential sediments in a wide debris fan (after [53]

Both rock avalanche dams at Lamabagar and Gath-Chaunrikharka display long-term stability [41], which suggests that the large volume, the cohesiveness of the deposited material, and rapid siltation in the impoundment (Fig. 8.18) are the most important (Fig. 8.19) factors for the stability and life span of these natural rock blockages. Lamabagar, with its boulder-armed spillway, has been stable for more than 3 generations and is completely silted up. The remnants of the postglacial Chaunrikharka dam, and the sediments of the formerly dammed Gath-Phakting Lake behind it, indicate a period dam stability of at least 120 years. After a first outbreak, a period of silting up with periodical downcutting of the dam followed. This is documented by alluvial terraces within the outflow channel [53].

## *3.5 General Description of Lake-Damming Rock Avalanches in the Qin Ling Mountains of Northern China*

Rock avalanches are common in China and seismic activity is their main trigger [57]. As some of them have occurred within Mesozoic granites of the Qin Ling

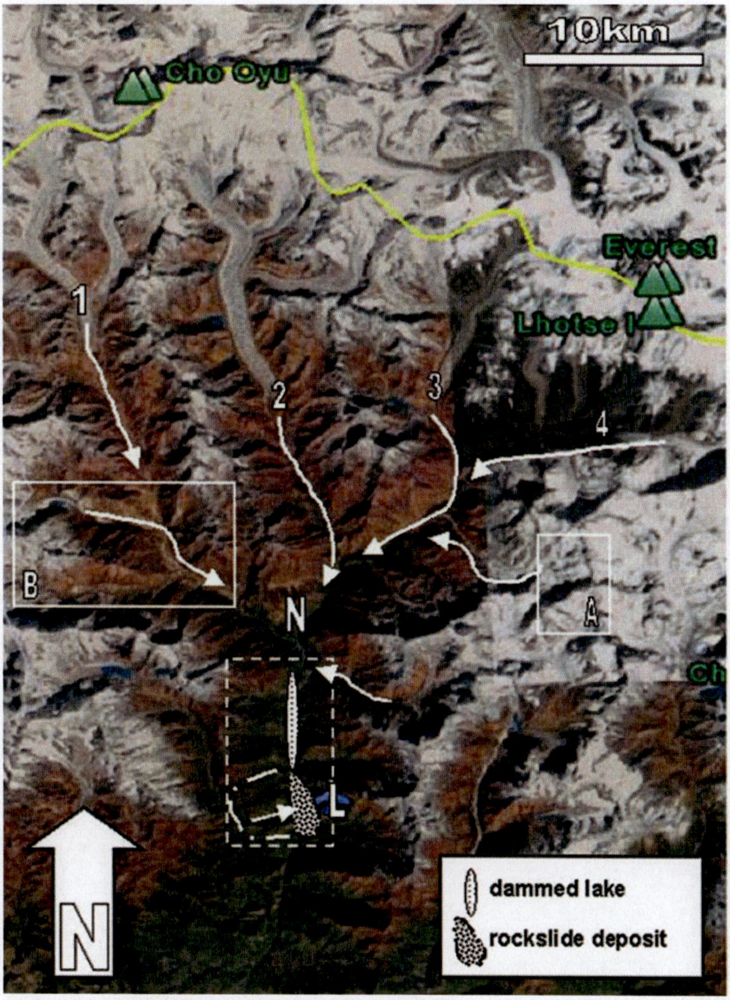


**Fig. 8.19** Overview of the catchment area of Gath-Chaunrikharka rockslide dam (source: Google Earth): *1* – Nangpa glacier, *2* – Ngozumpa glacier, *3* – Khumbu glacier, *4* – Lhotse glacier. *L* – Rockslide deposit close to Lukla, scarp and formerly dammed lake (shown in Fig. 8.18), *N* – Namche Bazar village; inferred tracks of prehistoric and recent GLOFs in the Khumbu area (*white arrows*), which may have been contributed to infilling of the former dammed lake: *A* = Nare Dranga from Ama Dablang South Glacier (6,856 m a.s.l), which occurred in 1977, *B* = Dig Tsho from Langmoche glacier, which occurred in 1985, and other possible sources in upper tributary valleys (after [53]

Mountains (Fig. 8.1), an orogenic belt of the collision type [23] comparable to litho-tectonic conditions to the Himalayas, results from that area are included in this study.

Our two rock avalanche dams in the Cui Hua Mountains (Fig. 8.20) 30 km south of Xi' an (Dong Cha Valley, Shaanxi) were triggered by an earthquake in 780 BC [51]. The lower of them – the composite dam of the three Cui Hua rock avalanches–

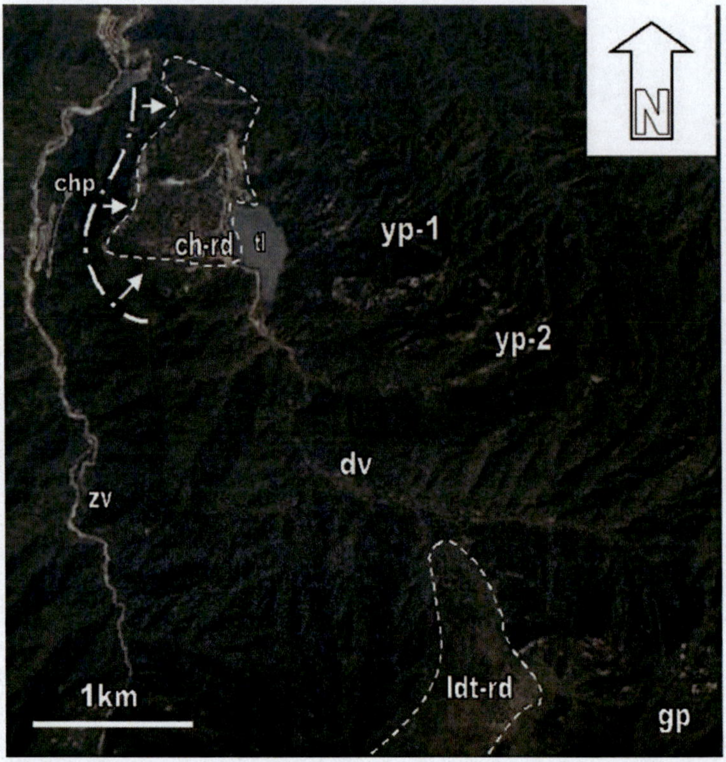


**Fig. 8.20** Satellite image of Cui Hua- (*ch-rd*) and Liu Dshe Tse (*ldt-rd*) rock avalanches (source: Google Earth, position: 109°00'E, 33°58'N); *zv* = Zhengza valley, *dv* = Dong Cha valley, *chp* = Cui Hua Peak (1,414 m), *yp-1* and *yp-2* = Yunan Peak I (1,668 m) and II (1,928 m), *gp* = Gangiu peak (2,045 m)

dammed Tianchi Lake. On the shore of this "Lake of Heaven" (alt. 1,210 m) local people have founded a village called Chui Tsho Tshe close to an ancient spillway of the lake to enjoy the benefits of this natural water storage. The upper Liu Dshe Tse rock avalanche dam, situated about 5 km south of Tianchi Lake, did not dam a lake because of permeability on the bottom of the impoundment created by tectonic structures.

## 3.6 Stability Conditions of Rock Avalanche Dams and Life Span of a Dammed Lake in the Qin Ling Mountains of China

The upper surface of Cui Hua rock avalanche – a giant fan-like cone up to the main scarp in the west – exhibits extremely huge blocks and boulders (Fig. 8.21). Instead of piping inside the rock avalanche, seeping waters have given a stabilizing input by filling voids between the blocks with fine-grained sediments. Nevertheless, in

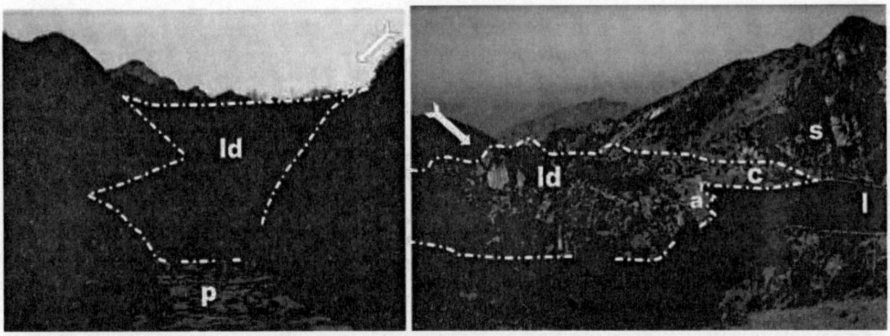


**Fig. 8.21** *Left*: view from the pathway (*p*) towards SSE, to the dam of Cui Hua rock avalanche (*ld*); ↓ = direction of movement. *Right*: Chui Tsho Tshe village (*c*) on top of Cui Hua rock avalanche (*ld*); *s* = spillway of Tianchi Lake (*l*) regulated for agriculture; *a* = artificial sealing of lake

the upper part of the deposit (carapace) water can still easily flow through the big boulders and the coarse-grained material. That is why the northern shore of Tianchi Lake has been sealed by the locals with artificial cementation, preventing the water from seeping through the material of the rock avalanche.

Due to this stabilizing input, the rockslide dammed waters of Tianchi Lake are kept at a higher level, thus functioning as a water reservoir. And with the isometric shape of the deposit and a volume of more than 100 times that of the dammed waters behind it, the barrier of the Cui Hua rock avalanche is considered safe from breaching.

### *3.7 Results, Conclusions and General Discussion on Rock Avalanche Dams*

The mode of rockslope failure is the main reason for the stability of these natural dams. The primary deposited materials of the landslides were mainly poorly sorted blocks and boulders, while secondary fine-grained sediments that were introduced by seeping river and lake waters have been reducing leakage and increasing the resistance of the dam. Additionally, these dammed lakes were filled with alluvial and lacustrine sediments before the dams could breach.

Although erosion and sedimentation has left only Tianchi Lake, each of these examples had a long life span.

## 4 Lake-Damming Landslides in the Lesser Himalayas of India and Nepal

Landslide dams made of shattered fine debris pose a frequent and serious problems to life and infrastructure; their dammed lakes are always of very short life span as these dams tend to breach by overtopping within short time. This special kind of an in-situ collapsed debris stream as the mechanism of a landslide is typical of the Lesser Himalayas.

## 4.1 General Description of Landslides in the Lesser Himalayas of Nepal and India

The rock avalanche of Ghatta Khola Valley [43, 44] is located in Western Nepal's Lower Himalayas (Fig. 8.1) [11]. It has dammed the main river, which descends from NNW to SSE, up to an altitude of 3,010 m.

In the Lower Himalayas of Kumaon (Uttarakhand, India) [14], one of the rare landslide-dammed lakes on the southern slopes of the Himalayas, the Gohna Tal, has existed throughout the twentieth century. Five km north of Chamoli village its water surface was situated at an elevation of 1,680 m, on the northern banks of the river Birahi Ganga, a tributary of the Alakananda (easternmost origin of Ganga River). A landslide that occurred in September 1893 during heavy rainfalls, dammed the main river, forming a lake that increased in size for a year until its first outburst (Fig. 8.22).

From the Buri Gandhaki Valley (Manaslu Himal, Central Nepal) Jacobsen [28] reported two rock avalanche deposits that have dammed the main river during recent times. One of them is located 1.5–3 km south of the village of Jagat, 40 km SE of

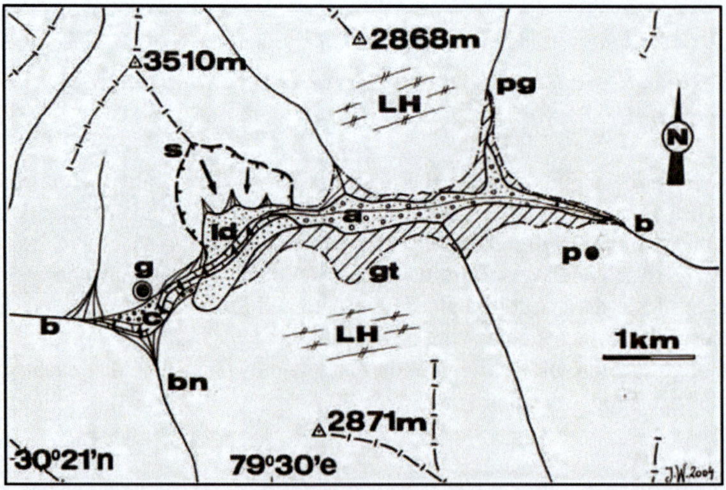


**Fig. 8.22** Geologic sketch-map of Gohna Tal rockslide; *LH* = Lesser Himalayas, *s* = scarp (2,900 m) (↓) = direction of movement, *ld* = landslide deposit, *gt* = max. extent of the lake, *a* = final extent and alluvial filling of the lake, *c* = terraced outbreak channel along the Birahi Ganga River (*b*), *g* = Gohna village (1,600 m), *p* = Pagna village, *pg* = Pul Gadera River, *pn* = Begar Nala River

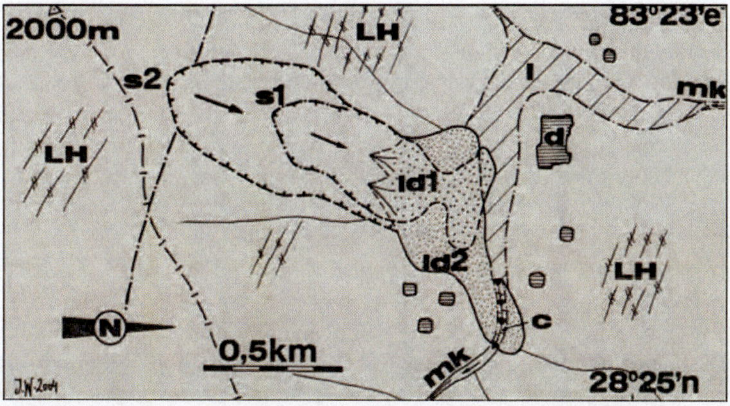


**Fig. 8.23** Geologic sketch map of Dharbang landslide; *LH* = Lesser Himalayas, *s1/2*, *ld1/2* = scarp and landslide deposit from 1926 (1,400 m) respectively from 1988 (1,750 m), ↓ = direction(s) of movement, *l* = temporary dammed lake 1988, *c* = outbreak channel along the Myagdi Khola River (*mk*), *d* = Dharbang village, houses (1,150 m)

Manaslu Peak (8,162 m); the other one is near the village of Labubesi, just 30 km NE of Gorkha.

The landslide of Dharbang in the Myaghdi Khola valley (western tributary of the Kali Gandaki River) is located within the Lesser Himalayas of Western Nepal (Fig. 8.1) [12]. The special geotectonic location [54, 56] and the climatic circumstances during monsoon where the preparatory causal factors and the triggers of the initial landslide that occurred in 1926, and the reactivation of the 1926 debris in 1988 [24, 44]. The S-N falling masses destroyed the former villages on the right bank of the river and dammed the river temporarily at an altitude of 1,100 m (Fig. 8.23).

The 1998 landslide of Tatopani in the lower Kali Gandhaki Valley (Nepal) represents another example of the collapse of a mountain flank triggered by increased pore water pressure during long periods of heavy monsoon precipitation. It occurred in the Lesser Himalayas in an area of a gigantic rock mass that is creeping towards the valley [42]. After initial movements on September 7 1998 and the opening of the main phase of failure on September 10, 400,000 $m^3$ of rock that originated from about 250 m above the Kali Gandhaki River, fell down from the eastern side of the valley on September 26, and accumulated in the form of a huge debris fan that functioned as a barrier, damming the river temporarily.

## *4.2 Stability Conditions of Landslide Dams in the Lesser Himalayas and Life Span of Temporary Dammed Lakes*

The boulders of Ghatta Khola landslide are flattened and cracked along pre-existing joints. Huge ones reach diameters of 4–5 m, although the bulk of the material is much smaller. Due to the avalanching mechanism of the landslide, the material was not heavily shattered. Seeping waters through the deposit have caused stabilization due to sedimentation between the boulders. The sediment-filled younger lake basin behind the dam and terraced lake sediments on the basin margin give evidence that an older lake with higher level must have had an active spillway and/or unspectacular outbreaks. Simultaneously Ghatta Khola River, originating from Ghurchi Lagna pass (alt. 2,500 m) and draining an area of 25–30 $km^2$, has silted up the entire lake basin.

Much more spectacular was the catastrophic outburst of the lake Gohna Tal, [see Chap. 1 by Evans et al., this volume] which devastated the valley of Alakananda River over a long distance downstream, when the dam failed on the August 26 1894, most probably because of overtopping. Subsequent catastrophic outbreaks of the lake are indicated by at least four river terraces of different ages within the eroded flanks of the landslide deposit and downstream in the valley, which were created during and after flood events (Fig. 8.24). Later, local eyewitnesses observed that the water level of Gohna Tal was lowered to a depth of 120 m. This water volume (80 $Mm^3$) was silted up with lake and alluvial sediments within a period of 78 years. This high rate of sedimentation was a function of the monsoon climate and the huge catchment area (200 $km^2$) in the vicinity of the Nanda Ghunti massif (6,500 m). In the 1972 monsoon Gohna Tal disappeared completely after a massive slide within the landslide deposits near its spillway.

Jagat and Labubesi lakes display a similar behaviour as Gohna Tal. Both lakes were dammed in recent times; both had a very short life span. One of the main reasons is the high average rate of sedimentation of the river. Jacobsen [28] calculated

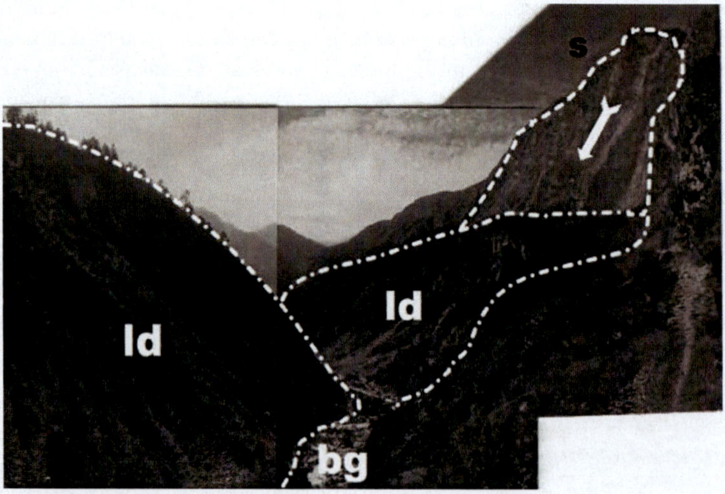


**Fig. 8.24** View from the western end of the former dammed Gohna Lake (alt. 1,660 m) towards the W; *s* = scarp of the landslide, ↓ = direction of movement, *ld* = deeply eroded deposit of the landslide on both sides of the Birahi Gang River (*bg*)

the flux of sediments into the former dammed lake at Labubesi at more than 16,000 t/day. The entire impoundment was filled up with lake and alluvial sediments within less than 17 years, before 1985. Similar depositional rates must be taken under consideration for the Jagat landslide, as in 1979 the sediments of the impoundment had already reached and partly buried an older landslide deposit upstream. Simultaneously the spillways of the lakes easily eroded and cut through the barriers of a totally crushed and fine-grained landslide material until the lakes were geological history (Fig. 8.25).

Because of the rapid breaching of a natural rock blockage, the Dharbang landslide of 1988 maintains an extraordinary position among all those investigated in the Himalayas so far. The source landslide material was saturated during the monsoon season before the event. Heavy rainfalls during late September caused the rock failure resulting in the rapid filling of a lake. Morphology and internal structure of the remaining deposition area after the lake's outburst (Fig. 8.26) show that the movement of the slide was an in-situ collapse due to increasing pore water pressure.

The fine-grained gravel and debris of Tatopani landslide provides a similar example of a dammed lake with a very short life span. Its level of water reached the height of Tatopani village, a thermal water health resort and tourist site located 1 km valley upstream on an alluvial terrace of the right side of the river. The rockslide dam failed due to overtopping. The landslide material was washed away, and the lake level dropped to a depth of 5 m. Recently even this remaining water disappeared by the accumulation of alluvial sediments in the lake basin and the erosion of the spillway.

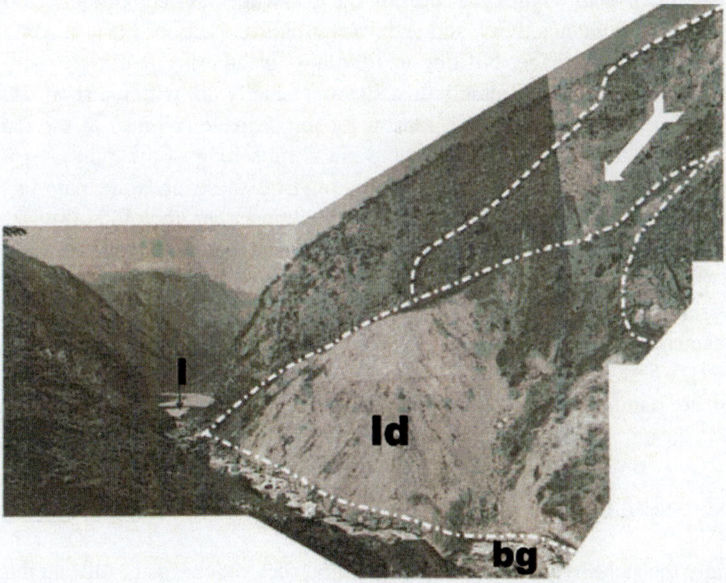


**Fig. 8.25** N-view towards the remnants of Labubesi landslide (*ld*); *bg* = Buri Gandhaki River, *l* = silted up impoundment, the former dammed lake. Note scarp and ↓ = direction of movement. Observation point at 1,150 m [28]

**Fig. 8.26** Panoramic view over the remnants of Dharbang landslide dam (*ld*): Note the morphology of the cone of landslide debris); *s2* = scarp (1988), ↓ = direction of movement, *mk* = Myagdi Khola River

## 4.3 Results, Conclusions and General Discussion on Recent Landslide Dams

Giant creeping rock masses associated with deglaciation, postglacial erosion and deep pre-existing joints and fissures have preconditioned the slopes and mountain flanks of the Lesser Himalayas to massive rock slope failure. In addition, sliding planes parallel to foliation and strata of rocks with intensive tectonic history have contributed to destabilization. In addition the landscape has been influenced by erosion processes, human activity, and extreme climatic conditions [25]. In some cases these slides dam drainage, but due to lithology, weathering and high pore water content, the barriers are extremely unstable and usually fail within a short time.

Pulverized rocks are the main reason for the seepage (piping) of the dammed water through the landslide dam and the initial breaching of the dam after a short period. That is why mountain waters originating from these landslide materials show higher electrical conductivity than those from compact hardrock [35, 48].

As the lithologies and geo-tectonics of the Lesser Himalayas are similar to areas in the Alps, the results of this study are comparable with landslides [16] there, e.g., the river damming rock avalanche of Val Pola 1987 [22]. Also mixtures of rock and ice avalanches such as those from Huascarán (6,654 m) in the Cordillera Blanca in Peru [32], where the Rio Santa River was dammed up to a lake of 2 km in length for 30 min, are comparable.

# 5 Discussion and Conclusions

Generally the stability of lake-damming natural rock blockages is attributed to their morphology [3, 17, 29], volume of debris, composition of the material including different lithological behavior [6, 37] as well as climatic conditions and the rates of sedimentation and/or transport of alluvial sediments in the lake. The analysis of twenty case studies presented in this paper shows that there are other additional important factors affecting the stability of landslide dams;

- *Movement-type* such as (i) en masse translational rock sliding, (ii) rock avalanching with rolling and bouncing components, (iii) in-situ collapse landslides with crushed material partly supersaturated with water.
- *Disintegration grade* of the landslide mass, which depends on (i) the volume of the mass, (ii) the mechanism of sliding and (iii) the distance of displacement.
- *Shattering grade and cataclastic reactions* depending on (i) the involved lithologies and (ii) their intensity of weathering.
- *Grain-boulder block-size distribution* of the landslide debris, e.g., (i) non disintegrated and interlocked blocks, (ii) boulders and (iii) shattered and crushed material, which is influenced by the type and distance of dislocation [7].
- *Secondary cementation/compaction* (which again influences the grain, boulder and blocksize) not only by seeping waters and/or transported sediments but also by natural mineralized and/or thermal springs in the landslide areas.
- *Mixture of landslide and moraine material*, as morainic material is highly susceptible to erosion by overtopping and piping of a dam.
- *Size of the catchment area, climatic conditions and rates of sedimentation* of the landslide-dammed river, as the availability of lake and alluvial sediments and their rapid transport is most important for the silting up of the lake basin before the breaching of the dam.

## 6 Blockages and the Life Span of Their Dammed Lakes

The above conclusions led to a diagram (Fig. 8.27), a tool for estimating potential hazard, which should help in determining whether a lake-damming landslide is stable. The diagram correlates the grain, boulder and block size of landslide material and the stability of a dam (life span of the dammed lake). Figure 8.27 shows that the greater the average diameter of the components, the longer the life of the dam and the lake. This relationship is suggested by the field data analyzed in this study, although it was not possible to determine quantitatively the grain size of the landslide materials [4]. The average grain, boulder and block size was visually estimated by analyzing different outcrops. En masse translational dislocated blocks, which were shattered by mechanical stress during sliding, are counted as single pieces.

The scientific message of Fig. 8.27 for the life span of landslide dammed lakes is the following: not only very large, en masse dislocated blocks (dozens of meters in diameter) are stabilized and recompacted by secondary infilling, but also boulders of meter-size as seeping mountain waters give a stabilizing input by filling the interspaces with fine-grained sediments (transported by the water) instead of piping

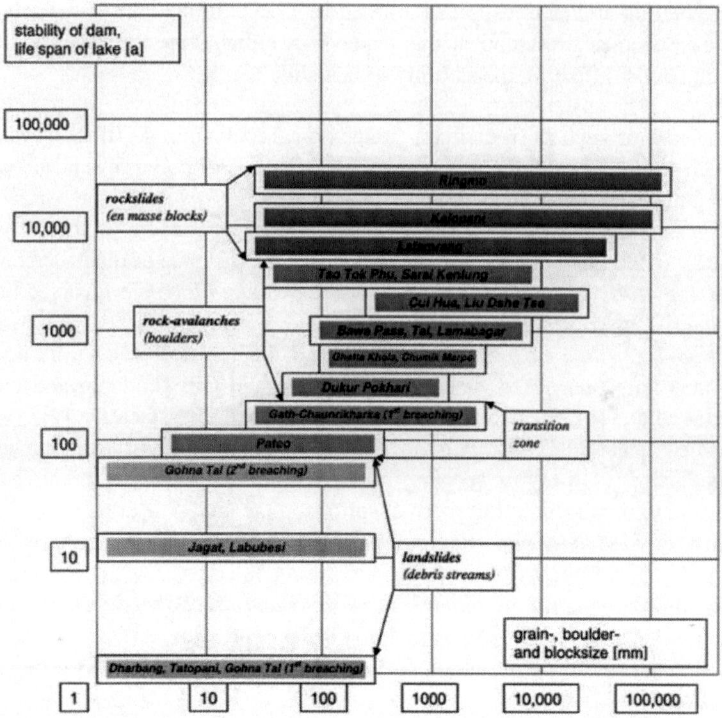


**Fig. 8.27** Grain boulder block size stability diagram of lake-damming rockslides, rock avalanches and debris stream landslides from India, Nepal and China

inside the deposit. The critical element of this system is defined by the average grain size of the deposit that can easily be transported out by seepage waters. The weakest stability conditions are represented by debris-stream dams that have been shown to fail rapidly by overtopping and/or piping. These results correlate very well with observations recently made on the behavior of landslide dams triggered by the 2008 Wenchuan earthquake in China [57].

**Acknowledgements** Field trips to India and Nepal were financially supported by the Austrian Science Foundation, Vienna (grant No. P 9433-Geo). The study in the Qin Ling Mountains was sponsored with a habilitation grant by the Foreign Office of the University of Salzburg and the Northwest University (Xibei Daxue) of Xi'an (China). The author is grateful to Prof. Jiading Wang, Prof. Naixi Ma and Prof. Zhousu Li for scientific discussions during the field work. Financial support for the field trip around the Annapurnas in autumn 2003 was provided by the Stiftungs- und Förderungsgesellschaft of the University of Salzburg. Analysis of the field data was done in the Department of Geography and Geology, Salzburg University as well as Erkudok© Institute/K-Hof Museums of Gmunden (Upper Austria). Last but not least, special thanks to the organizers of the Nato ARW 2004 in Bishkek for their invitation, Dr. Oliver Korup (formerly of SLF, Davos, Switzerland) and an anonymous reviewer for carefully checking the manuscript and inspiring comments.

## References


1. Baade, J., Lang, A., Mäusbacher, R. and Wagner, G.A. (1998) Quaternary lake deposits in the Thakkhola Graben. Mustang, Nepal. *13th Himalaya-Karakorum-Tibet International Workshop, April 20–22, 1998, Peshawar, Pakistan, Abstract Volume,* pp. 22–23.
2. Bookhagen, B., Thiede, R.C. and Strecker, M.R. (2005) Late Quaternary intensified monsoon phases control landscape evolution in the northwest Himalaya, *Geology* **33**, 149–152.
3. Casagli, N. and Ermini, L. (1999) Geomorphic Analysis of Landslide Dams in the Northern Apennine, *Transactions of the Japanese Geomorphological Union* **20**, 219–249.
4. Casagli, N., Ermini, L. and Rosati, G. (2003) Determining grain size distribution of the material composing landslide dams in the Northern Apennines: Sampling and processing methods, *Engineering Geology* **69**, 83–97.
5. Colchen, M., Le Fort, P. and Pêcher, A. (1986) *Geological Researchs in the Nepal's Himalaya: Annapurna – Manaslu – Ganesh Himal. Notice of the geological map on 1:200.000,* Editions du Centre National de la Recherche Scientifique, Paris.
6. Costa, J.E. and Schuster, R.L. (1988) The formation and failure of natural dams, *Geological Society of America Bulletin* **100**, 1054–1068.
7. Davies, T.R. and McSaveney, M.J. (2004) Dynamic fragmentation in landslides: Application to natural dam stability, in K. Abdrakhmatov, S.G. Evans, R. Hermanns, G. Scarascia Mugnozza, and A.A. Strom (eds.), *NATO ARW Bishkek: Security of Natural and Artificial Rockslide Dams Abstract Volume*. pp. 28–34.
8. Dortch, J.M., Owen, L.A., Haneberg, W.C., Caffee, M.W., Dietsch, C. and Kamp, U. (2008) Nature and timing of large landslides in the Himalaya and Transhimalaya of northern India, *Quaternary Science Reviews* **28**, 1037–1054, DOI 10.1016/j.quascirev.200805.002.
9. Ermini, L. and Casagli, N. (2003) Prediction of the behaviour of landslide dams using a geomorphological dimensionless index, *Earth Surface Processes and Landforms* **28**, 31–47.
10. Fort, M. (2000) Glacier and mass wasting processes: Their influence on the shaping of the Kali Gandaki Valley (higher Himalaya of Nepal), *Quaternary International* **65/66**, 101–119.
11. Fuchs, G. (1977) The Geology of the Karnali and Dolpo Regions, Western Nepal, *Jb Geol B-A Vienna* **120**, 165–217.
12. Fuchs, G. and Frank, W. (1970) The Geology of West Nepal between the Rivers Kali Gandaki and Thulo Bheri, *Jb Geol B-A Vienna* **Sb. 18**, 1–103.
13. Fuchs, G. and Linner, M. (1995) Geological Traverse across the Western Himalaya – a Contribution to the Geology of Eastern Ladakh, Lahul, and Chamba, *Jb Geol B-A Vienna* **138**, 655–685.
14. Fuchs, G. and Sinha, A.K. (1978) The Tectonics of the Garhwal-Kumaun Lesser Himalaya, *Jb Geol B-A Vienna* **121**, 219–241.
15. Hanisch, J. (1995) Large-scale valley damming of Kali Gandaki (Nepal) – Lacustrian sediments upstream and debris flow deposits downstream, in D.A. Spencer, J.-P. Burg, and C. Spencer-Cervato (eds.), *10th Himalaya-Karakorum-Tibet Workshop, Abstract Volume. Mitt Geol Inst ETH Zürich* **298**, 294–297.
16. Heim, A. (1932) *Bergsturz und Menschenleben*. Fretz und Wasmuth, Zürich.
17. Hermanns, R.L., Niedermann, S., Gonzalez Diaz, F.E., Fauque, L., Folguera, A., Ivy-Ochs, S. and Kubik, P. (2004) Landslide dams in the argentine andes, in K. Abdrakhmatov, S.G. Evans, R. Hermanns, G. Scarascia Mugnozza, and A.A. Strom (eds.), *NATO ARW Bishkek: Security of Natural and Artificial Rockslide Dams Abstract Volume*. pp. 79–85.
18. Heuberger, H. and Weingartner, H. (1986) Die Ausdehnung der Letzteiszeitlichen Vergletscherung an der Mount-Everest-Südflanke, *Nepal Mitt Österr Geograph Ges* **127**, 71–80.
19. Hewitt, K. (1998) Catastrophic landslides and their effects on the upper indus streams. Karakorum Himalaya, northern Pakistan, *Geomorphology* **26**, 47–80.
20. Hewitt, K. (2002) Styles of rock avalanche depositional complexes conditioned by very rugged terrain, Karakoram Himalaya, Pakistan, *Geological Society of America Reviews in Engineering Geology* **XV**, 345–377.

21. Hodges, K.V., Wobus, C., Ruhl, K., Schildgen, T. and Whipple, K.X. (2004) Quaternary deformation, river steepening, and heavy precipitation at the front of the higher Himalaya ranges, *Earth and Planetary Science Letters* **220**, 379–389.
22. Huber, A. (1992) Der Val Pola Bergsturz im oberen Veltlin vom 28. Juli 1987, *Eclogae Geologicae Helvetiae* **85**, 307–325.
23. Hsu, K.J., Wang, Q., Li, J., Zhou, D. and Sun, S. (1987) Tectonic evolution of the Qinling mountains, China, *Eclogae geol Helv* **80**, 735–752.
24. Ibetsberger, H.J. and Weidinger, J.T. (2000) Role of extreme meteorological anomalies in initiating the darbang landslide, Dhaulagiri Himal, Western Nepal, *Journal of Nepal Geological Society* **21**, 35–40.
25. Ives, J.D. and Messerli, M. (1987) *The Himalayan Dilemma – reconciling development and conservation*. Routledge, London, New York, NY.
26. Iwata, S., Yamanaka, H. and Yoshida, M. (1982) Glacial landforms and river terraces in the Thakkohla region, central Nepal, *Journal Nepal Geological Society* **2**, 81–94.
27. Iwata, S., Sharma, T. and Yamanaka, H. (1984) A preliminary report on geomorphology of central Nepal and Himalayan uplift, *Journal Nepal Geological Society* **4**, 141–149.
28. Jacobsen, J.-P. (1990) Die Vergletscherungsgeschichte des Manaslu Himalayas und ihre klimatische Ausdeutung. *Geo aktuell 1-terrimago-Verlag Pachnicke, Göttingen*, 1–76.
29. Korup, O., Strom, A.L. and Weidinger, J.T. (2006) Fluvial response to large rock-slope failures: Examples from the Himalayas, the Tien Shan and the Southern Alps in New Zealand, *Geomorphology* **78**, 13–35.
30. Mitchell, W.A., Dunning, S. and Taylor, P.J. (2001) Preliminary investigations of rock avalanches in the Indian Himalaya, *Journal of Asian Earth Sciences* **19**, 45.
31. Mitchell, W.A., McSaveney, M.J., Zondervan, A., Kim, K., Dunning, S.A. and Taylor, P.J. (2007) The Keylong Serai rock avalanche, NW Indian Himalaya: Geomorphology and palaeoseismic implications, *Landslides* DOI 10.1007/s10346-007-0085–0.
32. Patzelt, G. (ed.) (1983) *Die Berg- und Gletscherstürze vom Huascaran, Cordillera Blanca, Peru*. University Press Wagner, Innsbruck.
33. Pratt-Sitaula, B.A., Burbank, D.W., Heimsath, A. and Ojha, T. (2004) Landscape disequilibrium on 1,000–10,000 years scales, Marsyandi River, Nepal, central Himalaya, *Geomorphology* **58**, 223–241.
34. Schneider, J.F. (2004) Risk Assessment of remote Geohazards in central and Southern Pamir/Tajikistan, GBAO, Tajikistan, in K. Abdrakhmatov, S.G. Evans, R. Hermanns, G. Scarascia Mugnozza, and A.A. Strom (eds.), *NATO ARW Bishkek: Security of Natural and Artificial Rockslide Dams Abstract Volume*. pp. 164–169.
35. Schramm, J.-M. and Weidinger, J.T. (1996) Distribution of electrical conductivity at Tsergo Ri landslide, central-north Nepal, in *Proceedings of the 7th International Symposium on Landslides*, Balkema, Rotterdam, pp. 889–894.
36. Schramm, J.-M., Weidinger, J.T. and Ibetsberger, H.J. (1998) Petrologic and structural control on geomorphology of prehistoric Tsergo Ri slope failure, Langtang Himal, Nepal, *Geomorphology* **26**, 107–121.
37. Schuster, R.L. (1986) *Landslide Dams, Processes, Risk and Mitigation*. Geotechnical Special Publication 3, ASCE, New York, NY.
38. Shaller, P.J. (1991) Analysis and implications of large martian and terrestrial landslides. Ph. D. thesis, Calif. Inst. of Tech., Pasadena.
39. Strom, A. (1999) The morphology and internal structure of large rockslides as indicators of their formational mechanisms, *Doklady Earth Sciences* **369**, 1079–1081.
40. Tapponnier, P., Peltzer, G. and Armijo, R. (1986) On the mechanics of the collision between India and Asia, in M.P. Coward and A.C. Ries (eds.), *Collision Tectonics. Geological Society of London, Special Publication* **19**, 115–157.
41. Uhlir, C.F. (1998) Landslide dammed lakes: A case study of the Lamabagar and Ghat-Chaurikharka landslide deposits, Dolakha and Solukhumbu districts, eastern Nepal, *Journal Nepal Geological Society* **18**, 329–334.

42. Völk, H.R. (2000) The 1998 Tatopani Landslide in the Kali Gandaki Valley of Western Nepal: Cause and relation to mass rock creep, *Journal of Nepal Geological Society* **22**, 405–412.
43. Weidinger, J.T. (1997) Case history and hazard analysis of two lake-damming landslides in the Himalayas, *Journal of Asian Earth Sciences* **16**, 323–331.
44. Weidinger, J.T. and Ibetsberger, H.J. (2000) Landslide dams of Tal, Latamrang, Ghatta Khola, Ringmo and Dharbang in the Nepal Himalayas and related hazards, *Journal of Nepal Geological Society* **22**, 371–380.
45. Weidinger, J.T., Ibetsberger, H.J. and Nuschej, F. (2002) Hazard and Risk in the Areas of the Rock Avalanches of Darcha, Pateo and Sarai Kenlung (Manali-Leh-Road, Himachal Pradesh, India), *Geoöko* **23**, 251–267.
46. Weidinger, J.T. and Schramm, J.-M. (1995) A Short Note on the Tsergo Ri Landslide, Langtang Himal, Nepal, *Journal of. Nepal Geological Society* **11**, 281–287.
47. Weidinger, J.T. and Schramm, J.-M. (1995) Tsergo Ri (Langthang Himal, Nepal) – Rekonstruktion der "Paläogeographie" eines gigantischen Bergsturzes, *Geology Paläont Mitt Innsbruck* **20**, 231–243.
48. Weidinger, J.T., Schramm, J.-M. and Madhikarmi, D.P. (1995) Electrical conductivity in a landslide-area with uniform lithology (Tsergo Ri Landslide, Langtang, Nepal). Spatial trends and application, *Journal of Nepal Geological Society* **12**, 50–51.
49. Weidinger, J.T., Schramm, J.-M. and Nuschej, F. (2002) Ore Mineralization Causing Slope Failure in a High-Altitude Mountain Crest – On the Collapse of an 8,000 m Peak in Nepal, *Journal of Asian Earth Sciences* **21**, 295–306.
50. Weidinger, J.T., Schramm, J.-M. and Surenian, R. (1996) On preparatory causal factors, initiating the prehistoric Tsergo Ri landslide (Langthang Himal, Nepal), *Tectonophysics* **260**, 95–107.
51. Weidinger, J.T., Wang, J. and Ma, N. (2002) The earthquake-triggered rock avalanche of Cui Hua, Qin Ling Mountains, P.R. of China – the benefits of a lake-damming prehistoric natural disaster, *Quaternary International* **93/94**, 207–214.
52. Weidinger, J.T. (2006) Pre-design, failure- and displacement-mechanisms of large rockslides in the Annapurna Himalaya, Nepal, *Engineering Geology* **83**, 201–216.
53. Weidinger, J.T. and Korup, O. (2009) Frictionite as evidence for a large Late Quaternary rockslide near Kanchenjunga, Sikkim Himalayas, India - Implications for extreme events in mountain relief destruction, *Geomorphology* **103**, 57–65.
54. Yagi, H. (1992) Hazard mapping on large scale landslides in Lower Nepal Himalayas. 7th Int. Conf. and Field Workshop on Landslides in Kathmandu, Abstract Volume, 111–116.
55. Yagi, H. (1997) Origin of the Phoksundo Tal (lake), Dolpa district, wetern Nepal, *Journal of Nepal Geological Society* **15**, 1–7.
56. Yagi, H., Maruo, Y., Saijo, K. and Nakamura, S. (1990) The Sept. 1988 large landslide in the vicinity of MCT, Darbang, Nepal, *Journal of Japan Landslide Society* **26**, 45–49.
57. Yin, Y.P., Wang, F.W. and Sun, P. (2009) Landslide hazards triggered by the 2008 Wenchuan earthquake, Sichuan, China, *Landslides* DOI 10.1007/s10346-009-0148–5.

# Chapter 9
# Volcanic Natural Dams Associated with Sector Collapses: Textural and Sedimentological Constraints on Their Stability

**Lucia Capra**

## 1 Introduction

The occurrence of a sector collapse and debris avalanche at Mount St. Helens in 1980 brought to light the recognition of uniquely hummocky morphology of many analogous avalanches at volcanoes around the world [4, 16, 39, 40, 44, 47]. Subsequent studies revealed the occurrence of edifice collapses at several of the better-known volcanoes of the world. Since a debris avalanche is a very rapid gravitational flow, its emplacement generally changes the volcano morphology, e.g. by blocking a river network. The emplacement of the Mount St. Helens debris avalanche produced two main natural dams (Coldwater, Castle Creek) and raised the level of Spirit Lake by 60 m (Fig. 9.1). Artificial stabilization by spillways obviated subsequent rupture. Volcanic debris avalanche deposits are not the only process responsible for dam formation in volcanic settings. In fact, if the geomorphologic characteristics of the surrounding area are appropriated, other types of volcanic flows such as pyroclastic flows or non-cohesive debris flows can block drainage [25, 45]. Generally, in these cases, the resulting dams are smaller in volume and of shorter duration. In this paper a bibliographic survey of volcanic natural dams is presented, and based on it, the parameters controlling dam formation, duration and type of rupture are analyzed and discussed. In particular, ASTER images are used to illustrate the best studied cases and, where sedimentological data are available, discrimination diagrams are used to illustrate the relation between dam dimension/duration vs. the type of volcaniclastic material involved in its formation. Finally, the best-studied examples are used to illustrate the different aspects of volcanic dam formation, duration and failure.

L. Capra (✉)
Centro de Geociencias, UNAM, Queretaro, México
e-mail: lcapra@dragon.geociencias.unam.mx

S.G. Evans et al. (eds.), *Natural and Artificial Rockslide Dams*, Lecture Notes in Earth Sciences 133, DOI 10.1007/978-3-642-04764-0_9, © Springer-Verlag Berlin Heidelberg 2011 

**Fig. 9.1** ASTER image of Mount St. Helens showing the two new lakes (Coldwater and Castle) formed by the emplacement of the debris avalanche deposit during the May 18, 1980, eruption. The pre-existing Spirit Lake underwent an increase in water level of approximately 60 m

## 2 Terminology

A volcanic debris avalanche is a rapidly moving, incoherent, and unsorted mass of rock and soil mobilized by gravity [33]. Two end member facies define the deposit texture [19]: block facies and mixed (or matrix) facies. The block facies is composed mainly of debris avalanche blocks (unconsolidated pieces of the volcano transported to their place of deposition) with practically no matrix. The mixed facies is a mixture of clast and interclast matrix that may contain clasts of all rock types and sizes from millimeters to meters. This differentiation is fundamental because these facies control the stability of a natural dam.

In proximal areas, the surface of volcanic debris avalanche deposits generally consists of mounds (hummocks or *cerrillos*) consisting of single debris avalanche blocks; in distal areas the surfaces are generally flat, with fewer mounds but with lateral levees and an abrupt front.

A debris flow is a flowing mixture of debris and water with a sediment concentration between 70 and 90% by weight [30]. When the content of clay fraction is >3–5% by weight, they are defined as cohesive debris flows [35]. A lahar is a general term for a rapidly flowing water-saturated mixture of rock debris and water from a volcano, which emplaces different types of debris-flow deposits [41].

Granulometric distribution are based on phi values, where phi is the – *log(d)* and *d* is the diameter of the clast. Statistical parameters are calculated based on Folk [15], where Md is mean grain-size and $\sigma_I$ is sorting, a measure of the uniformity of the grain-size distribution, which increases as $\sigma_I$ approach to 0.

## 3 Origin of Volcano Instability

The instability of a volcanic edifice is dependent on many factors directly related to volcanic activity as well as exogenous processes such as weathering. The volcanic factors include direct magmatic intrusion into the edifice (Bezymianny-type activity [21]) or subvolcanic crust [12, 14], deposition from voluminous pyroclastic deposits on steep slopes [27], hydromagmatic processes [13], and phreatomagmatic activity (Bandai-type activity [28]).

The structural setting of the volcano may influence the direction of collapse and, in some cases, faulting may trigger collapse [24, 27, 39]. In addition, the volcano mass may induce isostatic flexure, compactation and deformation that can lead directly to collapse [1, 48, 49].

Although simple gravitational failure may occur in response to progressive weakening of an edifice, discrete triggering mechanisms are commonly independent of the processes producing edifice instability. Keefer [23] established that numerous large landslides during historic times were triggered by earthquakes. Schuster and Crandell [33] determined that approximately 35% of landslides causing natural dams were seismogenic. The most recent example of a catastrophic seismogenic flow from volcanic terrain occurred in 1994 at Nevado del Huila volcano (Colombia) in response to an earthquake of M6.4 [35]. Here, saturated by recent rainfall, more than 3,000 shallow failures coalesced with flank collapses to produce a single wave of cohesive debris flow that traveled over 100 km and caused as many as 1,000 fatalities. Other triggering mechanisms include phreatic explosions and precipitation. A hurricane-induced rainfall-triggered flank collapse at Casita volcano in Nicaragua in 1998 killed 2,500 people [36, 38].

Progressive weakening of a volcanic edifice by hydrothermal alteration is an indirect factor leading to collapse [5, 6, 17, 46]; e.g. cohesive debris flows directly mobilized from flank collapses of saturated, highly altered and clay-rich material [11, 34, 37, 46]. In general, cohesive debris flows may also originate in other ways, such as: (1) transformation during transport of the distal portion of a water-saturated debris avalanche [29] (2) post-depositional remobilization of water-saturated portions of a debris avalanche [20], and (3) rupture of natural dams produced by debris avalanches [2, 3, 9, 10]. This last case is very important because dam rupture associated with cohesive-debris flows may be more devastating than the initial debris avalanche that caused the river-blockage.

## 4 Volcanic Natural Dams

The emplacement of a debris avalanche generally modifies the topography of the area around the volcanic edifice and provokes the interruption of the drainage. Therefore, the formation of natural impoundments by blocking the river systems, and their subsequent rupture, should be a common process in volcanic terrains. Schuster and Crandell [33] determined that, of all the known cases of natural dams, 9% of the causalities were associated with volcanic collapse. A summary table of all known cases of volcanic natural dams associated with volcanic collapse is reported here (Table 9.1). One of the best examples of such phenomenon is represented by the 18,500 year BP collapse of the Nevado de Colima volcano (México), first described by Stoopse and Sheridan [42] that resulted in the largest debris avalanche deposit in the world, which travelled more than 120 km from its source. Subsequently, Capra and Macias [3] reinterpreted this debris avalanche to be a cohesive debris flow originating from a dam break. They determined that the edifice collapse generated a debris avalanche flow that traveled just 45 km from the source and obstructed the Naranjo river drainage forming a 1 km$^3$ natural dam (Fig. 9.2). The deposit interpreted by Stoopes and Sheridan [42] as a debris avalanche is actually a cohesive debris flow originating from the dam break. At the same Colima Volcanic Complex, approximately 3,600 years ago, a debris avalanche from the collapse of the western sector of the younger Volcán de Colima obstructed the Armeria river forming a ~0.41 km$^3$ temporary lake [8] (Fig. 9.2).

In the literature it is possible to find other examples of volcanic natural dams not directly associated with a volcano collapse, including: (1) the 1982 eruption of the Chichón volcano (Mexico) where block-and-ash flow deposits blocked the Magdalena river forming a 26*10$^6$ m$^3$ dam which failed after approximately 2 months [25]; (2) the 1992 Pinatubo eruption, where several post-eruption lahars obstructed the Mapanuepe river [45]; (3) the 1980 rain-induced Polallie Creek debris flow that formed a dam on the east Fork Hood River which broke after just 12 min [18]. All of these cases demonstrate that different types of volcaniclastic deposits can produce natural dams and, depending on the type of material and deposition rate, the longevity of the dam and the characteristics of secondary flows associated with its failure will change.

Schuster and Costa [32] showed that the duration of a natural dam depends on: (1) volume of the dam; (2) type of material; and (3) inflow rate. These authors present an analysis of the duration time and he found that the 90% of the dams that failed during the first year. This scenario is probably not appropriate for volcanic natural dams since they generally last longer. In fact, the volume of the mass produced by the volcanic collapse is generally more than 1 km$^3$ and produces a dam higher than

**Table 9.1** Summary of known cases of volcanic natural dams associated with volcano collapse

| Volcano | Age of the event type of activity | Type and dimensions of the dam | Time to fill | Lake dimensions | Dam "life" Mechanism of rupture | References |
|---|---|---|---|---|---|---|
| **Bandai** Japan | July 15, 1888 phreatic | ? V: 1.5 $km^3$ | Several months | Five lakes A: 1.5 $km^2$ | Partial overflowing – still existing | [39, 43] |
| **Mt. St. Helens** USA | May 18, 1980 Criptodome intrusion (Bezymianny type) | Block facies Andesite lava | Spirit lake 3.7 years Coldwater and Castle lakes 1.7 years | Spirit lake: A: 1,4 $km^2$, D: 69 m, V: 618 $10^6$ $m^3$ Coldwater lake: A: 0.4 $km^2$, D: 71 m, V: 140 $10^6$ $m^3$ Castle lake: A: 0.1 $km^2$, D: 37 m, V: 30 $10^6$ $m^3$ | Three main dams are controlled by artificial spillway. Estimated peak flow discharge 16,240 $m^3/s$ | [32] |
| **Mt Spurr** Alaska | 10,000–3,800 year BP | Block facies 17 km × 150 m | ND | V: $1.2 \times 10^9$ $m^3$ | Overflowing associated with erosion of the downstream end of the dam face during knick point retreat. | [50] |
| **Iriga** Philipine | 1628 A.D. tectonic-induced | Block facies 1.5 $km^3$ | ND | ND | Still existing | [24] |
| **Nevado del Huila** Colombia | 46,000–200,000 yr. BP tectonic-induced | Mixed facies 1.5 $km^3$ | 3 days | V: 0.5 $km^3$ | Overflowing | [31] |
| **Nevado de Colima** México | 18,500 yr. BP Bezymianny type | Mixed facies 1.5 $km^3$ | 9 days to 1 year | V: 1 $km^3$ D: 250 m | Overflowing | [3] |
| **Volcán de Colima** México | 3,600 yr. BP Bezymianny type | Mixed facies 1 $km^3$ | 11 days | V: 0.4 $km^3$ D: 200 m | Overflowing | [8] |
| **Parinacota** Chile-Boliva | 8,000 yr. BP tectonic-induced | Block facies, Andesite lava V 6 $km^3$ | ND | A: 15 $km^2$ | Still existing | [7] |

Abbreviations are: V, volume; A: area; D: depth. ND: not determined.
Note: yr. is conventional abbreviation when reporting geological ages.

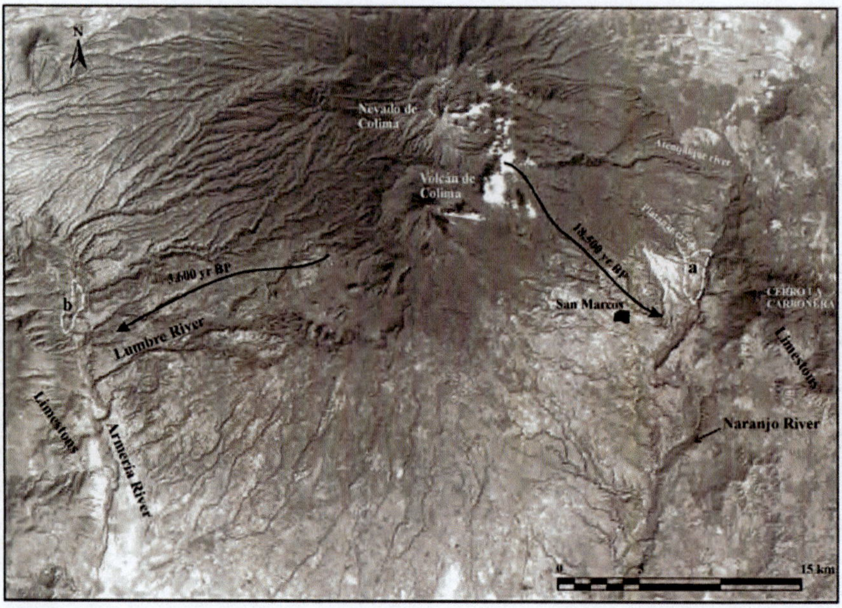


**Fig. 9.2** ASTER image of the Colima volcanic complex. *Black arrows* indicate the path of debris avalanches that produced the obstruction of both the Naranjo River (**a**) and the Armeria river (**b**), 18,500 yr. BP and 3,600 yr. BP respectively

100 m. In addition, because of the greater cohesive strength of the volcanic material, it can resist failure longer. In the best studied cases of this type (i.e. Volcán de Colima [8], Nevado de Colima (Mexico) [3] and Nevado del Huila (Colombia) [31]) 10–20 m thick lacustrine sequences are found in the infilling basin indicating a longer duration for the dams.

Valley morphology is fundamental in determining the formation and dimension of natural dams. As noted by Costa and Schuster [10] volcanic natural dams correspond to type III of Swanson et al. [43] where the dam fills the valley from side to side and fills considerable distances upvalley and downvalley, blocking other tributary rivers to the main valley. Natural dams generated by volcanic collapses are some of the most voluminous, with a reservoir capacity that may range from $30{*}10^6$ $m^3$ to 1.2 $km^3$ (Table 9.1). Different geometries of the reservoir are possible. In most cases, volcanoes are surrounded by deep valleys and high relief, favouring the formation of deep lakes, and elongated along river drainage (e.g., Mount St. Helens, Nevado de Colima and Volcán de Colima; Figs. 9.1 and 9.3).

In contrast, when the river network of the surrounding area is not well developed and the volcano is surrounded by a subdued topography, the relationship between the volume of the dam, its height and the lake depth is more expanded and shallow, around 20 m deep (e.g., Parinacota and Iriga volcanoes, Figs. 9.4 and 9.5). The presence of geomorphological obstacles favouring the formation of a natural

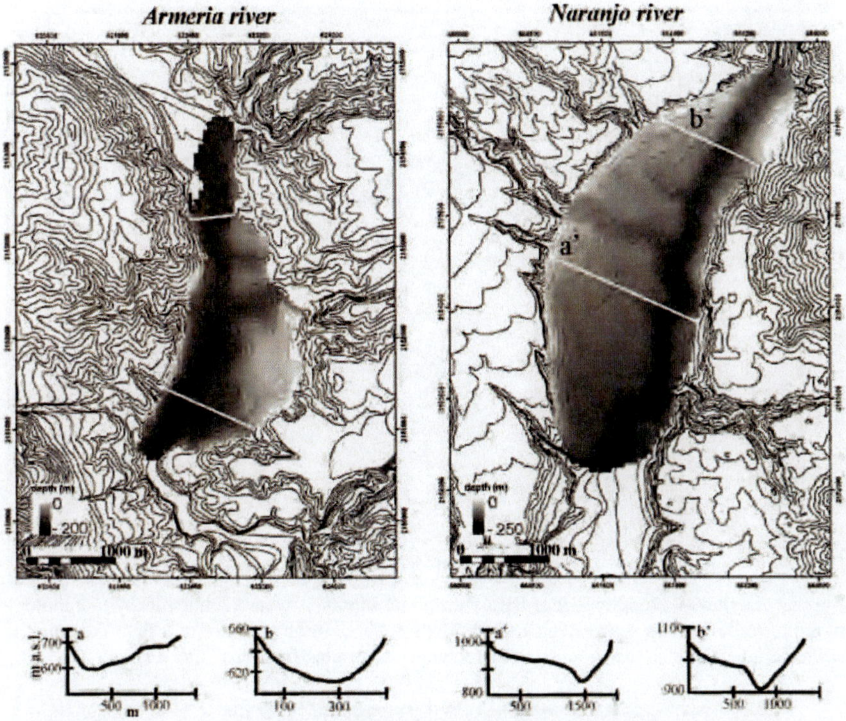


**Fig. 9.3** Reconstruction of the extension and depth of lakes formed at the Colima volcanic complex after the 3,600 year BP collapse in the Armeria River (*left*) and the 18,500 year BP event in the Naranjo River (*right*). In these cases, lakes were narrow with depths up to 250 m

impoundment is thus not necessary because if the volume of the displaced mass is extremely large it will be enough to fill in the drainage system.

The obstruction generally occurs in the proximal zone, at the base of the volcanic edifice, where the debris avalanche may consist of block facies only (with no matrix). In particular, it appears that these types of dams resist longer than other types by reaching equilibrium in their hydrological balance. For example, debris avalanches at Parinacota (Chile) and Iriga (Philippines) volcanoes produced natural lakes that have lasted since the Early Holocene and the seventeenth century, respectively (Figs. 9.4 and 9.5).

This may also be explained by the relatively high permeability of the debris due to the absence of matrix, and the low internal and/or superficial erosion because the fragments that constitute the deposit surface are too large (10–100 m in diameter) to be removed by water. Figure 9.6 shows the texture of the Parinacota debris avalanche where it is possible to observe a large number of voids between blocks. This texture gives a high permeability to the material, leading to establishment of a hydrological balance between the infilling rate and outflow via a natural spillway.

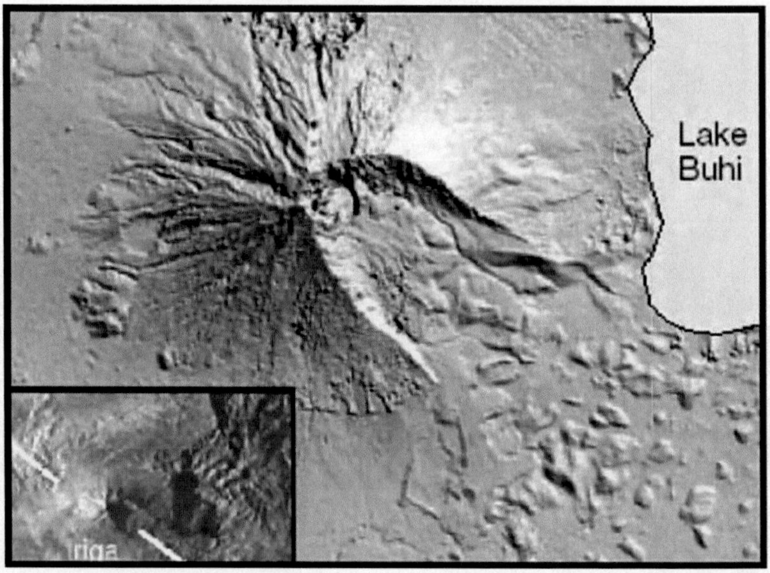


**Fig. 9.4** Digital elevation model of Iriga volcano, Philippines. Note the hummocky morphology in the SE sector, where Lake Buhi formed. The inset shows the presence of a fault crossing the edifice, suggesting that the collapse was structurally controlled [modified after 24]

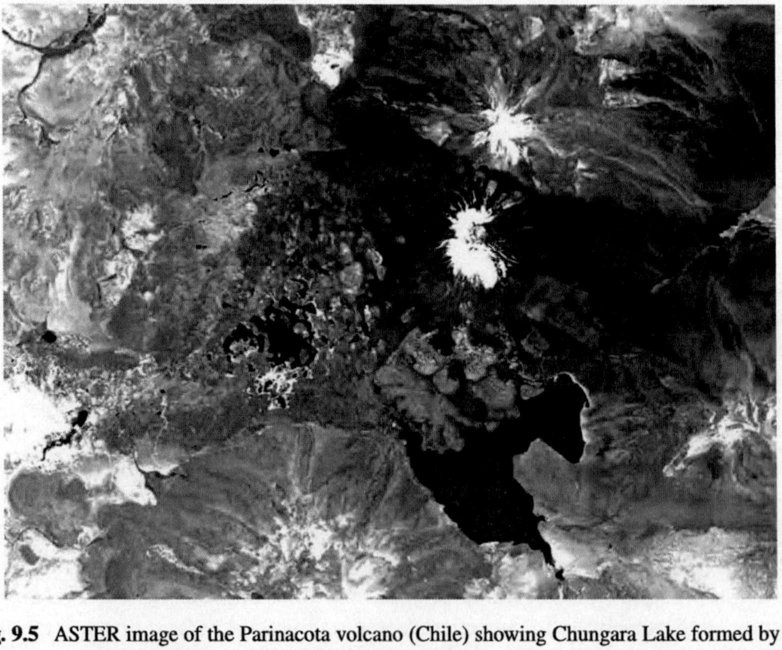

**Fig. 9.5** ASTER image of the Parinacota volcano (Chile) showing Chungara Lake formed by the emplacement of the ~8,000 year BP debris avalanche. Based on the stratigraphic study of the debris avalanche deposit [7] the dam consists of block facies

**Fig. 9.6** Picture showing the texture of the block facies of a debris avalanche deposit, Parinacota volcano, Chile (Photo by L. Capra)

In most observed cases, debris avalanche deposits responsible for damming rivers consist of mixed facies, and therefore the dam contains clasts and megaclasts (that can be more that 10 m in diameter) with a muddy-cohesive matrix filling all voids (Fig. 9.7). This sedimentological "architecture" produces a low permeability dam that is only removable by superficial erosion, when dam overflowing occurs. Similarly, Scott [34] stated that cohesive volcanic debris flows may form stable dams that last for thousands of years.

## 5 Sedimentology of Volcanic Dams

Very few papers report detailed data about the sedimentology of volcanic deposits at sites were natural dams have formed. As the grain-size distribution of a volcanic or volcaniclastic deposit is highly variable, one cannot use data of samples collected from other sites downstream. This limits the data to the few cases reported in the diagrams below. By analyzing the sedimentological characteristics of different volcaniclastic deposits that formed natural dams, and by considering their duration, results show that low sorting and coarse grain-size (but significant clay content) equate with a longer duration natural dam. In particular, by plotting mean grain-size (Md) against sorting ($\sigma_I$) (Fig. 9.8), the value of –1 phi for Md represents an important dividing line between volcanic natural dams formed by pyroclastic and/or

Fig. 9.7 Photograph showing the texture of the matrix facies of a debris avalanche deposit at Nevado de Colima Volcano, Mexico (Photo by L. Capra)

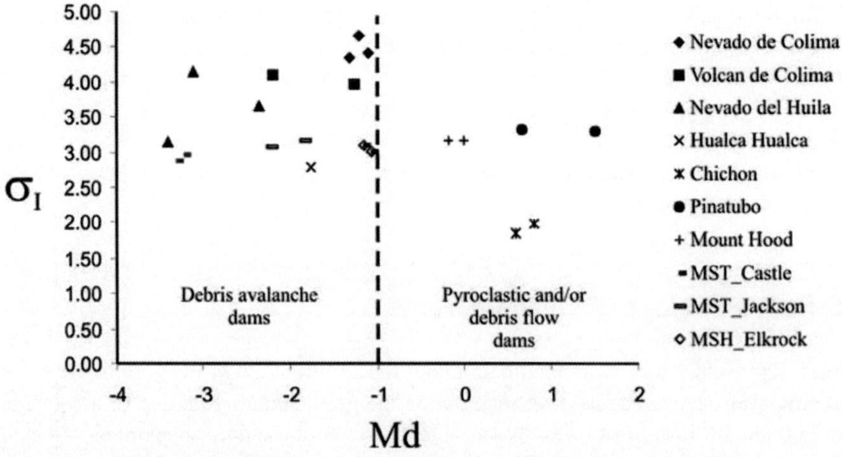


Fig. 9.8 Mean grain size (Md) vs. sorting ($\sigma_I$) diagrams. Note that the value of –1 phi separates debris avalanche dams from other dams formed by other types of volcanic and volcaniclastic deposits (Modified from Capra [2])

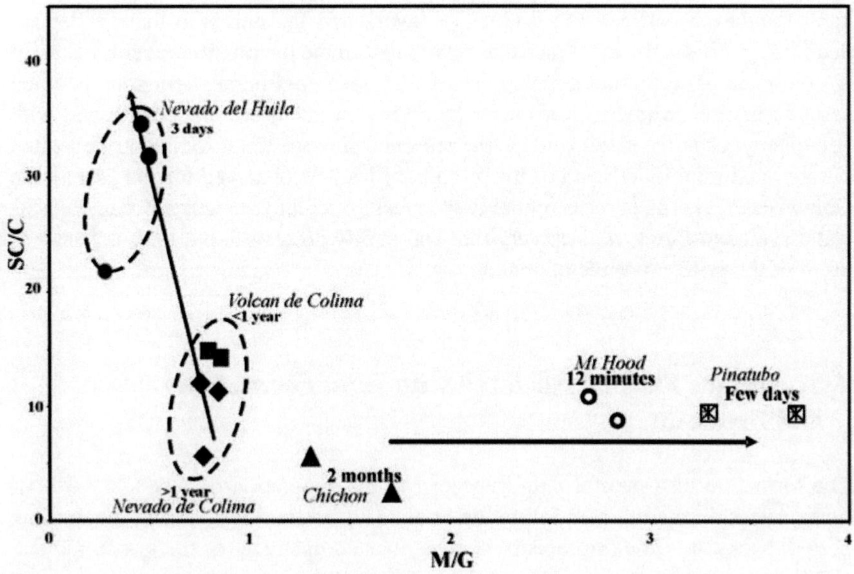


**Fig. 9.9** Matrix/Gravel (M/G) vs. Silt+Clay/Clay (SC/C) diagram showing the relation between the dam grain-size and duration of the impoundment

debris flow deposits that last just few days or months (Md > –1 phi) from those produced by debris avalanche deposits, that last more that 1 year or are still existing (Md < –1 phi).

For debris avalanche dams, the relation between M/G (M: matrix, G: gravel) and S+C/C (S: silt; C: clay) indicates that for M/G lower than 1, there is an inverse relation between mud content (S+C) and dam longevity. But for dams formed exclusively of fine material (phi less than –1, with a comparable fine fraction), dam duration decreases as matrix content increases (Fig. 9.9). In fact, with a higher percentage of silt and clay fractions in the mass damming the river, the dam will have a lower permeability, favoring a rapid rise of the lake level leading to overflow. The El Chichón Volcano dam (which lasted longer then the Hood and Pinatubo dams) was dominated by sand fraction, with only 1.8% of silt and was practically barren of clay. Despite these observations it is clear that other factors, such as the volume of the lake and the infilling rate play an important role in determining volcanic dam stability.

## 6 Identification of Natural Volcanic Dams

Although natural volcanic dams are common in volcanic terrains, their recognition is more difficult where the dams have breached especially for those of Holocene or older age. The important factors in the identification of a natural volcanic dam are: (1) relation between the drainage network and the debris avalanche deposit distribution (this is the most intuitive way to determine the possible river blockage); (2) presence of lacustrine deposits in a valley lake environment (in some studied cases lacustrine sediments have been found in elevated areas that correspond with lateral terraces of the river), and (3) the presence of slack-water sediments deposited during maximum inundation of the temporary lake, as observed for the Nevado de Colima case [3]. The latter is probably the most difficult to recognize because slack-water sediments are generally very thin and poorly preserved, but their presence is proof of the prior existence of a lake.

## 7 Secondary Flows Associated with Dam Formation and Breakout

The formation of a natural dam may lead to the generation of secondary debris-flows, either during the dam formation or after its failure. Thus, during the infilling period, backwater flooding occurs in areas located upstream of the impoundment. Debris flows can also originate from the initial emplacement of the dam or from lateral drainage filled by the same material which obstructed the main river. When the lake reaches equilibrium, lacustrine sedimentation can take place, with the deposition of fine, muddy layers, possibly interbedded with coarser layers. But, when the dam fails, devastating debris-flows occur downstream in the main drainage. As stated before, the most common process of failure is overflow of the dam with its

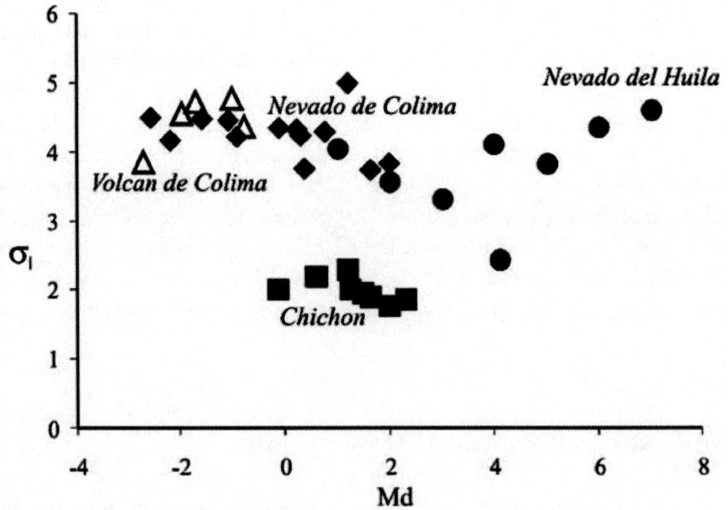


**Fig. 9.10** Mean grain-size vs. sorting of some debris flow deposits originated from a volcanic dam breakout

subsequent superficial erosion. In this sense, after an initial formation of small debris flows from the over-saturated zones, or from the remobilization of the more external portions of the dam, the main rupture of the dam give rise to several debris-flows that can coalesce into one.

Figure 9.10 shows the difference between sedimentological characteristics of the well – sorted, coarse sandy hyperconcentrated flow originated from the failure of the dam formed by block- and ash-flow deposits at El Chichón volcano during the 1982 eruption, from the coarser, poor sorted, cohesive debris flow deposits originating from the failure of the dam constituted by the debris avalanche deposit from the Colima Volcanic Complex (México) and Nevado del Huila volcano (Colombia) [3, 8, 31].

Based on these examples, the failure of a dam constituted by pyroclastic flows or non-cohesive debris flow deposits, generally produce non-cohesive debris flows that rapidly transform to more dilute hyperconcentrated flows. In contrast, the failure of a dam formed by debris avalanche deposits can produce non-cohesive to cohesive debris-flows that, instead of going through a process of down-flow dilution, they experience an increase in volume (bulking process) up to six times the original volume, with a discharge up to 4 $Mm^3/s$ or even more (i.e. Nevado de Colima and Nevado del Huila events [3, 31]). Eventually, they may transform into hyperconcentrated debris flow after more than 100 km of travel from the source.

## 8 Conclusions

The collapse of a volcanic edifice may result in several secondary effects, including the formation of one or more natural dams. The duration of the dam depends in the first case on the volume of the obstructing mass and on its textural characteristics. The block facies of a debris avalanche produces durable dams, where its texture leads to the establishment of a hydrological balance between the infilling rate and outflow through a natural spillway. In contrast, the less permeable mixed facies is easily eroded after overflowing. The other types of volcanic natural dams show a slight relationship between texture and longevity, which decreases as mud content increases.

However the most devastating effect is related to the generation of a secondary debris flow, in particular cohesive debris flows, which can originate from dam failure. These secondary flows are generally more voluminous and extensive than the primary debris-avalanche deposit.

It is very important to evaluate the potential interaction between the emplacement of gravity-driven flows and the morphology of the surrounding area as these secondary effects can threaten areas a long distance from the volcano. Recent experiences, such as the 1980 Mount Saint Helens eruption, have demonstrated that engineering work can mitigate such secondary effects, but they can still represent a hazard in volcanic areas where this phenomenon it is not yet well understood. If it is not feasible to prevent the dam failure, and if the volume of past flows is known or the flow discharge is calculated using a dam-break model (e.g. DAMBRK) [26, 51], the limit of the inundation area of secondary debris-flows can be delineated by using a GIS-based model (e.g. LAHARZ [22]), to prepare a hazard map for the local authorities.

**Acknowledgements** I would like to acknowledge all the organizing committee of the NATO advanced workshop on "Security Of Natural And Artificial Rockslide Dams" who gave me this great opportunity for an incredible scientific experience. José Luis Macías and Duncan Keppie provided useful suggestions that improved this manuscript. All ASTER images were acquired form glovis.usgs.gov web site.

## References

1. Borgia, A. (1994) Dynamic basis of volcanic spreading, *Journal of Geophysical Research* **99**, 17791–17804.
2. Capra, L. (2007) Volcanic natural dams: Identification, stability and secondary effects, *Natural Hazards* **43**, 45–61.
3. Capra, L. and Macías, J.L. (2002) The cohesive Naranjo debris flow deposit (10 $km^3$): A dam breakout flow derived from the pleistocene debris-avalanche deposit of Nevado de Colima volcano (México), *Journal of Volcanology and Geothermal Research* **117**, 213–235.
4. Capra, L. and Macias, J.L. (2000) Pleistocene cohesive debris flows at Nevado de Toluca Volcano, central Mexico, *Journal of Volcanology and Geothermal Research* **102**, 149–167.
5. Capra, L. (2000) Colapsos de Edificios Volcánicos: Transformación de Avalanchas de Escombros en Flujos de Escombros Cohesivos. Los casos del Monte Santa Elena (EEUU), Nevado de Toluca y Nevado de Colima (México). *Universidad Nacional Autónoma de México, Mexico City,* PhD Thesis, 180 pp.
6. Carrasco-Nuñez, G., Vallance, J.W. and Rose, W.I. (1993) A voluminous avalanche-induced lahar from Citlaltepetl volcano, Mexico: Implications for hazard assessment, *Journal of Volcanology and Geothermal Research* **59**, 35–46.
7. Clavero, J.E., Sparks, R.S.J. and Huppert, H.E. (2002) Geological constraints on the emplacement mechanism of the Parinacota debris avalanche, northern Chile, *Bulletin of Volcanology* **64**, 40–54.
8. Cortes, A., Macías, J.L., Capra, L. and Garduño, V.H. (2009) Sector collapse of the SW flank of Volcán de Colima, México: The 3,600 yr BP La Lumbre-Los Ganchos debris avalanche and associated debris flows, *Journal of Volcanology and Geothermal Research* **197**, 52–66.
9. Costa, J.E. (1988) Floods from dam failures, in V.R. Baker, R.C. Kochel and P.C. Patton (eds.), *Flood Geomorphology*. Wiley, New York, NY, pp. 439–463.
10. Costa, J.E. and Shuster, R.L. (1988) The formation and failure of natural dams, *Geological Society of America Bulletin* **100**, 1054–1068.
11. Crowley, J.K. and Zimbelman, D.R. (1997) Mapping hydrothermally altered rocks on Mount Rainier, Washington, with Airbone Visible/Infrared Imaging Spectrometer (AVIRIS) data, *Geology* **25**, 559–562.
12. Day, S.J. (1996) Hydrothermal pore fluid pressure and the stability of porous, permeable volcanoes, in W.J. McGuire, A.P. Jones and J. Neuberg (eds.), *Volcano Instability on the Earth and Other Planets*. Geological Society Special Publication, London, pp. 77–93.
13. Dzurisin, D. (1998) Geodetic detection of inflating stratovolcanoes: A potential breakthrough for mitigating volcanic hazards in the twenty-one century, *EOS* **79**, 973.
14. Elsworth, D. and Voight, B. (1996) Dike intrusion as a trigger for large earthquakes and the failure of volcano flanks, *Journal of Geophysical Research* **100**, 6005–6024.
15. Folk, R.L. (1980) *Petrology of Sedimentary Rocks*. Hemphill, Austin, TX, 182 p.
16. Francis, P.W. and Wells, G.L. (1988) Landsat thematic mapper observation of debris avalanche deposits in Central Andes, *Bulletin of Volcanology* **50**, 258–278.

17. Frank, D. (1983) Origin, distribution and rapid removal of hydrothermally formed clay at Mount Baker, Washington, *US Geological Survey Professional Paper* **1022-E**, 31.
18. Gallino, G.L. and Pierson, T.C. (1985) Polallie creek debris flow and subsequent dam-break flood of 1980, East Fork Hood River Basin, Oregon, *US Geological Survey Water Supply Paper* **2273**, 22.
19. Glicken, H. (1991) Sedimentary architecture of large volcanic-debris avalanches, in: G.A. Smith and R.V. Fisher (eds.) *Sedimentation in Volcanic Settings*, SEPM, Tulsa, Special Publication **45**, 99–106.
20. Glicken, H. (1998) Rockslide-debris avalanche of May 18, 1980, Mount St. Helens Volcano, Washington, *Bulletin of the Geological Survey of Japan* **49**, 55–106.
21. Gorshkov, G.S. (1962) On the classification and terminology of Pelée and Katmai type eruption, *Bulletin of Volcanology* **24**, 155–165.
22. Iverson, R.M., Schilling, S.P. and Vallance, J.W. (1998) Objective delineation of lahar-inundation hazard zones, *Geological Society of America Bulletin* **110**, 972–984.
23. Keefer, D.K. (1984) Landslides caused by earthquakes, *Geological Society of America Bulletin* **95**, 406–421.
24. Lagmay, A.M.F., Van Wyk Vries, B., Kerle, N. and Pyle, D.M. (2000) Volcano instability induced by strike-slip faulting, *Bulletin of Volcanology* **62**, 331–346.
25. Macias, J.L., Capra, L., Scott, K.M., Espindola, J.M., Garcia-Palomo, A. and Costa, J.E. (2004) The 26 May 1982 breakout flows derived from failure of a volcanic dam at El Chichón, Chiapas, Mexico, *Geological Society of America Bulletin* **116**, 233–246.
26. Manville, V., White, J.D.L., Houghton, B.F. and Wilson, C.J.N. (1999) Paleohydrology and sedimentology of a post-1.8 ka breakout flood from intracaldera Lake Taupo, North Island, new Zealand, *Geological Society of America Bulletin* **111**, 1435–1447.
27. McGuire, W.J. (1996) Volcano instability: A review of contemporary themes, in W.J. McGuire, A.P. Jones and J. Neuberg (eds.), *Volcano Instability on the Earth and Other Planets*, Geological Society of London Special Publication **110**, 1–24.
28. Moriya, I. (1980) Bandaian Eruption and landforms associated with it. Collection of articles in memory of retirement of Prof K Nishimura, Tohoku University, Tokyo **66**, 214–219.
29. Palmer, B.A. and Neall, V.E. (1989) The Murimotu Formation-9500 year old deposits of a debris avalanche and associated lahars, Mount Ruapehu, North Island, New Zealand, *New Zealand Journal of Geology and Geophisics* **32**, 477–486.
30. Pierson, T.C. and Costa, J.E. (1987) A rheologic classification of subaerial sediment-water flows, in J.E. Costa and G.F. Wieczorek (eds.), *Debris flows/Avalanches: Process, Recognition, and Mitigation.* Geological Society of America Reviews in Engineering Geology, Boulder, CO, pp. 1–12.
31. Pulgarin, B., Capra, L., Macias, J.L. and Cepeda, H. (2001) Depósitos de flujos de escombros gigantes (10 $km^3$) asociados a colapsos de edificios volcánicos: Los casos de los volcanes Nevado del Huila (Colombia) y Nevado de Colima (México), *Revista de Geofísica del Instituto Panamericano de Geografía e Historia* **55**, 51–75.
32. Schuster, R.L. and Costa, J.E. (1986) A perspective in landslide dams, in R.L. Schuster (ed.), *Landslide Dams: Processes, Risk, and Mitigation*, Geotechnical Special Publication No. 3. American Society of Civil Engineers, New York, NY, 1–20.
33. Schuster, R.L. and Crandell, D.R. (1984) Catastrophic debris avalanches from volcanoes, IV International Symposium on landslides proceedings, Vol. 1, Toronto, 567–572.
34. Scott, K.M. (1985) Lahars and flow transformations at Mount St. Helens, Washington, USA, International Symposium on Erosion, Debris Flow and Disaster Prevision, Tsukuba, Japan, 209–214.
35. Scott, K.M., Macias, J.L., Vallance, J.W., Naranjo, J.A., Rodriguez-Elizarrarás, S.R. and McGeehin, J.P. (2002) Catastrophic debris flows transformed from landslides in volcanic terrains: Mobility, hazard assessment, and mitigation strategies, *US Geological Survey Professional Paper* **1630**, 61.
36. Scott, K.M., Vallance, J.V., Kerle, N., Macias, J.L., Strauch, W. and Devoli, G. (2005) Catastrophic precipitation-triggered lahars at Casita Volcano, Nicaragua, flow bulking and transformation, *Earth Surf Process Landforms* **30**, 59–79.

37. Scott, K.M. and Vallance, J.W. (1995) Debris flow, debris avalanche, and flood hazards at and downstream from Mount Rainier, Washington, *US Geological Survey Hydrologic Investigation Atlas*, 729.
38. Sheridan, M.F., Bonnard, C., Carreno, C., Siebe, C., Strauch, W., Navarro, M., Calero, J.C. and Trujilo, N.B. (1999) Report on the 30 October 1998 Rock Fall/Avalanche and breakout flow of Casita Volcano, Nicaragua, Triggered by Hurricane Mitch, *Lanslide News* **12**, 2–4.
39. Siebert, L. (1984) Large volcanic debris avalanches: Characteristics of source areas, deposits and associated eruptions, *Journal of Volcanology and Geothermal Research* **22**, 163–197.
40. Siebert, L., Glicken, H. and Ui, T. (1987) Volcanic hazards from Bezymianny- and Bandai-type eruptions, *Bulletin of Volcanology* **49**, 435–459.
41. Smith, G.A. and Fritz, W.J. (1989) Volcanic influences on terrestrial sedimentation, *Geology* **17**, 375–376.
42. Stoopes, G.R. and Sheridan, M.F. (1992) Giant debris avalanches from the Colima Volcanic complex, Mexico: Implication for long-runout landslides (>100 km), *Geology* **20**, 299–302.
43. Swanson, F.J., Norio, O. and Masaki, T. (1986) Landslide dams in Japan, in R.L. Schuster (ed.), *Landslide Dams – Processes, Risk and Mitigation*, Geotechnical Special Publication No. 3. American Society of Civil Engineers, New York, NY, pp. 131–145.
44. Ui, T. and Glicken, H. (1986) Internal structural variations in a debris-avalanche deposit from ancestral Mount Shasta, California, USA, *Bulletin of Volcanology* **48**, 189–194.
45. Umbal, J.V. and Rodolfo, K.S. (1996) The 1991 lahars of southwestern Mount Pinatubo and evolution of the lahar-dammed Mapanuepe lake, in C.G. Newhall and R.S. Punongbayan (eds.), *Fire and Mud; Eruptions and Lahars of Mount Pinatubo*, Philippine Institute of Volcanology and Seismology. University of Washington Press, Seattle, WA, London, 951–970.
46. Vallance, J.V. and Scott, K.M. (1997) The Osceola Mudflow from Mount Rainier: Sedimentological and hazard implication of a huge clay-rich debris flow, *Geological Society of America Bulletin* **109**, 143–163.
47. Vallance, J.V., Siebert, L., Rose, W.I., Girón, J.R. and Banks, N.G. (1995) Edifice collapse and related hazard in Guatemala, *Journal of Volcanology and Geothermal Research* **66**, 337–355.
48. Van Wyk Vries, B. and Borgia, A. (1996) The role of basement in volcano deformation, in M.J. McGuire, A.P. Jones, and J. Neuberg (eds.), *Volcano Instability on the Earth and Other Planets*, Geological Society of London Special Publication **110**, 95–110.
49. Van Wyk Vries, B., Self, S., Francis, P.W. and Keszthelyi, L. (2001) A gravitational spreading origin for the Socompa debris avalanche, *Journal of Volcanology and Geothermal Research* **105**, 225–247.
50. Waythomas, C.F. (2001) Formation and failure of volcanic debris dams in the Chakachatna River Valley associated with eruptions of the Spurr volcanic complex, Alaska, *Geomorphology* **39**, 111–129.
51. Waythomas, C.F., Walder, J.S., McGimsey, R.G. and Neal, C.A. (1996) A catastrophic flood caused by drainage of a caldera lake at Aniakchak Volcano, Alaska, and impplications for volcanic hazards assessment, *Geological Society of America Bulletin* **108**, 861–871.

# Chapter 10
# Formation and Treatment of Landslide Dams Emplaced During the 2008 Wenchuan Earthquake, Sichuan, China

**Peng Cui, Yongshun Han, Dang Chao, and Xiaoqing Chen**

## 1 Introduction

The Wenchuan Earthquake, had its epicenter at Wenchuan County in the Sichuan Province of China (30.40°N/103.47°E), occurred at 14:28 local time (Beijing time) on 12 May 2008, and was measured at $M_S$8.0 according to the China Earthquake Administration (CEA). The earthquake brought overwhelming destruction to eight provinces and cities, including Sichuan, Gansu, Shanxi, Chongqing, Yunnan, Shaanxi, Guizhou and Hubei. The most highly affected areas were along the Central and Frontal faults of the Longmenshan fault system in the west of Sichuan, a 300-km-long belt-shaped zone. With the strong shock, along the main rupture zone, Modified Mercalli Intensities ranged up to XI (Fig. 10.1). Distinct surface displacement appeared widely along the ruptures and some of the vertical scarps were as high as 6.2 m; measurements also indicated as much as 4.9 m in horizontal displacement (Fig. 10.2 from [21]).

The quake completely destroyed houses, roads, water and electricity projects, and "life-lines" such as transportation and communication facilities. According to Chinese State Officials, as of 19 June 2008, the quake caused 69,200 known deaths, with 18,195 people missing and 374,216 injured; 5,362,500 houses collapsed, 21,426,600 houses were badly damaged, and more than 5 million people were made homeless. With 80~90% of houses collapsed, Beichuan County, Yingxiu Town, Xuankou Town and others were almost completely destroyed by the earthquake.

The Wenchuan earthquake occurred in an area with great topographical relief and consequently produced many earthquake-induced geo-hazards. The strong and intensive main quake and aftershocks, as well as large co-seismic surface displacement, triggered numerous rockslides, rock falls, other landslides, and debris flows [7]. The relatively huge amount of debris delivered by landslide or debris-flow blocked river channels as landslide dams and formed many landslide-dammed

P. Cui (✉)
The CAS Key Laboratory of Mountain Hazards and Earth Surface Process, Institute of Mountain Hazards and Environment, Chengdu, China
e-mail: pengcui@mde.ac.cn

S.G. Evans et al. (eds.), *Natural and Artificial Rockslide Dams*, Lecture Notes in Earth Sciences 133, DOI 10.1007/978-3-642-04764-0_10, 

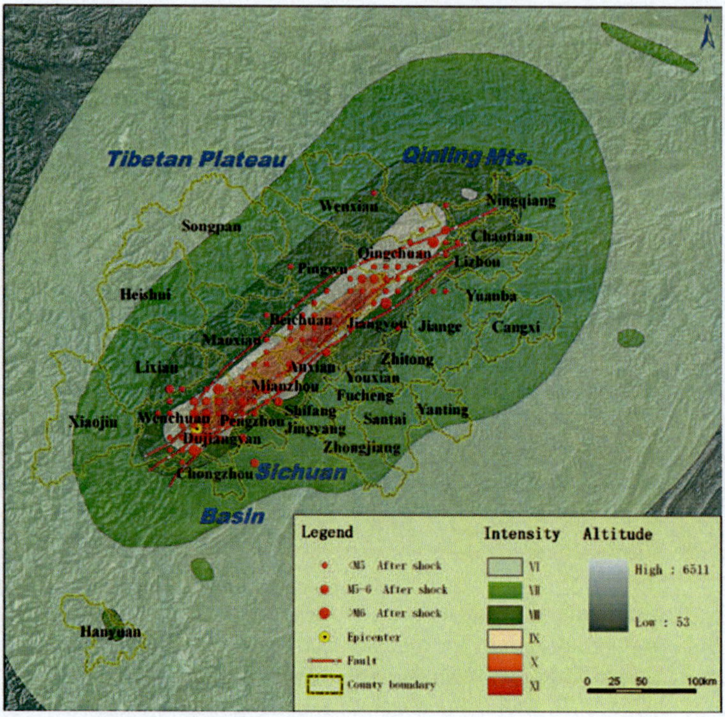


**Fig. 10.1** Map showing faults, aftershocks, major shock and intensity for the 2008 Wenchuan Earthquake (courtesy of China Earthquake Administration)

lakes. Most of the landslide dams were assessed as being unstable due to the fact that they were composed of large volumes of unconsolidated material and fragmented rock. Sudden dam failure and resulting outburst floods were considered to be common scenarios for these landslide lakes because the dams formed spontaneously as a result of the earthquake, in a short time-span [6]. According to the authors' investigation, 257 landslide dams in total were emplaced in the area hit by the Wenchuan earthquake. The distribution, emergency risk evaluation and risk mitigation treatments for those dams will be discussed in this chapter.

## 2 Distribution of the Landslide Dams

### *2.1 Investigation Techniques*

The earthquake triggered large-scale landslides that formed dams in channels and blocked rivers in the narrow valleys of the region. Lakes rapidly formed behind the debris dams as increasing amounts of runoff occurred at the start of the rainy season.

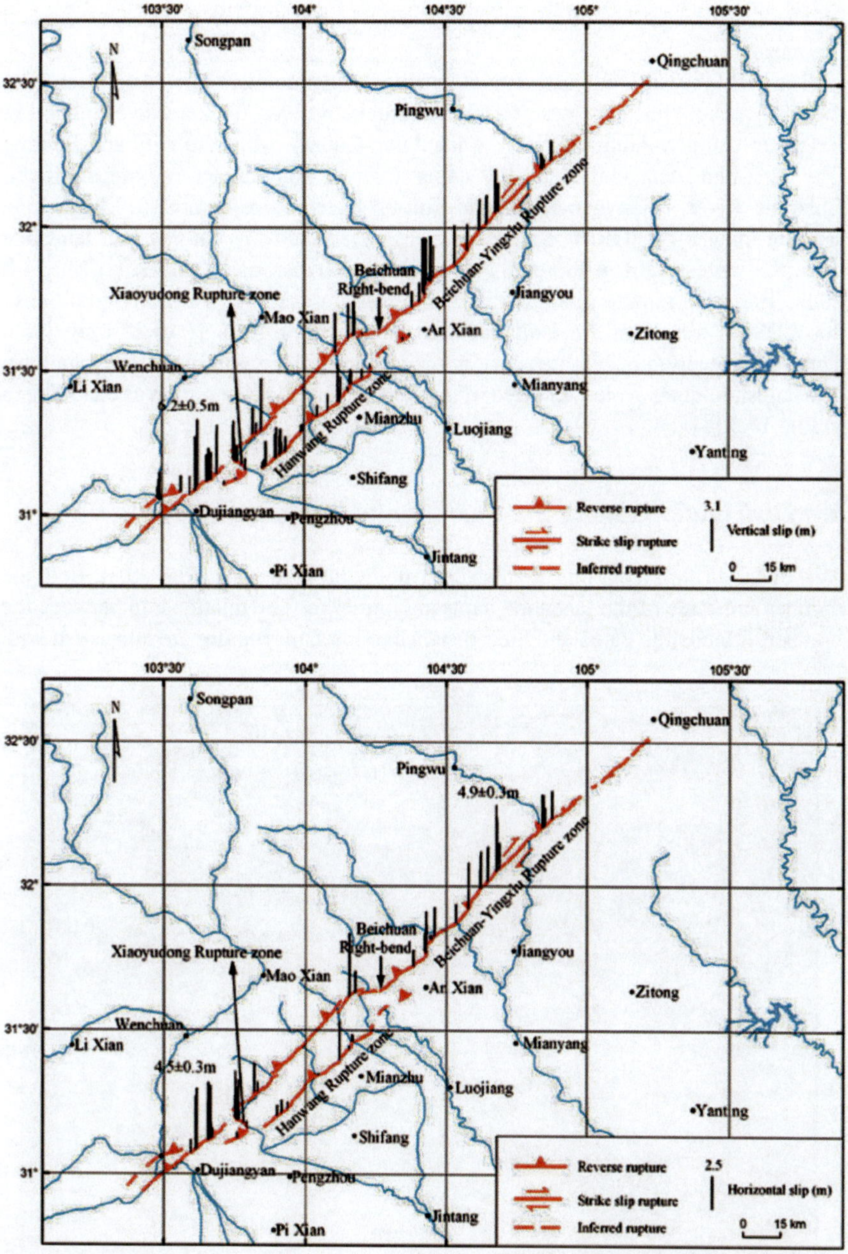


**Fig. 10.2** Maps of the parallel distribution of measured coseismic vertical (upper map) and horizontal displacements (lower map) along the Beichuan, Pengguan and Xiaoyudong rupture zones (from [21])

This called for rapid action to survey the dams for emergency risk assessment and treatment.

Both field investigation and remote sensing inspection were carried out soon after the earthquake. Our team went to the earthquake hit-area for field investigation in Beichuan County, Anxian County, Qingchuan County, Mianzhu City and Shifang City. First-hand data on the landslide dams along the Tongkou River, Qingzhu River, Chaiping River, Mianyuan River and Shiting River, were compiled. Meanwhile, satellite images of FORMOSAT-2 were inspected, and the Tangjiashan landslide dam, the largest dam in the earthquake affected-area, was identified on May 14, 2008. Then, the remote images with 2 m resolution taken by Aerial Digital Senso 40 (ADS40) were used for landslide dam identification from 16 to 28 May 2008. Through inspection of remote sensing imagery, combined with field investigations, 257 landslide dams were identified in the earthquake-affected region as can be seen in Fig. 10.3 [8].

## 2.2 Distribution Along the Earthquake Rupture Zone

The landslide dam locations are clustered along the fault zone (Fig. 10.3). Both the number and scale of the landslide dams are immense. The relationship between the number of landslide dams and their distance to the fault rupture are consistent with

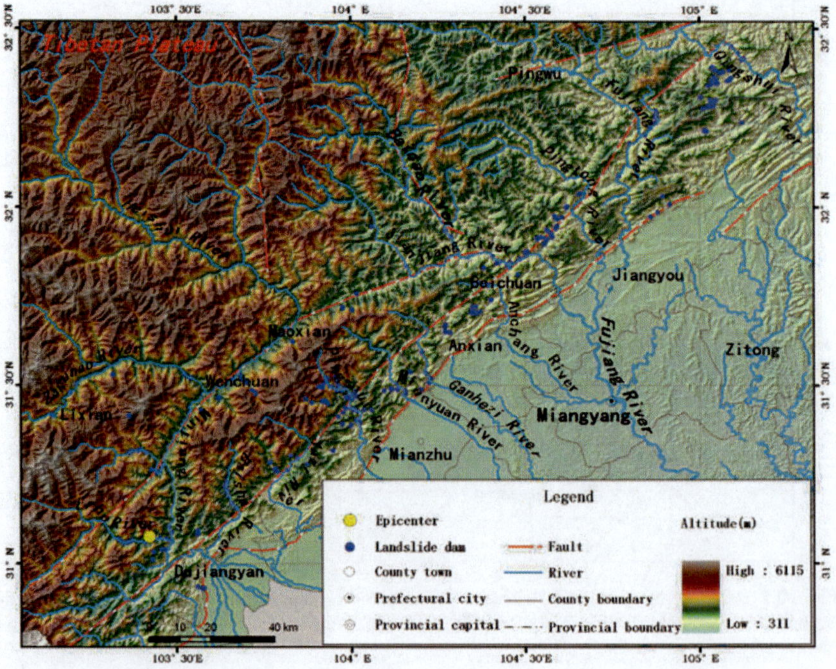


Fig. 10.3 The distribution of landslide dams caused by the 2008 Wenchuan Earthquake

the logarithm attenuation principle with a negative correlation coefficient of 0.9699. Approximately 85.6% of the dams are clustered within 10 km of the three strands of the Longmenshan fault system. There are 176 landslide dams distributed within 10 km of the main Central fault, also called the Beichuan-Yingxiu fault, accounting for 68.5% of the total (Fig. 10.3). Examination of the landslide dams located near the Beichuan-Yingxiu fault reveals that the dams are increasingly abundant towards the surface projection of the major fault. There are 109 landslide dams within 5 km of the Central fault, which is 42.4% of the total number of dams generated. Within 5–10 km of the Central fault are 67 dams, 26.1% of the total; within 10–15 km are 29 dams, 11.3% of the total; within 15–20 km are 15 dams, 5.8% of the total; within 20–25 km are 11 dams, 4.3% of the total; within 25–30 km are 10 dams, 3.9% of the total; within 30–35 km are 4 dams, 1.6% of the entire number; within 35–40 km are 4 dams, accounting for 1.6%; within 40–45 km are 2 dams, accounting for 0.8% (Fig. 10.4). These data indicate a strong correlation between the location of the surface trace of the main fault and the earthquake-generated landslides.

By comparing the distribution of landslide dams in Fig. 10.3 and the seismic intensity in Fig. 10.1 as well as vertical and horizontal displacements along the Beichuan, Pengguan and Xiaoyudong rupture zones in Fig. 10.2, it is revealed that the spatial location of the landslide dams is consistent with the areas of high seismic intensity and great surface displacement. Most landslide dams occurred in the area where Modified Mercalli Intensities were XI, X and IX. The highest density of landslide dams appeared in the places with strongest surface displacement such as the adjacent rupture areas in Wenchaun County, Shifang City, Mianzhu City and Beichan County where the vertical displacement is generally larger than 3 m and the horizontal displacement larger than 2.5 m.

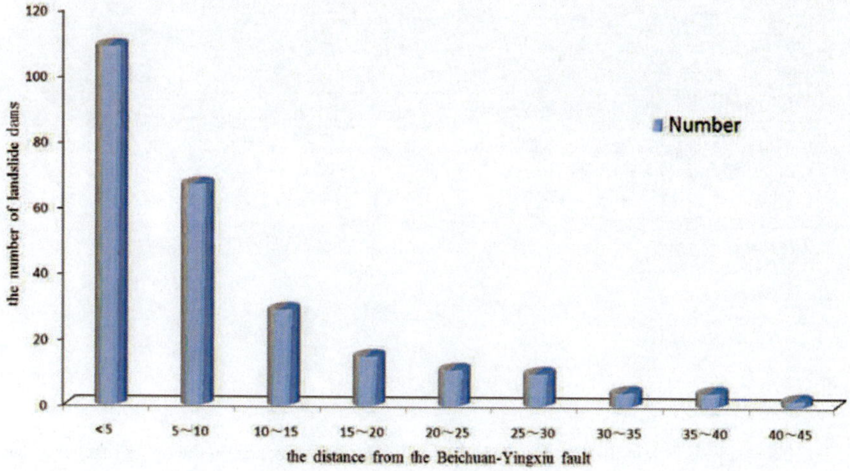


**Fig. 10.4** The relationship between number of landslide dams and distance from Beichuan-Yingxiu fault

## 2.3 Distribution Along River Valleys

Another feature of the landslide dams is that they are distributed in clusters along river valleys like a string of beads (Fig. 10.5). Within 30 km of the Tongkuo River in Beichuan County there are nine strings of landslide dams, and within 10 km of the Mianyuan River are four strings, with a density of 0.4 dams/km. There are seven within 8 km of the Shiting River, with a density of 0.88 dams/km; four within 7.5 km of the Anzi River, with a density of 0.53 dams/km; three within 5 km of the Qing River, with a density of 0.6 dams/km; and eight within 14 km upreaches of Tongkou River (a part of which is shown in Fig. 10.6), with a density of 0.57 dams/km.

The Tangjiashan Dam, located on the upper reaches of the Tongkou River (tributary to the Fujiang River) is the largest of the earthquake-created landslide dams. Along the channel of the Tongkou River downstream of the Tangjiashan Dam in Beichuan, landslide dams occur every two kilometers on average, according to the image from the Chinese Mapping Bureau (Fig. 10.6). If the Tangjiashan landslide dam had collapsed, it was likely that the dams downstream would have collapsed in a chain reaction due to the released flood waters. The damage caused by this and

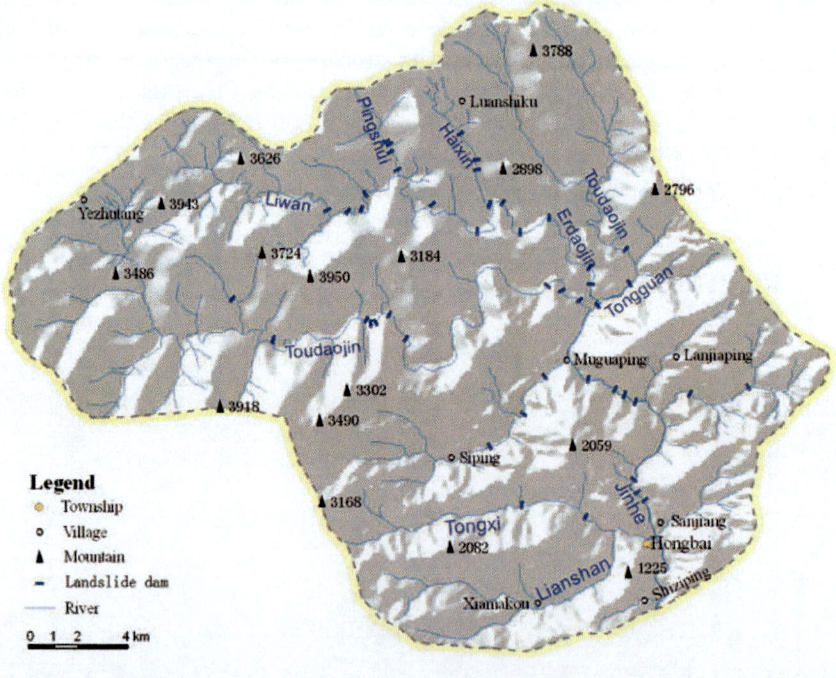


Fig. 10.5 Landslide dam distribution along rivers like a string of beads, in the upper reaches of the Shiting River basin, Shifang city. There are 47 landslide dams which cluster along both tributary and main channels and account for 18.29% of the total dams emplaced in the area affected by the Wenchuan earthquake

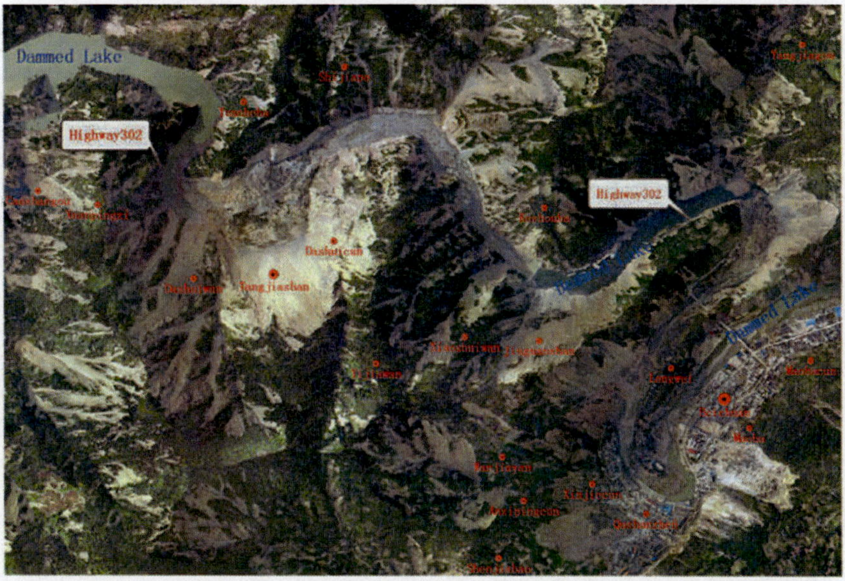


**Fig. 10.6** The cluster of landslide dams along the major Tongkuo River Valley upstream from Beichuan County to the Tangjiashan rockslide dam. From *left* to *right* are the Tangjiashan dam, the Kuzhu landslide dam and the New – Street village landslide dam (Image from the Chinese Mapping Bureau)

other rivers' outburst chain reactions would have been enormous, and this problem was addressed with investigation and mitigation.

## 3 Rapid Assessment of Landslide Dam Breaching Potential

### 3.1 *Main Features of the Landslide Dams*

According to preliminary investigations, 257 landslide dams formed in total, including 34 dams presenting a potential of outburst flooding. By analyzing the information obtained by field investigation, remote sensing interpretation and reports by local people, the major features of the 34 dams are described in Table 10.1. As the rainy season began, normally in June and lasting until October, the water levels behind the many landslide dams rose, increasing the risk of uncontrolled outburst and greatly threatening the lives and property of more than 1.3 million people downstream. The Tangjiashan Dam in Beichuan County was judged the most dangerous one on the basis of its location and size, potentially impacting over one million people downstream in many towns as well as Mianyang City. The landslide dam, consisting of 20.4 $Mm^3$ of landslide debris, was located at Tangjiashan 3.2 km upstream from Beichuan township, and before its engineered partial breaching on June 10, had an estimated water storage capacity of 315 $Mm^3$, and a submerged area over 23 km upstream along the river thalweg (Figs. 10.6 and 10.7).

**Table 10.1** The main features of the barrier dams

| No | Dam | Location | River | Height (m) | Capacity ($10^4$ $m^3$) | Width-height ratio | Volume ($10^6$ $m^3$) | Composition |
|---|---|---|---|---|---|---|---|---|
| 1 | Zhicheng | Beichuan | Tongkou River | 10 | 120 | 3 | 1.2 | Block and soil |
| 2 | Tangjiashan | Beichuan | Tongkou River | 82–124 | 31,600 | 9.8 | 20.37 | Block and soil |
| 3 | Kuzhuba | Beichuan | Tongkou River | 60 | 200 | 3.3 | 1.65 | Block and soil |
| 4 | Xinjiecun | Beichuan | Tongkou River | 20 | 200 | 10 | 2 | Soil |
| 5 | Beiguocun | Beichuan | Tongkou River | 10 | 80 | 20 | 0.4 | Block and soil |
| 6 | Sunjiayuanzi | Beichuan | Tongkou River | 50 | 560 | 3.6 | 1.6 | Block and soil |
| 7 | Yanyangtan | Beichuan | Tongkou River | 20 | 400 | 7.5 | 1.6 | Block and soil |
| 8 | Guanzipu | Beichuan | Fuxing River | 60 | 585 | 6.5 | 1.8 | Block and soil |
| 9 | Tangjiawan | Beichuan | Fuxing River | 40 | 1,520 | 13.3 | 4 | Clastics and soil |
| 10 | Wenjiaba | Pingwu | Shikan River | 25 | 686 | 23.2 | 5.32 | Clastics and soil |
| 11 | Ma'anshi | Pingwu | Shikan River | 67.6 | 1,150 | 14 | 5.8 | Clastics and soil |
| 12 | Shibangou | Qingchuan | Qingzhu River | 70 | 1,000 | 11.4 | 10.5 | Soil and stone |
| 13 | Hongshihe | Qingchuan | Hongshi River | 50 | 300 | 10 | 4 | Soil and stone |
| 14 | Donghekou | Qingchuan | Qingzhu River | 20 | 300 | 15 | 10 | Soil and stone |
| 15 | Xiaojiaqiao | Anxian | Chaping River | 73 | 2,230 | 4.3 | 2 | Clastics |
| 16 | Laoyingyan | Anxian | Suishui River | 130 | 1,010 | 2.3 | 3 | Clastics |
| 17 | Guantan | Anxian | Suishui River | 60 | 1,000 | 3.3 | 4.8 | Soil and block |
| 18 | Heidongya | Mianzhu | Mianyuan River | 50 | 180 | 2.4 | 0.4 | Block and soil |
| 19 | Xiaogangjian 1# | Mianzhu | Mianyuan River | 62 | 1,100 | 2 | 0.6 | Block and soil |
| 20 | Xiaogangjian 2# | Mianzhu | Mianyuan River | 30 | 700 | 5 | 0.34 | Blocks |
| 21 | Yibadao | Mianzhu | Mianyuan River | 25 | 50 | 4 | 0.1 | Blocks |
| 22 | Ganhekou | Shifang | Shiting River | 10 | 50 | | 0.1 | Block and soil |
| 23 | Macaotan 1# | Shifang | Shiting River | 40 | 200 | 4 | 1 | Block and soil |
| 24 | Macaotan 2# | Shifang | Shiting River | 40 | 250 | 2.3 | 0.2 | Block and soil |
| 25 | Macaotan 3# | Shifang | Shiting River | 30 | 100 | 3.3 | 0.12 | Block and soil |
| 26 | Muguaping | Shifang | Shiting River | 15 | 50 | 6.7 | 0.2 | Block and soil |
| 27 | Yanziyan | Shifang | Shiting River | 10 | 10 | 4 | 0.01 | Block and soil |
| 28 | Hongcun | Shifang | Shiting River | 40 | 150 | 2 | 0.24 | Block and soil |
| 29 | Xiejiadianzi | Pengzhou | Shajin River | 10 | 100 | 7 | 0.12 | Block and soil |

**Table 10.1** (continued)

| No | Dam | Location | River | Height (m) | Capacity ($10^4$ $m^3$) | Width-height ratio | Volume ($10^6$ $m^3$) | Composition |
|---|---|---|---|---|---|---|---|---|
| 30 | Fengmingqiao | Pengzhou | Shajin River | 10 | 180 | 8 | 0.14 | Block and soil |
| 31 | Zhugenqiao | Chongzhou | Wenjing River | 90 | 450 | 0.8 | 1.53 | Blocks |
| 32 | liudinggou | Chongzhou | Wenjing River | 60 | 300 | 0.8 | 0.75 | Blocks |
| 33 | Huoshigou | Chongzhou | Wenjing River | 120 | 150 | 0.3 | 1.2 | Blocks |
| 34 | Haiziping | Chongzhou | Wenjing River | 8 | 300 | 8.8 | 0.34 | Blocks |

**Fig. 10.7** The Tangjiashan landslide dam (Provided by Water Resources Bureau of Mianyang city)

The landslide dams also brought tremendous threat to rescue and reconstruction work in the earthquake-damaged areas. It was, and continues to be, critical and urgent to assess the risks and mitigate the hazards posed by the Tangjiashan case and the numerous other landslide dams.

Confronted with numerous landslide dams, it is critical to perform a rapid analysis of the possibility for each dam's potential breaching in order to focus emergency response on the most important situations. Unfortunately, this is not only critical, but it is difficult within the required time framework. Damaged, destroyed and unstable roads, traffic delays, unstable hillsides, and frequent aftershocks all made it difficult to get close to the landslides for investigation and analysis. The challenge of rapidly determining the types of materials present at depth in the landslide dams added to the difficulty of analyzing their outburst flooding risk.

## 3.2 Risk Assessment for Dam Breaching

The components used to make a rapid assessment of the risk caused by the landslide dam failure were (1) interpretation of remote sensing imagery, (2) basic information from field investigations, (3) the length, width, height and composition of the dam determined from comprehensive topographic analysis, and (4) the backwater length of the dams. By selecting the dam height, dam composition and the water storage capacity as the main indices, we established a statistical method of assessing the risk of each dam (Table 10.2), and categorized its risk into four types: extremely high, high, medium and low risk.

**Table 10.2** The risk assessment indices with respect to outburst flood

| Index \ Grade | Extreme high risk | High risk | Medium risk | Low risk |
|---|---|---|---|---|
| Dam height (m) | >100 | 50–100 | 25–50 | <25 |
| Water storage capacity ($\times 10^4$ $m^3$) | $10^4$ | $10^3$–$10^4$ | $10^2$–$10^3$ | <$10^2$ |
| Composition of dam materials | Soil | Soil with some boulders | Boulders with some soil | Boulders |

After excluding the 12 already-overtopped dams which occurred before 25th May (they were determined to be less dangerous than others due to the decrease in dam height due to vertical erosion of the spillway elevation during overtopping), it was concluded that the Tangjiashan Dam was extremely dangerous. Seven dams demonstrated a high risk, five dams presented a medium risk, and eight dams were low risk (Table 10.3). While determining the risk, we also made judgments on the ways that the dams might collapse from instantaneous full collapse, partial instantaneous collapse, progressive collapse, or a gradual collapse (Table 10.3). These modes of failure were assumed for analyzing the possible impact on the downstream environment.

We further analyzed the high- and extremely high-risk dams by considering their width to height ratio and the dam's probable mode of failure. The eight dams with the highest risk of breaching, in descending order, were: Tangjiashan, Laoyingyan, Nanba, Upstream of Xiaogangjian, Xiaojiaqiao, Tangjiawan, Guzipu and Yanyangtan.

The risk evaluation results for dam failure reported by this paper were accepted by the Earthquake Disaster Relief Office. The 17 dams with medium risk up were being planned for emergency treatment at the end of May 2008. All of the mitigation treatments recommended for reducing potential risk have subsequently been successful.

## *3.3 Risk Analysis for Potential Outburst Flood Inundation Area – The Tangjiashan Rockslide-Dammed Lake*

This involves three exercises in estimating the discharge, travel time and flood inundation area of a potential outburst flood. The largest landslide dam, the Tangjiashan Dam, identified as the dam posing the highest risk is taken as a case study and is discussed below.

### 3.3.1 Calculation of Peak Discharge for Outburst Flood

Estimation of discharge from a landslide dam failure involves the calculation of the peak flow at the toe of the dam and the simulation of the evolution of the peak

**Table 10.3** The parameters and risk assessment results for key landslide dams

| Location | County (city) | Drainage basin | Dam height (m) | Water storage capacity ($\times 10^4$ $m^3$) | Width to height ratio | Composition of dam | Risk category | Probably style of dam failure |
|---|---|---|---|---|---|---|---|---|
| Tangjiashan | Beichuan | Jianjiang River (Tongkouhe) | 82–124 | 30,200 | 9.8 | Boulders with some soil | Extremely high risk | Progressive collapse |
| Downstream of Kuzhuba | Beichuan | Jianjiang River (Tongkouhe) | 60 | 200 | 3.3 | Boulders | Medium | Partial collapse |
| Xinjiecun | Beichuan | Jianjiang River (Tongkouhe) | 20 | 200 | 10.0 | Soil | Medium | Instantaneous collapse |
| Baiguocun | Beichuan | Jianjiang River (Tongkouhe) | 10 | 80 | 20.0 | Boulders with some soil | Low | Gradual collapse |
| Yanyangtan | Beichuan | Jianjiang River (Tongkouhe) | 50 | 560 | 3.6 | Soil with some boulders | High | Gradual collapse |
| Sunjiayuanzi | Beichuan | Jianjiang River (Tongkouhe) | 20 | 400 | 7.5 | Boulders with some soil | Medium | Gradual collapse |
| Guanzipu | Beichuan | Jianjiang River (Tongkouhe) | 60 | 585 | 6.5 | Boulders with some soil | High | Gradual collapse |
| Tangjiawan | Beichuan | Fuxinghe – branch of Jianjiang river | 40 | 1,520 | 13.3 | Small boulders with soil | High | Progressive collapse |
| Nanba | Pingwu | Shikan River | 25 | 686 | 23.2 | Small boulders | High | Progressive collapse |
| Xiaojiaqiao | An | Chaping River | 64 | 2,000 | 4.3 | Small boulders | High | Partial collapse |
| Xiaogangjian 1# | Mianzhu | Mianyuan River | 62 | 1,100 | 2.0 | Boulders with some soil | High | Instantaneous collapse |
| Xiaogangjian 2# | Mianzhu | Mianyuan River | 30 | 700 | 5.0 | Boulder s | Medium | Progressive collapse |
| Yibadao | Mianzhu | Mianyuan River | 25 | 50 | 4.0 | Large boulders | Low | Progressive collapse |
| Ganhekou | Shifang | Shitingjiang River | 10 | 50 | | Boulders with some soil | Low | Partial collapse |

**Table 10.3** (continued)

| Location | County (city) | Drainage basin | Dam height (m) | Water storage capacity ($\times 10^4$ $m^3$) | Width to height ratio | Composition of dam | Risk category | Probably style of dam failure |
|---|---|---|---|---|---|---|---|---|
| Upstream of Macaotan | Shifang | Shitingjiang River | 40 | 200 | 4.0 | Boulders with some soil | Low | Progressive collapse |
| Middle of Macaotan | Shifang | Shitingjiang River | 40 | 250 | 2.3 | Boulders with some soil | Medium | Gradual collapse |
| Downstream of Macaotan | Shifang | Shitingjiang River | 30 | 100 | 3.3 | Boulders with some soil | Low | Partial collapse |
| Muguaping | Shifang | Shitingjiang River | 15 | 50 | 6.7 | Boulders with some soil | Low | Partial collapse |
| Yanziyan | Shifang | Shitingjiang River | 10 | 10 | 4.0 | Boulders with some soil | Low | Partial collapse |
| Hongcun Power Station | Shifang | Shitingjiang River | 40 | 150 | 2.0 | Boulders with some soil | Low | Partial collapse |
| Laoyingyan | An | Jushui River | 130 | 1,010 | | Small boulders | High | Progressive collapse |

flow downstream [11]. Three methods have been applied: (1) use of an empirical relationship between the known outburst parameters and peak flows – an approach involving large uncertainty [5, 15, 17], (2) use of the geometry of the breaches and observed durations of failure to form a physical model as applied by Fread [12], (3) calculation of peak flow from analysis of the breach-forming processes and duration [13, 14, 20].

For the Tangjiashan dam, because of the lack of previous engineering surveys and unknown boundary conditions, and the time-crucial need for an expedited analysis, we proposed the following approach, for scenarios of 20, 25, 33, and 50% failure of the entire rockslide dam structure.

Peak flood discharge at the toe of a dam composed of loose granular material can be determined from:

$$Q_{\max} = \frac{8}{27}\sqrt{g}\left(\frac{B}{b}\right)^{\frac{1}{4}} bH_0^{\frac{3}{2}} \tag{1}$$

where $Q_{max}$ is the maximum flow at the dam ($m^3/s$), $B$ is the width of the dam top (m), $H_o$ is the upstream water depth before failure (m), $g$ is the acceleration of gravity (9.8 m/s), and $b$ is the width of the breach (m) [23]. The parameters applied for the Tangjiashan dam are listed in Table 10.4.

### 3.3.2 Downstream Progression of Flood

Peak flow along individual reaches of a river downstream from a failed dam can be determined [18] as:

$$Q_{LM} = \frac{W}{\frac{W}{Q_M} + \frac{L}{VK}} \tag{2}$$

where $Q_{LM}$ is the peak discharge along a given reach with distance $L$ (m) from the dam, $W$ is the channel capacity, $Q_M$ is the maximum peak discharge at the dam, and $VK$ is an empirical coefficient (7.15 for mountain rivers, 3.13. for rivers on plains, and 4.76 for rivers flowing through terrain of intermediate relief [18]). The value of

**Table 10.4** The parameters applied for estimating peak discharge resulting from dam breaching

| Percent of dam failure | Average width of over – topping mouth (B/m) | Height of water head ($H_0$/m) | Discharge during peak flood flow ($Q_{max}/m^3/s$) |
|---|---|---|---|
| 20 | 125.2 | 16.53 | 8,726.9 |
| 25 | 119.0 | 20.66 | 12,089.2 |
| 33 | 109.2 | 27.27 | 17,175.7 |
| 50 | 88.0 | 41.33 | 27,269.2 |

*VK* used here is the intermediate value, believed appropriate based on the average slope of the main channel of the Tongkou River (< 1%).

The time ($t$) is the time for the peak discharge of the flood wave to reach a distance ($L$) downstream:

$$t = k\frac{L^{1.4}}{W^{0.2}H_0^{0.5}h_m^{0.25}} \tag{3}$$

where $k$ is a coefficient, generally 0.8–1.2, and $h_m$ is the water level at maximum discharge at a control section after dam failure [16].

Based on our past experience with case histories involving similar materials, gradual collapse of 25–33% of the structure was considered to be the most likely scenario. Once breaching occurs, water level will fall, discharge will decline, and the breach will stabilize temporarily by progressive armoring with coarse material. With the beginning of the rainy season, however, rapid inflow may trigger renewed breach incision. For a conservative estimate, we apply equations (2) and (3) in a numerical simulation to analyze four possible scenarios of failure equivalent to 20, 25, 33 and 50% of the total volume of the dam. Because post-earthquake conditions made flow and stage measurements difficult, we applied the Manning formula to estimate flows at the section in Mianyang, 74.9 km downstream from the Tangjiashan rockslide dam (Table 10.5).

### 3.3.3 The Determination of Areas of Potential Risk

The evolution of a potential outburst flood, peak discharges, cross-sectional areas and water levels in each section downstream of Tangjiashan Dam, was calculated by equation (2) and equation (3). Based on the digital elevation model (DEM) in this area, and the combining of calculated hydrological parameters along the stream, the four delineated inundation areas were identified as low risk of 3.35 $km^2$ for 20% for dam-failure, medium risk of 3.84 $km^2$ for 25%, high risk of 4.22 $km^2$ for 33%, and very high risk of 4.65 $km^2$ for 50% respectively. The results of the four possible dam-failure scenarios provided information necessary for planning emergency evacuation.

**Table 10.5** The calculated peak discharges at the section in Mianyang for dam-failure percentages of 20, 25, 33, and 50%

| Breach geometry (%) | Maximum discharge ($m^3$/s) | Cross-sectional area ($m^2$) | Average depth (m) | Width (m) | Arrival time (h) |
|---|---|---|---|---|---|
| 20 | 6,106.4 | 1,588.8 | 4.6 | 424.5 | 5.02 |
| 25 | 7,396.6 | 1,803.7 | 5.1 | 441.5 | 4.38 |
| 33 | 9,061.8 | 2,071.4 | 5.7 | 457.8 | 3.71 |
| 50 | 11,260.9 | 2,248.0 | 6.2 | 521.7 | 2.95 |

# 4 Emergency Treatment-Case Studies of Tangjiashan and Xiaogangjian #1 Rockslide Dams

## 4.1 Emergency Risk Reduction Treatment of Tangjiashan Dam

### 4.1.1 Morphology and Geology of the Dam

Tangjiashan rockslide dam stands at 6 river-km upstream from the Beichuan township along the Tongkou River and 3.2 km in a direct line [4, 10]. The dam, 803.4 m long and 611.8 m wide, possessed a maximum possible reservoir capacity of 315 $Mm^3$ before the excavation of the sluiceway described below. The drainage area of 3,550 $km^2$. The elevation at the highest point of the landslide dam was 793.9 m a.s.l. and 669.5 m a.s.l. at the lowest. Weakly weathered base rock is present at the level of 638.0 m a.s.l. The maximum height of the dam was 124.4 m and lowest point of the dam surface being 82.7 m high (Fig. 10.7). The dam was formed in what was an asymmetrical V-shaped valley before the landslide, with the right bank steeper (60°) than the left (30°) (Fig. 10.8).

The relief varies on the dam surface. Three gullies developed on the right, middle, and the left, respectively. The gully on the right, which stretches from the upper to the lower part, is 20–40 m wide and ends at elevation 752.2 m; the other two gullies formed on the lower part of the dam and are about 400 m long and 10~20 m wide.

The slope of the upper part of the dam is about 20° while the lower flank is steep in the upper section (50 m high with gradient of 1:0.7) and lower section (20 m high with gradient of 1:0.5); slope is gentle in the middle, with a gradient of 1:2.5.

Precambrian rocks of the Qingping Formation underlie the dam. They consist of intercalated hard and soft rocks (sandstone, marlite, phyllite, and muddy rocks). The strata strike at N60ºE/NW∠60º. The dip slope is on the right bank and the anti-dip slope on the left. Cracks develop intensively in the basal rock, including a strongly weathered zone of 5–10 m. Alluvial deposits, up to 15 m in depth, are distributed

**Fig. 10.8** The landform of Tangjiashan before Wenchuan Earthquake (Provided by Water Resources Bureau of Mianyang City)

on both banks. The composition of the deposits is silty loam (60%), rock debris in size of 5~20 cm (30–35%), and crushed rocks (5–10%).

### 4.1.2 The Emergency Treatment for Risk Reduction

The drainage area of the Tongkou River, a tributary of the Fujiang River and ultimately the Yangtze River, receives an average annual precipitation of 1,355.4 mm at Beichuan weather station. Rainfall between May and October accounts for 86.3% of that total. The Tangjiashan Dam impounded between 8 and approximately 13.4 $Mm^3$ of water daily [2, 22], beginning on 12 May 2008 after the river was blocked. By 19 May, a week after the earthquake, the water level submerged the Zhicheng hydropower station upstream, having risen to 676 m elevation, with a lake volume of about 4 $Mm^3$. By 25 May the water level reached 729.2 m elevation, with a volume of about 173 $Mm^3$. The water storage behind the dam reached 203.7 $Mm^3$ on May 31. As a consequence of the approaching rainy season and the occurrence of frequent rainfall after the earthquake, authorities, concerned with the rapid rate of the rise of the lake surface, and especially with the risk posed to the downstream population in Mianyang City and over 70,000 residents in over 10 villages and towns downstream, decided to reduce risk by excavating a drainage sluiceway to release water from the dammed lake [4, 10].

The sluiceway was designed to make use of the low point on the upper right side of the dam, in a curve with the convexity facing the right bank (Fig. 10.9; also refer to Fig. 10.7). The designed channel has a trapezoid section with slope of 1:1.5 on both sides, the bottom is 8 m wide, the elevation of the entrance was 742 m a.s.l. and the length is 695 m. The channel gradient is 0.6% in the upper part; and it is 24 and 16% in the middle and the lower part respectively (Fig. 10.10).

After 6 days of construction from May 26 to May 31, a channel with trapezoid cross sections had been created on the surface of the rockslide dam (Figs. 10.11 and 10.12). Before drainage took place, the sluiceway was 50 m wide, the base of the inlet was 7 m wide, and increases up to 10 m at the outlet; the maximum depth was 13 and 12 m on average. The sluiceway channel was 475 m long, the highest elevation at the base of the entrance was 740 m a.s.l. (2 m lower than designed) and the outlet was 739 m a.s.l. The total excavated debris was 135,550 $m^3$ in volume and 4,200 $m^3$ of stone-filled gabions were placed to stabilize the channel. The project also included 17 km of road construction and 35,000 $m^3$ of wood clearance.

### 4.1.3 The Overflow Process

The drainage process of Tangjiashan landslide lake began at 7:08 a.m. on June 7 when the lake water level reached the base of the entrance of the sluiceway and water began to flow in the excavated channel over the debris. The controlled breach was initiated when the water level rose to el. 743.10 m a.s.l. at 1:30 a.m., June 10. At 11:30 that day, the discharge achieved the peak of 6,500 $m^3/s$, which only moderately exceeded the flood discharge with a 100-year return period (6,040 $m^3/s$). From 8:00 am on June 7 to 8:00 am on June 11, about 47 $Mm^3$ of water had poured into

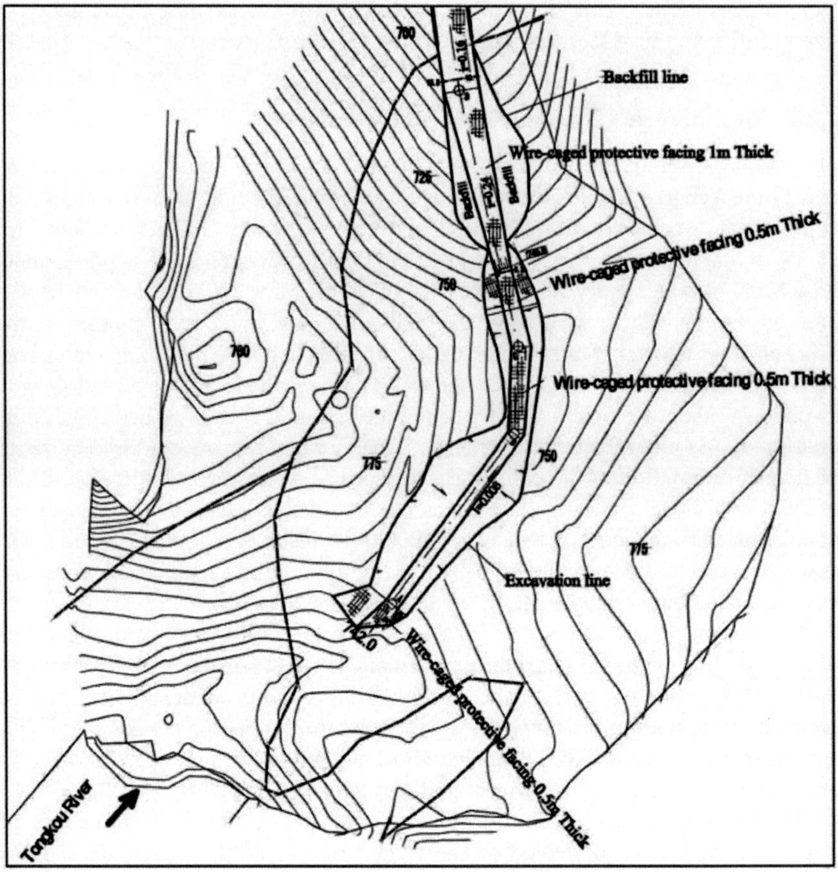


**Fig. 10.9** The layout of the sluiceway excavated on the debris surface of the Tangjiashan rockslide dam (from [3])

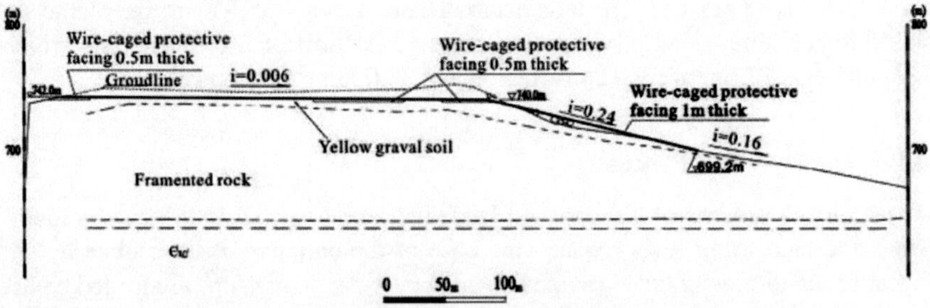


**Fig. 10.10** Long profile of the Tangjiashan sluiceway (Provided by Chen Zuyu)

Fig. 10.11 The Tangjiashan sluiceway under construction (Provided by Water Resources Bureau of Mianyang city)

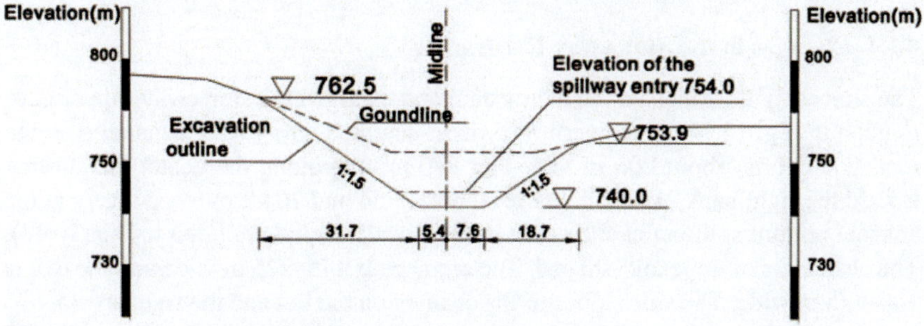


Fig. 10.12 Cross section of the Tangjiashan sluiceway (from [3])

the lake from the upper reach; 186 $Mm^3$ had been delivered through the overflow channel. By 8:00 a.m June 11, the water level had fallen to 714.62 m, and the impounded water decreased from 246.6 to 86.1 $Mm^3$ (Fig. 10.13). This reduced the peak discharge 4 times in comparison with that of the overflow under natural conditions (i.e. without treatment) of the dam, which was calculated as 27,627 $m^3/s$ according to Cenderelli [1].

Comparing the drainage peak discharge of 6,500 $m^3/s$ and the calculated peak discharges for scenarios of 20, 25, 33, and 50% failure of the entire dam in Table 10.4, the actual peak discharge is 43.26% less than calculated one in case of 20% failure and 4 times less than that of 50% failure. This comparison documents that the emergency treatment was successful [4, 10].

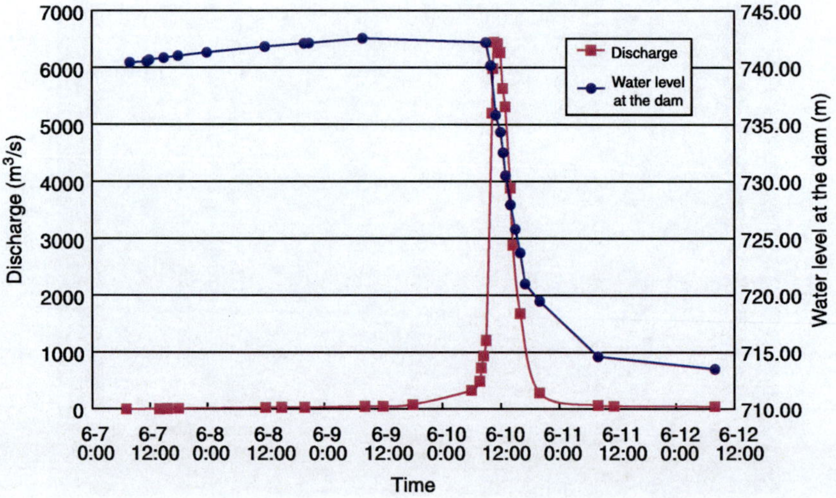


**Fig. 10.13** The drainage-discharge process of Tangjiashan rockslide dam showing water level and discharge at the dam (June 7–12, 2008)

### 4.1.4 Changes in the Sluiceway Excavation

The sluiceway flow incised the debris dam and enlarged the sluiceway (upper photograph in Fig. 10.14). The overflow process sculpted a new river channel over the rockslide debris, about 200 m wide and 890 m long along the centerline curving toward the right bank, with a water level of 713.54 and 704.2 m respectively at the channel bottom at the inlet and outlet of the dam (lower photograph in Fig. 10.14). The channel is of trapezoid-shaped. The entrance is 145–225 m wide and the exit is 100–145 m wide. The side slopes of the channel on the left and the right are 35–50° and 45–60°, respectively; both slopes are between 10–60 m in height. About $5\times 10^6$ $m^3$ of dam material, roughly a quarter of the total volume of the rockslide, has been eroded during the channel incision.

At present, the slope of the channel is higher in the middle reach and lower in the upper and lower reaches; the maximum scoured depth is 60 m. The material composition is similar for both bank slopes, which are yellow rock detritus 20–40 m in thickness on the upside, 10–20 m gravels on the middle, and black grey silica rich rocks outcropping on the foot. The right bank has been intensely eroded and failures appear much more extensive than the left bank. The channel bed is well armed by boulders (Fig. 10.15).

The toe of the dam has also experienced strong erosion, but without large collapses. The body of the dam mainly occupies the left of the river channel, arch-like in plan. It is 80–120 m thick, 600 m long along the river and 300 m wide. Its volume remains 15 $Mm^3$ (lower photograph in Fig. 10.14).

**Fig. 10.14** Tangjiashan rockslide-dammed lake during and after partial drainage on June 10, 2008. *Upper picture* (view downstream) shows the incised channel over the debris caused by water release from Tangjiashan lake through the sluiceway (photograph from Prof. Liu Ning). *Lower photograph* (view upstream) is an overview of new channel after water release from Tangjiashan landslide lake (photograph is provided by Water Resources Bureau of Mianyang city)

**Fig. 10.15** Tangjiashan Lake: upstream view of the armoured spillway channel bed 15 months after risk reduction operation (Photograph taken on September 22, 2009)

## *4.2 Emergency Risk Reduction Treatment of Xiaogangjian Dam*

### 4.2.1 Topography and Geology of the Dam

The Xiaogangjian #1 rockslide dam (Table 10.3), is 300 m upstream from the Xiaogangjian hydropower station on the Mianyuan River (Fig. 10.16), 6 km downstream of the confluence with the Qingpingxiang River. The river at the rockslide dam flows from south to north and the streambed is narrow with steep side slopes. The geology consists of the Devonian $D_3^{tn}$ interbedded layers of dolomite and limestone with orientations of N10-20°W/NE∠70–80° (Fig. 10.17) [7, 9].

The dam blocking the Mianyuan River is formed by a rocky landslide that slid from the right bank. The top surface of the dam is lower on the left and higher on the right. The left side of the dam is 50 m wide with a surface slope is 2–5°. Near the left cliff there is an eroded 5 m wide channel at an altitude of 884.7 m a.s.l. in the middle, the dam is 120 m wide and the surface slope is 0–5°. The right side consists of stones and bedrock fragments. The surface slope of the rockslide dam parallel to river flow is between 25–30°. The dam is 250 m wide across the river, 300 m long along the thalweg, and 62–120 m high above the streambed. The total volume of the rockslide is $2\times10^6$ m$^3$. The maximum volume of impounded water is about $1.050\times10^7$ m$^3$. The normal discharge of Mianyuan River in June is about 15 m$^3$/s at Qingping.

Boulders in the dam are as large as 10 m diameter and 2–4 m on average, accounting for 50% of the debris; rocks 0.3–0.6 m in diameter account for 25%; gravels are 0.1~0.2 m in diameter for 10%; soils make up the rest of the debris (15%). Isolated

**Fig. 10.16** Aerial view of Xiaogangjian #1 rockslide dam. View is upstream

stones are mainly concentrated on the left side of the dam. These form a complex openwork structure.

The river banks near the dam are also of loose structure. The slope is steep and is overlain with thick colluvium. The channel on the left bank can be easily enlarged by flooding. Much debris, deposited on the slopes opposite the dam and induced by earthquakes, may slide down and block the drainage channel. The risk assessment results in Table 10.3 shows this dam had a high risk for dam failure.

### 4.2.2 Emergency Treatment for Risk Reduction

The sluiceway is designed to reduce the risk [7, 9]. The channel is 25~30 m wide so as to lower and maintain the water level down to elevation 842 m a.s.l. (Figs. 10.17 and 10.18). Explosions were used to dig the channel because of the inaccessibility of the site to engineering machinery. The sluiceway discharge reached 1,300 $m^3/s$, which is equal to a flood with a 10-year return period (Fig. 10.19).

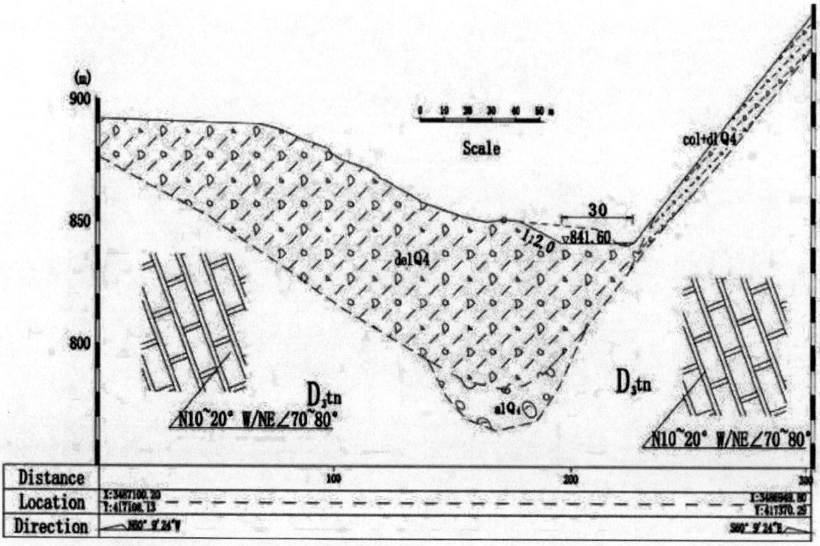


Fig. 10.17 Cross-section of the Xiaogangjian #1 rockslide dam and the location of designed sluiceway (Provided by Chen Zuyu)

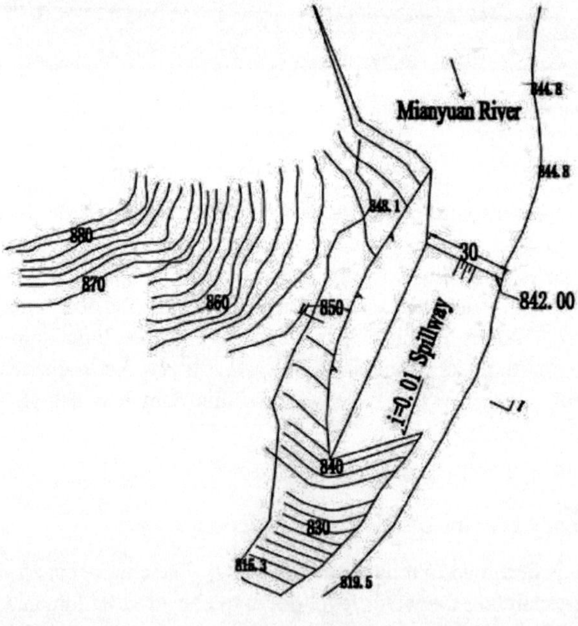


Fig. 10.18 The design of sluiceway for emergency operations at Xiaogangjian #1 rockslide dam (Provided by Chen Zuyu)

**Fig. 10.19** The peak discharge of the spillway flood from Xiaogangjian #1 rockslide-dammed lake passing the Hanwang township, Mianzhu city (Provided by Chen Zuyu)

## 5 Conclusions

The M8.0 Wenchuan earthquake of 12 May 2008 caused strong surface motion in the mountainous area surrounding Longmenshan fault zone, triggering over 20,000 landslides in the quake-hit area. Many landslides blocked river channels and formed debris dams and barrier lakes. By means of field investigation and remote sensing inspection, 257 landslide-dams have been identified. The dams were mainly distributed in a belt along the rupture zone and in clusters along the river valleys. Most of the landslide-dams were the result of rockslides.

These dammed lakes threatened hundred thousands of inhabitants downstream. Due to the risk associated with the frequent rainfall after the earthquake and the onset of the rainy season, an urgent emergency risk assessment was carried out. Under the emergency circumstances, the dam height, dam composition and storage capacity of dammed lake were used as indices of the dam's risk of collapse. Four risk levels were assigned and 21 landslide dams were assessed – 1 dam had an extremely high danger risk, 7 dams had a high danger, 5 dams medium danger and 8 dams were of low danger. The Earthquake Disaster Relief Office adopted all of the risk evaluation results and arranged the emergency mitigation operations for potential risk reduction according to the assessment results.

Two cases, Tangjiashan Dam and Xiaogangjian #1 Dam, are evaluated in detail in this chapter. For the case of Tangjiashan Dam, the scenarios were for 20, 25, 33, and 50% of the dam failing, yielding, respectively, flood heights of 4.6, 5.1, 5.7, and 6.2 m, and discharges of 6,106, 7,397, 9,062, and 11,260 $m^3/s$, at the

Fujiang Bridge in Mianyang, the second largest city in Sichuan Province. The inundation areas for these four scenarios were estimated by GIS as 3.35, 3.84, 4.22, and 4.65 $km^2$, respectively, values useful for planning emergency management and evacuation downstream.

Sluiceways (spillways) were designed to lower water levels of the dammed lakes and to reduce the risk of rockslide dam failure. Emergency mitigation operations for risk-elimination proved to be effective and successful. No further collapses of any rockslide dam occurred subsequently during the rainy season of 2008.

On the other hand, numerous landslides and collapses have provided plenty of source material for potential large-scale debris flows which are likely to pour into rivers and form new dams in the coming 5–10 years [7, 19]. Therefore, risk evaluation for potential outburst flooding from debris dam overtopping or collapse is still needed, along with implementation of appropriate countermeasures. The methods described in this chapter can be applied to that task. At the same time, it is of prime importance to take measures and prevent the debris flows from damming the major river channels of the region and posing new threats to downstream areas.

**Acknowledgements** The work was supported by the State Key Fundamental Research Program (No. 2008CB425802), Natural Science Foundation of China (No. 40671025) and the Knowledge Innovation Project of the Chinese Academy of Sciences (No. KZCX2-YW-302). The authors acknowledge Prof. Chen Zuyu, Prof. Liu Ning, Ministry of Land and Resources, and the Water Resources Bureau of Mianyang City, for providing photographs and layouts. The authors also appreciate Dr. Kevin M. Scott, USGS, for his kind help in editing the manuscript.

## References

1. Cenderelli, D.A. (2000) Floods from natural and artificial dam failures, in E.E. Wohl (ed.), *Inland Flood Hazards: Human Riparian, and Aquatic Communities*. Cambridge University Press, New York, NY.
2. Chen, J.C., Fan, K.X., Li, Z.P. and Zhang, H.G. (2008) An analysis of hydrological characteristics of Tangjiashan landslide lake[J], *Yangtze River* **39**, 26–28 (in Chinese).
3. Chen, W.Y., Zheng, J.X., Tan, J. et al. (2008) The proposals of emergency and comprehensive treatment for Tangjiashan Barrier Lake, *China Power* **34**, 10–14 (in Chinese).
4. Chen, X-Q., Cui, P., Li, Y. and Zhao, W.Y. (2010) Emergency response to the Tangjiashan landslide-dammed lake resulting from the 2008 Wenchuan Earthquake, China, *Landslides*, doi 10.1007/s10346-010-0236-6.
5. Costa, J.E. and Schuster, R.I. (1985) *Floods from Dam Failures*. US Geological Survey Open-File Report 85–560, 54 p.
6. Costa, J.E. and Schuster, R.I. (1988) The formation and failure of natural dams, *Geological Society of America Bulletin* **100**, 1054–1068.
7. Cui, P., Chen, X.Q., Zhu, Y.Y. et al. (2009) The Wenchuan Earthquake (12 May 2008), Sichuan Province, China, and resulting geohazards, *Natural Hazards* doi 10.1007/s11069-009-9392–1.
8. Cui, P., Zhu, Y., Han, Y., Chen, X. et al. (2009) The 12 May 2008 Wenchuan earthquake-induced landslide lakes: Distribution and preliminary evaluation, *Landslides* **6**, 209–223.
9. Cui, P., Chen, X.Q. and Chen, Z.Y. (2009) The barrier lakes created during the Wenchuan Earthquake and its mitigation works. *Proceedings of SINOROCK 2009 – ISRM International Symposium on Rock Mechanics, 19–22 May 2009, Hong Kong*.
10. Cui, P., Dang, C., Zhuang, J-Q., You, Y., Chen, X-Q. and Scott, K.M. (2010) Landslide-dammed lake at Tangjiashan, Sichuan province, China (triggered by the Wenchuan

Earthquake, May 12, 2008): risk assessment, mitigation strategy, and lessons learned, *Environmental Earth Sciences*, doi 10.1007/s12665-010-0749-2.

11. Fread, D.L. (1977) The development and testing of a dam-break flood forecasting model, in *Proceedings of Dam-Break Flood Routing Model Workshop* held in Bethesda, MD, Water Resources Council, Washington, DC, pp. 164–197.
12. Fread, D.L. (1980) Capabilities of NWS model to forecast flash floods caused by dam failures, in *Proceedings of Second Conference on Flash Floods*. Am. Meteorol. Soc, Boston, MA, pp. 171–178.
13. Fread, D.L. (1988) *BREACH: An Erosion Model for Earthen Dam Failures*. National Weather Service (NWS) Report, NOAA, Silver Spring, MD.
14. Fread, D.L. and Lewis, J.M. (1998) *Theoretical Description and User Documentation of NWS FLDWAV Model*. Hydrologic Research Laboratory: Office of Hydrology, National Weather Service (NWS), NOAA, Silver Spring, MD, 20910.
15. Froehlich, D.C. (1995) Peak outflow from breached embankment dam, *Journal of Water Resources Planning and Management (ASCE)* **121**, 90–97.
16. Hydrology Office of Sichuan Province (1979) Hydrological manual of Sichuan Province, Sichuan Science and Technique House.
17. Kirkpatrick, G.A. (1977) Evaluation guidelines for spillway adequacy. In *Proceedings of the Engineering Foundation Conference, The Evaluation of Dam Safety*. American Society of Civil Engineers, New York, NY, pp. 395–414.
18. Li, W. (1980) *Handbook of Hydraulic Calculations*. Water Publication, Beijing, pp. 612–644 (in Chinese).
19. Luo, J., Wang, S. and Zhang, J. et al. (1987) The projects for controlling debris flow at the lower reaches of the Jiangjia ravine and the improving suggestions, *Journal of Mountain Research* **5**, 14–18.
20. Singh, V.P. (1996) *Dam Breach Modeling Technology*. Kluwer, Dordrecht.
21. Xu, X., Wen, X., Ye, J. et al. (2008) The MS 8.0 Wenchuan earthquake surface ruptures and its seismogenic structure, *Seismology and Geology* **30**, 597–629 (in Chinese).
22. Yang, Q.G. (2008) Key technologies of emergency treatment of Tangjiashan dammed lake, *China Water Resources* **16**, 8–11 (in Chinese).
23. Zhao, Z.X. and He, J.J. (2005) *Hydraulics*. Tsinghua University Press, Beijing (in Chinese).

# Chapter 11
# The Importance of Geological Models in Understanding and Predicting the Life Span of Rockslide Dams: The Case of Scanno Lake, Central Italy

**G. Bianchi-Fasani, C. Esposito, M. Petitta, G. Scarascia-Mugnozza, M. Barbieri, E. Cardarelli, M. Cercato, and G. Di Filippo**

## 1 Introduction

The Scanno rockslide-avalanche dammed the Tasso river and caused the impoundment of the Scanno Lake which is one of the most famous examples of a naturally dammed lake in the Central Apennines (Fig. 11.1); it is located at an elevation of 922 m a.s.l. and has an area of approximately 1.2 km$^2$, a perimeter of about 5 km and a maximum depth of 33 m [27]. Far more significant is the presumed age of the lake. According to reports and documents by Roman historians the event could date back to 217 B.C [22]. Further, radiocarbon dating of a paleosoil sample collected in the accumulation area just below the rockslide debris yielded an age of about 12,800 years, thus giving a lower bound for the time of occurrence [4]. These datings, together with considerations of the landslide debris distribution and upstream morphology, would suggest an age between 12,000 and, at least, 2,300 years. Thus the life span of the dam is remarkable, since there is no historic record of catastrophic breaching during the past 2,300 years.

To explain such a long life span of the rockslide dam and to get an indication of its future stability conditions, it was considered fundamental to build a representative geological model specifically aimed at defining:

- geometry of the boundary surface between the landslide debris and the bedrock;
- characteristics of the debris with special reference to hydraulic behaviour;
- flownet within the rock-avalanche deposit taking into account the complex geological and hydraulic boundary conditions imposed by the palaeovalley morphology and by the numerous springs downstream of the dam which are fed by different aquifers.

G. Scarascia-Mugnozza (✉)
Department of Earth Sciences, University of Rome "La Sapienza", Piazzale Aldo Moro 5, 00185 Rome, Italy
e-mail: gabriele.scarasciamugnozza@uniroma1.it

S.G. Evans et al. (eds.), *Natural and Artificial Rockslide Dams*, Lecture Notes in Earth Sciences 133, DOI 10.1007/978-3-642-04764-0_11, © Springer-Verlag Berlin Heidelberg 2011

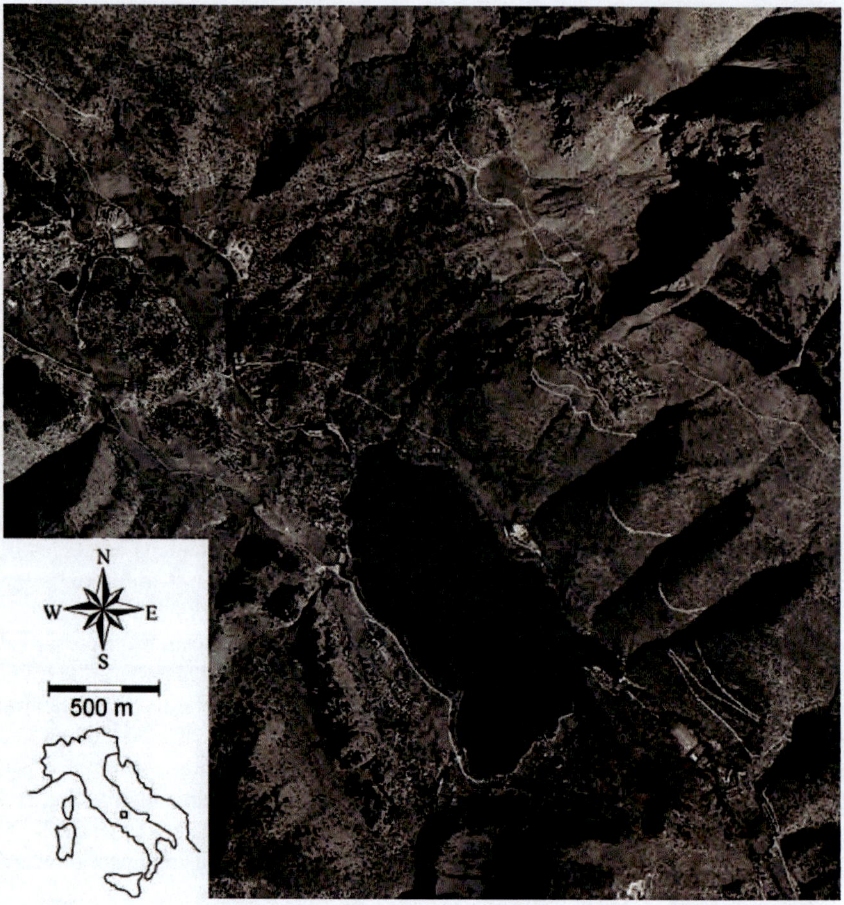


**Fig. 11.1** Aerial view of the rockslide-dammed Scanno Lake. Inset shows location of Scanno Lake in Central Italy

According to the above mentioned objectives, a multidisciplinary research project was planned. The performed activities encompass geological and geomorphological survey, borehole logging, geophysical surveys based on both electrical tomography and seismic refraction techniques, pumping tests and hydrogeological measurements, chemical and isotope analyses. To obtain further indications about the stability conditions of the Scanno rockslide dam, previously proposed geomorphological indexes [16] were computed. In the following sections, after an outline of the geological and hydrogeological setting of the Scanno area, the research methodology and the results so far acquired are presented and discussed.

## 2 Geological Setting

The study area is located in the central Apennines, a NE-verging imbricate, fold-thrust belt developed under a Neogene, ENE-oriented, compressive thrust system. Beginning in the Pliocene-Lower Pleistocene, this phase of thrusting was followed by extensional tectonic movements [8]. The latter were compounded by regional uplift processes that have affected the entire Italian peninsula since Lower Pleistocene [14] up to Present. As a result, the general morphostructural setting is characterized by a pattern of ridges bounded by high angle, normal faults on the western side and by NW-SE to NNW-SSE thrust fronts on the eastern one. These ridges, mainly made up of Triassic-Middle Miocene carbonate platform, ramp, and intraplatformal facies [7], are often separated by valleys usually filled with synorogenic, siliciclastic deposits and continental, clastic covers.

The Tasso-Sagittario valley is located in the eastern Marsica region within a complex geological-structural setting, where different geological-structural units belonging to different palaeogeographic domains (the Latium-Abruzzi and Apulian carbonate platforms and the Molise pelagic basin) are present. The contacts among these units are represented by important tectonic lineaments predominantly N-S and NNW-SSE oriented [21].

In particular, the Scanno Lake area is bounded on its western side by the Mt. Argatone ridge, an homoclinal sequence of Meso-Cenozoic dolomitic limestones, micritic and detritic limestones ascribable to the Latium-Abruzzi carbonate platform palaeogeographic domain (Fig. 11.2). The contact between the Mt. Argatone eastern slope and the Tasso-Sagittario river valley is marked by a tectonic element, corresponding to a NNW-SSE oriented and eastwards dipping thrust surface.

The eastern flank of the valley is represented by the Mt. Genzana ridge, where a homoclinal sequence of rocks belonging to the transition zone between the Apulian carbonate platform and an ancient basin then erased during the orogenesis [13] crop out. These deposits consist of Meso-Cenozoic limestone and marly-limestone with chert nodules. Mt. Genzana is bordered on its western slope by an important tectonic lineament (Genzana Fault in [10]) NNW-SSE oriented dipping 60–80° westwards. The kinematic indicators point out at least a first strike-slip activity and a subsequent extensional one [21].

Clayey and sandy-clayey, synorogenic flysch deposits outcrop within the valley bottom: their contacts with the local pre-Miocenic sequences are exclusively represented by tectonic elements.

Another geological-structural unit (Mt. della Rovere-Piana Malvascione Unit) crops out between the Mt. Genzana and Mt. Argatone ridges: it is composed of micritic and detritic limestones with interlayered clayey beds. As for the flysch deposits, this unit has tectonic contacts with the surrounding geological units marked by NNW-SSE oriented, high angle faults.

The local stratigraphic framework is completed by Pleistocene up to Recent continental, clastic deposits such as alluvial fans, debris cones, toe-of-slope debris, alluvial and colluvial deposits.

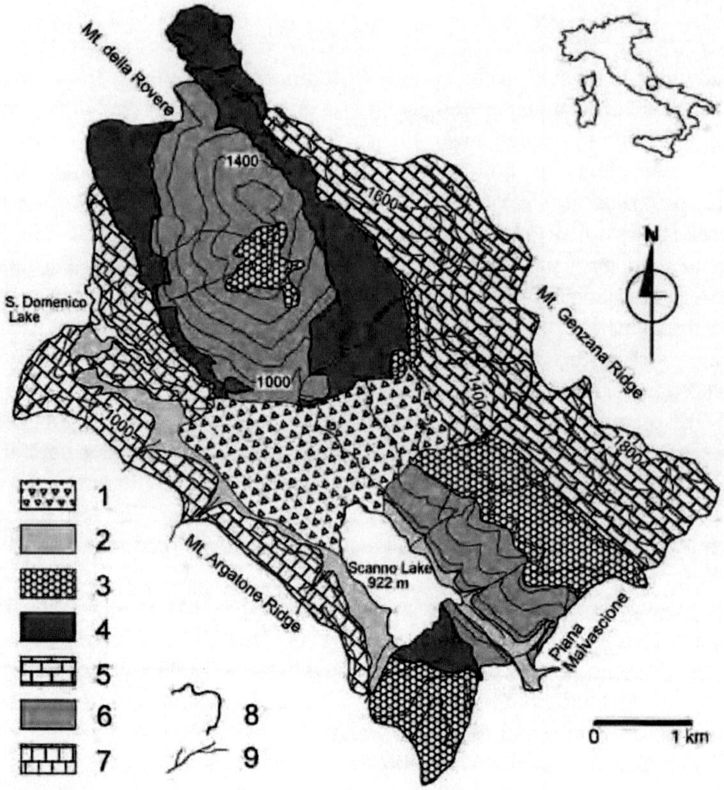


**Fig. 11.2** Simplified geological map. Key to legend: *1* – Rockslide debris; *2* – Quaternary deposit; *3* – Quaternary cemented breccias; *4* – Turbiditic clay and sandstone; *5* – Mt. Genzana unit; *6* – Malvascione unit; *7* – Mt. Argatone unit; *8* – Rockslide scar; *9* – Surficial hydrography

# 3 Hydrogeological Setting

The Tasso-Sagittario basin consists of mesozoic carbonate aquifers, with very complicated structural setting (Fig. 11.3). The aquifers belonging to the carbonate platform domain have an effective infiltration of about 900 mm/year; the aquifers belonging to the transitional domain have an effective infiltration which has been estimated about 600 mm/year. Effective infiltration recharges large karst aquifers which feed huge springs with discharges in the order of some $m^3$. In addition, some streambed springs (natural discharge of groundwater emerging directly into the streambed) have been detected in the area [5, 26].

The head of the Sagittario River is located at elevations of about 1,300 m a.s.l., where it is called Tasso River. This river flows into the Scanno Lake (922 m a.s.l.) with a minimum discharge of 0.4 $m^3$/s (base flow), including La Marca Spring (0.1 $m^3$/s) flowing directly into the Scanno Lake [25, 28]. Scanno Lake has no

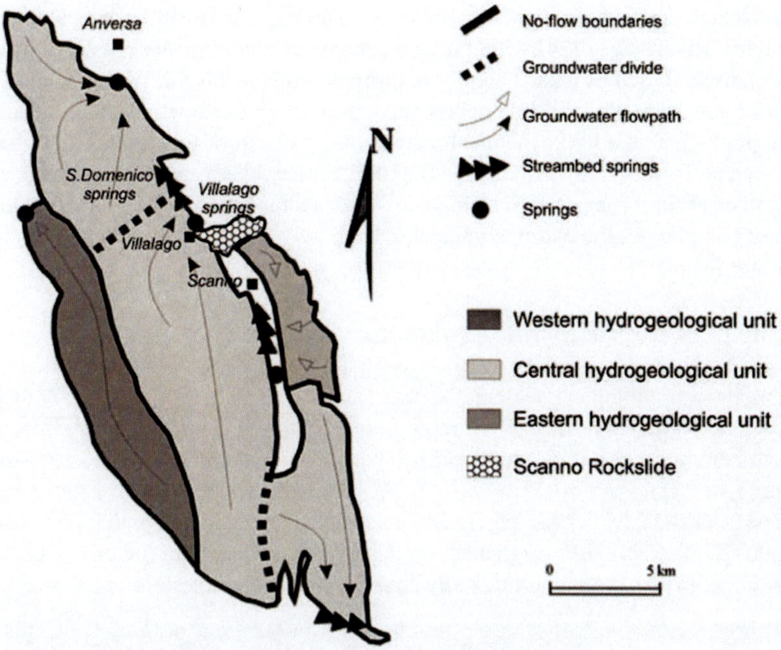


**Fig. 11.3** Hydrogeological setting of the Scanno area (modified after [28])

outlet during dry periods (summer and autumn) and frequently during the entire year. The lake retains a steady water level. Only during intense precipitation periods do waters flow into the overflow inducing the formation of an ephemeral lake at the Prato site where infiltration into the rockslide-avalanche deposit occurs (Fig. 11.4). When discharge is not observed at the overflow, the inflow is assumed to be lost by evaporation and/or by infiltration into the natural dam. Taking into account the climatic conditions, a preliminary water budget shows high infiltration values into

**Fig. 11.4** Prato site within the rockslide debris; in the right photograph the ephemeral lake perimeter during the wet period is visible in the debris

the rockslide deposits [6]. Downstream of the dam there are many springs along the Sagittario River valley [24] with a total discharge around 1 $m^3/s$ (Villalago Spring group), partially fed by a karst aquifer outcropping on the left side of the valley. The contribution from the landslide debris was up to today unknown, because detailed hydrogeological and hydrochemical studies to the origin of this groundwater have not been performed. Downstream the Sagittario valley, there are further springs and streambed springs (Sega and S. Domenico), with a total discharge of 1.8 $m^3/s$, which further complicates the hydrogeological setting.

## 4 The Scanno Rockslide-Avalanche

The main morphological features of the Tasso-Sagittario valley sector between Scanno and Villalago villages are represented by the Scanno rockslide-avalanche detachment and deposition areas and by the rockslide-dammed Scanno lacustrine basin [3, 4, 22].

The rockslide, which as previously mentioned could date back in the range 12,000–2,300 years BP originated on Mt. Rava, a shoulder of Mt. Genzana (Fig. 11.5); in plan the scar is markedly concave within the limestone and the rupture

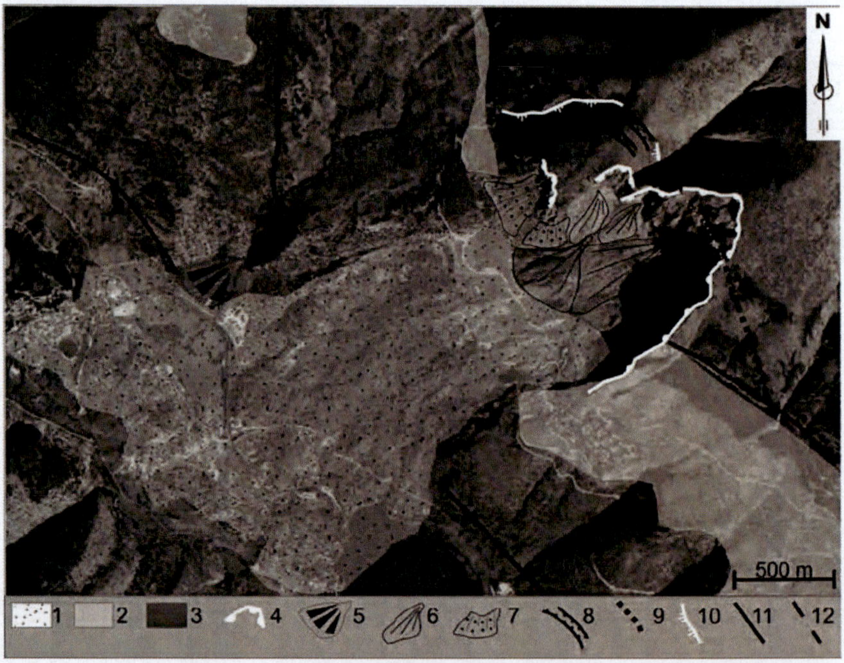


**Fig. 11.5** Aerial photo with the main geomorphological elements of the Scanno rockslide-avalanche area. Key to legend: *1* – Scanno rockslide-avalanche debris; *2* – Cemented continental breccias; *3* – Detached limestone block; *4* – Scanno rockslide-avalanche detachment area; *5* – Alluvial fan; *6* – Debris cone; *7* – Talus; *8* – Trench; *9* – Structural lineation; *10* – Gravity-induced deformation; *11* – Fault; *12* – Presumed fault

surface along the slope has a chair shape in section. The S-shaped geometry of the rupture surface is clearly related to the geological-structural setting, and in particular to the bedding attitude, which favoured a sliding mechanism in the early stage of failure [4]. The landslide involved a rock mass volume of about $80 \cdot 10^6$ m$^3$ that slid onto the valley bottom and partially ran up on the opposite slope, with a 3.6 km total run-out. After the collapse onto the valley bottom the debris dammed the Palaeo-Tasso River and formed Scanno Lake.

The landslide debris now covers a T-shaped area of about 2.6 km$^2$, mainly developed on the flysch deposits of the valley bottom, and has a typical hummocky topography with a marked frontal ridge. Some road cuts locally show an inverse sorting of the debris, that is composed by a sandy-gravel matrix with interspersed blocks and boulders with volumes up to tens of m$^3$. These morphological and sedimentological features together with the large volume involved and the evidence of a partial run-up on the opposite slope, indicate a rock-avalanche mechanism of emplacement [4, 22].

The dam resulting from the portion of the landslide debris that impounds the valley bottom, shows a considerable bulk and a relatively flat form. Such features, added to the presence of a pronounced frontal ridge as well as to the peculiar T shape of the debris, allow us to ascribe the Scanno rockslide dam to type III of classification by [11].

## 5 Methodology

The core goal of this paper is to understand the factors bearing on the long life span of the Scanno rockslide-avalanche dam and to assess its overall stability, in relation to its geometry and characteristics of the debris with particular reference to groundwater circulation.

The geometry and the geotechnical characteristics of the debris are a function of the pre-rockslide-avalanche Sagittario River valley morphology, of the mechanism of landslide debris emplacement and of the source rock mass outcropping along the western slope of Mt. Genzana.

In order to obtain information on these subjects, the following site investigations were carried out:

- borehole core logging down to the flysch bedrock to determine the thickness of the landslide deposit which filled the bottom of the Sagittario river valley;
- geophysical prospecting based on vertical electrical sounding, resistivity tomography and seismic refraction methods.

The groundwater transfer role played by the rock-avalanche debris, in the framework of the hydrogeological setting described above, can be outlined as follows: (a) the lake reservoir receives waters from both watershed and hydrogeological sources (inflow); b) the rock-avalanche debris transfers groundwater from the lake to the Villalago Spring (transfer); (c) the Villalago Spring group, receives contributions from both the rock-avalanche debris aquifer and regional carbonate aquifers (outflow).

Based on this scheme, hydrogeological investigations were performed to obtain the following information:

- water budget of the Scanno Lake, with the evaluation of the infiltration losses into the rockslide-avalanche deposit; the analysis was undertaken for a long time period (climatic conditions in 50 years);
- evaluation of the contribution from the landslide deposit to the Villalago Spring group, recording outflow and spring discharge values and their variations in different seasons of 2004;
- evaluation of the permeability of the landslide debris, acquiring data on the hydraulic gradient based on lake water table and spring elevations as well as through pumping tests in appositely drilled wells;
- separation of the contribution from carbonate aquifers and from the landslide deposit to the Villalago Springs, by hydrochemical coupled sampling from the inlet and the outlet, from the lake itself and from different springs, in different periods during the year;
- groundwater flowpaths modeling by stable isotope ($^{18}O/^{16}O$) investigations, to distinguish actual flowpaths into the carbonate aquifers and into the rockslide deposits as well as to obtain information about the recharge area elevations [9, 12].

# 6 Reconstruction of the Natural Dam Geometry

## *6.1 Borehole Investigations*

A couple of boreholes, used respectively as a well and a piezometer, were drilled in a depressed area within the rockslide-avalanche debris in order to characterise the stratigraphy of the debris, to find the contact with the local flysch bedrock and, finally, to calibrate the successive geophysical investigations (Fig. 11.6).

As regards the stratigraphy, after about 5 m of alluvial-colluvial deposits the drilling passed through a thickness of about 55 m of rock-avalanche debris (calcareous and marly-calcareous blocks within a sandy-gravel calcareous matrix). The clayey flysch bedrock was reached at a depth of about 60 m.

## *6.2 Geophysical Investigations*

To investigate the geometry of the shallow soil deposits, their physical properties, thickness and volume of the landslide body, a multi-parametric geophysical survey was performed in this area.

As shown in Fig. 11.6, three ERT (Electrical Resistivity Tomography) lines (ERT1, ERT2 and ERT3) and two seismic lines (L1 and L2) were run striking NW-SE, along the valley floor, while the ERT4 line is oriented roughly NE-SW.

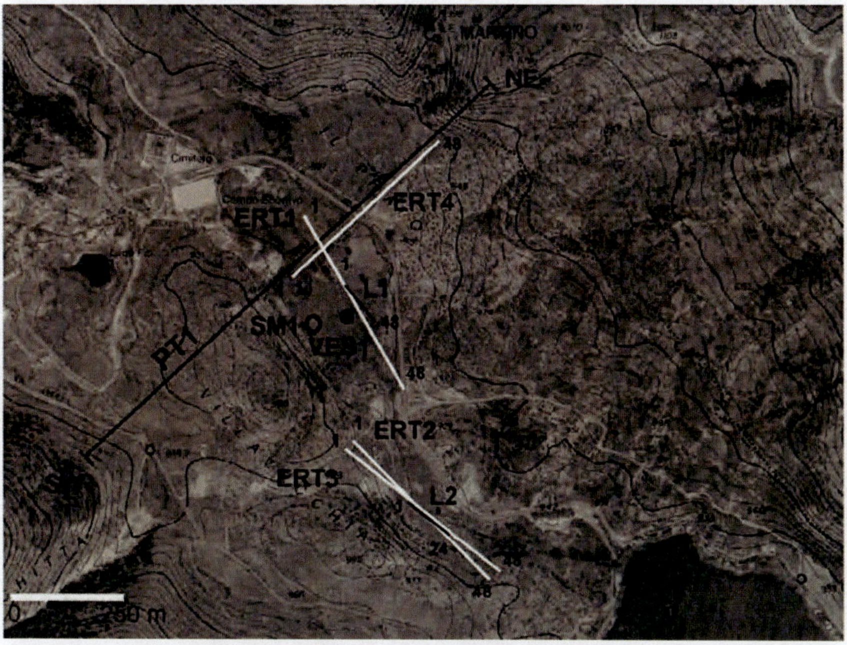


**Fig. 11.6** Location of the geophysical surveys. ERT1, ERT2, ERT3, ERT4: electrical resistivity tomography. L1, L2: seismic *lines*. VES1: vertical electrical sounding. SM1: boreholes location. PT1: topographic profile (see Fig. 11.15)

The acquisition parameters for these surveys are summarized in Table 11.1.

Electrical resistivity data were measured using the IRIS Syscal device with 48 electrodes spaced 10 m apart and were processed using the inversion algorithm proposed by [19].

Only the inverted models obtained with the pole-dipole configuration, which penetrated deeper into the subsoil [15], are presented. As ERT2 and ERT3 are approximately coincident, only the ERT3 section will be shown here.

An additional Vertical Electrical Sounding VES1, performed to reach deeper targets, completed the electrical prospecting.

Seismic data were acquired using a 24 or 48 channels device (GEODE by Geometrics). Joint acquisition of seismic refraction and Multichannel Analysis of Surface Wave or MASW data [23] were possible with such a device using the acquisition parameters in Tables 11.1 and 8 Hz vertical geophones. Seismic refraction data (first arrivals) were processed using the delay-time method [17] for shallow targets [2].

The processing applied to surface wave data is described as follows; after balancing the traces for cylindrical spread correction, the records were transformed in f-k domain and picking of maxima was performed to detect the dispersion curve values. We used the algorithm proposed by [18], to invert the dispersion curves. The running variables of the inversion process are the S-wave velocity values, while, in

**Table 11.1** Acquisition parameters for the geophysical measurements

| Electrical prospecting | | | | |
|---|---|---|---|---|
| **Survey** | **Electrodes** | **Quadripole** | | **AB/2 max** |
| VES1 | 4 | Schlumberger | | 250 |
| **Survey** | **Electrodes** | **Configuration** | | **Spacing (m)** |
| ERT1 | 48 | dipole-dipole, pole-dipole | | 10 |
| ERT2 | 48 | dipole-dipole | | 10 |
| ERT3 | 48 | Pole-dipole | | 10 |
| ERT4 | 48 | Dipole–dipole, pole-dipole | | 10 |
| Seismic prospecting | | | | |
| **Survey** | **Δt (ms)** | **ns** | **Δx (m)** | **Receivers** |
| L1 | 0.125 | 16,384 | 3 | 48 |
| L2 | 0.125 | 16,384 | 5 | 24 |

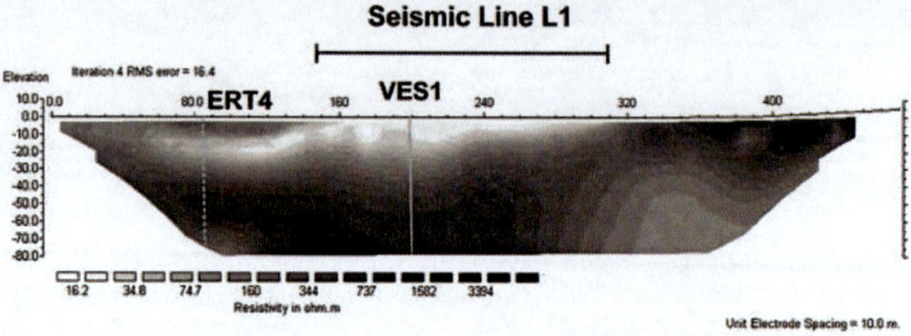


**Fig. 11.7** Resistivity section of ERT1 *line*. The vertical *dashed line* indicates the intersection with ERT4 *line* (see Fig. 11.9), while the location of L1 is indicated on the top of the section. VES1 is also located on the resistivity section

order to reduce the number of unknowns, P-wave velocities and density values are assigned as a-priori information.

The resistivity model ERT1, displayed in Fig. 11.7, shows the presence of a low resistivity overburden (in the first 300 m of the line), which slightly dips toward the NNW. This overburden is characterized by resistivity values in the 20–70 Ωm range, and by a maximum thickness of about 20 m (corresponding to the first electrodes of the line). The underlying layer is marked by higher resistivity values, up to 1,500 Ωm.

On the other hand, the right part of the section (distance from first electrode higher than 320 m, i.e. from electrode 33), which is the one closer to the lake, shows high resistivity for the outcropping layer, which forms in this area a moderate topographic high.

The outcome of the VES1 is shown in Fig. 11.8. Although the 1D distribution of resistivity, which was assumed for VES data interpretation, poorly represents the subsoil features in this geological context, it was possible to estimate the thickness of the high-resistivity layer as about 60 m, with a maximum depth for the bottom of this layer of approximately 90 m.

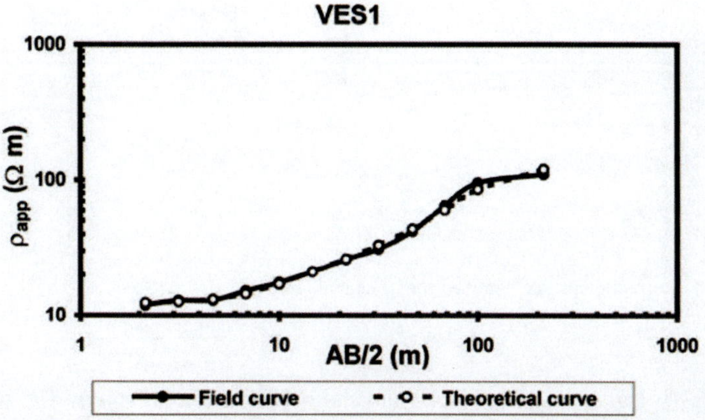


| Layer | Resistivity (Ω*m*) | Depth layer bottom (*m*) |
|---|---|---|
| 1 | 12 | 5.5 |
| 2 | 39 | 31.5 |
| 3 | >1500 | 90 |
| 4 | 30 | inf. |
| | | |
| residual (r.m.s.) 5.4 % | | |

**Fig. 11.8** VES1. *Top*: display of the measured and the inverted apparent resistivity curves. *Bottom*: inverted model

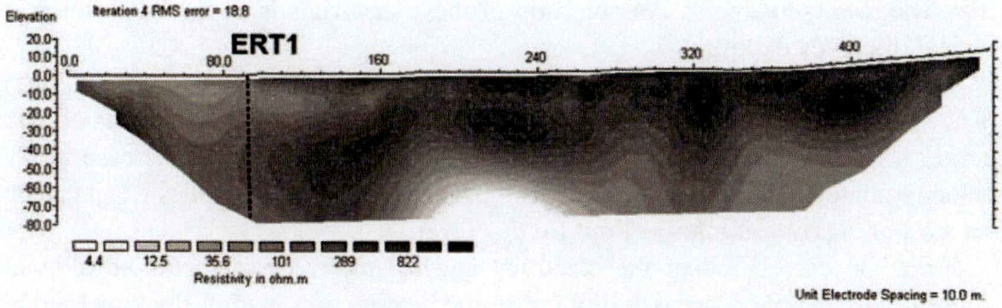


**Fig. 11.9** Resistivity section of ERT4 *line*. The vertical *dashed line* indicates the intersection with ERT1 line (see Fig. 11.7)

The ERT 4 (Fig. 11.9), which is roughly orthogonal to the ERT1, shows high resistivity values outcropping in the northern part of the array, which was deployed on the slope. On the other hand, the part of the resistivity line reaching the valley floor agrees with the result of the ERT1 investigation, as the resistivity properties fit well where the two lines are intersecting each other (dashed lines in Figs. 11.7 and 11.9).

The seismic line L1 was used to characterize the overburden and the top of the resistive layer in terms of propagation velocities of body waves. The refraction

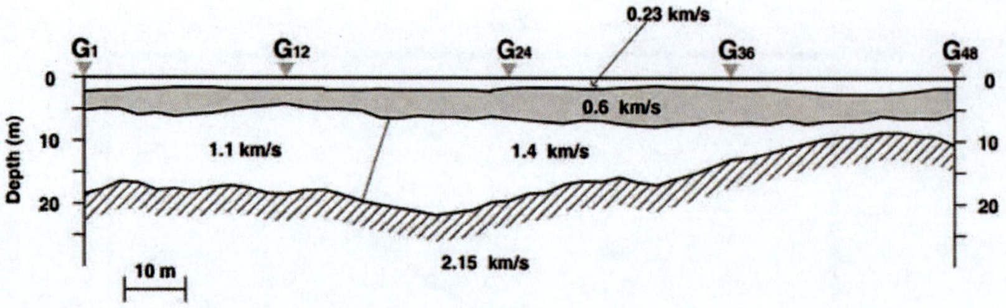


**Fig. 11.10** Refraction profile obtained for L1 seismic line

profile L1 is reported in Fig. 11.10 and shows four different layers. The weathering is characterized by a Vp of about 230 m/s and a thickness of roughly 2 m. The underlying layer is marked by a low P-wave velocity of 600 m/s in the first 5–6 m of depth. Below, there is a layer where Vp ranges from 1,100 to 1,400 m/s with slight lateral heterogeneities. The deeper refractor located by first arrival interpretation, marked by a P-wave velocity of about 2,100 m/s, corresponds to the top of the high resistivity layer in Fig. 11.7.

As the refractor is dipping with moderate slope in the N direction, using 1D assumption for the subsoil model in Rayleigh-wave dispersion interpretation, it was possible to characterize unambiguously the overburden in terms of $V_S$. Figure 11.11 shows the results of the MASW data inversion for two symmetric shots. The overburden shows $V_S$ values typical of poorly consolidated sediments, lower than 250 m/s. The average value of Poisson ratio of these materials is about 0.45, which is typical of clayey deposits.

The Vs value of the deeper refractor of the refraction profile ($Vp \approx 2{,}100$ m/s) is not well resolved by Rayleigh-wave inversion, as derived by the analysis of data kernel resolution. In spite of that, the dispersion curves show an increase in Vs velocity related to the deeper interface indicated by refraction seismics L1: it can be set a Vs of 500 m/s as a lower limit for this layer.

From the correlation of the electrical and seismic measurements an alluvial deposit was identified, weakly dipping in the N direction, with a thickness along the line in the range 10–20 m, corresponding to low resistivity values and seismic velocities typical of poorly consolidated materials.

The underlying layer is marked by higher resistivity and Vp velocities: these results, if compared to the geological model described above, point out the presence of the limestone debris of the landslide body, whose thickness has been estimated by VES sounding to be about 60 m in this area.

The ERT3 displayed in Fig. 11.12 depicts a completely different scenario if compared to the ERT1 section (Fig. 11.7). In this case, a heterogeneous resistive material is outcropping. The resistivity ranges from more than 1,000 Ωm at shallow depths to a few hundreds Ωm in depth. The average thickness of this resistive heterogeneous deposit can be estimated along the line as about 45–55 m. Below this material a very low resistivity deposit is identified.

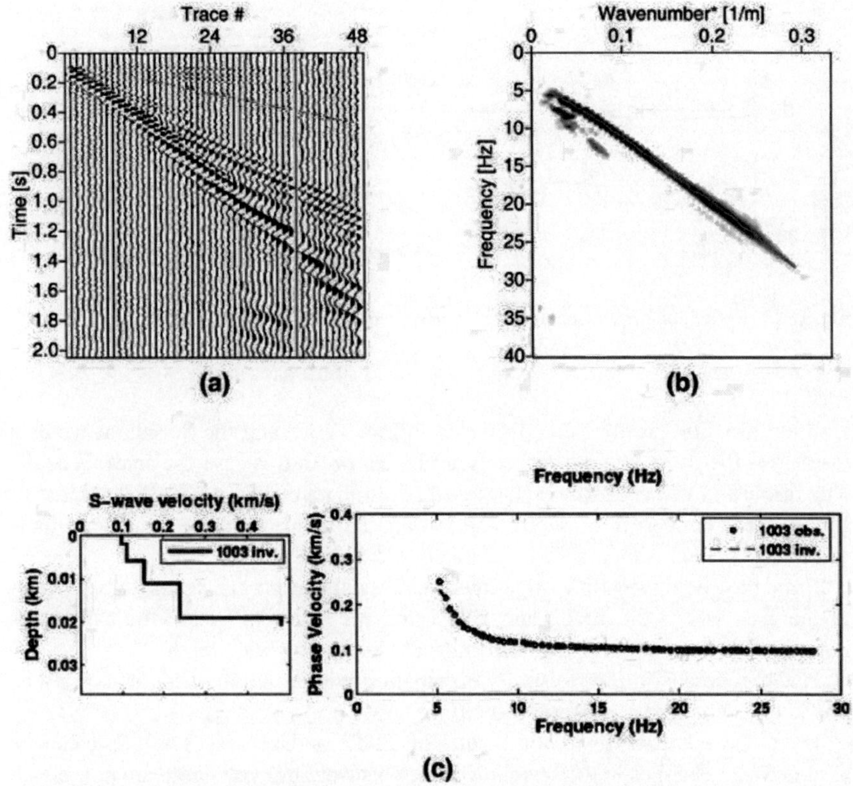


**Fig. 11.11** MASW data interpretation for seismic line L1. (**a**) Example shot gather (20 m offset from geophone 1). (**b**) *f-k* panel with picked dispersion curves. (**c**) Inverted resistivity profile (*left*) and comparison between experimental and theoretical data (*right*)

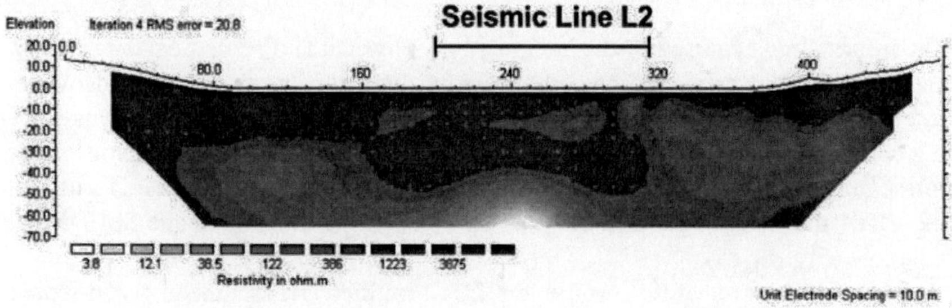


**Fig. 11.12** Resistivity section of ERT3 line. The location of L2 is indicated on the top of the section

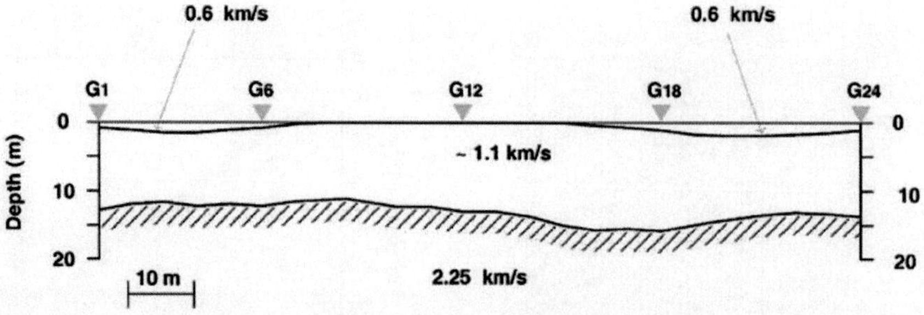


**Fig. 11.13** Refraction profile obtained for L2 seismic line

The refraction profile L2 is displayed in Fig. 11.13, and the Surface-wave data results for the same line are reported in Fig. 11.14. In this case the analysis of the data resolution kernel in Rayleigh-wave dispersion inversion indicates data reliability down to about 15 m of depth. The average Poisson's ratio for these deposits is about 0.33, which is typical of unsaturated rock debris.

From the geophysical investigations, no alluvial overburden seems to be present in the area where the ERT2 and ERT3 and L2 lines are taken: the calcareous landslide debris is outcropping, with stiffness increasing with depth.

Heterogeneities in the resistivity distribution within the landslide mass may be due to different degree of water saturation.

From the comparison of the results of ERT3 and seismic line L2 it can be assumed that changes in the seismic velocity distribution with depth are not related to a stratigraphic contact, but derive from an increase of stiffness within the landslide body, which has been estimated as 50 m thick along the line.

## *6.3 Remarks*

The information acquired by the integrated geophysical and geological (site surveys and boreholes) investigation has been used for the reconstruction of the paleovalley morphology and, finally, for the assessment of the debris volume actually contributing to the stream damming. For this purpose a NE-SW oriented topographic section (orthogonal to the valley direction) has been considered, partially coincident with the ERT4 multielectrodic line and passing through the location of the SM1 drilled borehole (Fig. 11.15).

For the NE sector of this profile, the ERT4 multielectrodic line gives an indication of the morphology of the contact between the landslide debris and the flysch "bedrock": it has been assumed to be coincident with the trend of upper boundary of the resistivity layers with values < 100 Ωm. Additional information about the trend of this contact has been acquired not only by the SM1 results, but also by projecting the information obtained from other multielectrodic lines onto the selected section.

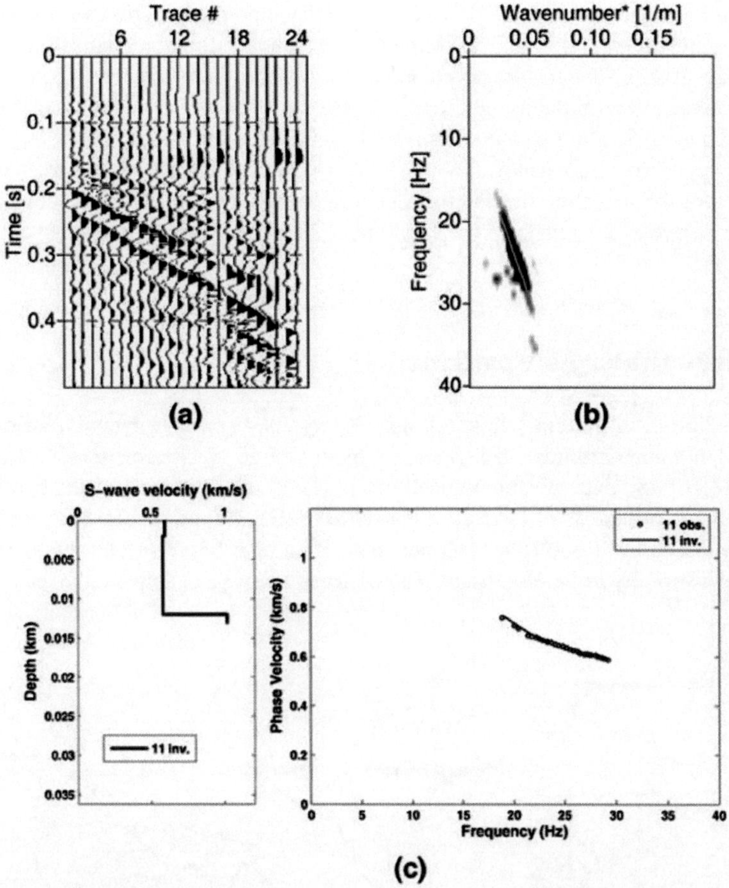


**Fig. 11.14** MASW data interpretation for seismic line L2. (**a**) Example shot gather (50 m offset from geophone 1). (**b**) *f-k* panel with picked dispersion curves. (**c**) Inverted resistivity profile (*left*) and comparison between experimental and theoretical data (*right*)

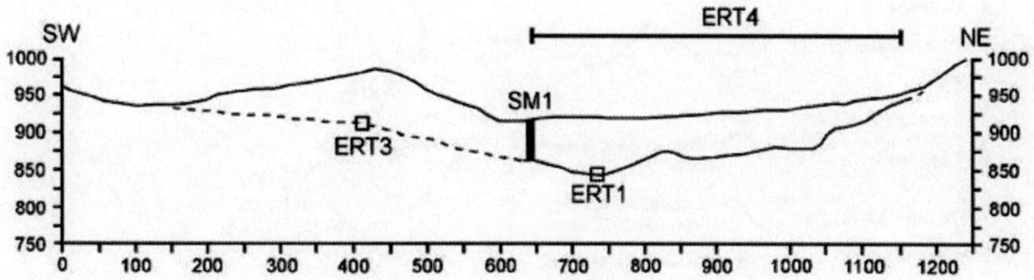


**Fig. 11.15** Reconstruction of the probable paleovalley topography along the profile PT1 (for location of ERT1, ERT3, ERT4, SM1 and PT1 refer to Fig. 11.6)

It should be observed how the derived paleovalley topographic cross section is consistent with some morphological elements: the frontal ridge of the landslide debris, resulting from a sudden loss of kinetic energy during emplacement, is located in correspondence with the abrupt change in the slope of the paleovalley, that is the site where deceleration is imposed on the landsliding mass.

The reconstructed palaeovalley morphology (Fig. 11.15) allowed us to calculate the area of the landslide debris cross-section (about $41 \cdot 10^3$ $m^2$): it was considered representative of the length of the dam (about 2 km), thus giving a volume of about $82 \cdot 10^6 m^3$.

## 7 Groundwater Flowpath Model

The hydrological system was monitored during 2004 (once in January, 3 times in March, and once in May and September) by discharge measurement along streams (Fig. 11.16) and data are summarised in Fig. 11.17. Contribution of the two inlets of the Scanno Lake (La Marca spring and Tasso River) shows a seasonal variation, mainly due to the Tasso River regimen, collecting both baseflow and runoff from a high-elevation hydrographic basin. The outlet of the lake is active only seasonally,

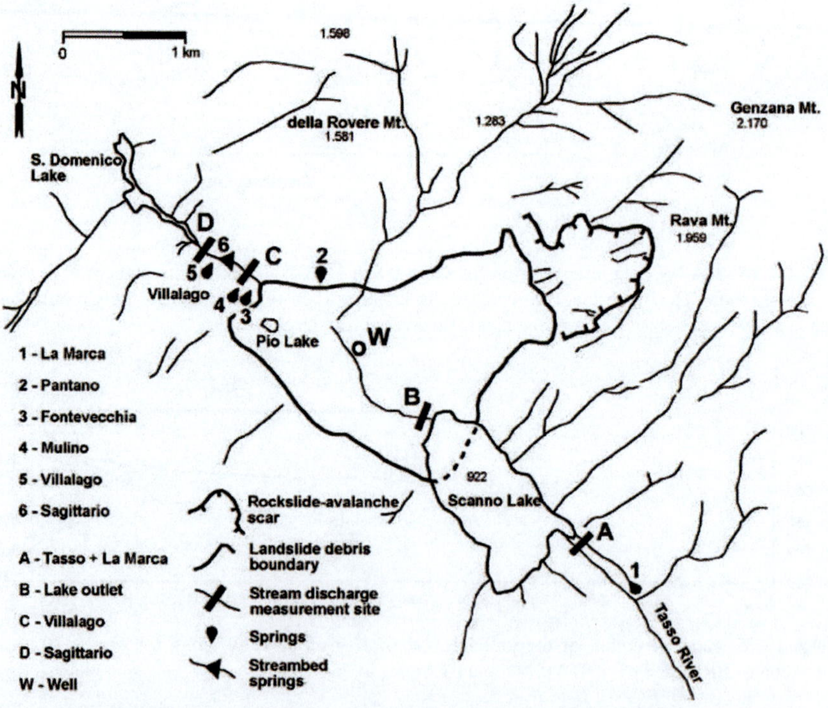


**Fig. 11.16** Sagittario valley sketch map with location of measurement site and main springs

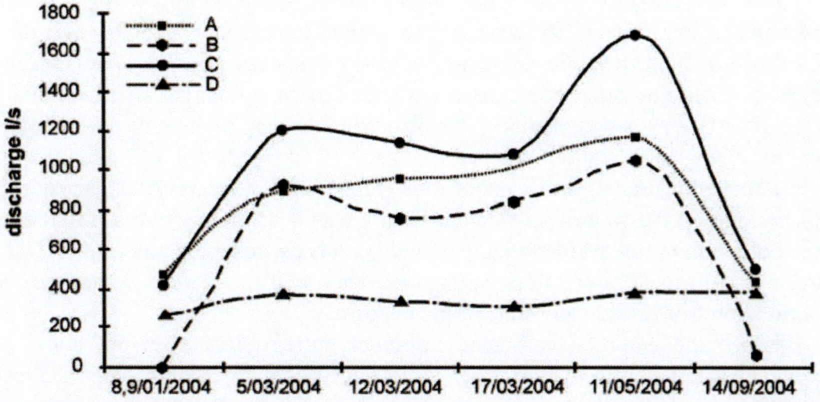


**Fig. 11.17** Comparison between the measured discharges along the Tasso-Sagittario river valley (letters refer to locations in Fig. 11.16)

when high discharge of the inflow streams is coupled with runoff water coming from the lake slopes. Evaporation from the lake has been estimated [29] as being very close to the direct rainfall on the lake surface (1,200 versus 1,050 mm of rainfall). When the outlet is active, its discharge spreads out on the landslide deposit surface, creating a seasonal lake. In a few 100 m, the outlet becomes completely exhausted and water seeps into the landslide deposits. Discharge of the outlet is very similar to the inlet discharge, but with lower values (Fig. 11.17).

The Villalago group of springs has been divided into two subgroups:

1. Fontevecchia and Mulino springs, located immediately downstream of the landslide deposit; these show a similar discharge regimen to the inlet-outlet system, but with higher discharge and with a positive contribution also during dry season (when outlet flow does not contribute to the discharge)
2. Villalago spring and Sagittario streambed spring, located along the Sagittario valley, where the river receives other contributions from the western ridge of Montagna Grande: total contribution of these springs has been evaluated by discharge measurements of the Sagittario River immediately before the S. Domenico artificial lake (Fig. 11.16). Its regimen and values are completely different from other springs and stream discharge (D in Fig. 11.17).

Other minor contribution to the discharge comes from eastern ridge (20 l/s from Pantano spring and 0–50 l/s variable discharge from Giordano stream (Fig. 11.16)).

A first water budget of the studied area shows strong correlation between the lake system (inlets-outlet) and spring discharge of the upper Sagittario valley (Fontevecchia and Mulino springs, 3 and 4 in Fig. 11.16), suggesting a contribution through the landslide deposits, which is evident even when the outlet is not active.

The two boreholes into the landslide deposits (W in Fig. 11.16), acting as a well and a piezometer respectively, were used to monitor the groundwater table over time and to obtain hydrodynamic parameters of the groundwater hosted by the landslide deposits. From pumping tests carried out with 7l/s of discharge, a transmissivity value of 4·10–2 $m^2/s$ was obtained, demonstrating the very high permeability of the deposits acting as an aquifer.

Hydrochemical analyses of major ions (Fig. 11.18) show very similar characteristics and all the waters are calcium-bicarbonate based. Only minor differences between groundwater are identifiable from this analysis, and these relate to the Tasso stream and to the Eastern slope springs (Pantano and La Marca), which receive contribution from clastic and siliciclastic deposits.

Because the common carbonate origin of springwaters does not allow us to obtain detailed information from basic hydrochemistry, two isotope surveys ($^{18}O/^{16}O$ and D/H) have been conducted on groundwater. Results show high correlation ($R^2 = 0.93$), identifying a trend similar to other regional trends previously reported [1, 20], with very little scatter due to the limited area under investigation (Fig. 11.19). The trend clearly identifies a different origin of groundwater feeding the Tasso river (A in Fig. 11.19) with respect to the other waters.

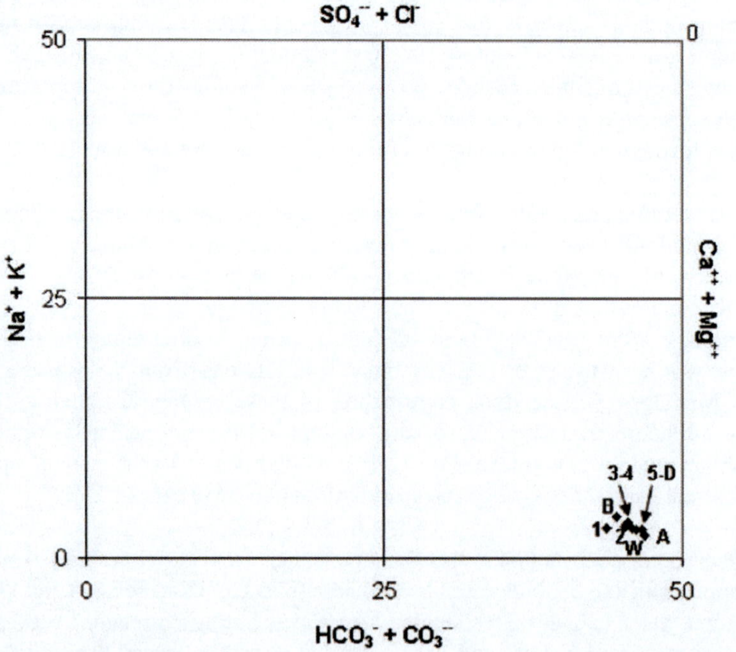


Fig. 11.18 Chebotarev diagram of hydrochemistry of springs and streams (letters and numbers refer to Fig. 11.16)

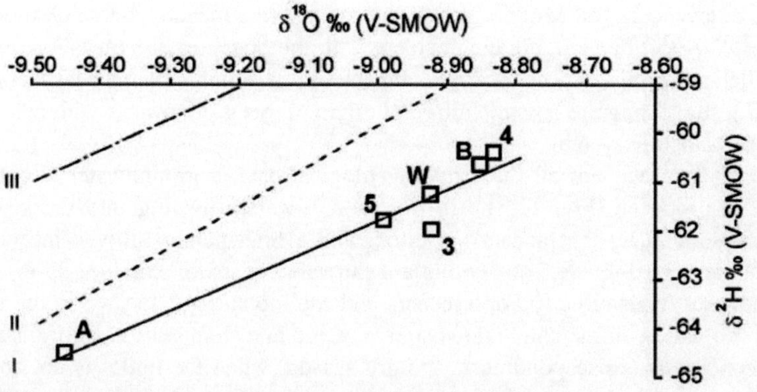


**Fig. 11.19** Correlation between the isotopic values δ $^{18}$O and δ $^{2}$H (I) for the collected samples and comparison with the regional trend elaborated by [1] (II) and [2] (III). Letters and numbers refer to Fig. 11.16

Starting from the hypothesis that outlet waters are related to the elevation of the lake surface, receiving direct rainfall, it was possible to make a first-attempt at correlation between $\delta^{18}O/^{16}O$ content and mean recharge elevation (Fig. 11.20), corresponding to the mean altitude of infiltration of groundwater fed by the studied spring, to separate groundwater coming from different flowpaths. This correlation shows a vertical isotope gradient of 0.15‰/100 m and a mean recharge elevation for Tasso river of around 1,300 m a.s.l., confirming the hydrogeological model proposed by [5] and validating our proposed hypothesis. All other waters (outlet,

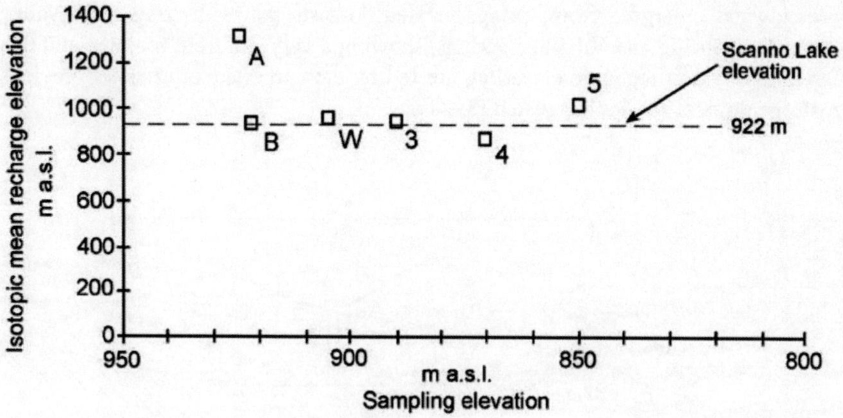


**Fig. 11.20** Calculated isotopic mean recharge elevation vs. sampling elevation. *Dashed line* represent lake elevation, corresponding to sample B (surface lake waters). Letters and numbers refer to Fig. 11.16

well, Fontevecchia and Mulino springs) are related to a mean recharge elevation of around 930–950 m a.s.l., quite coincident with the Scanno Lake elevation, except for Villalago springs, whose mean recharge elevation is higher than the lake (> 1,000 m a.s.l.), suggesting the possible different origin of this groundwater with respect to the lake-landslide system.

Taking into account all these hydrogeological data, a groundwater circulation model is proposed (Fig. 11.21). In this model, waters flowing into the lake are directly drained by the landslide aquifer, with a high permeability value (calculated around 1·10–2 m/s, considering a mean value of its thickness of 25 m). This groundwater feeds directly Fontevecchia and Mulino springs, located in the upper Sagittario valley, immediately downstream of the landslide debris. During steady-state conditions (corresponding to the dry season, when the outlet is not active), it is possible to calculate a hydraulic gradient of 20‰, considering the lake elevation (922 m a.s.l.), the water level in the well (–7 m from 919 m a.s.l.), spring elevation (895 m a.s.l.) and their distances. Groundwater discharge of Fontevecchia and Mulino springs is higher than 0.4 $m^3$/s, corresponding in the model to.the underground losses from the lake that coincide with the inflow, as rainfall and evaporation on the lake are exactly balanced.

During the wet season, the outlet is active and the landslide aquifer receives additional widespread contribution, which determine transient conditions of groundwater flow. The water table rises very close to the surface (maximum value of 0.5 m b.g.l. is registered in the piezometers) and its subsequent decrease shows the same gradient of the outlet discharge, confirming their relationship. The discharge from the landslide aquifer to the Fontevecchia and Mulino springs increases up to 1,600 l/s, with very rapid and impulsive response of the system. The calculated hydraulic gradient between the well and Fontevecchia and Mulino springs (respectively W, 3 and 4 in Fig. 11.16) is higher than 30‰. In the model, groundwater emerging from springs located downstream in the Sagittario Valley (streambed spring and Villalago spring), showing a very different regimen and differences in mean recharge elevation, are fed by the carbonate aquifer, as previous hydrogeological studies suggested [5, 28].

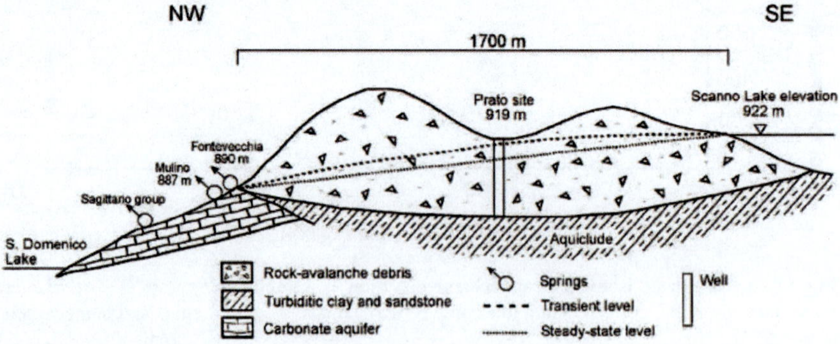


Fig. 11.21 Groundwater circulation model for the Scanno rockslide

Summing up, the landslide deposit represents an aquifer with high hydraulic conductivity which has an active groundwater circulation with an important transfer role of groundwater, from the surficial Scanno lake system to the springs located in the upper Sagittario valley, constituting a "tank" modulating different seasonal regimens of the lake, with different responses during dry and wet periods [25].

## 8 Conclusions

The Scanno rockslide dam stability conditions were assessed considering both the morphometric parameters inferred by geological and geophysical investigations, and the flownet conditions derived from hydrogeological investigations [3].

With reference to the morphometric parameters, according to the method proposed by [16], the volume ($Vd = 82 \cdot 10^6$ $m^3$) and height ($Hd = 60$ m) of the dam and the catchment area of the Tasso river upstream the lake ($Ab = 80$ $km^2$) were obtained. These values have been used to calculate the Dimensionless Blockage Index (DBI) [16] for the Scanno dam. As a result, the calculated DBI value (1.84)

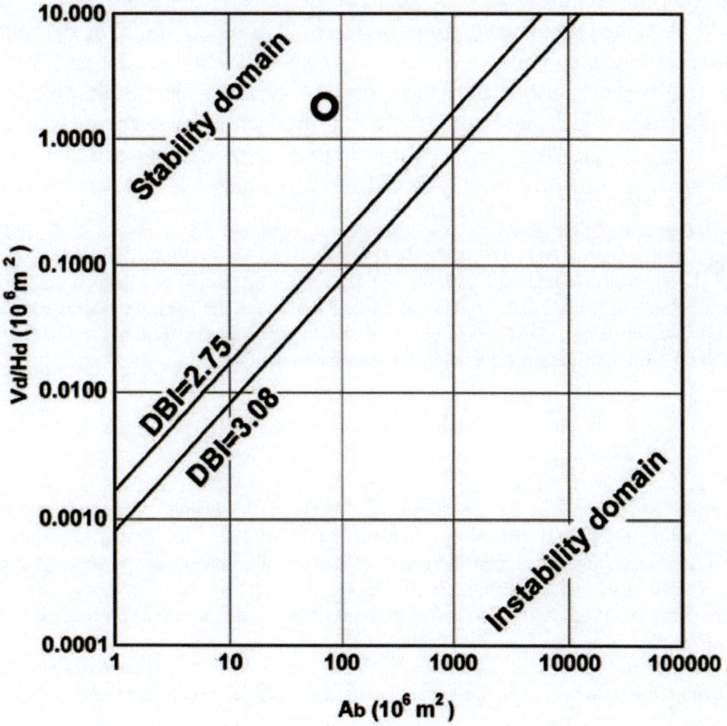


**Fig. 11.22** Scanno rockslide dam stability evaluation by means of morphometric parameters (DBI) (modified after [16])

is much lower than the threshold between stability and instability domains (2.75), thus indicating good stability conditions for Scanno rockslide dam [3] (Fig. 11.22).

In particular, the dam stability is favoured by two main geometric features:

(1) the high value of the ratio of the dam volume to the impounded valley width (1 km), gives to the natural dam a behaviour similar to that of an artificial earth dam;
(2) the small value of the ratio of the dam height to the length (about 2 km) indicates a markedly flat form that is another condition favourable for dam stability, especially if coupled with the low angle of the upstream slope.

Possible erosion and/or overtopping of the dam is related to the lake level variations. On a yearly basis, this variation during the period 1951–1994 was limited to 2.5 m, corresponding to $2 \cdot 10^6$ $m^3$ of water. The highest daily discharge registered by Tasso River (1929–1958) was 12.3 $m^3/s$, corresponding to $1 \cdot 10^6$ $m^3$ of lake water, which increased the lake level around 1.2 m. These values did not result in intense erosion processes and overtopping, both over short and long-term, taking into account the presence of the seasonal outlet, which becomes active when the lake level increases up to 0.5 m.

The hydraulic conductivity of the dam deposit is very high, as shown by hydrodynamic characterization and groundwater flow is active both in dry and wet conditions. Flowpaths show a hydraulic gradient of between 20 and 30‰; the springs fed by the landslide deposit represent a constant-head limit, able to drain higher discharge values and limiting hydraulic gradient increase during wet periods. These hydrogeologic characteristics contribute to dam stability and consequently, dam failure by piping processes is considered very unlikely.

**Acknowledgements** This research was financially supported by University of Rome "La Sapienza" grants (respectively G. Scarascia-Mugnozza and M. Petitta). Thanks to Prof. Paola Iacumin (Laboratorio Geochimica Isotopica of University of Parma) for isotope analyses. The authors wish also to thank Diana Violante, Emiliano Andreoni and Fabrizio Marini who cooperated in this project (both in field measurements and data processing) as part of their M.Sc. Thesis. Special thanks to S.G. Evans for the helpful discussion in the field.

## References

1. Barbieri, M., D'Amelio, L., Desiderio, G., Marchetti, A., Nanni, T., Petitta, M., Rusi, S. and Tallini, M. (2003) Gli isotopi ambientali ($^2H$, $^{18}O$ e $^{87}Sr/^{86}Sr$) nelle acque sorgive dell'Appennino abruzzese: considerazioni sui circuiti sotterranei negli acquiferi carbonatici, *Atti 1° convegno AIGA, Chieti, febbraio*, 69–81.
2. Bernabini, M. (1965) Alcune considerazioni sui rilievi sismici a piccola profondità, *Bollettino di Geofisica Teorica ed Applicata* **7**, 106–119.
3. Bianchi Fasani, G., Cercato, M., Esposito, C. and Petitta, M. (2005) Il Lago di sbarramento di Scanno: considerazioni riguardo le condizioni di stabilità, *Giornale di Geologia Applicata* **2**, 45–50.
4. Bianchi Fasani, G., Esposito, C., Maffei, A. and Scarascia Mugnozza, G. (2004) Geological controls on slope failure style of rock avalanches in Central Apennines (Italy). in *Landslides: Evaluation and Stabilization*, Lacerda, Ehrlich, Fontoura & Sayao (eds.), pp. 501–507. International Symposium Landslide. Rio de Janeiro 2004.

5. Boni, C. and Ruisi, M. (2005) *Carta idrogeologica della Marsica Orientale e note illustrative, scala 1:50,000*. Pubbl. GNDCI-CNR n.2866.
6. Casale, M. (1995) *Idrodinamica sotterranea e valutazione delle risorse idriche dell'area del Lago di Scanno (Abruzzo),* Quaderni di Geologia Applicata, IV Conv. Naz. Giov. Ric, 319–326.
7. Cavinato, G.P., Cosentino, D., De Rita, D., Funiciello, R. and Parotto, M. (1995) Tectonic-sedimentary evolution of intrappenninic basins and correlation with the volcano-tectonic activity of central Italy, *Memorie Descrittive Servizio Geologico Italiano* **39**, 44–59.
8. Cavinato, G.P. and De Celles, P.G. (1999) Extensional basins in the tectonically bimodal central Apennines fold-thrust belt, Italy: Response to corner flow above a subducting slab in retrograde motion, *Geology* **27**, 955–958.
9. Clark, I.D. and Fritz, P. (1997) *Environmental Isotopes in Hydrogeology*. Lewis Publishers, New York, NY.
10. Colacicchi, R. (1967) Geologia della Marsica Orientale, *Geologica Romana* **6**, 189–316.
11. Costa, J.E. and Schuster, R.L. (1988) The formation and failure of natural dams, *Geological Society of America Bulletin* **100**, 1054–1068.
12. Craig, H. (1961) Isotopic variation in meteoric water, *Science* **133**, 1702–1703.
13. D'Andrea, M., Miccadei, E. and Praturlon, A. (1991) *Rapporto tra il margine orientale della piattaforma Laziale-abruzzese e quello occidentale della piattaforma Morrone-Rotella, Studi Geologici Camerti* **spec. 1991/2**, 389–395.
14. Dramis, F. (1992) *Il ruolo dei sollevamenti a largo raggio nella genesi del rilievo appenninico, Studi Geologici Camerti* **spec. 1992/1**, 9–15.
15. Edwards, L.S. (1977) A modified pseudosection for resistivity and IP, *Geophysics* **42**, 1020–1036.
16. Ermini, L. and Casagli, N. (2003) Prediction of the behaviour of landslide dams using a geomorphological dimensionless index, *Earth Surface Processes and Landforms* **28**, 31–47.
17. Gardner, W.L. (1939) An areal plan of mapping subsurface structure by refraction shooting, *Geophysics* **4**, 247–259.
18. Herrmann, R.B. (2002) Computer Programs in Seismology.Website: http://www.eas.slu.edu/People/RBHerrmann/ComputerPrograms.html (Accessed December 2005).
19. Loke, M.H. and Dahlin, T. (2002) A comparison of the Gauss-Newton and quasi-Newton methods in resistivity imaging inversion, *Journal of Applied Geophysics* **49**, 149–162.
20. Longinelli, A. and Selmo, E. (2003) Isotopic composition of precipitation in Italy: A first overall map, *Journal of Contaminant Hydrology* **270**, 75–88.
21. Miccadei, E. (1993) Geologia dell'area Alto Sagittario-Alto Sangro (Abruzzo, Appennino centrale), *Geologica Romana* **29**, 463–481.
22. Nicoletti, P.G., Parise, M. and Miccadei, E. (1993) The Scanno rock-avalanche (Abruzzi south – central Italy), *Bollettino della Società Geologica Italiana* **112**, 523–535.
23. Park, C.B., Miller, R.D. and Xia, J. (1999) Multichannel analysis of surface waves, *Geophysics* **64**, 800–808.
24. Perrone, E. (1900) Carta Idrografica d'Italia – Aterno Pescara, *Roma* 1.
25. Petitta, M., Scarascia Mugnozza, G., Barbieri, M, Bianchi Fasani, G. and Esposito, C. (2010) Hydrodynamic and isotopic investigations for evaluating the mechanisms and amount of groundwater seepage through a rockslide dam, *Hydrological Processes* **24**, 3510–3520.
26. Pianelli, A. and Boni, C. (1995) Groundwater resources assessment in Sagittario River Basin – Central Apennines (Italy), Proceedings 2nd International Meeting Young Researchers in Applied Geology. Peveragno (Italy), 361–366.
27. Riccardi, R. (1929) Il lago di Scanno (Abruzzo), *Bolletino Reale Società Geografica Italiana, Ser.* **IV**, 162–182.
28. Salvati, R. (2002) Natural hydrogeological laboratories: A new concept in regional hydrogeology studies. A case history from central Italy, *Environmental Geology* **41**, 960–965.
29. Visentini, M. (1937) L'evapotraspirazione degli specchi liquidi, Memorie e Studi idrografici, pubblicazione n.2 del Servizio Idrografico. Istituto Poligrafico dello Stato, Roma.

# Chapter 12
# Formation, Characterisation and Modeling of the Val Pola Rock-Avalanche Dam (Italy)

**G.B. Crosta, P. Frattini, N. Fusi, and R. Sosio**

## 1 Introduction

Landslide dams (i.e., the natural blockage of river channels by hillslope-derived mass movements [6]) are common natural phenomena with great relevance for geomorphological conditions and for the safety of people living close to them. Landslide dams can induce significant geomorphic hazards in the adjacent areas. Flooding by impounded water can create large lakes in the upstream areas, causing damages to human activities. Catastrophic outburst floods and/or debris flows can be triggered by a rapid dam failure with exceptional rates of sediment erosion and deposition along the downstream part of the valley.

Different types of landslides (e.g. rock avalanche, debris avalanche, debris flow, rock fall, rockslide, rotational failure, rock-block slide, deep-seated complex failure, earthflow) can originate a landslide dam by damming the valley bottom and the river channel with their accumulation. The damming of the valley bottom can occur with different modalities [6] as a function of different factors as the landslide type and velocity, its size and geometry, the forming material, the size and conditions of the valley bottom, the river discharge, the geomorphological conditions at the landslide dam site and of the catchment upstream of the dam.

Fast moving and long run out landslides are among the main causes of damming because the large amount of material, rapidly transported and piled up at the valley bottom, can be eroded only through very large water discharge. Occurrence of the erosion and down-cutting of the accumulation is more difficult because of armouring by the presence of a carapace of large to very large blocks.

The internal part of the accumulation is often considered as homogeneous but its properties must be related to the complex motion and set of mechanisms occurring during the put in place. Both translation or sliding and internal deformation participate in the motion and deposition of flow-like landslides. At the same time the term flow-like landslides implies that the material behaves as a complex fluid.

---

G.B. Crosta (✉)
Dipartimento di Scienze Geologiche e Geotecnologie, Università degli Studi di Milano – Bicocca, 20126 Milano, Italy
e-mail: giovannib.crosta@unimib.it

S.G. Evans et al. (eds.), *Natural and Artificial Rockslide Dams*, Lecture Notes in Earth Sciences 133, DOI 10.1007/978-3-642-04764-0_12, 

Translation or sliding implies the resisting action at the base of the flowing material along the infinitesimal thickness of the failure/sliding surface, but the internal deformation or the interaction with a rough basal topographic surface imply the resisting action of the flowing material. Furthermore, flow-like landslides (rock and debris avalanches, rock falls, debris flows, earth flows, dry grain flows) could entrain or deposit material while moving, allowing for important changes in mass and volume.

The particular features and the complexity of the motion of these landslides generated the formulation of many different theories. The partial knowledge or description of the deposits has been frequently at the origin of different theories to explain observed exceptional runout.

We present and discuss data from the 1987 Val Pola rock avalanche. Field surveys, in situ and laboratory tests allowed a detailed reconstruction of the landslide accumulation and of the successive events following the damming of the river channel.

## 2 The Val Pola Rock Avalanche

During mid-July 1987, the central Alps experienced a period of exceptional rainfall and warm temperatures. More than 600 mm of rain fell between 15th and 22nd July 1987, representing half of the mean annual precipitation of 1,200 mm for the Valtellina area (central Italian Alps) [1, 5, 8, 10, 20]. The 0°C isotherm remained up to the elevations of 3,500–4,000 m a.s.l. for days, allowing abundant glacier melting which strongly increased runoff and river discharge in the Alps. Severe flooding and hundreds of landslides occurred between 15th and 22nd July. The upper Valtellina presents strong relief and very deep slopes developed in intrusive and metamorphic rocks, such as diorite, gabbro, ortho-quartzite and gneiss. The main valley trend is north-south, being controlled by a Late Alpine fault system.

The Val Pola landslide, that occurred in the upper Valtellina on the eastern slope of Mount Zandila, was by far the largest and most damaging event that occurred in 1987. The upper sector of the eastern slope (right hand flank of the Adda river valley, with average slope angle of 32°) of Mount Zandila was part of a large prehistoric landslide. The landslide was located at the intersection of two major joint sets steeply dipping (45° and about 90°) into the Adda River valley. Bedrock in the area consists of tight northward-dipping isoclinal folded and highly fractured and tectonized diorite and gabbro (Gabbro di Sondalo), ortho-quartzite and amphibolite (Mte. Tonale Formation) and fine-grained gneiss (Val Grosina Formation) overlain by thin glacial and colluvial deposits. The prehistoric landslide was clearly identified by a large and continuous head scarp, up to 700 m in length and 100 m in relief, which merged laterally with two sub-vertical E-W trending joints (Fig. 12.1a). The total relief of the landslide was about 700 m, between 1650 and 2350 m a.s.l.

The Val Pola creek, a tributary of the Adda River, marked the left hand flank of the landslide and deeply incised the metamorphic and igneous rocks. Before the landslide event on 28th July 1987, on 18–19 July intense and continuous rainfall

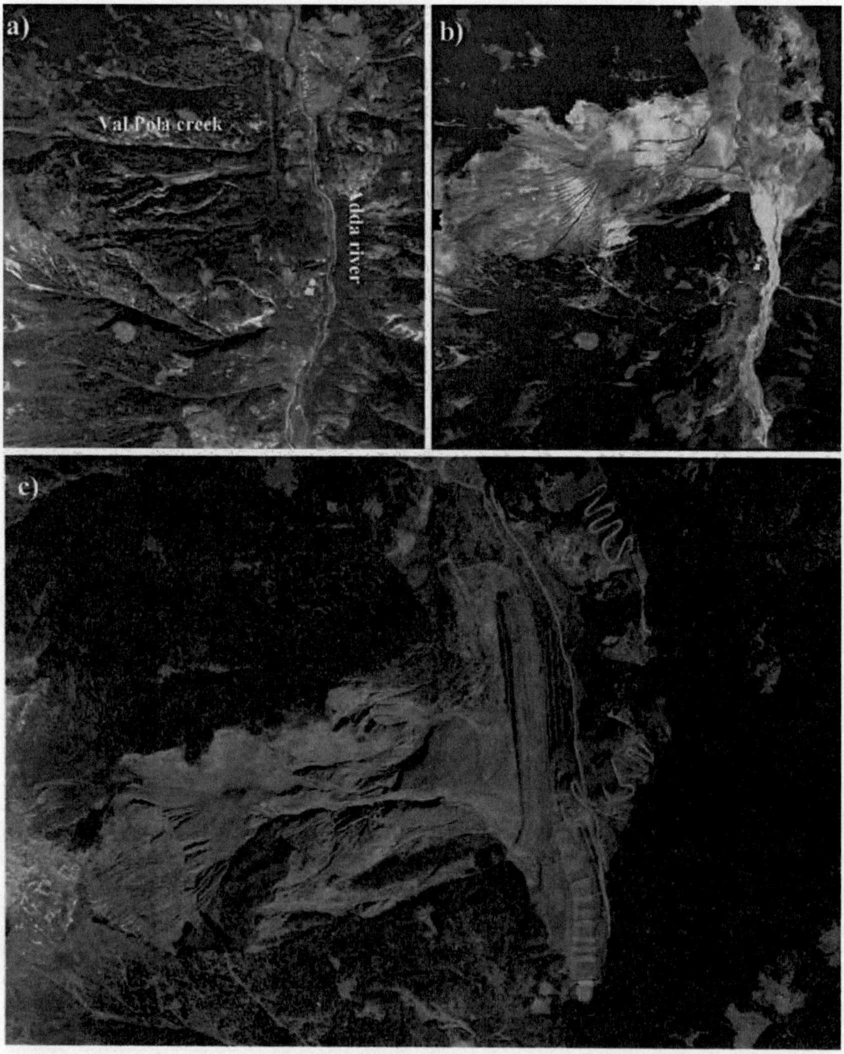


**Fig. 12.1** Aerial photos of the Val Pola rock avalanche area. **(a)** pre-failure slope conditions with the old landslide head scarp; **(b)** post-failure conditions just after the overtopping of the landslide dam; **(c)** post-failure conditions after the removal of the accumulation, construction of the large embankment dam at the slope toe, a series of check dams along the downstream toe of the deposit, two by-pass tunnels and new roads in the valley bottom and along the runup area

caused flooding and numerous superficial rockfalls and debris flows along the Val Pola creek. Slope instabilities along the southern flank of the Val Pola creek continuously fed debris flows which formed a large alluvial fan that dammed the Adda River valley. A large shallow lake started to form and reached a volume of ca. 50,000 $m^3$ and a depth of 1–5 m.

On July 25, a 600 m long fracture developed at the base of the old head scarp, and progressively reached a length of 900 m. On July 28 at 7:23 AM, a volume initially estimated at 34–43 millions of cubic meters detached from the slope, moved down the slope to reach the valley bottom, rising 300 m on the opposite valley side and flowed both upstream and downstream (Fig. 12.1b, c) along the valley. We recomputed the total volume by generating two DTM from topographic maps at 1:10,000 and 1:5,000 scale showing the pre and post failure conditions. The landslide entrained 5–8 million cubic meters of debris along its track. The rock avalanche debris is thought to have bulked by about 30% during deposition. After the event the compound multi-planar wedge shaped geometry of the initial failure, controlled by the presence of major fault, joint and schistosity planes, was recognized [1, 8] (Figs. 12.1 and 12.2). This peculiar geometry caused the detached mass initially to move slight with a northward component, transverse to the slope direction and toward Pola Creek. After impacting against a rocky spur on the north side of Val Pola Creek, the rockslide turned rapidly eastward and transformed into a rock avalanche. The estimated maximum runout velocity of the avalanche ranged from 76 up to 108 m/s, based on potential energy conversion [5, 8] and from 50 to 75 m/s, based on numerical modelling [9, 10]. The downstream run of the mass stopped at 1.5 km from the down-slope axis initially followed by the landslide (Figs. 12.1 and 12.2). The upstream movement was characterized by a large mud wave, up to 35 m high, that reached a distance of 2.7 km. A mud wave up to 95 m high was generated by the mass impacting the lake. It decreased to 15–20 m at a distance of about 1.3 km but continued travelling 2.7 km upstream. The landslide and the mud wave damaged the village of Aquilone and claimed 27 lives [5, 20].

The energy released by the landslide during its motion has been recorded by Italian and Swiss seismograph stations located at distances of 37–248 km from the site [9]. The seismic records last 70–120 s and the computed earthquake magnitude was 3.3 on the Richter scale.

## 3 Geomorphologic Features of the Accumulation

The avalanche deposits were poorly sorted, consisting of angular, coarse blocks several meters in diameter to fine rock debris. Ridges and furrows were largely parallel to the valley axis (Figs. 12.1 and 12.3), near the toe of the slope and close to downstream of the river valley [9–11]. Secondary slides, scarps and flow-like features were observed within the accumulation. These were originated by the return flow descending from the opposite valley side after the debris reached the maximum runup.

Coarse blocks and gravel were mainly distributed at the toe of the slope and around the valley central axis. Poorly sorted hummocky surfaces were also observed. Lobate landforms had surface ridges and coarse clasts. Fine-grained materials were concentrated on the lee-side. Alluvial and lacustrine sediments scoured by the moving mass from the valley bottom were found within the deposit.

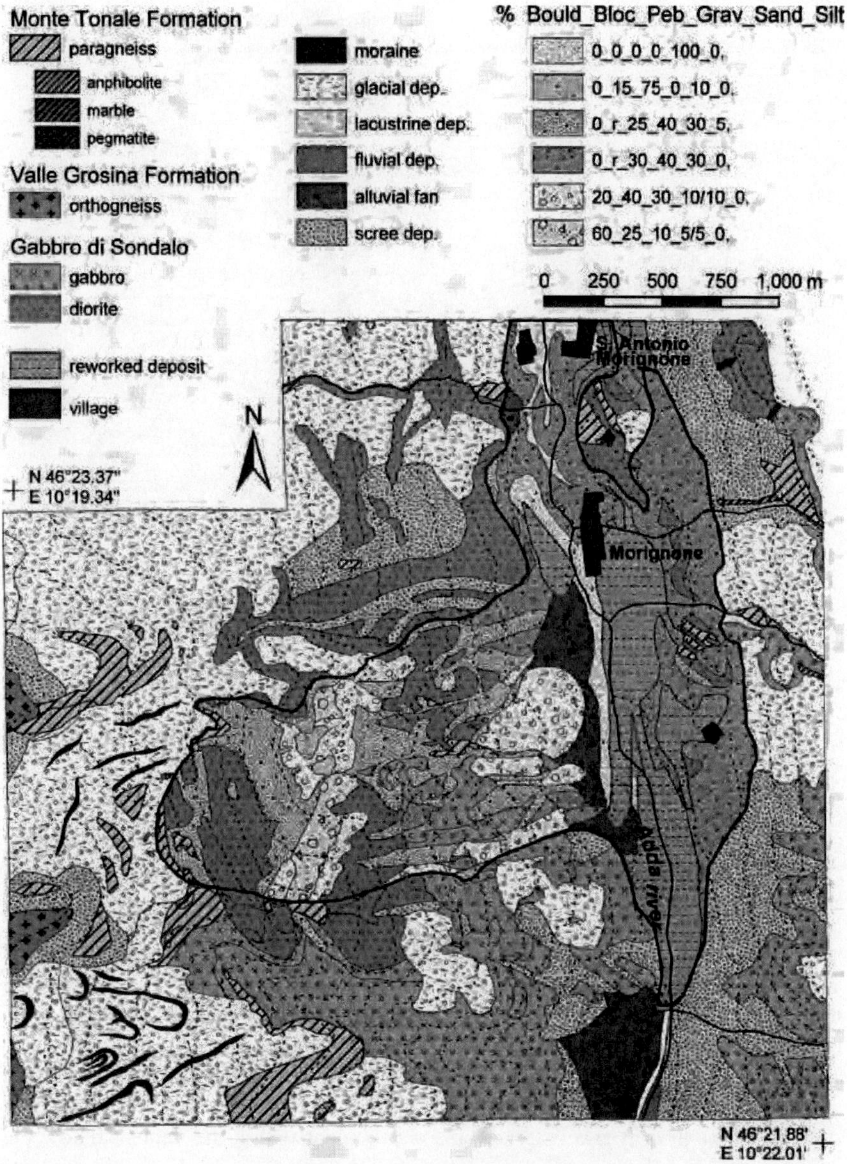


**Fig. 12.2** Geologic map of the Val Pola rock avalanche area with detailed description of the superficial deposits and of the grain size distribution within the landslide accumulation. Grain size distribution is described with % of boulders, blocks, pebbles, gravel, sand, silt. r: rare

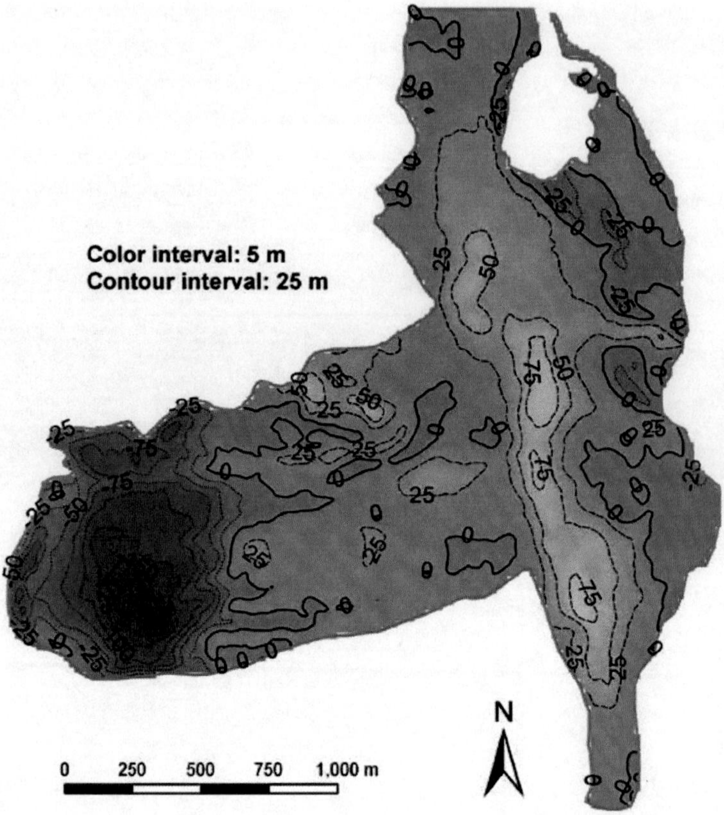


**Fig. 12.3** Map of the changes in elevation as computed from pre – (1:10,000) and post-failure (1:5,000) topographic maps. Maximum thickness of the deposit is reached close to the failed slope and southward from the landslide axis. Values of contour lines are positive for deposition and negative for erosion-depletion

Concerning the stratigraphic and textural characteristics of the debris deposits, large blocks up to tens of meters in a sandy matrix at the deposit surface were distinctly coarse and better sorted than were the subsurface particles [11] (Fig. 12.4). The internal texture appeared to be relatively massive. No significant difference in the internal texture of the deposit was observed between ridges and intermediate troughs. A rough decrease of grain size was evident within the deposit below the surface.

## 4 Deposit Characteristics

The quantitative evaluation of the volumetric contributions of geotechnically significant blocks and matrix of fine texture helps to characterize the physical properties and consequent behavior of the coarse to very coarse materials [11]. In mixed

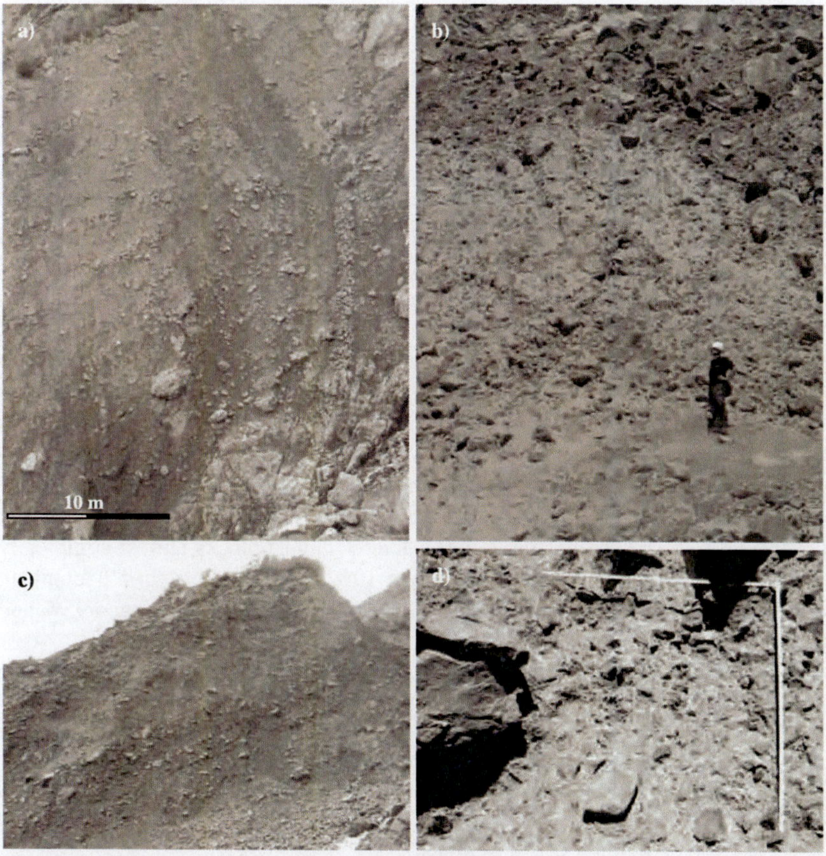


**Fig. 12.4** Rock avalanche deposits at the toe of the source slope. (**a**) Evidence of discontinuous and irregular layering are visible, together with inverse gradation in the upper part, subdivision in sectors with different lithologic characteristics (change in *colour* in figure **b**), and presence of large blocks at different depths (**c**). Figure (**d**) (scale 1 m × 1 m) shows the distribution of pebbles and blocks immerged in a fine matrix. Matrix separates fragments with bands of centimetric thickness. Fragments looks sub-angular to sub-rounded, equant or slightly elongated

grained sediments, the geomechanical behavior of the material is a function of the block volumetric proportions. A 20–30% fine matrix volumetric content was defined [21, 25] as the limit to pass from a behavior controlled by the coarse fraction to that controlled by the finer matrix. The grain size distribution of the material can be used to describe and understand segregation processes, comminution and fragmentation of landslide material, and to verify the feasibility of certain mechanisms of transport and deposition.

Shortly after the Val Pola event, 61 debris samples from 1 kg–20,000 Kg were collected at depth of 0–7 m below the surface within the deposition area (Fig. 12.4).

Four more samples were collected within the source area where 8 field surveys for soil description were also performed. 21 boreholes were drilled in the accumulation and 109 deep samples were collected (0–5 m: 13 samples; 5–10 m: 13 samples; 10–15 m: 13 samples; 15–20 m: 9 samples; 20–30 m: 12 samples; 30–40 m: 20 samples; 40–50 m: 13 samples; 50–70 m: 16 samples). Photo-analytical techniques have been used to collect more data both from the detachment zone and the deposit [11].

Grain size analyses were conducted on representative samples, both on site and in the laboratory, according to the type of material and the relative percentages of blocks, boulders and finer particles. Sample dimension ranged between few kilograms up to about 20,000 kg. Blocks and gravels were generally more abundant at the surface of the deposit especially in the valley bottom. Blocks were sized up to tens of meters with the mean values of 2–5 m. The remaining part of the materials roughly ranged between few millimeters and as much as 40 cm. Azzoni et al. [1] describe an average grain size value of about 50 mm, but the difficulty in determining such a value will be discussed in the following. Blocks and gravel parts at the deposit surface (Fig. 12.5) range from 8 to 95%, sand from 10 to 75%, and silt and clay from 6 to 60%. Within the 61 superficial samples, blocks almost completely form 10 samples. Silts are usually less than 10% in the deposits (in 52 samples). Fine matrix (< 0.075 mm) reaches in few cases a content between 50 and 95% and

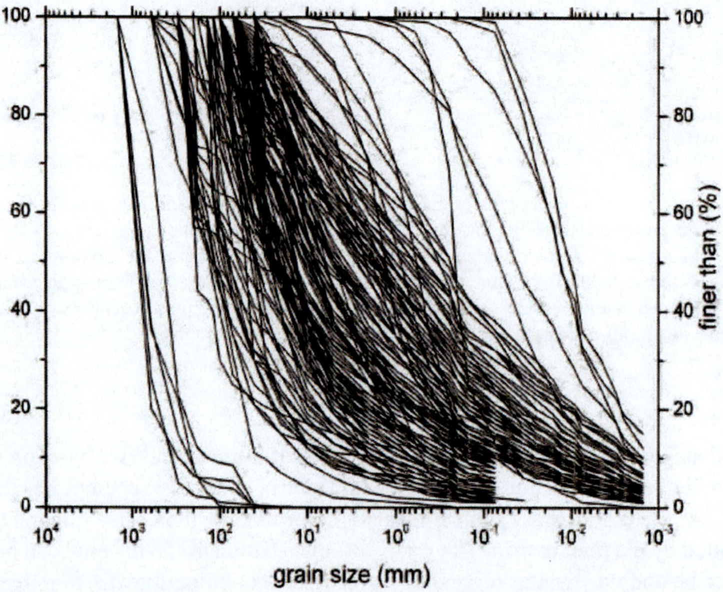


**Fig. 12.5** Grain size curves for 61 samples collected at the surface of the landslide accumulation. Samples were collected at different points within the landslide and include both landslide debris and alluvial material scoured from the valley bottom. Grain size analyses have been carried out both in the laboratory and in situ

mainly in coincidence with alluvial deposits scoured from the valley bottom or with washed out deposits. The coefficient of uniformity ($U = D_{60}/D_{10}$) is usually low.

Samples (finer than 100 mm) taken from within the bowl-shaped source area are well graded with a low degree of sorting. Gravel ranges between 30 and 60%, sand between 40 and 70% and finer than 0.075 mm are usually less than 5%. Blocks and pebbles are present but they are not much more abundant than in the main deposit along the valley bottom. Pebbles and gravel elements have been examined to determine their shape by analyzing the length of principal axes [11]. The elements pertain prevalently (Fig. 12.6) to the equant-spherical and to the oblate – disk-shaped classes with relatively few elements of the triaxial-blade and prolate-blade types. Grain size of samples coming from boreholes have been analyzed. We observe (Fig. 12.7) a slight decrease in average grain size with depth and an increasing value of the uniformity coefficient, U, and also the presence of well-graded materials within the deposit (Figs. 12.7 and 12.8). This increase is continuous up to about 30 m where the value is changing irregularly. The reason for this trend can be found in the entrainment of alluvial deposits from the valley bottom and the drilling and sampling below the landslide debris accumulation. Block content within the deposit range from a few percent to 40% with a clear decrease between the surface of the accumulation and 30 m below it. They are generally more abundant close to the surface, at the toe of the failed slope and at the extreme terminal tips of the deposit.

Gravel content follows the same trend within the initial 30 m with a progressive increase in the sandy grain size fraction. Silt and clay content remains below 10%

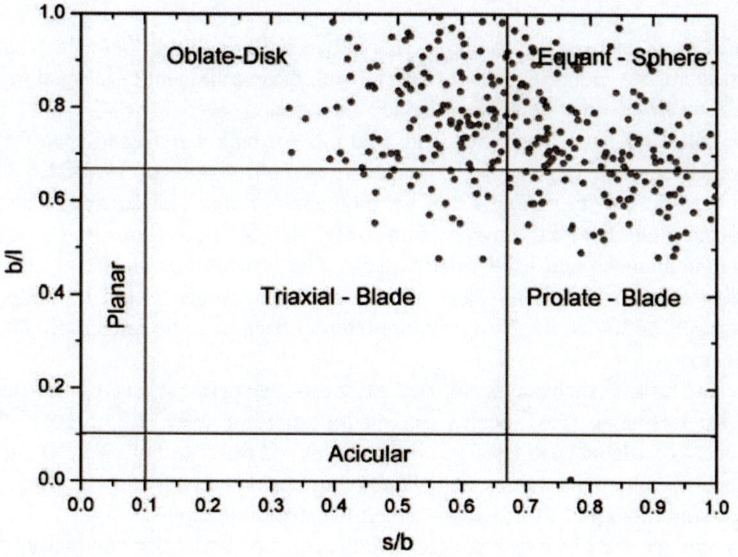


**Fig. 12.6** Zingg's plot for pebbles and gravels (size: 3–17 cm) sampled at the surface of the landslide accumulation

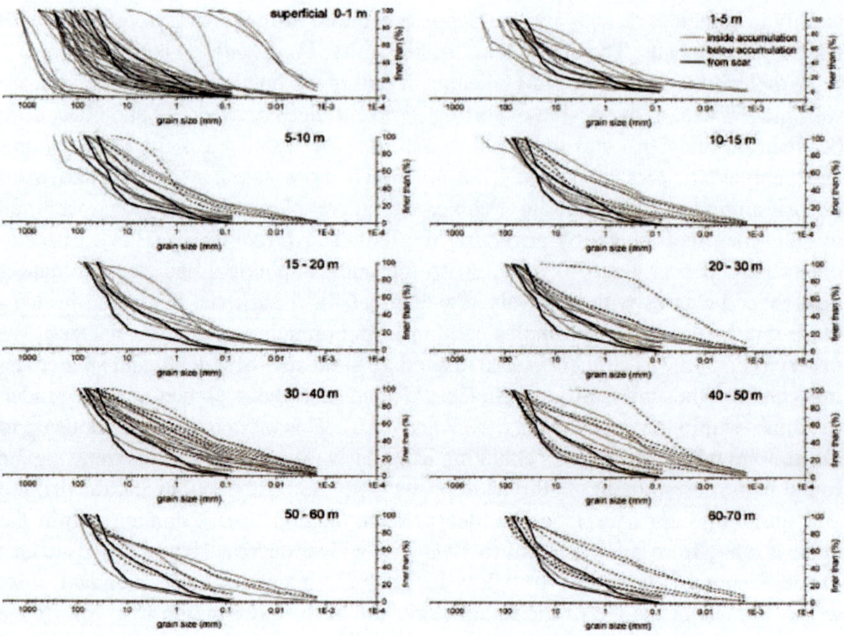


**Fig. 12.7** Grain size curves for the 109 samples from the 21 boreholes drilled through the debris accumulation. Samples are plotted in groups for different depth intervals (0–5 m, 5–10 m, 10–15 m, 15–20 m, 20–30 m, 30–40 m, 40–50 m, 50–70 m). Curves for samples from the landslide source scar are shown (between thick black lines)

up to 30 m, to increase up to about 37% between 30 and 70 m below the surface. We associate this increase in finer material with the entrainment of alluvial material or the sampling below the base of the accumulation.

We collected 7 more samples from trial pits for bulk weight and specific gravity evaluations and large-scale triaxial tests. A bulk weight of 19.8–21.6 kN/m$^3$ and a specific gravity of 2.75–2.88 were obtained. Large-scale triaxial tests were conducted under the confining pressures of 0.1–0.6 MPa on reconstituted samples, 500 mm in diameter and 1,000 mm in height. The maximum grain size of the reconstituted material was 100 mm, with a bulk weight that ranged 21.0–22.5 kN/m$^3$ and the porosity of 20–24.5%. The resultant internal friction angle ranged 45-47° with no cohesion.

In situ tests have been performed to assess the permeability of the deposit. Different techniques have been used: pit infiltrometric tests ($1.2 \times 10^{-4}$ –$1.1 \times 10^{-3}$ m s$^{-1}$), variable head tests within boreholes and tracer tests ($2.7 \times 10^{-3}$ m s$^{-1}$). Hydraulic conductivity values (Fig. 12.9) ranges between $10^{-6}$ and $10^{-2}$ m s$^{-1}$ with a dispersion of about two orders of magnitude at similar depths.

On average we observe a progressive decrease of hydraulic conductivity with depth, from $10^{-5}$ m s$^{-1}$ close to the surface of the deposit up to $10^{-6}$ m s$^{-1}$ at

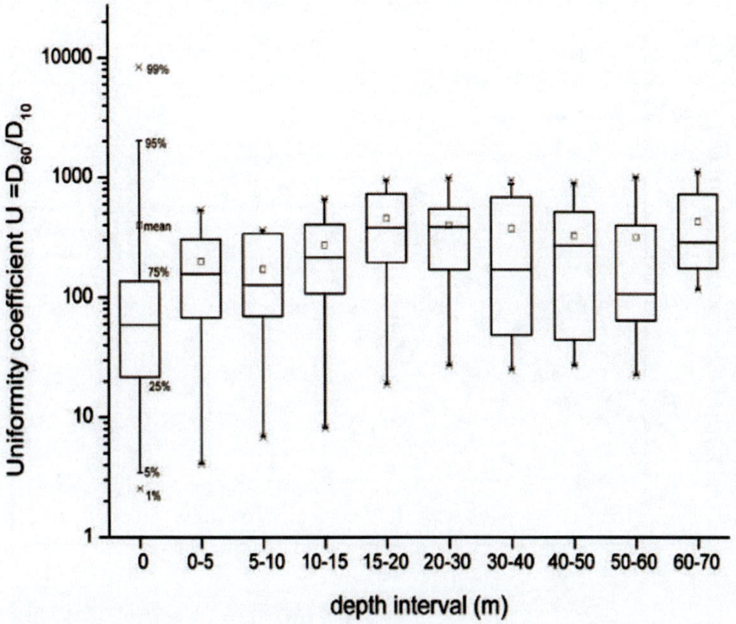


**Fig. 12.8** Distribution of the values for the Coefficient of Uniformity ($U = D_{60}/D_{10}$) for different depth intervals. Legend for the box plot is shown for the leftmost case

70 m. This is reasonable considering the mechanism of transport and deposition, the change in grain size and gradation, the thickness of the deposited material and the entrapment of alluvial material.

Geophysical surveys by the superficial shear waves (SSW) technique have been performed at different sites. Three main layers have been recognized: between 0 and 4 m with shear wave velocity, Vs, between 200 and 300 m s$^{-1}$; between 4 and 15 m, with Vs = 600–1,000 m s$^{-1}$; and below 15 m with Vs = 1,700 m s$^{-1}$.

## 5 Hydrogeological Data

The hydrogeologic characteristics of the landslide debris are among the most important in controlling the stability of landslide dams. Beyond the physical parametrization of the debris accumulation it is fundamental to observe its behavior during water level rising and perturbation of flow conditions.

21 piezometers were drilled within the debris (Figs. 12.10 and 12.11) and monitored during lake level rising, pumping and dam overtopping. The landslide dam closed a catchment of 533 km$^2$ containing two large artificial lakes for hydropower

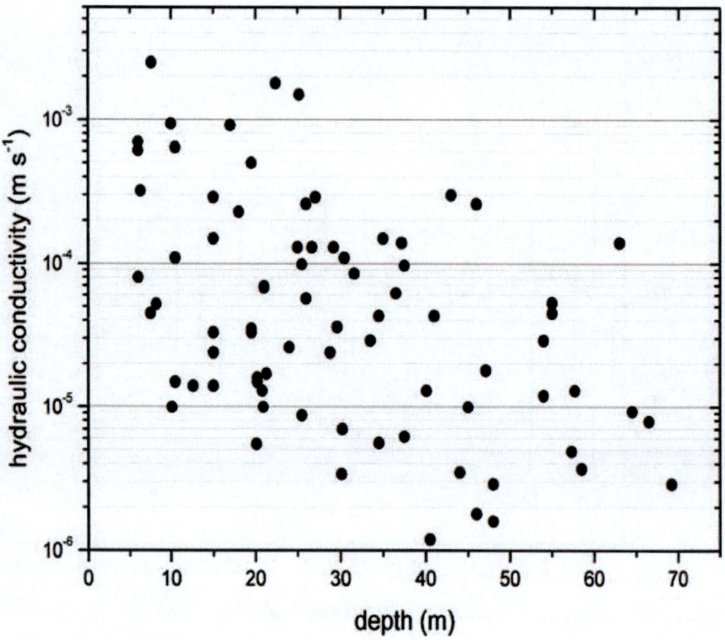


**Fig. 12.9** Plot of hydraulic conductivity vs depth for the debris of the Val Pola landslide from pit infiltrometric tests, borehole and tracer tests. The progressive decrease in hydraulic conductivity with depth can be recognized together with the dispersion of the measured values and partially explained by the heterogeneity of the landslide material

production. Between July 28th and August 23rd the rockslide-dammed lake level rose about 0.5 m $d^{-1}$ up to 1,089.65 m a.s.l. for a total area of 570,000 $m^2$. 200 mm of rain fell between August 24–25th raising the water level up to 1,097.62 m a.s.l (761,000 $m^2$). A 2 km long rip-rapped spillway channel, at 1,102 m a.s.l., was excavated between August 27th and 30.

The downstream slope of the landslide dam was graded to reach an average slope ranging between 8 and 12% to remain stable for an overflow discharge of 40 $m^3s^{-1}$. On August 29th, the lake level reached the crest of the landslide dam and flowed across it (Figs. 12.10 and 12.11). The water discharge at the inlet of the channel excavated on the landslide dam was 35 $m^3$ $s^{-1}$ for 3 days. 25,000 people were evacuated from the downstream area of Valtellina. No outflow from the spillway was observed because of the total infiltration of the water discharge. Overtopping elevation was about 1,107 m a.s.l. coincident with an impounded lake volume of about $22 \times 10^6$ $m^3$. Overtopping started on August 31st, when the landslide dam body was probably almost completely saturated as shown by lake level and piezometric level measurements. A water pumping system was built in 32 days. It was started on September 17th, 1987 with one pump (2 $m^3$ $s^{-1}$) to which two more were added on September 21st (5 $m^3$ $s^{-1}$) and October 2nd (6 $m^3$ $s^{-1}$). A temporary lake, with a volume of about 40,000 $m^3$, formed on the landslide debris at 1,072 m a.s.l. but

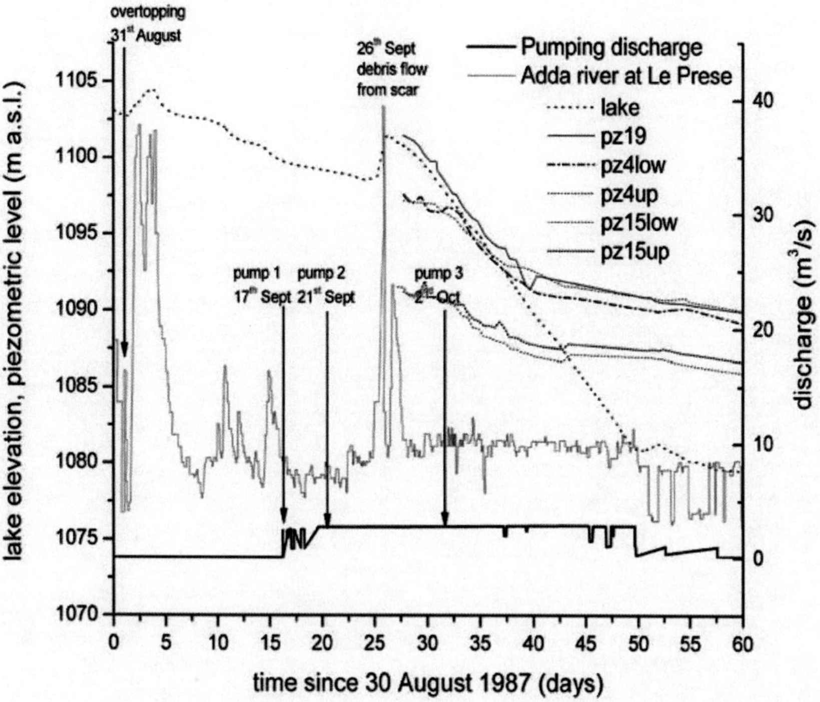


**Fig. 12.10** Lake level and piezometric levels, measured in some of the 21 piezometers, since August 30, 1987. Discharge of the Adda river (as measured at Le Prese) and of the pumping system are also represented. For the location of the piezometers see Fig. 12.11

dried up 20 days after the initiation of pumping. Following the overtopping, the downstream slope of the dam was deeply incised by the overflow with erosion and subsequent deposition of a large amount of debris.

A major irregularity in the trend of the lake level is represented by the September 26th peak (Fig. 12.10). This was originated by a large debris flow which dammed the spillway channel causing a rapid increase in the lake level.

The piezometric level (Fig. 12.11) within the landslide dam reacted in different ways to the lowering of the lake level. The groundwater level changed immediately with the lake level in piezometers close to the lake itself (piezometers #4, 15, 17, 19, 21 and just uphill #16, 18, 20) at a similar rate for all piezometers. Piezometers in the downstream sector of the dam (#2, 5, 7, 11, 12) reacted only 10 days later at similar rates but with different stabilization times. This behaviour can be the result of: seepage through the dam, water leakage from the spillway channel within the middle and lower sectors, as suggested by the total infiltration of the water discharge during the initial phases of the overtopping. Finally, the heterogeneity of the rock-slide debris could have played a role in controlling the seepage through it and the asynchronous lowering and stabilization of the groundwater level.

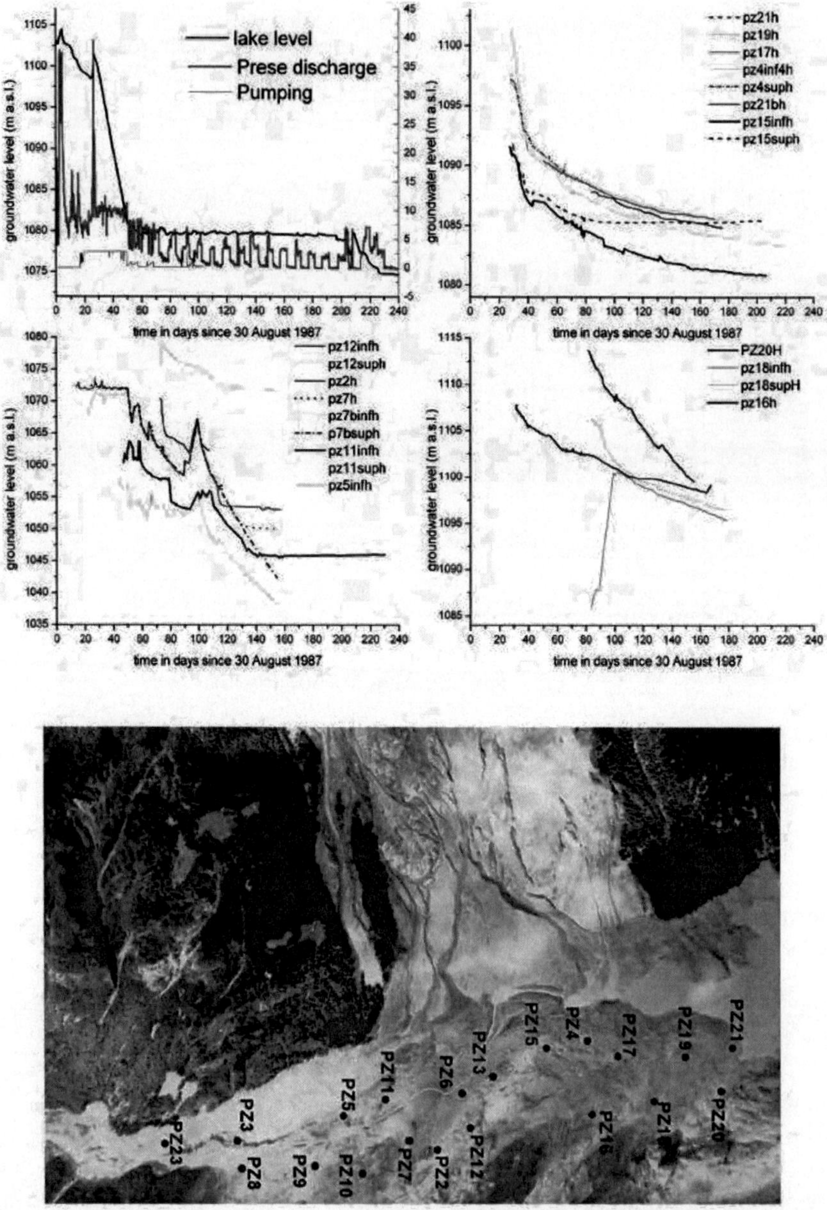


**Fig. 12.11** Location of the piezometers installed within the landslide deposit. Piezometers are plotted in different groups according to their behaviour. In the photograph the lake and the spillway channel are visible as well as the temporary lake in the middle part of the accumulation. Plots above show the lake level, piezometric levels, discharge at Le Prese, as well as pumping discharge since the end of August 1987

# 6 Dam Break Flood Assessment

Floods generated by failures of natural dams represent a relevant hazard to people and property. As a consequence, the stability of the dam and the peak discharge associated with the dam break are of major interest. The low degree of stability of landslide dams compared to earth fill and concrete dams is represented in Fig. 12.12 in terms of failure frequency. Dams with no formal compaction of the core material are about 7 times more likely to experience slides than the average and about 25 times more likely than dams with good compaction of the core [13].

The figure also shows the relevance of different mechanisms of dam failure (piping, sliding and overtopping). We must remember that overtopping is the most frequently observed mechanism of failure for natural dams. This can be partially due to the difficulty in performing direct observations both for the very short life of these features (about 37% of failures occur in less than 1 day, 90% fail in the first year), their remoteness and complexity. Furthermore, we must stress that frequency

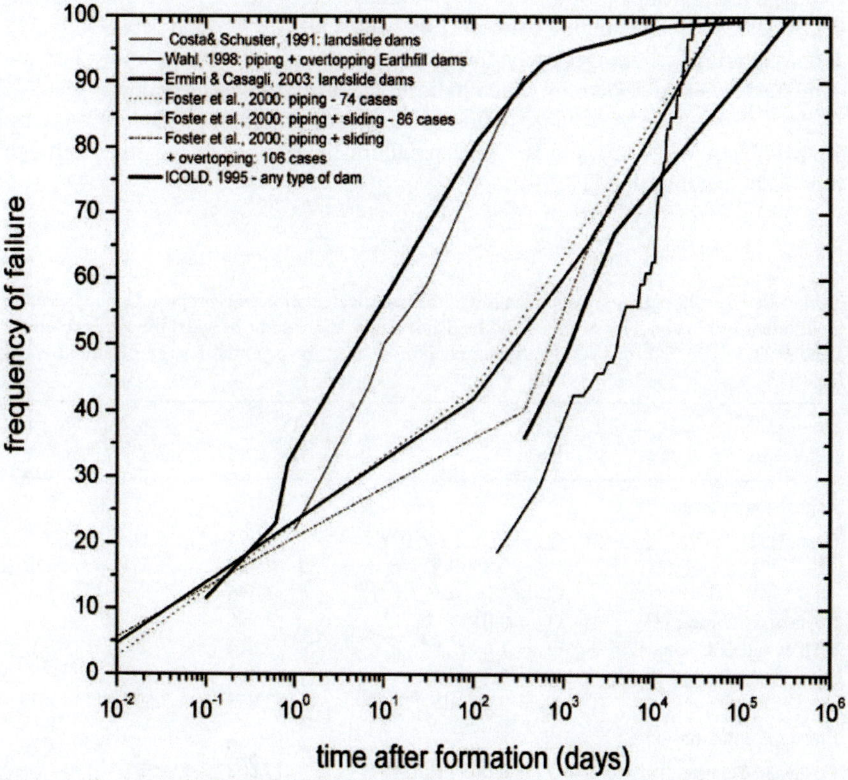


**Fig. 12.12** Frequency of failure for different times after formation/construction for different types of dams: natural and artificial, as consequence of different mechanisms of failure (piping, overtopping, sliding) and according to different datasets [7, 12, 13, 22, 27, 28]

curves for failure of landslide dams, like those in Fig. 12.12, are probably subjected to a sampling bias due to problems in mapping all existing landslide dams. This can result from the location of landslide dams in remote areas, burial below alluvial material, or simply by the fact that available datasets include only landslide dams that are still recognizable in the field or from aerial-photographs.

At the time of the Val Pola landslide, an initial emergency assessment of the peak discharge (16,500 $m^3 s^{-1}$) was estimated using Costa's relationship [4, 5]. We tried to re-examine this subject by using a new set of collected data and considering new approaches developed since 1987. The main methods that have been developed are empirical regression relationships, parametric methods, dimensionless analysis and physically based models. The volume of the lake, the height of the dam or the depth of the lake, the breach width and its development time are the most important factors controlling peak discharge.

Empirical relationships, in the form of regression or envelope curves, are a rapid method of determining the order of magnitude of peak discharge, but they are characterised by a high degree of uncertainty. Using some of the regression and envelope relationships proposed in the literature (see Table 12.1), we obtained values of peak discharge ranging between 2,492 and 4,487 $m^3 s^{-1}$, and between 97,341 and 133,946 $m^3 s^{-1}$, respectively. Table 12.1 also shows the mean values and the computed standard deviations which indicate the dispersion of the values.

We computed the breach size (mean width ca. 161.5 m; basal width. 100–118 m; top width: 204–222 m) and the breach failure time ($t_f$ = ca. 46 min) through Froehlich's relationship [17, 19].

**Table 12.1** Results obtained by application of the empirical relationships proposed in the literature by different authors both for artificial and landslide dams. $V_o$: volume of water released (complete failure: $22 \times 10^6$ $m^3$); d: lake depth (max depth: 43 m); Ep: potential energy of stored water ($\rho_w \cdot g \cdot d \cdot V_o$)

| References | Eq | $Q_p$ $m^3 s^{-1}$ | Mean $m^3 s^{-1}$ | St. Dev. $m^3 s^{-1}$ |
|---|---|---|---|---|
| Regression equations | | | | |
| Costa [11] | $Q_p = 672 \cdot (V_o/10^6)^{0.56}$ | 3,794 | | |
| | $Q_p = 6.3 \cdot d^{1.59}$ | 2,492 | | |
| | $Q_p = 181 \cdot (dV_o/10^6)^{0.43}$ | 3,446 | | |
| Costa and Schuster [1] | $Q_p = 0.0158 \cdot E_p{}^{0.41}$ | 3,276 | | |
| Walder and O'Connor [20] | $Q_p = 1.6 \cdot V_o{}^{0.46}$ | 3,816 | | |
| | $Q_p = 6.7 \cdot d^{1.73}$ | 4,487 | | |
| | $Q_p = 0.99 \cdot dV_o{}^{0.40}$ | 3,855 | 3,595 | 618 |
| Envelope equations | | | | |
| Costa and Schuster, [1] | $Q_p = 0.063 \cdot E_p{}^{0.42}$ | 17,608 | | |
| Walder and O'Connor [20] | $Q_p = 46 \cdot V_o{}^{0.46}$ | 109,717 | | |
| | $Q_p = 200 \cdot d^{1.73}$ | 133,946 | | |
| | $Q_p = 25 \cdot dV_o{}^{0.40}$ | 97,341 | 89,653 | 50,379 |

The methods proposed by Fread [14, 15] and Wetmore and Fread [31] (SMPDBK) were used for the determination of peak discharge by taking into account the computed breach dimension and failure time. The peak discharge values ranged between 2,360 and 19,731 $m^3\ s^{-1}$. Available data can be also used to compute the instantaneous (i.e. by instantaneous breach failure) peak discharge through adoption of broad-crested weir equations [16, 24]. With these approaches we obtained peak discharges ranging between 2,570 and 3,310 $m^3\ s^{-1}$.

A series of Monte Carlo analyses has been performed using some of the models (SMPDBK [18, 29, 30, 31]) and by introducing different values for the main variables (dam height, final breach width, elevation of breach base) sampled from assumed normal distributions. The results (Fig. 12.13) can be useful in showing the probability of occurrence for a certain value of peak discharge. They illustrate the variability of the results due to the different approaches, to the sensitivity of the approach itself to the change in value of the controlling variables, and uncertainty in parameter assessment. This has been done also for other methods such as BREACH [15, 16] (Fig. 12.14) and by Coleman et al. [3] for overtopping breaching of non cohesive homogeneous embankments.

We were particularly interested in the sensitivity of the models to the geometric and physical parameters.

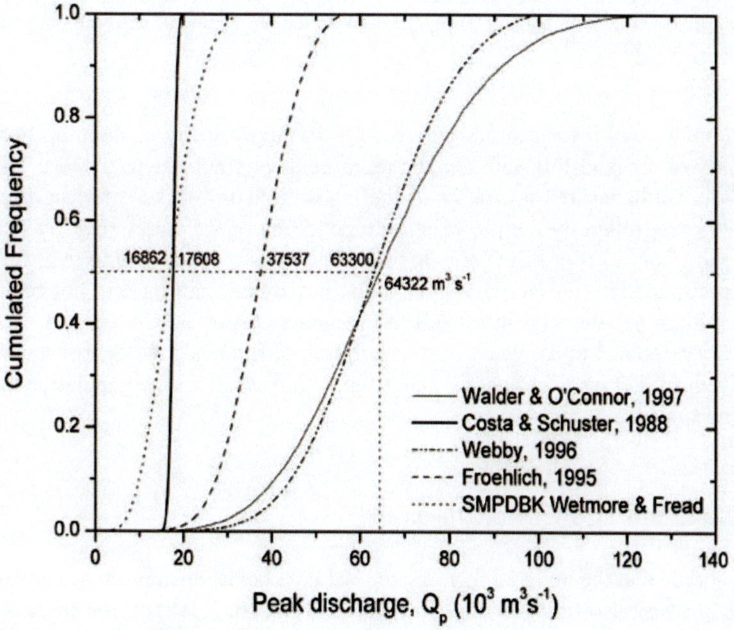


**Fig. 12.13** Plot of the results of Monte Carlo simulations (10,000 runs, Latin-Hypercube sampling) performed for different approaches. Range for each variable: dam height: 30–90 m; breach height: 15–55 m; mean breach width: 50–221 m; lake area: 1.42 $km^2$; lake volume: $22 \times 10^6\ m^3$

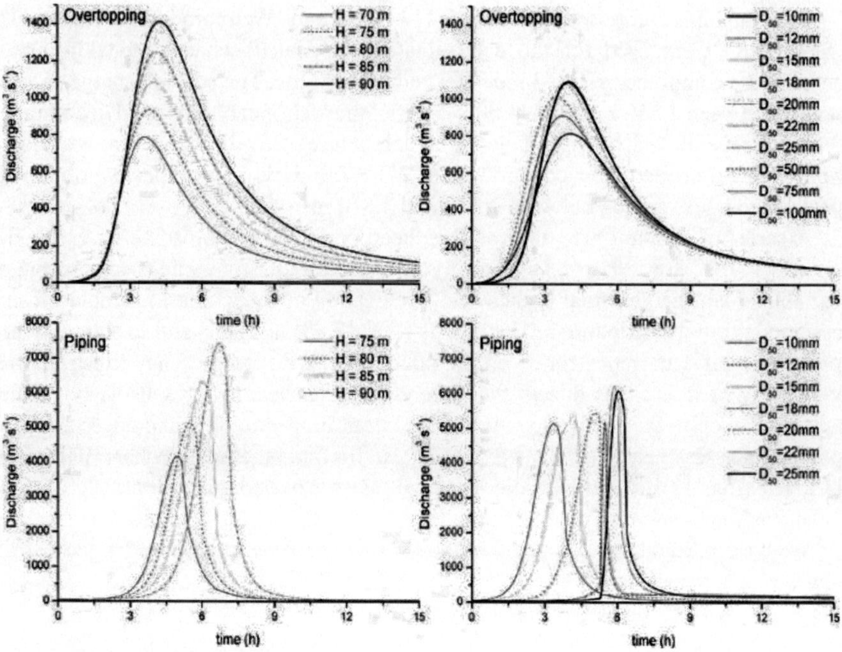


**Fig. 12.14** Sensitivity of BREACH model for overtopping and piping failure to changes in values of D50 (*right*) and to dam height (H) (*left*). Origin of the plots (time 0) has been put at a fixed discharge value (40 $m^3\ s^{-1}$)

The main reason for such an interest lies in the difficulty of defining both the geometry of the landslide dam and the representative grain size (e.g. $D_{50}$). The difficulty in evaluating the geometry of the landslide dam results from: its irregular geometry controlled by the pre-existing topography of the valley (e.g. by longitudinal and transversal profiles), the downstream sloping valley profile (e.g. for very large accumulations in alpine valleys a flat horizontal dam base is not realistic), and the main geometric features of the debris accumulation. Concerning the $D_{50}$, we make reference to the grain size description of the landslide debris, and to the described difficulties in collecting data from within the debris accumulation as well as at the surface.

## 7 Discussion and Conclusions

We observed that the material forming the Val Pola landslide deposits was relatively coarse but was also fining downward and well graded. It is common to pass from a grain-supported texture to a matrix-supported one by simply moving vertically within the debris accumulation or approaching the surface. The elements were subangular with spherical to tabular shape and the general angle of internal friction was relatively high. Our analyses allow us to raise some major questions concerning the data collected and to present some ideas about mechanisms of movement and deposition.

Grain size analyses (Figs. 12.5 and 12.7) show a certain homogeneity of two major sectors within the debris: the carapace, or the coarse sorted superficial zone, and the finer inner well-graded part which forms most of the mass. This latter part shows a slight fining downward (or inverse grading, Figs. 12.7 and 12.8) trend. Segregation is a frequently observed process in granular materials and can result from different mechanisms. The existence of a thin basal low-density and high particle velocity basal shear layer is a common assumption in modelling of granular avalanches down a steep incline. This behaviour has been observed in laboratory tests and numerical models on mono-sized materials, under low stress levels, with no particle breakage, in motion along very smooth and regular surfaces. These conditions seem quite different from those characteristic of very large rock- and debris-avalanches. Phenomena observed in nature are generally characterised by: breakable particles, very irregular superficial morphologies, and extremely high stress levels. As a consequence we can infer that many particular and local conditions can control both external and internal structures of the landslide debris accumulation, and that local or point-like descriptions are not sufficient to give a complete picture of the deposit.

Our observations also suggest that the difference in grain size between debris forming the main accumulation and debris present in large quantities within the bowl shaped scarp of the Val Pola rock avalanche are quite small. This contrasts with the assumption that the material undergoes a strong decrease in grain size during its motion down the slope. The change in grain size seems less important than usually assumed and could exclude the feasibility of some particular mechanisms or minimize their relevance in controlling landslide mobility. Furthermore, some recent numerical models of rock fragmentation both under static and dynamic stresses [23, 32] demonstrate a strong dependence of fragmentation on the confining stresses which are usually relatively low in rock avalanches, considering their average thickness. On the contrary, local stresses and shock waves, originated by motion along irregular geometries of the topographic surface, and the impact against large natural obstacles, can cause breakage of the particles.

Numerical models of the transportation and runout phases of a rock avalanche [9, 10] have also shown that different material behaviours can involve completely different strain distributions within the moving mass. As a consequence a different degree of shearing and grinding can be experienced by the material within the moving mass.

The most important factors controlling landslide dam stability and breaching (e.g. size of the breach, time of breach formation, maximum discharge) [3, 14, 15, 27, 28, 29] are its geometrical and physical characteristics (grain size, porosity, compaction, etc.). It has been suggested in the literature [2] that deposits originating from mass movements are not influenced by the emplacement process and that grain size and texture of debris in a landslide dam could be interpreted as a function of the parent rock (e.g. lithology, joint spacing, porosity). We believe that these two assumptions are not completely applicable to rock avalanche deposits. Rock- and debris-avalanches are usually characterized by long runout and particular mechanisms of transport and deposition typical of granular movements. Grinding, shearing, fragmentation, segregation, soil, water and ice entrapment, can occur in these instabilities.

Texture controls the stability of landslide dams by influencing the geotechnical and hydrogeological properties of the dam material. Dams showing a higher degree of stability, with no catastrophic failure, are those characterised by coarse grain sizes and grain-supported textures [2]. As a consequence, it becomes relevant to verify the presence of bimodal grain size frequency distributions (matrix-supported or grain-supported). Rock and debris-avalanche deposits show vertical and lateral changes in grain size. Parts of the deposits can work sometimes as a filter with respect to other sectors. One of the main factors controlling internal erosion (i.e. piping), of natural and artificial dams, is some effective grain-size ratio between the filter and the soil needing filter protection. The grain-size ratio serves as an index of whether or not erosion will occur (e.g. ratio $D15_{filter}/D85_{soil}$). Nevertheless, confining pressure may influence stability against erosion [26]. A filter material placed in a loose state will be characterized by interstitial spaces larger than for well packed materials. Under increasing confining pressure the filter particles will tend to consolidate into a denser state. At the same time, confining pressure may act as a destabilizing phenomenon. Under low-stress conditions interparticle friction may cause small soil particles to form arching bridges across interstitial voids, decreasing the propensity for erosion. Large confining pressures increase stresses on these bridges, which may overcome friction, causing particles to dislodge and arches to fail. As mentioned, dams with no formal compaction of the inner material are more likely to experience sliding and failure. Then landslide dams may be unstable but because of high energy during deposition some compaction can thus occur increasing their relatively low degree of stability (Fig. 12.12).

We have also shown how the uncertainty in the definition of representative values for the controlling parameters (e.g. mean particle size, dam height, gradient of the base of the dam, length and width) can strongly influence the computed values of the peak discharge and the flood duration (Figs. 12.13 and 12.14, Table 12.1). We used a Monte Carlo approach (Fig. 12.13) to consider the uncertainty in parameters definition and to compute the probability of reaching a specific value of peak discharge.

No breaching but only channel erosion was eventually observed in the case of the Val Pola landslide dam (Fig. 12.1b). The main engineering works carried out to guarantee the stability of the landslide dam and to avoid future damming of the valley bottom were (Fig. 12.1c) (1) two 3.5 km long by-pass tunnels were excavated, 4.2 and 6.0 m in diameter, with a discharge capacity of 150–350 $m^3\ s^{-1}$, (2) check dams, a rip-rapped spillway channel, 12 m above the lake level, and the by pass inlet were constructed along a 2 km sector to allow discharge of the expected 200–300 yr return period flood (500 $m^3\ s^{-1}$), and (3) a large embankment debris retention dam has been, constructed in two successive stages to entrap material delivered at the toe of the Mount Zandila slope by debris flows and small landslides.

**Acknowledgements** The research has been funded by the EC Project LessLoss and by the Ministry of Education (MIUR and FIRB projects). The Regione Lombardia, Ufficio Rischi Geologici, and Ismes – Enel New Hydro are thanked for making data available. The authors wish to thank the reviewers for their comments.

## References

1. Azzoni, A., Chiesa, S., Frassoni, A. and Govi, M. (1992) The Val Pola landslide, *Engineering Geology* **33**(1), 59–70.
2. Casagli, N., Ermini, L. and Rosati, G. (2003) Determining grain size distribution of the material composing landslide dams in the Northern Apennines: Sampling and processing methods, *Engineering Geology* **69**, 83–97.
3. Coleman, S.E., Andrews, D.P. and Webby, M.G. (2002) Overtopping breaching of noncohesive homogeneous embankments, *Journal of Hydraulic Engineering* **128**(9), 829–838.
4. Costa, J.E. (1985) Floods from Dam Failures, US Geological Survey Open-File Report 85–560, Denver, Colorado, 54 p.
5. Costa, J.E. (1991) Nature, mechanics, and mitigation of the Val Pola landslide, Valtellina, Italy, 1987–1988, *Zeitschrift fur Geomorphologie* **35**, 15–38.
6. Costa, J.E. and Schuster, R.L. (1988) The formation and failure of natural dams, *Geological Society of America Bulletin* **100**, 1054–1068.
7. Costa, J.E. and Schuster, R.L. (1991) Documented historical landslide dams from around the world, *US Geological Survey Open File Report* 91–239, 486 p.
8. Crosta, G. (1991) Studio di movimenti in massa. Modellizzazione teorica e sperimentale con osservazioni e rilievi di campagna. PhD Thesis, University Studi di Milano, 137 p.
9. Crosta, G.B., Imposimato, S. and Roddeman, D.G. (2003) Numerical modelling of large landslides stability and runout, *Natural Hazards and Earth System Sciences* **3**(6), 523–538.
10. Crosta, G.B., Chen, H. and Lee, C.F. (2004) Replay of the 1987 Val Pola Landslide, Italian Alps, *Geomorphology* **60**(1–2), 127–146.
11. Crosta, G.B., Frattini, P. and Fusi, N. (2007) Fragmentation in the Val Pola rock avalanche, Italian Alps, *Journal of Geophysical Research* **112**, F01006, doi: 10.1029/2005JF000455.
12. Ermini, L. and Casagli, N. (2003) Prediction of the behaviour of landslide dams using a geomorphological dimensionless index, *Earth Surface Process Landforms* **28**, 31–47.
13. Foster, M.A., Fell, R. and Spannangle, M. (2000) The statistics of embankment dam failures and accidents, *Canadian Geotechnical Journal* **37**(5), 1000–1024.
14. Fread, D.L. (1984) *DAMBRK: The NWS Dam-Break Flood Forecasting Model*. National Weather Service, Office of Hydrology, Silver Spring, MD.
15. Fread, D.L. (1988) *BREACH: An Erosion Model for Earthen Dam Failures*. National Weather Service, National Oceanic and Atmospheric Administration, Silver Spring, MD, 35.
16. Fread, D.L. (1996) Dam-breach flood, in V.P. Singh (ed.), *Hydrology of Disasters*. Kluwer, Dordrecht, 85–126.
17. Froehlich, D.C. (1987) Embankment-dam breach parameters, in *Proceedings of 1987 National Conf. on Hydraulic Engineering*, R. M. Ragan (ed.), Williamsburg, Virginia, August 3–7, 1987, Am. Soc. Civ. Eng., New York, 570–575.
18. Froehlich, D.C. (1995) Peak outflow from breached embankment dam, *Journal of WaterResources Planning and Management* **121**(1), 90–97.
19. Froehlich, D.C. (1996) Closure to discussion of "Peak outflow from breached embankment dam" (1995a), *Journal of Water Resources Planning and Management* **122**(4), 317–319.
20. Govi, M. (1989) The 1987 landslide on Mount Zandila in the Valtellina, Northern Italy, *Landslide News* **3**, 1–3.
21. Hawkins, A.B. and Pinches, G.M. (1992) Engineering description of mudrock, *Quarterly Journal of Engineering Geology* **25**, 17–30.

22. ICOLD (1995) Dam failures – statistical analysis. Bulletin 99 of the International Committe on Large Dams (ICOLD), Paris, France, ISSN 0534-8293, 73 p.
23. Kou, S.Q., Liu, H.Y., Lindqvist, P.A., Tang, C.A. and Xu, X.H. (2001) Numerical investigation of particle breakage as applied to mechanical crushing. Part II: Interparticle breakage, *International Journal of Rock Mechanics & Mining Sciences* **38**, 1163–1172.
24. MacDonald, T.C. and Langridge-Monopolis, J. (1984) Breaching characteristics of dam failures, *Journal of Hydraulic Engineering* **110**(5), 567–586.
25. Medley, E. and Lindquist, S.S. (1995). The engineering significant of the scale independence of some Franciscan melanges in California, USA. In: Proceedings of the 35th Symposium on Rock Mechanics. Balkema, Rotterdam, 907–914.
26. Tomlinson, S.S. and Vaid, Y.P. (2000) Seepage forces and confining pressure effects on piping erosion, *Canadian Geotechnical Journal* **37**, 1–13.
27. Wahl, T.L. (1998) Prediction of embankment dam breach parameters, a literature review and needs assessment. *Report No. DSO-98-004*, Bureau of Reclamation, US Department of the Interior, Denver, 67 p.
28. Wahl, T.L. (2004) Uncertainty of predictions of embankment dam breach parameters, *Journal of Hydraulic Engineering* **130**(5), 389–397.
29. Walder, J.S. and O'Connor, J.E. (1997) Methods for predicting peak discharge of floods caused by failure of natural and constructed earth dams, *Water Resources Research* **33**(10), 2337–2348.
30. Webby, M.G. (1996) Discussion of "Peak outflow from breached embankment dam" (Froehlich, 1995a), *Journal of Water Resources Planning and Management* **122**(4), 316–317.
31. Wetmore, J.N. and Fread., D.L. (1991). The NWS Simplified Dam Break Flood Forecasting Model. Federal Emergency Management Agency (FEMA). 36 p.
32. Zhu, W.C., Tang, C.A., Huang, Z.P. and Liu, J.S. (2004) A numerical study of the effect of loading conditions on the dynamic failure of rock – SINOROCK2004 CD-ROM, Elsevier Ltd. Symposium *Int. J. Rock Mech. Min. Sci.* 41, 3, Paper 2A 06.

# Chapter 13
# The 1786 Dadu River Landslide Dam, Sichuan, China

**C.F. Lee and F.C. Dai**

## 1 Introduction

Natural damming of rivers by landslides may cause significant hazards. Firstly, it presents serious threats to people and property due to possible upstream flooding as the impounded lake water level rises. Secondly, downstream flooding may occur due to dam breach and rapid release of the impounded water. Thus, it has been the subject of many research studies in recent years. Of these, Costa and Schuster [9] presented the most comprehensive inventory and bibliography of 463 landslide dams throughout the world. Clague and Evans [7] described 16 existing and 22 historical landslide dams in the Canadian Cordillera. Casagli and Ermini [2] presented an inventory of 68 present and historic landslide dams in the Northern Apennine. Hewitt [15] provided an excellent guideline on how to reconstruct landslide-damming events from geomorphic evidence.

In China, Chai et al. [4] presented an inventory of 147 present and historic landslide dams. It is by no means complete due to the remote nature of landslide-damming events and the vagaries of reporting such incidents. This inventory and more recent landslide-damming events suggest that at least 5,500 deaths have been caused by landslide dams in China since 1900. A number of these incidents are well documented. For example, on August 25, 1933, an earthquake of $M = 7.5$ near Diexi, Sichuan Province caused the formation of three landslide dams with a maximum height of 160 m on the Min River and nine landslide dams on its tributaries. The three lakes behind the landslide dams on the Min River merged into a large lake because of the continuous inflow of water from upstream and the higher elevation of the most downstream dam. The dam was overtopped 45 days later and a flood of water rushed down the valley for a distance of 250 km, killing at least 2,500 people [5, 24]. On June 8, 1967, a huge landslide with a volume of $68 \times 10^6$ m$^3$ occurred at Tanggudong on the Yalong River, Sichuan Province, forming a natural dam with a height of 175 m and a reservoir with a capacity of $680 \times 10^6$ m$^3$. Nine days

C.F. Lee (✉)
Department of Civil Engineering, The University of Hong Kong, Hong Kong, China
e-mail: leecf@hkucc.hku.hk

S.G. Evans et al. (eds.), *Natural and Artificial Rockslide Dams*, Lecture Notes in Earth Sciences 133, DOI 10.1007/978-3-642-04764-0_13, © Springer-Verlag Berlin Heidelberg 2011

later, the dam break by overtopping lasted 13 h and resulted in a maximum flood discharge of 53,000 $m^3/s$ [6], but no loss of life occurred because of precautionary evacuations. More recently, the Yigong River in southeastern Tibet was dammed on April 9, 2000 by a huge landslide with a volume of $300 \times 10^6$ $m^3$. This natural dam, about 130 m high and 1.5 km long, was created in 8 min. The dam partially failed on June 10, 2000 [22]. The resultant flood travelled more than 500 km downstream, damaged many bridges, and created numerous new landslides along both banks of the river (Chap. 22 by Evans and Delaney, this volume). The flooding also resulted in 30 deaths, more than 100 missing people, and more than 50,000 homeless in the five districts of Arunanchal Pradesh, India [33].

Historic records also revealed that the 1786 Kangding-Luding earthquake in Sichuan, southwestern China triggered a large landslide that dammed the Dadu River, which breached suddenly 10 days later, resulting in up to 100,000 deaths due to downstream flooding [11]. This incident may thus constitute the world's most disastrous landslide-damming event ever recorded. Although studies have been carried out to determine the location, magnitude and intensity of the 1786 Kangding-Luding earthquake [27], relatively little is known about the nature of the landslide dam. This paper presents an attempt to address the location and nature of this landslide dam and to reconstruct the peak discharge at the dam breach, based on historical documents and geomorphic evidence.

## 2 Geographic and Geological Setting

The Kangding-Luding area in Sichuan, southwestern China is located in the convergence of the NE-SW trending Longmenshan, NW-SE trending Xianshuihe and N-S trending Anninghe active fault zones (Fig. 13.1). This area is one of the most seismically active areas in China, although seismic activity, in terms of magnitude, at the south extremity of the Longmenshan fault zone is considered to be moderate [27]. The spatial and temporal characteristics of historic earthquakes along the Anninghe fault zone indicate that its seismic activity tends to decrease northward. The seismic activity of the Kangding-Luding area is controlled primarily by the Xianshuihe active fault zone [28]. The zone is one of the most active fault systems in the world, and can be divided into five segments: the Moxi fault, Selaha fault, Zheduotang fault, Yalahe fault, and northwest segment [1, 32]. As the Selaha and Zheduotang faults are, at most, only 10 km apart, and they converge toward their ends, it is possible that they dip steeply toward each other and merge at depth [1, 32]. The northwestern end of the Xianshuihe fault overlaps with the southeastern end of the Ganzi-Yushu fault, with a left en echelon stepover of about 40 km (Fig. 13.1). The southeastern extremity of the Xianshuihe fault zone connects with the Anninghe fault. Based on a detailed study of fault landforms, historical and paleo-earthquakes, co-seismic slip and geochronology, Zhou et al. [32] found that the average horizontal slip rates of the Yalahe, Selaha, Zheduotang, and Moxi faults during the late Quaternary are $2.0 \pm 0.2$, $5.5 \pm 0.6$, $3.6 \pm 0.3$, and $9.9 \pm 0.6$ mm/a, respectively, and that the

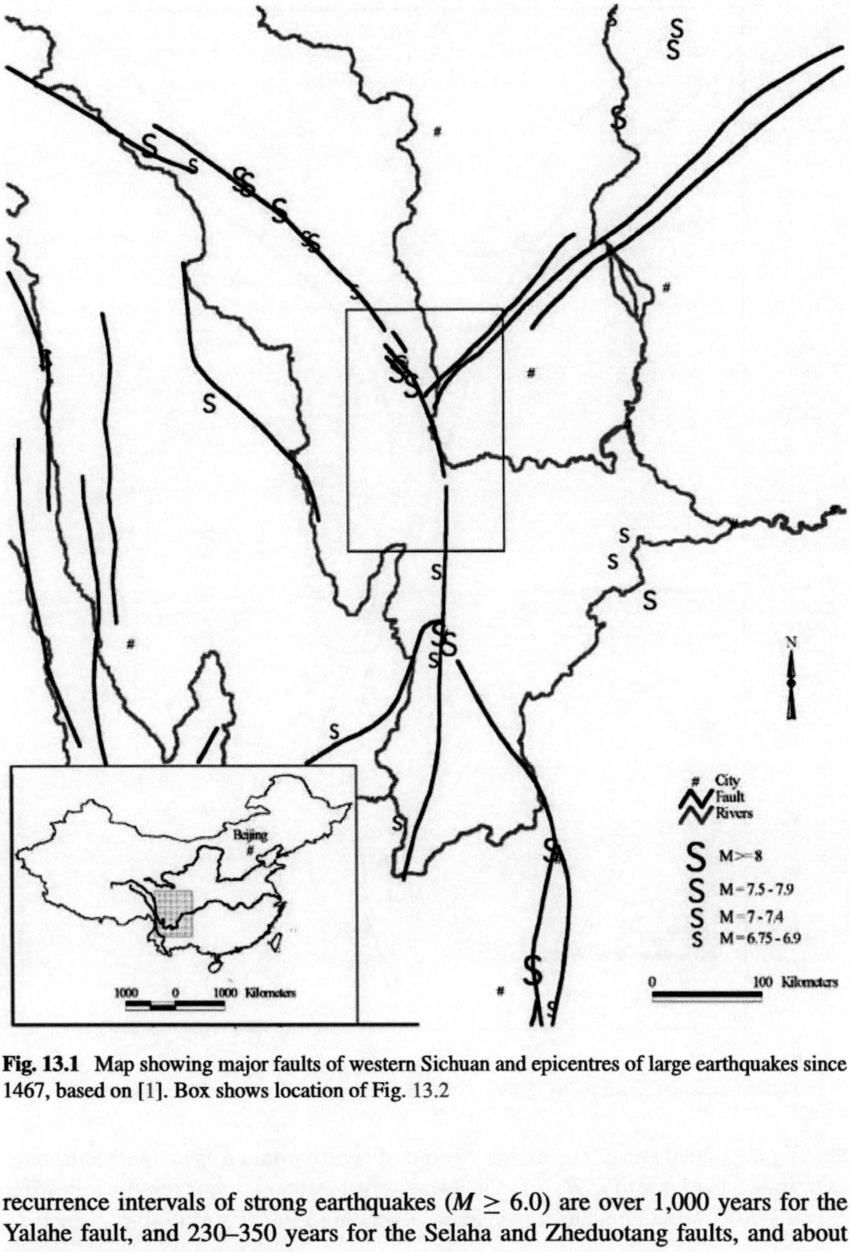


**Fig. 13.1** Map showing major faults of western Sichuan and epicentres of large earthquakes since 1467, based on [1]. Box shows location of Fig. 13.2

recurrence intervals of strong earthquakes ($M \geq 6.0$) are over 1,000 years for the Yalahe fault, and 230–350 years for the Selaha and Zheduotang faults, and about 300 years for the Moxi fault.

Figure 13.1 shows large historic earthquakes on the Xianshuihe and associated faults of southwestern China. It can be seen that the major earthquakes tend to occur on the major faults. It should be noted that the historical seismic records in this sparsely populated part of China are very short as compared to those of the more

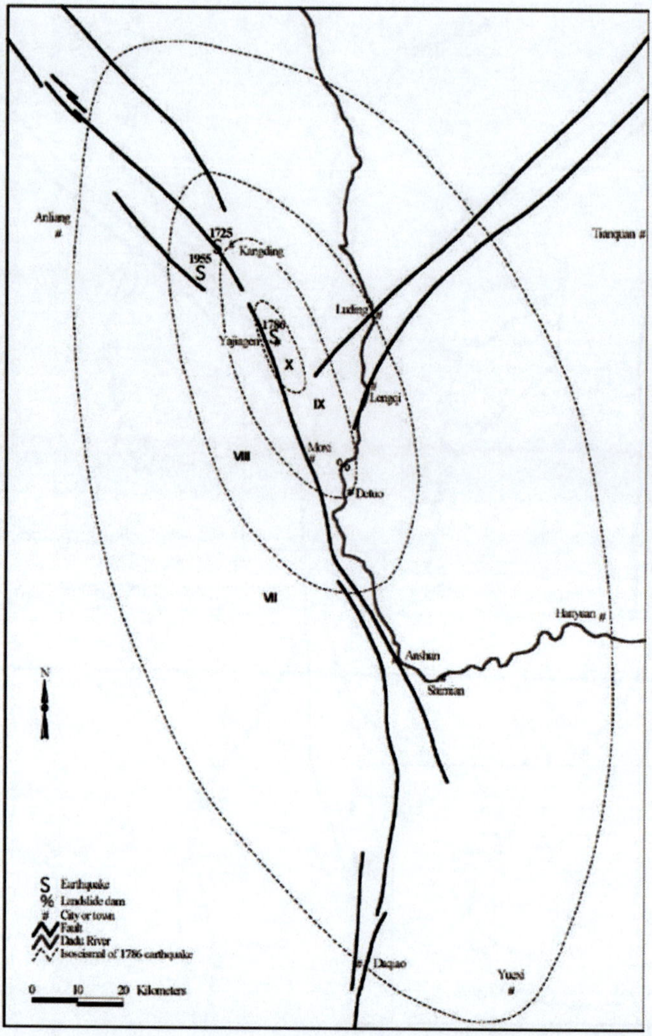


**Fig. 13.2** Isoseismal map of the 1786 Kangding-Luding earthquake, based on [27]. Modified Mercalli Intensities assigned to the different isoseismal zones are shown with roman letters

densely populated areas. The earliest record of large earthquakes on the Xianshuihe fault zone dates back to 1725. Among the recorded major earthquakes, the following events are of particular significance to the Kangding-Luding area (Table 13.1):

- The 1725 Kangding earthquake;
- The 1786 Kangding-Luding earthquake
- The 1955 Zheduotang earthquake, which killed 70 people, injured 217 persons, and caused extensive damage to property [29].

**Table 13.1** Historic large earthquakes in the Kangding-Luding area, southwestern China

| Date | Name | Coordinates | Fault | MMI at the epicentre | Magnitude | Focal depth (km) |
|---|---|---|---|---|---|---|
| August 1, 1725 | Kangding earthquake | 30°03'N, 101°56' | Selaha fault | IX | 7.0 | – |
| June 1, 1786 | Kangding-Luding earthquake | 29°53'N, 102°01' | Moxi fault | ≥X | 7.75 | 20 |
| April 14, 1955 | Zheduotang earthquake | 30°00'N, 101°54' | Zheduotang fault | IX | 7.5 | 20 |

The bedrock of the region is made up primarily of quartzites, marbles, slates, and phyllites of the late Triassic Xikang group; diorite of the Archean to early Proterozoic age; and Mesozoic and Precambrian granites [23]. Quaternary deposits are limited to small basins, particularly along the fault, in addition to local fluvial terraces and glacial deposits. Rocks along the fault zone are intensely sheared, with many tight folds as well as abundant mylonite and gouge. The crushed zone, which is easily erodible, controls the drainage pattern of the region.

The terrain is typically steep and mountainous. The 1786 earthquake-affected region is situated in the transitional zone between the Longmenshan Mountains and the Qinghai-Tibetan Plateau. This area has a relief of over 3,000 m with rivers flowing at approximately 1,100–1,300 m a.s.l., while the mountain peaks rise to above 6,000 m. The mean annual rainfall at Luding is 637 mm. The major rivers flowing through the study area include the Dadu River and its tributaries. The Dadu River has a U-shaped and steep-sided valley. The maximum discharge of the river generally occurs during the summer season, between June and September. A maximum peak discharge of 5,500 $m^3/s$ had been recorded at Luding in 1992, while the average flow varies from 100 to 1,500 $m^3/s$. The valleys have five levels of paired river terraces, which support agriculture and the main villages.

## 3 Methodology

With the aim to determine the location and nature of the 1786 landslide dam and subsequent dam-break flood, the methodology used in the present study can be summarized as follows:

- Initially, the published materials associated with the 1786 earthquake were collected and analyzed to provide the background information on the epicentre, magnitude and seismic intensities.
- Next, historic records were collated to define the possible locations, the nature of the landslide dam and the consequence of dam-break flood.

- A reconnaissance fieldwork was carried out to identify the location of the landslide dam. Diagrammatic mapping was then conducted to determine the nature of the landslide dam from the geomorphic evidence, after identifying the landslide site.
- Based on the dam crest height estimated in the field, the former lake created by the landslide dam was reconstructed, and the peak discharge at the dam breach was estimated from empirical regression equations that relate the observed peak discharge to some measure of the landslide dam and the impounded water volume, and a physically-based model developed by Walder and O'Connor [26].

## 4 The 1786 Kangding-Luding Earthquake

At noon on June 1, 1786, the Kangding-Luding area was shocked by a strong earthquake, which caused extensive damage in the affected area. Historic documents showed that the main shock was followed by a series of aftershocks during the next 12 days. Damage to buildings was reported from the affected area to the Emperor Qianlong of the Qing Dynasty. This report was prepared by Bao Ning, the governor of the Shaanxi-Sichuan region at that time, and will be termed as the Bao Ning report hereafter. The Bao Ning report can be accessed from the Atlas of Historic Earthquakes in China for the Qing Dynasty Period [16]. Based on the damage description in the report, Wang and Pei [27] adopted the following classification system with regard to structures or buildings: A type – buildings simply constructed using mud and dry grass; B type – structures built of slabs of uniform or non-uniform stones piled upon each other with small chips filling the voids and supporting the large stones, or those of thick slabs of uniform stones cemented with mud plaster or woods of poor quality; and C type – other types of construction with either an integral wooden frame or with brick masonry, normally used for temples or museums. On the basis of a detailed analysis of the Bao Ning report, complemented with field surveys and visual interpretation of available aerial photographs and satellite images, a very recent surficial fault rupture that extended for at least 70 km from south of Moxi to northwest of Kangding was identified, and an isoseismal map (Fig. 13.2) was produced on the basis of modified Mercalli intensity (MMI) scale by Wang and Pei [27]. Each intensity contour in Fig. 13.2 shows a NW-SE trend for its major axis and a NE-SW trend for its minor axis. The tectonics of the region, the isoseismal trend and the pattern of ground fractures resulted from the earthquake indicate that the causative fault of this earthquake was trending NW-SE, which follows the regional trend of the Xianshuihe fault. The major and minor axes of the isoseismal for MMI IX or above are 20 and 6 km long, respectively. The zone with MMI X or above thus covers an area of approximately 95 km$^2$ and is confined to a relatively sparsely populated area. It was found that in this zone, the ground failure occurred mainly along the major faults, and was characterized by the distribution of numerous sag ponds, faulting scarps and secondary surface ruptures with subsidence along the faults. The MMI zone IX extends from Kangding in the north to Detuo in the south, and covers an area of approximately 720 $km^2$. In this zone, the majority of the A and B-type structures, and 24–26% of the C-type buildings collapsed. Within the MMI zone VIII, more than 50% of the A- and B-type structures were tilted or even damaged, and a minority of the C-type buildings were tilted. Within the MMI zone VII, only few of the A- and B-type houses were either collapsed or damaged.

The epicenter of this earthquake was located in Yajiagen along the Moxi active fault, and the focal depth was estimated to be 20 km (Wang and Pei, [27]). The magnitude of this earthquake was estimated to be $M = 7.75$, because the length of the major axis of the MMI zone VIII is equivalent to that of other earthquakes with $M = 7.5$–8.

Available historic records revealed that the violent, prolonged shaking from this earthquake gave rise to extensive small-scale rock/block falls and landslides in the MMI zones VIII to X, and the majority of these events were confined to intensely jointed and crushed rocks on steep slopes. Historic records also mentioned that a large landslide occurred and dammed the Dadu River.

# 5 Landslide Dam

## *5.1 Historical Records*

As a direct result of the 1786 earthquake, a large landslide occurred along the bank of the Dadu River. The sliding materials blocked the river and formed a dam. The Bao Ning report offered the following description of this landslide-damming event: "*... In Tiger Cliff, the mountain fractured and fell due to the June 1 earthquake and dammed the river, causing submergence of the majority of the river-side agricultural land behind the dam. The water level in the lake behind the dam exceeded 66 m above the original river level. On June 10, the dam breached suddenly, washing away the agricultural land...*" [16]. Apart from the Bao Ning report, a monument was created by local residents in memory of this event. The monument (Fig. 13.3), now kept at the Luding Seismological Office, is a stone tablet with the following inscription in Chinese [31]: "*In 1786, the ground shook and the rocks collapsed. The mountain at Jindongzi collapsed and a landslide dam blocked the Dadu River for 9 days. In the early morning on June 9, water in the lake overflowed the crest of the dam*".

Although this landslide-damming event has been cited to illustrate the catastrophic effects of landslide dams [1, 17–19], there has been little consensus on the location and nature of the landslide dam. In fact, it has been a subject of debate among some Chinese seismologists and geologists [27, 31]. Although Tiger Cliff and Jindongzi were mentioned as the landslide locations in the Bao Ning report and the monument, respectively, it is now difficult to find these precise locations. In all likelihood, the names of some locations might have changed with time.

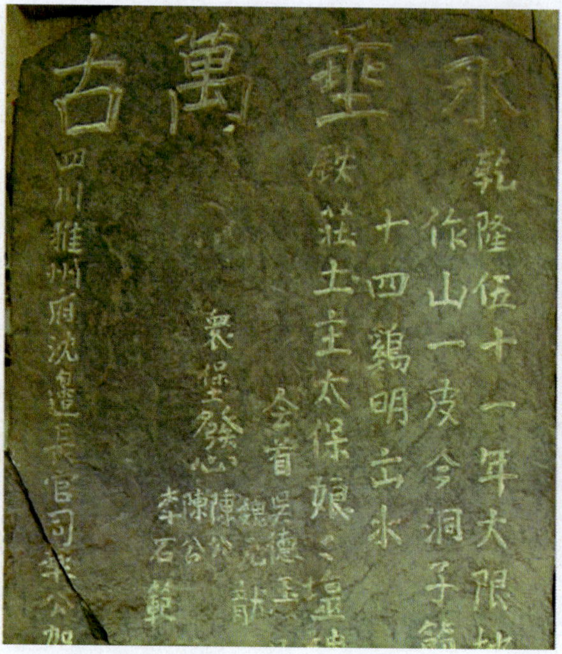


**Fig. 13.3** The monument created by local residents in memory of the 1786 landslide-damming event. For interpretation refer to the main text

However, the Bao Ning report stated that Tiger Cliff was located in the Shenbie administrative zone at that time. Wang and Pei [27] collected historic documents and presented a figure showing the geographical location of the Shenbie administrative zone (Fig. 13.4). This means that the landslide-damming events occurred on the Lengqi – Tianwan section of the Dadu River.

## 5.2 Geomorphic Features

We searched all published materials associated with the possible location of the landslide, and found a photograph taken in 1984, which was considered to represent the landslide site, in the Atlas of Historic Earthquakes in China for the Qing Dynasty Period [16]. The location as then photographed was found in the field in August 2003 (Fig. 13.5). A detailed walkover survey was then carried out, and it was found that the landslide could not have taken place there, on the grounds that (a) the terrace, composed possibly of glacial deposits [23] is relatively flat with slope angles less than 5°, although slope deposits originating from small-scale rockfalls could be observed on steep slopes at the back of the terrace (b) the bedrock, though heavily fractured, was exposed along the ephemeral streamlines; and (c) no remnants of the landslide debris can be found on the opposite bank.

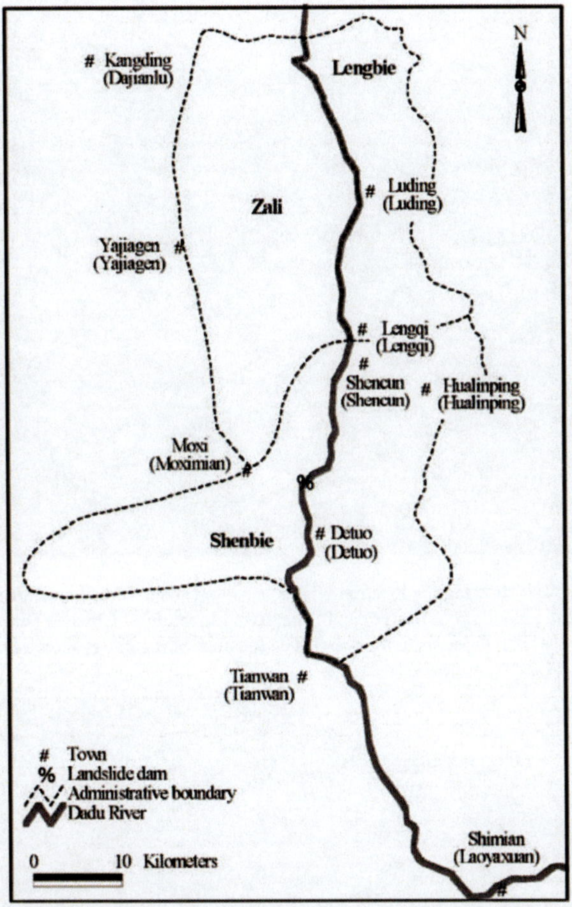


**Fig. 13.4** Location of the Shenbie administrative zone relative to the Dadu river in the period governed by the Emperor Qianlong of the Qing Dynasty. The name of towns used in the Qing Dynasty given in brackets (after [27])

We then checked all other possible locations, conducting a walkover survey from Lengqi to Qianwan along the Dadu River (Fig. 13.4). We found that, although numerous small-scale rockfalls and landslides could readily be observed on both banks of the Dadu River, a site on the right bank (Fig. 13.6) was the only sizable landslide that could dam the Dadu River. This landslide is adjacent to the site in Fig. 13.5 and lies within the MMI zone IX of the 1786 earthquake.

Geomorphic and geologic features of the landslide site are illustrated in a diagrammatic map of the landslide area (Fig. 13.7). Two adjoining hummocky slide masses can be seen above the steep right bank of the Dadu River. The landslide area is bush-covered, marked by an active debris avalanche scar. The landslide is bounded to the north and the south by two ephemeral drainage lines and separated

**Fig. 13.5** Photograph showing the location which was interpreted as the landslide site by Institute of Geophysics at the National Seismological Bureau and Institute of Chinese Historic Geography at Fudan University [16]. (View is to the west). The location of the site is shown in Fig. 13.7

**Fig. 13.6** Photograph showing the landslide site (view is to the west). The failure main scarp is marked with a *dashed line* and the location of the site shown in Fig. 13.4

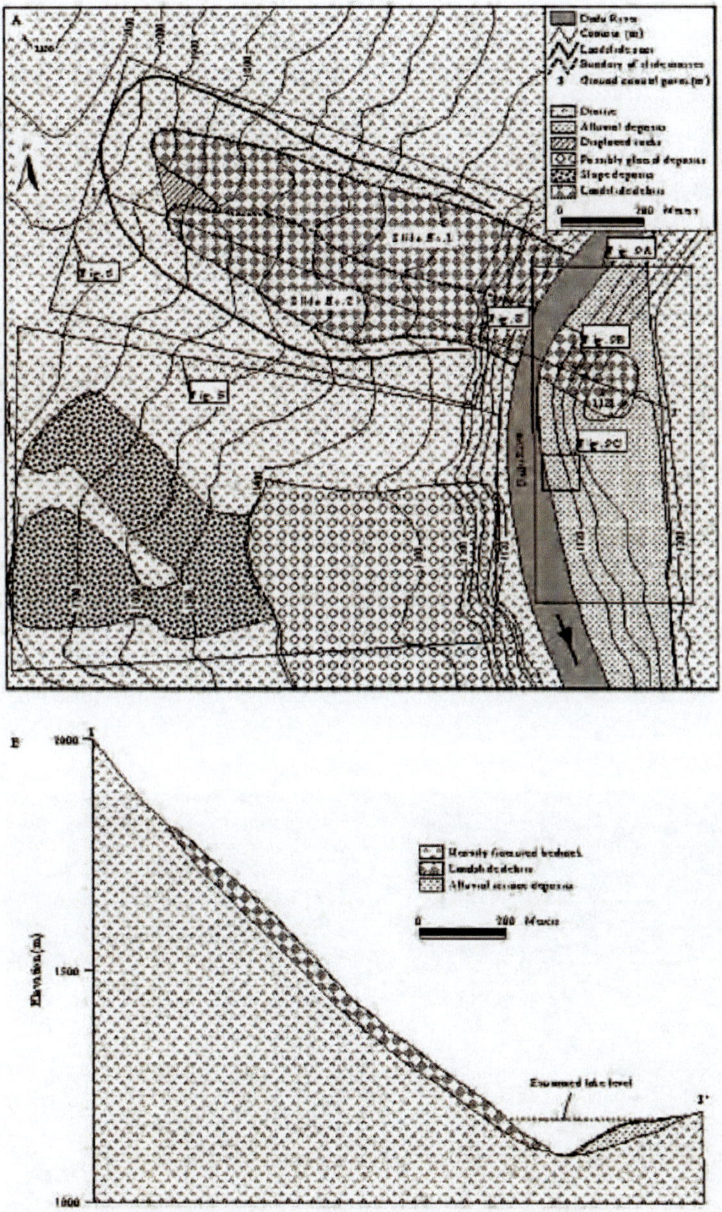


**Fig. 13.7** Sketch map showing the geomorphology and geology of the landslide site and the spatial relation of Figs. 13.5, 13.6, 13.8, and 13.9. **(a)** plan; **(b)** cross section of I–I'

into two parts by a drainage line along which the recent debris avalanche moved downslope (Fig. 13.6). It can be observed that the loose materials deposited along the drainage line were entrained during debris movement, exposing the underlying bedrock. It may be reasonable to divide the landslide into Slide No. 1 and Slide No. 2 as possibly two phases of movement (Fig. 13.7), because both slides are inferred to have a deep shear surface from the geomorphic features of the main landslide scar and the exposed bedrock on the drainage line. At the back of the landslide mass, a huge displaced rock block was sandwiched between these two slide masses. Apart from the minor erosion scar that constitutes the source area of the recent debris flow, no scarp shows evidence of active slope movements.

In the landslide area, the slope ranges from 35 to 40°, and the Dadu River has a gradient of about 1%, traversing a narrow and steep-walled valley. The width of the riverbed at the landslide site is estimated to be about 80 m. The geology of the landslide area is comprised mainly of diorite of the Archean – early Proterozoic age (Fig. 13.7). The bedrock is heavily fractured due to structural and weathering effects. On the left bank, two terrace levels composed of alluvial sands, gravels and boulders can be distinguished. The river turns right and erodes the toe of Slide No. 2. The intense undercutting and lateral erosion of the river water makes the riverbank very steep, with an estimated slope angle of 50–60°. The landslide debris consists predominantly of extremely crushed materials, ranging in particle size from clays to boulders and rock fragments (Fig. 13.8). It appears that both Slides No. 1 and No. 2 occurred within heavily fractured diorite and slid along a steeply inclined slip surface (Fig. 13.7b). The volume is approximately $5 \times 10^6$ $m^3$ for Slide No. 2, and

**Fig. 13.8** Photograph showing the landslide debris on the right bank (view is to the west)

$6 \times 10^6$ $m^3$ for Slide No. 1, if the average failure depth is assumed to be 35 m based on the topographic features of the main landslide scar.

From the geomorphic features of the landslide area, it appears that the distal part of Slide No. 2 released a significant volume of landslide debris into the valley of the Dadu River and dammed the river completely, whereas the remainder of the debris remained stuck in the valley bank, forming a hummocky accumulation. This can be verified by the fact that on the opposite bank of the moving direction of Slide No. 2, the remnants of landslide debris can be clearly observed, which can be distinguished from the upstream and downstream boulders and sands of alluvial origin (Fig. 13.9). This means that the landslide debris butted against the slope opposite from the failure site. On the terraces of the left bank, the remnants of landslide debris can be found. It can be inferred that the landslide debris ran up to the opposite terrace, implying that the landslide dam had a height equivalent to the elevation of the terrace. An official Ground Control Point (GCP) for surveying, at an elevation of 1,181 m, is located where the remnants of the landslide debris were observed (Fig. 13.7). Judging from the deposits, the landslide dam had a length of 320 m at the crest and 80 m at the river level. Schuster and Costa [21] concluded that in

**Fig. 13.9** (**a**) View of the opposite bank of the landslide (view is to the east). A small, fresh bank failure can be observed on the steep walled bank; (**b**) remnants of landslide debris on the opposite steep-walled bank, exposed due to the small, fresh bank failure (view is to the east); (**c**) alluvial deposits on the opposite steep-walled bank downstream the landslide dam site (view is to the east). There is no evidence to confirm that this slide completely blocked the Dadu River

steep-walled narrow valleys, there is no room for the landslide mass to spread out. The width of the landslide dam can, therefore, be estimated from the geometry of the landslide and the width of remnants of landslide debris deposited on the terraces of the left bank, and should be less than 150 m. In addition, it can be inferred from the topographical map that the riverbed has an elevation of 1,110 m. Therefore, this landslide dam had a height of about 70 m. This figure is in agreement with that recorded in the Bao Ning report. Since the landslide dam spanned the river valley floor and deposited materials on the opposite side of the valley, it belongs to Type II in the classification system proposed by Costa and Schuster [8] and the "main valley lake" type of Crozier and Pillans [10]. As for Slide No. 1, there is no evidence to confirm that this slide completely blocked the Dadu River.

It can be envisaged that after the landslide dam was formed, a reservoir began to develop behind the dam. The 1:50,000-scale topographic map covering the landslide area and its upstream was digitised and then used to generate the Digital Elevation Model (DEM) data with a resolution of 20 m so that the volume of the reservoir water can be reasonably estimated. Based on the ground control point in Fig. 13.7, it is reasonable to set the maximum elevation of the lake water as about 1,180 m a.s.l. Thus, the reservoir behind the landslide dam at this elevation was about 6.8 km long along the main valley; the volume and area were estimated to be about $50 \times 10^6$ $m^3$ and 1.7 $km^2$, respectively (Fig. 13.10).

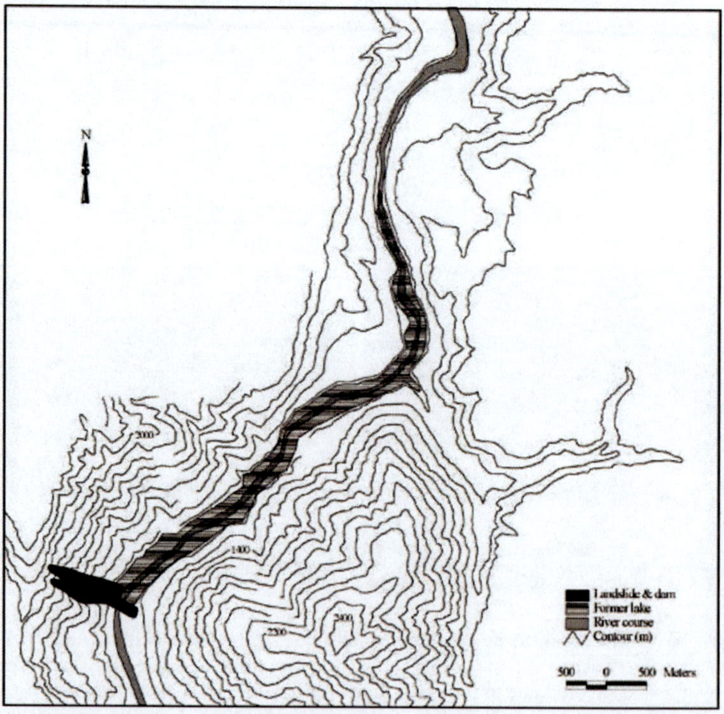


**Fig. 13.10** Map showing the 1786 landslide dam site and the lake behind the landslide dam. The lake was reconstructed for a dam crest height of 70 m

As described on the monument, the lake water overflowed the top of the landslide dam in the early morning of June 9. On June 10, the landslide dam breached suddenly, and the river valley downstream of the dam was flooded. This dam failure was triggered by an aftershock, rather than by overtopping or seepage and piping that are the most common mechanisms of natural dam failures, as described in the Memoirs of Tianquan Prefecture [12] "*... on June 10, a major aftershock triggered the dam failure...*" [12].

If compared to the size of many landslide dams observed elsewhere in the world, the size of this landslide dam on the Dadu River is not spectacular. What makes this landslide dam interesting is its catastrophic consequence downstream that will be subsequently described.

## 6 Consequences of the Dam Break

Based on the Bao Ning report, this landslide dam created damage due to two different types of flooding: (1) submergence of agricultural land due to upstream flooding as the reservoir impounded; and (2) downstream flooding as a result of the failure of the dam. The Bao Ning report provided the following description: "*... between Shimian and Hanyuan downstream of the landslide dam, the river levees, agricultural land, encamps and houses on both banks of the Dadu River were swept away by the flooding water...*". There are other historic documents kept in the local governmental archives that recorded the damage downstream caused by this dam failure flood. The Memoirs of Tianquan Prefecture (Editorial Group of Sichuan Earthquake Information, [12]) recorded that "*... the flooding caused by sudden dam breaching swept away the residents on both banks downstream. One day later, the flooding reached Leshan City. In the morning of June 11, a lot of people crowded on the east gate of the city wall and watched the flood. The wall collapsed suddenly and numerous people dropped into the flood water...*". The Memoirs of Leshan County (Editorial Group of Sichuan Earthquake Information, [12]) recorded that "*... the flood wave looked like a moving mountain; more than 10,000 people were drowned to death...*". A written document by Qi Zhong, a scholar in the Qing Dynasty, reported that "*... in Leshan, Yibin and Luzhou along the River, at least 100,000 people were drowned to death....*" [12, 20]. As a result 100,000 deaths were counted into the damage of the 1786 earthquake (p. 193 in [20])

It was impossible to record the exact number of deaths at that time. It can be concluded, however, that the number of deaths was already large in Leshan City, and that the number of deaths increased drastically, as the flood routed downstream and more people were drowned. Historic records on this flooding event in the Memoirs of Leshan County [12, 30] denied the possibility of additional flooding from the Min River, and thus the high number of fatalities should be attributed mainly to downstream flooding from this landslide dam break.

If it is accepted that 100,000 people were drowned by this landslide-dam break flood, this event should be the most disastrous in the world, in terms of the number of deaths caused by downstream flooding from landslide dam breaching.

## 7 Peak Discharge Estimate for the Dam-Break Flood

The peak discharge of a natural dam-break flood can be estimated by the following two methods. One of them relies on regression equations that relate the observed peak discharge to some measure of the impounded water volume: depth, volume or some combination thereof [9, 13]. Regression equations are expedient but generally provide no better than order-of-magnitude predictions of probable peak discharge [3, 9, 13, 26]. The other method employs computer implementation of physically based models [14, 25]. Such models have been developed to simulate breach formation, lake drainage, and the peak discharge at the dam breach based on lake volume, water depth, lake geometry, breach geometry, and the rate of erosion at the dam breach [14]. The models generally account for hydraulic constraints not reflected by regression equations but may be cumbersome to use, and often require detailed information on dam structure and material properties, which may be poorly known for natural dams [3].

Despite the limitations of the regression equations, they do provide useful information on general trends in the peak discharge at the dam breach for landslide dam failure floods. Cenderelli [3] provided the following regression equations for landslide dams:

$$Q_p = 24d^{1.73} \tag{1}$$

$$Q_p = 3.4\,V^{0.46} \tag{2}$$

$$Q_p = 1.9(Vd)^{0.4} \tag{3}$$

where $Q_p$ is the peak flood discharge ($m^3/s$), $d$ is the drop in lake level depth (m), and $V$ is the volume of water drained from the lake ($m^3$).

The peak discharge estimation generally involves a breach that erodes rapidly to the base of the dam to calculate the possible maximum discharge [26]. Therefore, we assume that the drop in lake level depth ($d$) is 70 m, and the volume of water drained from the lake ($V$) is $50 \times 10^6$ $m^3$. The peak discharge at breach is therefore 37,345 $m^3/s$ if Eq. (1) is applied, 11,831 $m^3/s$ if Eq. (2) is applied, and 12,485 $m^3/s$ if Eq. (3) is applied.

Walder and O'Connor [26] developed the following physically-based predictive equations for estimating the peak discharge at dam breach:

$$Q_p = 1.51(g^{0.5}d^{2.5})^{0.06}(kV/d)^{0.94} \qquad \textit{if } kV(gd)^{-0.5}d^{-3} < 0.6 \tag{4}$$

$$Q_p = 1.94\,g^{0.5}d^{2.5}(Dc/d)^{0.75} \qquad \textit{if } kV(gd)^{-0.5}d^{-3} >> 1 \tag{5}$$

where $g$ is gravitational acceleration ($m/s^2$), $k$ is the erosion rate at the breach (m/s), $V$ is lake volume drained ($m^3$), $d$ is the lake level decline during the flood (m), and $D_c$ is the height of the dam crest relative to the dam base (m). The erosion rate at the breach is determined by

$$k = d/t \tag{6}$$

where $t$ is the time for the breach to form (s).

Since the landslide dam failure in this study was triggered by a major aftershock, it may be reasonable to assume the time for the breach to form was very short and Eq. (5) should be applied. The peak discharge at the dam breach is estimated to be 248,977 $m^3/s$, if $D_c = 70$ m and $d = 70$ m.

The above analyses indicate that the peak discharges at the dam breach, estimated using the regression equations, and the physically-based predictive equation developed by Walder and O'Connor [26], are very variable at least on an order of magnitude. As shown in Fig. 13.7, the landslide dam with a 70 m deep flow has a cross-sectional area of about 9,800 $m^2$. Based on hydraulic continuity, the estimated discharges from Eqs. (1), (2) and (3), divided by the cross-sectional area, yield mean velocities of 3.5 m/s, 1.2 and 1.3 m/s, respectively. The value of 1.2 or 1.3 m/s seems unreasonably small. However, the estimate from Eq. (5) implies a mean velocity of 25.4 m/s, which is impossibly large and close to the critical velocity to achieve cavitation at that flow depth. Therefore, the values of 3.5 and 37,345 $m^3/s$ may be acceptable as the mean velocity and peak discharge at the dam breach, respectively.

The best way to estimate the peak discharge at the dam breach may be to collect information on the flood water levels downstream and then back analyse the peak discharge at the breach, based on flood routing models for unsteady flow [14]. This requires detailed information on downstream topographic data and hydraulic constraints. A preliminary study by Xie [30] concluded that the maximum flood level in Leshan City was 376.9 m a.s.l [30], but the maximum flood depth is not known and the accuracy of this estimation remains to be evaluated. Further study is needed in this regard.

## 8 Discussion

Landslide dams are significantly hazardous, especially in an area of high seismicity, because of their possible catastrophic damage on downstream settlements and infrastructures. A full development of the hydropower potential of the Dadu River is now on the agenda. However, the risks of extreme, if rare, hazards from future landslide-damming events may be seriously underestimated. The ability to identify the occurrence of such events and interpret the consequence is necessary for estimating future risks.

Although there is no geomorphic evidence indicating that this landslide is still active, it has not yet reached the position of minimum potential energy (Fig. 13.7). Hence, it is possible for the landslide to be reactivated under strong shaking conditions. There are the following four external and internal factors that may contribute to the future reactivation of the landslide. The first factor is the erosion of flowing water at the toe of the landslide. The Dadu River continuously and directly erodes the toe of Slide No. 2, removing the lateral support from an oversteepened slope. The second factor is that the shear strength of the sheared surface has been drastically reduced to the residual state due to past landsliding, increasing the possibility of future reactivation, as the heavily fractured bedrock has been extremely crushed or even powdered during the past movement. The third factor is that the headscarp of the landslide seems to show, albeit minor, symptoms of instability and the possibility of new rock falls from the cliffs should be considered, particularly under seismic conditions. The last factor is the possibility of future earthquakes. Based on seismological study, there are three near-site potential seismic sources that may produce strong earthquakes: (a) the Zheduotang and Selaha faults; (b) the Yalahe fault; and (c) the Moxi fault. For the Yalahe fault, the recurrence interval of strong earthquake with $M \geq 6.0$ exceeds 1,000 years and the last strong earthquake on this fault is estimated to have occurred in $1{,}380 \pm 60$ AD [32]. The possibility of strong earthquakes on this fault in the coming 100 years is thus very low. The same applies to the Zheduotang and Sehala faults because the strong earthquake ($M \geq 6.0$) recurrence interval of these faults is 230–350 years [32], and the last strong earthquake occurred in 1955. However, the Moxi fault will induce a strong earthquake in the coming 100 years, because the strong earthquake recurrence interval on this fault is about 300 years and the last strong earthquake occurred in 1786, which generated the landslide dam under study. In conclusion, it is possible for the landslide to be reactivated in the near future.

Reactivation of the landslide, if it does occur, will block the Dadu River again, as the oversteepened distal part of Slide No. 2 has already reached the riverbed and the river course is narrow. Elements at risk include the residents and buildings on the opposite bank as well as those in upstream and downstream areas. Detailed geotechnical investigations for evaluating the stability of the landslide are strongly recommended with due consideration for future seismic risk.

**Acknowledgements** This study was supported financially by the Research Grants Council of Hong Kong.

## References

1. Allen, C.R., Luo, Z., Qian, H., Wen, X., Zhou, H. and Huang, W. (1991) Field study of a highly active fault zone: The Xianshuihe fault of southwestern China, *Geological Society of America Bulletin* **103**, 1178–1199.
2. Casagli, N. and Ermini, L. (1999) Geomorphic analysis of landslide dams in the northern Apennine, *Transactions of the Japanese Geomorphological Union* **20**, 219–249.
3. Cenderelli, D.A. (2000) Floods from natural and artificial dam failures, in Wohl, E.E. (Ed.), *Inland Flood Hazards*. Cambridge University Press, New York, NY, pp. 73–103.
4. Chai, H.J., Liu, H.C. and Zhang, Z.Y. (1995) The catalog of Chinese landslide dam events, *Journal of Geological Hazards and Environment Preservation* **6**(1–9) (in Chinese).
5. Chai, H.J., Liu, H.C., Zhang, Z.Y. and Xu, Z.W. (2000) The distribution, causes and effects of damming landslides in China, *Journal of Chengdu University of Technology* **27**, 302–307.
6. Chen, Y.J., Zhou, F., Feng, Y. and Xia, Y.C. (1992) Breach of a naturally embanked dam on Yalong river, *Canadian Journal of Civil Engineering* **19**, 811–818.
7. Clague, J.J. and Evans, S.G. (1994) *Formation and failure of natural dams in the Canadian Cordillera, Geological Survey of Canada Bulletin* **464**, 35 p.
8. Costa, J.E. and Schuster, R.L. (1988) The formation and failure of natural dams, *Geological Society of America Bulletin* **100**, 1054–1068.

9. Costa, J.E. and Schuster, R.L. (1991) *Documented historical landslide dams from around the world*. U.S. Geological Survey Open-File Report 91–239, 486 p.
10. Crozier, M.J. and Pillans, B. (1991) Geomorphic events and landform response in south-eastern Taranaki, New Zealand, *Catena* **18**, 471–487.
11. Dai, F.C., Lee, C.F., Deng, J.H. and Tham, L.G. (2005) The 1786 earthquake-triggered landslide dam and subsequent dam-break flood on the Dadu River, southwestern China, *Geomorphology* **65**, 205–221.
12. Editorial Group of Sichuan Earthquake Information (1980) *Collections of Earthquake Information in Sichuan Province*. Sichuan People Publishing House, Chengdu.
13. Evans, S.G. (1986) The maximum discharge of outburst floods caused by the breaching of man-made and natural dams, *Canadian Geotechnical Journal* **23**, 385–387.
14. Fread, D.L. (1996) Dam-breach floods, in Singh, V.P. (Ed.), *Hydrology of Disasters*. Kluwer Academic Publishers, Dordrecht, pp. 85–126.
15. Hewitt, K. (1998) Catastrophic landslides and their effects on the upper indus streams, Karakoram Himalaya, northern Pakistan, *Geomorphology* **26**, 47–80.
16. Institute of Geophysics at the National Seismological Bureau and Institute of Chinese Historic Geography at Fudan University (1990) *Atlas of Historic Earthquakes in China for the Qing Dynasty Period*. China Cartographic Publishing House, Beijing.
17. Korup, O. (2002) Recent research on landslide dams – a literature review with special attention to New Zealand, *Progress in Physical Geography* **26**, 206–235.
18. Li, T. (1990) *Landslide Management in the Mountain Areas of China*. ICIMOD Occasional Paper 15, International Centre for Integrated Mountain Development. Kathmandu, Nepal.
19. Li, T., Schuster, R.L. and Wu, J. (1986) Landslide dams in south-central China, in Schuster, R.L. (Ed.), *Landslide Dams Processes, Risk and Mitigation*. Geotechnical Special Publication No. 3. American Society of Civil Engineers, pp. 146–162.
20. Lou, B.T. (1996) *Collections of Historic and Recent Earthquake damage*. Seismological Press, Beijing.
21. Schuster, R.L. and Costa, J.E. (1986) A perspective on landslide dams, in Schuster, R.L. (Ed.), *Landslide Dams Processes, Risk and Mitigation*. Geotechnical Special Publication No. 3. American Society of Civil Engineers, pp. 1–20.
22. Shang, Y.J., Yang, Z.F., Li, L.H., Liu, D.A., Liao, Q.L. and Wang, Y.C. (2003) A super-large landslide in Tibet in 2000: Background, occurrence, disaster, and origin, *Geomorphology* **54**, 225–243.
23. Sichuan Bureau of Geology and Mineral Resources (1991) *Regional Geology of Sichuan Province*. Geological Publishing House, Beijing (in Chinese).
24. Sichuan Seismological Bureau (1983) *The 1933 Diexi Earthquake*. Sichuan Science and Technology Press, Chengdu (in Chinese).
25. Singh, V.P. (1996) *Dam Break Modeling Technology*. Kluwer Academic Publishers, Dordrecht.
26. Walder, J.S. and O'Connor, J.E. (1997) Methods for predicting peak discharge of floods caused by failure of natural and constructed earthen dams, *Water Resources Research* **33**, 2337–2348.
27. Wang, X.M. and Pei, X.Y. (1987) The 1786 Kangding-Luding earthquake. Earthquake Research in Sichuan, 1–51 (Supplement Issue, in Chinese).
28. Wang, X.M. and Pei, X.Y. (1998) Major active faults and earthquake-induced cracks in Kangding-Luding area, *Earthquake Research in Sichuan* **1–2**, 46–56.
29. Wang, X.M., Yu, H.S. and Pei, X.Y. (1996) Application and results of statistical hazard data about the 1955 $M$=7.5 Kangding earthquake, *Earthquake Research in Sichuan* **3**, 57–64.
30. Xie, X.D. (1991) A preliminary study of the flood caused by the 1786 Luding-Kangding earthquake, in Guo, Z.J. (Ed.), *Collections of Historic Earthquake Research in China:Part 2*. Seismological Press, Beijing, pp. 169–173.

31. Zheng, Z.M. (1998) Interpretation of the Tiezhuang temple monument associated with the Luding earthquake, *Sichuan Wenwu* **4**, 59–60 (in Chinese).
32. Zhou, R.J., He, Y.L., Huang, Z.Z., Li, X.G. and Yang, T. (2001) The slip rate and strong earthquake recurrence interval on the Qianning-Kangding segment of the Xianshuihe fault zone, *Acta Seismologica Sinica* **23**, 250–261 (in Chinese).
33. Zhu, P.Y. and Li, T. (2001) Flash flooding caused by landslide dam failure. ICIMOD Newsletter No. 38.

# Chapter 14
# La Josefina Landslide Dam and Its Catastrophic Breaching in the Andean Region of Ecuador

**Galo Plaza, Othon Zevallos, and Éric Cadier**

## 1 Introduction

On March 29, 1993, a great landslide in the southern part of the Andean region of Ecuador, dammed the Paute river, creating an artificial lake of approximately 200 $Mm^3$ (Figs. 14.1, 14.2, and 14.3). The landslide of near 20 $Mm^3$, created a natural dam of 100 m height, 300 m width and 1.1 km in length parallel to the river channel. Approximately 1 month after the event, the rupture of the dam caused catastrophic damage along the river, upto 100 km downstream. Nearly 5,000 persons living downstream had to be evacuated.

Direct losses were estimated at 147 million dollars [9]. They were caused by the impact on the environment, flooding of almost 1,000 ha of fertile land, destruction of houses, the road and railway system and a thermoelectric plant [10]. This catastrophe could have put the country into a serious energy crisis, because the Daniel Palacios dam, which produced three quarters of the electric energy consumed by Ecuador, is located 50 km downstream from the slide.

In this work, the main phases of the catastrophe are described from the landslide occurrence until the flow wave created by the breaching [5, 7, 17, 28]. Then we describe the scenarios and compare the different predictive methods and models of earth dam breaching that were used. Finally, different aspects of the management of the crisis are analyzed, taking into account published reports [28].

Most data used or referred to in this work, come from measurements or observations carried out by INECEL (Ecuadorian Electricity Institute), and from reports prepared by the EPN (National Polytechnic School) technical team, both compiled in [6, 28].

G. Plaza (✉)
Escuela Politécnica Nacional, Quito, Ecuador
e-mail: higeodes@server.epn.edu.ec

S.G. Evans et al. (eds.), *Natural and Artificial Rockslide Dams*, Lecture Notes in Earth Sciences 133, DOI 10.1007/978-3-642-04764-0_14, © Springer-Verlag Berlin Heidelberg 2011

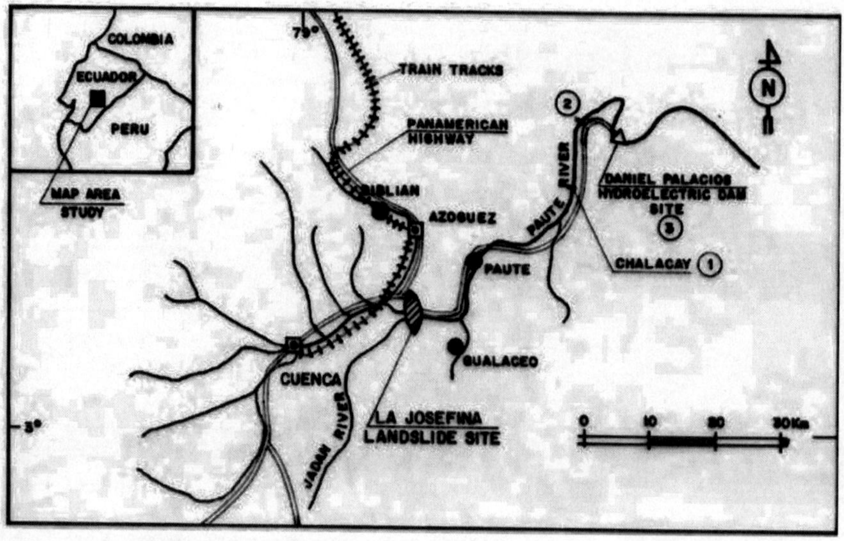


Fig. 14.1 Location and vicinity map of 1993 La Josefina Landslide dam, Ecuador. Dam located at 02°51′05″S/78°50′49″W

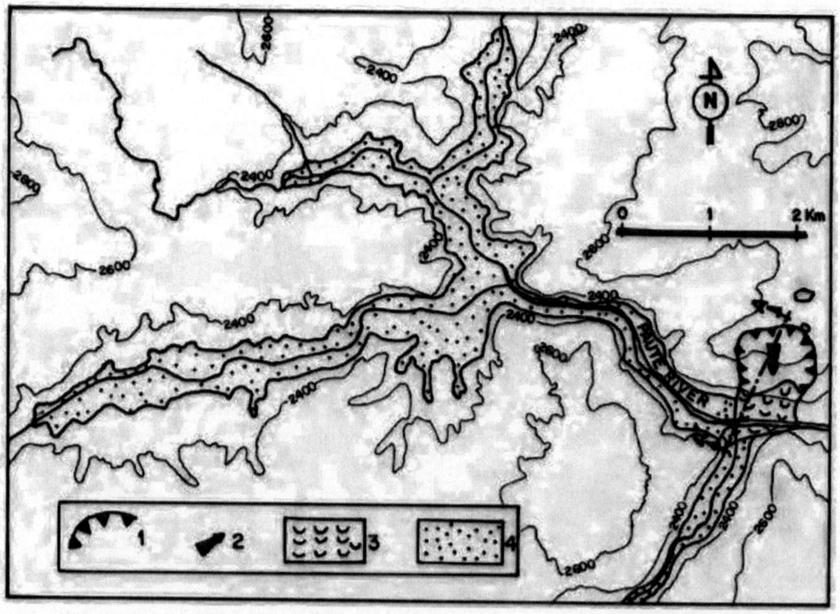


Fig. 14.2 Map of the landslide showing general morphology and the impounded lake. Key to legend: (*1*) Headscarp, (2) Direction of movement, (*3*) Area of landslide debris accumulation, (*4*) Impounded lake at its maximum extent

**Fig. 14.3** View of the landslide deposit and impounded lake from upstream, the day before the rupture. (Photo C. Menéndez). (**A**) Headscarp, (**B**) Direction of movement; (**C**) Partial debris accumulation, (**D**) Excavated channel, (**E**) Impounded lake

## 2 La Josefina Landslide: Dynamics, Causes and Composition

On the 29 March at 2030h, a great portion of the "Parquiloma" hill (locally known as "Cerro Tamuga"), located on the north side of the Paute River, about 20 km northeast of Cuenca, slid into the valley, killing approximately 35 people [9] and destroying the only road that connected the towns of Paute and Gualaceo, as well as the Daniel Palacios dam, with Cuenca and the rest of the country.

The landslide took place in the eastern part of the Inter Andean valley formed by Neogene sedimentary and volcanic rocks, overlain by colluvial deposit [18, 28]. This whole sequence overlies metamorphic basement rocks intruded with microdiorite dikes. The slide, along the direction of movement, was 1,500 m long and 600 m wide. The pre-failure slope of the ground was estimated to be between 25 and 45°, with a elevation difference of 500 m [17, 18].

First estimations of the landslide volume were highly variable; between 20 [18], 36 [2], and 44 million m$^3$ [13]. Accurate analysis carried out after the crisis, through photogrametry, resulted in a volume of 19.4 million m$^3$ [21, 22].

### *2.1 Dynamics of the Movement*

The landslide was composed of two violent successive mass movements [17, 28], (Figures 14.4 and 14.5). The first translational landslide took place along the failure plane located at the bottom of the hillslope and it was covered a short time later by a

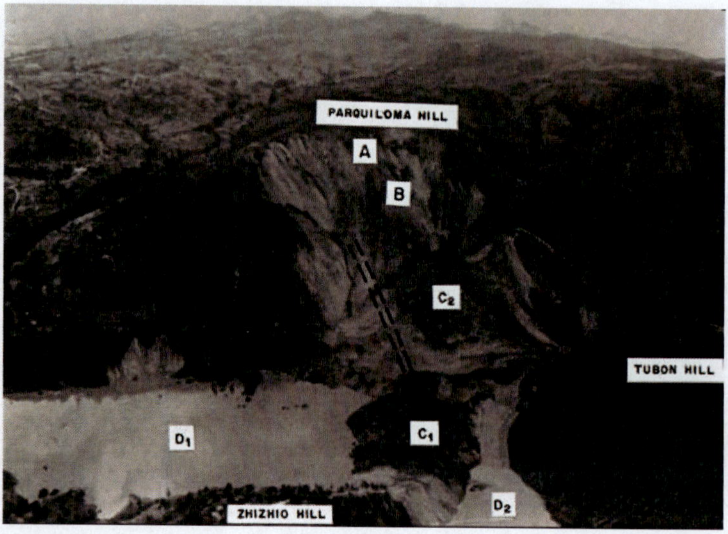


**Fig. 14.4** La Josefina Landslide dam 7 days after formation: (*A*) Headscarp, (*B*) Direction of movement, (*C1*) Debris acumulation of the first slide, (*C2*) Debris accumulation of the second slide, (*D1*) Impounded lake of the Paute River, (*D2*) Impounded lake of the Jadan River

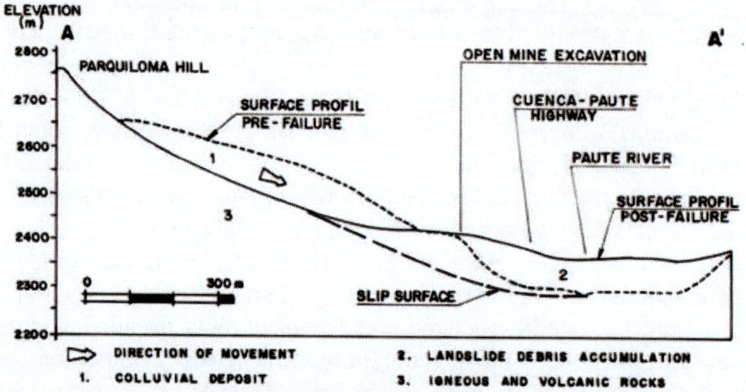


**Fig. 14.5** Cross section of the La Josefina Landslide along the first movement direction (A–A'), showing surface profiles pre-and post-failure (For location see Fig. 14.2)

second landslide. The velocity of the movement must have been very high, because some sections of the road located in the Paute River valley bottom, were found a few 100 m away on the opposite side of the river, testifying to the strong kinetic energy of the movement. In fact, the total vertical displacement of mass movement was 200 m.

The 1993 landslide occurred primarily within fractured igneous and volcanic rocks overlain by an ancient landslide deposit [17]. Microdiorites, ignimbrites, some andesites as well as all of the ancient landslide deposit were mobilized, creating a debris deposit with large size boulders in the western part, and highly fragmented rocks, of all sizes, within a granular and fine matrix, in the eastern part. During the research, two sets of fracture planes, with strikes parallel to the length of the slope and an ancient headscarp related to the ancient landslide deposit were identified. It's believed that the failure surface was formed by the ancient headscarp surface and the fractures in the rocks.

After the two main movements, some rockfalls occurred from the headscarp.

## 2.2 Landslide Factors

Some landslide factors triggered the movement. The hillslope was very steep, formed by tectonized and fractured rocks, overlain by a landslide deposit, testifying to an ancient mass movement with a failure plane. In fact, an analysis of aerial photographs of the slope obtained before March 1993, shows clear morphological evidence of an ancient landslide, whose headscarp had an approximate height of 80 m [18, 22]. The ancient debris accumulation covered an estimated surface area of 45.9 ha but the date of its occurrence is unknown.

On the other hand, the fracture planes in the rocks were steeply dipping toward the valley, contributing to instability. The quarry excavation at the foot of the slope had reached a dangerous depth of 160 m, with an estimated volume excavated of 2.31 million $m^3$ [28]. In addition, the magnitude of the rainfall in the preceding month was twice the average [1, 28].

## 2.3 Landslide Composition

The landslide dam material had a broad grain size distribution, extending between fine sand up to metric-size blocks. Within the deposit three layers were distinguished [28]. The top layer, between 10 and 15 m in thickness, corresponds to fine soil. The intermediate layer, between 20 and 30 m in thickness, contain blocks up to 1 m in diameter within a sandy and gravely matrix. The basal layer, with a thickness of between 45 and 75 m, was composed of large size blocks. Some observations carried out after the dam breaching, about 40 m under the blockage surface, showed that between 25 and 40% (in volume) corresponded to the matrix. The layers in the debris were identified by a seismic refraction survey during the crisis and were confirmed by excavation carried out in the material after the breaching of the dam [28].

# 3 Filling of the Impounded Lake and Upstream Flooding

Water accumulated behind the dam for 33 days, forming a natural lake of 191 million $m^3$. During the infilling, average inflow of the Paute River was 71 $m^3/s$. Maximum and minimum daily values were 131.7 and 28.4 $m^3/s$ [27]. The lake was 77 m deep, 9 km long and flooded an area of 800–1,000 ha [6] (Table 14.1).

**Table 14.1** Landslide dam and impounded lake characteristics

| | Length of dam at bottom (km) | Maximum width (m) | Maximum height (m) | Volume ($10^6$ $m^3$) |
|---|---|---|---|---|
| The landslide dam | 1.1 | 300 | 100 | 19.4 |
| The impouded lake | Volume ($10^6$ $m^3$) | Length of the reservoir (km) | Maximum depth (m) | Flooded area (ha) |
| | 191 | 9.0 | 77 | 800–1,000 |

## 4 The Construction of a Discharge Channel

In order to reduce flooding upstream of the natural dam and reduce catastrophic damage downstream, the Ecuadorian Army Corps of Engineers dug a channel in the right side of the dam, approximately 18 m deep, 6 m wide at the base and 407 m long (Fig. 14.6).

Using up to 23 heavy machines 165,000 $m^3$ of debris were removed during 13 days of continuous excavating [28]. Flow through this channel prevented the flooding of 250 additional hectares of land situated upstream. If the channel had not been dug, the volume of the reservoir would have reached 330 million $m^3$ and the peak discharge at the moment of the breaching would have been close to 30,000 $m^3/s$, with more serious consequences. These results were obtained based on calculations and the test of a physical scale model ran by Zevallos et al. [27].

Previously, some potential solutions were analyzed, which included the construction of a by-pass tunnel, pumping water from the lake and blowing up

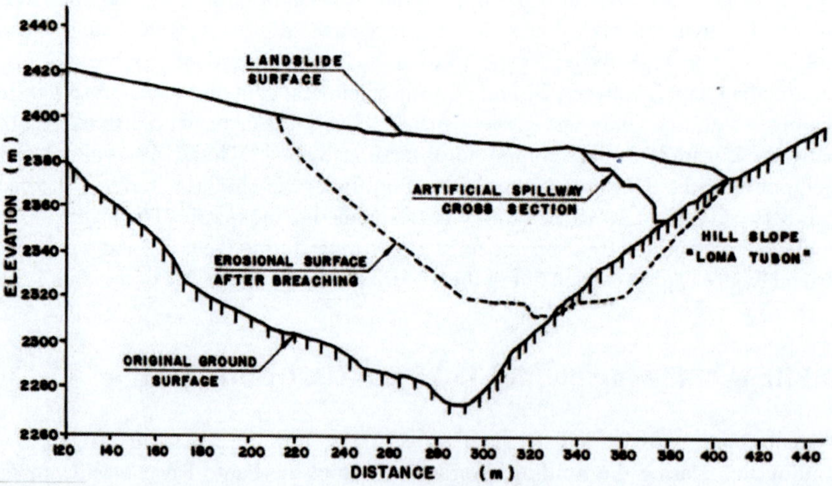


**Fig. 14.6** Cross section of the discharge channel and the erosional surface after breaching. Adapted from [2]

the dam with the use of dynamite. The stabilization and use of the deposit of the landslide in building a permanent dam was also considered.

## 5 Dam Breaching Process

The dam breaching process was difficult to predict; it was not possible to predict exactly when the breaching would begin, nor the duration of the breaching process.

### *5.1 Regressive Erosion Process in the Channel*

Twenty five days (April 23rd) after the landslide occurred water began to flow through the channel. It was estimated that between 1 and 3 days were needed for the erosion process to reach the upstream extreme of the channel, initiating the rupture of the dam. However, 7 days were necessary for this process, because the channel was continuously obstructed by material falling from the slopes of the channel sides.

### *5.2 Rupture of the Dam*

Thirty three days after the landslide occurred the dam breach took place when the water depth in the overflow channel reached 4 m. The previous day, the Ecuadorian Army had fired an anti-tank missile trying to break a large boulder that blocked the channel.

On May 1st at 05h30 in the morning, the discharge was estimated at about 200 $m^3/s$. After this, events happened very quickly and the dam break was a violent process. At 7h15, limnimetric measurements indicated that the level of the lake had began to decrease. Between 9h00 and 10h00, a lowering of the lake level of more than 4 m was registered, which corresponded to an outflow volume of 30 million $m^3$. The estimated maximum flow discharge rate occurred at around 9h40, varying within a range of 8,000 and 10,000 $m^3/s$. Flow velocities ranged between 15 and 20 m/s.

The total outflow volume was 175 million $m^3$. Between 08h00 and 11h00, 75 million $m^3$ were evacuated. As predicted by the models, a residual lake 40 m deep was left, with an approximated volume of 20 million $m^3$. Afterward, the level of this lake was lowered 7 m more by excavation.

Because the rapid down draw of the reservoir, some landslides took place upstream of the dam along the lake shoreline.

## 6 The Downstream Flood Wave Propagation

The flooding reached Paute town, located 20 km downstream from the landslide, in about 1.5 h, and reached the Amaluza Reservoir of de Daniel Palacios Dam in 3 h, as predicted. In the flooded areas, the water reached between 6 and 8 m in depth in the widest part, and between 25 and 30 m in the narrowest reaches. A depth of 24 m was reached at the Chalacay measuring in depth station, located in a very narrow valley downstream [6]. This station is located between the La Josefina landslide and the Daniel Palacios Dam (see Fig. 14.1).

Flooding took place during the day, and there were no casualties, since the evacuation of the inhabitants was previously organized. However, an important sector of Paute town and several kilometers of roads were destroyed, and a number of agricultural fields were swept away. The valley floor morphology was modified by deposits carried by the flood, which reached several meters in thickness. Numerous secondary landslides were induced by the passing flood that left important parts of the surrounding slopes without any support.

In order to prevent the overtopping of the Daniel Palacios Dam, some days before the La Josefina landslide dam rupture, the water level of the reservoir was lowered 31 m, to provide a storage volume of 51 million $m^3$ for the expected flood. For this reason the maximum flow registered through the spillway of the Daniel Palacios Dam was only 4,140 $m^3/s$, totaling 4,400 $m^3/s$ including base discharge. The Daniel Palacios Dam spillway was designed for a peak discharge flow of 7,700 $m^3/s$; thus the double arched concrete dam was not endangered.

Peak inflow into Daniel Palacios Dam of 9,300 $m^3/s$ was calculated with water level measurements carried out during the flooding. From this data a maximum discharge of 9,500 $m^3/s$ was back calculated at La Josefina [28]. Figure 14.7 shows the observed hydrograph at various sites.

In the Paute Hydroelectric Power Plant, two turbines were put to work 30 h after the passing of the flood, when the concentration of suspended sediments was still

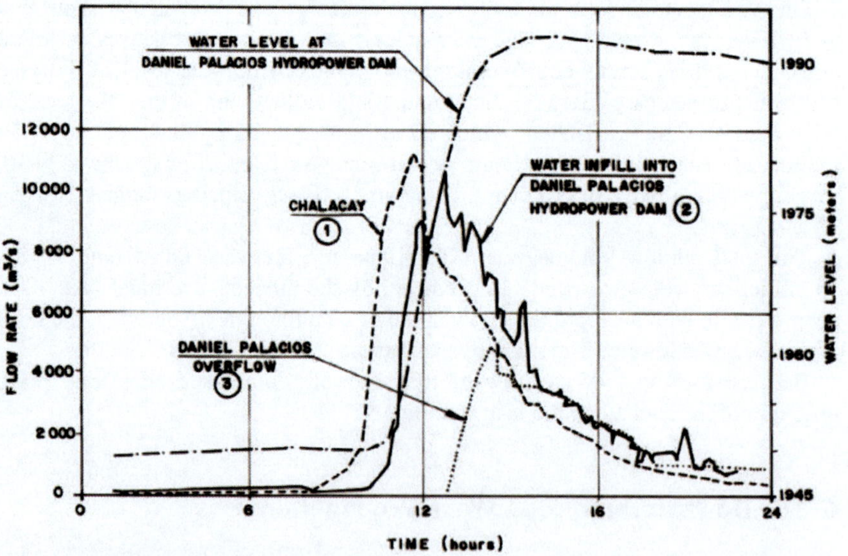


**Fig. 14.7** Flood hydrograph at various sites May 1st, 1993. INECEL data [28] (For location of sites see Fig. 14.1)

very high, which caused damages to the turbines. An estimated 1.2 million $m^3$ of sediments entered into the Amaluza Reservoir of the Daniel Palacios Dam during the flood [6, 28]. The following year, sedimentation in the reservoir suffered an increase of approximately 20% compared to the average of the previous year thus reducing the dam's design life.

## 7 Comparison of Dam Break Methods Used

Research goals at the Josefina landslide dam aimed to prevent the negative effects on the valley downstream and to evaluate certain scenarios, which would allow management of the risk to the hydroelectric Paute Project. The prediction of the flood hydrograph was also absolutely vital for ensuring the safety of the settlements downstream of the landslide dam.

### *7.1 Empirical Methods*

Applying the Schuster and Costa diagram [20] for this particular case, a peak discharge of 10,000 $m^3/s$ was obtained, for a 83 m dam height and 200 million $m^3$ of water, which reflected very closely the actual conditions (Table 14.2). Using the Hagen formula a peak discharge of 28,200 $m^3/s$ was obtained, which was an over estimation [6].

### *7.2 Numerical Modelling*

Natale [15] from the University of Pavia, in cooperation with the University of Cuenca, used an erosion model based upon the equations of continuity and movement of water and sediments. He evaluated five hypotheses and finally considered

**Table 14.2** Results of the different models used to predict the dam breaching peak discharge

| Model | Volume ($10^6$ $m^3$) | Peak discharge at La Josefina ($m^3/s$) |
|---|---|---|
| Empirical diagram due to Schuster and Costa | 200[a] | 10,000 |
| Natale's numerical model | 177 | 1,974 |
| DAMBRK model run by James et al. (Natale hypothesis; most unfavorable case) | | 2,698; 5,479 |
| BREACH model, run by Zevallos et al. (optimistic, intermediate and pessimistic hypothesis) | 75, 100, 117[a] | 7,100, 11,000, 16,500 |
| Physical scale model run by Zevallos et al (test 1, 2 and 3) | 310, 120, 184 | 30,000, 7,900, 16,000 |
| Physical scale model run by Rivero and Marin (test 1 and 2) | 90, 117 | 8,000, 6,500 |

[a]Adopted value.

that the rupture in 72 h constituted the most critical case. During this period, 177 million $m^3$ drained from the lake. According to this author, the peak discharge would reach 1,974 $m^3/s$, 22 h after the initial overflow (Table 14.2).

James et al. [14] in collaboration with the University of Cuenca technical team, ran the DAMBRK model using Natale's most critical case, starting from a layer 25 m thick (D50% = 10 cm, $n = 10\%$) situated upon a second layer with a different grain size distribution (D50% = 70 cm, $n = 50\%$). The initial elevation was 2,358 m, and the river discharge 100 $m^3/s$. The peak discharge 26 h after the initial flow would be 2,698 $m^3/s$.

James et al. [14] also ran the DAMBRK model with other parameters: a rupture time of 15 h, a final breach of 100 m, and with a lateral slope of 1:1. The resulting discharge in the most unfavorable case was 5,479 $m^3/s$. This model was also used to predict the peak flow flood levels downstream of the landslide site. For this purpose, the river channel was divided into 34 sections. As a security measure, the most pessimistic scenario with flooding levels between 7.8 and 19.3 m was adopted in order to design the evacuation plan [2].

Zevallos et al. [25–28] from the EPN technical team, in cooperation with INECEL, used the BREACH model in order to simulate the rupture of the dam. They adopted the following conditions: vertical erosion of the dam from 12 to 25 m; initial river discharge of 80 $m^3/s$; maximum breach width of 50 m at the base of the breach; river slope of 0.017 and a Manning's roughness coefficient of 0.05. For the flood wave routing in the river, they used the DAMBRK model in its dynamic wave version. For this purpose, the river channel was divided in 30 sections. For the flood routing process in the Amaluza reservoir, the HIDRO1 model developed by EPN was used.

The results of these simulations were (Table 14.2): a peak discharge at La Josefina landslide dam between 7,100 and 16,500 $m^3/s$; incoming peak discharge into the Amaluza reservoir between 5,500 and 12,000 $m^3/s$; peak overflow discharge leaving the Amaluza reservoir between 0 and 3,033 $m^3/s$; maximum depth in flooded zones between 8 and 16 m; filling time between 6 and 8 h, and an estimated arrival time to Paute town and the Amaluza reservoir of 1.5 and 3 h, respectively.

## 7.3 Physical Modelling

In order to improve the accuracy regarding the rupture of the landslide dam, Zevallos et al. [27, 28], tested an 1:200 undistorted physical scale model, using the Froude similitude. The material grain size was simulated with linear scale, using a specific weight similar to the prototype. This ensures that the Shields mobility parameter is the same in model and prototype, thus guaranteeing the reproduction of the erosion phenomenon.

Type, characteristics and distribution of material, in surface and depth, were the most important parameters chosen for modelling the rupture, but at the same time they were sources of discrepancies and uncertainty. The average values adopted for the modelling were the following:

Specific weight of the material: 2.76 t/m$^3$
Grain size distribution: D50% : 100 mm
D90/D30 : 30
Porosity: 15% (10–50%)
Friction angle: 45°
Cohesion: 0.5 t/m$^2$
Angularity: high.

Due to uncertainties with regards to the dam's composition three tests were completed [25]. The first test was run without an overflow channel and the others by lowering the dam crest with a dug channel. The first test assumed an homogeneous deposit (10 cm diameter fragments), resulting in 310 million m$^3$ of total volume discharge and a maximum discharge of 30,000 m$^3$/s. The second assumed a 20 m thick upper layer with a mean fragment diameter of 25 cm while underneath this material was a layer of large blocks. The third test was run assuming a homogenous deposit composed of fragments 10 cm in diameter. The maximum discharge registered in the second case was 7,900 m$^3$/s, and in the third case was 16,000 m$^3$/s, evacuating 120 million m$^3$ and 184 million m$^3$ of water in each case [28] (Table 14.2). Figures 14.8 and 14.9 allow for a comparison of the scale model and actual conditions.

Rivero and Marin [19, 28] ran two tests into a 1:150 scale model, considering two different levels at the bottom of the dug channel. Using a mean fragment diameter of

**Fig. 14.8** Actual erosion during the breaching of La Josefina landslide dam, May 1st, 1993; view downstream at 9 h 30 min, 10 min before the maximum outflow discharge

**Fig. 14.9** Erosion in the physical scale model, view upstream, tested by EPN technical team [28]

15 cm (D50% = 15 cm) for the landslide dam, the following results were obtained. For the first case the maximum discharge was 8,000 $m^3/s$, with a stored volume of 126 million $m^3$, and a final volume of 36 million $m^3$. For the second case the maximum discharge was 6,500 $m^3/s$, the stored volume 150 million $m^3$, and a final volume of 33 million $m^3$.

These tests were carried out before the rupture of the dam. The differences concerning the results demonstrate the complexity of the subject.

Results of the different methods and models (empirical methods, numerical modelling, and physical modelling) used to calculate the dam breaching peak discharge of La Josefina landslide dam are summarised in Table 14.2.

## 7.4 Some Aspect About Dam Breaking Methods Used

These aspects were pointed out by Cadier et al. [6].

(a) The graph proposed by Schuster and Costa [20], which was also very easy to use, gave a good approximation to peak discharge.

(b) The combination of physical scale models and mathematical models is recommended in order to simulate the rupture of the dam and to determine its possible effects downstream. Nevertheless, its use is delicate given that the results obtained are very sensitive to the parameters used, which are not always well known. Particularly, it is vital to know accurately the grain size distribution of the material in order to adequately simulate the rupture of the dam.

(c) The DAMBRK model was adequate to simulate the flood wave routing, but not the rupture of the dam itself, given that it is necessary to assume certain parameters such as the height and width of the breach and rupture time, which are unknown factors. Additionally, since the model does not take into account sediment transport, it did not reproduce accurately water level in the first few kilometers downstream of the artificial dam, where the riverbed has been elevated several meters due to the deposit of material originating in the rupture.

(d) The BREACH model is inadequate to simulate the dam breaking process, given that it requires that the final level of equilibrium of the dam be set, as well as the final width of the breach, which were unknown factors. Additionally, this model exhibits a lack of stability regarding variation of parameters such as the gradient of the side slope and the width of the dam. In this model, the average diameter of the material that makes up the dam has no influence upon the rupture process, which does not reflect actual conditions.

(e) The difficulties found when trying to make accurate predictions showed the need to improve the knowledge regarding the following aspects:

- Specific characteristics of the landslide damming site: river discharge, dam height, volume stored, elevation-volume curve, dimensions of the dam, dimensions of the evacuation channel, the type of material that makes up the dam.
- Dam erosion process: it is necessary to study the interaction between water flow and sediments under unsteady torrential flow conditions, the size of the rupture gap, sedimentation downstream of the site, the stabilization height of the dam and in consequence, the resulting flood hydrograph.

## 8 Lessons and Aspects Concerning Crisis Management

### *8.1 Real Perception of the Risk*

The instability of La Josefina was known before the 1993 landslide. Some large cracks on the top of the hillslope and continuous displacement along the road located at the bottom of the slope were clear preliminary signs of movement [18, 28]. Two years before the catastrophe, a field inspection was carried out by the local authorities accompanied by an engineer from the Ecuadorian Geological Survey, which was reported in the press ("El Tiempo", 03/19/91). The press article pointed out the potential danger of the site, describing an imminent landslide. Before this time, several reports by technicians from the Ecuadorian Geological Survey identified cracks and other instability features, and recommended the cessation of mining activities located at the foot of the slope [28].

Therefore, it was clear that there was technical information and geological knowledge about the potential catastrophe. It also indicates the lack of real consideration by the decision-makers and their institutions, about the hazard pointed out by technical teams.

## 8.2 Institutional Response and Organization for Crisis Management

Prevention and management of these catastrophes was normally handled by the "Junta Provincial de Defensa Civil" with cooperation of technical departments of various institutions (Army Forces, INECEL, CREA, Ecuadorian Geological Survey, etc.), as a regional unit of the National Civil Defense System [23, 24].

Considering the importance of the event, the President of Ecuador led the crisis management personally. At the time, two special organizations were created for this purpose [27]: (1) the "Crisis Management Committee" (CMC) under the responsibility of the Army, and (2) the "Emergency Works on the Paute River Watershed and Tributaries Programming Council" (CPO), under the responsibility of local civil institutions. The last organization coordinated the relationships with national and international aid organizations, like the Red Cross, the Church, the USAID Agency, the European Union and Non Governmental Organizations. Also, a coordinated international technical assistance team distributed aid for the population and organized the rebuilding of the affected areas.

In Ecuador, the creation of temporary advisory committees for disaster management is normal practice. They must be able to quickly organize and manage unexpected situations. Most of the time, they have to relegate and even take over the tasks of the National Civil Defense System. This situation has generated recurrent power struggles, ambiguous decisions and lack of effective response. On the other hand, this produces a delay in order to develop an optimal National Disaster Prevention and Management System.

## 8.3 Real-Time Crisis Management

Technical aspects during the crisis were managed by several independent working groups, lacking coordination between them. Each team did not have the same basic data, therefore the results were very different in terms of the peak discharge at the time of the dam breaking and the filling time of the reservoir. A group defended the position that the dam would not break, or that the rupture would take place within a period ranging between 3 and 15 days [16]. In spite of the difference in the results, all of them agreed upon the severity of the consequences of the landslide dam rupture.

This situation was generated by the lack of preparedness from the governmental organizations and from the advisory committee. Therefore, in the future the government should guarantee an organizational structure that manages these kinds of crisis.

In the case of a breaching landslide dam it must be considered that most landslide damming structures fail in a very few days. It is necessary to immediately identify the hazard areas and organize the evacuations plan and mitigations measures. For this purpose, there must be an available working group with optimal support.

## 8.4 Prevention Polices

In the Andean region of Ecuador the damming causing by the La Josefina landslide was the second phenomenon of this type. It occurred 3 years after the Pisque River event [3], close to Quito, capital of Ecuador. This type of hazard is common in the region and has been identified even in historical records [28]. After the crisis, a pilot project carried out in the Paute River watershed demonstrated the high risk of the zone due to the population density, and the magnitude and frequency of the phenomena [4].

On the other hand, several natural hazards occurring in the last decade had produced very important losses, causing negative effects on Ecuador's socioeconomic development. Therefore, the government should be aware that it is better to invest in prevention, mitigation and planning, rather than in reconstruction after the event. The current strategy for adequate risk management is to incorporate the risk parameter in investment and regional development projects.

## 8.5 Post-Crisis Activities

At this stage, there were a lot of participants. Ranging from Ecuadorian governmental institutions, various NGOs and even international organizations, whose help ranged from small, on-the-spot donations to the implementation of far reaching projects. In this way, an important reconstruction project with an estimated 14 million Euros was executed with the European Union. At the same time, a multidisciplinary pilot project was executed by Ecuadorian institutions and foreign organizations [4]. One of the goals of this project was to carry out the landslide hazard assessment of the Paute river watershed (3,600 km$^2$).

Post-crisis activities required an additional national budget, which was used for health, road system reconstruction, production, electricity, agriculture and housing projects. In the case of this Ecuadorian Andean region, the Catholic Church was a very important social group which helped to complete successful reconstruction.

### *8.6 Scientific Interest of This Catastrophe*

This event demonstrated that the knowledge concerning the process of a breaking landslide dam is still limited. The available models for calculating dam breaks, such as BREACH, or DAMBRK [12], are developed upon the basis of great simplifications of the dam erosion process. In consequence, the results obtained are not always reliable. Physical models of dam breaking, have been applied in some other cases such as the Sarras dam [11], and the 1974 Mayunmarca landslide in Perú [8].

During the Josefina case, it was observed that the breaching process stopped and created a residual dam. It would seem that the high percent of great-size blocks created a channel-bed armoring, stopping the erosive process. Therefore the problem resides in knowing the velocity of the erosive process, and when and at what height the stabilization of the residual dam will take place.

After the crisis, several scientific and technical forums about the analysis and risk management of La Josefina took place and were compiled and published. Among them, the book *Sin Plazo para la Esperanza*, compiled by Zevallos et al. [28], gathered the most important information concerning the disaster and its multiple features.

## 9 Conclusions

La Josefina Landslide and its effects demonstrated the importance of this kind of natural hazard for regional and local planning, in order to protect human lives, material goods and infrastructure.

This event also illustrated the high vulnerability to landslides in the Inter Andean Ecuadorian region, a hazard frequently increased by human actions. In these conditions and with abundant landslides, this area has a high risk and many potential disaster situations.

The need for an appropriate organization is evident as well as the preparedness of all social actors involved, whose objective must be to improve the risk management and reduce the impact of landslide hazards and damming effects.

By using available tools it is possible to obtain reasonable scenarios of dam rupture for crisis management. The mathematical and physical models must be used to predict the breaching process and identify hazard areas, but some variables cannot always be determined with accuracy during the crisis period.

La Josefina Landslide showed preliminary signs of instability which had not been taken in account from a real risk perception. So, it is necessary to continually assess landslide risk in order to prevent and mitigate this kind of natural disaster.

**Acknowledgements** We want to express our gratitude to Patricia Mothes and Remigio Galárraga for their contribution in improving the final English version of the manuscript. We want also to thank Giovanni Taco and O. Zevallos P. for their help in the typing of this essay, and Victor López for his contributions to the drawing of the graphs.

## References


1. Acosta, T. and Salazar, J. (1993) Hipótesis sobre el origen del deslizamiento de la Josefina y geomorfología del área de influencia, in Escuela Politécnica Nacional (eds.), *Memorias del Coloquio científico El deslizamiento de la Josefina*, Quito.
2. Aguilera, E. and Romo, M. (1993) Deslizamiento La Josefina: Planificación y decisión para el manejo de la crisis, in Escuela Politécnica Nacional (eds.), *Memorias del Coloquio científico El deslizamiento de la Josefina.*, Quito.
3. Asanza, M., Plaza-Nieto, G., Yepes, H., Shuster, R.L. and Ribadeneira, S. (1992) Landslide blockage of the Pisque River, northern Ecuador, *Proc. 6th International Symp. on Landslides*, Christchurch, New Zealand, 1229–1234.
4. Basabe, P., Almeida, E., Ramón, P., Zeas, R. and Alvarez, L. (1996) Avances en la prevención de desastres naturales en la cuenca del río Paute, Ecuador, *Bull. Ints. fr. études andines* **25**(3), 443–459.
5. Cadier, E., Zevallos, O. and Vacca, A. (1994) Le Glissement de terrain et les inondations catastrophiques de la Josefina en Équateur, in *Actes des 23émes journés de lhydraulique de la SHF*, Nimes, pp. 355–358.
6. Cadier, E., Zevallos, O. and Basabe, P. (1996) Le glissement de terrain et les inondations catastrophiques de la Josefina en Équaterur, *Bull. Ints. fr. études andines.* **25**(3), 421–441.
7. Canuti, P., Frasoni, A. and Natale, L. (1994) La Josefina rockslide and rio Paute landslide dam, Ecuador. Failure of the rio Paute landslide dam, *Landslide News* **8**, 6–7.
8. Chang, L. and Alva, J. (1991) Deslizamientos y Aluviones en el Perú, CISMID, Universidad Nacional de Ingeniería, Lima.
9. C.R.E.A. (1993) Evaluación preliminar de los daños por el deslizamiento de la Josefina.
10. Cruz, M. (1993) Evaluación de daños. Defensa Civil., in Escuela Politécnica Nacional (eds.), *Memorias del Coloquio científico El deslizamiento de la Josefina.*, Quito.
11. Escande, L., Nougaro, J., Castex, L. and Barthet (1961) Influence de quelques paramétres sur une onde de crue subite á l'aval d'un barrage. *La Houille Blanche*, No. 5, pp. 565–575.
12. Fread, D. (1984) *DMBRK: The NWS Dam-Break Flood Forecasting Model.* National Weather Services, Office of Hydrology, Silver Spring, MD.
13. James, M. (1993) Análisis de experiencias de roturas de presas para preparación de un plan de emergencia. in Escuela Politécnica Nacional (eds.), *Memorias del Coloquio científico El deslizamiento de la Josefina*, Quito.
14. James, M., Galarza, L., Cisneros, F., Cordero, P., de Brieve, B., Abril, B. and Martínez, F. (1993) Informe de resultados de los modelos matemáticos.
15. Natale, L. (1993) Reporte técnico de la misión italiana sobre simulación de la erosión en la presa sector la Josefina.
16. Penille, I. and Jácome, A. (1993) Informe. Deslizamiento Josefina ciudad de Cuenca-Ecuador, abril.
17. Plaza, G. and Zevallos, O. (1994) La Josefina rockslide and rio Paute landslide dam, Ecuador. The la Josefina rockslide, *Landslide News* **8**, 4–6.
18. Plaza, G. and Egüez, A. (1993) Consideraciones geológico-geotécnicas sobre el deslizamiento de la Josefina, in Escuela Politécnica Nacional (eds.), *Memorias del Coloquio científico El deslizamiento de la Josefina.*
19. Rivero, J. and Marin, L. (1993) Breve descripción del comportamiento del modelo hidráulico de la presa la Josefina, Guayaquil, 16 de abril.
20. Schuster, R. and Costa, J. (1986) A perspective on landslide dams: Processes, risk, and mitigation, *Geotechnical Special Publication ASCE* **3**, 1–20.
21. Sevilla, J. (1993) El deslizamiento La Josefina. Consideraciones geológicas, in Escuela Politécnica Nacional (eds.), *Memorias del Coloquio científico El deslizamiento de la Josefina.*, Quito.

22. Sevilla, J. (1994) The Josefina landslide and its implications in the electrical service for the Republic of Ecuador, in *Proceedings 7th International IAEG Congress*. Balkema, Rotterdam, pp. 1801–1810.
23. Zeas, R. (1993) El deslizamiento de la Josefina, tragedia nacional, in Escuela Politécnica Nacional (eds.), *Memorias del Coloquio científico El deslizamiento de la Josefina.*, Quito.
24. Zeas, R. (1994) El deslizamiento de la Josefina, tragedia nacional. Universidad de Cuenca, Instituto de Investigaciones técnicas.
25. Zevallos, O. (1993) El deslizamiento de La Josefina – Evaluación Hidráulica, in Escuela Politécnica Nacional (eds.), *Memorias del Coloquio científico El deslizamiento de la Josefina*, Quito.
26. Zevallos, O. (1994) Expériences de modélisation physique et mathématique pour la prevision de rupture de barrages, in Actes des 23émes journées de l'hydraulique de la SHF, Nimes, pp. 261–267.
27. Zevallos, O., Bastidas, L. and Valdospinos, N. (1993) Estudio del proceso de ruptura del embalse La Josefina río paute, Quito-Ecuador. Escuela Politécnica Nacional, abril 20.
28. Zevallos, O., Fernández, M., Plaza, G. and Klinkicht, S. (compilers) (1996) *Sin plazo para la esperanza.* Reporte sobre el desastre de la Josefina, Ecuador, 1993, Escuela Politécnica Nacional, Quito, 348 p.

# Chapter 15
# The Flims Rockslide Dam

**Andreas von Poschinger**

## 1 The Flims Rockslide Site

The Flims rockslide in the Swiss Alps (Fig. 15.1) is the largest landslide deposit known in the Alps and one of the largest world wide. Its volume has been estimated [16] to be 8 $km^3$. It covers a surface of about 52 $km^2$. Also larger volumes have been assumed; 15 $km^3$ by Heim [6], up to 13 $km^3$ by Abele [1] and 12 $km^3$ by Pollet [13]. Its long history of investigations was summarised by Poschinger et al. [16]. Although the huge dimensions impede its study, the deep level of erosion by the Vorderrhein and the Rabiusa rivers has revealed many good outcrops enabling quite easy access and making this site one of the most important for the study of large rockslides. Furthermore, it has not seen important changes in its morphology; Blumenthal [2] pointed out that the surface of the slide mass has been modified very little by erosion. Only along the Vorderrhein valley and its main tributaries has strong incision taken place. In almost all other places the landscape represents that of the early post event state. There was some settling of the freshly deposited masses, but due to its compact structure probably not more than some few meters. Only right along the linear erosion forms of the gorges have some funnel-shaped sinkholes formed. This excellent morphological preservation of the deposits gives many hints to its generation.

In recent years, new impetus has come into the study of the Flims rockslide [17]. Some studies were made for the subsurface investigation for a road tunnel, diverting traffic around the village of Flims. Furthermore, in the last few years several teams of Swiss, French and German scientists have revealed new aspects of the event [14–16, 18, 20]. One important aspect was the dating of the rockslide, thought to be late glacial until then [1, 6, 7, 10], to about 8,200–8,300 $^{14}C$ yr BP (uncalibrated) by the author [15, 16]. Early Holocene ages are also reported by [5, 8]. This means the slide occurred during a warm climate (Boreal) without any direct glacial influence as a trigger.

A. von Poschinger (✉)
Bavarian Environment Agency, D-80696 München, Germany
e-mail: Andreas.Poschinger@lfu.bayern.de

S.G. Evans et al. (eds.), *Natural and Artificial Rockslide Dams*, Lecture Notes in Earth Sciences 133, DOI 10.1007/978-3-642-04764-0_15, © Springer-Verlag Berlin Heidelberg 2011

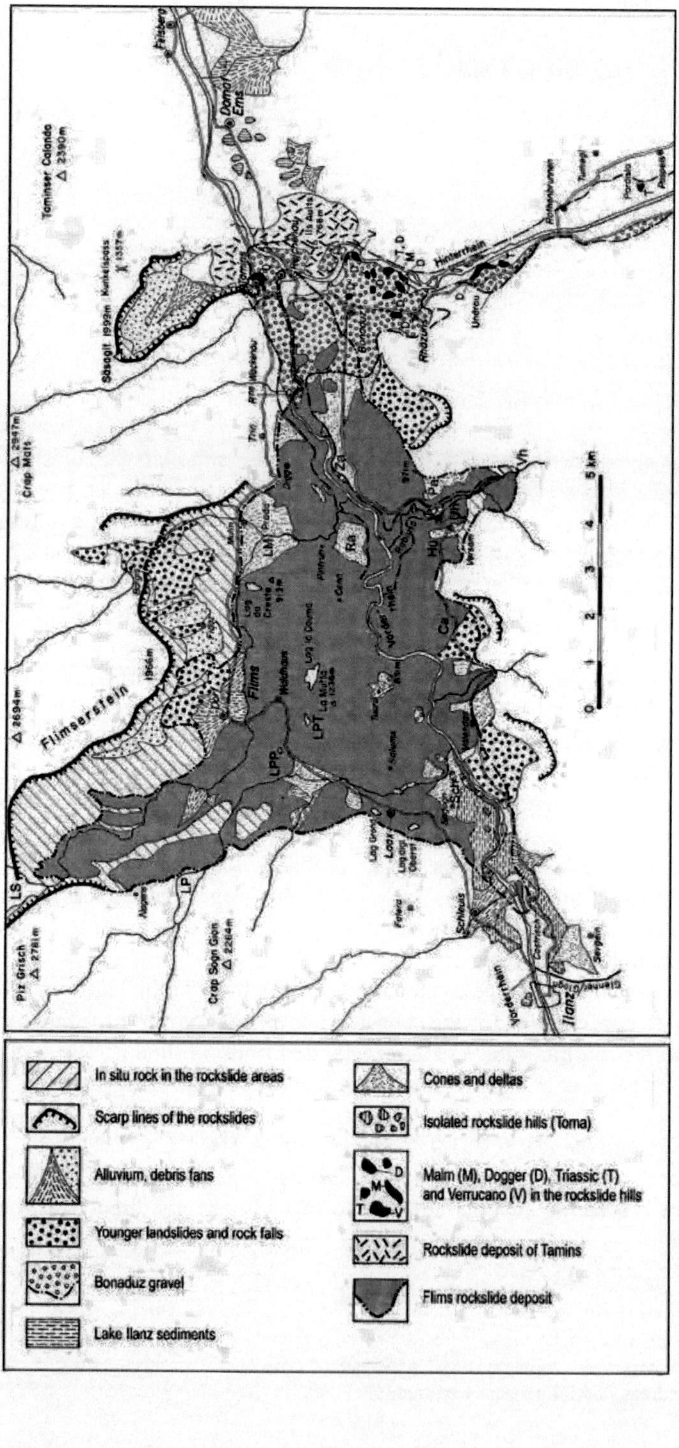
Flimserstein
Flims
Ilanz
In situ rock in the rockslide areas
Scarp lines of the rockslides
Alluvium, debris fans
Younger landslides and rock falls
Bonaduz gravel
Lake Ilanz sediments
Cones and deltas
Isolated rockslide hills (Toma)
Malm (M), Dogger (D), Triassic (T) and Verrucano (V) in the rockslide hills
Rockslide deposit of Tamins
Flims rockslide deposit

## 2 The Rockslide and Its Lakes

### *2.1 The Rockslide Event*

It must be assumed that the rockslide happened "in one great stroke", as already ascertained by Heim [6]. The sedimentary layering of the Jurassic and Cretaceous limestone is parallel to the slope and dipping southward at an angle of about 25–35°. The whole Flimserstein mountain, the source area of the rockslide, is not built up by a simple sequence of strata, but is tectonically shaped by several parallel folds and imbricate thrusts [10]. Within this, the general main structure is striking parallel to the slope. The limestone beds near the sliding surface are separated by fine synsedimentary clay-rich layers, assumed to be responsible for the slope failure. The whole sediment sequence has suffered very low-grade metamorphism.

During the slide event, the moving rock mass did not spread out far to both sides, but kept together its main body. So the movement was "en bloc" and not turbulent or flow-like. Accordingly, the term "sturzstrom", implying according to Heim [7] a flow structure, cannot be applied. Due to the en bloc movement the primary rock structure is still preserved in many places. Nevertheless, the rock is entirely crushed. The degree of fracturing differs from place to place.

The undulating form of the slide surface must have caused a first destruction of the rigid limestone block during the motion. Also the existence of steps, higher than 50 m, within the slide surface must have been responsible for further internal destruction of the sliding mass. The maximum deformation however happened during the impact of the rock mass with the opposite slope, a fact that recently has been described by Pollet [13]. Hence, the resulting facies differs according to the local stress situation attained during the event.

In many places a small-scale three-dimensional jigsaw has been produced, still preserving the original structure (Fig. 15.2). This general structure is cross-cut, more or less densely, by deformation planes and zones. Along these zones the limestone has been crushed to fine gouge. In some places these planes are sub-parallel to the former sedimentary layering as indicated by Pollet [13]. It seems that even more important are ramp-like deformation planes, cross-cutting the bedding and mostly dipping northward.

Only the upper portions of the rockslide mass show the typical facies of a rockslide, i.e. a chaotic blocky material with finer grained matrix and with reduced compactness. Pollet [13] and Wassmer et al. [20] called it "granular facies". This facies is attributed to the lack of confinement of the material and by the

**Fig. 15.1** Site map (previous page) and legend. In the map the following sites are indicated by letters: Ca, Carnifels; Hü, Hüschera; Pa, Parstogn; Ra, Ransun; Sch, Schiedberg; Uh, Unterhof; Vh, Vorderhof; Za, Zault. The lakes are: LM, Lag Mulin; LP, Lag Plaun; LPP, Lag Prau Pulta; LPT, Lag Prau Tuleritg; LS, Lag Segnes Sut

**Fig. 15.2** Rockslide material in "three-dimensional jigsaw-facies" in the Rabiusa gorge. The direction of movement was from *right* to *left*. At the *right* margin sub-horizontal former bedding is visible (parallel to the *white arrow*). Important deformation plains cross-cut it and follows upward ramps (*white line*)

disintegration of the top of the sliding mass. Also an effect of relaxation following the preceding shock wave is probable. The thickness of this facies is quite variable; in the outcrops in the frontal part it rarely exceeds 10 m.

The coarse grained facies is clearly distinguished from the compacted, dense rockslide masses and especially from the "jigsaw-facies". The latter shows a broad spectrum of the degree of crushing with all transitions.

With few exceptions the slide mass has not been cemented and so cannot strictly be called a "breccia". The rare exceptions are either matrix-supported breccias or grain-supported breccias. Real matrix-supported breccias occur within the internal shear zones. Rockslide fragments up to some centimetres are embedded within the finest detrital rockslide material. They are to be found mainly at the front and at the bottom of the slide mass. Grain-supported breccias are found on top of the slide mass. They can be rather coarse and not only include rockslide components, but sometimes also fluvial well-rounded pebbles. Their cementation is due to post-event processes.

The upper coarse grained facies has very high permeability and due to its high content of unconsolidated fines, it is easy to erode on the surface, as well as by subsurface erosion in the underground. More complicated is the estimation of the stability of the "jigsaw-facies". In any degree of fracturing it is completely dissected by dense joints. The permeability of this facies may be restricted by two facts. First, the gouge along the deformation planes is very fine and may act as an aquitard. Secondly, the particles have a close "grain to grain" contact with exactly fitting boundaries, deposited under pressure. So, the space between the particles is restricted. Nevertheless, under water pressure, the jigsaw-facies is, due to its high degree of fracturing, quite permeable. But it is very important, as also pointed out by Davies and McSaveney [4], that the special structure impedes significant internal erosion. This material is not easy to erode and especially for subsurface water there is almost no possibility to transport significant amounts of solids.

## *2.2 The Lakes*

### 2.2.1 Lake Ilanz

The Flims rockslide dammed two large lakes and several small lateral ponds [9]. The biggest lake was that of Ilanz. The maximum level of that lake is a crucial point of discussion. A level of about 820 m a.s.l. is demonstrated by lake sediments (Fig. 15.3), morphological traces and by delta sediments (Fig. 15.4). This evidence marks the level at which the lake existed for some time. As the floodplain sediments also reach to that level, a first breach must have affected a higher level and the lake must have been dammed at that higher level for only a short time without leaving any clear traces. The exact position of that maximum lake level has not been established

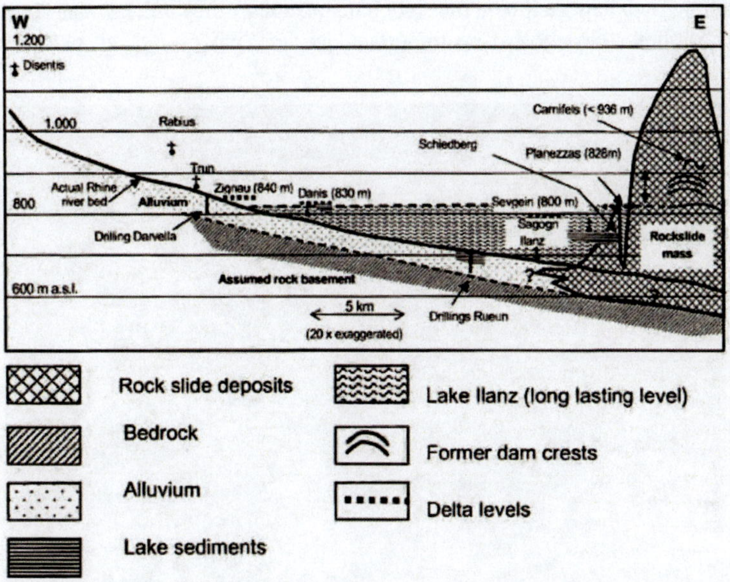


**Fig. 15.3** Section along Lake Ilanz. The distribution of rockslide material under the Rhine river bed is assumed only. The maximum lake level did not exceed the Carnifels pass at 936 m a.s.l. and probably reached 870 to 900 m only. The attribution of the Zignau delta to Lake Ilanz is in doubt

**Fig. 15.4** Sediments of Lake Ilanz near Schiedberg. The clearly and regularly layered silts and sands show a horizontal bedding surface. The origin of the bedding might have been seasonal changes, but, also, repeated single heavy rain events

and can only be assumed. It is evident that it was significantly higher than 820 m a.s.l., because the large flood plain must have been created by a large amount of water. It also obvious that the lake level did not exceed 936 m a.s.l. At that elevation, a morphological depression on the right bank near Carnifels – Erlacresta (Figs. 15.1 and 15.5) shows clearly that no important overflow has ever taken place. So, the

**Fig. 15.5** Delta sediments of the Laaxerbach at Planezzas south of Laax with topset and foreset beds. The top of the foreset beds has an elevation of about 820 m

maximum lake level was somewhere between 820 and 936 m a.s.l. with a higher probability in the range of 850–920 m. For levels lower than 850 m probably the amount of water to be released was too small to create the large flood plain. As the lake surface approached the level of 936 m the risk of seepage and a breach in that area increased. As no signs of such processes are visible, the level probably did not get close to the level of 936 m.

Between 820 and 936 m a.s.l. along the flanks of the slopes only indistinct traces can be found, but cannot be attributed without doubt to either the Lake Ilanz phase or to earlier, late glacial processes. The lack of any clear morphological hints or of sediments higher than 820 m indicates that the first breach occurred very soon after the rockslide event, probably in connection with the first overtopping.

The filling of the lake up to 820 m, considering the figures for the current mean discharge of the Vorderrhein and the rough volume taken from the 1:25,000 topographic map is calculated to take about 13 months [16]. For a further filling up to 870 or even to 920 m, neglecting seepage and in contrast to older estimates indicated by Poschinger et al. [16], a filling time of 2–3 years can be assumed.

The dimensions for Lake Ilanz for the different levels can be roughly estimated from the 1:25,000 topographic map. At the level of 820 m it had a length of about 23 km and a volume of about 1.5 $km^3$. Assuming a level of 870 m the length is 25 km. At 900 m a.s.l., the length is 27 km and the volume about 2.8 $km^3$. The surface area of the lake varies with the different water levels between 24 and 33 $km^2$. These rough figures are to be checked by DTM/GIS calculations in the near future.

In the literature there has been much speculation about the persistence of Lake Ilanz and especially about its break out. The typical lake sediments beneath the present Vorderrhein river-bed near Rueun, about 5 km west of Ilanz that are right in the centre of the Lake Ilanz basin, have a thickness of about 20 m. But also on the shoulders of the valley, near the documented lake level of about 820 m a.s.l., important outcrops of lake sediments are preserved. One site, with about 18 m of well-bedded lake sediments, is located near the Schiedberg ruin, south of the village of Sagogn (Figs. 15.1, 15.3 and 15.5). The layering is almost horizontal and indicates that these sediments are not only lateral remnants of a partial filling, but that the basin was perhaps even completely filled by those sediments.

Thus, the lake must have persisted at this level for many years. The origin of the varve-like layers is not clear. It may be due to yearly seasonal changes but it also can be due to episodic sediment inputs following heavy rainfall events in the catchment. Taking a mean thickness of each layer of about 2 cm, the sequence at Schiedberg consists of about 900 layers. This may indicate a duration of about 900 years. Even in the case of rainfall induced layers it suggests a period of sedimentation of several hundreds of years. And these figures refer only to the 18 m of sedimentation at Schiedberg, not for the probable filling of the whole lake, comprising more than 100 m.

Taking the large volume of the basin filled by sediments, a time span of 1,000 or more years is reasonable. As already mentioned above, these sediments are the only evidence of the longer lasting lake level at 820 m. An important flood after a breach at a higher level is documented by the morphology of the flood plain, described later on.

### 2.2.2 Lake Versam

Lake Versam and Lake Ilanz were dammed by the same event but filled and drained independently. The smaller Lake Versam was dammed to a level of about 870–880 m a.s.l. and had a length of about 5 km (Fig. 15.6). The steep gradient of the Rabiusa river and the steepness of its slopes are responsible for the comparatively small volume of the basin. As the Rabiusa valley has been strongly eroded, no assumption about the previous volume of the basin is possible. Lake sediments are found near Versam-Unterhof (Fig. 15.1) at an elevation of 860 m and at Parstogn at 870 m a.s.l. On the right bank a slightly inclined shoulder is visible, climbing southward up to about 900 m near Vorderhof. This shoulder is interpreted to be the former delta level. Furthermore, well-bedded laminated silty lake sediments that can be attributed to Lake Versam are found at the footpath from Parstogn down to the Rabiusa bridge at 850 m a.s.l. The lake level did not exceed the elevation of 908 m, because the morphological pass leading to the depression of Hüschera (Fig. 15.1) did not experience any major discharge. According to the gravel and sand layers found at Parstogn, Lake Versam was filled up completely with sediments before the incision of the gorge started.

### 2.2.3 Small Lakes

Several small lakes were created due to lateral damming by the rockslide mass (Fig. 15.1). Lac Grond and Lac digl Oberst still exist, Lac Plaun, Lac Segnes Sut and Lac Mulin have since been filled up with sediments. The basin of Hüschera near Versam was drained through the permeable rockslide mass and was not filled by

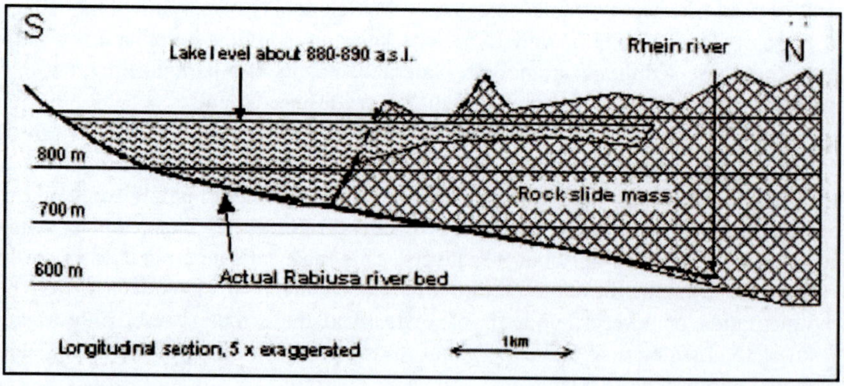


Fig. 15.6 Section along Lake Versam. The steep gradient of the Rabiusa river and the narrow valley are responsible for a relatively small basin

water immediately after the rockslide event. Due to subsequent debris flows of fine-grained sediments from the neighbouring Bündnerschiefer slopes, the basin became less permeable and then for a certain time bonded a small lake [16].

Also on top of the rockslide mass some lakes were created (Fig. 15.1). Lac Prau Pulta, Lag Prau Tuleritg, Lag la Cauma and Lag la Cresta are lakes or ponds with a very sophisticated hydrology. Analysis of the lake sediments in those lakes by Deplazes et al. [5] showed a younger age that did not contradict the radiocarbon dating of the rockslide mentioned above.

## 2.3 Reconstruction of the Events

### 2.3.1 Outbreak of Lake Ilanz

Several earlier authors [e.g. 16] assumed a complete and instantaneous break of the Ilanz dam with catastrophic circumstances. All refer to the huge inclined eroded floodway, stretching over more than 6 km from Ransun to Zault (Fig. 15.7), further on to Reichenau, and finally disappearing under younger sediments at Domat-Ems. The surface of the floodway has a strikingly constant inclination of about 2–2.5° (4%). Only the uppermost part at Ransun is flatter, indicating that in that part erosion of the barrier was prevalent. The floodway spread out laterally up to more than 1 km. The volume of the debris eroded is hard to estimate, because its thickness is only known in few outcrops and borings. Very roughly it is assumed to be about 30–50 × $10^6$ $m^3$.

**Fig. 15.7** View from Trin towards SW over the rockslide deposits (forested hills in the centre) to the Ilanz floodway (*arrows*) of Ransun (Ra), Zault and Dabi (see Fig. 15.1)

**Fig. 15.8** Sediments of the Ilanz outbreak flood at Zault showing clear layering. The fragments consist of angular rockslide material (limestone). In some layers the particles have the size of coarse gravel, in some they consist of sand or even silt only. Around the single large block no disturbance of the bedding is apparent

In several erosion gullies the vertical structure of the flood deposits is well exposed (Fig. 15.8). The rockslide deposits are covered by gravel sediments that have been bulldozed by the rockslide from their former situation in the valley floor to an elevated position [14, 15, 16]. Accordingly, they are part of the rockslide event, forming together with the rockslide debris an undulating surface. This relief was filled up by outbreak sediments. NW of Zault the sedimentation of the flood starts with a typical debris flow deposit, dominated by coarse, slightly rounded blocks. These blocks are concentrated in several layers, separated by finer material. This blocky sediment is about 15 m in thickness. Above it, gradually finer sediments (medium gravel size) occur, showing a layering (Fig. 15.8). The sequence of the layered gravel attains about 10 m in thickness, each single layer about 5–15 cm. All constituents are angular fragments of the rockslide. Some layers contain only sand or even silt. At first sight, the layered gravel indicates a low energy environment. However, single large blocks within these layers (Fig. 15.8) led to questions of origin. Manville (personal communication, 2004) has suggested that, a "hyperconcentrated flow", as described in Pierson and Costa [12], might be the origin for this sediment. A giant flood as assumed to have happened could have induced such a process of transport. Its deposition was the last phase of the flood event. The layered gravel beds with few boulders are parallel to the surface of the inclined floodway. After this sedimentation, the floodway was incised by fluvial erosion.

As no clear evidence for the maximum lake level (> 820 m) can be found, it must be assumed not to have existed for a very long time. Probably the first overtopping was responsible for the break. This might explain the observations of Wessels [21] in the sediments of Lake Constance, 80 km from Flims downstream in the Rhine river. He found two anomalies in the regular layering of the sediments, separated by several layers of normal sedimentation. Schneider et al. [19] attribute these anomalies to the outbreak of Lake Tamins/Bonaduz (first anomaly) and Lake Ilanz (second anomaly). According to the findings reported above it is more realistic to link these two layers to the Flims rockslide with the generation of the Bonaduz gravel (first anomaly) and to the outbreak of Lake Ilanz (second anomaly). They could also can have been generated by subaquatical slides, e.g. induced by earthquakes.

Uncertainty also prevails about the structure of the former rockslide barrier. Probably the break started in the surficial coarse upper facies. It must be assumed that subsurface erosion as a consequence of heavy seepage within this facies played an important role in the first failure. The main part of the dam was built up by the dense three-dimensional jigsaw facies that is evidently more resistant. Erosion also has cut into this facies, but with the breach duration getting longer and longer, and discharge reducing with the lowering lake surface, the breaching must have come to a stop.

Obviously the remaining barrier was stable until the remaining basin was filled, at least to a great part, with sediments. From then, the sediment load at the barrier increased and erosion became more active again. The further downcutting of the gorge from 820 m to the Vorderrhein level of 610 m, an additional 210 m, has not produced any important flood sediments. This indicates that it progressed slowly. The terraces observed indicate that the erosion happened in several steps. Abele [1] differentiated only three different levels of terraces. My own field investigations showed that at least 5 levels can be mapped.

### 2.3.2 The Flims/Tamins Relation

The Flims rockslide is closely related to its neighbour, the Tamins rockslide (Fig. 15.9). With a volume indicated by Abele [1] of "only" about 1–1.6 km$^3$ it always stood in the shadow of its big brother, the Flims slide. The interpretation of the succession of the events is important for the understanding of the mechanisms. Unfortunately, there is until now no radiocarbon dating for the Tamins slide. Preliminary results of recent surface exposure dating on $^{36}$Cl done by Ivy-Ochs (personal communication) indicate an age close to that of Flims. The range would be less than ± 1,000 years.

Several observations favour the assumption that the Tamins slide is older than the Flims event (Fig. 15.10):

(a) The Bonaduz gravels, a very special sediment assumed to have been formed by fluidisation by the impact of a great Films rockslide onto the alluvium [e.g. 11, 15], have straight contact to the Tamins deposits. The surface of the Tamins

**Fig. 15.9** The Tamins rockslide deposits, seen from the Toma hills of Domat-Ems (looking towards West). The slide mass forms a straight barrier across the Rhine valley. A distinct breach (*white line*), caused by the outbreak of Lake Bonaduz, was partly filled later by the outbreak sediments of Lake Ilanz. The inclined flood plain in the centre of the photo has a direct extension to that of Zault. Within the breach in the background the Flims deposits are visible

rockslide block masses shows about 10–20 cm of a well-cemented breccia. This hard crust is smoothened, polished and deeply scratched. Even if this surface is similar to a glacial-polished one, and it has been interpreted as such by some authors [e.g. 16], the direction of the surface gouges is inconsistent with a glacial origin. Furthermore no other glacial influence is evident. Instead, a polishing by the fluidized Bonaduz gravel during its transport is possible.

The contact between the Flims deposits and the Bonaduz graves is not sharp. In many places the rockslide masses and the gravel are mixed and interfingered. So, obviously the Bonaduz gravel were mobilized by the Flims masses and hit against the already existing Tamins deposits.

(b) The Bonaduz gravels are almost only found upstream of the Tamins rockslide barrier and far up the Hinterrhein valley (Fig. 15.1). If they were generated by the Tamins slide they should be found downstream as well. In the case of generation by the Flims slide, an already existing barrier at Tamins would have been the perfect barrier responsible for diversion of the debris up the Hinterrhein valley (Figs. 15.1 and 15.10/3).

(c) The flood sediments of the outbreak of Lake Ilanz are incised into the Bonaduz gravel. Accordingly, the flood event is younger than the Bonaduz gravel (Fig. 15.10/5).

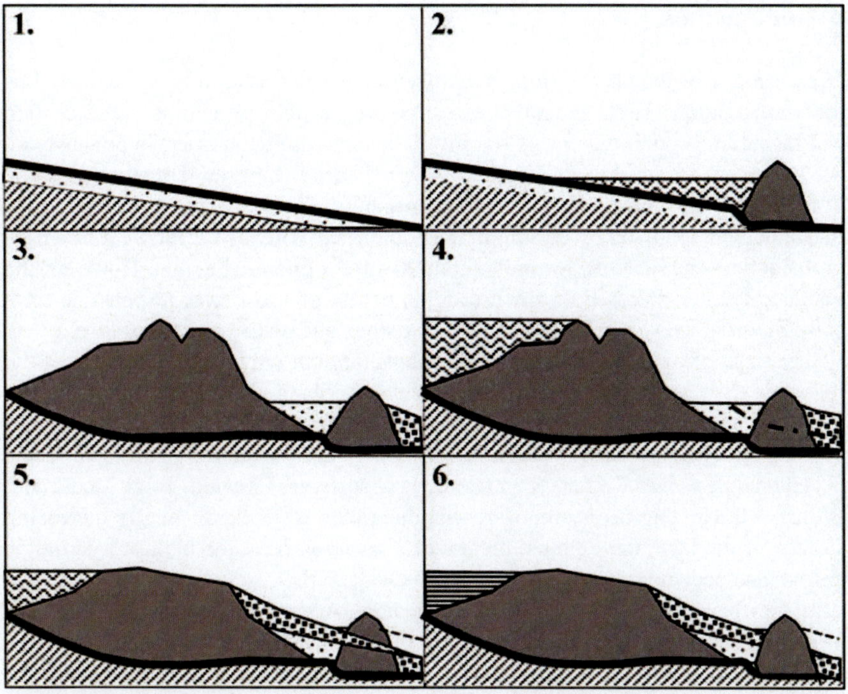


**Fig. 15.10** Sketch of the development of the Flims site: 1. Original state of the Vorderrhein river with alluvial sediments; 2. Deposition of the Tamins rockslide and damming of lake Bonaduz; 3. Deposition of Flims rockslide and mobilisation of alluvial sediments (= Bonaduz gravel), drainage of Lake Bonaduz with flood sediments downstream of the Tamins deposits; 4. Partial or even complete filling of Lake Ilanz, erosion in Tamins deposits; 5. Break of Flims barrier down to 820 m, flood sediments crossing the breach in Tamins deposits; 6. Filling of Lake Ilanz with sediments before further erosion to the present river bed

(d) The astonishingly straight and evenly inclined floodway of the Flims dam crosses the Tamins barrier without any disturbance. This indicates that not only the Tamins barrier existed during the flood event, but also the breach in it (Fig. 15.9). Lake Bonaduz, dammed by the Tamins barrier, had already drained through this breach. Remnants of a second flood sediment are to be found in an erosional terrace at the village of Felsberg-Altdorf (Figs. 15.1 and 15.10/3). They have a clearly higher elevated position with respect to the Flims flood sediments and have no extension upstream of the Tamins barrier. So, they are attributed by the author to the Tamins breach event.

All of these observations are an indication for an older age for the Tamins rockslide in relation to the Flims event.

## 3 Conclusions

Even if many of the observations mentioned above still have to be verified by further investigations, some general conclusions are possible. The Flims rockslide dam had favourable conditions for its stability. The length of the barrier (in downstream flow direction) favours a low hydrologic gradient. The jigsaw structure of some of the debris is relatively stable and prevents subsurface erosion. Nevertheless, the rockslide was followed by an important flood event. The lake level went down by probably less than 100 m, for only about 20–30% of its total height. However, the volume of water, released by the breach, in excess of 1 km$^3$ was responsible for a catastrophic flood wave that travelled 80 km downstream to Lake Constance.

The example of the Flims rockslide shows that not only the total breach of the rockslide dam may cause great flood disasters. Looking also to other rockslides in the Alps it is obviously not the exception to have a first breaching of the topmost part of the dam. Also the Köfels rockslide has experienced a similar history [3]. Abele [1] enumerates several others (e.g. Col de la Madeleine, Ehrwald, Totalp and In der Wöhr). All dam engineers are aware that due to the often exponentially increasing surface of the lake, the topmost meters of a reservoir have the highest hydrologic and so also economic potential. In the same way, in the case of a break of a natural dam, they have the highest catastrophic potential. Accordingly, measures to prevent an overtopping, and to reduce the lake level for only few meters, might increase the stability of rockslide dams significantly, and decrease the corresponding hazard.

## References

1. Abele, G. (1974) Bergstürze in den Alpen, *Wiss Alpenvereinshefte* **25**, 1–230.
2. Blumenthal, M. (1911) Geologie der Ringel-Segnesgruppe, *Beiträge zur Geologischen Karte der Schweiz* **NF 33**, 1–71.
3. Brückl, J. and Heuberger, H. (2001) Present structure and prefailure topography of the giant rockslide of Köfels, *Zeitschrift für Gletscherkunde und Glazialgeologie* **37**, 49–79.
4. Davies, T.R. and McSaveney, M.J. (2004) Dynamic fragmentation in Landslides: Application to Natural Dam Stability, *NATO ARW Bishkek, Kyrgyzstan, Extended Abstracts Volume*, 28–34.
5. Deplazes, G., Anselmetti, F.S. and Hajdas, I. (2007) Lake sediments deposited on the Flims rockslide mass: the key to date the largest mass movement of the Alps, *Terra Nova* **19**, 252–258.
6. Heim, A. (1883) Der alte Bergsturz von Flims, *Jb. d. Schweizer Alpenclubs* **18**, 295–309.
7. Heim, A. (1934) Bergsturz und Menschenleben, *Vjschr. D. Naturforsch. Ges. Zürich*, 1–218.
8. Ivy-Ochs, S., Poschinger, A.v., Synal, H-A. and Maisch, M. (2009) Surface exposure dating of the Flims landslide, Graubunden, Switzerland, *Geomorphology* **103**, 104–112.
9. Nabholz, W. (1975) Geologischer Überblick über die Schiefersackung des mittleren Lugnez und über das Bergsturzgebiet Ilanz-Flims-Reichenau-Domleschg, *Bulletin Ver Schweizerischen Petroleum Geology U Ing* **42**, 38–54.
10. Oberholzer, J. (1933) Geologie der Glarneralpen, *Beiträge zur Geology Karte der Schweiz* **28**, 626 p.
11. Pavoni, N. (1968) Über die Entstehung der Kiesmassen im Bergsturzgebiet von Bonaduz-Reichenau (Graubünden), *Ecl. Geologic. Helvetica* **61/2**, 494–500.

12. Pierson, T.C. and Costa, J.E. (1987) A rheological classification of subaerial sediment-water flows, *Geological Society of America, Reviews of Engineering Geology* **7**, 1–12.
13. Pollet, N. (2004) Mouvements gravitaires rapides de grandes masses rocheuses: Apport des observations de terrain à la compréhension des processus de propagation et dépôt. Application aux cas de La Madeleine (Savoie, France), Flims (Grisons, Suisse) et Köfels (Tyrol, Autriche), *Thèse, École Nationale des Ponts et Chaussées*, 252 p.
14. Pollet, N. and Schneider, J.-L. (2004) Dynamic disintegration processes accompanying transport of the Holocene Flims sturzstrom (Swiss Alps), *Earth and Planetary Science Letters* **221**, 433–448.
15. Poschinger, Av. and Haas, U. (1997) Der Flimser Bergsturz, doch ein warmzeitliches Ereignis? *Bulletin für angewandte Geologie* **2**, 35–46.
16. Poschinger, A.v., Wassmer, P. and Maisch, M. (2006) The Flims Rockslide: history of interpretation and new insights, in S.G. Evans, G. Scarascia-Mugnozza, A. Strom and R.L. Hermanns (eds.), *Landslides from Massive Rock Slope Failure*, NATO Science Series IV, v. 49, 329–356.
17. Poschinger, A.v. (2006) Weitere Erkenntnisse und weitere Fragen zum Flimser Bergsturz, *Bull. Angew. Geol.* **11**, 35–43.
18. Poschinger, A.v. (2009) Alluvial deposits liquefied by the Flims rock slide, *Geomorphology* **103**, 50–56.
19. Schneider, J.-L., Pollet, N., Chapron, E., Wessels, M. and Wassmer, P. (2004) Signature of Rhine valley sturzstrom dam failures in Holocene sediments of Lake Constance, Germany, *Sedimentary Geology* **169**, 75–91.
20. Wassmer, P., Schneider, J.-L., Pollet, N. and Schmitter-Voirin, C. (2004) Effects of the internal structure of a rock-avalanche dam on the drainage mechanism of its impoundment, Flims sturzstrom and Ilanz paleo-lake, Swiss Alps, *Geomorphology* **61**, 3–17.
21. Wessels, M. (1998) Late-glacial and postglacial sediments in Lake Constance (Germany) and their paleolimnological implications, *Archives of Hydrobiological Special Issue Advances in Limnology* **53**, 411–449.

# Chapter 16
# Usoi Rockslide Dam and Lake Sarez, Pamir Mountains, Tajikistan

**A.R. Ischuk**

## 1 Introduction

Landslides and rockfalls can provoke river damming, the creation of lakes upstream and in some cases, floods and debris flows after landslide dam breach [14]. Such phenomena are common in mountainous regions and have occurred often in Tajikistan both in the past and recent times. 1,300 lakes with a total area of 705 $km^2$ are known in Tajikistan, most of which have been created by landslide and rockfall damming, including those triggered by earthquakes, and by moraine deposits (glacial lakes). The majority of such lakes have a water surface area of less than 1 $km^2$ so that 97.5% of them cover only 9% of the total impounded area. However several large lakes exist such as the Karakul, Sarez, Yashilkul, Zorkul Lakes in Pamirs, the Iskanderkul Lake and the cascade of 7 Shing Lakes in the Southern Tien Shan. Most of these lakes are situated in the Pamirs and Alai Mountains at 3,500–5,000 m. a.s.l. elevation. They store 46.3 $km^3$ of water. Furthermore, there are 9 man-made reservoirs with 664 $km^2$ of total water surface area, which store 15.344 $km^3$ of water. These reservoirs have 7.63 $km^3$ net storage volume that is 13% of the average annual flow of all rivers of the Aral Sea basin [13].

The 670-m high Usoi rockslide dam, and the 300-m high Nurek embankment dam are located in Tajikistan; these are respectively, the highest natural and constructed dams in the world.

As a rule, large man-made embankment dams are safer than natural blockages, since they have special engineering design including spillways, special technology of construction and special monitoring systems. Nevertheless, numerous large natural dams in mountainous regions, most of which have been formed by slope failures triggered by strong earthquakes, exist for hundreds and even thousands of years. However the lifetime of other rockslide dams is rather short and they pose a threat

A.R. Ischuk (✉)
Institute of Earthquake Engineering and Seismology, Academy of Sciences of the Republic of Tajikistan, Dushanbe, Tajikistan
e-mail: a_ischuk@yahoo.com

S.G. Evans et al. (eds.), *Natural and Artificial Rockslide Dams*, Lecture Notes in Earth Sciences 133, DOI 10.1007/978-3-642-04764-0_16, © Springer-Verlag Berlin Heidelberg 2011

**Fig. 16.1** Satellite image of Lake Sarez in the Pamirs Mountains, Tajikistan

to people living downstream. Thus, the reduction of natural dam outburst flood risk is very important.

The study of Lake Sarez provides a good example of outburst flood risk assessment. The lake was formed by huge landslide (about 2.2 $km^3$ in volume) caused by strong (M 7.4) earthquake on February 18, 1911, which buried Usoi village and dammed the Murgab River (Fig. 16.1). Most of inhabitants of Usoi were killed (only 4 who were not in the village at that time survived). The lake started forming upstream from the dam and in Summer-October, 1912 the larger Sarez village (6 km upstream from the dam) was inundated. For this reason the lake was named Lake Sarez and the dam was named the Usoi rockslide dam. The lake water did not pass through the Usoi Dam until 1914 when, in the autumn, the filtration of water from the lake started and the Murgab River appeared again downstream.

## 2 Main Characteristics of the Lake Sarez Area

### *2.1 Geomorphology and Climate*

The Pamirs area is a mountain region located within the Euro-Asian and Indian tectonic plates collision zone. It is one of the most geologically active zones in the World [12]. Active tectonics, complex geological structure and high seismicity form the background favorable for a variety of geological hazards, especially large-scale bedrock slope instability.

From a geomorphic point of view, the Pamirs can be divided into Western and Eastern parts separated by an arbitrary North–South trending boundary that passes through the eastern part of Lake Sarez.

The Western Pamirs is characterized by rugged topography with river canyons up to 2–3 km deep and mountain ranges up to 6–7 km high. A major zone of glaciers plays an important role for river run-off in Central Asia. Main floods are due to snow and ice melting and occur in June–August. Winter discharge is much less. The difference between morning and evening water level in the rivers can reach 2 m.

The climate of Western Pamirs ranges from subtropical in the river valleys to gently warm climate in the middle parts of the ridges and cold climate in the highland. It is characterized by a big difference between day and night temperatures. The maximum temperature was recorded near Khorog, (+38°C) and minimum was –32°C at the same place. The annual rainfall is between 150–200 and 500–700 mm.

The Eastern Pamirs is quite different. It is a highland plateau at 3,500–5,000 m. a.s.l. with ridges that rise 500–1,000 m above river valleys. Climate of Eastern Pamirs is continental with negative annual temperature, very short warm period and long severe winter with strong winds, low precipitation usually in the form of snow. The thin blanket of snow and negative annual temperature provide strong freezing of the rocks and form permafrost up to 1.5 m deep. The maximum temperature was recorded in Murgab (+32°C) and the minimum was –47°C at the same place. Annual rainfall is only 60–100 mm and air humidity is lower than in the deserts of Central Asia.

## 2.2 *Geological Background*

According to geological structure, the Pamirs can be divided into Northern, Central, and Southern structural zones. Lake Sarez is located within the 20 km wide Rushan-Pshart thrust fault zone that forms the boundary between the Central and Southern Pamirs. Rocks within this fault zone are crushed and mylonitized, resulting in extensive fracturing and formation of shear zones and rock mass cleavage. The fault zone consists of several faults. At some localities the Southern Pamirs overthrust the Rushan-Pshart zone completely.

The Usoi landslide occurred on the southern slope of Muzkol Ridge. The landslide scar (Fig. 16.2) is cut into the Permo-Triassic deposits (dolomite, limestone, gypsum, anhydrite) in its upper part, while its main portion is in terrigenous-carbonate deposits (sandstone, schist, quartzite) of the Carboniferous Sarez Formation. These two units are divided by the Usoi Thrust dipping 60° to 80° towards SE. There is one more shear zone dipping 50° towards NW at the eastern part of the scar. Large-scale slope failure was caused by the collapse of the wedge delimited by these two fault planes. The bedding plains dip mainly 30 to 45° towards NNW, but with marked local variations due to local folding and faulting. A high degree of fracturing has led to the practically permanent occurrence of rockfalls and talus formation along the slopes on the right flank of Murgab Valley. The talus slopes have a surface angle between 35 and 40° and form rock glaciers in the upper part of the slopes (above 4,000 m. a.s.l.).

**Fig. 16.2** Scar of the Usoi landslide. View from the south (photo A. Ischuk, 2004)

The left bank of the Murgab River valley near the Usoi landslide is formed of the same types of rocks, also belonging to the Carboniferous Sarez Formation, but their tectonic conditions are less complex than on the right side of the valley. Thus, the location of the Usoi slope failure was determined, in general, by the combination of several tectonic factors [12]:

- High degree of rock mass fracturing formed by tectonic activity.
- Presence of the major Usoi thrust.
- Series of large shear zones that formed conditions favorable for a typical wedge failure.
- Presence of an active SW-NE trending wrench fault just in the innermost corner of the wedge.

The main part of the Usoi landslide body (Figs. 16.3 and 16.4) consists of quartzite, sandstone and schist of the Sarez Formation and, in its northern part, of marble and shale with subordinate gypsum, anhydrite and dolomite of Permo-Triassic age.

## *2.3 Seismicity of Lake Sarez Area*

The Pamirs is characterized by frequent strong earthquakes. Two groups of earthquakes occur: shallow-focus earthquakes (within Earth crust depth <70 km) and deep-focus earthquakes of the Pamirs-Hindu Kush zone at a depth of 100–300 km. Most of the Pamirs belong to the seismic zone with anticipated intensity of 9 points on the MSK-64 (MM) scale.

According to [3] the Lake Sarez area is located at the intersection of the Bartang-Pshart and Karakul-Sarez seismogenerating zones [4].

**Fig. 16.3** Usoi rockslide blockage. Fragment of high-resolution KFA-3000 space photograph

**Fig. 16.4** Usoi Dam and two rockslide-dammed lakes: Lake Sarez and Lake Shadau (photo A. Ischuk, 2002)

The Bartang-Pshart zone is associated with the Rushan-Pshart fault zone (Rushan is a village located at the Bartang River mouth). Besides the 1911 M 7.4 Sarez earthquake effects there are several other large rockslides in the region and Pleistocene-Holocene surface ruptures that are interpreted as geological evidences of strong past earthquakes. This zone is considered as a possible source of $M \leq 7.5$ earthquakes.

The Karakul-Sarez sub-meridional zone coincides with a transversal system of recent normal faults about 200 km long and up to 20 km wide, which conventionally separates the Eastern and Western Pamirs. Epicenters of the strong 1911 Sarez and 1974 M 7.2 Markansu earthquakes are located at the intersections of this zone with the Darvaz-Zaalay and Bartang-Pshart seismogenic zones. Taking into account the superposed character of this zone and its association with strong crustal and mantle earthquakes, it was assumed that earthquakes with magnitudes up to 8.0 can occur within this zone [3].

Seismic Hazard Assessment of the Lake Sarez area was carried out in June 2000 by GIBB Ltd. as a part of Lake Sarez Safety Assessment under IFAS Project [18].

The main results of PSHA are as follows.

- Mean peak ground accelerations of up to 0.1 g were obtained for the earthquakes with return periods of up to 40 years.
- Earthquakes with return period of around 100 years produce mean peak ground acceleration of 0.2 g.
- 0.15 g mean peak ground acceleration was estimated for the site during the 1911 Sarez earthquake. It corresponds to a return period of 80–85 years.
- A maximum credible earthquake of $M = 7.9$ with the same focal depth as the 1911 earthquake is likely to generate an average mean peak ground acceleration of 0.213 g at the site. This acceleration corresponds to an earthquake with a return period of just slightly above 130 years.
- Based on mean PGAs obtained it is likely that the earthquakes with return periods $> 100$ years and peak ground accelerations $> 0.20$ g are unlikely to occur close to Lake Sarez.
- The above mean PGA values are likely to increase by 50% if a standard deviation is taken into account.

## 3 The Usoi Dam

No direct data on the Usoi Dam internal structure are available. According to surface observations, significant and abrupt variations in grain-size composition of the dam body can be expected. Fragment size may range from sandy-silty material to blocks and boulders more than several cubic meters in volume. We can divide the dam body into three parts. The southern part is the highest, with a maximum height of about 250–270 m above the level of Lake Sarez (Fig. 16.5). Its surface is covered by blocks of various sizes from 2 to 20 m with no fines visible. Along the southern

**Fig. 16.5** Helicopter view of the scarp of the Usoi landslide from the highest point of the dam (photo A. Ischuk, 2004)

limit of the dam body there is a unit composed of granite pebbles, with a fine loam "matrix", typical of moraine deposits that rest on the Carboniferous bedrock in the valley left slope. We interpret them as moraine deposits bulldozed in front of the rockslide. Similar deposits are also clearly visible at the upstream side of the dam body near Shadau Lake. The central part of the dam body lies 100 m above the lake level, on average, and has a sharp escarpment at its right side, where the surface is lower. The latter is composed of strongly comminuted material without large blocks. However, at some locations in this zone one can see remnants of the pre-landslide natural slope surface that indicate that a huge block moved intact (see Fig. 16.3).

The northern part of the dam (see Fig. 16.5) is the lowest part with a minimum freeboard of approximately 38–45 m. This part is covered by large blocks of sandstone and schist from 2 to 20 m in size without fines on the surface. The part that abuts the right flank of the Murgab River valley is covered by secondary debris flows

**Fig. 16.6** Helicopter view of the downstream side of the Usoi Dam, debris flow and canyon (photo A. Ischuk, 2004)

and mudflows deposits (Figs. 16.5, 16.6, 16.7 and 16.8) composed of limestone, marble, gypsum debris and fines from moraine deposits that rest in the upper part of the scar and in the glacier zone above it. Mudflows from the Usoi Landslide scar passed towards the downstream face up to 1949 when their direction was changed by rockfall from the headscarp wall. After that they have flowed towards the lake (see Fig. 16.3). The volume of sediments from these mudflows is not too big and does not influence the seepage flow through the dam due to void space clogging [9].

Three topographic steps are clearly visible in the central and southern parts of the dam (see Fig. 16.3). The first (upstream) step is a ridge with a very steep (60–70°) western slope. The crushed and broken surface of the next step is clearly visible from this ridge. Along the downstream face of the dam (close to the western scar wall) a large cone of fine debris flow material has been deposited since dam formation. They rest partly on blocky dam material and this depositional mass is terminated by

**Fig. 16.7** Usoi dam. View from the right bank slope. Source area is to the right (photo A. Ischuk, 2002)

**Fig. 16.8** Helicopter view of the proximal (northern) part of the Usoi Dam (photo A. Ischuk, 2002)

a local canyon on the downstream face of the dam, which has developed by erosion caused by seepage flow through the dam (Fig. 16.6).

Analysis of the Usoi Landslide body and headscarp suggest the following mechanism of slope failure. Due to earthquake a huge wedge of rocks slid downslope like a single block. The speed of movement was very high and when this block hit the valley bottom it started separating lengthwise into three large parts, which, in turn, fragmented into smaller blocks. Several smaller rock falls from the western wall of the scarp also occurred during and after the earthquake. Later, several small landslides formed within the northern part of the dam body where rockslide debris is overlain by the debris flow deposits. Their bow-shape headscarps are visible on Figs. 16.3, 16.5 and 16.8.

## 4 Lake Sarez Hydrology and Permeability of the Dam's Body

The main parameters of the dam are as follows: maximum height of the dam – 620 m; minimum recorded height of the lowest part of the dam above the lake level – 38 m, the maximum height – 288.8 m; dam volume – 2.2 km$^3$; dam length – 5 km; dam average width – 3.2 km; dam area – 10.8 km$^2$; maximum altitude of the dam's crest – 3,469.1 m a.s.l.; minimum elevation of the dam's foot – 2,850 m. a.s.l.; average elevation of the lower part of the dam's crest – 3,290–3,300 m a.s.l.; difference between lake level and the level where seepage of lake water appears in downstream face – 148.2 m; number of the springs in the canyon – 57.

The main parameters of Lake Sarez are: length – 55.8 km; average elevation of the lake level – 3,263 m a.s.l.; maximum width of the lake – 3.3 km; maximum depth of the lake – 500 m; average recorded water inflow – 47.1 m$^3$/s (1,487 million m$^3$/year); average recorded water outflow – 47.7 m$^3$/s (1,505 million m$^3$/year); maximum seasonal oscillation of the lake level – 12 m; maximum volume of water in the lake – 16.074 km$^3$. We should note that estimates of water inflow and outflow contradict the fact that water level increases gradually. This contradiction can be explained as follows: outflow discharge is measured at the gaging station 20 km downstream from the dam, near Barchidiv village, and includes water that comes from the small Khurmo-Hatts and Vavzid Rivers that flow into the Murgab downstream from the blockage. Furthermore, the Lake Sarez inflow measurements are not accurate and do not take into account the underground water inflow as well as discharge of some small tributaries that flow into the Lake.

57 springs on the downstream slope of the dam are all located in the canyon at the nearly the same level – about 3,100 m. a.s.l., 140–150 m below present Lake level (Figs. 16.9 and 16.10). Seepage started when the Lake level rose to 3,050 m a.s.l.; indicating that the dam below this elevation is impervious. The cumulative discharge of all springs is 35 m$^3$/s in winter and 75 m$^3$/s in summer (season variation) and depends on the lake level, which varies annually ±6 m on average.

The lake level increased very rapidly during the 1911–1914 period and when it reached the 3,050 m a.s.l. elevation, the filtration started in the spring of 1914 with

**Fig. 16.9** The downstream canyon (photo A. Ischuk, 2002)

**Fig. 16.10** Springs in the downstream canyon (photo A. Ischuk, 1999)

about 2 $m^3/s$ discharge. In 1925 discharge reached 78 $m^3/s$ in the summer, and after decreased up to 50 $m^3/s$ in 1940. Later, the discharge varied from 32 $m^3/s$ (in winter) to 75 $m^3/c$ (in summer). The highest level of 3,270 m a.s.l. was recorded in 1994 when maximum depth reached 520 m and the freeboard of the dam was only 38 m. The perennial observations between 1949 and 1989 indicate very slow increase of the seepage outflow (Fig. 16.11) and attenuating growth of the average, maximal and minimal lake levels (Fig. 16.12). These processes are interrelated due to the higher permeability of the dam's upper part so that the lake level increase leads to outflow discharge increase.

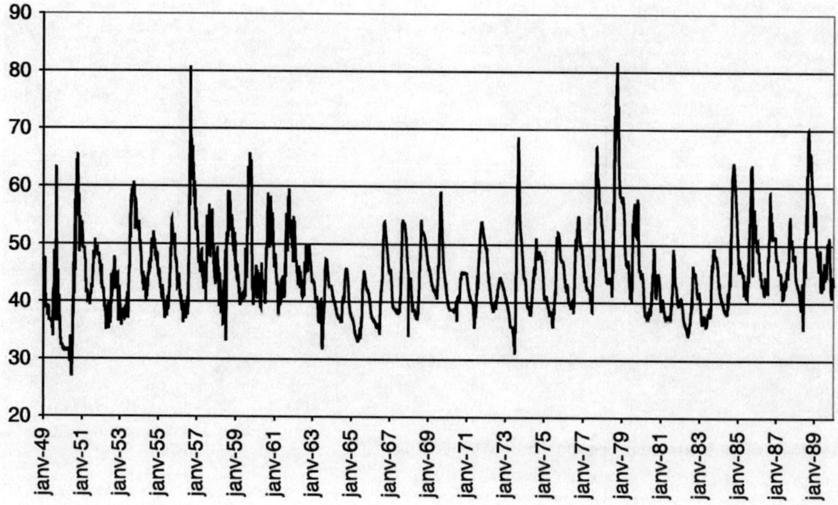


**Fig. 16.11** Evolution of seepage outflow $Q_{out}$ ($m^3/s$) from 1949 to 1989 ($m^3/s$) [after 5]

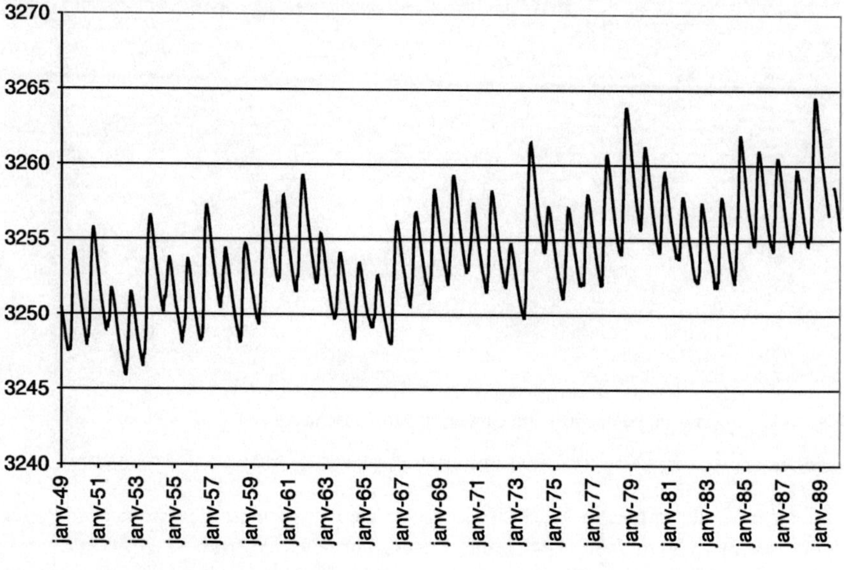


**Fig. 16.12** Evolution of the Lake Sarez level (m a.s.l.) from 1949 to 1989 [after 5]

## 5 Dam Stability Assessment

There are two alternative viewpoints on the Lake Sarez problem concerning its stability. The first one is that the Usoi Dam is an unstable feature and overtopping can occur after some time causing a giant outburst flood that will devastate the Bartang and Pyandj River valleys. The opposite point of view is that the Usoi Dam is a stable

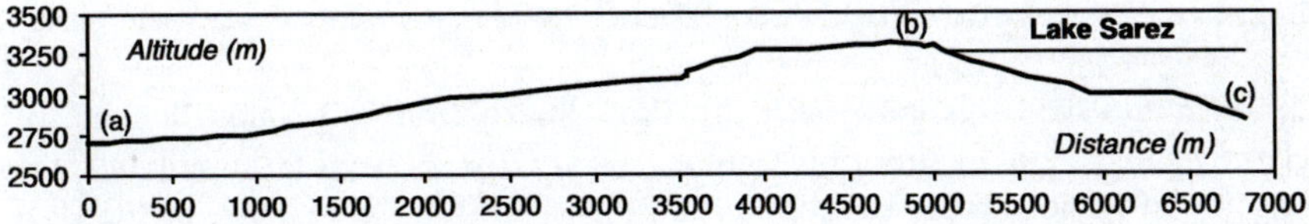


**Fig. 16.13** The Usoi Dam cross-section [after 5]

natural dam and Lake Sarez will exist for a long time like other similar lakes such as the Yashilkul Lake in Pamirs, Iskanderkul Lake in Central Tajikistan and many others [7].

This author holds the second opinion. The stability of the dam with reference to its very compact structure and the huge dimensions of its body (Fig. 16.13) ensures a high safety factor, though dam morphology indicates that some secondary movements have taken place. As a matter of fact a settling of up to 75 m is mentioned in various reports. The origin of such deformations is not known. Hypothetically they could be due to debris compaction (earthquake induced or not), that were more intensive during the first decades of the dam's lifetime, since the latest investigation showed that contemporary surface subsidence does not exceed 5–10 mm per year [7, 18].

The water flow from the lake towards the downstream springs is estimated to be very rapid (2–5 m/s), though measurements performed in the past are not completely consistent. The "intake zone", about 1,000-m long, was identified on the upstream slope of the dam. The head of the erosional canyon is about 1,500 m from the lake, at 3,110 m. a.s.l. The average hydraulic gradient through the dam is 10%.

Taking into account the average hydraulic gradient of 10%, and a Darcy coefficient K corresponding to a flow through large-size gravel without fines ($K = 10^2$ cm/s), it was found that the maximum velocity of water through the dam should not exceed some decimeters per second, i.e. 15–60 times less than the measured values [5]. "These observations lead to the conclusion that the structure of the Seepage Zone consists of a vast inter-block system through which preferential hydraulic pathways run. The high-recorded seepage velocities and the springs are related to the preferential hydraulic ways. Presently, the exact rate of water flowing through the preferential hydraulic ways and through the inter-block system is not known. However, we know that none of these two phenomenon is negligible" [5].

The canyon eroded by the filtrating water will progress in its present direction, along the contact between the collapsed Carboniferous material and the debris flow and mud flow sediments (see Figs. 16.5 and 16.9). In this case the head of the canyon would progressively approach the lowest part of the dam crest, but significantly slower than in the past.

Though the detailed internal structure of the Usoi Dam is unknown, all data available indicate that there is no danger of a general dam failure due to the water pressure of the Lake against the upstream face. Moreover, "even under an extremely heavy earthquake with horizontal acceleration of 0.5 g, the estimate of the safety factor still results in a safe value" [2].

## 6 Lake Sarez Shoreline Slope Stability

In 1967, the field mission headed by Prof. Sheko from All-Union Institute of Hydrogeology and Engineering Geology (VSEGINGEO) [16] showed that the ongoing exogenous processes such as landslides, erosion, debris flows on the Usoi Dam body can not endanger the stability of the Usoi Dam. At the same time they pointed out that the dam's body can be destroyed by huge waves that could be produced by two large landslides from the right and left shorelines very close to the Usoi Dam, which could be triggered by a strong earthquake. Landslide-triggered waves could destroy the dam body causing water overflow and a catastrophic flood downstream in the Bartang and Pyandj River valleys.

From this point the study of slope stability of the Lake Sarez area was carried out by the Tajik Geological Survey under the direction of Prof. Sheko [15]. From 1981 untill the late 80s these investigations were directed by Prof. Fedorenko of Moscow State University.

Analysis of practically all research reports on Lake Sarez area completed during 1967–1990, of old reports and papers since the Shpilko investigation made in 1913 [17] showed that there were different points of view on the possibility of dam breach by a Vaiont-like landslide triggered wave.

During 1981–1984 engineering-geology mapping of the Lake Sarez area at 1:25,000 scale was carried out by Tajik Geological Survey [10, 11]. Numerous landslides that could pose a threat to Usoi Dam were revealed on the shorelines. The most dangerous is the so-called "Right Bank landslide" (RBL) located about 4 km upstream from the dam on the right shoreline (see Figs. 16.8 and 16.14). It

**Fig. 16.14** The Right Bank landslide (RBL)

was proposed that the Right Bank landslide has been forming inside the body of a more ancient and bigger bedrock slope failure, named the Murgab landslide [6].

The next cycle of investigations in 1985–1987 [1] was focused on detailed engineering-geological, geodetic, geophysical investigation of the RBL site.

In 1986 a 243-m deep borehole was drilled on the right shoreline in the lower part of the Right Bank landslide. According to the core samples, it was concluded that the sliding surface of the ancient Murgab landslide is present at a depth of 211.4–235 m. However, this interpretation has been open to dispute.

The whole slope was divided into two parts: the south-eastern mass 0.3 $km^3$ in volume that is quite active and is collapsing step by step in relatively small portions (see lower part of the RBL on Fig. 16.14), and the north-western mass 0.6 $km^3$, the instability of which seems to be doubtful (see upper part of the outlined slope on Fig. 16.14). The total volume of the two blocks is 0.9 $km^3$. Unfortunately, no data that allow us to estimate the thickness of the sliding blocks are available. The average thickness of 200 m that was obtained on the basis of slope reconstruction was used in [1]. At the same time we find in the same report the following sentence: "On the basis of geophysical research (seismic profiling) the depth of the sliding surface does not exceed 100 m". Furthermore, it has been indicated that it is unlikely that both defined blocks will move simultaneously to form one block 0.9 $km^3$ in volume even during strong earthquake shaking.

Nevertheless, the collapse of a 0.9 $km^3$ block was used for the triggered wave calculation and modeling. Such results indicated that the speed of the moving landslide will reach 25–35 m/s and the height of the wave on the Usoi Dam would reach 60–150 m. In this case the volume of the overflow water can reach $30 \times 10^6$ $m^3$ [9].

Results of the geophysical investigation [1, 9] showed very contradictory information related to the RBL structure. The seismic profiling provided the most complete information, but the analysis indicated only a crushed layer 100 m thick (locally up to 250 m thick) with very low reliability, and did not indicate the presence of any sliding surface (Fig. 16.15).

No cogent evidence of both large massifs moving along deep-seated sliding surface have been found in surface geodetic data. Practically all survey benchmarks are located at a depth of 1.5–1.8 m from the surface in the loose deposits (as a rule represented by loam with schist) and, thus, indicate only near-surface movements of these deposits. The reported 10–15 cm per year average rate of movement relates only to the surface layers (up to 60–100 m). There is no evidence that these movements involve deeper parts of the rock mass. There is only a relatively small (about 0.3 $km^3$) active block in the southern-east part of this slope, clearly visible on Fig. 16.14, where failures occur, but they are not dangerous for Usoi Dam stability.

Interpretation of the Quaternary deposits and their location within the RBL provided in [1, 9] is also considered doubtful.

As for numerous linear features on the slope that were mapped as landslide cracks [1, 9], two groups can be identified from our point of view: fractures caused by sliding and those due to sagging. The first group forms on the steep sections of the slope, i.e. on the sidewall of the erosion channels, hollows, and steep ledges. These cracks have variable orientation on the slope – lateral, transversal, and diagonal, but

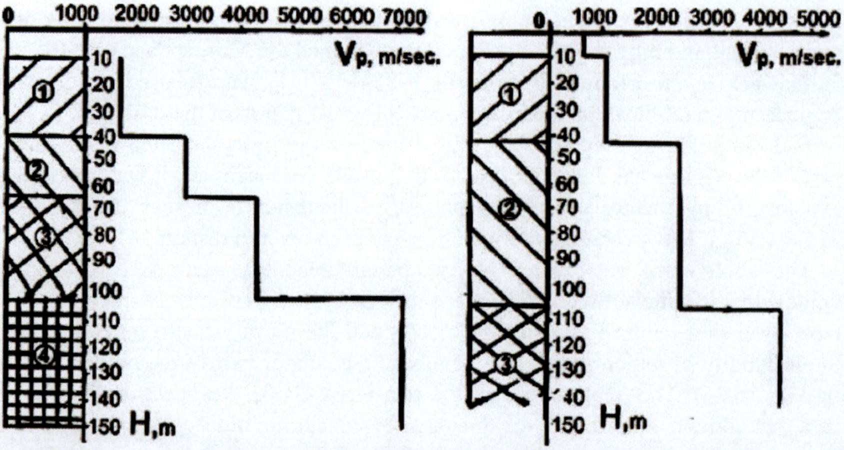


**Fig. 16.15** Examples of the RBL seismic profiling. *1* – loose deposits, Vp = 1,000–2,000 m/s; *2* – crushed bedrock, Vp = 2,000–3,000 m/s; *3* – weakly crushed bedrock, Vp = 4,000–5,000 m/s; *4* – bedrock, Vp > 6,000 m/s

their offsets always correspond with the local direction of gravity force. They are localized in the loose sub-surface deposits and do not affect deeper layers.

The elongated geomorphic features of the second group are located on the prominent sections of the slope, composed of moraine (fluvial-glacial) deposits with large content of sandy-loam fill. They are up to 1.5–2.0 m deep and their sides are not displaced vertically. It is important that they are located on the any side of the slope prominence, but not only toward to the slope bottom. Most likely they are cryogenic features.

According to [1, 9] the biggest landslide crack is located in the upper part of the slope. However, most likely it is not a fracture but traces of contacts between glacial body and the valley slope located at 4 levels (Fig. 16.16).

These 4 terraces were erroneously interpreted as a 2 km long sliding surface crack with 200 m difference of the elevation between its headcrown (corresponding to line 4 on Fig. 16.16) and top of sliding block (corresponding to lines 1 and 2). However it seems to be incorrect, though there are some fractures associated with landslide (arrows on Fig. 16.16).

Thus, the so-called "Right Bank Landslide" slope is interpreted to be a typical slope in an ancient glacial valley with remnants of fluvial-glacial deposits and the traces of the glacial levels. Taking into account the formation of Pamirs valleys, the thickness of the loose deposits on the Right bank slope is probably not more than 50–60 m.

The ongoing post-glacial evolution of the slope results, in particular, in the downslope motion of the loose deposits only. They reflect cryogenic processes rather than deep-seated landslide processes. We can find many such slopes in other Pamirs valleys.

**Fig. 16.16** Linear features on the upper part of the RBL. *Dotted lines* 1–4 mark along-slope traces of contacts between glacial body and valley slope, *arrows* mark landslide crack

# 7 Conclusions

Though the internal structure of the Usoi Dam and the form of seepage water table are not known with engineering accuracy, it can be concluded that only the upper portion of the dam body (not more than 150 m below lake level) is subjected to seepage through the dam. Due to the high permeability of the dam's uppermost part, the rise of the water level is compensated by increasing seepage. This process should lead to more intensive erosion in the canyon and its propagation towards the lower part of the dam crest. It seems, however, that these processes do not pose a real hazard at present or in the near future. At present the Usoi Dam is stable and can not be completely destroyed either by internal erosion or by overtopping because of its huge dimensions, composition and structure [8].

There is not sufficient evidence suggesting that a Right Bank Landslide (0.9 $km^3$ in volume) could collapse simultaneously. This slope is undergoing typical post-glacial evolution and relatively small failures can occur. There can be either sub-surface landslides in the loose deposits or rockslide in crushed bedrock in the southern-east part of the slope up to 0.3 $km^3$ in volume. Either of them could not cause the destruction of the Usoi Dam.

The most unfavourable scenario is the rise in Lake Sarez level caused either by dam body compaction due to strong earthquakes resulting in the decrease of seepage outflow, or by increasing inflow due to climate changes. It may result either in dam overtopping or in intensive propagation of the erosional canyon. However, in both cases it should not lead to immediate catastrophic consequences, since the composition of the dam surface, covered by the blocky carapace, prevents its rapid destruction.

## References


1. Akdodov, Yu. and Kazakov, Yu.M. (1987) *Complex engineering-geological investigation in the Lake Sarez area for the slopes stability assessment and forecast of geological processes in the scale 1:25000*. Tajikglavgeologia unpublished report (in Russian).
2. Alford, D. and Schuster, R.L. (eds.) (2000) *Usoi Landslide Dam and Lake Sarez: An Assessment of Hazard and Risk in the Pamir Mountains, Tajikistan*, United Nations ISDR Prevention Series No. 1,113 p.
3. Babaev, A., Mirzoev, K. and Koshlakov, G. (1980) *Seismic Zoning of Tajikistan Territory*. Donish, Dushanbe (in Russian).
4. Babaev, A., Lyskov, L. and Achilov, G. (1990) *Map of seismogenerating zones of Tajikistan* Dushanbe (in Russian).
5. Biedermann, P. and Attewill, L. (2002) Evolution of seepage through landslide dams: the case of Usoi Dam in Tajikistan. Paper presented at IAHR Syposium, St. Petersburg, 8 p.
6. Fedorenko, V.S. (1988) *Rockslides and Rock Falls and Their Forecast*. Moscow State University Publishing House, Moscow (in Russian).
7. Ischuk, A.R. (2000) Critical analysis of the works executed on Lake Sarez – Geological aspects of the problem, in *Lake Sarez Risk Mitigation Project*, Dushanbe.
8. Ischuk, A.R. (2001) Lake Sarez – current knowledge about the problem, *Izvestia Academy of Sciences of the Republic of Tajikistan. Department of Physics, Mathematics, Chemical and Geological Sciences* **1**, 170–183 (in Russian).
9. Kazakov, Yu.M. and Akdodov, Yu. (1990) *Detail engineering-geological investigation of the landslide prone slopes in the Lake Sarez area*. Tajikglavgeologia unpublished report (in Russian).
10. Lim, V.V. and Akdodov, Yu. (1984) *Results of engineering geological investigation for future evolution forecast of landslides in the lower part of Lake Sarez*. Tajikglavgeologia unpublished report (in Russian).
11. Lim, V.V. and Akdodov, Yu. (1998) *Landslides of Lake Sarez Area*. Donish, Dushanbe (in Russian).
12. Lukk, A.A., Ynga, S.L. and Shevchenko, V.I. (1995) Earthquake focal mechanisms, deformation state, and seismotectonics of the Tien Shan region, Central Asia, *Journal of Geophysical Research* **100**(B10), 20321–20343.
13. Negmatoullaev, S.Kh., Vladut, T.J. and Ischuk, A.R. (2003) *Big Reservoirs in the Seismic Dangerous Regions in Tajikistan and Safety of Dams*, Second edition. TIEES, Dushanbe (in Russian).
14. Schuster, R.L. (1993) Landslide dams – a worldwide phenomenon: *In Proc., Annual Symposium, The Japan Landslide Society, Kansai Branch, Osaka*, 27 April, 1–23.
15. Sheko, A.I. (1968) The Usoi blockage stability and the Lake Sarez breach assessment, *Bulletin of Moscow Nature Investigation's Society Geological Section*, Issue 4 (in Russian).
16. Sheko, A.I. and Lehatinov, A.I. (1970) *The Usoi blockage stability assessment*, Tajikglavgeologia unpublished report (in Russian).
17. Shpilko, G.A. (1914) The 1911 earthquake in Pamirs and its effects, *Proceedings of Russian Geographical Society* **3**, 69–94 (in Russian).
18. Stucky Consulting Engineers (2001) *Design report – Monitoring and Early Warning Systems/Studies on Long Term Solutions: Lake Sarez Risk Mitigation Project*, Renens, Switzerland.

# Chapter 17
# Rock-Avalanche Size and Runout – Implications for Landslide Dams

**T.R. Davies and M.J. McSaveney**

## 1 Introduction

### *1.1 Factors Affecting Landslide-Dam Stability*

A common and widespread hazard associated with landslides is river damming. If the dam fails, it releases impounded water as a potentially destructive dambreak flood, which sweeps the debris of the dam and any accumulated reservoir sediment to the river downstream, causing severe aggradation and progressive channel instability. Landslide-dam stability depends to a large extent on the dam size relative to the flow of the river it blocks, and on the dam material. These in turn depend on the nature of the triggering event, the source material and on processes that modify the material in its travel. The rock fragmentation that occurs during rock-avalanche motion results in a distinctive grading in the interior of a deposit, while the surface and upper few metres of the deposit is much coarser, less intensely comminuted rock. Landslide dams formed by blockslides are composed of relatively large, intact blocks of material; the fragmentation that influences the motion of the block is confined to a thin basal layer. Such dams are less likely to fail than those formed by rock avalanches.

### *1.2 Landslide Size*

The impact of a landslide on society is determined by many factors but principally its location, speed, and size. Location is with reference to position in the landscape, and in relation to people and their infrastructure; it is not a fundamental landslide property. Size and speed convey information relating to a landslide's impact on its surroundings. They are useful information to know in advance of a landslide's

T.R. Davies (✉)
Department of Geological Sciences, University of Canterbury, Christchurch, New Zealand
e-mail: tim.davies@canterbury.ac.nz

S.G. Evans et al. (eds.), *Natural and Artificial Rockslide Dams*, Lecture Notes in Earth Sciences 133, DOI 10.1007/978-3-642-04764-0_17, © Springer-Verlag Berlin Heidelberg 2011 

occurrence. Landslide size can mean any of the quantities of mass, volume, and area that indicate how big it is, or might be, before and after failure. For landslide dams, volume is a crucial factor; can this be determined in advance of occurrence?

In 1991, the summit and northern flank of New Zealand's highest peak fell in a huge rock avalanche. What determined the size of this, or any other historical landslide? Why was it not larger (or smaller)? Mt Cook has had previous rock avalanches and the shape of this and other peaks derives from episodic gravitational release of rock masses. What is the statistical distribution of such landslide sizes? Are rock avalanches larger end-members of a rockfall probability-density distribution, or are rockfalls and rock avalanches fundamentally different? Mount Cook is episodically shaken in great earthquakes, but not in historical time. What is the likely release volume at Mount Cook in the next great earthquake? This event has a probability of 10% in the next 20 years!

In March 1929, a $M_L$ 6.9 earthquake triggered a release of $55 \times 10^6$ m$^3$ from Falling Mountain, on the main divide of New Zealand's Southern Alps. What determined this volume, and not some other? If the earthquake had been larger, would more have fallen? Why is the volume of this rock avalanche such an anomaly in the region's rockfalls and landslides of March 1929? We cannot yet answer any of these "size" questions, but we wish to address whether such questions have answers.

### *1.3 Effects of Dynamic Grain Fragmentation*

In recent papers [3, 4] we proposed that the remarkably long travel distance of large rock avalanches results from internal dispersive stresses caused by dynamic grain fragmentation during runout. We also suggested that grain fragmentation at the base of large low-angle blockslides explains their extraordinarily long travel [4]. Knowing the mechanical modifications that occur in rock avalanches and blockslides is critical in determining the amount of motion and in assessing landslide-dam stability; this paper summarizes our understanding. We then consider how the presence of finely fragmented rock affects dam stability.

## 2 Regional Landslide Hazard Assessment

### *2.1 Probabilistic Landslide Hazard Analysis – A New Zealand Example*

What is the volume of the largest landslide that might be expected in the Kyrgyz Republic in an average year? This is a most readily answered type of "size" question. It can be answered only through probabilistic analysis of Kyrgyz landslide data. Unless a site has a long record of repeating landslides, such landslide data are likely to be for a large area, and will not be site specific. If we had this Kyrgyz data, we could estimate from it the probability of a landslide of a particular volume occurring within the susceptible area of the Kyrgyz Republic, but we would not be able to predict precisely where or when it might occur.

The New Zealand large landslide inventory (www.geonet.org.nz) records location and area for all known large landslides in New Zealand, but few ages are known. The New Zealand landslide catalogue has fewer data, and less precise locations, but it records time of occurrence. For North Island, the inventory contains data on 5,773 landslides. Analysis provides information on relative sizes of landslides, but nothing on rate of occurrence.

Systematic data collection for the catalogue started in August 1996. It records landslides as they occur, compiled by daily monitoring of news media and other sources, and special mapping for events that trigger many landslides. The catalogue is "complete" for landslides with footprint areas >10,000 $m^3$, and provides useful data on landslide rate despite the short (7 year) record.

North Island landslide-area probability-density distributions can be modelled as either inverse-gamma [13] or double Pareto [24] distributions (Fig. 17.1); the larger-area portion is fractal, with a fractal dimension of $2.56 \pm 0.05$ [6]. Catalogue data have statistically the same fractal dimension, and combining the two analyses allows exceedence probabilities of given landslide areas over time to be estimated (Fig. 17.2) [6].

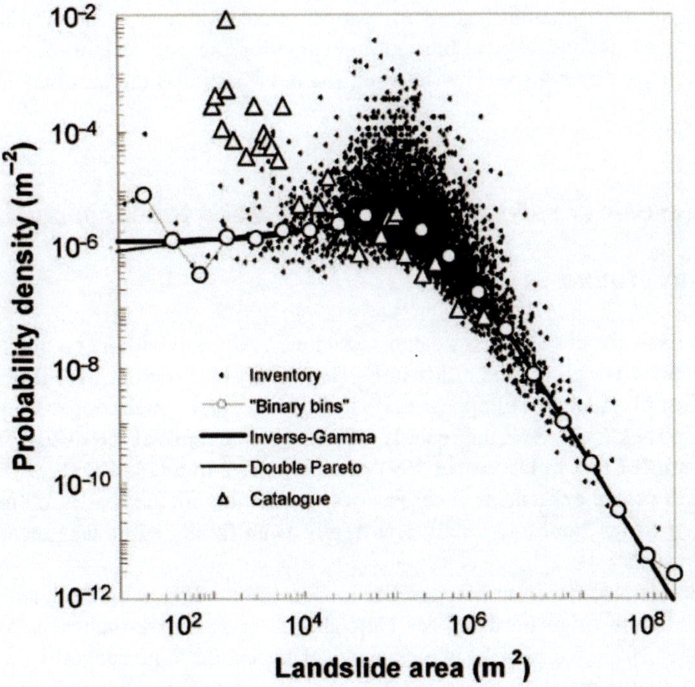


**Fig. 17.1** Probability-density distributions of landslide areas for North Island in the New Zealand large landslide inventory (5,773 landslides) and landslide catalogue (53 landslides) [6]

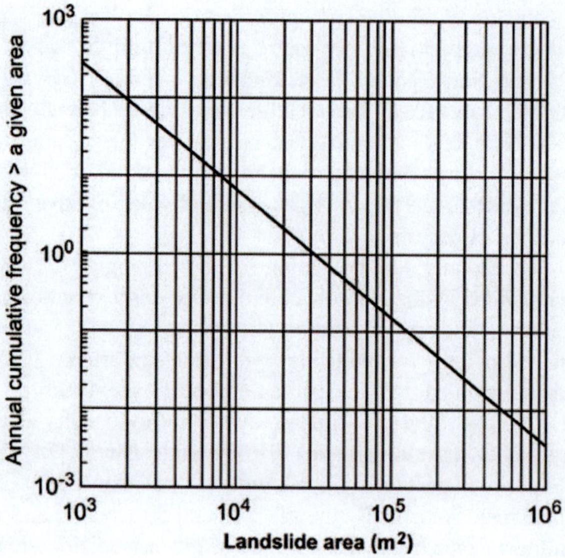


**Fig. 17.2** Probabilistic landslide analysis for North Island landslides answers the question of what sizes of landslides might be expected over some specified interval (1 year) [6]

Probabilistic landslide hazard evaluation is achievable with appropriate data. There are no methodological barriers in estimating the size of landslide expected in a given area over a specified interval; the only barrier is the chronic paucity of data.

# 3 Endogenic ("Self-Induced") New Zealand Rock Avalanches

## *3.1 Background*

Many rock avalanches in New Zealand's Southern Alps fall without recognized triggers, or prior warning. One at Mt Cook (Fig. 17.3) [14], two at Mt Fletcher [14], and one at Mt Adams [10] appear to be induced by static gravitational stress in the mountain rock mass. Seismic records indicate that the bulk of the energy from the release at Mt Cook in December 1991 dissipated in 2 min [14, 17] (Fig. 17.4 top). Witnesses describe the main release as over in 15 min, but the roar of falling rocks continued for an hour and a half, and tremor from falling rocks was recorded for most of the day.

Rockfalls also began at Mt Fletcher in December 1991, and continued sporadically until the ridge north of Mt Fletcher fell in rock avalanches in May and September 1992. On May 2, witnesses watched several large rockfalls during the day, and a large piece of the slope fell that night. Seismic records indicate that the bulk of the energy in May dissipated in 3 min, and similarly in September, but with different phasing and lesser amplitude (Fig. 17.4 centre). There had been numerous

**Fig. 17.3** The east face of Mount Cook collapsed in December 1991 releasing an avalanche of $12 \times 10^6$ $m^3$ of rock and ice. Based on seismogram (Fig. 17.4) interpretation, the bulk of the rock mass broke from the source in the first 20 s of the rapidly escalating collapse, and almost all of the mass was in simultaneous motion. Photo by Lloyd Homer

**Fig. 17.4** Vertical component seismograms of four large rock avalanches in New Zealand's Southern Alps showing evidence of different phasing of mass release in the various rock avalanches [17]. All from the one station (Berwen) and displayed at equal gain

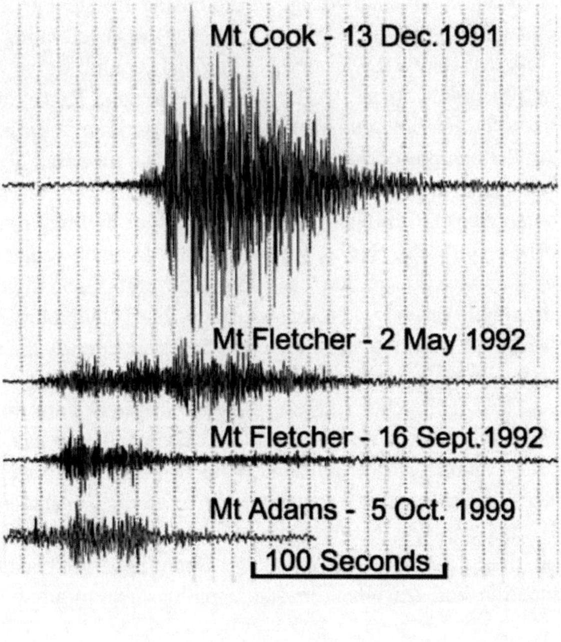

rockfalls from Mt Fletcher over at least the last 50 years, with some of them being many thousands of cubic metres in volume, such that the ablation area of Maud Glacier has been kept covered by rock debris. On 6 October 1999, a rock avalanche of some 10–13 × $10^6$ $m^3$ fell from near the summit of Mt Adams and blocked Poerua River [10]. It was a more abrupt seismic event than the other three examples (Fig. 17.4 lower) indicating one dominant release, as at Mt Cook.

### 3.2 Development of Release Volumes

To model a deep-enough release surface at Mt Cook, McSaveney [14] invoked rock-mass inhomogeniety – a stress-release joint system subparallel to the face of the mountain, but dipping slightly more steeply. Because of the low factor of safety calculated for small failures near the base of a buttress, he suggested that the avalanche did not instantaneously release on one surface, but was a rapid cascade of releases that extended up the slope. This analysis is too simplistic. The seismic record comes from the highest rate of energy dissipation in a cataclysmic phase of failure. Preceding this was an unrecorded phase of accelerating creep. Stress redistribution during creep is likely to have progressively developed a single release surface. If creep began at the most highly stressed location, it would have begun near the surface in the lower slopes of a rock buttress, and propagated from there up and into the slope, intersecting numerous stress-release joints. Creep did not lead to any external change that could be recognized as unusual by untrained observers scaling the summit in the early hours of the previous day, or in passing aircraft during the latter parts of the day until sunset. Accelerating creep in the lower slope is likely to have shown as an increasing frequency of rockfalls, but no frequency change was reported. We surmise that the accelerating creep phase was brief, possibly less than a few hours. No unusual rockfall or ice-avalanche activity was noted until people were alerted by the main collapse. Retrogressive failure appears to have been "minor" and due to stress-release cracking in the remaining "intact" rock mass. An "interrupted" slump of ~0.5 × $10^6$ $m^3$ remains on the right upper flank of the release surface to this day. Spalling of rocks from the release surface continued for some weeks, diminishing exponentially from a maximum in the moments after the primary release. Hence, we reason that a cataclysmic release of the rock avalanche took away a 700-m buttress and the summit in one flowing surge of rapidly collapsing rubble on a single release surface. The principal release volume (~90% or more) was in the main event, followed by an exponentially declining series of stress-release rockfalls that diminished in volume and frequency over weeks.

The Mt Fletcher release sequence differed from that at Mt Cook. First, the Mt Fletcher releases were not isolated events. There had been sporadic rockfalls and rock avalanches over decades, with two causally related avalanches within months. Second, the seismic records of the three rock avalanches are different (Fig. 17.4). If Mt Cook was a single-phased release, then the two Mt Fletcher events cannot be single phased. The May and September rock avalanches fell from the same location and travelled the same path; seismogram differences can only be from differing styles of release of the source masses. The May seismograms have been read [17] as indicating that the May event escalated "slowly" and had one main but "extended" phase. The September release escalated more rapidly but was not as large as in May. As the first phase in September began to die away, a late, smaller release phase began, and was prolonged relative to the first. The releases from Mt Fletcher may be retrogressive failures. The whole sequence of rockfalls and rock avalanches can be interpreted as a series of retrogressive failures. The underlying cause of the sequence appears to be rock-mass creep consequent on melting of ice formerly buttressing a structurally weak rock mass on the right lateral margin of Maud Glacier [14]. The structural weakness arises because Mt Fletcher is formed in the hanging wall of the low-angle, reverse Main Divide fault, and its rock mass is very closely jointed. Stress redistribution within Mt Fletcher is likely to have been over the long term during the thinning of Maud Glacier over more than 100 years. The creep phase has been ongoing for decades, during which, the rock mass has dilated. A protracted phase of accelerating creep on the lower slopes of the ridge led to an increasing rockfall frequency. There were several large falls in the hours before the rock avalanche in May 1992. Diminishing stress-release spalling of rocks from the failure-surface scar continued for months after both major rock avalanches. The principal mass of the release sequence was split between two main events in a ratio of 2.6:1, with perhaps as little as 10% in pre- and post-rock-avalanche rockfalls [17].

## *3.3 Prospects for Prediction*

In the above failure scenarios, the initial Mt Cook rock-avalanche mass was created by propagation of a release surface during creep prior to cataclysmic failure. Hence static 3-D stability analysis of the slope, with appropriate generalized rock-strength parameters would have given a useful estimate of the release volume. The summit flanks of Mount Cook appear to have a modal release volume of ~3.6–36 $\times$ $10^6$ $m^3$ under static loading. The location and volume of the next endogenic failure is determined by the current topography and generalized in situ rock-mass strength and might be found by searching the existing topography for the site causing the highest local stress concentrations in the rock mass. A great earthquake on the nearby Alpine fault could trigger some of these impending "self-induced" failures, but also could trigger a much-more deeply seated failure (see Sect. 4). In the Mt Fletcher scenario, the rock avalanches are only parts of a lengthy sequence of progressive failure. Static 3-D stability analysis of the slope with appropriate generalized rock-strength parameters can be expected to give only an estimate of the total volume, and not of the individual components. Nevertheless, we infer that the flanks of Mt Fletcher have a gross modal failure volume similar to that at Mt Cook, ~3.6–36 $\times$ $10^6$ $m^3$ under static loading. To determine the volume of the next failure in the sequence will require specific knowledge of the thickness and extent of the dilated, creeping rock mass of the larger failure. The obvious rockfall hazard of the site precludes any prospect of obtaining in situ geotechnical parameters for the mountainside.

# 4 Earthquake-Triggered Rock Avalanches and Rockfalls

## 4.1 New Zealand's 1929 Falling Mountain Rock Avalanche

On 6 March 1929, a landslide of $55 \times 10^6$ $m^3$ of closely jointed rock was released from Falling Mountain (Fig. 17.5) [16]. Although it is the only historical rock avalanche from this location, it was not the first release from the local valley wall. An adjacent valley was truncated prehistorically at Tarn Col by cataclysmic loss of its headwaters [16].

At the top of the source area, the main scarp is about 900 m wide. The main scarp and flanking scarps of the release surface bite deeply into the mountainside. The slide was thicker (up to 100 m) in the crown area. There are no visible crown cracks.

## 4.2 Controls on Volume

The remnants of Falling Mountain show evidence of "ridge renting" or sackung, and it is tempting to attribute the size of the failure to the spacing of specific defects associated with the ancient sackungen. However, there are many mountainsides in the

**Fig. 17.5** Falling Mountain (FM) on the main divide of New Zealand's Southern Alps. The distinctive deep release surface indicates an earthquake trigger. The release volume is almost as predictable as the earthquake location and magnitude. Location of the truncated valley at Tarn Col (TC) is indicated. A sackung ridge crosses the *right* mid-flank of the mountain. Photo by Lloyd Homer

vicinity with similar sackungen and no rock avalanches, and Sackungen still survive on Falling Mountain. The geometry of surviving defects does not suggest that the landslide release surface was controlled by pre-existing defects; rather, favourably located defects were exploited. Major defects remain. Small rockfalls occurred from surrounding slopes in the area, but no other long-runout landslides occurred in that event. The Falling Mountain rock-avalanche volume is an extreme outlier among landslides triggered by the March 1929 earthquake. The controls on the volume appear to have been the precise location and magnitude of the triggering earthquake. We suggest that the now-destroyed mountain mass resonated at a frequency that was strongly represented in the earthquake shaking, and shook itself apart during the earthquake.

### *4.3 Prospects for Prediction*

There is prospect for a specific estimate of earthquake-triggered landslide volume when earthquake location and magnitude are predictable. Meanwhile, there are other challenges for seismic slope stability.

There is a family of deep, cirque-like, earthquake-induced rock-avalanche scars that differ markedly from the "thin-slab" failures that occur between earthquakes. These deep-seated failures arise because the rock mass under seismic loading behaves as if it were a weak rock under static loading. It is not in fact weak; failure arises because the triggering force is large in comparison to the rock-mass strength. Typically, the rock strength is exceeded by a large amount. Important factors in seismic triggering of landslides are local resonance and topographic amplification of wave energy, including double-amplitude displacements at reflection of waves. These are not only surface phenomena, they also occur at internal rock-mass defects and velocity contrasts.

A question arises in earthquake-induced landsliding: "If the earthquake were larger, or closer, would the landslide have been larger?" This question cannot be answered empirically: it is impossible to test the same slope to failure under different loading conditions. Among our colleagues who claim that landslide size is determined by pre-existing defects, are some who claim that a larger or closer earthquake would have triggered a larger landslide. How is this possible if pre-existing defects are the control? Do the defects move? An explanation of the role of rock-mass defects comes from failure theory. When materials break, they always break at pre-existing defects, but the breakage is not defect controlled – other, sometimes-weaker defects do not break. As stress increases in the rock mass, pre-existing defects that are appropriately aligned to the stress tensor grow, while others are inhibited. Failure occurs when sufficient defects coalesce. In this model of failure, a larger or closer earthquake causes a larger landslide from the same site if it activates different and deeper defects. The mechanism of seismic triggering of landslides in brittle rock, however, is not well understood and a number of plausible hypotheses are available. This limits the practicality of numerical modelling to accurately determine the volume of failure. One method of overcoming the inability to investigate the question empirically is through physical modelling. We currently are developing apparatus to investigate this problem.

## 4.4 Rockfalls from New Zealand's 2003 Fiordland Earthquake

The New Zealand landslide catalogue includes 457 landslides that were triggered by a $M_W$ 7.2 shallow earthquake in Fiordland on August 21, 2003 [20]. Volumes of these landslides were visually assessed during several flights over the area, and most volumes are expressed as a range (e.g. 250,000–500,000 $m^3$) or as maxima (e.g. <1,000 $m^3$) or minima (e.g. >100,000 $m^3$), which present some difficulty in numerical analysis. To overcome the difficulty, values were arbitrarily assigned to the indeterminate volumes according to a few simple rules. Initial inspection of the data suggested a fractal distribution of rockfall volumes, and so indeterminate volumes were randomly distributed within their assessed range according to the fractal distribution of the population. Volumes assessed as <1,000 $m^3$ were assumed to lie within the range 10–1,000 $m^3$, and those assessed as >100,000 $m^3$ were given a range of 100,000–125,000 $m^3$. To ensure that any particular randomly selected volume under these rules did not bias the analysis, the procedure was repeated 5 times, and the 5 values for each landslide were averaged. The entire process was iterated until there was no significant change in the power-law exponent (slope of the distribution in log-log space), indicating that the indeterminate volumes were distributed within their ranges in the same distribution as the total. The probability density distribution of rockfall volumes >100 $m^3$ from the August 2003 Fiordland earthquake appears to be fractal over the entire range of assessed volumes, with a power-law exponent of $-1.5 \pm 0.3$ (Fig. 17.6a).

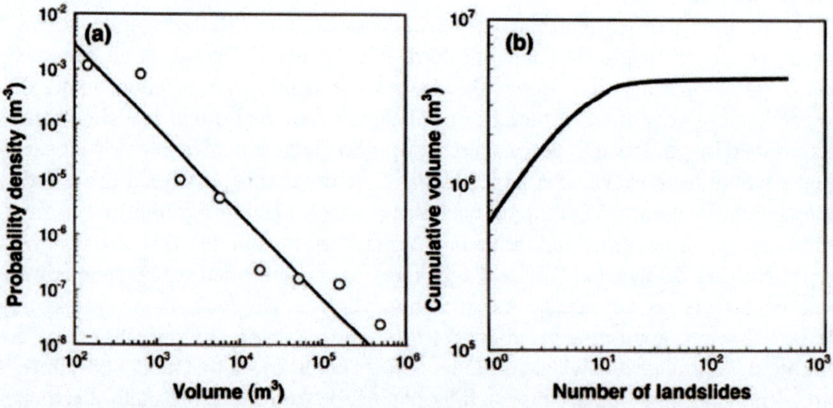


**Fig. 17.6** (**a**) Probability density distribution of rockfall volumes in the August 2003 Fiordland NZ $M_L$ 7.1 earthquake. Line represents a power law with an exponent of –1.5. (**b**) Cumulative rockfall volume in the Fiordland earthquake showing that most of the volume is in the few largest rockfalls

The cumulative rockfall volume (Fig. 17.6b) is very revealing of the prospects for prediction of the total volume of landslides in a given earthquake. The largest 1% of the Fiordland rockfalls contains 52% of the total rockfall volume, and the largest 10% contains 96%. The few large landslides that contributed most of the volume are likely to have had specific and unusual reasons for being so large, and so they were not amenable to useful statistical prediction. In the 1929 Arthur's Pass earthquake that gave rise to the $55 \times 10^6$ m$^3$ Falling Mountain rock avalanche, the total volume of landslides is likely to be only a little more than the volume of that one landslide, and we have already discussed the difficulties in predicting the size of this landslide prior to the earthquake.

## 5 Summary – Landslide Size

A long-recognized fundamental landslide problem is predicting the reach for hazard assessment. Let us consider in a flight of fancy that we have solved the problem of the physics of landslide runout and have an infallible numerical model for predicting landslide reach. We can be certain that a key input parameter in the model will be information on the landslide mass or volume. We have to know how big a landslide will be in order to predict its potential runout, or the size of the dam it may form. There are people (probably many people) among us who do not believe that the size of landslides can be determined in advance without a great deal of costly geotechnical investigation. Is it possible that they are mistaken? There is an analogue computer out there that gets the answer right every time. If more effort is put into researching this "big" problem will we obtain useful answers in the intermediate or longer time? One starting point might be to identify failure mechanisms where there might be "simple" solutions, and those where there might not. When we know how big a landslide is going to be, then the calculation of its shape at the end of its runout will have much more meaning.

Edifice shape appears to be much more important than specific, weakest defects in determining the shapes of release surfaces, and so the probability-density distribution of potential failure sizes from a site is intrinsically estimable from topography, general knowledge of the rock-mass characteristics, and the probability-density distributions of potential triggering events. When a landslide will occur is quite another question.

## 6 Stresses Within a Fragmenting Rock Avalanche

### *6.1 Grain Bridges, Elastic Limits and Super Stress*

Of particular importance in the flow of granular materials is the presence within the shearing grain mass of "bridges" (or arches), which are filamentous assemblages of grains aligned in the direction of maximum compressive strain of the mass (at 45° to

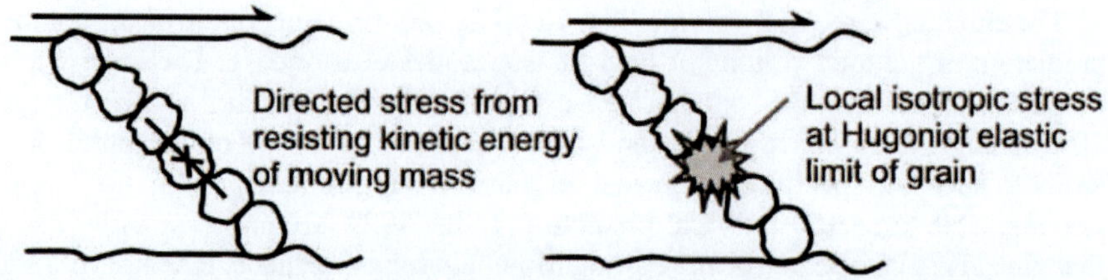


**Fig. 17.7** Idealized sketch of a grain bridge where confining pressure inhibits slip and rotation of grains and the weakest grain in the bridge fails by fragmentation, creating a local dispersive pressure of at least the Hugoniot elastic limit of the grain material

the flow direction; Fig. 17.7). Such bridges are a well-recognized hazard in industrial grain flows [19]; they also have been observed in experimentally deformed fault gouge [21]. Shear strain forces grains against adjacent grains, and stresses are transmitted between grains across bridges of grains in compressional contact. Shear ceases if the grain bridges do not fail. They mostly fail by slip between adjacent grains, and by rolling of grains, and these processes control the rate of grain-mass deformation.

A less frequent grain-bridge-failure mechanism is grain breakage: when stresses within a bridge are sufficiently high, the weakest grain can break. A characteristic of breaking brittle rock is a delay between stress application and rupture while micro-cracks form and propagate [22, 23]. When the stress in a grain bridge exceeds the strength, or elastic limit of the weakest grain, a degree of "super-stressing" is inevitable and the grain explosively disintegrates into fragments – a process we term "fragmentation". In the explosion, an isotropic dispersive force is transferred to all adjacent grains, and while fragmentation is taking place, the shear resistance within the volume occupied by the fragmenting grain is vanishingly small or zero.

## 6.2 Magnitude of Stresses

A one-dimensional numerical model [12] has been used [3] to show that the time- and depth-averaged internal longitudinal dispersive stress needed to simulate the runout of two rock avalanches in New Zealand is of the order of five times the average geostatic stress of the deposit; in the case of the $55 \times 10^6$ $m^3$ Falling Mountain event [3] this is about 2.5 MPa. If the stress generated by a single fragmenting grain were 2.5 MPa, then every grain in the avalanche would need to fragment simultaneously to develop the required stress, and the isotropic dispersive stress at all points within the translating debris mass would exceed the geostatic stress by about a factor of five. This would cause the whole mass to explode in every direction at about 5 g (J.N. Hutchinson, Imperial College, London, *pers. comm.* 2003). There is no field evidence for such behaviour. It therefore cannot be the case that all (or even the majority) of grains are fragmenting simultaneously, so the local stress due to a grain fragmenting must be much greater than the spatially averaged dispersive stress.

With a small proportion of grains fragmenting at any time, the upward vertical component of the dispersion caused by any particular fragmentation event is fully reversed by gravity before the next event occurs in the vicinity. No net work is accomplished in the vertical direction by this mechanism. Hence local, intermittent high isotropic dispersive forces do not necessarily cause general upward motion of a granular mass. By contrast, instantaneous longitudinal dispersions are not reversed by gravity, but persist until the next fragmentation event in the vicinity. The absence of general upward explosion of rock avalanches thus is not evidence for the absence of locally high isotropic dispersive stresses during runout.

Experimental evidence of the magnitude of the dispersive stress generated by rock failure comes from early inelastic unconfined compression tests on a variety of rock types at low shear rates ($10^{-5}$ $s^{-1}$) [25]. Two modes of failure are reported; one (Type I) in which failure envelopes showed decreasing stress with increasing strain, and one (Type II) in which axial strain briefly reversed against the applied compressive stress. In the latter case, if the failure stress path follows the envelope during rapid stress application, the dispersive stress generated by failure exceeds the applied compressive stress. Such Type II behaviour, once initiated, requires no further input of energy to complete the fragmentation, and failure proceeds rapidly and explosively; this occurs often in dynamic grain fragmentation [15]. Because Type I behaviour does not result in a reversal of strain at failure, we make the conservative assumption that the local dispersive stress generated by a fragmenting rock particle is approximately equal to the unconfined compressive strength of the particle.

In the Falling Mountain example, the local fragmentation-induced dispersive stress is of the order of 0.5 GPa [4]; to obtain a spatially averaged stress of 2.5 MPa then requires that one grain in about 200 is fragmenting at any instant.

### *6.3 Direction of Stresses*

Fragmentation of a rock particle in the shear field of a large rock avalanche delivers non-isotropic forces to its surrounding grains if it breaks into only a few pieces. However, most break into many pieces, and the number of fragmentation events occurring in any sizable aggregate of grains is large enough that the integrated effect of even non-isotropic individual stresses will be that of a uniformly isotropic stress. In addition, only isotropic stresses – not shear stresses – can be transmitted in materials stressed above their Hugoniot elastic limits. The ubiquitous presence of finely comminuted rock shows that rock-avalanche materials have been repeatedly stressed beyond their elastic limits.

### *6.4 Grain-Size Distribution of Rock Avalanche Material*

Grain-size distributions of rock avalanches and fault gouges from published and unpublished data from New Zealand and elsewhere (Fig. 17.8) all have a fractal dimension close to 2.58, a value corresponding to a three-dimensional geometry

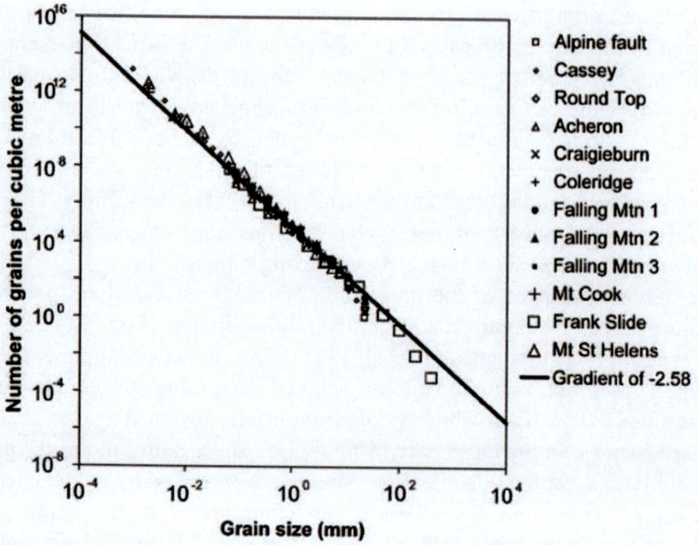


**Fig. 17.8** Rock avalanche grain-size distributions (Alpine fault and Cassey are fault gouges). Frequency is expressed as number of grain per $m^3$ assuming spherical grains of density 2,650 kg $m^{-3}$

that equalises and minimises the probability of fracture of all particles. Thus all grains present in a rock avalanche are equally likely to fragment at any given time.

## *6.5 Fragmentation Occurs Throughout the Runout*

It is fundamental to our explanation that fragmentation occurs throughout the runout, and we have much supporting evidence:

- shattered undisaggregated clasts are often found close to the distal margin at depth within rock-avalanche deposits. Such clasts could not have been shattered at the onset of motion and retained their close proximity and arrangement for the duration of the avalanche.
- the deposit of the Vaiont landslide, which travelled only a relatively short distance before halting, is not fragmented [8].
- the Köfels landslide deposit is not completely fragmented; in particular, one component of the mass that was halted by an obstruction is fairly intact [8]

The reason for fragmenting lies in strain-induced fracture of grains during grain-bridge failure, and it must continue so long as shear deformation continues. Dispersive stresses due to fragmentation are therefore present throughout rock-avalanche fall and runout.

# 7 Conceptual Model of Fragmentation

Our concept of rock-avalanche fragmentation (Fig. 17.9) is that particles of all sizes are fragmenting at any one time, each generating a local isotropic dispersive pressure, the integrated longitudinal component of which causes increased spreading during translation. In the Falling Mountain rock avalanche, the local dispersive stress generated with each fragmentation is assumed to be ~0.5 GPa. The grain-size distribution (and therefore the fractal dimension) does not vary with depth below the carapace [7], suggesting that the spatial probability of occurrence of fragmentation events is the same at all levels below the carapace.

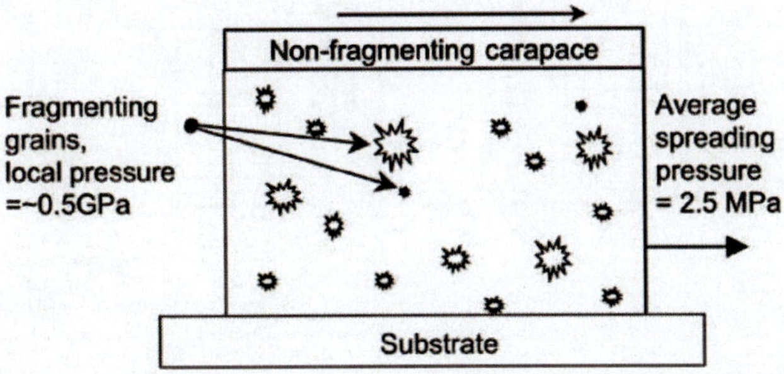


**Fig. 17.9** Diagram of fragmentation in rock avalanches. Fragmentation of individual grains is represented by stars of various sizes; particles of all sizes are fragmenting at any one time, each generating a high local isotropic dispersive pressure, the integrated longitudinal component of which causes increased spreading during translation. Pressures refer to those in the Falling Mountain rock avalanche discussed in Sect. 4.1

# 8 Laboratory Confirmation

We have designed and built a shallow cone-and-plate rheometer for measuring the direct and shear stresses developed in shearing granular rock material that is under sufficient direct stress that shearing is accompanied by fragmentation (Fig. 17.10). Briefly, the stationary roof is loaded with the chosen direct stress. The sample container is rotated about its axis at a chosen rate; the geometry ensures that the shear rate is uniform across the width and depth of the sample. The sample is sheared and, due to the high direct stress, grains fragment. A load cell attached to the torque-measuring annulus records the shear stress transferred from the rotating base to the roof. The apparatus and its ancillaries are described in greater detail elsewhere [2].

At low shear rates (~10 $s^{-1}$) similar to those found in rock avalanches, with a maximum grain diameter less than 20% of the shearing material depth, shearing of fragmenting granular materials (coal, argillite and limestone) gave a ratio of shear to direct stress (or friction coefficient) that was always similar to that of the same material when not fragmenting.

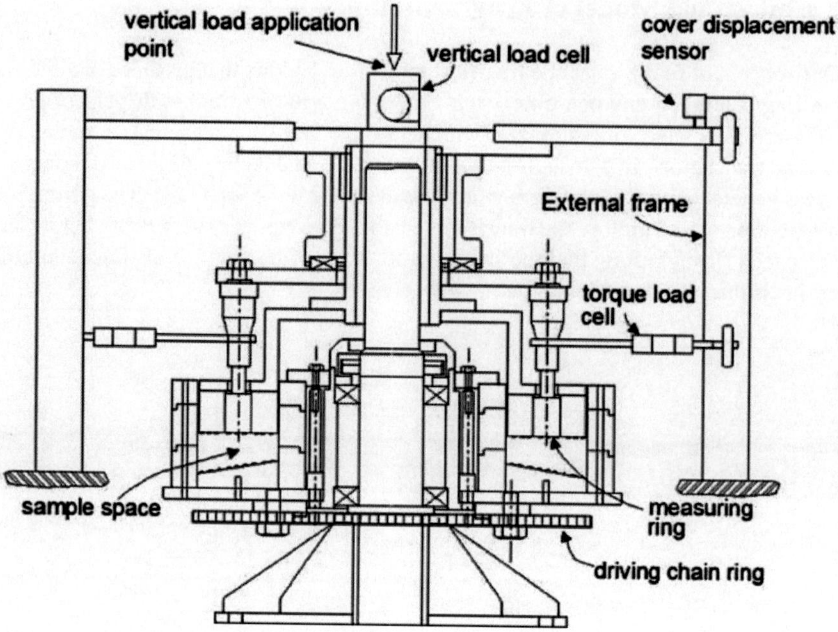


**Fig. 17.10** Principle of fragmenting rheometer in which test data (Fig. 17.11) were measured

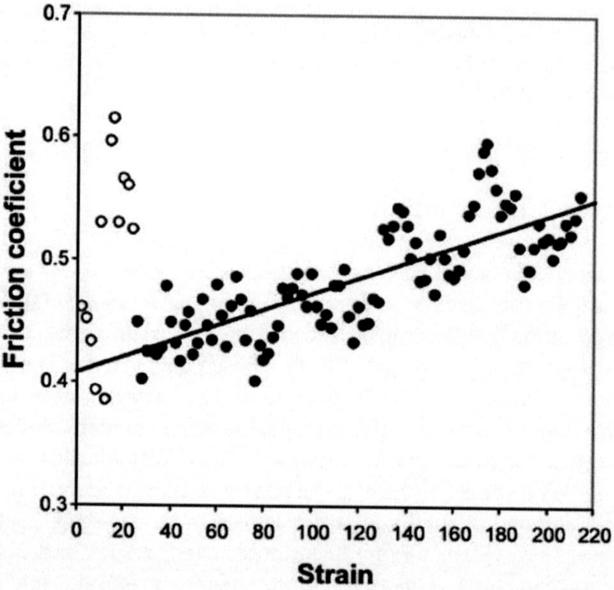


**Fig. 17.11** Sample test data for fragmenting coal. Shear rate is ~100 $s^{-1}$, overburden is 300 kPa. Trend line (slope 0.00065 ± 0.00005/unit of strain) is fitted to data of filled circles. Maximum reduction in friction coefficient is 20%

However, when we loaded the shear cell with large particles (up to the same size as the shear cell depth) and subjected them to greater shear rates (~100 $s^{-1}$), the measured friction coefficient was significantly lower than that for non-fragmenting shearing material. This was because the large fragmenting grains exerted a dispersive direct stress on the upper and lower cell boundaries, reducing the effective direct stress on the remaining non-fragmenting grains and thus reducing the frictional force they could apply to the torque-measuring annulus. Typical test data (Fig. 17.11) show that the friction coefficient increased with shear strain, as grain sizes in the cell reduced through fragmentation, reducing the ability of fragmenting particles to deliver direct stress simultaneously to both upper and lower cell boundaries. The reduction in friction coefficient allows the reduction in effective direct stress to be calculated, which is the value of the stress generated by fragmenting grains; thus the fragmentation stress is measured by this experiment. For coal grains, the fragmentation-generated stress was about 0.15 MPa, of the order of 1% of the unconfined compressive strength of the coal, indicting that about 1 grain in 100 was fragmenting at any time.

## 9 Application to Blockslides

A remarkable prehistoric landslide deposit at Waikaremoana, North Island, New Zealand is interpreted as a low-angle rapid blockslide [1, 5], in which a 1.2 $km^3$ intact block of sandstone moved about 2 km on a sandstone surface sloping at about 6–8° to the horizontal at a speed of up to 30–50 $ms^{-1}$ (Fig. 17.12). The long, accelerating travel on a very low-angle surface requires that there was very low frictional

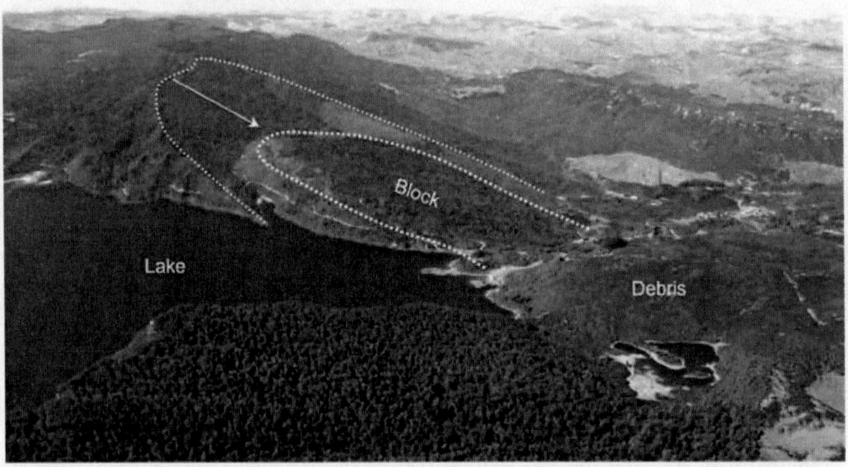


**Fig. 17.12** Waikaremoana blockslide and landslide-dammed Lake Waikaremoana. Photo by Lloyd Homer

resistance to motion of the block on the sandstone substrate. We suggest that this low friction was due to fragmentation at the sliding interface, given the inevitable initial rugosity expected of any failure surface within intact rock [5].

A 0.3 m thick layer of finely ground rock at the base of the displaced block recently has been proven by drilling (R. Beetham, GNS, Lower Hutt, N.Z. personal communication), and is interpreted as the fragmented material. The vertical dispersive pressure generated by fragmentation of individual clasts, if applied directly to the underside of the detached intact block, provides sufficient vertical force to support most of the weight of the block during motion (Fig. 17.13). This reduces the normal stress between the block and its substrate, and hence the frictional resistance to motion; a fragmenting rock at the substrate-block interface cannot transmit any shear force because it is stressed above its Hugoniot elastic limit. Analysis of the stresses involved shows that fragmentation of one particle in about 35 is required [5] to explain the acceleration of the block to the inferred velocity [1, 5].

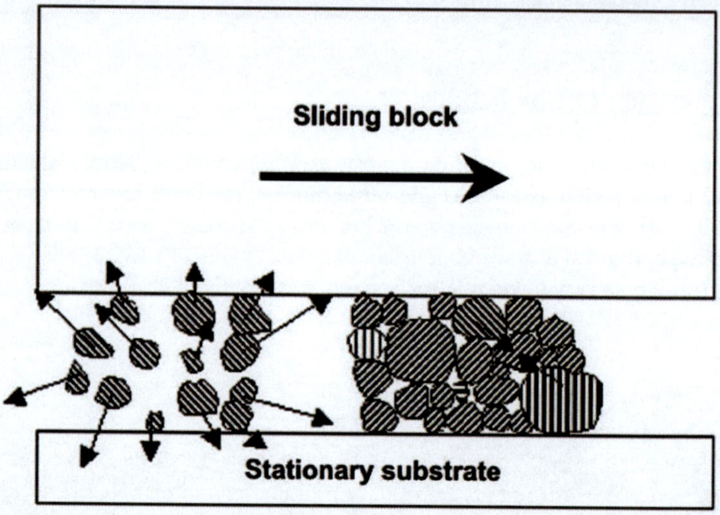


**Fig. 17.13** Fragmenting grain supports weight of block reducing direct stress on adjacent shearing non-fragmenting debris

## 10 Application to Landslide Dams

Fragmented rock-avalanche debris dominates the material properties of landslide dams, and the dams' ability to resist failure due to overtopping, slope failure or piping. A large rock-avalanche deposit is a relatively unstructured mass of all grain sizes, apart from the top several metres, which has a higher proportion of larger clasts. The surface layer of large clasts often gives rock-avalanche deposits an appearance of being comprised completely of angular boulders (Fig. 17.14).

**Fig. 17.14** Mystery Creek (Green River) rock-avalanche deposit, British Columbia, Canada showing uniformly large surface clasts forming a carapace on the rock avalanche. The carapace is formed by collapse of the rock mass along joints. Its mass is its only contribution to the dynamics of the rock-avalanche runout process. Note figure, trees and power pylons for scale

However, the material beneath the boulder carapace is finely fragmented (Fig. 17.15), and matrix – rather than clast-supported. If the carapace is disrupted by water flow, the interior is highly erodible. The finer subsurface material, however, is likely to be of low permeability due to its wide grading and consequent very small average void size. It is therefore likely that seepage rates will be low following emplacement of the dam. Slumping of the downstream dam face due to daylighting of the phreatic surface is likely to require a long time to occur, making this mode of failure relatively unlikely prior to overtopping. This was the case with the Poerua landslide dam in Westland, New Zealand [10, 11]. Piping failure also has a lower likelihood than overtopping, because the widely graded material of the dam acts as an efficient filter. Even when seepage has developed it is unlikely that void spaces will be sufficiently large to allow particle migration.

Not all overtopping events cause failure; the Poerua dam initially overtopped about 2 days after emplacement, but did not fail until 5 days later, in a higher flow event [11]. This could be due to saturation of the downstream slope by seepage from the overflow water, or to translation of the phreatic surface to the downstream face during the interval.

Alternatively, the failure could have been caused simply by the greater and more erosive flow. Some landslide dams do not fail. Recent attempts to identify the factors that determine whether or not a landslide dam will fail [9, 18] conclude that dam-material size is a dominant factor in determining the long-term stability of the dam. Landslide dams caused by blockslides (e.g. Waikaremoana) usually comprise intact bedrock with a few large fractures but no general disaggregation into fines. Seepage through a blockslide dam is often much greater than through a rock avalanche dam,

**Fig. 17.15** Section through Mystery Creek (Green River) rock-avalanche deposit, showing the finely crushed interior. Dynamic grain fragmentation in the failure of grain bridges is the cause of the comminution. Grain fragmentation imparts a dispersive stress of the magnitude of the Hugoniot elastic limit of the rock material appropriate for the local strain rate and confining stress

and the probability of failure is much lower. The Waikaremoana dam is now the site of a hydropower development. Prior to this development, the seepage through the debris was large, discharging all the lake inflow except in major storms.

## 11 Conclusions

- Edifice shape appears to be more important than specific, weakest defects in determining release-surface shapes, and so the probability-density distribution of potential failure sizes from a site is intrinsically estimable from topography, general knowledge of the rock-mass characteristics, and the probability-density distributions of potential triggering events.
- The dispersive stress generated by fragmenting rock particles is equal to or greater than the rock's Hugoniot elastic limit under the prevailing conditions. In order to explain the runout of the Falling Mountain rock avalanche, about one particle in >~200 would need to be fragmenting at any time.
- Fragmentation takes place throughout the runout phase of rock avalanches.
- The grain-size distribution of rock-avalanche debris has a fractal dimension corresponding to a particle arrangement that equalises and minimises the probability of fragmentation of each individual particle.
- Low-angle rapid blockslides result from low frictional resistance to block motion, caused by intense fragmentation under the block generating forces sufficient to support most of the block weight.

- The stability of a landslide dam is related to the grain-size distribution within it. The relative rarity of landslide dam failures due to seepage, slumping or piping can be explained wide grain-size distribution and very small average void size of the dam material.
- Landslide dams formed by blockslides are less likely to fail than those formed of fragmented debris. Seepage through substantial cracks in the block is likely to prevent significant overtopping.

**Acknowledgement** This research was supported by the New Zealand Foundation for Research, Science and Technology, through the Public Good Science Fund. We are grateful to our colleagues who attended the NATO Advanced Research Workshops in Celano in 2002 and in Bishkek in 2004 for their stimulating feedback on some of the ideas expressed here.

## References

1. Beetham, R.D., McSaveney, M.J. and Read, S.A.L. (2002) Four extremely large landslides in New Zealand, in J.Rybar, J. Stemberk and P. Wagner (eds.), *Landslides*. A.A. Balkema, Rotterdam, pp. 97–102.
2. Davies, T.R., Deganutti, A.M. and McSaveney, M.J. (2003) A high-stress rheometer for fragmenting rock, in L. Picarelli (ed.) *Proceedings of the International Conference on Fast Slope Movements: Prediction and Prevention for Risk Mitigation*. Naples, May 11–13, 2003 Vol. 1, pp. 139–141.
3. Davies, T.R.H. and McSaveney, M.J. (2002) Dynamic simulation of the motion of fragmenting rock avalanches, *Canadian Geotechnical Journal* **39**, 789–798.
4. Davies, T.R.H. and McSaveney, M.J. (2003) Runout of rock avalanches and volcanic debris avalanches, in L. Picarelli (ed.) *Proceedings of the International Conference on Fast Slope Movements: Prediction and Prevention for Risk Mitigation*. Naples, May 11–13, 2003 Vol. 2.
5. Davies, T.R., McSaveney, M.J. and Beetham, R.D. (2006) Rapid block glides: slide-surface fragmentation in New Zealand's Waikaremoana landslide, *Quarterly journal of Engineering Geology and Hydrogeology* **39**, 115–129.
6. Dellow, G., McSaveney, M.J., Stirling, M.W., Lukovic, B., Heron, D.W., Berryman, K.R. and Peng, B. (2004) Probabilistic landslide hazard in North Island, New Zealand, from landslide magnitude-frequency distributions, Unpublished file report. Institute of Geological & Nuclear Sciences Ltd, Lower Hutt, New Zealand.
7. Dunning, S.A. (2004) Rock avalanches in high mountains – A sedimentological approach, Unpublished PhD Thesis University of Luton, UK.
8. Erismann, T.H. and Abele, G. (1999) *Dynamics of Rockfalls and Rockslides*. Springer, Heidelberg, 316p.
9. Ermini, L. and Casagli, N. (2003) Prediction of the behaviour of landslide dams using a geomorphological dimensionless index, *Earth Surface Processes and Landforms* **28**, 31–47.
10. Hancox, G.T., McSaveney, M.J., Davies, T.R.H. and Hodgson, K.A. (1999) Mt Adams rock avalanche of 6 October 1999 and the subsequent formation and breaching of a large landslide dam in Poerua river, Westland, New Zealand, *Institute of Geological and Nuclear Sciences Science Report* **99/19**, 33p.
11. Hancox, G.T., McSaveney, M.J., Manville, V.R. and Davies, T.R. (2005) The October 1999 Mt. Adams rock avalanche and subsequent landslide dam-break flood and effects in Poera River, Westland, New Zealand, *New Zealand Journal of Geology and Geophysics* **48**, 683–705.
12. Hungr, O. (1995) A model for the runout analysis of rapid flow slides, debris flows, and avalanches, *Canadian Geotechnical Journal* **32**, 610–623.

13. Malamud, B.D., Turcotte, D.L., Guzzetti, F. and Reichenbach, P. (2004) Landslide inventories and their statistical properties, *Earth Surface Processes and Landforms* **29**, 687–711.
14. McSaveney, M.J. (2002) Recent rockfalls and rock avalanches in Mount Cook National Park, New Zealand. in S.G. Evans, and J.V. DeGraff (eds.). *Catastrophic landslides: Occurrence, mechanisms and mobility*. Geological Society of America Reviews in Engineering Geology, **15**, 35–70.
15. McSaveney, M.J. and Davies, T.R.H. (2006) Rapid rock-mass flow with dynamic fragmentation:inferences from the morphology and internal structure of rockslides and rock avalanches, in S.G. Evans, G. Scarascia-Mugnozza, A. Strom and R. Hermanns (eds.), *Landslides from Massive Rock Slope Failure*. NATO Science Series IV, Earth and Environmental Sciences v. 49. Springer, Dordrecht, pp. 285–304.
16. McSaveney, M.J., Davies, T.R.H. and Hodgson, K.A. (2000) A contrast in deposit style and process between large and small rock avalanches, in E. Bromhead, N. Dixon and M.-L. Ibsen (eds.), *Landslides in Research, Theory and Practice*. Thomas Telford Publishing, London, 1053–1058.
17. McSaveney, M.J. and Downes, G. (2002) Application of landslide seismology to some New Zealand rock avalanches, in J. Rybar, J. Stemberk and P. Wagner (eds.), *Landslides*. Balkema, Lisse, pp. 649–654.
18. Nash, T. (2004) Engineering geological assessment of selected landslide dams formed from the 1929 Murchison and 1968 Inangahua earthquakes, Unpublished M.Sc. (Eng Geol) thesis, University of Canterbury, 230p.
19. Nedderman, R. (1992) *Statics and Kinematics of Granular Materials*. Cambridge University Press, Cambridge.
20. Reyners, M., McGinty, P., Cox, S., Turnbull, I., O'Neill, T., Gledhill, K., Hancox, G., Beavan, J., Matheson, D., McVerry, G., Cousins, J., Zhao, J., Cowan, H., Caldwell, G. and Bennie, S. the GeoNet team (2003) The Mw7.2 Fiordland earthquake of August 21, 2003: Background and preliminary results, *Bulletin of the New Zealand Society for Earthquake Engineering* **36**, 233–248.
21. Sammis, C., King, G. and Biegel, R. (1987) The kinematics of gouge formation, *Pure and Applied Geophysics* **125**, 777–812.
22. Scherbakov, R. and Turcotte, D.L. (2004) Damage and self-similarity in fracture, *Theoretical and Applied Fracture Mechanics* **39**, 245–258.
23. Scherbakov, R. and Turcotte, D.L. (2004) A damage mechanics model for aftershocks, *Pure and Applied Geophysics* **161**, 2379–2391.
24. Stark, C.P. and Hovius, N. (2001) The characterization of landslide size distributions, *Geophysical Research Letters* **28**, 1091–1094.
25. Wawersik, W.R. and Fairhurst, C. (1970) A study of brittle rock fracture in laboratory compression experiments, *International Journal of Rock Mechanics and Mining Sciences* **7**, 561–575.

# Chapter 18
# Prospects for Prediction of Landslide Dam Geometry Using Empirical and Dynamic Models

**O. Hungr**

## 1 Introduction

Some 50% of those landslide dams that fail, do so within 10 days of emplacement [22], producing dangerous flooding downstream. Risks deriving from such sudden, catastrophic scenarios could be substantially reduced, if it were possible to predict in advance the geometry of a dam resulting from a given potential slope instability. Often, the location and volume of the landslide can be anticipated with reasonable accuracy. The remaining crucial question is how the unstable mass will travel after detaching from its source area and how it will deposit.

## 2 Runout Prediction

### *2.1 Empirical Models*

Empirical methods useful for estimating landslide runout have been reviewed in [8]. The first type of methods is based on the concept of fahrböschung ("travel angle") introduced by Heim [5] as a measure of landslide mobility. The fahrböschung is the vertical angle between the crown of the source area and the toe of the deposit, measured along the centreline of the movement path. Heim [5], Scheidegger [21] and others noted that the angle decreases approximately in inverse proportion with the logarithm of the landslide volume. As shown in Fig. 18.1, the correlation tends to be highly scattered, even with the use of logarithmic scales. Several authors attempted to reduce the scatter by sorting empirical data sets with regard to the morphology of the path, e.g. unconfined, elongated or transverse to the valley [3, 18]. However, such attempts generally have not been highly succesful, as there are other factors controlling the spreading of the landslide mass, which are responsible for the scatter.

O. Hungr (✉)
Earth and Ocean Sciences Department, University of British Columbia, Vancouver, BC, Canada V6T 1Z4
e-mail: ohungr@eos.ubc.ca

S.G. Evans et al. (eds.), *Natural and Artificial Rockslide Dams*, Lecture Notes in Earth Sciences 133, DOI 10.1007/978-3-642-04764-0_18, © Springer-Verlag Berlin Heidelberg 2011

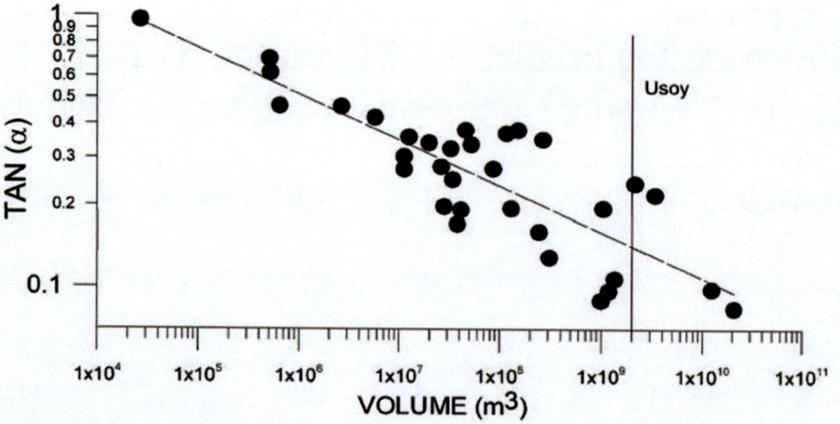


**Fig. 18.1** Correlation between tangent of the fahrböschung angle (α) and rock avalanche volume proposed by Scheidegger [21]

An example application of the fahrböschung concept is shown in Fig. 18.2. The example is the Usoy Dam in Tadjikistan [4], generated by a 1911 earthquake-triggered landslide with an estimated volume of 2 km$^3$. The two limiting values of the fahrböschung angle have been taken from the diagram in Fig. 18.1. It is apparent that in this case, the average empirical value of the travel angle would produce a good estimate of the position of the toe of the landslide deposit, which could lead to a reasonable prediction of the shape of the landslide crest, provided that the average thickness of the deposit can also be estimated.

Correlations between the deposit thickness and volume provide another means of estimating the dimensions of a potential landslide dam. A correlation between deposit volume and plan area was produced by Li [15], based on rock avalanche data compiled by [1, 5]. Li's Equation 2 can be re-written as:

$$h = 0.013\, V^{0.433} \tag{1}$$

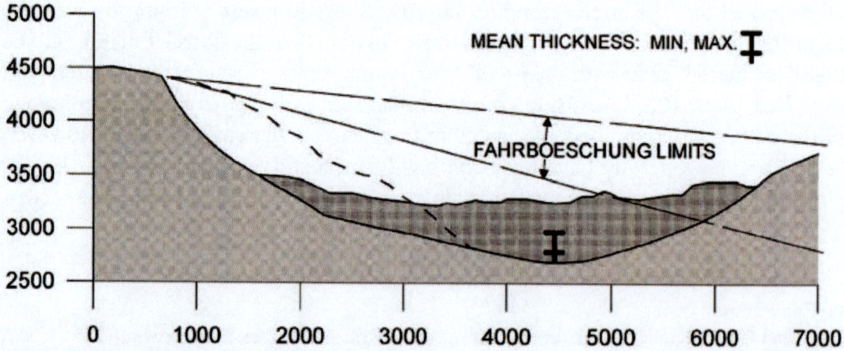


**Fig. 18.2** Application of two empirical methods of runout prediction to the Usoy landslide dam (Data courtesy A.Ischuk (unpublished) and [4, 12])

where V is the deposit volume and h the mean thickness of the debris. A similar equation was proposed by Hungr ([6], p. 309), although it is based on a less detailed database. Li's relationship has a standard deviation of about ±60%. The range of mean thicknesses obtained from this relationship for the Usoy landslide dam is also shown in Fig. 18.2. Apart from dealing with the large scatter of the correlation, another difficult step is to estimate the thickness along the axis of the dam from the mean value given by the equation. In case of the Usoy example, as shown in Fig. 18.2, the ratio between the mean thickness estimate and the true maximum thickness of the dam at its crest is about 1:4.

While these empirical methods can roughly predict the overall travel distance of the landslide mass, or its areal extent, they can give no indication of the distribution of debris in the deposition area, information that is needed for planning protective measures such as a spillway or bypass tunnel, or even the all-important prediction of the pool elevation behind the dam. The empirical methods also suffer from great scatter of data, making even the limited predictions very unreliable. Recent observations indicate that entrainment of surficial materials covering the landslide path plays an important role in determining both the quantity and distribution of the deposits, as shown later in this article.

## 2.2 Dynamic Models

In recent years, a number of numerical dynamic models for simulation of landslide motion have been developed (for a review, see [8]). Most of the models are based on depth-integrated St.Venant solutions of the equations of motion and compatibility. Some models are resolved in a one-dimensional framework, to produce two-dimensional (2D) representations of flow depth and displacement (e.g. [13, 20]). One model uses a combination of a user-defined path width outline and the equation of continuity to provide an approximate three-dimensional representation of flow [7]. Several new models resolve the governing equations in two dimensions, to derive three-dimensional (3D) flow solutions [2, 14, 16].

Most existing three-dimensional algorithms rely on the St. Venant theory of shallow flow. The potential landslide mass is expanded in volume to represent bulking due to fragmentation and represented by a digital thickness (isopach) map. The source area of the landslide and the path is represented by a digital terrain model. The analysis divides the moving mass into columns, with boundaries normal to the path surface [14, 16], or vertical [2]. Each column is then allowed to flow over the path surface under the influence of gravity forces, basal resisting forces and internal tangential stress resultants based on a modified theory of stress in a deforming frictional mass. All of the existing models have been tested against laboratory experiments involving flow of granular material and real landslide cases.

The constitutive relationships used in these models to set the resisting forces at the base of the flow include Bingham flow [23], frictional flow [20], frictional flow with pore-pressure, combined frictional and turbulent resistance (the Voellmy model [7, 24], and other relationships. Some models can include more than one rheological component within the moving mass [13, 14]. Others allow the choice of several alternative rheologies, suitable for modelling a range of material flow characteristics [2, 14, 16].

Predictions of landslide deposits can be strongly influenced by the type of model used and the value of the input parameters, as illustrated in the following paragraphs. Unfortunately, we have no means at present of determining these controlling aspects beforehand. The most promising approach is to use the semi-empirical concept of "equivalent fluid", proposed by [7]. The material used in the flow solution is not a real material, but an "equivalent fluid" whose rheology is such that the bulk dynamic behaviour of the landslide will be reasonably approximated by the model. The success of the simulation is judged in terms of overall displacement, distribution of deposits and any available velocity data. The parameters controlling the rheology of the equivalent fluid are selected by back-analysis of landslide examples similar to the case under consideration. The fluid should have a simple rheology, so that only a few parameters need to be constrained. Recently, attempts have been made to change the rheology along the path of motion, to simulate the important process of entrainment of saturated path material [10].

# 3 Examples

## *3.1 2D Versus 3D*

Three-dimensional solutions are preferred for dam geometry predictions, as there is often a large degree of lateral spreading. A sample back-analysis was carried out of the 1987 Val Pola Slide in Valtellina, Northern Italy (Fig. 18.3). The analysis, carried out using the DAN 3D model described by [11, 16] is shown in Fig. 18.4. The analysis uses the frictional model with pore pressure. The basal frictional resistance is characterized by a single parameter, the "bulk friction angle", defined by [7]:

$$\phi_b = \arctan[(1 - r_u)\tan(\phi)] \tag{2}$$

where $\phi$ is the friction angle of the material and $r_u$ is the constant pore pressure ratio (the ratio between the pore pressure and total stress at the base of the moving mass). Thus, even though pore pressure is included, the resisting stress on the base of the landslide is controlled only by total normal stress, similarly to dry sliding. In the present case, $\phi_b = 14°$ was used. The internal friction angle, $\phi_i$, used in a manner suggested by [20] was 35°. A comparative analysis carried out with the two-dimensional model DAN [7, 11] using the same input parameters and a path width estimated from a map is shown in Fig. 18.5. This is a difficult test for the pseudo three-dimensional solution, due to the great degree of spreading of the slide and the uneven topography of the valley slope opposite the source. However, the distribution of the debris across the valley corresponds very well to the 3D model

**Fig. 18.3** The Val Pola slide deposit. Note substantial runup on the right-hand slope opposite the source slope (Photo: Banco Popolare Sondrio)

**Fig. 18.4** Val Pola Slide, Valtellina, Italy, analysed using DAN 3D [16]. Frictional rheology, $\phi_b = 14°$, $\phi_i = 35°$. (**a**) before slide (**b**) after slide. Light gray areas mark terrain overrun by the slide, but not covered by deposits. Compare B with Fig. 18.3 (Digital terrain models kindly provided by Professor G. Crosta, Università degli Studi di Milano-Bicocca)

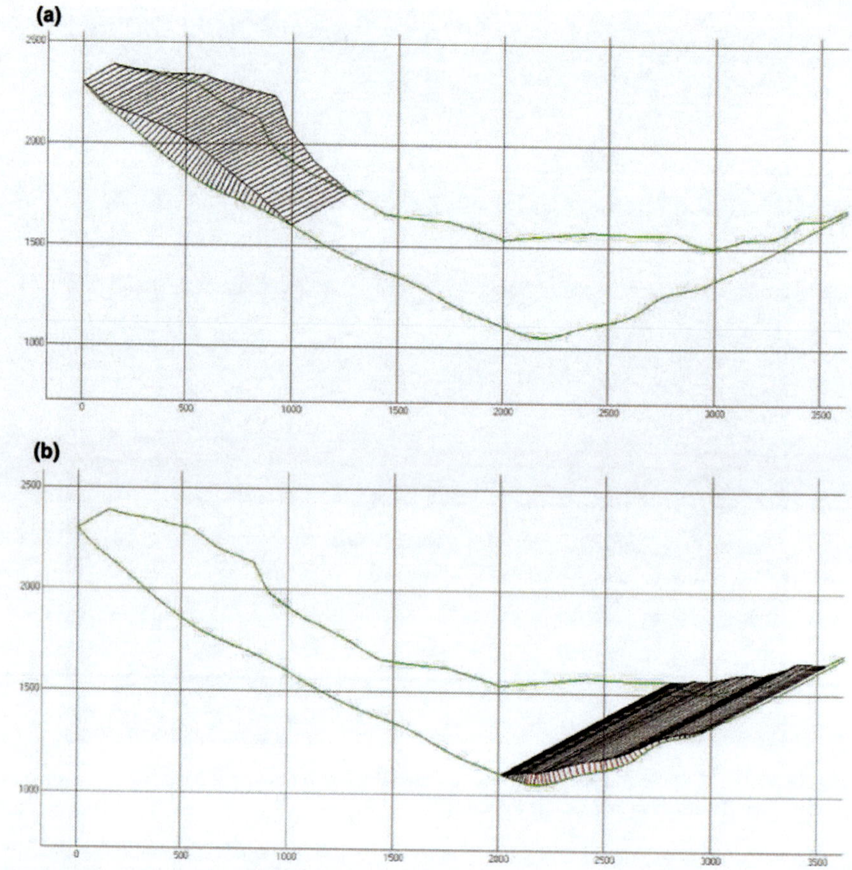


**Fig. 18.5** ValPola slide analysed with the 2D program DAN [7], using the same frictional parameters as in the 3D model shown in Fig. 18.4. The isometric view represents only the left half of the slide path. In the forefront is the scaled cross-section of the centerline of the slide path, showing the mean thickness of the debris. The upper line is the left margin of the slide, measured from a map of the slide **(a)** Source volume **(b)** Final deposit, 123 s after the beginning of motion

and the 2D solution represents a reasonable preliminary analysis of the situation (It is to be noted that the frictional model does not represent the optimal solution in this case, as shown in Sect. 3.2).

## *3.2 Influence of Rheology*

Figure 18.6 shows the results of the 2D Val Pola analysis using DAN with three alternative rheological models. In all three cases, the model parameters were fitted so as to obtain the correct maximum runout distance and height on the bank opposite

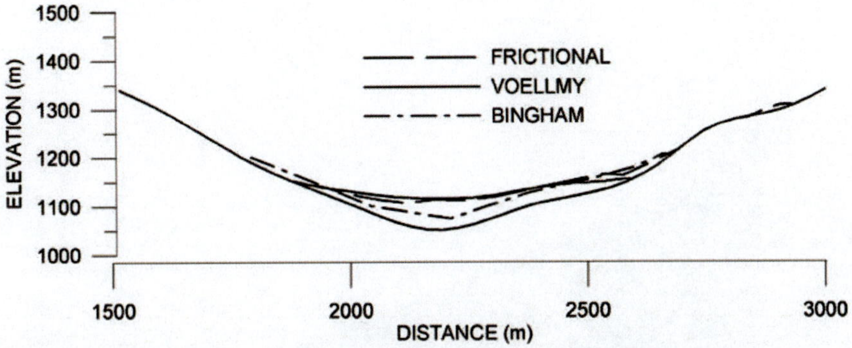


**Fig. 18.6** Three alternative mean deposit thickness cross-sections of the Val Pola rock avalanche produced by the program DAN [7]. The full line is the pre-landslide ground profile

the failure, as well as a realistic velocity profile (cf [9]). Each of the three rheologies produces a different distribution of debris across the valley floor. The frictional run produced runup onto the bench on the opposite side of the valley, followed by a fall-back, leaving a small part of the deposit on the bench. The Voellmy deposit is similar, but more compact at the front, so that the bench crest was reached, but no deposit was left. The Bingham result left a large part of the mass on the proximal slope and did not predict any fall-back. Its deposit is of a fairly uniform ("critical") thickness. The Voellmy model yields the best comparison with the actual distribution of deposits, both in terms of the maximum thickness of the debris and the shape of the "dam" crest (compare with Fig. 18.3).

### *3.3 Influence of Pore-Pressure*

Analyses presented in the earlier paragraphs presume that the landslide mass turns into a fluid on failure. However, many landslides with limited displacements remain relatively coherent. Furthermore, given a the soil material that has no cohesion and is not contractant, neither the frictional strength nor pore-pressure at the rupture surface may change substantially during failure. In such cases, landslide motion may simply be a response to natural transient increase of pore-pressure along the rupture surface.

A real case of a landslide involving a sudden re-activation and displacement of a landslide in clayey glacial till in southern British Columbia, Canada, is illustrated in Fig. 18.7. The landslide has been in existence for at least 50 years. Recently, as a result of unusually high infiltration caused by combined precipitation and snowmelt, the landslide moved several metres, causing a slight constriction of a stream channel at its toe. Fortunately, the magnitude of this movement episode was not sufficient to cause damming of the stream. In course of a hazard assessment for a community located on an alluvial fan short distance downstream of the landslide, the question

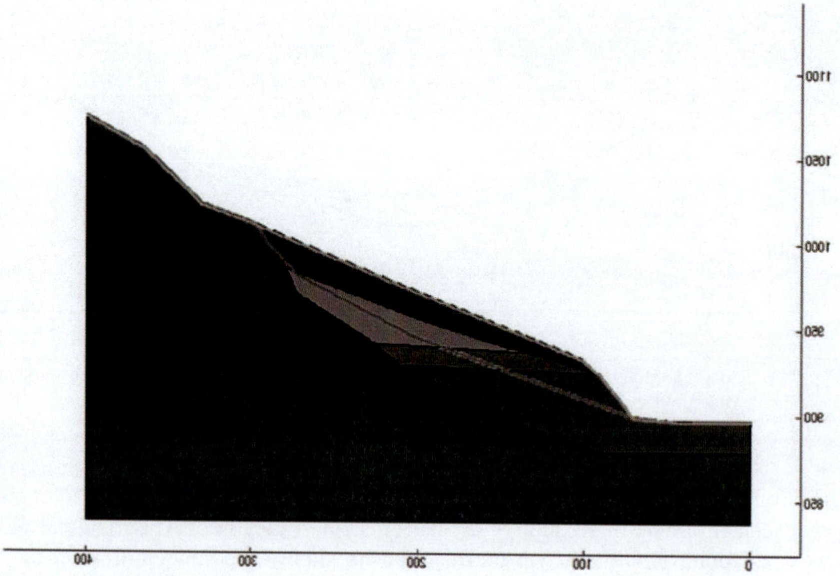


**Fig. 18.7** Schematic cross-section of an existing landslide in clayey glacial till. The light-coloured oblique line is the rupture surface, interpreted from inclinometer observation and the kinematics of observed movement

was asked whether a greater transient increase might not propel the landslide front across the creek channel. It became necessary to estimate the maximum height of a potential landslide dam, to serve as input into a dam break analysis. The pseudo-three-dimensional model DAN [7] was used to establish the relationship between pore-pressure increase and slide displacement. One of the analyses is depicted in Fig. 18.8.

Figure 18.9 summarizes the results of the dynamic analyses, in terms of horizontal displacements of the slide mass as function of the position of the mean pore-pressure head on the rupture surface relative to the ground surface. Under existing conditions, the piezometric surface is 3 m beneath the ground surface (i.e. +3 on Fig. 18.9). The original version of the model [7] produced estimates of the displacement of the front, uphill end ("tail") and the centre of gravity of the deforming mass. The model predicts displacements even under existing conditions, as a result of internal deformation of the slide mass. This is a conservative result which neglects internal cohesion of the dense glacial till material. A second analysis was carried out with a modified version of the model, in which longitudinal deformation of the slide mass was prevented, allowing the mass to slide as a flexible block.

These results, combined with the knowledge of the site groundwater conditions, were used to provide a range of landslide dam dimensions, corresponding to estimated extremes of pore-pressure increase. This in turn provided the required input into dam break analyses forming a part of the hazard assessment.

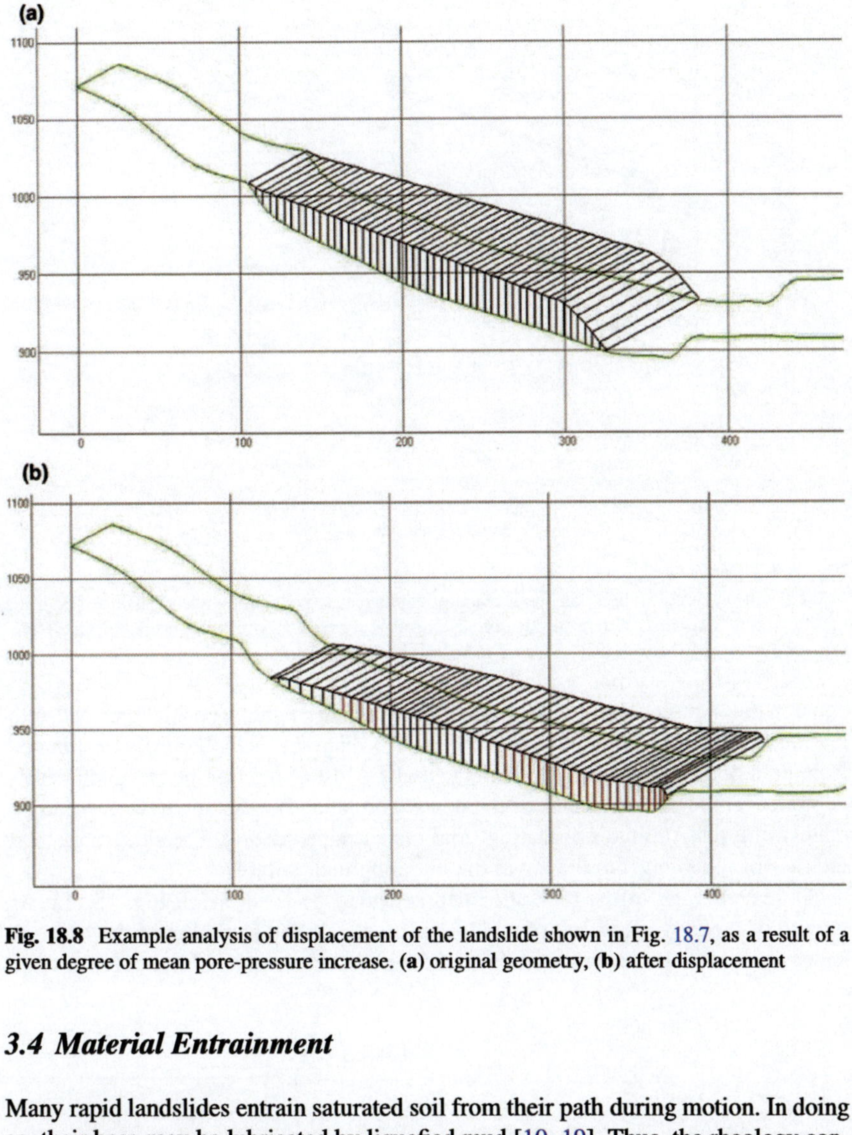


**Fig. 18.8** Example analysis of displacement of the landslide shown in Fig. 18.7, as a result of a given degree of mean pore-pressure increase. (**a**) original geometry, (**b**) after displacement

### *3.4 Material Entrainment*

Many rapid landslides entrain saturated soil from their path during motion. In doing so, their base may be lubricated by liquefied mud [10, 19]. Thus, the rheology controlling the landslide motion may change dramatically, and with it the mode of emplacement of the debris and the dimensions of a potential dam. A striking example of such behaviour is the 1999 rock slide debris avalanche on Nomash River near Zeballos, Vancouver Island, Canada, described by [10]. The event started as a rock slide in marble and andesite, with a fragmented volume of approximately 400,000 $m^3$.

The slide debris collapsed onto an apron of relatively fine-grained colluvial and glacial soils accumulated at the foot of the source slope. A volume approximately

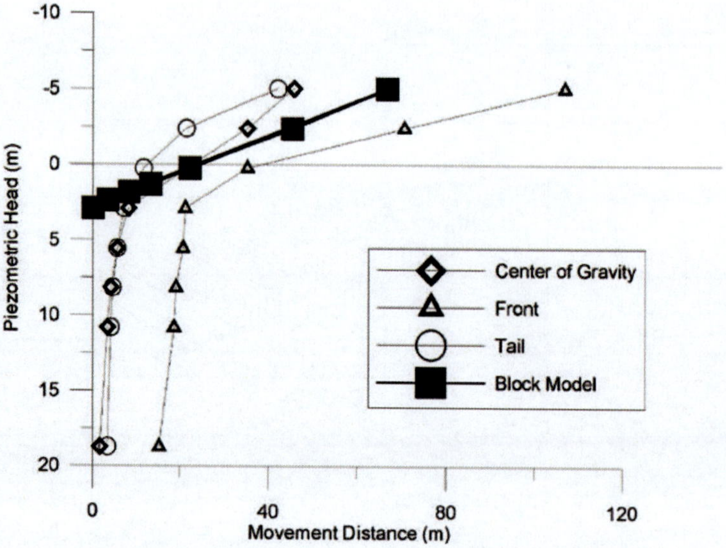


**Fig. 18.9** Result of a parametric study of motion of the landslide shown in Fig. 18.7, as a function of the mean piezometric head on the surface of rupture, relative to the ground surface (i.e. zero on the vertical axis means that the piezometric head is coincident with the ground surface.) The present piezometric head is at +3 m, or 3 m below the ground surface

equal to the source volume was entrained from this apron and incorporated into the moving mass. A volume balance ("mass") diagram of the landslide estimated from field observations is shown in Fig. 18.10. Most of the fragmented rock debris deposited in the proximal part of the deposition area. The distal part was made up mainly of liquefied mud originating from the entrained debris. The slide ran a total distance of 2,000 m from the toe of the initiating rock failure.

An overall view of the rock slide debris avalanche is shown in Fig. 18.11a. As a result of the long spreading of the debris, practically no damming of the Nomash River occurred. The upstream margin of the slide debris was only about 5 m thick

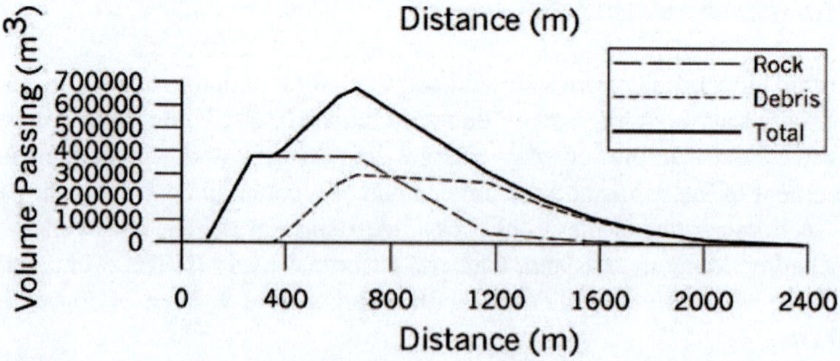


**Fig. 18.10** Volume balance diagram of the Nomash rock slide debris avalanche (after [10])

and only a small pond accumulated above it. The river filtered through the coarse debris and cut a poorly defined channel in the fine part.

Figure 18.11b shows a dynamic analysis of the landslide produced using the model DAN 3D [17]. The input parameters are equivalent to those used in a 2D DAN analysis by [10]. The initial slide is frictional and dry, with a friction angle of 30°. The slide entrains the requisite amount of material from the path in accordance with Fig. 18.10. At the same time, the rheological model is changed to Voellmy, with a friction coefficient of $f = 0.05$ and $\xi = 400\ \mathrm{m/s^2}$. These are the same parameters as used in the 2D analysis and the resulting distribution of debris is also very similar, corresponding approximately to the actual distribution observed in the field. Thus,

(a)

**Fig. 18.11** Analyses of the Nomash River rock slide debris avalanche using the program Dan3D [17]. **(a)** An aerial view of the slide. **(b)** Analysis using the Voellmy fluid model as described by Hungr and Evans [10]. **(c)** Analysis using the frictional model, with a bulk friction angle of 20°. The depth contour interval is 2 m. The thin line is the actual path outline. Time intervals 25 s

**Fig. 18.11** (continued)

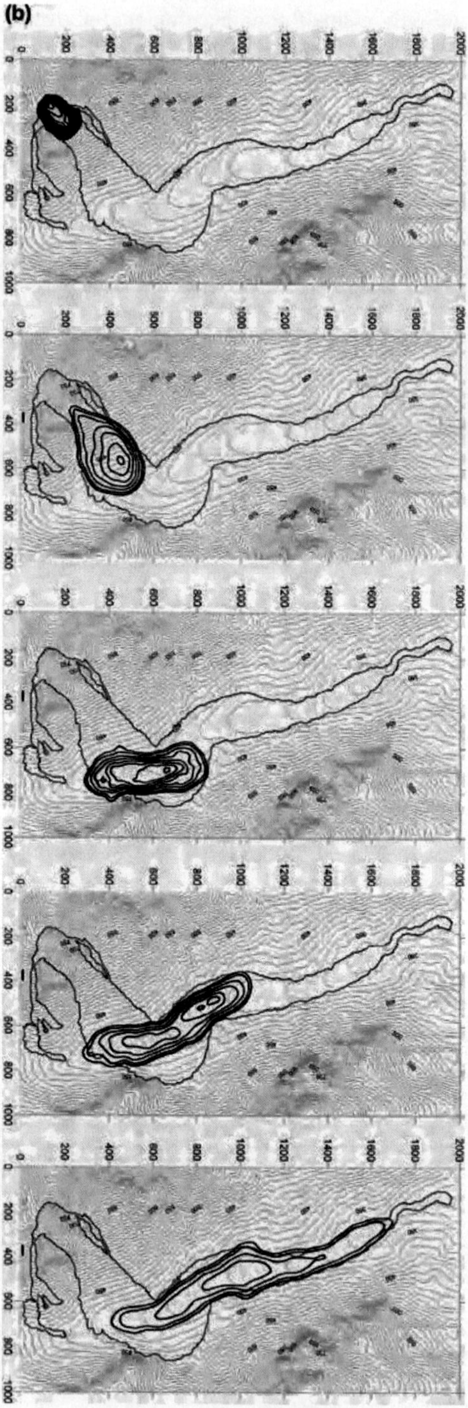

(c)

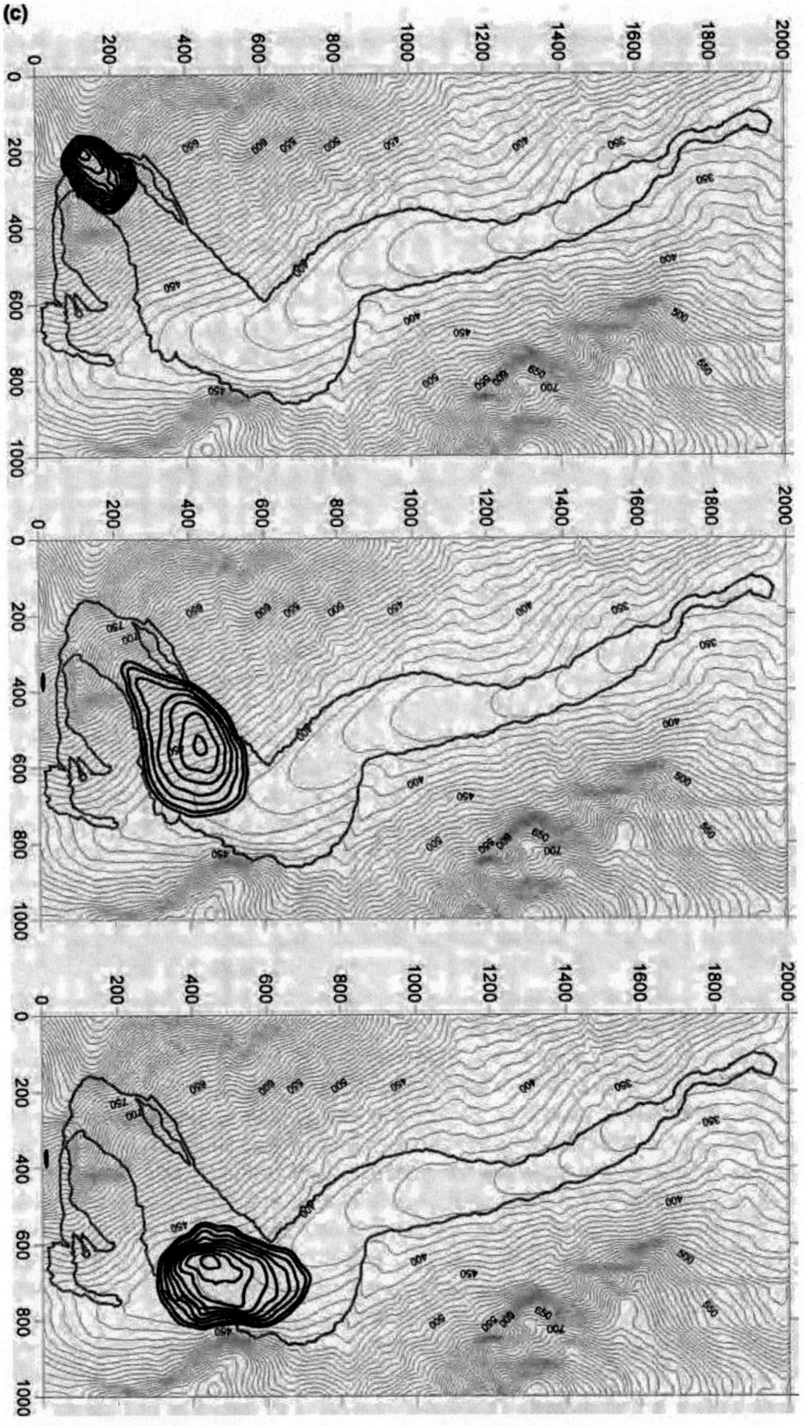

Fig. 18.11 (continued)

both the 2D and 3D models give good agreement with field observations, using the same parameters.

A second analysis was carried out, in which the uniform frictional model was used, with an average bulk friction angle, $\phi_b$ of 20°. As shown on Fig. 18.11c, this produces a very different distribution of debris and forms a landslide dam 16 m high. Thus, neglect of the rheological change due to liquefaction and entrainment of saturated substrate would lead to a very unrealistic result. Models capable of simulating material entrainment have only recently become available. Considerable amount of calibration is required, before these models can be used to make predictions for cases such as the Nomash slide described here.

## 4 Conclusion

Powerful tools exist for analysis of landslide dynamics. These tools have the potential to produce predictions of the geometry of landslide dams. In some cases, the dynamic behaviour of potential landslides can be constrained sufficiently to provide prediction at least of credible ranges of dam geometry. Three-dimensional analyses, which have only become available relatively recently, show particular promise. However, much work remains in calibrating the models so as to facilitate reliable choice of the rheology and allow for the important influence of material entrainment.

**Acknowledgements** The 3D analyses presented here were carried out by Scott McDougall, formerly a Ph. D. student at the University of British Columbia, working under a National Science and Engineering Research Council of Canada Scholarship.

## References

1. Abele, G. (1974) *Bergstürze in den Alpen*. Wissenschaftliche Alpenvereinshefte, Vol. 25. Munich.
2. Chen, H. and Lee, C.F. (2000) Numerical simulation of debris flows, *Canadian Geotechnical Journal* **37**, 146–160.
3. Corominas, J. (1996) The angle of reach as a mobility index for small and large landslides, *Canadian Geotechnical Journal* **33**, 260–271.
4. Gaziev, A. (1984) Usoy landslide dam and Lake Sarez, Tadjikistan, *Proceedings, 4th International Symposium on Landslides*, **1**, pp. 511–515, Toronto.
5. Heim, Albert. (1932) *Bergsturz und Menschenleben*, Zurich, Fretz and Wasmuth Verlag, 218 p.
6. Hungr, O. (1981) *Dynamics of Rock Avalanches and Other Types of Mass Movements*, Ph.D. Thesis. University of Alberta, Edmonton, AB.
7. Hungr, O. (1995) A model for the runout analysis of rapid flow slides, debris flows and avalanches, *Canadian Geotechnical Journal* **32**, 610–623.
8. Hungr, O. (2006) Rock avalanche occurrence, process and modelling, in S.G. Evans, G. Scarascia-Mugnozza, A.L. Strom and R.L. Hermanns (eds.), *Landslides from massive rock slope failure.* NATO Science Series IV, v. 49, Springer, Dordrecht, 243–266.
9. Hungr, O. and Evans, S.G. (1996) Rock avalanche runout prediction using a dynamic model, *Proceedings, 7th International Symposium on Landslides*, Trondheim, Norway **1**, 233–238.

10. Hungr, O. and Evans, S.G. (2004) Entrainment of debris in rock avalanches; an analysis of a long run-out mechanism, *Geological Society of America Bulletin* **116**, 1240–1252.
11. Hungr, O. and McDougall, S. (2009) Two numerical models for landslide dynamic analysis, *Computer & Geosciences* **35**, 978–992.
12. Ischuk, A. (2006) Usoy natural dam: problem of security (Lake Sarez, Pamir Mountains, Tadjikistan), *Journal of Engineering Geology and Environment*, Special Issue **1**, 189–192.
13. Iverson, R.M. (1997) The physics of debris flows, *Reviews of Geophysics* **35**, 245–296.
14. Iverson, R.M. and Denlinger, R.P. (2001) Flow of variably fluidized granular masses across three-dimensional terrain. 1. Coulomb mixture theory, *Journal of Geophysical Research* **106**, 537–552.
15. Li, T. (1983) A mathematical model for predicting the extent of a major rockfall, *Zeitschrift fur Geomorphologie* **27**, 473–482.
16. McDougall, S. and Hungr, O. (2004) A model for the analysis of rapid landslide runout motion across three- dimensional terrain, *Cahadian Geotechnical Journal* **41**, 1084–1097.
17. McDougall, S. and Hungr, O. (2005) Dynamic modelling of entrainment in rapid landslides, *Canadian Geotechnical journal* **42**, 1437–1448.
18. Nicoletti, G. and Sorriso-Valvo, M. (1991) Geomorphic controls of the shape and mobility of rock avalanches, *Geological Society of America Bulletin* **103**, 1365–1373.
19. Sassa, K. (1985) The mechanism of debris flows. *Proceedings, 11th International Conference on Soil Mechanics and Foundation Engineering*, San Francisco vol. 1, pp. 1173–1176.
20. Savage, S.B. and Hutter, K. (1989) The motion of a finite mass of granular material down a rough incline, *Journal of Fluid Mechanics* **199**, 177–215.
21. Scheidegger, A.E. (1973) On the prediction of the reach and velocity of catastrophic landslides, *Rock Mechanics* **5**, 231–236.
22. Schuster, R.L. and Costa, J. (1986) A perspective on landslide dams, in R.L. Schuster (ed.) *Landslide Dams: Processes, Risk, and Mitigation*, American Society of Civil Engineers Geotechnical Special Publication No. 3, 1–20.
23. Sousa, J. and Voight, B. (1991) Continuum simulation of flow failures, *Geotechnique* **41**, 515–538.
24. Voellmy, A. (1955) Über die Zerstorungskraft von Lawinen, *Schweiz. Bauzeitung* **73**, 212–285.

# Chapter 19
# The Grain-Size Distribution of Rock-Avalanche Deposits: Implications for Natural Dam Stability

**Stuart A. Dunning and P.J. Armitage**

## 1 Introduction

Rock avalanches are a catastrophic mass movement derived from the failure of a commonly agreed upon minimum volume of $1 \times 10^6$ m$^3$ of bedrock and associated cover material. The initial failure may initiate as a rockslide, topple or poly-phase event, but transitions to a rock avalanche leaving behind a characteristic granular deposit. The mechanisms of motion that allow rock avalanches to achieve runout distances that are considered excessive are at present unknown, but has been, and still is, the focus of intense research providing a multitude of possible solutions. Historically, in this search for a global mechanism for the runout of rock avalanches the sedimentology of rock-avalanche deposits has often been overlooked at best, poorly described at worst yet the field features are now beginning to drive theories of how they move [7].

As "natural", non blast triggered (cf. Chap. 25 by Adushkin, and Chap. 26 by Korchevskiy et al., this volume) rock avalanches cannot be predicted, monitored or instrumented and are rarely witnessed, it is the resulting rock-avalanche deposit that can provide the best insight into the final moments of motion and the possible processes acting during "flow".

Rock avalanches primarily occur in the high mountains of the world [17], and, after failure travel to valley floors, often impacting upon the drainage network over human [8, 9] and geological timescales [14]. Rock-avalanche deposits are capable of creating high and wide, valley-blocking dams by virtue of their volume and capability to run-out for considerable distances across and down valley. Of the reported landslides that create dams, rock-avalanche deposits are one of the most common forms [3]. Of concern is that is that *if* a landslide dam does fail, it usually does so rapidly; it has been calculated that of those that fail, 50% will fail within 10 days of formation [3]. For those that last above these short-term thresholds, the time to

S.A. Dunning (✉)
Discipline of Geography and Environment, Northumbria University, Newcastle Upon Tyne NE1 8ST, UK
e-mail: stuart.dunning@northumbria.ac.uk

S.G. Evans et al. (eds.), *Natural and Artificial Rockslide Dams*, Lecture Notes in Earth Sciences 133, DOI 10.1007/978-3-642-04764-0_19, © Springer-Verlag Berlin Heidelberg 2011

**Fig. 19.1** Preserved banding sub-facies at the Falling Mountain rock-avalanche deposit consisting of highly fragmented, angular, fine, bands of dark argillite and highly fragmented, angular, coarse, light coloured greywacke. Highly fractured but relatively undisaggregated clasts are common; a large greywacke example is highlighted. The boundaries between bands are sharp with no evidence of mixing. Height of river-cut exposure is approximately 4 m

failure may be measured in years, decades, or even centuries [13] and may never "fail" but slowly erode. Of the factors relating to the stability of a rock-avalanche dam the material properties are deemed critical [3], sensibly based on the history of engineering experience of earth-fill dams.

It appears that as with the mechanisms of motion for rock avalanches, landslide-dam stability study has much to gain from studies of the sedimentology of the resulting deposits. The current state of knowledge adheres to commonly described features; crude inverse grading [5, 13], highly fractured but relatively undisaggregated clasts (Figs. 19.1 and 19.2) [6], preservation of source stratigraphy in the final deposit (Fig. 19.1), and, interaction with the deposit substrate [16].

## 2 Methodology

Recent sedimentological studies have attempted to plug the gaps in current knowledge with a variety of techniques, based on either direct sampling [2, 4] or an observational, facies based approach [20]. Numerous problems persist in the attempt to characterise rock-avalanche sedimentology, not least of which is the issue of how truly representative any sampling regime can be. The presumed minimum volume above which characteristic rock-avalanche sedimentology can be observed is $1 \times 10^6$ m$^3$; to sample just 1% of this minimum mass would involve sieving of 10,000 kg of debris, at the Flims rock-avalanche deposit, more than 1,000,000 kg of debris for a point sample. Any direct sampling regime has to accept this inherent

**Fig. 19.2** Close up view of a highly fractured but undisaggregated greywacke clast at the Falling Mountain rock-avalanche deposit. Distinct "impact" marks are preserved with fractures radiating away from them. These features must have either been transported without the fractures opening and allowing matrix infill and further fracture as observed in Fig. 19.1; or, this represents a fracture event immediately before the cessation of motion of the rock avalanche

problem and be appropriate to the features observed. Rock-avalanche deposits are not simple, structureless, breccias; they preserve original source stratigraphy in the final deposit, often many kilometres from that source. Another common observation is that of crude inverse grading of deposits from a finely fragmented interior to a blocky openwork surface. For these reasons, a random sampling regime should not be applied to any direct sampling of rock-avalanche deposits. Significant variations in sedimentology must be related directly to the features observed or measured, this requires biased sampling, be that towards preserved lithological band types and their boundary effects, grading, height in deposit or distance along flow.

For this research, data from five rock-avalanche deposits have been identified for biased direct sampling; Acheron [21], Falling Mountain [18], Poerua [11, 12], and Round Top [25] in New Zealand, and the Flims rock-avalanche deposit [20] in the Swiss Alps. The volumes involved vary by orders of magnitude but all can be described as valley-confined (TYPE II, III of Costa and Schuster [3],) or down-valley directed. Sampling used a combination of field and laboratory sieving along with laser granulometry for a least errors method [24]. Individual sample sizes were in the order of 15–20 kg and deemed sufficient to determine the grain-size distribution (GSD) of the site-specific features studied and to yield comparative data between and within deposits. A maximum clast size of 256 mm and minimum of 0.002 mm is characterised using this technique; no clasts in the *interior* of deposits sampled exceeded this maximum size. The raw GSD data obtained has been entered into a sedimentological analytical package to calculate the standard suite of descriptive statistics [1].

Further analysis utilised model fitting methods, using the techniques of [15] to test for a fractal distribution of the data and of [23] for a Weibull (Rosin-Rammler) type distribution as described previously for rock avalanche GSDs [22].

## 3 The Grain Size Distribution of Rock-Avalanche Deposits

It rapidly becomes apparent from field observations and data analyses that the preserved stratigraphy found in the interior of deposits is a key control on the GSD. Individual lithologies have GSDs that are significantly different from one another yet remarkably consistent between layers of the same rock type at a site. This can be illustrated with GSD data derived from the Flims rock-avalanche deposit (Fig. 19.3) composed of distinct, alternating preserved bands of dark and light coloured calcareous Jurassic Malm limestone. The GSDs segregates on preserved lithological band type over any other tested variable, this includes height in deposit and distance along flow path (although this is complicated by topographic confinement). It is the lithological and stratigraphic composition of the source region for a deposit that dominates the final GSD and overprints any down-flow fining trends. This lithological control is found to be true for all sampled deposits where the source stratigraphy allows for recognition in the final deposit of distinct preserved rock types.

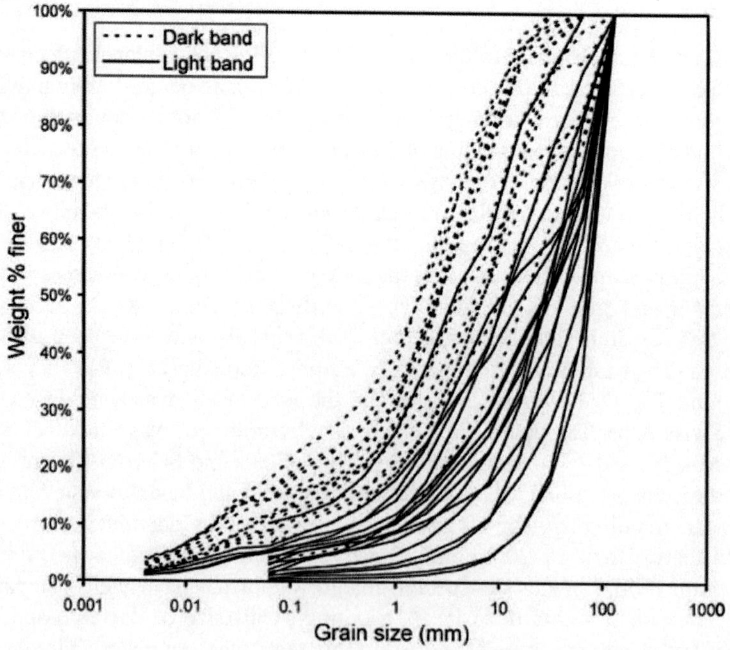


**Fig. 19.3** Grain size distributions for samples taken from the Flims rock-avalanche deposit by preserved band type

**Table 19.1** The range of values for selected descriptive statistics based on five rock-avalanche deposits of varied lithology

| | N | Minimum | Maximum | Mean | SD |
|---|---|---|---|---|---|
| Mean (Phi) | 89 | −0.13 | −5.69 | −2.63 | 1.31 |
| Median (Phi) | 89 | −0.82 | −6.19 | −3.24 | 1.36 |
| Sorting (Phi) | 89 | 1.46 | 3.69 | 2.60 | 0.48 |
| Gravel (%) | 89 | 46.97 | 98.65 | 75.85 | 0.10 |
| Fractal dimension | 89 | 1.95 | 3.04 | 2.44 | 0.20 |

All of the deposits sampled display finely skewed GSDs and can be classed as poorly to very-poorly sorted. A range of descriptive statistics derived from the GSD can be summarized as data ranges (Table 19.1) and as a gravel-sand-clay plot showing the possible and most likely GSD assemblages (Fig. 19.4). Each sample GSD has been tested using both a Weibull approach and a fractal approach. All samples tested prove to be fractally distributed, the range of values are shown in Table 19.1. The fractal dimension of samples increases with decreasing grain size and achieves a maximum value of 3.04 approaching values of both natural and simulated fault gouges [19]. Application of a Weibull distribution [23] has not proven successful and only fits over an extremely limited portion of the GSD of a sample. Examples of both methodologies applied to a single sample are shown in Figs. 19.5 and 19.6.

The notion of inverse grading of rock-avalanche deposits can be tested at several of the sampled sites. Data from the Falling Mountain rock-avalanche deposit (Fig. 19.7) is presented as an example of the findings. The Falling Mountain exposure presented is located 3 km from the source region and is a near full basal to

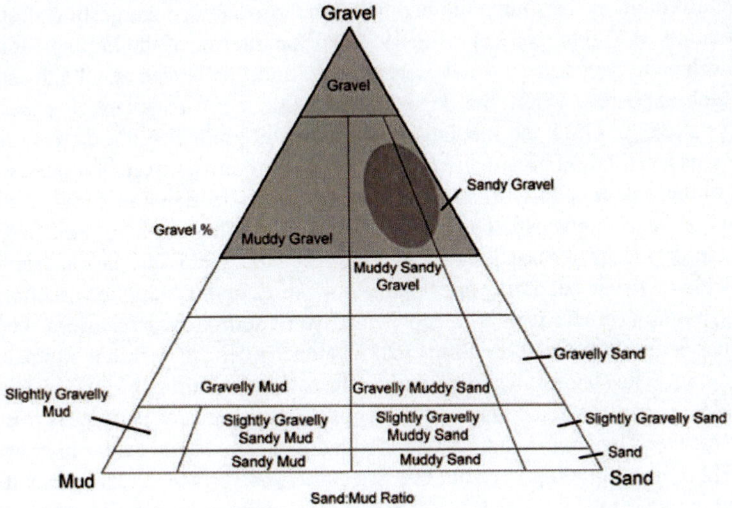


**Fig. 19.4** Gravel-sand-mud plot for the possible (*light*) and usual (*dark*) grain size assemblages for rock-avalanche deposits based on 89 samples from 5 rock-avalanche deposits

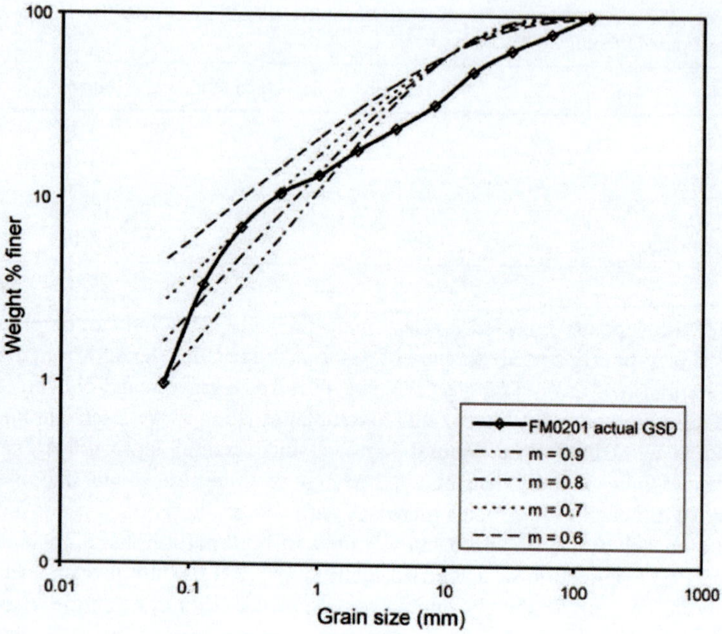


**Fig. 19.5** Application of the Weibull distribution [15] to a sample from the Falling Mountain deposit varying the exponent *m*

surface exposure composed of fine dark argillite bands and coarser, light coloured greywacke, preserved sub-horizontal bands, similar to those observed in the source region, although in the source the bedding is near to vertical suggesting stretching and rotation. It is apparent from Fig. 19.7 that the interior of the Falling Mountain rock-avalanche deposit is not inversely graded, and this is true of all sites studied with such exposures where the deposits have spread and fragmented extensively. The variations in GSD and median, mean or modal grain size are directly related to the preserved lithological band type a sample is removed from. Comparisons of similar lithological band types at varied heights in the section also show no indication of coarsening upwards. The only instances of grading are within thicker units, for example the near-basal greywacke unit (Fig. 19.7); this grading is normal in nature, that is to say becoming finer grained upwards, and it cannot be ascertained if this is an emplacement feature or representative of source rock features. The same trend has been observed at the Flims rock-avalanche deposit within the thicker preserved units. The abnormally coarse unit at the top of the Falling Mountain exposure (Fig. 19.7) is argillite that composes the deposit surface and near surface at this point as a carapace. This unit is sharply bounded below against another, far finer argillite unit. There is no grading between the two units; the surface layer is separate and distinct in its GSD.

Variations in grain size, unrelated to source rock strength/lithology are also observed for the distance from source at Falling Mountain. Proximal exposures

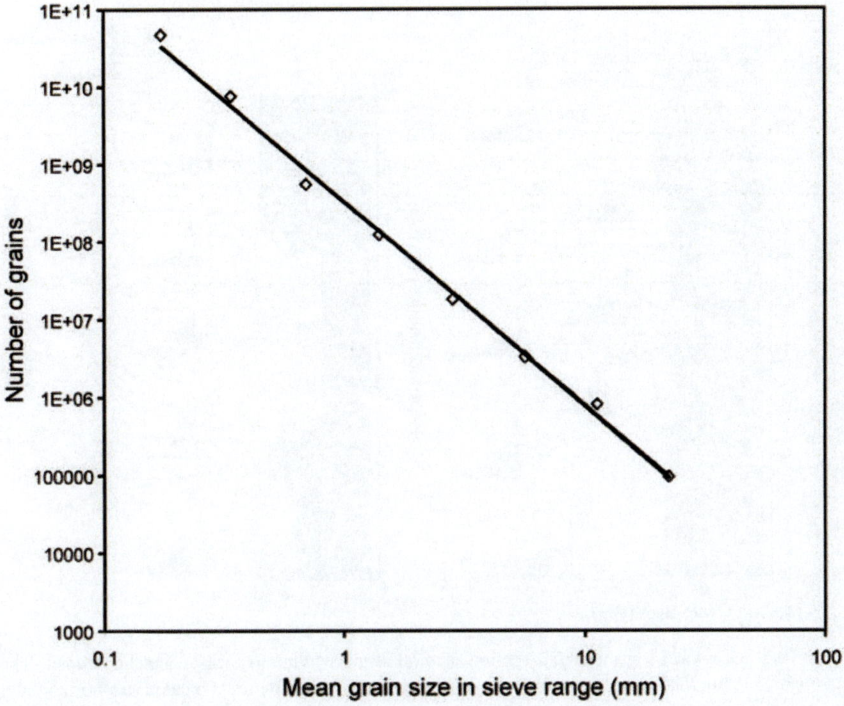


**Fig. 19.6** Application of the fractal method to the same sample as Fig. 19.5. The resulting fractal dimension yielded by the line of best fit is 2.64 with an $r^2$ value of 0.99

show coarse GSDs dominated by very coarse gravel and larger, more distal exposures show a fining of this GSD for the same rock types with a decreasing gravel content (absolute loss of coarse grades). This is a complex phenomena, the results are clearest in the simple case of a rock avalanche leaving the source region and travelling, confined, along a valley such as at Falling Mountain. Deposits of considerable size, or those that spread across and along wide valleys in multiple directions, show far more complicated relationships such as at Flims.

## 4 A Simplified Facies Model for Rock-Avalanche Deposits

Detailed field investigations in combination with GSDs from the interior of the deposits presented above reveal that the deposits investigated are not inversely graded but at their simplest show three distinct facies; a surface and near surface carapace facies, the main interior body facies, and a basal facies that includes the regions of the interior that have incorporated substrate material or are affected by it. A detailed description of the facies is beyond the scope of this paper and for the purposes of this study only a brief summary is required.

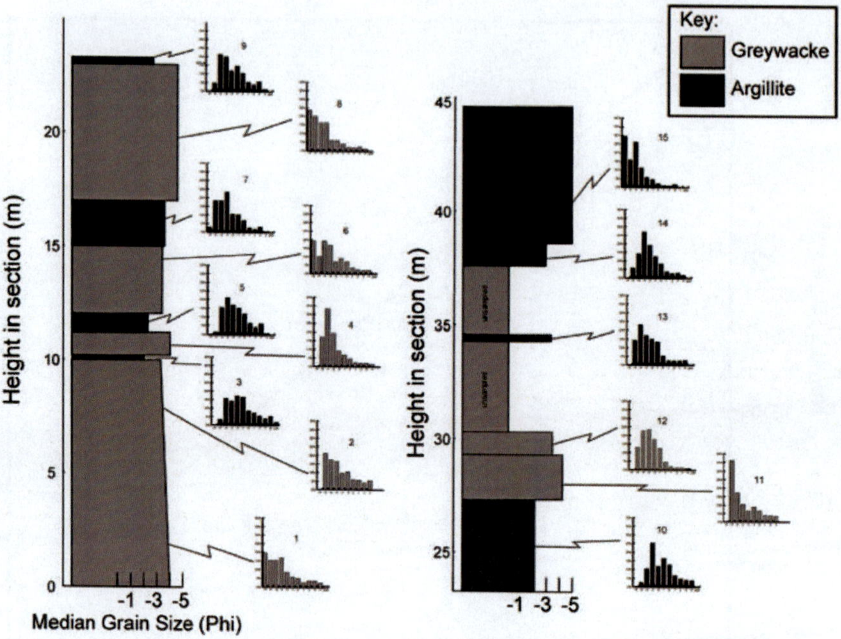


**Fig. 19.7** Graphical log through a section of the Falling Mountain rock-avalanche deposit. The exposure is composed of coarser greywacke units and finer argillite units banded sub-horizontally

## *4.1 The Carapace Facies*

The carapace facies is the coarsest unit of a rock-avalanche deposit and is the material composing the surface and near surface, observations suggest a depth of up to 30% of total-deposit thickness as a maximum in deposits that have not been prematurely stalled by topography. The facies is clast supported and retains source stratigraphy as discreet bands without mixing. This is often obscured as the surfaces of rock-avalanche deposits, particularly proximal regions, can exhibit lobes of material orientated toward the source, interpreted to represent "pulses" of debris in the final stages of failure. The key point of relevance is that the facies is sharply bounded below (Fig. 19.7), with the boundary defined as the line below which all material is intensely fragmented, matrix supported and showing fragmentation derived features – the body facies. Observational and experimental evidence suggest that the carapace facies may show inverse grading but this is a near surface phenomena and not representative of the entire depth of deposit. Deposits that are relatively thin, either through runout of a low volume of material (Acheron for example), or through unconfined spreading, may consist almost entirely of the carapace facies. For reasons like this it is, entirely understandable that observations/measurements have been made on the "inverse grading" of deposits, since it is often either an exposure primarily of a graded carapace or an erroneous observation made on the basis of a coarse near surface, a fine interior and qualitative assumptions of grading between the two. The carapace facies is of critical importance to rock-avalanche dam stability because it is the material forming the dam crest, and is of course the only material available for observation on arrival at a recent event requiring urgent hazard assessment.

### 4.2 The Body Facies

The body facies forms the main body of rock-avalanche deposits and is usually the most voluminous in valley-confined deposits that have undergone significant runout. Features of note include, highly fractured but relatively undisaggregated clasts, matrix support, impact marks with radial fracture preserved on clasts (Fig. 19.2), and as described, preserved source stratigraphy, although orientations and thicknesses are considerably altered. A number of sub-facies can be identified within the body facies but they are outside of the scope of this paper and more related to variations in fragmentation and topographic confinement.

### 4.3 The Basal Facies

This base of a rock-avalanche deposit is rarely exposed, and never in the case of hazard assessment planning at recent dam forming deposits. The basal facies not only includes rock-avalanche material that is interpreted to have interacted with the substrate, but also the substrate material altered by passing rock avalanche, including that bulldozed in front of the rock avalanche. This includes erosion and entrainment of substrate material into the base of the moving debris, and also subsequent deposition of modified mixes of substrate and rock-avalanche material. This is of course substrate dependant, be it soft erodable valley fills and surface vegetation that can be carried for some distance, or regions of bedrock that undergo superficial erosion as the mass passes – a good example being the bedrock gorge at Falling Mountain, subsequently re-exhumed by erosion of the rock-avalanche deposit. The boundary between the basal facies and the body facies is often indistinct and variable in its height above the assumed deposit base. The facies is, however, interpreted to be the smallest by volume within a rock-avalanche deposit, although in deposits that have spread and thinned completely unrestricted over deformable substrates it may only be the basal and carapace facies that remain.

## 5 Discussion

The results of a sedimentological study provide a number of new advances in the field as well as interpretations that take the study beyond the simply descriptive. Detailed description has been required for the sedimentology of rock-avalanche deposits, be they dam forming or not. Many past, and current models of their mechanisms of motions ignore these first principles of field study; models should fit with quantified deposit features rather than simply being a "black box" approach that may yield the required travel distance or dam-overtopping time, without consideration of the deposit features and sedimentology. It can be hoped that usage of a facies approach and the GSD data presented allows a brief glimpse into the sedimentology of rock-avalanche deposits. The data raise questions on how to relate initial source rock strength to final deposit GSD, and the effect of multiple lithological units. Deposits such as Falling Mountain show clear fragmentation segregation, fragmentation is far more intense along the weaker argillite layers that have correspondingly higher fractal dimensions resembling fault gouge. Investigation will reveal if this is simply a response to the relative strengths and initial fracture distribution of the source rock, or if in fact the fragmentation does localise in the weaker units. All interior rock-avalanche GSDs prove fractal in this study and those selected and tested from the literature where sampling methodology allows. This in essence shows that rock-avalanche deposit interiors are self-similar at scales of observation. The values of the fractal dimension of the rock-avalanche GSDs vary (Table 19.1) but all show an average of ~2.44. This is below the value of 2.58 interpreted to represent equal probability of fracture across all size grades, a configuration representing maximal spacing of same size clasts [19]. Values above 2.58 are interpreted to represent configurations with an excess of fines [19]. It can be interpreted for the rock-avalanche deposits studied that there is an excess of coarse material, that motion has stopped before maximal "cushioning" of same-size clasts could be achieved. It also follows that the deposits studied were, immediately prior to the cessation of motion, preferentially fragmenting these coarser clasts. This is backed by the fining of rock-avalanche deposits with distance from the source (an absolute loss of coarse grades) and a significant statistical relationship showing that as grain size becomes finer, the fractal dimension increases, becoming closer to the 2.58 value. Observations show that maximal "cushioning" has not been achieved, i.e., there are numerous fragmented but relatively undisaggregated clasts at all distances from the source. These can be interpreted to either represent transported fragmentation events that occurred near the source; or, they represent fragmentation events immediately prior to stopping. Those with significant matrix infill in the fractures represent may slightly older fragmentation events.

## 6 Application of the Data to Rock-Avalanche Dams

### *6.1 Prediction of Parameters*

The data set presented is important for the study of rock-avalanche dams in a number of ways. Firstly, and possibly most importantly, it provides the GSD and basic properties of the material forming the dam, something that has previously been reported as critical [3]. The distinction of a carapace facies and body facies allows for an

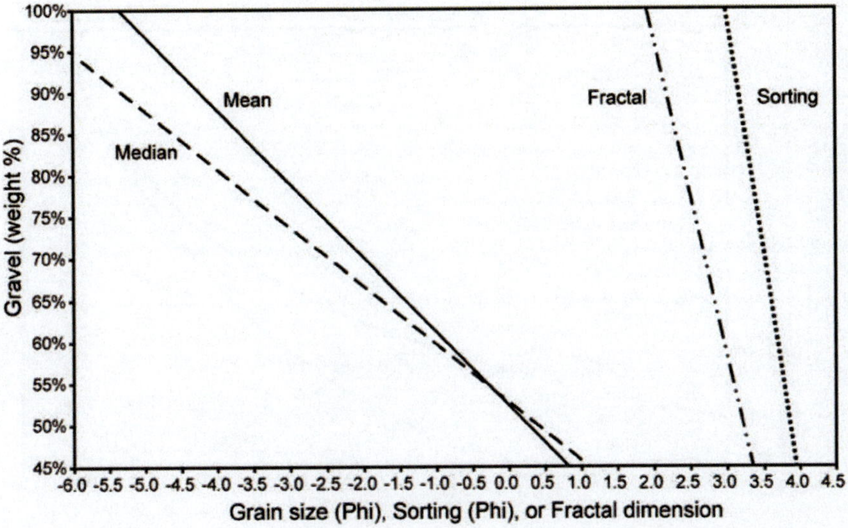


**Fig. 19.8** Predictive plot generated after study of rock-avalanche deposit interiors to calculate mean and median grain sizes (Phi), fractal dimension, or sorting (Phi) based on statistical relation to weight percent gravel. The plot can be used to calculate each parameter from any single measurement; mean, median or weight percent gravel are the recommended inputs

estimation of the properties of the material unobserved on arrival at a recent rock-avalanche dam forming deposit. Preliminary results for similar data in the carapace facies, taken in a relatively fine grained carapace formed from argillite at Falling Mountain, allow for conservative values to compare against the main interior values. For estimation of the properties of any particular rock-avalanche dam, predictive plots have been generated (Figs. 19.8 and 19.9) based on a simple measure or estimation of weight percent gravel. From these, values of mean and median grain size, sorting, fractal dimension, and permeability can be estimated. Deposits vary in their relative coarseness by lithology. Although the relationships to weight percent gravel are applicable for all lithologies, more discreet zones of lithology are required so that even the simple measure/estimation of gravel is no longer required for rapid assessment.

## 6.2 Data Used for Modelling

Grain size data is a key component in various dam break/dam breach software packages. These models use inputs such as median grain size of the material composing the dam interior and exterior to calculate the rate of dam crest breaching via overtopping – the most common cause of landslide dam failure [3]. However, the sedimentology presented above, with its distinct differences between the dam crest formed of the carapace facies and the main bulk formed of the body facies

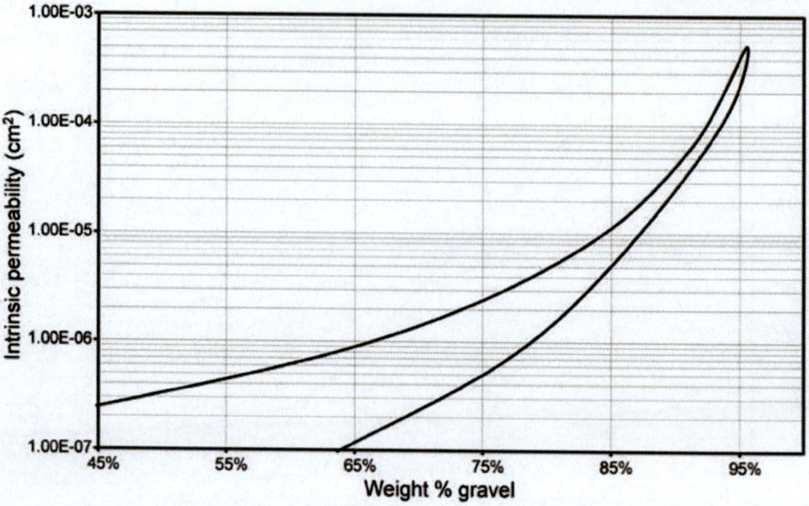


**Fig. 19.9** Predictive plot showing the range of intrinsic permeability for rock-avalanche deposits based upon a measure of weight percent gravel. Alternative plot for hydraulic conductivity is also available based on the data set

raise some interesting questions. From the GSD it is possible to calculate simple approximations of the hydraulic conductivity and intrinsic permeability (Fig. 19.9) and, from density measurements and wetting/drying experiments, it is possible to calculate porosity of the debris assemblage. These properties, along with several that are more calculable from case-study examples, allow us to begin to assess the role of sedimentology on dam stability. Although rock-avalanche dams are commonly assumed to fail during overtopping, it is interesting to study the process of the impounded lake filling against the dam until that overtopping since the final failure mechanism may actually be face-failure rather than overtopping (which should be confined to the headward erosive flow of water over a dam). This should include the effect of the sedimentology described above as opposed to a simple heterogeneous debris mass. In this preliminary modelling, using both Geo-Slope 2003 [10] and its incorporated packages, SEEP/W and SLOPE/W, an idealised rock-avalanche dam based upon the 1999 failure into the Poerua River, Westland, New Zealand involving 10–15 $Mm^3$ of schistose bedrock and colluvium [11, 12] is used. The Poerua rock-avalanche dam overtopped within 48 h but remained intact with flow confined in a relatively stable spillway with no obvious signs of erosion [11, 12]. The dam did, however, breach with an associated flood 6 days after emplacement after heavy rainfall. Preliminary modelling focuses on the emplacement of a dam that is relatively stable under normal flow conditions when overtopped, such as described initially for Poerua [11, 12]. Under these conditions the sedimentology of the dam becomes even more critical as the impounded lake fills as will be illustrated below. The modelling uses a 2-D section of a rock-avalanche dam at right angles to the inferred runout direction of a rock avalanche that has crossed a narrow valley and completely blocked it. The section is at the lowest point of the dam crest, approximately 120 m high, with the lake impounded filling within 48 h to the overtopping point and then remaining at that maximum level. The dam is founded on schist bedrock with a valley gradient of 6°; the dam face is 24° downstream and 12° upstream, as at Poerua [11, 12]. The material properties of the dam are based upon GSD sampling carried out at Poerua and results in values of hydraulic conductivity for the carapace facies of $1.24 \times 10^{-1}$ and $3.69 \times 10^{-4}$ m/s for the body facies both equal in the x and y cartesian directions. For the preliminary modelling a comparison between a dam made entirely of the body facies and one consisting of the observed, but simplified sedimentology of a carapace and body facies is presented. Further details on the modelling conditions such as fitting data to a conductivity curve, the node spacing, strength, friction angles, volumetric water content function, and iteration conditions are to be discussed further in work in preparation.

An obvious exclusion to the modelling presented here is that of a dam formed only by the carapace facies. Preliminary results indicate that such a sedimentology is not suitable for forming a dam that can fill to overtopping since flow is able to move through the mass rapidly. However, assumptions for this form of sedimentology are common based on the surficial material presented to first investigators, and the assumption is dangerous that a lake will not form [9].

### *6.3 Seepage Analyses*

Seepage analysis has been carried out using Geo-Slope (2003) SEEP/W [10], a finite element product to model movement and pore-water distribution within porous materials such as soil and rock. SEEP/W is able to model both saturated and unsaturated flow, the inclusion of unsaturated flow in groundwater modelling is important for obtaining physically realistic analytical results. SEEP/W accounts for the drainage of water from soil pores, or water filling soil pores, and the changes in hydraulic conductivity function that occur in a transient flow system. The computed head distribution can then be used in SLOPE/W slope stability analysis, particularly powerful in the case of transient systems.

### *6.4 Slope-Stability Analyses*

Slope-stability analyses have been carried out using Geo-Slope (2003) SLOPE/W [10], a limit equilibrium theory based programme that calculates the factor of safety (FoS) for earth and rock slopes. Using limit equilibrium, it has the ability to model heterogeneous soil types, complex stratigraphic and slip surface geometry, and variable pore-water pressure conditions using a large selection of soil models. Analyses can be performed using deterministic or probabilistic input parameters. Porewater pressure conditions can be specified in SLOPE/W in several ways, including finite element computed porewater pressures. SLOPE/W has the ability to import porewater pressure data computed by SEEP/W, VADOSE/W or SIGMA/W, three of GEO-SLOPE's finite element programs. This capability is especially useful for performing slope stability analyses where the groundwater flow conditions are transient and/or significantly affected by the stress state within the soil. In this case the computed head distribution for each time increment in SEEP/W has been imported into SLOPE/W to determine the FoS as a function of time as the impounded lake fills to overtopping and beyond, for the two sedimentological models. The style of failure in use is interpreted to be of a form that would remove the dam crest allowing for a catastrophic breach. Larger failures that remove the entire dam in one failure are deemed an unlikely occurrence. A brief summary of conditions derived, and through approximation with known materials, yields cohesion of zero for both facies, a unit weight of 20 $kN/m^2$ and a friction angle close to the maximum angle of repose at the dam, 24° as the "worst credible" value.

More complex modelling has been carried using FLAC (produced by Itasca) in which the failure surface evolves naturally, and in all cases is curved but FoS values and the depths of failure approximate well to the GEO-SLOPE values for the simplified purposes of this paper and so these results are presented.

## *6.5 Modelling Results*

### 6.5.1 Single Facies Model

Seepage analysis shows a typical progressive saturation front for a homogenous earth dam with a steadily filling impounded lake. Figure 19.10 shows the position of the phreatic surface at 6 h intervals from the commencement of lake filling until 12 h after the lake had reached crest level. Progression is initially rapid, due to the filling of the lake and associated rising head on the upstream face. As the lake level

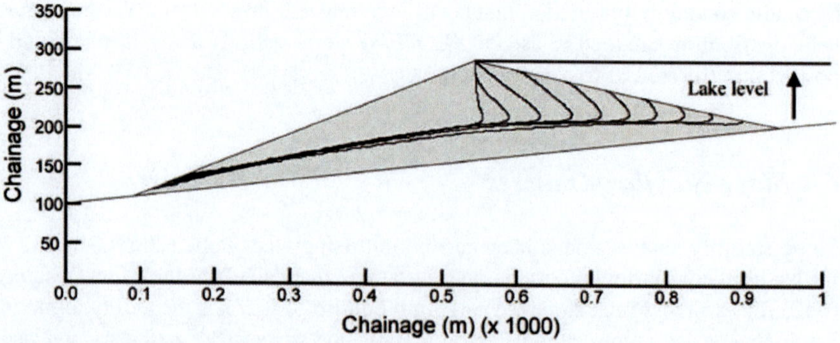


**Fig. 19.10** Six hourly positions of the phreatic surface within a single facies dam from 0 to 60 h. Note the steady progression of the phreatic surface to overtopping and the presence of a low seepage point on the downstream dam face

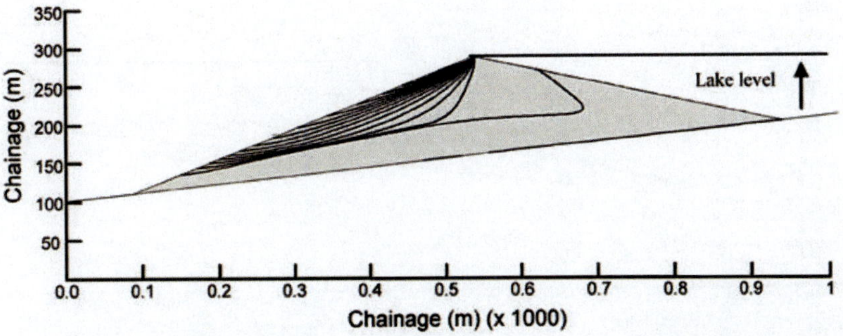


**Fig. 19.11** Forty hourly positions of the phreatic surface within a single facies dam from 0 to 720 h. Note the steadily increasing position of the seepage point on the downstream dam face

reaches the dam crest, the rate of progression of the phreatic surface lowers as the driving head reduces within the embankment with distance from the downstream face to the phreatic surface. After 720 h, the seepage has not yet reached steady state, however the phreatic surface has progressed sufficiently to cause a reduction of the FoS to below unity. Figure 19.11 shows the progression of the saturation front at 40 h intervals to 720 h.

### 6.5.2 Simplified Observed Facies Model

In this case seepage analysis shows an identical progressive saturation front to 38 h. Figure 19.12 shows the positions of the phreatic surface at 2 h intervals from the commencement of lake filling until the lake has reached crest level. After 38 h, the lake level has reached the carapace facies. As the lake level rises above the carapace, the phreatic surface progresses more rapidly through the carapace than through the less permeable body facies. This results in a downstream "tonguing"

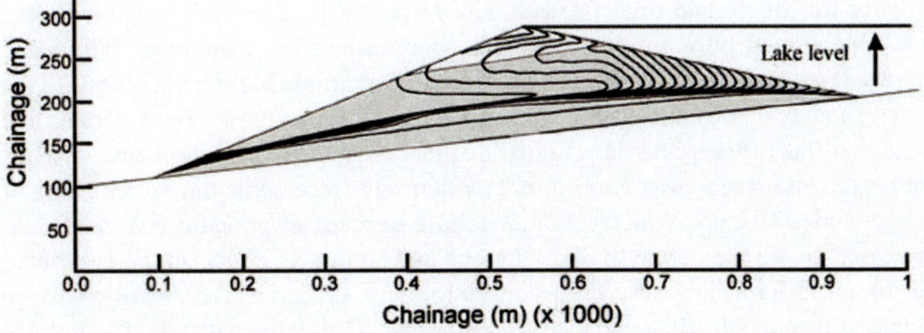


**Fig. 19.12** Two hourly positions of the phreatic surface within a two facies dam from 0 to 48 h. Note the rapid advance of the surface once the lake level reaches the carapace facies and the high point on the downstream dam face that seepage occurs

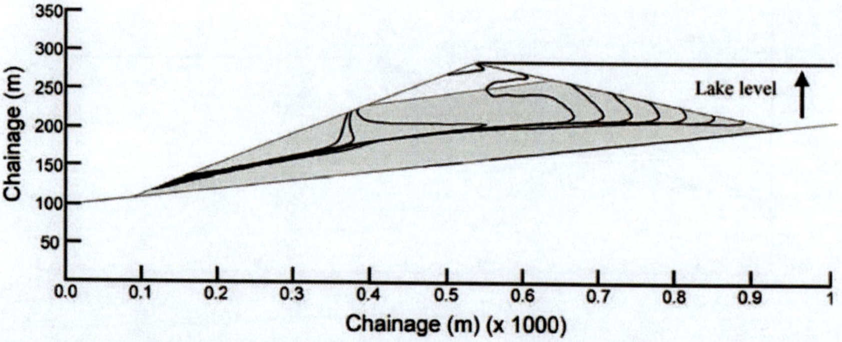


**Fig. 19.13** Six hourly position of the phreatic surface within a two facies dam from 0 to 60 h. The carapace facies offers virtually no resistance to flow in comparison with the body facies and the phreatic surface reacts accordingly

projection of the phreatic surface along, above, and slightly below the boundary between the body facies and the carapace facies. As the lake level approaches the dam crest, the saturation front progresses almost entirely through the carapace, and partially through the body. Figure 19.13 shows the positions of the phreatic surface at 6 h intervals from the start of the impounded lake filling until 12 h after the lake had reached crest level. After 48 h have passed, the carapace becomes entirely saturated, taking just 2 h to do so, and the saturation front continues to progress through the body up to 60 h. After 60 h, the seepage has not yet reached steady state, however the phreatic surface had progressed sufficiently to cause a reduction of the FoS to well below unity.

### 6.5.3 Slope Stability

Computed head distribution for time increments in the SEEP/W analyses were used in SLOPE/W slope stability analyses to determine the Factor of Safety as a function of time and therefore lake level. The resulting FoS are presented in Figs. 19.14 and 19.15 for the single and dual facies model respectively. The FoS computed were for a translational failure surface using a Morgenstern-Price analyses. The same failure surface has been used in each analysis, and approximately corresponds to the carapace facies – body facies contact, a likely failure level based on observational evidence of lake filling and dam failure sequences [9]. For both the 1 and 2 facies systems the FoS remains as 3.03 from 0 to 38 h. This represents the initial filling of the impounded lake up to the level of the failure surface, the phreatic surface has not progressed above the defined failure surface and so has no effect on the computed FoS. From 39 h onwards, the FoS reduces for each system as the phreatic surface progresses through the dam and also above the specified failure surface. The FoS for the two facies system reduces at a higher rate than for the 1 facies system, falling below unity after 49 h, the one facies system falls below unity after approximately 450 h.

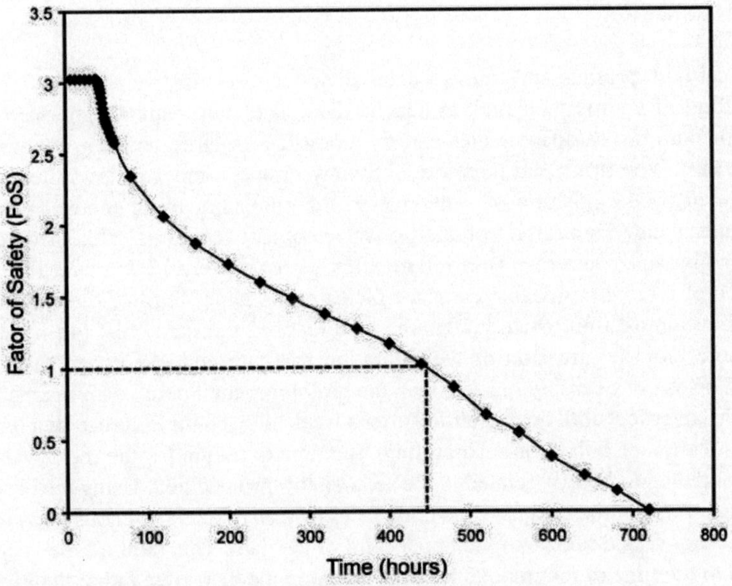


**Fig. 19.14** Variation of the FoS of the rock-avalanche dam with time (lake level increase) using a specified block slide geometry and a Morgenstern-Price analysis method for a heterogeneous dam

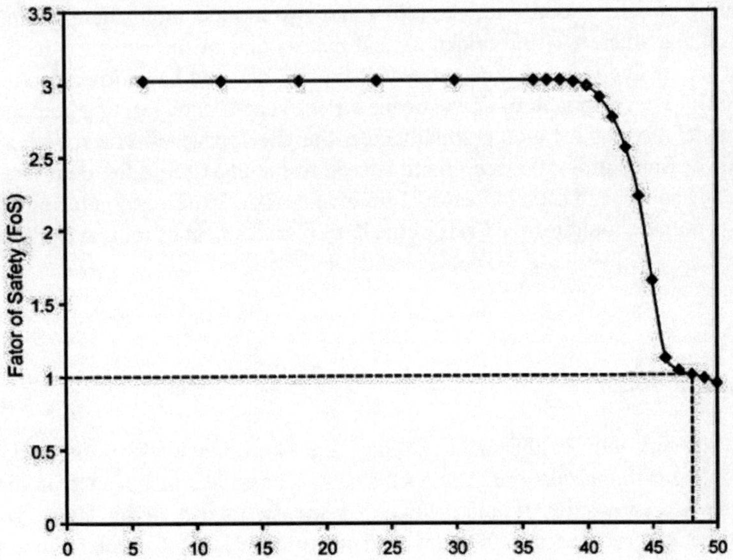


**Fig. 19.15** Variation of the FoS of the rock-avalanche dam with time in hours (lake level increase) using a specified block slide geometry and a Morgenstern-Price analysis method for a dam showing a developed carapace and body facies

### *6.6 Discussion*

Identical dam profiles and rates of external water level rise were used within the modelling of a simplified rock-avalanche dam. The same unit weight was specified for both the carapace facies and the body facies. Differences in the position of the saturation front, and therefore FoS between the 1 and 2 facies systems were controlled by the position of water within the embankment as controlled by the hydraulic conductivity and volumetric water content functions. The specification of a single facies dam results in failure after approximately 18.75 days. The introduction of a more permeable carapace facies in the upper third of the dam results in failure almost immediately after the lake reached the dam crest (48 h), due to the more rapid progression of the saturation front through the upper part of the dam. This is considerably quicker than the prototype; the Poerua dam maintained a "steady" overflow until heavy rainfall after a week allowed for failure, either through downstream face failure or head-cutting. The time to failure for the rock-avalanche dam modelled is directly related to the sedimentology specified. Using the observed sedimentology of the carapace and body faces instead of a homogenous mass results in a far more rapid failure of the upper third of the dam. This failure time is closely linked to the time of the impounded lake reaching the dam crest rather than just the body-carapace boundary. It is possible that this sort of failure mechanism is responsible for the observational evidence of rock-avalanche dams failing rapidly due to overtopping [3]. Future work will utilise more complex sedimentological models for modelling of dam response to lake filling and also more complex dam geometries. Of particular interest is the thickness and distribution of the carapace facies that appears to be key in the failure sequence. From a risk and hazard perspective, the location and development of downstream surface seeps is crucial. In a homogenous dam model the seep location gradually creeps up the downstream face; with a more realistic sedimentology the seep point is seen to rapidly rise up the dam face with increasing speed, and is an indicator of imminent risk of failure. Monitoring of such seepage points should form a basis of the hazard assessment of rock-avalanche dam failure.

## 7 Conclusions

This paper has outlined the sedimentology of rock-avalanche deposits, a common natural dam forming material. The results show a considerable advance on the current accepted knowledge of both the main fragmented interior of the deposit and the overall zonal structure; this includes evidence for the lack of crude inverse grading. This detailed sedimentology should form a vital part of any theory for the mechanism(s) of motion that allow(s) rock avalanches to travel for such excessive distances. As an example of the use of such sedimentological data, preliminary finite element and limit equilibrium modelling has been presented. Although simplified at present, the modelling shows the significant role that sedimentology can play in the time of failure of a rock-avalanche dam. This variation in failure time has major consequences for hazard assessment and subsequent evacuation procedure. It has been shown that the monitoring of seepage fronts in terms of its progression up the downstream dam face could be a key indicator of internal sedimentology of the dam and lead to subsequent predictions of its stability. Numerous possibilities exist for future work related to this study. We must better constrain the variation in both rock-avalanche dam geometry and its effect, if any, on dam sedimentology. The pro-forma approach to sedimentology and stability using plots such as those presented in Figs. 19.8 and 19.9 can be developed to the point of lithological zonation reducing the need to sieve material and cut trenches.

In final summation, the sedimentology of rock-avalanche deposits should form a vital part of models for the processes during transport, and for the assessment of the post-emplacement behaviour of the resulting debris.

**Acknowledgements** This research and attendance of the NATO Advanced Research Workshop in 2004 from which this paper has developed was funded by the International Landslide Centre, University of Durham. The work was carried out in association with Dr. Tim Davies, Canterbury, New Zealand, and Dr. Mauri McSaveney, Geological and Nuclear Sciences, New Zealand, their help is much appreciated. The 2004 comments of the editorial committee when an earlier version of this paper was submitted are gratefully acknowledged.

## References

1. Blott, S.J. and Pye, K. (2001) Gradistat: A grain size distribution and statistics package for the analysis of unconsolidated sediments, *Earth Surface Processes and Landforms* **26**, 1237–1248.
2. Casagli, N., Ermini, L. and Rosati, G. (2003) Determining grain size distribution of the material composing landslide dams in the Northern Apennines: Sampling and processing methods, *Engineering Geology* **69**, 83–97.
3. Costa, J.E. and Schuster, R.L. (1988) The formation and failure of natural dams, *Geological Society of America Bulletin* **100**, 1054–1068.
4. Crosta, G.B., Frattini, P. and Fusi, N. (2007) Fragmentation in the Val Pola rock avalanche, Italian Alps, *Journal of Geophysical Research – Earth Surface* **112**, F01006.
5. Cruden, D.M. and Hungr, O. (1986) The debris of the Frank Slide and theories of rockslide-avalanche mobility, *Canadian Journal of Earth Sciences* **23**, 425–432.
6. Davies, T.R., McSaveney, M.J. and Hodgson, K.A. (1999) A fragmentation-spreading model for longrunout rock avalanches, *Canadian Geotechnical Journal* **36**, 1096–1110.
7. Davies, T.R. and McSaveney, M.J. (2009) The role of rock fragmentation in the motion of large landslides, *Engineering Geology* **109**, 67–79.
8. Dunning, S.A., Mitchell, W.A., Rosser, N.J. and Petley, D.N. (2007) The Hattian Bala rock avalanche and associated landslides triggered by the Kashmir earthquake of 8 October 2005, *Engineering Geology* **93**, 130–144.
9. Dunning, S.A., Rosser, N.J., Petley, D.N. and Massey, C.I. (2006) Formation and failure of the Tsatichhu landslide dam, Bhutan, *Landslides* **3**, 107–113.
10. Geo-Slope (2003) Geo-Slope Office (SLOPE/W, SEEP/W, QUAKE/W). Geo-Slope International, Calgary, AB.
11. Hancox, G.T., McSaveney, M.J., Davies, T.R., Hodgson, K. and Daniel, R. (2000) The October 1999 landslide dam in Poerua River, Westland, New Zealand., New Zealand Society on Large Dams (NZSOLD) Symposium, November 2000.

12. Hancox, G.T., McSaveney, M.J., Manville, V.R. and Davies, T.R. (2005) The October 1999 Mt. Adams rock avalanche and subsequent landslide dam-break flood and effects in Poera River, Westland, New Zealand, *New Zealand Journal of Geology and Geophysics* **48**, 683–705.
13. Hewitt, K. (1998) Catastrophic landslides and their effects on the Upper Indus streams, Karakoram Himalaya, northern Pakistan, *Geomorphology* **26**, 47–80.
14. Hewitt, K., Clauge, J.J. and Orwin, J.F. (2008) Legacies of catastrophic rock slope failures in mountain landscapes, *Earth Science Reviews* **87**(1–2), 1–38.
15. Hooke, R.L. and Iverson, N.R. (1995) Grain-size distribution in deforming subglacial tills: Role of grain fracture, *Geology* **23**, 57–60.
16. Hungr, O. and Evans, S.G. (2004) Entrainment of debris in rock avalanches: An analysis of a long runout mechanism, *Geological Society of America Bulletin* **116**, 1240–1252.
17. Korup, O., Clague, J.J., Hermanns, R.L., Hewitt, K. and Strom, A.L. (2007) Giant landslides, topography, and erosion, *Earth and Planetary Science Letters* **261**(3–4), 578–589.
18. McSaveney, M.J. and Davies, T.R. (1999) The Falling Mountain rock avalanche of 1929, Arthur's Pass National Park, New Zealand, Institute of Geological and Nuclear Sciences Science Report, 15 pp.
19. Sammis, C.G., White, P., Osborne, R.H., Anderson, J.L. and Banerdt, M. (1986) Self-similar cataclasis in the formation of fault gouge, *Pure and Applied Geophysics* **124**, 53–78.
20. Schneider, J.-L., Wassmer, P. and Ledésert, B. (1999) The fabric of the sturzstrom of flims (Swiss Alps): Characteristics and implications on the transport mechanisms, *Earth and Planetary Sciences* **328**, 607–613.
21. Smith, G.M., Davies, T.R. and McSaveney, M.J. (2006) The Acheron rock avalanche, Canterbury, New Zealand – morphology and dynamics, *Landslides* **3**(1), 62–72.
22. Strom, A.L. and Pernick, L.M. (2004) Utilisation of the data on rockslide dams formation and structure for blast-fill dams design, in K. Abdrakhmatov, S.G. Evans, R. Hermanns, G. Scarascia Mugnozza, and A.L. Strom (eds.) Security of natural and artificial rockslide dams, extended abstracts, NATO Advanced Research Workshop, Bishkek, Kyrgyzstan, June 8–13, 2004, 177–182.
23. Weibull, W. (1951) A statistical distribution function of wide applicability, *Journal of Applied Mechanics* **18**, 837–843.
24. Wen, B., Aydin, A. and Duzgoren-Aydin, N.S. (2002) A comparative study of particle size analyses by sieve-hydrometer and laser diffraction methods, *Geotechnical Testing Journal* **25**, 1–9.
25. Wright, C.A. (1998) The AD 930 long-runout Round Top debris avalanche, Westland, New Zealand, *New Zealand Journal of Geology and Geophysics* **41**, 493–497.

# Chapter 20
# Incorporating the Effects of Groundwater and Coupled Hydro-Mechanical Processes in Slope Stability Analysis

**E. Eberhardt and D. Stead**

## 1 Introduction

The presence of groundwater in a rock or soil slope mass can have several destabilizing effects including those related to freeze-thaw, weathering, and/or erosion (either at the slope's toe or internally through piping). Although some of these processes may be treated in a slope stability analysis, they generally require the use of complex techniques that are applied in a highly conceptualized fashion (e.g. weathering/strength degradation [8]; seepage erosion/piping [6]). In contrast, it is the counterbalancing effect of pore pressures described through the effective stress relationship that is taken into consideration in most deep-seated slope stability analyses. Slope stability analyses involve three main procedures depending on the importance and complexity of the slope geology – kinematic, limit equilibrium and numerical. The first stage should always be a kinematic assessment of the feasibility of varied slope failure mechanisms; this is especially important in the case of rock slopes. Such analyses are usually undertaken using stereographic techniques that check for daylighting structures and whether sliding is possible assuming a specified friction angle. Pore and/or joint water pressures are not normally considered as kinematic analyses are meant to focus on the spatial configuration and alignment of persistent discontinuities.

Having identified the potential failure mechanism, limit equilibrium and numerical modelling codes may then be employed to investigate the influence of groundwater on the stability state of the slope. Variations in the mode of application of these methods generally depend on the objectives of the analysis and the quality of the field observations and in situ monitoring data available to constrain the analysis. For most analyses, the key data constraint is the location of the groundwater table as determined through piezometer readings or field observations of surface springs.

E. Eberhardt (✉)
Geological Engineering/Earth and Ocean Sciences, University of British Columbia, Vancouver, BC, Canada V6T 1Z4
e-mail: eeberhar@eos.ubc.ca

S.G. Evans et al. (eds.), *Natural and Artificial Rockslide Dams*, Lecture Notes in Earth Sciences 133, DOI 10.1007/978-3-642-04764-0_20, © Springer-Verlag Berlin Heidelberg 2011

**Table 20.1** Conventional and advanced methods of groundwater-based slope analysis

| Method | Groundwater input | Advantages | Limitations |
|---|---|---|---|
| Limit equilibrium | Multiple water tables/piezo-lines. Pore water pressure ratio (Ru and Hu coefficients). Total/pressure head and pore pressure grid points. Saturated and anisotropic permeability. Unsaturated permeability models. User-defined permeability models. Nodal inflow/outflow. Ground water boundary conditions. Steady state or transient groundwater analysis. | Minimal input requirements. Solutions for several modes of failure (rotational, planar and wedge, etc.). Rapid sensitivity analysis to show influence of water pressure. Can import/incorporate data from finite element groundwater analysis. Probabilistic analyses possible. Unsaturated conditions may be considered. Can be readily incorporated with GIS. | Assumed failure mechanism. Pore pressure assumptions simplistic for jointed rock. Deformation and intact failure not allowed for. Influence of water pressures on failure mechanism not considered. Temporal effects ignored or difficult to analyse (e.g. drawdown). Groundwater flow analysis needed to adequately consider influence of seepage forces. |
| Continuum modelling (*e.g. finite-element, finite-difference*) | Water table and boundary conditions (e.g. flux, pore pressure, etc.). Saturated or unsaturated analysis. Fluid bulk modulus and density. Moisture content and negative pressure head. Permeability tensor. Permeability – volumetric strain relation. Poroelastic parameters (α, β). Steady state or transient coupled analysis. | 2-D and 3-D, uncoupled or fully coupled hydro-mechanical modelling possible. Single phase or two phase flow. Useful for modelling effect of seepage forces. Can incorporate dynamic analysis. Output includes effective stress, pore pressures, displacements, failure zones, seepage vectors, and particle tracking. | Users must be well trained and experienced. Need to be aware of model limitations (e.g. boundary effects, mesh aspect ratios, etc.). Availability of groundwater input data generally poor with scale dependency problems. Required input parameters not routinely measured. Can be difficult to perform sensitivity analysis due to run times. |
| Discontinuum modelling I (*e.g. distinct-element*) | Water table and boundary conditions (pore pressure/flux/etc.). Fluid pressure and gradients, fluid bulk | Allows for movement of blocks relative to each other. Can model complex behaviour and mechanisms | General limitations similar to those listed above. Must be aware of scale effects. Requires data on |

**Table 20.1** (continued)

| Method | Groundwater input | Advantages | Limitations |
|---|---|---|---|
| | modulus and viscosity. Intact yield strength and elastic constants. Joint contact hydraulic aperture, initial saturation and pore pressure. Joint spacing, trace length and gap length. Joint shear strength and stiffness properties. | (combined material and discontinuity deformation, coupled with hydro-mechanical and dynamic analysis). No flow, steady state and transient coupled flow. Output includes joint fluid flow and velocity, flow vectors and pore pressure. | discontinuity network (spacing, persistence, etc.). Limited data available for required joint properties. Limited experience with fully coupled solutions. Only fracture permeability considered, intact rock assumed impermeable. |
| Discontinuum modelling II *(e.g. particle flow codes)* | Circular particles or spheres representing solid impermeable media. Porosity, fluid bulk modulus, permeability and residual aperture. Particle and wall geometries and joint sets for permeability anisotropy. Inter particle bond strength and stiffness properties. | Fluid particle interaction able to consider uniform or hydrostatic pressure, saturated particle assembly with weak pressure gradients, fluid flow in notional cracks. Failure mechanisms such as piping, channelling and surface erosion can be simulated. Simulation of bond breakage between particles due to water forces possible. | Models require careful calibration. Experience with coupled particle flow code modelling limited. Models are computationally intensive. Considerable development needed before solutions can be applied to practical problems. |

Table 20.1 presents a selection of the common and newly developed research techniques used in slope stability analysis, highlighting the required input data, and the advantages and limitations of the methods with respect to the incorporation of groundwater.

This paper will review the current understanding of groundwater flow in natural slopes, and the means used to analyze the influence of groundwater and coupled hydro-mechanical processes on slope failure mechanisms. Three case histories are presented illustrating the treatment of groundwater in a slope stability analysis employing one of the key methodologies used in slope stability analysis: limit equilibrium, finite-element (continuum mechanics) and distinct-element (discontinuum mechanics).

## 2 Groundwater in Natural Slopes

The groundwater table/pore pressure distribution within a natural slope reflects the existing steady state or transient groundwater flow system [11]. Often, some form of steady state condition is assumed to enable a straight-forward means to calculate the pore pressure distribution along one or more potential deep-seated slip surfaces. This is usually performed in the context of the effective stress relationship, where the total stress is assumed to remain constant so that any increase in pore pressure results in a reduction in effective stress, which in turn lowers the frictional component of strength available for shear resistance:

$$\tau = c' + (\sigma_n - \mu) \tan \phi' \tag{1}$$

Where:

$\tau$ = shear strength
$c'$ = effective cohesion
$\sigma_n$ = normal stress acting across the potential shear/sliding plane
$\mu$ = pore pressure
$\phi'$ = effective friction angle

In reality the groundwater conditions existing in most natural slopes are rarely simple. The hydrologic response of a slope to a precipitation, snow melt event, etc., involves a complex, transient, saturated–unsaturated interaction that leads to some change in the pore pressure distribution (e.g. a water-table rise). The magnitude and duration of such a response, and the time lag between a rainfall event, for example, and any resulting rise in pore pressures, will vary widely depending on the slope geology and topography, the rainfall duration and intensity, and the saturated and unsaturated hydrogeologic conditions present in the slope [11, 23].

To deal with this uncertainty, the water table in most slope analyses is taken at its highest likely position or coincident with the ground surface. Although the latter is usually assumed to be the most conservative case, Freeze and Cherry [11] note that several geometrical and/or geological conditions may be present for which anomalously high pore pressures may exist. As will be later shown for one of the case histories presented (Campo Vallemaggia), this was indeed the case and consideration of the regional groundwater flow system was necessary. Another complexity involves the near surface unsaturated zone. Collins and Znidarcic [5] suggest that a rising water table with increased precipitation should not always be assumed, and that the actual infiltration profile may lead to perched water zones near the surface of the slope with an unsaturated profile below. Accordingly, they incorporated the effect of both positive and negative pore pressures into a coupled infinite slope analysis to present a predictive formulation of slope failures triggered by rainfall.

Such treatments, however, are generally more applicable to shallow failures or debris-flow initiation [9, 18]. Paddingtom [21] successfully combined saturated/

unsaturated groundwater flow models with both limit-equilibrium and finite-element codes in the analysis of debris flows in gentle-over-steep terrain. Debris flow runout analyses incorporating the effects of complex saturation/pore pressure profiles on flow rheology have also been explored by Hungr [15] and Iverson [16].

In deep-seated failures, pore pressures are usually treated assuming an equivalent homogeneous continuum (Fig. 20.1a). Heterogeneity resulting from permeability differences between material types is only of real significance with respect to flow analysis and/or the transient response of the system; it is less significant with respect to pore pressure distributions when steady state conditions are assumed. A notable exception is when artesian conditions are present within a confined aquifer, or where the flow regime is dictated by fracture permeability (for example, in a jointed, crystalline rock slope; Fig. 20.1d).

In joint-controlled slope stability problems, the same effective stress principles as described in Eq. 1 are applied to the discontinuities along which shear failure may occur. However, the definition of a groundwater table in these situations is much more questionable. Patton and Deere [22] point out that in the case of fracture permeability, the pore pressure distribution in a jointed rock slope is likely very irregular due to the controlling influence of individual structures (Fig. 20.2). They also note that the porosities of jointed rock are extremely small (0–10%) relative to that for soils (20–50%). This can lead to large, rapid, water-table fluctuations in a rock slope in response to a precipitation event, with larger pore pressure increases than those experienced in soil slopes.

Faults further complicate this picture as they can either act as impermeable barriers or to concentrate groundwater flow depending on their genesis and configuration relative to the host rock (Fig. 20.2). In massive, creeping rock slopes, this complexity

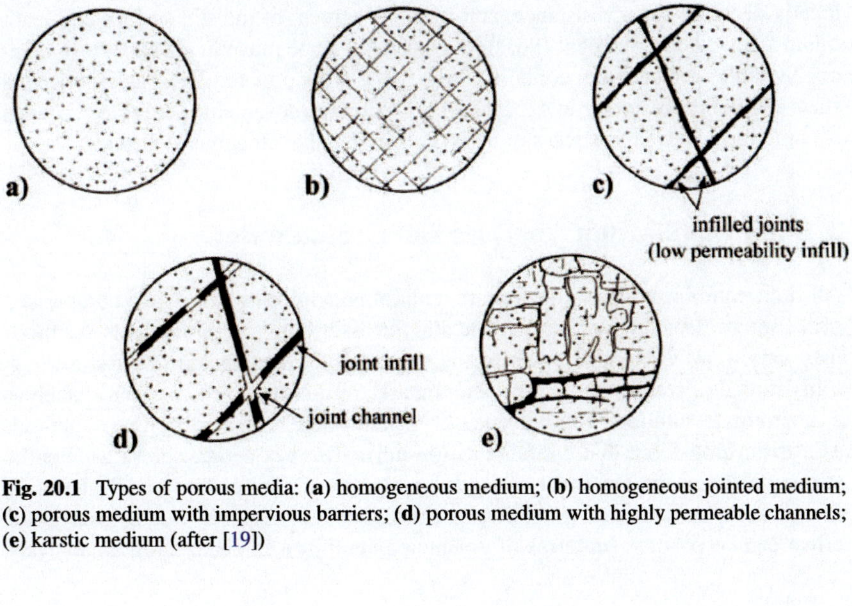


**Fig. 20.1** Types of porous media: (**a**) homogeneous medium; (**b**) homogeneous jointed medium; (**c**) porous medium with impervious barriers; (**d**) porous medium with highly permeable channels; (**e**) karstic medium (after [19])

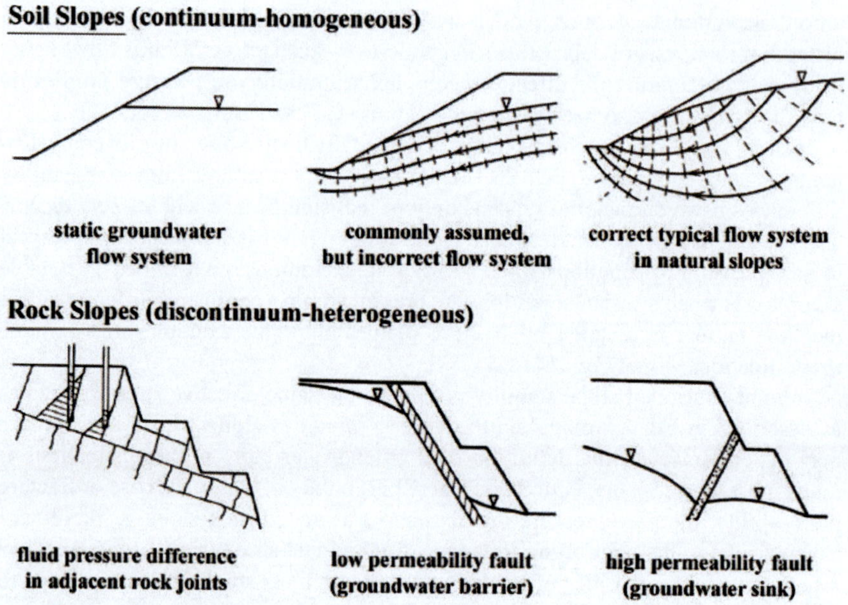


**Fig. 20.2** Groundwater flow systems in natural slopes, showing difference between soil slopes treated as a homogeneous continuum, and rock slopes, where fracture permeability presents a major control (after [22])

may evolve into a dynamic ever-changing system where rock slope movements work to generate and open cracks in the bedrock and colluvium, facilitating groundwater flow [20]. Pore water pressure changes in these zones of enhanced permeability may locally decrease shear resistance, activating a segment of the slip surface and leading to a period of movement (i.e. slip). Similarly, slope movements may close other cracks and/or groundwater conduits, thereby lowering pore pressures and increasing shear resistance, resulting in a localized cessation of movement (i.e. "stick" in what is commonly viewed as stick-slip behaviour in massive creeping rock slopes).

## 3 Limit Equilibrium Analyses and Groundwater

All limit equilibrium techniques share a common approach based on a comparison of resisting forces/moments mobilized and the disturbing forces/moments. Methods may vary, however, with respect to the slope failure mechanism in question (e.g. translational or rotational sliding), and the assumptions adopted in order to achieve a determinate solution. The presence of groundwater is treated by these methods as a disturbing force that counteracts the normal forces acting across a potential slip surface and the frictional strength enabled by means of these normal forces; in the case of unsaturated groundwater conditions, i.e. negative pore pressures, the effect can be positive (in terms of enhancing stability). In performing an analysis, pore pressures are calculated as a function of the location of a specified water table relative to the sliding plane. Alternatively, a pre-specified pore pressure distribution may be defined. Recent developments in commercial limit equilibrium codes have worked towards combining finite-element derived groundwater analyses with a "method of slices" solution, to more accurately calculate pore pressure distributions and magnitudes throughout the landslide mass.

In soil slopes, most analyses assume drained conditions, although undrained soil slope conditions may also be treated using a "$\phi = 0$" analysis (e.g. [25]). It should be noted that undrained soil slope conditions are generally only applicable to temporary short term engineered slopes, where the total stress state may be rapidly increased, for example during the construction of an embankment. In natural slopes, rapid loading causing an undrained condition may arise during an earthquake, leading to liquefaction. Rigorous treatments of liquefaction potential are now available through commercial codes that couple finite element derived dynamic stress conditions and generated pore-water pressures with limit equilibrium method of slices analyses [12].

In rock slopes, limit equilibrium analyses may provide a useful "first step" where failure occurs along well-defined discrete discontinuities [4]. The assumption of an assumed water table in rock slopes without due consideration of the flow properties of the individual joint sets must however be treated with caution and used only as an indicator or in sensitivity analyses.

## *3.1 Example Limit Equilibrium Application – Usoi Rockslide Dam*

The application of a limit equilibrium analysis generally involves the objective of testing the sensitivity of a slope to fluctuations in the groundwater table, for example those that occur following a heavy rainfall event, and the potential for such fluctuations to trigger a catastrophic failure. Several different incrementally increasing water tables may be analyzed producing a different factor of safety for each. The water level above which an acceptable factor of safety is not attained is then deemed the critical threshold

Similarly, the fluctuations in water level behind an earth dam or landslide dam may be tested for its influence on stability of the dam. Such is the concern for the case of the Usoi Landslide Dam in the Pamir Mountains of Tajikistan (Fig. 20.3), although overtopping due to a landslide-generated tsunami is also of great concern. The Usoi Landslide dam was created in 1911 following an earthquake that triggered a massive landslide that blocked a valley creating a lake, Lake Sarez, 60 km long and more than 500 m deep [1, Chap. 16 by Ischuk, this volume]. Due to high regional seismicity and the large volume of water behind the dam, questions have been raised regarding the potential for catastrophic collapse – an event that would threaten between 7,000 and 35,000 inhabitants immediately downstream from the dam in the Panj River Valley [17], as well as destroy villages and infrastructure in the Amu Darya River basin inhabited by more than 5 million people.

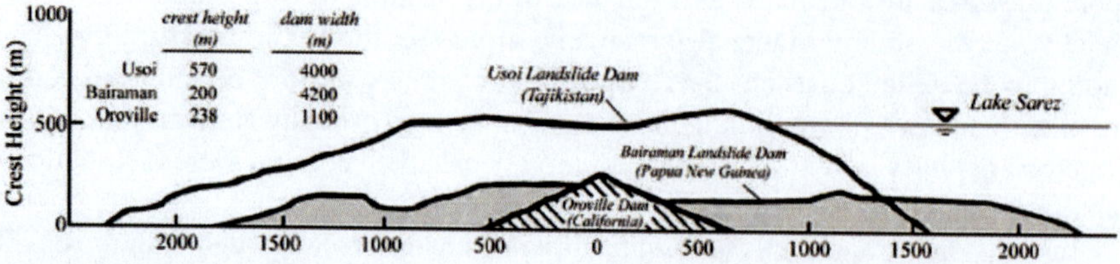


**Fig. 20.3** Comparative cross sections of the 1911 Usoi and 1986 Bairaman landslide dams, and the Oroville Dam in California, one of the world's largest embankment dams (after [24])

Initial limit equilibrium analysis by Hanisch and Söder [13] showed that given the enormous mass of the landslide dam, a Factor of Safety of 9 could be calculated against the dam sliding on its base due to the horizontal acting hydrostatic load on the upstream face of the dam. Their analysis showed that this value may drop by up to 50% in the event of a strong earthquake, but would still be sufficiently high. Instead, a catastrophic breach of the dam would more likely arise if one or more rotational slides developed within the body of the landslide dam. Figure 20.4 shows a method of slices limit equilibrium analysis of the Usoi Dam employing a Morgenstern-Price/General Limit Equilibrium solution (see [10] for a description of the formulation). Pore pressures for the analysis were imported from a finite element analysis assuming a lake level 536 m above the base and toe of the slope (or 34 m below the maximum crest of the dam)

Little is known about the internal structure of the landslide dam and therefore several critical assumptions are required regarding the dam's strength and flow characteristics. For the analysis shown in Fig. 20.4, it is assumed that the dam foundation is more impervious than the overlying landslide debris and that the composition of the landslide dam would primarily involve finer-grained material overlain by a carapace of coarser rockslide debris. The material properties used are given in Fig. 20.4 and are based on those reported by Hanisch and Söder [13].

Two different slip circles and their respective Factors of Safety are shown for slope failures ranging in size from the smallest but most critical slip surface, to one that would breach the entire dam (Fig. 20.4). In the case of the latter, the Factor of Safety is significantly higher, suggesting that the failure of the dam through a single large sliding event is unlikely. Instead, several smaller retrogressing circular slips, developing in series behind the smaller, most critical slip circle, provides a more likely threat to catastrophic breach (with respect to unstable slope activity). Another possible scenario is the development of a non-circular slip plane, for which an active-passive block scenario may develop [26].

Figure 20.5 contains the results for a sensitivity analysis in which the influence of a major earthquake is accounted for (equated in terms of horizontal acceleration) with respect to the most critical slip surface as determined in the analysis shown in Fig. 20.4. Here the left boundary of the plot coincides with the Factor of Safety initially calculated (FoS = 2.4; see Fig. 20.4), i.e. for the case of a horizontal seismic coefficient of zero (or no earthquake). As the seismic coefficient increases, the

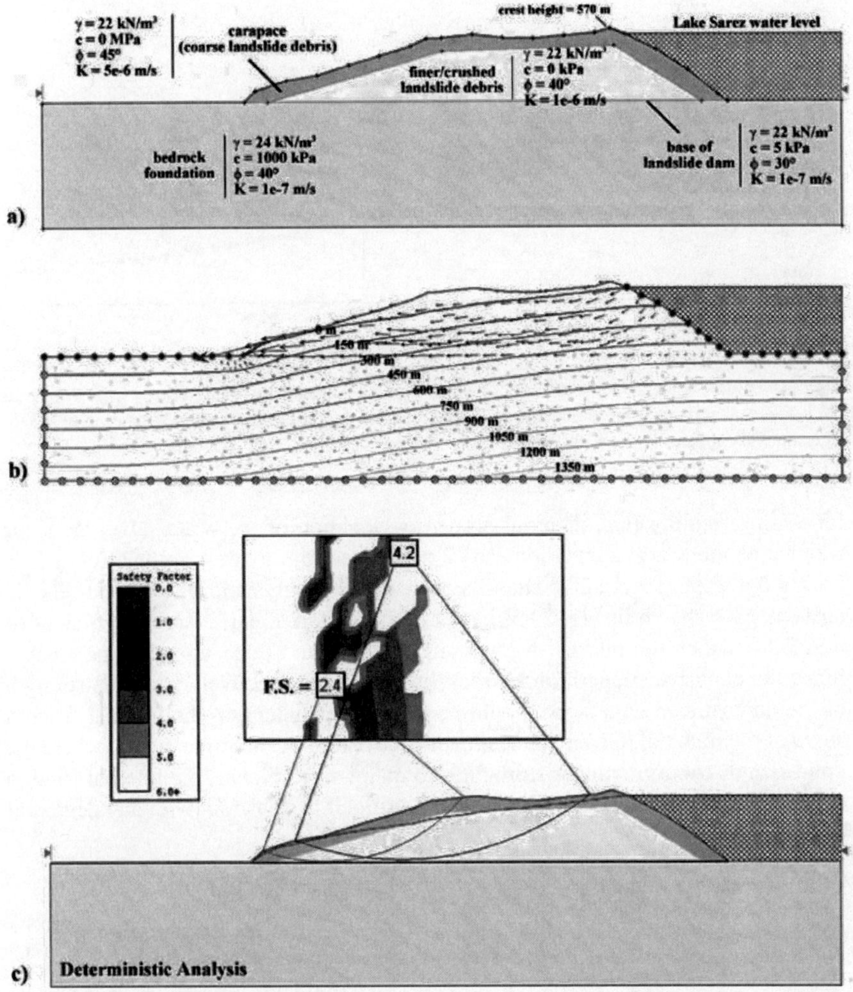


**Fig. 20.4** Limit equilibrium analysis of the Usoi landslide dam, showing from *top* to *bottom*: (**a**) the problem geometry and material assumptions; (**b**) finite element derived hydraulic head contours and flow vectors imported into the limit equilibrium analysis; and (**c**) deterministic analysis adopting a Morgenstern-Price solution

Factor of Safety along this slip surface decreases reaching 1.0 for a horizontal seismic coefficient of 0.325 g. In terms of the influence of pore pressures for this case, the Factors of Safety calculated would be 25% higher than those provided if pore pressures were neglected (i.e. if the slide mass was assumed to be dry).

The efficiency of computer-aided limit equilibrium analyses, where thousands of calculations can be performed in a matter of seconds, allows for uncertainty associated with different input parameters to be dealt with using probabilistic techniques. Statistical distributions of the different input parameters may be defined to account

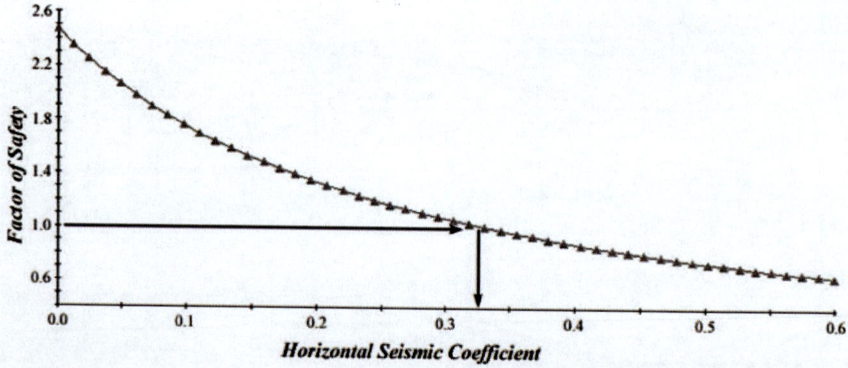


**Fig. 20.5** Limit equilibrium sensitivity analysis of the Usoi landslide dam plotted with respect to the horizontal seismic coefficient, showing the potential for failure for values greater than 0.325. This analysis applies to the most critical slip surface (initial FoS = 2.4), as shown in Fig. 20.4

for their variability (e.g. material properties, location of the water table, etc.), for which a Monte Carlo simulation can be performed to provide a probability of failure for the slope. Figure 20.6 shows a probabilistic analysis for the critical seismic analysis case shown in Fig. 20.5, i.e. assuming a horizontal seismic coefficient of 0.325, for which the internal friction angle is held to be the uncertain parameter. Since the cohesive strength of the rockslide debris is effectively zero, the strength of the downstream dam slope would be largely dependent on the internal friction angle of the material. Given the heterogeneous nature of rockslide debris, where the grain size distribution ranges from fines to metre-sized blocks, the internal friction angle of the slide debris may be somewhat difficult to constrain with any degree of certainty.

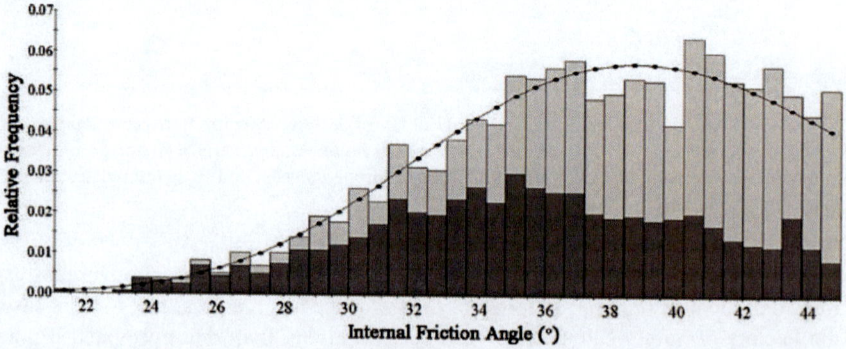


**Fig. 20.6** Limit equilibrium probability analysis of the Usoi landslide dam, showing the statistical gamma distribution of the internal friction angles used in the Monte Carlo simulation (*light grey bars*) and the resulting Factors of Safety (*dark grey bars*). The analysis corresponds to the critical horizontal seismic coefficient of 0.325 as calculated in Fig. 20.5 and the critical slip surface with Factor of Safety = 2.4 as shown in Fig. 20.4

Assuming a mean internal friction angle of 40° [13] and applying a gamma distribution that favours higher friction angles, as would be expected for broken rock, a probabilistic range in Factor of Safety can be obtained as shown in Fig. 20.6. Based on these results, a probability of failure of 44% is calculated for a seismic event with a horizontal seismic coefficient of 0.325. The probability of failure is simply equal to the number of analyses with a Factor of Safety less than 1, divided by the total number of samples analyzed by the Monte Carlo simulation. The reliability index for the analysis, a measure of the number of standard deviations which separate the mean from the critical (<1.0) Factor of Safety, is 0.3.

## 4 Numerical Approaches

Limit equilibrium analyses are limited by their inability to account for material deformation or yield, factors which help to provide insights into potential slope failure mechanisms [7]. In cases where the consideration of complexities relating to geometry, material anisotropy, non-linear stress-strain behaviour, in situ stresses and/or the presence of coupled processes (e.g. pore pressures, seismic loading, etc.) are deemed important, numerical modelling methods should be employed. There exists a wide range of numerical codes for the analysis of slopes, ranging from continuum techniques (e.g. finite element), ideally suited for soil slopes, to discontinuum techniques (e.g. distinct element), which are more applicable to rock slopes where both discontinuities and intact material deformation contribute to the development of an instability.

Numerical techniques incorporate an array of means for assessing the importance of groundwater in a slope stability analysis. Many of these, especially in the case of a continuum analysis, are similar to those previously discussed for a limit equilibrium analysis. In rock slopes, high-risk situations may require the characterization of discontinuity-controlled groundwater flow. This may either be done using a continuum groundwater flow code (assuming equivalent rock mass approaches and varying permeability) or preferably discontinuum codes with varying discontinuity properties such as joint permeability, spacing, persistence and aperture. Analyses may be undertaken using either an uncoupled, partially coupled or fully coupled flow-mechanical models.

### *4.1 Example Coupled H–M Continuum Application – Lutzenberg Shallow Landslide*

When multiple conditioning and triggering factors (heavy rain, weakened slope, anthropogenic factors, etc.) influence the failure of a slope, it is often difficult to pinpoint the primary trigger. The 2002 Lutzenberg landslide in eastern Switzerland was one such case in which various potential causes could be inferred [27]. The landslide occurred during a period of heavy precipitation and involved a shallow soil

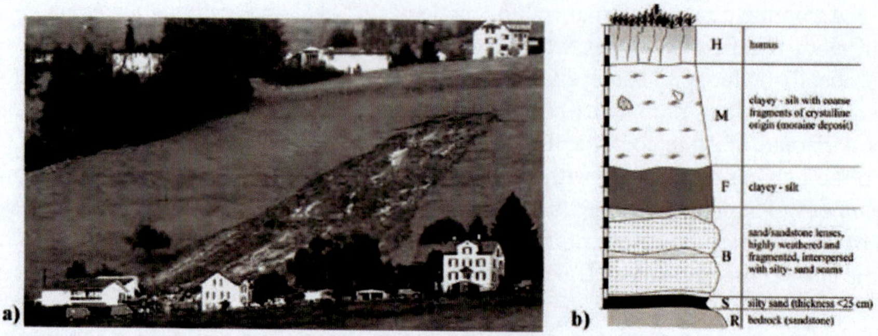


**Fig. 20.7** (**a**) Lutzenberg landslide, and (**b**) soil profile along the slide scarp (after [27])

slip of 2,500 $m^3$ along a weathered rock bedding plane resulting in three fatalities (Fig. 20.7). Such slopes are extremely common throughout Switzerland and have a high potential for repeat occurrence.

Results from an integrated study, combining geological and geotechnical in situ investigations with numerical modelling, showed that failure may not have occurred solely through the effects of heavy precipitation. A series of fully-coupled hydro-mechanical 2-D finite-element models were used to ascertain the cause of failure by testing different scenarios with increasing degrees of complexity, from dry slope conditions, to fully saturated slope conditions (i.e. coincident with heavy precipitation events), to a final scenario where the influence of a leaking water pipe was added. Material properties were determined through a series of in situ shear box tests and laboratory testing. From these test results, upper and lower bounds were placed on the cohesive and frictional strengths of the controlling soil units making up the shear surface and slide body.

Depending on the fit of the modelling results to those observed in the field, or the assumptions required to achieve a fit, the model scenario was then either validated or rejected (Fig. 20.8). Based on this scenario-approach, results from the finite element analysis suggested that failure under dry conditions would be highly improbable. When a fully saturated slope was considered, representing the conditions expected following a period of heavy precipitation, slope failure was seen to be possible but only if the lower bound strength properties were used. Even then, the models could not reproduce the location of the head and toe of the slide, favouring an instability forming lower down on the slope [27].

Models were then assembled to reproduce the effect of a leaking water pipe on the pore pressure distribution (Fig. 20.9). These results produced the overall best fit (Fig. 20.10). As such, and being further corroborated by observations made in the field, the coupled finite-element results suggest that a leaking water pipe was responsible for creating an unstable slope situation, and that heavy precipitations then triggered the fatal landslide. These results demonstrate the value of performing an integrated analysis for which each of the different components of the study is

**Fig. 20.8** Modelling procedure and scenarios tested to determine cause of Lutzenberg slide (after [27])

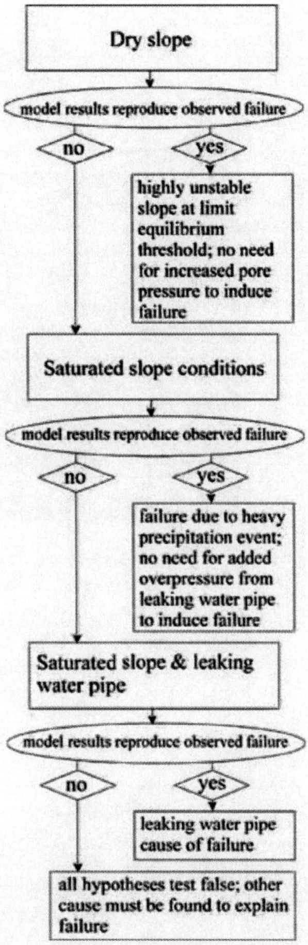


planned with the other components in mind [27]. In doing so, this allowed for a better comprehension of the landslide failure mechanism, which can be carried forward in planning future mitigation measures to be applied to similar slopes in the region.

## *4.2 Example Coupled H–M Discontinuum Application – Campo Vallemaggia*

Discontinuum methods, like the distinct-element method [14], treat the problem domain as an assemblage of deformable blocks for which complex non-linear interaction between blocks are solved (i.e. slip and/or opening/closing along discontinuities). When applying the methodology to coupled hydro-mechanical problems, mechanical joint deformations can be equated to changes in fracture conductivity

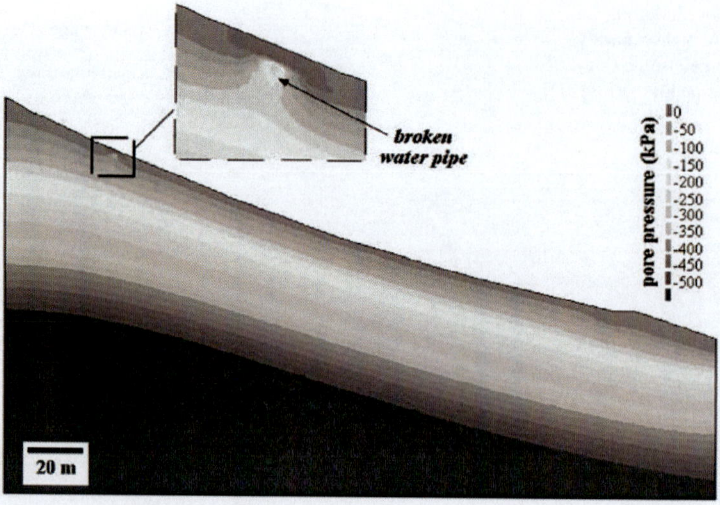


**Fig. 20.9** Finite-element pore pressure distribution; *inset* shows localized elevated pore pressures due to a simulated broken water pipe

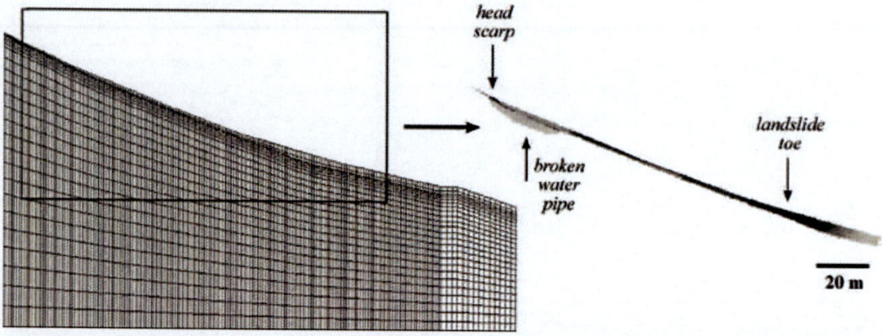


**Fig. 20.10** Coupled hydro-mechanical finite-element model, assuming a fully saturated slope with a leaking water pipe. Results show plastic yielding of slope (as used to indicate outline of failure). Superimposed on these results are *arrows* showing the mapped boundaries of the landslide head scarp and toe

(by means of a cubic law relationship), and conversely, joint water pressures can be viewed as affecting the mechanical behaviour/aperture of the joints (Fig. 20.11).

The Campo Vallemaggia landslide is a deep-seated creeping landslide in the south-central Swiss Alps, and involves a 300 m deep, complex structure consisting of schists and gneisses sub-divided into distinct blocks by a series of extensive faults ([2]; Fig. 20.12). The hydrogeology of the landslide is dominated by strongly anisotropic flow through discontinuities and especially along high conductivity fault zones. When combined with regional topographical controls, these structures result

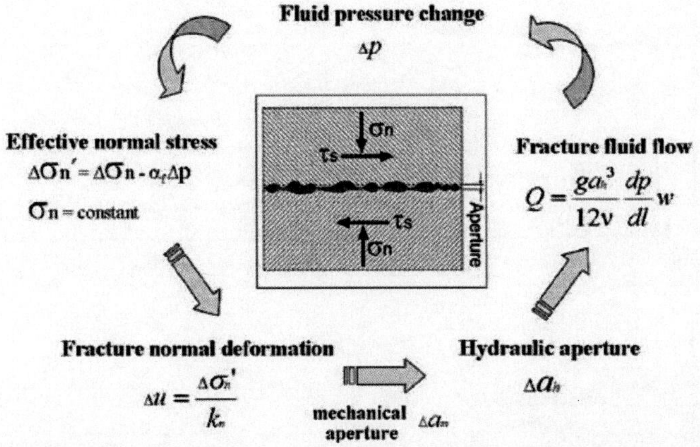


**Fig. 20.11** Hydro-mechanical coupling of the fluid/joint interaction solved for by the distinct-element method

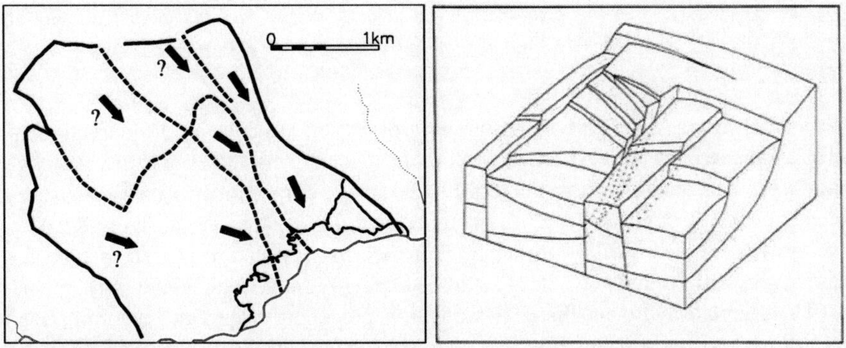


**Fig. 20.12** Plan view and block model of the Campo Vallemaggia slide mass (after [3])

in upward directed seepage forces at the base of the slide body. Instrumentation records showed that when pore water pressures exceeded an apparent threshold, sudden accelerations of the slide mass occurred; velocities returning to background levels as the pore pressures dissipated (Fig. 20.13).

Given the risk to communities located on the unstable slide mass, mitigative measures were undertaken. This involved the construction of a deep drainage adit within the undisturbed rock below the creeping slide mass, from which drainage boreholes were drilled into the slide base. Total discharge on completion of the adit, at the end of 1995, was approximately 50 l/s decreasing to 30 l/s by 1998. This resulted in an immediate reduction in water head of 150 m and the almost complete cessation of downslope movement (Fig. 20.13). Despite this apparent success, scepticism arose as to the true effectiveness of the deep drainage solution due to the relatively small water outflows measured from the drainage gallery and competing arguments that

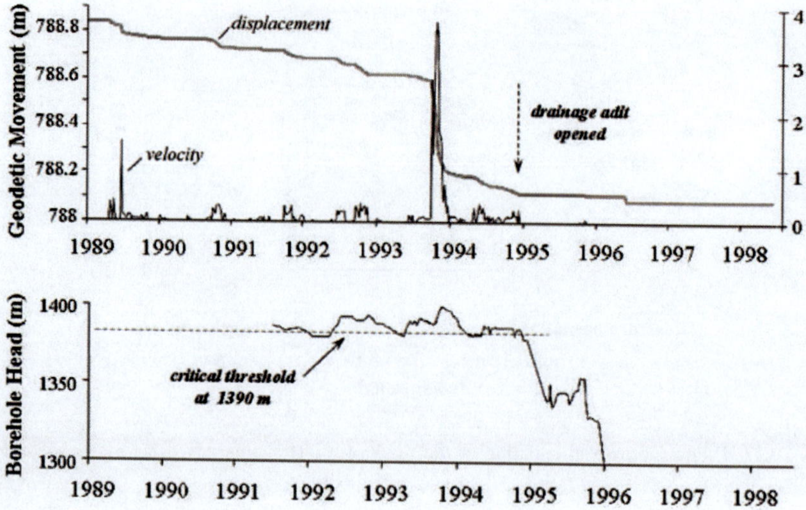


**Fig. 20.13** Plots of slide block movement and velocity versus time (as measured using an automated geodetic station), and corresponding borehole pore water pressures (after [3])

the instability was the result of continuous erosion at the slide's toe. Numerical modelling was therefore undertaken to explore the mechanisms and effects that drainage may have had on stabilizing the slope. The study was performed using a distinct element code through which a fully coupled hydro-mechanical analysis was undertaken where the controlling influence that rock mass structure played in both the kinematics and groundwater flow could be explicitly modelled (Fig. 20.14).

The coupled hydro-mechanical modelling procedure used can be summarized as follows: (i) pore water pressures were introduced to the model using a water table based on site investigation data; (ii) the model was stepped to a steady state

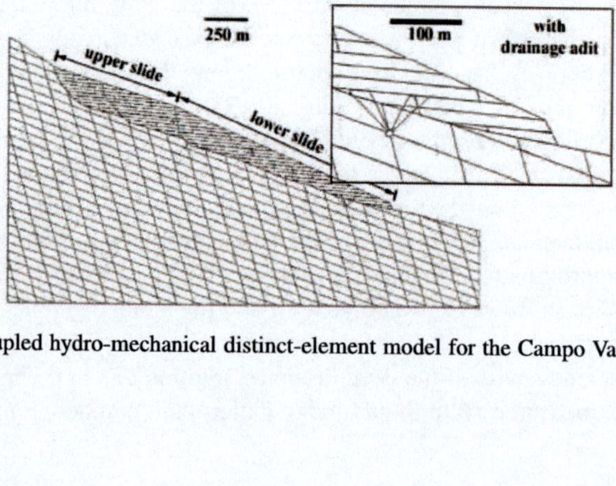


**Fig. 20.14** Coupled hydro-mechanical distinct-element model for the Campo Vallemaggia landslide

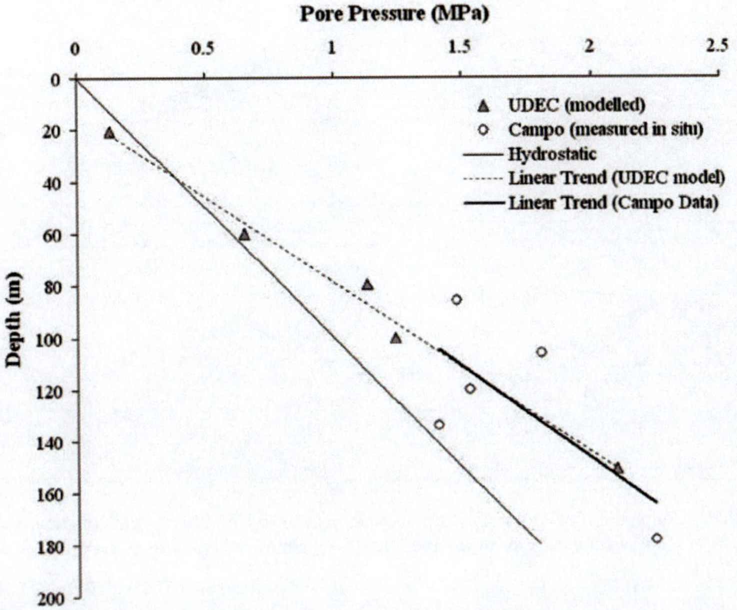


**Fig. 20.15** Comparison of modelled pore pressures with depth and those measured in situ

and modelled pore water pressure histories for varied locations compared to those recorded in the field (Fig. 20.15); (iii) strength properties along the failure surface were varied, where together with the addition of pore water pressures, an unstable slope state was created and movements could be compared to those recorded in the field; (iv) drainage of the slide mass is enabled through the modelled opening of the drainage adit.

Opening of the adit resulted in an abrupt termination of the modelled downslope velocities (Fig. 20.16). With respect to tunnel inflows the model showed that very little drainage is required for stabilization to be achieved (approximately 10–20 l/s). These magnitudes were in agreement with the observed tunnel inflows. In terms of the underlying contributing mechanisms, the distinct element models show that in the case of fracture permeability, where storativities are low, large water inflows through drainage are not necessary to achieve significant reductions in head.

## 5 Conclusions

Groundwater is recognized to be the critical factor in many, if not most, slope failures. Considerable research has been undertaken on the use of limit equilibrium and numerical methods in slope design and analysis. A concerted research effort, however, is required to improve our understanding of groundwater flow in soil

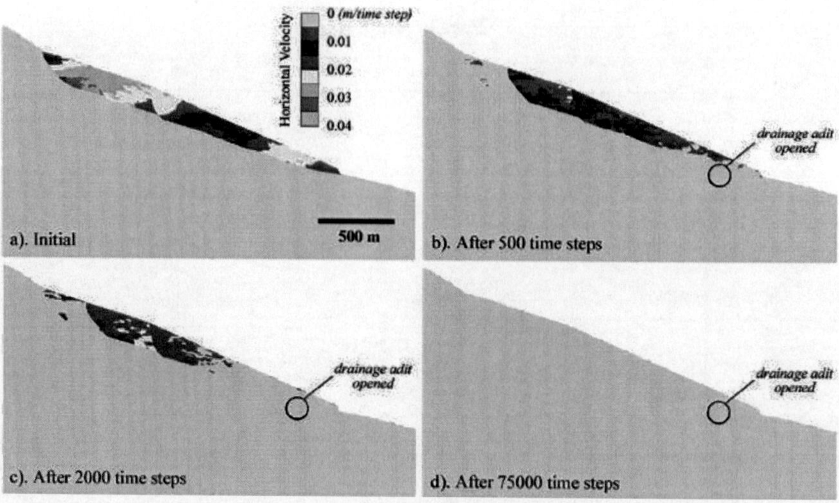


**Fig. 20.16** Coupled hydro-mechanical distinct-element model of Campo Vallemaggia, showing snapshots of slope velocities prior to (**a**) and after (**b–d**) simulated opening of the drainage adit

and rock slopes and its coupled influence on slope deformation. The authors argue that back-analysis of groundwater remedial drainage schemes, such as the Campo Vallemaggia landslide, may provide valuable information on natural slope groundwater characteristics, and would recommend that the success or failure of drainage measures be used as a constraint for groundwater models on a more frequent basis.

A wide range of options exist for the analysis of soil and rock slopes, from limit equilibrium to numerical modelling, each incorporating an array of techniques for assessing the importance of groundwater. Further progress in the application of these techniques, however, requires additional constraints with improved slope characterization and site investigations. For example, there is considerable scope for improved characterization of groundwater flow within jointed rock slopes; routinely used discontinuity survey sheets often do not provide much of the information needed in order to conduct coupled hydro-mechanical models. Similarly, improved instrumentation of groundwater in unstable slopes is needed in order to provide important constraints for numerical analysis. Current research is underway in order to simulate hydraulic fracturing mechanisms in the development of rock slope shear surfaces, as well as the effects of intact rock fracture on groundwater pressures and the subsequent transport characteristics of the rock failure debris (of major importance in any risk assessment).

**Acknowledgments** The authors would like to thank their collaborators in the various projects presented in this paper, especially Benoît Valley and Prof. Simon Loew (ETH Zurich, Switzerland), Dr. Luca Bonzanigo (Geolog.ch, Switzerland) and Prof. Kurosch Thuro (Technical University of Munich, Germany). This work has been supported in part by the National Science and Engineering Research Council of Canada (NSERC).

## References


1. Alford, D. and Schuster, R.L. (2000) Introduction and summary, in *Usoi Landslide Dam and Lake Sarez: An Assessment of Hazard and Risk in the Pamir Mountains, Tajikistan (ISDR Prevention Series Nº 1)*. United Nations, New York, NY, pp. 1–18.
2. Bonzanigo, L., Eberhardt, E. and Loew, S. (2000) Interpretation of the effects of a drainage adit in a deep creeping slide mass, in E. Bromhead et al.(eds.), *Proceedings of the 8th International Symposium on Landslides, Cardiff*. Thomas Telford, London, pp. 151–156.
3. Bonzanigo, L., Eberhardt, E. and Loew, S. (2001) Hydromechanical factors controlling the creeping Campo Vallemaggia landslide, in M. Kühne et al. (eds.), *International Conference on Landslides – Causes, Impacts and Countermeasures, Davos*. Verlag Glückauf, Essen, pp. 13–22.
4. Coggan, J.S., Stead, D. and Eyre, J.M. (1998) Evaluation of techniques for quarry slope stability assessment, *Transactions of the Institution of Mining and Metallurgy, Section B* **107**, 139–147.
5. Collins, B.D. and Znidarcic, D. (2004) Stability analyses of rainfall induced landslides, *Journal of Geotechnical and Geoenvironmental Engineering* **130**, 362–372.
6. Crosta, G. and di Prisco, C. (1999) On slope instability induced by seepage erosion, *Canadian Geotechnical Journal* **36**, 1056–1073.
7. Eberhardt, E. (2006) From cause to effect – Using numerical modelling to understand rock slope instability mechanisms, in Evans et al. (eds.), Landslides from *Massive Rock Slope Failure*. NATO Science Series IV, v. 49, Springer, Dordrecht, 85–101.
8. Eberhardt, E., Thuro, K. and Luginbuehl, M. (2005) Slope instability mechanisms in dipping interbedded conglomerates and weathered marls – The 1999 Rufi landslide, Switzerland, *Engineering Geology* **77**, 35–56.
9. Fannin, R.J. and Jaakkola, J. (1999) Hydrological response of hillslope soils above a debris-slide headscarp, *Canadian Geotechnical Journal* **36**, 1111–1122.
10. Fredlund, D.G. and Krahn, J. (1977) Comparison of slope stability methods of analysis, *Canadian Geotechnical Journal* **14**, 429–439.
11. Freeze, R.A. and Cherry, J.A. (1979) *Groundwater*. Prentice-Hall, Englewood Cliffs, NJ.
12. Geo-Slope (2003) *Geo-slope office: Slope/w, seep/w, quake/w*. Geo-Slope International Ltd., Calgary, AB.
13. Hanisch, J. and Söder, C.-O. (2000) Geotechnical assessment of the Usoi landslide dam and the right bank of Lake Sarez, in *Usoi Landslide Dam and Lake Sarez: An Assessment of Hazard and Risk in the Pamir Mountains, Tajikistan (ISDR Prevention Series Nº 1)*. United Nations, New York, NY, pp. 23–42.
14. Hart, R.D. (1993) An introduction to distinct element modeling for rock engineering, in J.A. Hudson (ed.), *Comprehensive Rock Engineering: Principles, Practice & Projects*, Vol. **2**. Pergamon Press, Oxford, pp. 245–261.
15. Hungr, O. (2000) Analysis of debris flow surges using the theory of uniformly progressive flow, *Earth Surface Processes and Landforms* **25**, 483–495.
16. Iverson, R.M. (1997) The physics of debris flows, *Reviews in Geophysics* **35**, 245–296.
17. Ives, J. and Pulatova, G. (2000) Human geography/demography, in *Usoi Landslide Dam and Lake Sarez: An Assessment of Hazard and Risk in the Pamir Mountains, Tajikistan (ISDR Prevention Series Nº 1)*. United Nations, New York, NY, pp. 83–86.
18. Johnson, K.A. and Sitar, N. (1990) Hydrologic conditions leading to debris-flow initiation, *Canadian Geotechnical Journal* **27**, 789–801.
19. Louis, C. (1976) Introduction à l'hydraulique des roches, Ph.D. Thesis, Paris.
20. Okunishi, K. and Okimura, T. (1987) Groundwater models for mountain slopes, in M.G. Anderson and K.S. Richards (eds.), *Slope Stability – Geotechnical Engineering and Geomorphology*. Wiley, New York, NY, pp. 265–285.
21. Paddington, S. (2004) The characterization of drainage related landslides on gentle-over-steep forest terrain in the Interior of British Columbia, M.Sc. Thesis, Simon Fraser University, Canada.

22. Patton, F.D. and Deere, D.U. (1971) Geologic factors controlling slope stability in open pit mines, in C.O. Brawner and V. Milligan (eds.), *Stability in Open Pit Mining*. American Institute of Mining Engineers, New York, NY, pp. 23–48.
23. Sangrey, D.A., Harrop-Williams, K.O. and Klaiber, J.A. (1984) Predicting groundwater response to precipitation, *Journal of Geotechnical Engineering* **110**, 957–975.
24. Schuster, R.L. (2000) A worldwide perspective on landslide dams, in *Usoi Landslide Dam and Lake Sarez: An Assessment of Hazard and Risk in the Pamir Mountains, Tajikistan (ISDR Prevention Series N° 1)*. United Nations, New York, NY, pp. 19–22.
25. Simons, N., Menzies, B. and Matthews, M. (2001) *A Short Course in Soil and Rock Slope Engineering*. Thomas Telford Publishing, London.
26. Sultan, H.A. and Seed, H.B. (1967) Stability of sloping core earth dams, *Journal of Soil Mechanics and Foundations Division, ASCE* **93**, 45–67.
27. Valley, B., Thuro, K., Eberhardt, E. and Raetzo, H. (2004) Geological and geotechnical investigation of a shallow translational slide along a weathered rock/soil contact for the purpose of model development and hazard assessment, in W.A. Lacerda et al. (eds.), *Proceedings of the 9th International Symposium on Landslides, Rio de Janeiro*. A.A. Balkema, Leiden, pp. 385–391.

# Chapter 21
# Paleohydrology of Volcanogenic Lake Break-Out Floods in the Taupo Volcanic Zone, New Zealand

**V. Manville and K.A. Hodgson**

## 1 Introduction

All floods result from the rapid release of water, stored either in the atmosphere or in terrestrial reservoirs (lakes, snow and ice), onto the landscape [74]. The magnitude of the flood therefore reflects the characteristic volume and release rates of the water source and the physiographic properties of the landscape that receives it, while the combination of geologic, climatic and physiographic factors make some areas of the Earth more flood-prone than others. Large-scale terrestrial impoundments of surface water, i.e. lakes, are an effective source of major floods because they can release huge volumes of water directly into a drainage route should their margins be breached. Conversely, a large meteorological flood requires that precipitation and run-off be integrated across a wide geographic area over a period of time. Temporary lake impoundments can develop behind barriers composed of ice, moraine material, landslides, fluvial deposits, and volcanic debris [21]. Globally, 16% of the 714 identified Holocene (last 10,000 years) volcanoes host one or more crater lakes [26], while 8% of naturally-dammed lakes have volcanic origins [21], being impounded by debris avalanche (volcanic landslide) deposits, lava flow, pyroclastic flows, or lahar deposits.

### *1.1 The Taupo Volcanic Zone*

The Taupo Volcanic Zone (TVZ) in the central North Island of New Zealand (Fig. 21.1) is an area of intense Quaternary silicic volcanism associated with extension and thinning of continental crust at the southern end of the Tonga-Kermadec

V. Manville (✉)
Institute of Geological and Nuclear Sciences, Wairakei Research Centre, Taupo, New Zealand
e-mail: v.r.manville@leeds.ac.uk

Current address: V. Manville, School of Earth and Environment, University of Leeds, Leeds LS2 9JT, UK, e-mail: v.r.manville@leeds.ac.uk

S.G. Evans et al. (eds.), *Natural and Artificial Rockslide Dams*, Lecture Notes in Earth Sciences 133, DOI 10.1007/978-3-642-04764-0_21, © Springer-Verlag Berlin Heidelberg 2011

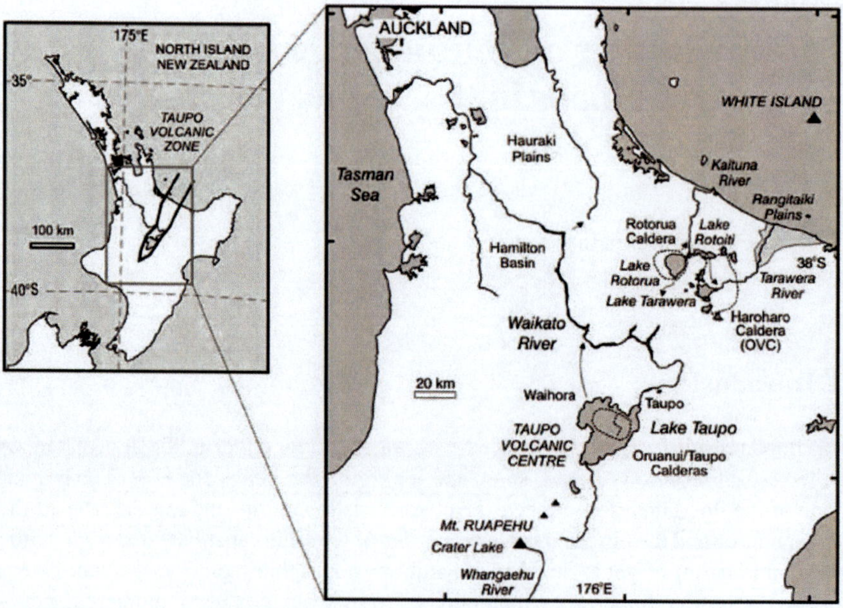


**Fig. 21.1** Location map of volcano-lake systems in the Taupo Volcanic Zone known to have produced break-out floods. Former drainage paths indicated by arrows

volcanic arc. The region is characterized by frequent caldera-forming explosive eruptions from multiple short-lived, nested and/or overlapping volcanic centres [40, 108]. Individual eruption volumes have ranged up to $10^2$–$10^3$ km$^3$, with primary fall deposition occurring over >$10^6$ km$^2$ and ignimbrites covering >$10^4$ km$^2$. The combined effects of volcanism and faulting in the central TVZ have led to the creation and destruction of numerous lakes spanning several orders of magnitude in scale and longevity developed in topographic lows formed by explosion craters, volcano-tectonic calderas, and depressions dammed by pyroclastic and lava flows, lahar deposits, and debris avalanches [48, 63, 90]. Amongst these, historic break-out floods have been recorded from the summit Crater Lake of Mt. Ruapehu as recently as March 2007 [14, 60], and from intracaldera Lake Tarawera [38]. Prehistoric break-out floods have also been identified from the geological record at Lake Tarawera [38] and intracaldera Lake Taupo [59, 61, 63], and inferred at Lake Rotorua [63, 65]. It is these events which form the subject of this paper.

## 2 Paleohydraulic Analysis

Typical paleohydraulic analyses consist of converting geologic stage-level indicators and channel geometry to estimates of peak discharge [74], with flow velocity and duration being additional important parameters. An additional factor in break-out floods is the outflow hydrograph from the impoundment, which is governed by such parameters as breach geometry and growth rate, and reservoir hypsometry [98].

## *2.1 Dam-Break Analysis*

Estimation of the potential flood hydrograph from a dam-breach is complicated by its dependence on four inter-related factors: the volume of the lake, the height of the dam (or depth of the lake), the average width of the breach, and the breach development time. Four main techniques [52], have been developed to assess the magnitude of potential dambreak floods from a range of natural (landslide, moraine, ice) and artificial dams (concrete, earthen): (i) empirical regression relationships between observed values of peak discharge and some dimensional characteristic of the lake, dam, or breach [e.g. 20, 32, 49, 98]; (ii) parametric models, based on time-stepping calculations of flow through a developing breach using published broad-crested weir-flow equations [e.g. 15, 81]; (iii) dimensional analysis, a method of grouping parameters of unknown significance together in order to simplify relationships and compare their relative importance [e.g. 98, 100]; and (iv) physically-based dam erosion models, often computer-implemented [e.g. 18, 31, 80]. Of these (i)–(iii) require the minimum input data and are hence the most useful for paleohydraulic analyses where information is limited.

### 2.1.1 Empirical Analysis

Regression relationships between observed values of peak discharge and some dimensional characteristic of the lake or dam, such as the total drop in lake level during the flood (often synonymous with the depth of the lake), and/or the volume of water released (also referred to as the excess lake volume) typically take a power-law form with empirically-derived coefficients [e.g. 20, 98]. However, the actual peak discharge of a dam-break flood is rarely a simple function of the volume of lake water or dam height/lake depth or some combination of the two. Furthermore, the data on which the relationships are based are often collected in different ways, and the actual data points themselves are highly scattered. Therefore, regression lines can be fitted almost at whim, and different published equations can produce estimates of peak discharge that vary by an order of magnitude for the same inputs [20].

An alternative set of empirical relationships has been developed based on the dimensional characteristics of the breach as well as the lake/dam system [e.g. 41, 49, 100]. However, these are difficult to apply unless the breach geometry is well-preserved.

### 2.1.2 Dimensional Analysis

Dimensional analysis, as applied to dambreak phenomena is a technique for investigating the relative importance of breach growth rate and lake volume and lake shape on the peak discharge of a dambreak flood [98], assuming that the fundamental physical mechanisms involved are the same for all dambreak flood events. Given that an ideal dam breach behaves as a broad-crested weir with critical flow through the outlet (i.e. Froude No., F = 1), breach discharge $\boldsymbol{Q_p}$ and lake volume $\boldsymbol{V_o}$ can be recast in dimensionless terms:

$$\textit{dimensionless peak discharge} \quad Q_p^* = \frac{Q_p}{g^{0.5} d^{2.5}} \tag{1}$$

$$\textit{dimensionless lake volume} \quad V_o^* = \frac{V_o}{d^3} \tag{2}$$

where $\boldsymbol{g}$ is gravity and $\boldsymbol{d}$ the drop in water level or breach depth. Analysis of a physically-based model of dam-breach formation shows that the dimensionless peak discharge is primarily a function of the dimensionless lake volume and the dimensionless breach erosion rate $\boldsymbol{k^*}$ $(=\boldsymbol{k/g^{0.5}d^{0.5}})$ so that only three parameters are required to predict the full outflow hydrograph: $\boldsymbol{d}$, $\boldsymbol{V_o}$, and the vertical breach erosion rate $\boldsymbol{k}$ (typically 10–100 m/h). The last two factors are most critical as these govern whether or not substantial drawdown of the lake occurs before the breach has reached its maximum size. Other influences on outflow rate such as breach geometry (width-depth ratio) and lake hypsometry are relatively well constrained and can be excluded for simplicity. The breach hydrograph therefore depends on a dimensionless parameter given by [98]:

$$\eta = k^* V_o^* = \frac{kV_o}{g^{0.5} d^{3.5}} \tag{3}$$

In practical terms, $\boldsymbol{Q_p}$ is only influenced by the breach erosion rate for $V_o^* < 10^4$, while $\eta > 1$ means that breaching of the dam is effectively instantaneous so that peak discharge is solely controlled by the dimensions and geometry of the outlet. In dimensional form the relationship becomes:

$$Q_p = C\, g^{0.5} d^{2.5} \tag{4}$$

The value of the coefficient $\boldsymbol{C}$ depends on the breach geometry and lake hypsometry. Solution of the relevant equations for a number of general and special geometric/hypsometric cases produces a series of graphs which enable dimensionless peak discharge to be rapidly determined from $\eta$: for example, $\boldsymbol{C}$ has the value 1.94 for overtopping failures where $\eta >> 1$ [98].

### 2.1.3 Parametric Methods

An alternative means of reconstructing the outflow hydrograph from a dam failure is to apply open-channel weir flow equations [e.g. 31, 81] in a time-stepping numerical model which balances the flow through an enlarging breach with the rate of reservoir drawdown and hence hydraulic head reduction. Although this method tends to overestimate discharge because it neglects backwater and flow resistance effects and assumes level-pool routing in the lake, it offers a very rapid means of determining potential peak discharges for a range of combinations of lake size and breach geometry. The parameters most commonly required for this technique are the final dimensions of the breach (width, depth and sidewall slope), the breach development time, and a hypsometric description of the lake (surface area vs. elevation).

## *2.2 Paleohydraulic Analysis*

While paleohydraulic analysis of breach development can give information about the maximum *potential* peak discharge at the outlet, actual evidence of high-magnitude discharges is better obtained from further downstream. Two methods of obtaining information about the size of a paleoflood are (i) analysis of the maximum size of particles that were transported, and (ii) analysis of the maximum stage heights and water surface profile of the flood, from which flow velocity and discharge can be reconstructed.

### 2.2.1 Flow Competence and Velocity

The existence of established relationships between hydraulic flow parameters (i.e. velocity, shear stress, stream power) and the maximum size of particles that can be transported [19, 46, 73] permit local discharges to be estimated. In an analysis of the Bonneville flood, O'Connor [73] compared boulder deposits with local flow conditions derived from step-backwater models (see below) to obtain a regression relationship between particle size $\boldsymbol{D_i}$ (the mean intermediate diameter in cm of the 5 largest boulders at a site) and flow velocity $\boldsymbol{v}$ (m/s):

$$v = 0.29\, D_i^{0.61} \tag{5}$$

In combination with estimates of flow cross-section from channel geometry and stage level markers, discharge can be derived from such velocity estimates.

### 2.2.2 Step-Backwater Modelling

Step-backwater analysis is based on the principle of conservation of mass and energy in a steady, gradually varying flow. In paleohydrologic applications, a series of flow cross-sections defined by channel geometry and paleostage height indicators can be used to calculate an energy-balanced water-surface profile that is a function of discharge, flow resistance, and channel geometry [74]. Calculated profiles can be iteratively fitted to the geologic data to estimate peak discharge at various points along the flow path. However, the method is very sensitive to channel geometry, including cross-section stability (i.e. no erosion or deposition during the flood), spacing, and areas of ineffective flow such as channel embayments and flood plains. Hydraulic jumps arising from rapidly varying flow conditions, for example along steep irregular channels can cause numerical instabilities in the model, while the key assumption of steady, slowly varying flow is often untenable for dam-break flood, which are characterized by a rapid rising limb. The technique has however been successfully applied to bedrock gorge-confined reaches of paleofloods such as the Missoula [75] and Bonneville events [73].

## 3 Lake Taupo

Lake Taupo is the largest lake in the modern TVZ, with an area of 620 $km^2$ and a volume of c. 60 $km^3$ [48]. It partially occupies a volcano-tectonic collapse structure whose present configuration largely reflects caldera collapses during the 26.5 ka Oruanui and 1.8 ka Taupo eruptions and faulting and downwarping on regional structures [25, 105]. The lake overflows with a mean discharge of c. 130 $m^3/s$ (including a 60 $m^3/s$ contribution from hydro-electric diversions) into the 320 km long Waikato River, the North Island's longest and largest. The river descends 340 m in the first 180 km downstream of the outlet, through a steep, narrow valley confined between late Pleistocene fluvial terraces and ignimbrite gorges, before crossing the lowlands of the Hamilton Basin. Extensive hydroelectric development has turned the first 180 km of the Waikato River into a chain of eight reservoirs, which, coupled with two thermal power plants further downstream, produced ~25% of New Zealand's power. Meteorological floods in the Waikato basin are buffered by the storage capacity of Lake Taupo and the thick permeable pyroclastic sequences that blanket much of the central North Island.

### *3.1 Post-26.5 ka Oruanui Flood*

The c. 530 $km^3$ (magma-equivalent) Oruanui eruption from the Taupo volcanic centre at 26.5 ka [105] had a profound impact on the geography of the central North Island, destroying a long-lived Pleistocene lake system, creating a new landscape around the vent site where nonwelded pyroclastic flow deposits accumulated to thicknesses of hundreds of metres, and producing a major new closed topographic depression through caldera collapse and ignimbrite emplacement [59].

#### 3.1.1 Intracaldera Shoreline Terraces

Post-eruption, Lake Taupo accumulated to eventually reach a highstand level of c. 500 m a.s.l., as marked by a poorly preserved shoreline terrace [34, 58–59, 63]. Initial overspill occurred through the Waihora Bay area, 20 km west of the present outlet, resulting in a drop in lake level of c. 20 m to an elevation controlled by a resistant horizon of welded Whakamaru Ignimbrite. This stable stillstand persisted for decades to centuries, allowing formation of a better-developed shoreline terrace

**Fig. 21.2** Post-Oruanui shoreline terraces in the Western Bays of Lake Taupo (photo L. Homer, GNS)

before the ignimbrite barrier was eventually compromised near the present outlet in the Taupo area (Fig. 21.2).

Tributaries downstream of Lake Taupo are inferred to have re-established and sapped their way back to the lake, possibly aided by drainage of an ephemeral lake developed at c. 360–400 m a.s.l. in the Reporoa Basin [53] that lowered base levels and accelerated erosion. Breaching of the rim of Lake Taupo triggered a drop in its level from c. 480 to c. 405 m a.s.l. and established the course of the upper Waikato River in essentially its modern form. Significantly, post-Oruanui fluvial terraces along the upper Waikato River grade to this 405 m lowstand level, not the 480 m a.s.l. stillstand. Drawdown is inferred to have been rapid, based on the absence of any intermediate shoreline terraces or tributary deltas surrounding the lake, or any resistant horizons in the ignimbrite barrier. Downcutting below 405 m a.s.l. was hindered by a broad surface of indurated Huka Falls Formation sediments and the accumulation of a several metre-thick lag of dense lithic gravel at the base of the breach [59].

### 3.1.2 Extracaldera Flood Features

A c. 80 m drop in the level of a water body of the magnitude of Lake Taupo corresponds to the release of c. 60 km$^3$ of water, enough to inundate the entire North Island of New Zealand to a depth of 0.5 m. Intracaldera evidence suggests that this release was relatively rapid, but the best indication of a catastrophic break-out flood is the Smythe's Quarry Boulder Member [94], a massive bouldery unit (Fig. 21.3).

**Fig. 21.3** (**a**) Boulder transported by the post-Oruanui break-out flood from intracaldera Lake Taupo, 75 km downstream of the outlet. (**b**) Matrix facies of the Smythe's Quarry Boulder Bed, metre stick for scale

This unit can be traced from 68 to 88 km downstream of Lake Taupo and rests unconformably on fluvially reworked gravels and sands that lie with erosive contact on primary Oruanui pyroclastics that fill a pre-eruptive paleovalley. It is buried by the main aggradational phase of post-Oruanui volcaniclastic sedimentation and was hence deposited relatively early in the post-eruptive chronology [59]. The member comprises boulders between 1 and 10 m in intermediate diameter, often arranged in imbricate cluster bedforms and bar forms separated by anastomosing braid channels, in a matrix of finer gravels and coarse sands. Clast lithology varies downstream, reflecting local erosion and transport. Distinctive clasts of hydrothermally altered pumiceous breccia outcropping at Atiamuri have been transported at least 8 km from Ohakuri. This flood deposit is taken as marking the base of the Hinuera C Formation [59], the main phase of post-Oruanui volcaniclastic resedimentation coinciding with the peak of the Last Glacial Maximum [79]. Massive fluvial aggradation along the Waikato River eventually enabled it to overtop a wind-gap and avulse into the Hamilton Basin, where it became confined by incision following climatic amelioration at c. 17.6 ka [72].

## 3.2 *Post-1.8 ka Taupo Flood*

The most recent explosive eruption from the Taupo volcanic centre occurred at 1.8 ka [106]. During a complex and multiphase eruption c. 35 km$^3$ of rhyolitic magma was expelled as a series of Plinian and phreatomagmatic fall deposits more than 0.1 m thick over an area of 30,000 km$^2$ east of the vent, and as a climactic radially-distributed pyroclastic flow that devastated an area of 20,000 km$^2$ [104, 106], triggering a complex and prolonged sedimentary response in impacted catchments [54, 62, 64, 87]. The pyroclastic flow deposited two geomorphic variants: a thin veneer layer that drapes the pre-eruption landscape and a thicker but still

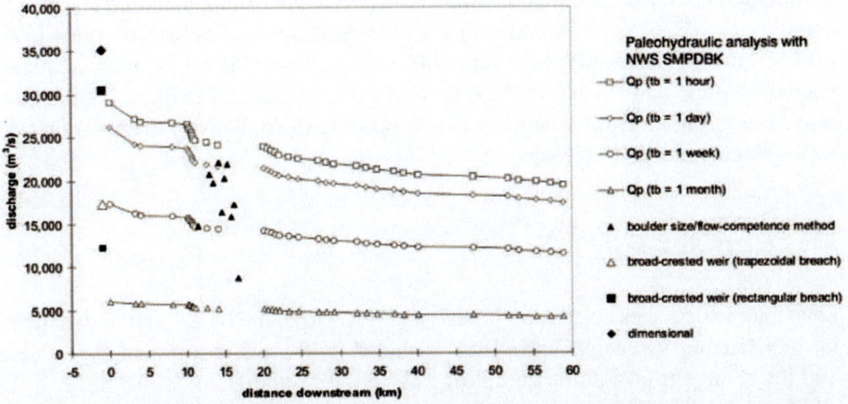


**Fig. 21.4** Summary of paleohydraulic analyses of the post-1.8 ka Taupo flood [61]. Breach peak discharge estimates from dimensional and parametric methods, downstream estimates from boulder flow competence data (Eq. 5), and hydraulic flow routing using US NWS programme SMPDBK [102]

unwelded valley-filling facies which infilled the Waikato River valley to depths of 40–70 m for 120 km downstream of the outlet. At the climax of the eruption, much of pre-eruption Lake Taupo was expelled, evaporated, or drained into the sub-rectangular caldera collapse structure beneath the lake floor [25]. The post-eruption highstand shoreline is marked by a semi-continuous, tectonically warped and offset, wave-cut bench and shoreline deposits at elevations of 28–43 m above the modern lake level of 357 m a.s.l [83, 107]. Refilling of Lake Taupo to the mean highstand level of +34 m is estimated to have taken 15–20 years based on modern precipitation and run-off rates and basin hypsometry [89, 106], before breaching of the unconsolidated barrier of ignimbrite choking the outlet channel was initiated by overtopping [61, 63].

Failure is inferred to have been rapid, based on the absence of well-developed intermediate shoreline terraces between the +34 m highstand level and a semi-continuous wave-cut bench at +2–5 m, indicating rapid drawdown during a single phase that released c. 20 $km^3$ of water [61, 63]. A variety of paleohydraulic techniques have been used to constrain the magnitude of the post-1.8 ka flood [61]: estimates of peak discharge at the breach fall in the range of 17,000–30,000 $m^3/s$ (Fig. 21.4).

### 3.2.1 Field Evidence for Flooding

Evidence of the catastrophic nature of the sudden release of 20 $km^3$ of water from Lake Taupo can be traced downstream along the Waikato River for over 220 km [54, 61]. Features associated with a large magnitude flood include: (1) a 12 km long, vertical walled spillway immediately downstream of the outlet that is floored with thin downstream-dipping fans of coarse lithic gravel and boulder lags; (2)

streamlined landforms sculpted from older deposits and exhumed river terraces; (3) bouldery fan deposits or expansion bars downstream of valley constrictions [57]; (4) fine-grained slackwater deposits in off-channel areas; (5) valley-wide erosional unconformities; and (6) buried forests in distal areas. Many of these features are also recognised, albeit at a larger scale, in other terrestrial environments impacted by major paleofloods [4, 73].

# 4 Lake Tarawera

Lake Tarawera covers an area of 41 $km^2$ and lies within the 64 ka Haroharo caldera in the Okataina Volcanic Centre [68], bounded by the western rim of the caldera and the resurgent lava dome complexes of Tarawera and Haroharo to the east. In addition to lying on the former drainage path from the adjacent Rotorua caldera, it also represents a remnant of a Haroharo intracaldera water body modified by caldera resurgence and dome development [63, 68].

## *4.1 SYN-700a Kaharoa Flood*

The 5 $km^3$ DRE 700a Kaharoa rhyolite eruption formed much of the Tarawera dome complex and deposited plinian tephra falls over much of the North Island [70]. Primary block-and-ash flows and remobilized pyroclastic debris blocked the outlet to the lake, infilling a narrow channel cut through an older fan structure (Fig. 21.5). Lake Tarawera rose c. 30 m above its pre-eruption level (298 m a.s.l.) before tsunami

**Fig. 21.5** Oblique aerial view of the outlet area of Lake Tarawera showing geomorphic elements of the pyro/volcaniclastic fan dam emplaced by the 700a Kaharoa eruption and the subsequent break-out flood (Photo DL Homer/GNS)

generated by late-stage pyroclastic flows overtopped the dam and triggered its catastrophic failure [38]. Approximately 1.7 $km^3$ of water was released into the head of the Tarawera valley as the lake level fell by >40 m, excavating a 300 m wide and 3 km long spillway before overtopping the 70 m high Tarawera Falls [38]. Flood deposits including boulders up to 13 m in diameter and giant bars extend up to 40 km from the lake, approximately 700 $km^2$ of the Rangitaiki Plains was resurfaced, and the shoreline advanced by c. 2 km [82]. Peak discharge at the outlet was estimated at c. $1.5 \times 10^5$ $m^3/s$ assuming instantaneous breach development, while boulder flow-competence relations further downstream indicate flows in the range of $10^4$–$10^5$ $m^3/s$ [38].

### *4.2 Flood*

In the aftermath of the 1886 basaltic plinian eruption of Tarawera volcano the level of Lake Tarawera rose by 12.8 m, before a rain-triggered break-out in November 1904 dropped it back by 3.3 m, generating a flood that was estimated to peak at c. 780 $m^3/s$ 24 km downstream [103]. The post-eruptive rise has been attributed to construction of a small alluvial fan across the outlet channel by flash-flood induced remobilisation of 1886 pyroclastic and older material in the Tapahoro gully [38].

## 5 Crater Lake, Mt. Ruapehu

The summit of Mt. Ruapehu, New Zealand's largest and most active onshore andesitic stratovolcano, normally hosts a hot acidic Crater Lake (Fig. 21.6a) at an elevation of c. 2,530 m a.s.l. with a volume of c. $9 \times 10^6$ $m^3$ [16]. The lake has

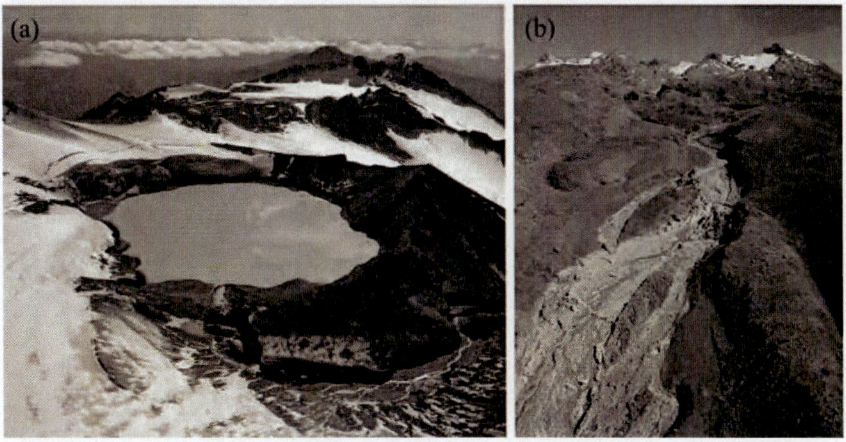


**Fig. 21.6** (**a**) Oblique aerial photo of Crater Lake, Ruapehu, after the break-out flood showing the breach in the tephra dam at lower centre. (**b**) Break-out lahar deposits and highwater marks in the lower Whangaehu Gorge c. 7 km downstream of Crater Lake (photos DL Homer, GNS)

existed for c. 2,000 years, based on the Holocene phreatomagmatic tephra and lahar records preserved on the eastern ring plain [23, 27, 39, 47], although it only became known to European settlers in 1879. Primary lahars have accompanied all large historic eruptions due to the explosive ejection of Crater Lake water over the rim and volumetric displacement [24, 36, 69], with secondary lahars being triggered by heavy rain on ash deposits [24, 37].

Lahars caused by break-outs from the Crater Lake include the 18 March 2007 flow [14, 60], the 1953 Tangiwai disaster detailed below, and a flood in the Whangaehu River in January 1925 that was estimated at 430 $m^3$/s at Tangiwai [92] and accompanied by an unexplained drop in the level of Crater Lake [77]. The cause of the 1861 lahar [22, 93], the largest historical event at Ruapehu is unknown, but may have been a lake break-out as no eruption was reported [33]. A number of pre-historic break-out events related to deep-seated rim failures have also been inferred from the geological record [39, 47].

## *5.1 March 2007 Break-Out Lahar*

The most recent volcanogenic break-out flood in New Zealand occurred on 18 March 2007 [14, 60], when Ruapehu's Crater Lake breached (Fig. 21.6a) a fragile barrier of unconsolidated tephra and scoria layers deposited over the former outlet area by the 1995–1996 eruption sequence after refilling over the preceding 11 years [12]. The 400 m diameter lake fell by c. 6 m in a little over 90 min as the barrier failed, resulting in the release of 1.3 million $m^3$ of warm, highly mineralised water into the steep upper gorge of the Whangaehu River at a peak discharge of 540 $m^3$/s. Entrainment of glacial ice, older laharic alluvium, talus fans, and syn-lahar landslide debris caused the flow to rapidly bulk by a factor of >3 into a transitional hyperconcentrated/non-cohesive debris flow within the gorge (Fig. 21.6b). Ten km from source, the flood debouched onto the low gradient, unconfined Whangaehu Fan where it anastomosed into multiple braided distributary channels before coalescing back into a single-thread channel at 17 km, continuing downstream confined within an increasingly deeply incised meandering channel cut into uplifted marine sediments of the Wanganui Basin. The lahar reached the coast, 215 km downstream of Crater Lake c. 16 h after the onset of dam failure, flowing as a series of partially overlapping and intermixing waves of kinematic form, comprising displaced ambient river water, highly mineralised Crater Lake water, and entrained sediment, each with their own celerity.

In contrast to the disastrous 1953 break-out lahar (see below), the 2007 dam breach was predicted a decade in advance. An interagency governmental task force put together a NZ$7 million response plan to associate risks through the installation of an early warning system and engineered countermeasures [35, 44, 45]. As a result, no lives were lost and infrastructural damage was minimal, while in parallel, a comprehensive research programme was able to maximise the scientific data and observational opportunities presented by this event [60].

### 5.2 The 1953 Tangiwai Lahar

On Christmas Eve 1953 the summit Crater Lake of Mount Ruapehu breached an unstable barrier composed of volcanic material deposited during the 1945 eruption sequence [77], and buttressed by the Crater Basin Glacier. Approximately 1.8 million cubic metres of water was released into the headwaters of the Whangaehu River, where it rapidly entrained snow, ice, and volcanic debris to form a lahar that reached the Tangiwai railway bridge c. 39 km downstream in a little over 2 h [77, 92]. The flow critically damaged the bridge moments before passage of the Wellington-Auckland express: unable to stop in time the train fell into the lahar-swollen river with the loss of 151 lives, making it New Zealand's worst volcanic disaster.

Paleohydraulic analysis of the 1953 event indicates that the former summit glaciers did not impede the outflow of hot Crater Lake water, enlargement of the ice tunnel keeping pace with growth of a trapezoidal breach at the outlet to generate a peak clear-water discharge in the 300–400 $m^3/s$ range [55]. By c. 10 km downstream, flow bulking through the entrainment of particulate material along the Whangaehu Gorge had increased this to c. 2,000 $m^3/s$, before rapid attenuation on the Whangaehu Fan, calibrated against the 1995 lahar sequence [24], reduced the peak flow by c. 60% at 23 km downstream. The crest of the lahar reached Tangiwai railway bridge 39.0 km downstream 2.1–2.3 h after the onset of breaching as a debris – to hyperconcentrated flow with a peak discharge of c. 590–647 $m^3/s$ [92]. Paleohydraulic analysis has highlighted both the dependence of bulking factors on lahar magnitude and the role of infiltration losses on the Whangaehu Fan, and how these interact to produce a lahar attenuation curve for the Whangaehu River.

## 6 Lake Rotorua

The 20 km-diameter Rotorua caldera, formed at 220 ka by eruption of the voluminous Mamaku Ignimbrite [40, 67], is currently partially occupied by the shallow intracaldera Lake Rotorua. Extensive suites of highstand lacustrine terraces and shoreline benches fringe much of the Rotorua basin [34, 43], evidence of a complex history of lake level oscillations influenced by volcanic activity at the adjacent Okataina Volcanic Centre [68].

The highest lacustrine terrace (387–414 m a.s.l.) within the Rotorua caldera corresponds to a post-220 ka highstand level associated with filling of the newly created basin in the immediate aftermath of the Mamaku Ignimbrite eruption [65]. Considerable uncertainty surrounds the direction of overflow of this level, but the lake potentially breached southwards through the Hemo Gorge, a v-shaped notch in the topographic rim of the caldera through the Ngakuru Graben into the Waikato River drainage. At some time however, a northeasterly outlet became established at a lower level. Eruption of the Rotoiti Breccia at 64 ka from the adjacent Okataina Volcanic Centre (OVC) blocked this drainage route, causing the lake to rise by

c. 90 m above its modern level. Lake level stabilised at c. 370–380 m a.s.l. for long enough for well-developed highstand lacustrine terraces to form around much of the basin [43, 65], before a northerly overspill was superseded by a north-easterly break-out through the Haroharo caldera (formed by the Rotoiti eruption and probably also temporarily host to a post-eruption intracaldera lake) into the Rangitaiki Plains. A deeply incised channel buried beneath the floor of Lake Rotoiti is inferred to represent the path of this flood. Resumption of volcanic activity in the Haroharo caldera during eruption of the Mangaone Sub-group [68] again blocked this drainage path at c. 36 ka, causing Lake Rotorua to rise to 348–353 m a.s.l. before breaking out again [63, 65]. Growth of the Haroharo resurgent dome complex at 9–7.5 ka ultimately blocked this north-easterly route out of the Rotorua caldera, forcing Lake Rotorua to rise to c. 285–290 m a.s.l. before capture of the headwaters of the Kaituna River established the current outlet channel to the north [68].

## 7 Potential Future Volcanic Lake Break-Out Floods in the TVZ

The potential for future dam-break floods from the numerous volcanic lakes in the central North Island depends in large part upon the role of future volcanic activity on their outlets. Historic events at Ruapehu have demonstrated that lake break-out floods are a recurrent hazard at this volcano. Lake Tarawera is particularly vulnerable to any resumption of activity at the adjacent Tarawera volcano: indeed, the lake remains c. 9.5 m higher than before the 1886 eruption due to incomplete breaching of the volcaniclastic fan dam during the 1904 flood [38]. On a geological timescale, volcanic modification and damming of the largest extant water bodies and rivers in the TVZ is inevitable, with serious implications for downstream areas [cf. 64].

## 8 Discussion and Conclusions

The Taupo Volcanic Zone hosts multiple indicators of dam-break flooding (Table 21.1), including: (i) sources of water, such as a lakes that have experienced a temporary highstand; (ii) a mechanism for generating a flood, such as breaching of the rim of the lake by overtopping or headward erosion; and (iii) corroborating evidence for the rapid release of large volumes of water, both from within the lake basin in the form of markers of rapid drawdown, and along the outlet channel in the form of indicators of large magnitude flows, above and beyond what would be expected from a purely meteorological event.

In addition to the 1953 and 2007 events in New Zealand, historic crater lake rim failures have occurred at Agua in Venezuela in 1541 [71], Kelut in Indonesia in 1875 [88], and Santa Ana in El Salvador in 2005 [9]. Notable historic break-out floods from other classes of volcanic impoundment have followed damming of rivers by pyroclastic flows at El Chichón in 1982 [50]; pyroclastic flows and lahars at Pinatubo in 1991 and subsequently [86, 96]; and by a debris avalanche

**Table 21.1** Paleohydraulic data for volcanic lake break-out flood in the Taupo Volcanic Zone

| Lake | Date | Dam type | Dam height/lake drop $H_o$ (m) | Excess lake volume $V_o$ ($\times 10^6$ $m^3$) | Peak discharge $Q_p$ ($m^3$/s) |
|---|---|---|---|---|---|
| Rotorua | Post-220 ka | Caldera rim | 27 | 4,500 | 23,000[a] |
| | Post-64 ka | Pyroclastic flow | 90 | 9,600 | 50,000–470,000[a] |
| | Post-36 ka | Pyroclastic flow | 75 | 8,100 | 50,000–300,000[a] |
| Taupo | Post-26.5 ka | Caldera rim | 80 | 60,000 | 350,000[a] |
| | Post-1.8 ka | Pyroclastic flow | 32 | 20,000 | 35,000[a] 25,000[b] |
| Tarawera | Syn-700a | Pyroclastic flow/lahar | 40 | 1,700 | 19,000–60,000[a] 75,000[b] |
| | 1904 | Lahar | 3.3 | 139 | 780[c] |
| Crater | 1953 | Pyroclastic fall and lava | 7.9 | 1.8 | 110–1,100[a] |
| Lake, Ruapehu | 2007 | Pyroclastic fall | 6.3 | 1.35 | 400[d] 540[e] |

[a]Peak discharge calculated using dimensional method [98]. Range is for breach erosion rates of 1–100 m/h.
[b]Peak discharge calculated using broad-crested weir equations.
[c]Gauged 24 km downstream of outlet.
[d]Calculated from BREACH model, broad-crested weir equations, and coupled thermal/hydrodynamic model.
[e]Calculated from breach erosion model and measured rate of lake-level drawdown, constrained by time-lapse images of dam-failure sequence.

at Mount St. Helens in 1981 [66]. Such initially clear-water break-outs frequently entrain abundant pyroclastic material from the dam and channel, rapidly increasing in discharge to generate major lahars (volcano-hydrologic mass flows and floods) in downstream catchments, often resulting in major loss of life [71]. Outside of New Zealand prehistoric volcanogenic floods have been recognised from breaching of debris avalanche deposits at Nevado de Colima [13] and Mount St. Helens [85], lava flow dams on the Colorado River [29, 30], and pyroclastic flow valley dams in Europe [2, 78] and Japan [42]. Intracaldera lake break-out floods have occurred historically at Pinatubo, Philippines by overtopping [6], Ilopango, El Salvador, associated with subaqueous dome emplacement [51], Karmsky, Kamchatka, following blocking of the outlet by an emergent tuff cone-forming eruption [8, 28] and Okmok, Alaska [1, 7]. Pre-historic intracaldera break-out floods have been recognised in many volcanic arcs, including Alaska (Aniakchak [99], Fisher [91], and Okmok [1, 7], China/North Korea (Baitoushan/Tianchi [101]).

A plot of peak discharge versus lake volume discharge (Fig. 21.7) shows that break-outs from intracaldera lakes, in New Zealand and worldwide, are amongst the largest known floods on Earth, only being exceeded by Late Pleistocene glacial lake outbursts and continental-scale meteorological events [5, 76]. Break-outs from other volcanogenic impoundments are comparable with floods from the breaching of other

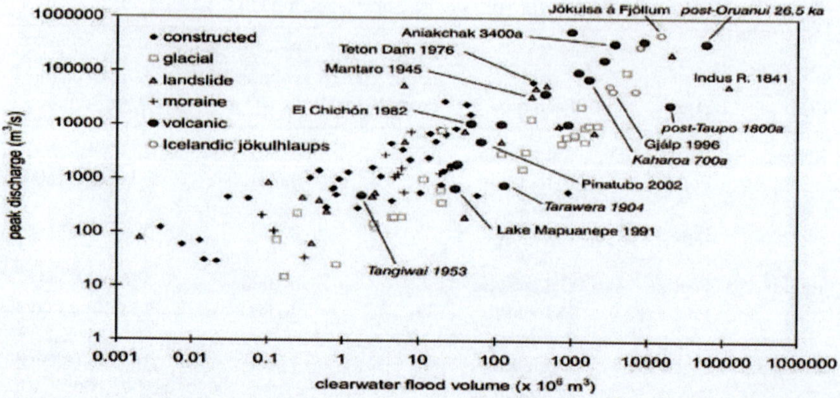


**Fig. 21.7** Comparison plot of break-out floods from volcanic dams, versus landslide and artificial dam breaches. Data compiled from [20, 76, 84, 98]

types of natural dam such as landslides and moraines [21, 84]. Jökulhlaups [e.g. 10, 95], although influenced by sub-glacial volcanism as a mechanism for melting ice, are functionally a class of glacier-lake breakout as the outflow hydrograph is controlled by the glacier, not the volcano [17, 97].

Crater lakes appear to be particularly ephemeral, being vulnerable to explosive ejection or drainage through breaching or collapse of the crater rim, while volcanic dams in general appear to predominantly fail by overtopping and surface erosion rather than piping or gravitational slope failure. This is apparently a function of the geometric, granulometric, and mechanical properties of the volcanic dam and channel. Unwelded pyroclastic dams appear more failure-prone than debris avalanche deposits, possibly because they are typically composed of unconsolidated, fine-grained, non-cohesive, and low density (vesiculated) material. There also appears to be a correlation between the silica content of the caldera-forming eruption (i.e. explosivity) and the likelihood that the caldera will develop a subaerial (as opposed to a subglacial, as in the case of Icelandic basaltic calderas [11]) water body large enough to potentially overspill and trigger a break-out flood [56]. Calderas developed by magma withdrawal at basaltic volcanoes tend to be dry or host only small ponds due to the permeability of the volcanic pile: felsic calderas such as Taupo tend to be lined and fringed by fine-grained relatively impermeable ignimbrite resulting in over-filling. Intermediate composition calderas may host a large lake that achieves a stable level without a surface outlet through seepage, such as Crater Lake, Oregon [3], or overfill and breach like Baitoushan/Tianchi [101]. The dominance of these effects over other variables such as climatic conditions and catchment water balance has yet to be explored. The frequency of events in the Taupo Volcanic Zone appears to relate to the intense geological scrutiny of this area rather than any special feature of the actual geological environment. Similar detailed research in other volcanic areas is expected to show that volcanic lake break-out floods are a ubiquitous feature, and hence represent a significant new class of hazard that should be incorporated into hazard assessments.

**Acknowledgements** Research for this study developed out of the Catastrophic Flooding and Lahars programme founded by B.F. Houghton. Funding was provided by the Foundation for Research, Science and Technology, New Zealand under contracts C05516 and C0005X6. The manuscript benefited from reviews by BJ Scott and M McSaveney.

## References

1. Almberg, L.D., Beget, J.E. and Larsen, J.F. (2003) Magmatic production rate determined by volcaniclastic lithofacies variations in a rapidly fluctuating caldera lake, Okmok volcano, Alaska. Proceedings, IAVCEI General Assembly, Sapporo, Japan, pp. A519.
2. Baales, M., Jöris, O., Street, M., Bittman, F., Wninger, B. and Wiethold, J. (2002) Impact of the Late Glacial eruption of the Laacher See volcano, central Rhineland, Germany, *Quaternary Research* **58**, 273–288.
3. Bacon, C.R., Gardner, J.V., Mayer, L.A., Buktenica, M.W., Dartnell, P., Ramsey, D.W. and Robinson, J.E. (2002) Morphology, volcanism, and mass wasting in Crater Lake, Oregon, *Geological Society of America Bulletin* **114**, 675–692.
4. Baker, V.R. (1985) Large-scale erosional and depositional features of the Channeled Scabland, in V.R. Baker and D. Nummedal (eds.), *The Channeled Scabland.* NASA, Washington DC, pp. 81–115.
5. Baker, V.R. (2002) High-energy megafloods: Planetary settings and sedimentary dynamics, in I.P. Martini, V.R. Baker and G. Garzón (eds.), *Flood and Megaflood Processes and Deposits: Recent and Ancient Examples*. International Association of Sedimentologists Spec. Publ. 32. Blackwell Science, Oxford, pp. 3–15.
6. Bornas, M.A., Tungol, N., Maximo, R.P.R., Paladio-Melosantos, M.L., Mirabueno, H.T., Javier, D.V., Corpuz, E.G., Dela Cruz, E.G., Ramos, A.F., Marilla, J.D. and Villacorte, E.U. (2003) Caldera-rim breach and lahar from Mt. Pinatubo, Philippines: Natural breaching and resulting lahar, Proceedings, IUGG 2003 Congress, Sapporo, Japan., p. A558.
7. Begét, J.E., Larsen, J.F., Neal, C.A., Nye, C.J. and Schaefer, J.R. (2005) Preliminary volcano-hazard assessment of Okmok volcano, Umnak Island, Alaska. Alaska Department of Natural Resources, Division of Geological and Geophysical Surveys, Fairbanks, Alaska, Report of Investigations 2004-3, 32 pp.
8. Belousov, A. and Belousov, M. (2001) Eruptive process, effects and deposits of the 1996 and the ancient basaltic phreatomagmatic eruptions in Karymskoye lake, Kamchatka, Russia, in J.D.L. White and N.R. Riggs (eds.), *Volcanogenic Sedimentation in Lacustrine Settings*. International Association of Sedimentologists Spec. Publ. 30. Blackwell Science, Oxford, pp. 35–60.
9. BGVN (2005) Santa Ana, El Salvador – eruption and crater lake break-out lahar, *Bulletin of the Global Volcanism Network* **30**, 14–16.
10. Björnsson, H. (1975) Subglacial water reservoirs, jökulhlaups and volcanic eruptions, *Jökull* **25**, 1–14.
11. Björnsson, H. (2003) Subglacial lakes and jökulhlaups in Iceland, *Global and Planetary Change* **35**, 255–271.
12. Bryan, C., Christenson, B.W., Hodgson, K.A., Houghton, B.F., Johnston, D.M., Jurado-Chichay, Z., Lindsay, J., Manville, V., Nakagawa, M., Nairn, I.A., Otway, P.M., Sherburn, S., Scott, B.J., Thordarson, T., Wood, C.P. and Wilson, C.J.N. (1996) Volcanic eruption at a New Zealand ski resort prompts re-evaluation of hazards, *Eos Transaction AGU* **77**, 189–191.
13. Capra, L. and Macías, J.L. (2002) The cohesive Naranjo debris-flow deposit (10 $km^3$): A dam breakout flow derived from the Pleistocene debris-avalanche deposit of Nevado de Colima Volcano (México), *Journal of Volcanology and Geothermal Research* **117**, 213–235.

14. Carrivick, J.L., Manville, V. and Cronin, S.J. (2009) A fluid dynamics approach to modelling the 18th March 2007 lahar at Mt. Ruapehu, New Zealand, *Bulletin of Volcanology* **71**, 153–169 doi:10.1007/s00445-008-0213-2.
15. Chow, V.T. (1959) *Open channel hydraulics*. McGraw-Hill, New York, NY.
16. Christenson, B.W. and Wood, C.P. (1993) Evolution of a vent-hosted hydrothermal system beneath Ruapehu Crater Lake, New Zealand, *Bulletin of Volcanology* **55**, 547–565.
17. Clarke, G.K.C. (2003) Hydraulics of subglacial outburst floods: New insights from the Spring-Hutter formulation, *Journal of Glaciology* **49**, 299–313.
18. Coleman, S.E. and Andrews, D.P. (2000) Overtopping breaching of noncohesive homogeneous embankments. Department of Civil and Resource Engineering, University of Auckland, Auckland, Report 589, 339 pp.
19. Costa, J.E. (1983) Paleohydraulic reconstructions of flash-flood peaks from boulder deposits in the Colorado front range, *Geological Society of America bulletin* **94**, 986–1004.
20. Costa, J.E. (1988) Floods from dam failures, in V.R. Baker, R.C. Kochel and P.C. Patton (eds.), *Flood geomorphology*. Wiley, New York, NY, pp. 439–463.
21. Costa, J.E. and Schuster, R.L. (1988) The formation and failure of natural dams, *Geological Society of America bulletin* **100**, 1054–1068.
22. Crawford, J.C. (1870) On the geology of the province of Wellington, *Transactions of the New Zealand Institute* **2**, 342–360.
23. Cronin, S.J. and Neall, V.E. (1997) A late quaternary stratigraphic framework for the north-eastern Ruapehu and eastern Tongariro ring plains, New Zealand, *New Zealand Journal of Geology and Geophysics* **40**, 185–197.
24. Cronin, S.J., Neall, V.E., Lecointre, J.A. and Palmer, A.S. (1997) Changes in Whangaehu River lahar characteristics during the 1995 eruption sequence, Ruapehu volcano, New Zealand, *Journal of Volcanology and Geothermal Research* **76**, 47–61.
25. Davy, B.W. and Caldwell, T.G. (1998) Gravity, magnetic and seismic surveys of the caldera complex, Lake Taupo, North Island, New Zealand, *Journal of Volcanology and Geothermal Research* **81**, 69–89.
26. Delmelle, P. and Bernard, A. (2000) Volcanic lakes, in H. Sigurdsson, B. Houghton, S. McNutt, H. Rymer and J. Stix (eds.), *Encyclopedia of Volcanoes*. Academic Press, New York, NY, pp. 877–895.
27. Donoghue, S.L., Neall, V.E., Palmer, A.S. and Stewart, R.B. (1997) The volcanic history of Ruapehu during the past 2 millenia based on the record of Tufa Trig tephra, *Bulletin of Volcanology* **59**, 136–146.
28. Fazlullin, S.M., Ushakov, S.V., Shuvalov, R.A., Aoki, M., Nikolaeva, A.G. and Lupikina, E.G. (2000) The 1996 subaqueous eruption at Academii Nauk volcano (Kamchatka) and its effects on Karymsky lake, *Journal of Volcanology and Geothermal Research* **97**, 181–193.
29. Fenton, C.R., Webb, R.H. and Cerling, T.E. (2006) Peak discharge of a Pleistocene lava-dam outburst flood in Grand Canyon, Arizona, USA, *Quaternary Research* **65**, 324–335.
30. Fenton, C.R., Poreda, R.J., Nash, B.P., Webb, R.H. and Cerling, T.E. (2004) Geochemical discrimination of five Pleistocene lava-dam outburst flood deposits, Western Grand Canyon, Arizona, *Journal of Geology* **112**, 91–110.
31. Fread, D.L. (1996) Dam-breach floods, in V.P. Singh (ed.), *Hydrology of Disasters*. Kluwer, Dordrecht, pp. 85–126.
32. Froehlich, D.C. (1987) Embankment-dam breach parameters, in R.M. Ragan (ed.), *National Conference on Hydraulic Engineering*. ASCE, Williamsburg, VA, pp. 570–575.
33. Graettinger, A.H., Manville, V. and Briggs, R.M. (2009) Depositional record of historic lahars in the Whangaehu valley, Ruapehu, New Zealand: Implications for trigger mechanisms, flow dynamics, and lahar hazards, *Bulletin of Volcanology* doi:10.1007/s00445-009-0318-2.

34. Grange, L.I. (1937) The geology of the Rotorua-Taupo subdivision, *New Zealand Geological Survey bulletin* **37**, 138 pp.
35. Hancox, G.T., Keys, H. and Webby, M.G. (2001) Assessment and mitigation of dam-break lahar hazards from Mt Ruapehu Crater Lake following the 1995-96 eruptions, in K.J. McManus (ed.), *Engineering and Development in Hazardous Terrains*, Proceedings, New Zealand Geotechnical Society/NZ Institute of Professional Engineers conference, Christchurch, New Zealand, pp. 385–409.
36. Healy, J., Lloyd, E.F., Rushworth, D.E.H., Wood, C.P., Glover, R.B. and Dibble, R.R. (1978) The eruption of Ruapehu, New Zealand on 22 June 1969, *New Zealand Department of Scientific and Industrial Research Bulletin* **224**, 80 pp.
37. Hodgson, K.A. and Manville, V. (1999) Sedimentology and flow behaviour of a rain-triggered lahar, Mangatoetoenui stream, Ruapehu volcano, New Zealand, *Geological Society of America Bulletin* **111**, 743–754.
38. Hodgson, K.A. and Nairn, I.A. (2005) The ~AD 1315 syn-eruption and AD 1904 post-eruption breakout floods from Lake Tarawera, Haroharo caldera, North Island, New Zealand, *New Zealand Journal of Geology and Geophysics* **48**, 491–506.
39. Hodgson, K.A., Lecointre, J. and Neall, V.E. (2007) Onetapu formation: The last 2000 yr of laharic activity at Ruapehu volcano, New Zealand, *New Zealand Journal of Geology and Geophysics* **50**, 81–99.
40. Houghton, B.F., Wilson, C.J.N., McWilliams, M.O., Lanphere, M.A., Weaver, S.D., Briggs, R.M. and Pringle, M.S. (1995) Chronology and dynamics of a large silicic magmatic system: Central Taupo Volcanic Zone, New Zealand, *Geology* **23**, 13–16.
41. Johnson, F.A. and Illes, P. (1976) A classification of dam failures, *International Water Power and Dam Construction* **28**, 43–45.
42. Kataoka, K.S., Urabe, A., Manville, V. and Kajiyama, A. (2008) Large-scale breakout flood from an ignimbrite-dammed river: Aftermath of the Numazawako eruption (BC 3400), Northeast Japan, *Geological Society of America Bulletin* **120**, 1233–1247.
43. Kennedy, N.M., Pullar, W.A. and Pain, C.F. (1978) Late Quaternary land surfaces and geomorphic changes in the Rotorua Basin, North Island, New Zealand, *New Zealand Journal of Science* **21**, 249–264.
44. Keys, H.J.R. (2007) Lahars of Ruapehu volcano, New Zealand: Risk mitigation, *Annals of Glaciology* **45**, 155–162.
45. Keys, H.J.R. and Green, P.M. (2008) Ruapehu lahar New Zealand 18 March 2007: Lessons for hazard assessment and risk mitigation 1995–2007, *Journal of Disaster Research* **3**, 284–296.
46. Komar, P.D. (1989) Flow-competence evaluation of the hydraulic parameters of floods: An assessment, in K. Beven and P.A. Carling (eds.), *Floods: Hydrological, Sedimentological and Geomorphological Implications*. Wiley, Chichester, pp. 107–134.
47. Lecointre, J., Hodgson, K.A., Neall, V.E. and Cronin, S.J. (2004) Lahar-triggering mechanisms and hazard at Ruapehu volcano, New Zealand, *Natural Hazards* **31**, 85–109.
48. Lowe, D.J. and Green, J.D. (1992) Lakes, in J.M. Soons and M.J. Selby (eds.), *Landforms of New Zealand*. Longman Paul Ltd, Auckland, pp. 107–143.
49. MacDonald, T.C. and Langridge-Monopolis, J. (1984) Breaching characteristics of dam failures, *Journal of Hydraulic Engineering* **110**, 567–586.
50. Macías, J.L., Capra, L., Scott, K.M., Espíndola, J.M., García-Palomo, A. and Costa, J.E. (2004) The 26 May 1982 breakout flows derived from failure of a volcanic dam at El Chichón, Chiapas, Mexico, *Geological Society of America Bulletin* **116**, 233–246.
51. Mann, C.P., Stix, J., Vallance, J.W. and Richer, M. (2004) Subaqueous intracaldera volcanism, Ilopango Caldera, El Salvador, Central America, in W.I.J. Rose, J.J. Bommer, D.L. López, M.J. Carr and J.J. Major (eds.) *Natural Hazards in El Salvador*, Geological Society of America Special Paper 375, Boulder, CO, pp. 159–174.
52. Manville, V. (2001) *Techniques for evaluating the size of potential dam-break floods from natural dams*. Institute of Geological and Nuclear Sciences, Wellington. Science Report 2001/28. pp. 70.

53. Manville, V.R. (2001) Sedimentology and history of Lake Reporoa: An ephemeral supra-ignimbrite lake, Taupo Volcanic Zone, New Zealand, in J.D.L. White and N.R. Riggs (eds.), *Volcanogenic Sedimentation in Lacustrine Settings*. International Association of Sedimentologists Spec. Publ. 30. Blackwell, Oxford, pp. 109–140.
54. Manville, V. (2002) Sedimentary and geomorphic responses to a large ignimbrite eruption: Readjustment of the Waikato River in the aftermath of the A.D. 181 Taupo eruption, New Zealand, *Journal of Geology* **110**, 519–542.
55. Manville, V. (2004) Palaeohydraulic analysis of the 1953 Tangiwai lahar: New Zealand's worst volcanic disaster, *Acta Vulcanologica* **XVI**, 137–152.
56. Manville, V. (2010) An overview of break-out floods from intracaldera lakes, *Global and Planetary Change* **70**, 14–23.
57. Manville, V. and White, J.D.L. (2003) Incipient granular mass flows at the base of sediment-laden floods, and the roles of flow competence and flow capacity in the deposition of stratified bouldery sands, *Sedimentary Geology* **155**, 157–173.
58. Manville, V. and Wilson, C.J.N. (2003) Interactions between volcanism, rifting and subsidence: Implications of intracaldera palaeoshorelines at Taupo volcano, New Zealand, *Journal of the Geological Society of London* **160**, 3–6.
59. Manville, V. and Wilson, C.J.N. (2004) The 26.5 ka Oruanui eruption, New Zealand: A review of the roles of volcanism and climate in the post-eruptive sedimentary response, *New Zealand Journal of Geology and Geophysics* **47**, 525–547.
60. Manville, V. and Cronin, S.J. (2007) Break-out lahar from New Zealand's Crater Lake, *Eos Transaction AGU* **88**, 441–442.
61. Manville, V.R., White, J.D.L., Houghton, B.F. and Wilson, C.J.N. (1999) Paleohydrology and sedimentology of a post-1800a breakout flood from intracaldera Lake Taupo, North Island, New Zealand, *Geological Society of America Bulletin* **111**, 1435–1447.
62. Manville, V., Newton, E.H. and White, J.D.L. (2005) Fluvial responses to volcanism: Resedimentation of the 1800a Taupo ignimbrite in the Rangitaiki river catchment, North Island, New Zealand, *Geomorphology* **65**, 49–70 doi: 10.1016/j.geomorph.2004.07.007.
63. Manville, V., Hodgson, K.A. and Nairn, I.A. (2007) A review of break-out floods from volcanogenic lakes in New Zealand, *New Zealand Journal of Geology and Geophysics* **52**, 131–150.
64. Manville, V., Segschneider, B., Newton, E.H., White, J.L.D., Houghton, B.F. and Wilson, C.J.N. (2009) Environmental impact of the 1.8 ka Taupo eruption: Landscape responses to a large-scale explosive rhyolitic eruption, *Sedimentary Geology* **220**, 318–336 doi: 10.1016/j.sedgeo.2009.04.017.
65. Marx, R.S., Manville, V. and White, J.D.L. (2009) Sedimentology and allostratigraphy of post-220 ka to pre-26.5 ka lacustrine terraces at intracaldera Lake Rotorua, Taupo Volcanic Zone, New Zealand, *Sedimentary Geology* **220**, 349–362 doi: 10.1016/j.sedgeo.2009.04.025.
66. Meyer, W., Sabol, M.A. and Schuster, R.L. (1986) Landslide dammed lakes at Mount St. Helens, Washington, in R.L. Schuster (ed.), *Landslide Dams – Process, Risk, and Mitigation*. American Society of Civil Engineers Geotechnical Spec. Publ. 3, pp. 21–41.
67. Milner, D.M., Cole, J.W. and Wood, C.P. (2003) Mamaku Ignimbrite: A caldera-forming ignimbrite erupted from a compositionally zoned magma body in Taupo Volcanic Zone, New Zealand, *Journal of Volcanology and Geothermal Research* **122**, 243–264.
68. Nairn, I.A. (2002) *Geology of the Okataina Volcanic Centre*, Institute of Geological & Nuclear Sciences, geological map no. 25, 1 sheet and 156 pp., Lower Hutt, New Zealand.
69. Nairn, I.A., Wood, C.P. and Hewson, C.A.Y. (1979) Phreatic eruptions of Ruapehu: April 1975, *New Zealand Journal of Geology and Geophysics* **22**, 155–173.
70. Nairn, I.A., Self, S., Cole, J.W., Leonard, G.S. and Scutter, C. (2001) Distribution, stratigraphy and history of proximal deposits from the ~AD 1305 Kaharoa eruption of

Tarawera volcano, New Zealand, *New Zealand Journal of Geology and Geophysics* **44**, 467–484.
71. Neall, V.E. (1996) Hydrological disasters associated with volcanoes, in V.P. Singh (ed.), *Hydrology of Disasters*. Kluwer, London, pp. 395–425.
72. Newnham, R.M., Eden, D.E., Lowe, D.J. and Hendy, C.H. (2003) Rerewhakaaitu Tephra, a land-sea marker for the last termination in New Zealand, with implications for global climate change, *Quaternary Science Reviews* **22**, 289–308.
73. O'Connor, J.E. (1993) *Hydrology, Hydraulics, and Geomorphology of the Bonneville Flood*, Geological Society of America, Special Paper **274**, 83 pp.
74. O'Connor, J.E. and Webb, R.H. (1988) Hydraulic modelling for paleoflood analysis, in V.R. Baker, R.C. Kochel and P.C. Patton (eds.), *Flood Geomorphology*. Wiley, New York, NY, pp. 393–402.
75. O'Connor, J.E. and Baker, V.R. (1992) Magnitudes and implications of peak discharges from glacial Lake Missoula, *Geological Society of America Bulletin* **104**, 267–279.
76. O'Connor, J.E., Grant, G.E. and Costa, J.E. (2002) The geology and geography of floods, ancient floods, modern hazards: Principles and applications of paleoflood hydrology, in P.K. House, R.H. Webb, V.R. Baker and D.R. Levish (eds.), *Ancient Floods, Modern Hazards: Principles and Applications of Paleoflood Hydrology*. Water Science and Applications **5**. American Geophysical Union, Washington, DC, pp. 359–385.
77. O'Shea, B.E. (1954) Ruapehu and the Tangiwai disaster, *New Zealand Journal of Science and Technology* **B36**, 174–189.
78. Park, C. and Schmincke, H.-U. (1997) Lake formation and catastrophic dam burst during the Late Pleistocene Laacher See eruption (Germany), *Naturwissenschaft* **84**, 521–525.
79. Pillans, B., Mcglone, M., Palmer, A., Mildenhall, D., Alloway, B. and Berger, G. (1993) The last glacial maximum in Central and Southern North Island, New Zealand – A paleoenvironmental reconstruction using the Kawakawa Tephra formation as a chronostratigraphic marker, *Palaeogeography Palaeoclimatology Palaeoecology* **101**, 283–304.
80. Ponce, V.M. and Tsivoglou, A.J. (1981) Modeling gradual dam breaches, *Journal of the Hydraulics Division* **107**, 829–838.
81. Price, J.T., Lowe, G.W. and Garrison, J.M. (1977) *Unsteady flow modelling of dam-break waves*, Proceedings, Dam-break flood routing workshop, U.S. Water Resources Council, Hydrology Committee, Bethesda, Maryland, Report PB–275–437.
82. Pullar, W.A. and Selby, M.J. (1971) Coastal progradation of Rangitaiki Plains, New Zealand, *New Zealand Journal of Science* **14**, 419–434.
83. Riggs, N.R., Ort, M., White, J.D.L., Wilson, C.J.N. and Clarkson, R. (2001) Marginal lacustrine sedimentation and the post-eruptive rise and fall of Lake Taupo, in J.D.L. White and N.R. Riggs (eds.), *Volcanogenic Sedimentation in Lacustrine Settings*. International Association of Sedimentologists Spec. Publ. 30. Blackwell, Oxford, pp. 151–177.
84. Schuster, R.L. (2000) Outburst debris-flows from failure of natural dams, in G. Wieczorek and N. Naeser (eds.), *Debris-Flow Hazards Mitigation: Mechanics, Prediction, and Assessment*. Balkema, Rotterdam, pp. 29–42.
85. Scott, K.M. (1988) *Origin, behavior, and sedimentology of prehistoric catastrophic lahars at Mt. St. Helens, Washington*, Geological Society of America, Special Paper **229**, pp. 23–36.
86. Scott, K.M., Janda, R.J., De La Cruz, E.G., Gabinete, E., Eto, I., Isada, M., Sexon, M. and Hadley, K.C. (1996) Channel and sedimentation responses to large volumes of 1991 volcanic deposits on the east flank of Mount Pinatubo, in C.G. Newhall and R.S. Punongbayan (eds.), *Fire and mud: Eruptions and lahars of Mount Pinatubo, Philippines*. University of Washington Press, Seattle, WA, pp. 971–988.
87. Segschneider, B., Landis, C.A., White, J.D.L., Wilson, C.J.N. and Manville, V. (2002) Resedimentation of the 1.8 ka Taupo ignimbrite in the Mohaka and Ngaruroro river catchments, Hawke's Bay, New Zealand, *New Zealand Journal of Geology and Geophysics* **45**, 85–102.

88. Simkin, T. and Siebert, L. (eds.) (1994) *Volcanoes of the world.* Geoscience Press, Tucson, AZ, 349 pp.
89. Smith, R.C.M. (1991) Landscape response to a major ignimbrite eruption, Taupo Volcanic Center, New Zealand, in R.V. Fisher and G.A. Smith (eds.), *Sedimentation in Volcanic Settings.* Society of Economic Paleontologists and Mineralogists Spec. Publ. 45, pp. 123–137.
90. Smith, R.C.M., Smith, I.E.M., Browne, P.R.L. and Hochstein, M.P. (1993) Volcano-tectonic controls on sedimentation in the Taupo Volcanic Zone, New Zealand, in P.F. Ballance (ed.), *South Pacific Sedimentary Basins.* Elsevier Science Publishers B.V, Amsterdam, pp. 143–156.
91. Stelling, P., Gardner, J. and Beget, J.E. (2005) Eruptive history of Fisher Caldera, Alaska, USA, *Journal of Volcanology and Geothermal Research* **139**, 163–183.
92. Stilwell, W.F., Hopkins, H.J. and Appleton, W. (1954) *Tangiwai Railway Disaster: Report of Board of Inquiry.* Government Printers, Wellington, 31 pp.
93. Taylor, R. (1861) *Journal of Rev. Richard Taylor 1833–1873.* Typescript in Alexander Turnbull Library, Wellington.
94. Thompson, B.N. (1958) The geology of the Atiamuri dam site, *New Zealand Journal of Geology and Geophysics* **1**, 275–306.
95. Tómasson, H. (2002) Catastrophic floods in Iceland, in Á. Snorrason, H.P. Finnsdóttir and M.E. Moss (eds.), *The Extremes of the Extremes: Extraordinary Floods.* IAHS, Reykjavik, pp. 121–126.
96. Umbal, J.V. and Rodolfo, K.S. (1996) The 1991 lahars of southwestern Mount Pinatubo and evolution of the lahar-dammed Mapanuepe Lake, in C.G. Newhall and R.S. Punongbayan (eds.), *Fire and Mud, Eruptions and Lahars of Mount Pinatubo, Philippines.* University of Washington Press, Seattle, WA, pp. 951–970.
97. Walder, J.S. and Costa, J.E. (1996) Outburst floods from glacier-dammed lakes: The effect of mode of drainage on flood magnitude, *Earth Surface Processes and Landforms* **21**, 701–723.
98. Walder, J.S. and O'Connor, J.E. (1997) Methods for predicting peak discharge of floods caused by failure of natural and constructed earthen dams, *Water Resources Research* **33**, 2337–2348.
99. Waythomas, C.F., Walder, J.S., McGimsey, R.G. and Neal, C.A. (1996) A catastrophic flood caused by drainage of a caldera lake at Aniakchak Volcano, Alaska, and implications for volcanic – hazards assessment, *Geological Society of America Bulletin* **108**, 861–871.
100. Webby, M.G. and Jennings, D.N. (1994) Analysis of dam break flood caused by failure of Tunawaea landslide dam, Proceedings, International Conference on Hydraulics in Civil Engineering: Hydraulics working with the environment, Institute of Engineers, Brisbane, Australia, pp. 163–168.
101. Wei, H., Sparks, R.S.J., Liu, R., Fan, Q., Wang, Y., Hong, H., Zhang, H.W., Chen, H., Jiang, C., Dong, J., Zheng, Y. and Pan, Y. (2003) Three active volcanoes in China and their hazards, *Journal of Asian Earth Sciences* **21**, 515–526.
102. Wetmore, J.N. and Fread, D.L. (1984) The NWS simplified dam break flood forecasting model for desk-top and hand-held microcomputers, Federal Emergency Management Agency, 122 pp.
103. White, J.D.L., Houghton, B.F., Hodgson, K.A. and Wilson, C.J.N. (1997) Delayed sedimentary response to the A.D. 1886 eruption of Tarawera, New Zealand, *Geology* **25**, 459–462.
104. Wilson, C.J.N. (1985) The Taupo eruption, New Zealand. II. The Taupo ignimbrite, *Philosophical Transactions of the Royal Society of London* **A314**, 229–310.
105. Wilson, C.J.N. (2001) The 26.5 ka Oruanui eruption, New Zealand: An introduction and overview, *Journal of Volcanology and Geothermal Research* **112**, 133–174.
106. Wilson, C.J.N. and Walker, G.P.L. (1985) The Taupo eruption, New Zealand. I. General aspects, *Philosophical Transactions of the Royal Society of London* **A314**, 199–228.

107. Wilson, C.J.N., Riggs, N.R., Ort, M.H., White, J.D.L. and Houghton, B.F. (1997) An annotated atlas of post-1.8 ka shoreline features at Lake Taupo, New Zealand. In Institute of Geological and Nuclear Sciences, Lower Hutt, New Zealand, pp. 35.
108. Wilson, C.J.N., Houghton, B.F., McWilliams, M.O., Lanphere, M.A., Weaver, S.D. and Briggs, R.M. (1995) Volcanic and structural evolution of Taupo Volcanic Zone, New Zealand: A review, *Journal of Volcanology and Geothermal Research* **68**, 1–28.

# Chapter 22
# Characterization of the 2000 Yigong Zangbo River (Tibet) Landslide Dam and Impoundment by Remote Sensing

**Stephen G. Evans and Keith B. Delaney**

## 1 Introduction

Detailed case histories of landslide dams provide key data on their formation and behaviour, longevity, volumes of impounded water, outburst volumes in the case of landslide dam breaching, as well as input data for the possible simulation of downstream outburst floods (e.g. [7]). Together this data provides insight into the physical processes involved in the damming of major rivers and the behaviour of subsequent landslide dams. They also contribute to global data on landscape effects of catastrophic landslides as well as contributing to the development of a hazard assessment framework for the affected region and landslide dam hazard mitigation strategies worldwide.

This paper concerns the damming of the Yigong Zangbo River by the Yigong rock avalanche which occurred in the mountains of Tibet at 30° 10′ 39″N/94° 56′ 25″E (Fig. 22.1) in the evening of April 9, 2000 [10, 14]. The landslide debris completely blocked the Yigong Zangbo River, a major river that is a tributary of the Yarlung Zangbo, which in turn is a major tributary of the Brahmaputra River (Fig. 22.2). The dam failed 62 days later on June 10, 2000 when the debris dam breached following overtopping through an excavated spillway, generating a major outburst flood downstream in the Yarlung Zangbo (Tibet) and Dihang (India) Rivers.

The Yigong landslide dam case history has global significance because, if the volumes reported in the literature are correct, (a) it formed one of the largest landslide-dammed impoundments documented in history, (b) the reported volume (3 $Gm^3$) of the outburst in the most detailed report on the event [10] is the largest outburst flood from a landslide-dammed lake ever documented, (c) discharges of the outburst flood reported 17 km downstream approach the highest reported for an outburst flood, and (d) the 2000 event occurred at the same location as an even larger

S.G. Evans (✉)
Landslide Research Programme, Department of Earth and Environmental Sciences,
University of Waterloo, Waterloo, ON, Canada N2L 3G1
e-mail: sgevans@sciborg.uwaterloo.ca

S.G. Evans et al. (eds.), *Natural and Artificial Rockslide Dams*, Lecture Notes in Earth Sciences 133, DOI 10.1007/978-3-642-04764-0_22, © Springer-Verlag Berlin Heidelberg 2011 

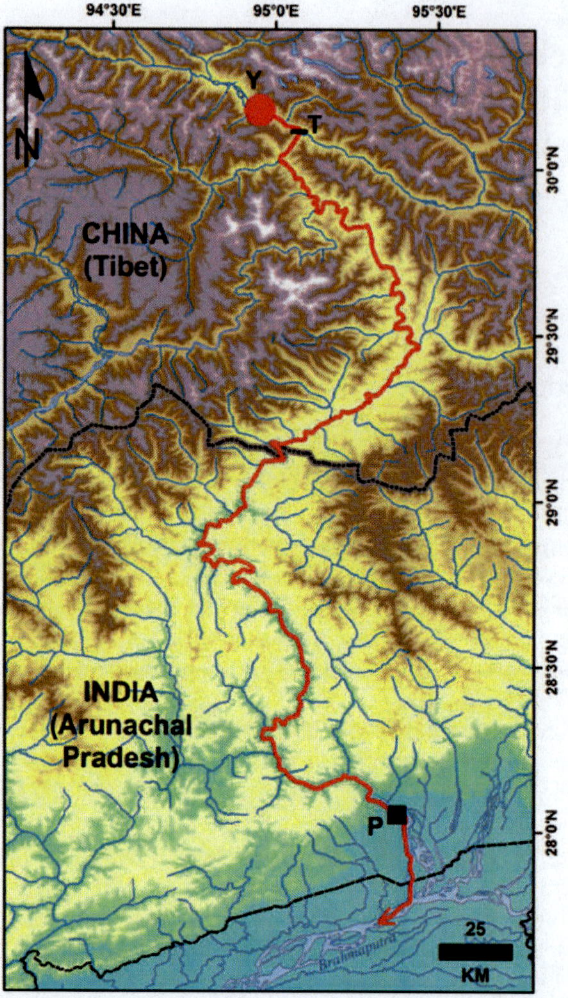


**Fig. 22.1** Location of 2000 Yigong landslide dam (Y) and path of outburst flood (*red line*) resulting from the breach of the dam 62 days after emplacement. The flood followed the Yarlung Zangbo (*Tibet*) and Dihang (*India*) Rivers, and entered the main Brahmaputra below Pasighat (*P*) which is located 462 km downstream from the landslide dam. Tongmai Bridge (*T*) is a locality discussed in text

landslide dam and landslide-dammed lake which formed and failed, resulting in a massive outburst flood, in 1900 (e.g. [10]).

The impetus for the present paper is the existence of notable inconsistencies in the body of published literature on the Yigong landslide and the landslide-dammed Yigong lake. These inconsistencies concern the geometry and volume of the 2000 damming landslide and the filling and maximum volume of the impounded Yigong lake.

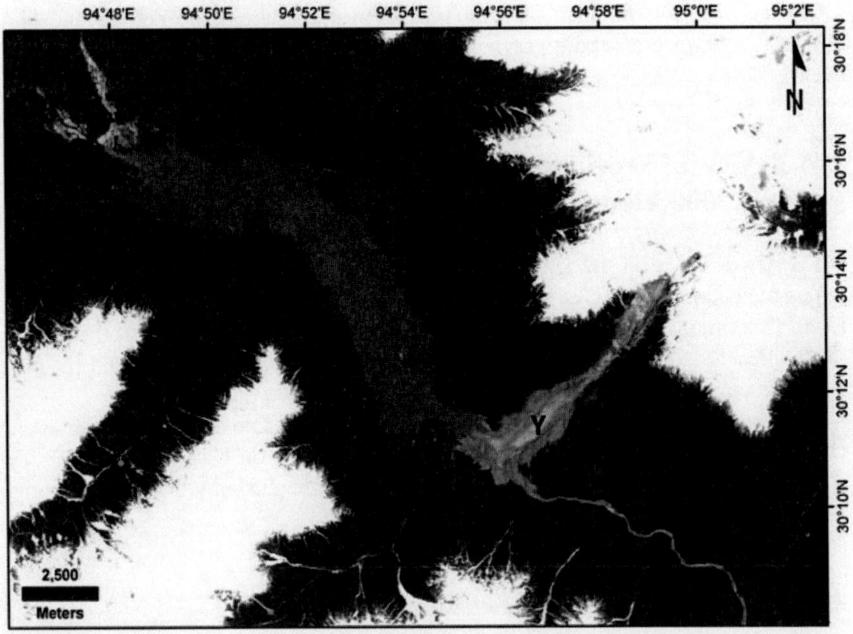


**Fig. 22.2** LANDSAT-7 image obtained on May 4, 2000 (Day 25 of impoundment) showing Yigong landslide (*Y*) and landslide-dammed Yigong lake forming upstream. Lake pool elevation is estimated at 2,225 m a.s.l. and the lake had a volume of approximately 313 $Mm^3$

Further, important discrepancies were noted in the most-cited paper on the event by Shang et al. [10] published in *Geomorphology*. These discrepancies concern such key data as the geometry of the damming landslide path, the volume of the source rock mass, the volume of material entrained by the landslide, the volume of the landslide debris and the geometry of the dam itself, the rate of filling of the landslide-dammed lake, and the final volume of the impoundment which breached on June 10, 2000. The issues concerning the occurrence and behaviour of the damming landslide are examined in detail elsewhere; here we focus on the landslide dam and the resulting filling and eventual outburst of Yigong Lake.

The objective of this paper is thus to attempt to resolve some of these issues by the analysis of remote sensing data available for the Yigong area. These include (1) digital topographic data acquired during the Shuttle Radar Topography Mission (SRTM) in February 2000 [5], only months before the valley-damming Yigong landslide occurred, and (2) LANDSAT-7 satellite images acquired at various times before the landslide dam formed, during the filling of the landslide-dammed lake and after the June 10 breach and outburst flood. We also document the extent of the outburst flood (which extended over 500 km downstream into neighbouring India and into the main Brahmaputra river valley) and compare our estimate of the Yigong Lake outburst volume, to other high-magnitude landslide-dammed lake outburst events worldwide.

Before we report our analysis of the formation and failure of the Yigong landslide dam, we undertake a detailed review of previously published work on the Yigong event.

## 2 A Review of Previously Published Work (2001–2008) on the 2000 Yigong Events

We review articles on the 2000 Yigong events published between 2001 and 2008 with a focus on, (1) the volume of the damming landslide, (2) the rate of lake filling, (3) the maximum pre-breach volume of the landslide-dammed lake, (4) the timing of the breach, (5) the nature and magnitude of the downstream flood, and (6) pre-landslide conditions.

The landslide dam was emplaced on April 9, 2000. Our review uses an event chronology beginning as Day 1 on April 10 and ending on Day 62 (June 10), the date of the breach. For consistency we refer to the watershed in which the damming landslide occurred as the Zhamulong Gully.

### *2.1 Volume of Damming Landslide*

Ren et al. [9], Zhou et al. [18], and Wang and Lu [13] report that the landslide occurred at 20:00 h on April 9, 2000. It had its source in a rock slope high up in the head of Zhamulong Gully. Ren et al. [9] and Xu et al. [16] are precise about the time of initial failure (19:59 h) based on landslide-generated seismogram traces at nearby seismic stations. The landslide was also reported by the USGS National Earthquake Information Centre (http://neic.usgs.gov) as two shallow earthquakes (M3.5 and 4.6) approximately 12 s apart with estimated epicentres 5.5 and 2.5 km, respectively, from the source area.

Ren et al. [9], Zhou et al. [18], Wang and Lu [13] report the landslide volume as 300 $Mm^3$. Other authors distinguish initial and final volume of debris but differ in their estimates of initial and entrained volumes. Both Han [6] and Xu et al. [16] suggest that the landslide initiated as a 30 $Mm^3$ rockfall from the source area, with the final deposit of 300 $Mm^3$ apparently due to the entrainment of 270 $Mm^3$ of material along its path. Ren et al. [9] present an account based on GPS-controlled traverses of the landslide debris and estimates the total volume of the 2000 debris to be more than 380 $Mm^3$.

In the most-cited paper on the 2000 Yigong event, Shang et al. [10] estimate that the damming landslide had a volume of 300 $Mm^3$ and report that the initial wedge failure involved a rock mass of about 100 $Mm^3$ that increased its volume by entrainment to 300 $Mm^3$ in its nearly 10 km of travel. Wen et al. [14] report that the initial failure was 200 $Mm^3$ in volume and entrained 100 $Mm^3$ in its travel down Zhamulong Gully for a final total volume of 300 $Mm^3$.

Wang [12] reports a detailed remote sensing and topographic analysis. With respect to the source volume a comparison of before and after imagery and digital elevation data indicates that 91.18 $Mm^3$ of rock was involved in the initial failure. However, Wang [12] also calculates the volume of the deposit to be 95.5 $Mm^3$, roughly the same as the source volume, an estimate apparently not reflecting (a) bulking of the source rock mass due to fragmentation, and (b) entrainment (which, based on published photographs of the debris may have been considerable) in Zhamulong Gully. In recognising that previous work overestimated the volume of the landslide (ca. 300 $Mm^3$), Wang [12] mentions that this may have been due to the inclusion of the 1900 landslide debris in estimates of the 2000 landslide debris volume (see below).

## 2.2 The Filling of the Landslide-Dammed Lake

The landslide debris blocked the Yigong River forming a landslide-dammed lake (Yigong Lake) which is reported to have flooded several villages and the Yigong tea plantation [15]. Zhou et al. [18] report a satellite image analysis of the filling of Yigong Lake, expressed as an increase in lake surface area as follows; April 14 (Day 5) – 15 $km^2$, May 4 (Day 25) – 34.7 $km^2$, and May 9 (Day 30) – 37.1 $km^2$. However, no additional data is reported for the lake filling between May 9 (Day 30) and the day of the breach, June 10 (Day 62), reported incorrectly by Zhou et al. [18] as occurring on June 11.

Based on the interpretation of a SPOT image on April 19 (Day 10), Han [6] gives the area of the landslide dammed lake as 18.6 $km^2$. On May 4 (Day 25) and on May 9 (Day 30) the areas of the lake were 34.7 $km^2$ and 37.1 $km^2$ respectively. The latter two estimates correspond to the areas reported by Zhou et al. [18]. On June 7 the area of the lake had reportedly increased to 70 $km^2$ ([6], p. 23). The volume of the lake was calculated as 2.92 $Gm^3$ and the lake had reportedly risen 48.34 m in the 60 days since the landslide dam was emplaced [6], a rise in water level presumably measured at the landslide dam itself. If we take the elevation of the river channel directly upstream of the dam as 2,210 m (see Fig. 5 in Zhou et al. [18]) this suggests a lake level of 2,259 m a.s.l. on June 7 (Day 59).

Wang and Lu [13] also outline a satellite image analysis of the Yigong events and they provide data on the most detailed and systematic study of lake filling ([13]; Table 22.1). These are plotted in Fig. 22.3. Topographic control was provided by (1) measurement of post-event ground control points by GPS on the surface of the landslide, and (2) selection of pre-event ground control points from a 1:100,000 scale topographic map. As seen in Table 22.1, Wang and Lu [13] examined 5 sets of satellite imagery obtained during the impoundment; the last satellite image examined is dated May 20 (Day 41), 21 days before the breach. Pre-event contour data was apparently derived from the 1:100,000 topographic map.

Shang et al. [10] report that the landslide dam was 2,500 m in length along the thalweg of the Yigong River. No elevations of the water level during lake filling are provided by Shang et al. [10] but, noting that the lake was 15 $km^2$ in area on April 14 (Day 5; cf. [18]), Shang et al. [10] report the lake volume as 70 $Mm^3$ on that day. No additional volume estimates are given during filling and lake areas given by

**Table 22.1** The filling of Yigong rockslide-dammed lake reported by Wang and Lu [13] (Data reproduced from Table 2 in Wang and Lu [13]). Pre-lake datum is given as 2,210 m a.s.l

| Date | Day of impoundment | Elevation of rockslide-dammed lake (m a.s.l.) | Lake area ($km^2$) | Lake volume ($Mm^3$) |
|---|---|---|---|---|
| 2000-4-13 | 4 | 2,214 | 18.909 | 85.4 |
| 2000-5-4 | 25 | 2,225 | 33.659 | 514.3 |
| 2000-5-9 | 30 | 2,228 | 36.32 | 706.2 |
| 2000-5-12 | 33 | 2,229 | 37.979 | 770.7 |
| 2000-5-20 | 41 | 2,234 | 43.121 | 1,234.5 |
| 2000-6-10 | 62 | 2,264[a] | 52.855 | 2,259.0 |

[a] Elevation given in original table is 2,244 m a.s.l. which was a typographical error. Correct value is 2,264 m a.s.l (Z. Wang, personal communication to the authors, 2009).

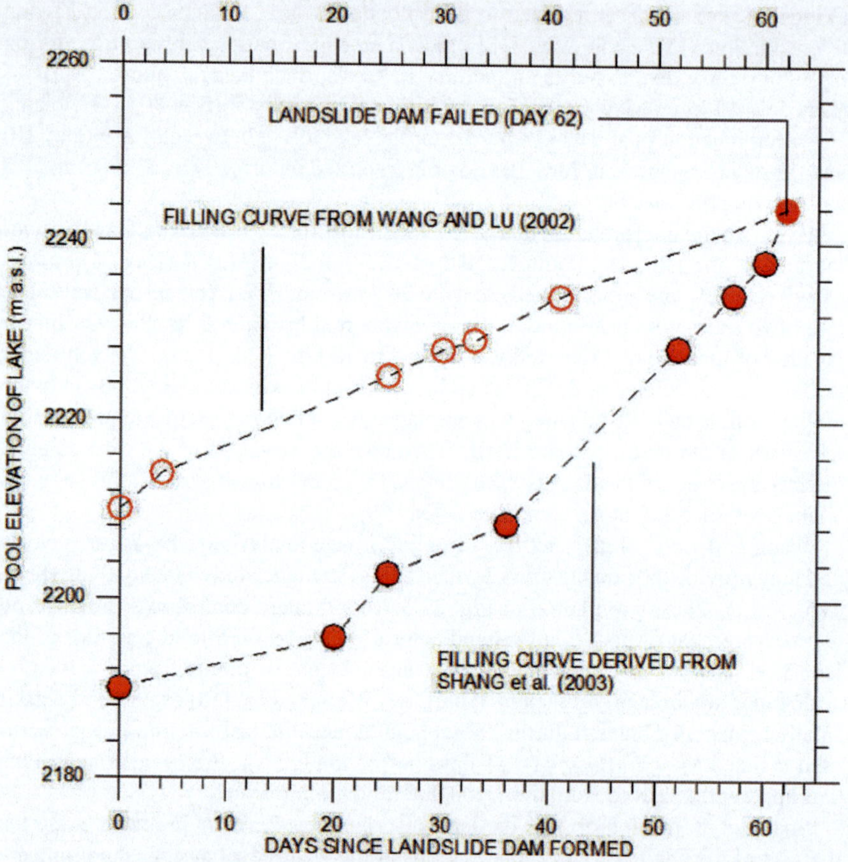


**Fig. 22.3** Filling curves for rockslide-dammed Yigong lake based on data of Wang and Lu [13] in Table 22.1 and data derived from Fig. 8 in Shang et al. [10]. Note that despite the fact that they begin at different datums the final pool elevation (2,244 m a.s.l.) is the same in both cases

Shang et al. [10] are reproduced from Zhou et al. [18] up to May 9 (Day 30), 32 days before the breach. Shang et al. [10] note that during the 62 days of impoundment the water rose by about 54 m in the landslide-dammed lake. Pre-event elevations in the paper are apparently taken from a report on the area completed in the mid-1980s (see Fig. 3 in [10]) and the elevation of the Yigong River channel directly upstream of the landslide is given as 2,190 m a.s.l (see Fig. 6 in [10]), 20 m below that reported by Zhou et al. [18] and Wang and Lu [13]. Data given by Shang et al. [10] thus suggest a final pool elevation of 2,244 m a.s.l. However, if we assume that the pre-landslide river bed had an elevation of 2,210 m a.s.l. as given in Zhou et al. [18] and Wang and Lu [13], then the depth (54 m) reported by Shang et al. [10] is equivalent to a full pool elevation of 2,264 m a.s.l. Shang et al. [10] also present a time-line diagram (Fig. 8 in Shang et al. [10]) which indicates that the lake level increased by 45 m in the 37 days between May 4 (Day 25) and June 10 (Day 62), the day of the breaching. This data does not correspond with that in Table 22.1 which tabulates the data of Wang and Lu [13].

We compared the filling data from Table 22.1 (Wang and Lu [13]) with the filling data we derived from Fig. 8 in Shang et al. [10]. The filling curve for both data sets is plotted in Fig. 22.3. Quite different filling curves result; however, although beginning at a different datum, the final pool elevation (2,264 m a.s.l.) is the same in both cases.

## 2.3 The Maximum Pool Elevation and Pre-breach Volume of Yigong Lake

Zhou et al. [18] do not report the maximum lake level before breaching nor estimate the pre-breach/outburst volume. However, they estimate the lowest height of the landslide dam above the river channel at 55 m (pp. 328, 331). The elevation of the river channel directly upstream of the dam is given as 2,210 m (Fig. 5 in [18]) suggesting a maximum pool elevation of the landslide dammed lake as ca. 2,265 m a.s.l.

Wang and Lu ([13]; Table 22.1) report the immediate pre-breach volume as 2.259 $Gm^3$ for a landslide dammed lake 52.855 $km^2$ in area and a maximum water level elevation of 2,244 m a.s.l. The sources of this final estimate are not given.

Zhu et al. [20] and Shang et al. [10] both estimate outburst volume as 3 $Gm^3$. Wang [12] estimates that 2.4 $Gm^3$ of water outburst from Yigong Lake after reaching a maximum pool elevation of 2,280 m a.s.l., the highest of the maximum published pool elevation estimates.

## 2.4 The Timing of the Breach

The timing of the breach was linked to the excavation of a spillway over the debris by Chinese soldiers using heavy machinery [6, 15]; also see time-line diagram in

Fig. 8 in [10]. According to Shang et al. [10] excavation of the spillway began on May 3 (Day 24) and was completed as designed on June 3 (Day 55) after the excavation of 1.36 $Mm^3$ of rock and soil. The spillway had a maximum depth of 24.1 m but was 18 m deep at the dam crest [17]. No data is reported on the elevation of the spillway. Han [6] reports that water began to flow in this channel at 0600 h on June 8 (Day 60) and that the dam failed catastrophically on June 10 (Day 62).

Wang and Lu [13] give the time and date of the breach as 21:30 h on June 10, 2000 (Day 62). Xu et al. [16] note that following the excavation of a man-made spillway the lake drained catastrophically on June 10 (Day 62) at 21:00 h.

## 2.5 The Outburst Flood and Downstream Effects

Zhu et al. [20] estimate the outburst involved a volume of 3 $Gm^3$ of water and lasted 6 h. This is equivalent to an average discharge of 138,888 $m^3/s$; they also estimate the maximum breach discharge as 30,250 $m^3/s$. Discharge calculated at an unspecified point downstream (probably the Tongmai Bridge) was 126,400 $m^3/s$, 36 times greater than the normal river flood [20].

With respect to the downstream flood, Shang et al. [10] report a maximum instantaneous discharge of about 120,000 $m^3/s$ at the Tongmai Bridge, 17 km downstream from the dam (T in Fig. 22.1) and they note that the outburst volume totalled 3 $Gm^3$ of impounded water.

## 2.6 The 1900 Landslide Dam

The pre-event Yigong River already had a complex channel pattern due to the fact that a similar landslide from Zhamulong Gully had blocked the Yigong River in 1900 (see below). The 2000 blockage occurred at the same site as a previous landslide in Zhamulong Gully (the Jazhong landslide) in July 1900 which Zhou et al. [18] estimate had a volume of 1 $Gm^3$. Zhu et al. [20] also give details of the 1900 landslide dam (reported as being emplaced in 1902). According to Shang et al. [10] the 1900 landslide was about 500 $Mm^3$ in volume and impounded a lake with an area of 51.9 $km^2$. After 1 month the 1900 landslide dam breached releasing a reported 2.7 $Gm^3$ of water [20]. The 1900 damming and the development of debris flow fans from the south, had resulted in a constriction in the Yigong valley, aggradation upstream and the development of a braided channel which during high river flows formed a lake (Lake Yigong).

## 2.7 Summary of Review

Thus in the published works reviewed above (1) the reported initial failure volume of the Yigong landslide varies by an order of magnitude from 30 to 300 $Mm^3$, (2) the final debris volume of the landslide reported varies from 95 to 300 $Mm^3$, and (3) the entrained volume reported varies from 4 to 270 $Mm^3$.

With respect to the landslide-impounded Yigong Lake, (1) the maximum pool elevations reported vary from 2,244 to 2,280 m a.s.l., and (2) reported impounded volumes varied from 2.259 to 3 $Gm^3$, estimates that also correspond to the range in reported outburst volumes.

## 3 Data Used in the Analysis and Its Precision

We attempt to resolve the discrepancies in previously published reports identified above by the analysis of a range of remotely sensed data. These data consist of Landsat 7 imagery obtained on September 23, 1999 (before impoundment), May 4 and May 20, 2000 (during impoundment) and June 17, 2000 (after impoundment). The May images allowed us to delimit the extent of Yigong Lake as it was filling. In conjunction with the optical images, we also utilised digital terrain data from the Shuttle Radar Topography Mission (SRTM) obtained in February 2000 [5], only months before the impoundment, to construct a digital elevation model of the Yigong area. We used SRTM-3 (3 Arc Second) data. This data has an estimated vertical error of approximately 10 m in the general region of the Yigong rockslide dam (Fig. 14 in [5]).

The vertical precision of the SRTM-3 data in the region was examined in more detail and appears to be high. The elevation of the Yigong river at the upstream margin of the landslide dam prior to the event is given as 2,210 m a.s.l. by Wang and Lu [13] and Zhou et al. [18] as noted above. SRTM data also gives an elevation of 2,210 m a.s.l. for the same location.

As a further test of SRTM precision in the region we used the approach of Becek [1] which involves the comparison of known airport runway elevations with SRTM data for the same locations. We examined data for two Tibetan airports; Gonggar Airport (29° 17′ 52″N/90° 54′ 43″E – 400 km to the west of Yigong), and Qamdo Bangda Airport (30° 33′ 13″N/97° 06′ 31″E – 213 km to the east of Yigong). The newly constructed Nyinchi Airport (29° 18′ 10″N/94° 20′ 02″E – 113 km to the south–south west) did not exist at the time of the SRTM mission. The two airports that existed in 2000 are in mountain valleys similar in setting to that of the Yigong landslide dam. SRTM elevations at both ends and the middle of the runways were obtained and the mean error was found to be +2.0 m. This data suggests that SRTM elevations in mountain valleys in Tibet are well within the error range specified for the region in Farr et al. [5] and superior to the vertical resolution, suggested by the contour interval, of a 1:100,000 scale topographic map.

Chinese topographic maps of greater than 1:1,000,000 scale of the Yigong area were not available to us and field work in Tibet was not carried out.

## 4 The Yigong Rock Avalanche: Geometry of Initial Failure Mass, Entrainment and Volume of Debris

We estimated the extent, area and volume of the source rock mass of the Yigong rock avalanche by (1) examining pre- and post-event Landsat-7 imagery and (2) analysing digital terrain data from the pre-event SRTM datasets (3) importing and re-constructing in GIS, the isopleths of the depletion zone reported by Wang [12; Fig. 8].

By examining the SRTM data and satellite imagery, Wang's [12] estimate of the initial failure volume (91 $Mm^3$) appears plausible but may be overestimated. We calculate the missing volume as 75 $Mm^3$. If correct, our volume estimate would produce ca. 90 $Mm^3$ of rock debris on fragmentation, assuming a 20% increase in volume due to bulking (e.g. [2]).

In estimating the volume of the debris we imported and re-constructed in GIS the isopleths of debris thickness in Wang [12; Fig. 13]. The isopleths represent a debris volume of only 95 $Mm^3$, a volume roughly equivalent to the bulked initial failure volume estimated above but a volume that does not reflect any entrainment from Zhamulong Gully. We also compared the SRTM and LANDSAT-7 imagery to obtain a rough estimate of debris volume. The comparison is complicated by the fact that the imagery reflects considerable post-event incision in the debris particularly along the main channel of the Yigong River (Fig. 22.2). Using GIS, we estimated the volume of the debris to be in the order of 105 $Mm^3$.

Entrainment during movement down Zhamulong Gully is evident on post-event imagery (Fig. 22.2) and is remarked on by several authors (e.g. [14]). The volume of entrainment is estimated by subtracting the bulked initial failure volume from the total volume of the debris; this yields ca. 15 $Mm^3$ of material entrained from Zhamulong Gully.

Thus despite increases in the volume of the initial failure mass due to bulking and entrainment, the estimated total volume of the landslide (105 $Mm^3$) is almost one third of the maximum estimate of 300 $Mm^3$. On the other hand, our estimate is over three times greater than the minimum estimate (30 $Mm^3$) cited above.

## 5 Landslide Dam Geometry

The measurement of the 2000 landslide dam geometry is complicated by the presence of the dissected remnants of the 1900 landslide dam discussed above. This is evident in the longitudinal profile of the Yigong River in the vicinity of the 2000 landslide dam (Fig. 22.4). The 2000 dam is approximately 2.8 km long along the thalweg of the Yigong. The lowest point on the crest of the upstream face is in the order of 2,282 m a.s.l. and the highest point of the dam is 2,306 m a.s.l. The spillway reported to have been dug was 18 m deep at the dam crest [17] which effectively constrains a maximum pool elevation at about 2,265 m a.s.l. This compares with the estimate of maximum lake level at 2,264 m a.s.l. reported by Wang and Lu [13] and

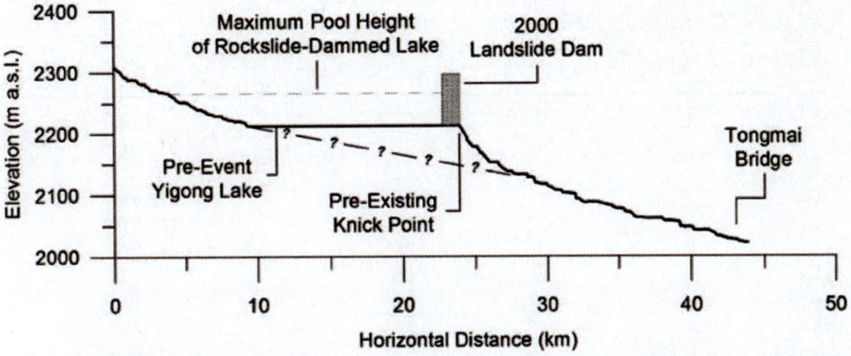


**Fig. 22.4** Profile (Vertical exaggeration = 41.7 × horizontal) of Yigong River derived from SRTM-3 data showing location of 1900 rockslide dam at pre-existing knick point, upstream aggradation formed behind this dam, the location of the 2000 landslide dam and the maximum elevation of the 2000 rockslide-dammed lake. Note also location of Tongmai Bridge mentioned in text

suggested by the data of Shang et al. [10], but is much below a later estimate by Wang [12] of 2,280 m a.s.l.

## 6 The Filling of Yigong Lake and Estimate of Maximum Lake Volume

In order to track the filling of Yigong Lake and to estimate its maximum volume before it breached on June 10, 2000 (Day 62) we constructed a digital elevation model (DEM) of the topography forming the Yigong Lake basin using SRTM-3 digital terrain data obtained in February 2000.

As noted above, SRTM data and elevations given by Wang and Lu [13] and Zhou et al. [18] establish that the elevation of the Yigong River valley at the upstream face of the landslide dam is 2,210 m a.s.l. This is also taken as the pre-event datum by Wang and Lu [13] and assumed as the datum in our calculations for relative water level measurements reported in Shang et al. ([10]; Fig. 8).

LANDSAT-7 images from May 4 (Day 25) and May 20 (Day 41) showing the extent of Yigong Lake on those dates were imported into the SRTM DEM and the lake levels were visually matched at 2,225 and 2,240 m a.s.l. respectively on these dates. Data obtained from Table 2 in Wang and Lu [13], reproduced above in Table 22.1 in this paper, and derived from Fig. 8 in Shang et al. [10] give lake elevations for May 4 (Day 25) of 2,225 and 2,220 m a.s.l. respectively, assuming the datum at 2,210 m a.s.l., and for May 20 (Day 41) of 2,234 and 2,233 m a.s.l. respectively. For May 4, the elevation of Wang and Lu [13] corresponds precisely with the SRTM data. The estimate derived from Shang et al. [10] is 5 m below the lake level given by SRTM data based on the extent of Yigong Lake obtained from LANDSAT-7 images; for May 20 the differences are 6 and 7 m below the

**Table 22.2** Area and volume of rockslide-dammed Yigong Lake at successive pool elevations based on DEM generated by SRTM-3 data. Maximum pool elevation attained was 2,265 m a.s.l

| Elevation of lake surface (m a.s.l.) | Lake area ($km^2$) | Lake volume ($Mm^3$) |
|---|---|---|
| 2,220 | 24.072 | 175.404 |
| 2,225 | 28.010 | 312.615 |
| 2,230 | 31.939 | 469.282 |
| 2,240 | 38.241 | 850.688 |
| 2,250 | 43.371 | 1,286.914 |
| 2,255 | 45.251 | 1,511.389 |
| 2,260 | 47.064 | 1,759.248 |
| 2,265 | 48.926 | 2,014.946 |
| 2,270 | 50.578 | 2,280.128 |
| 2,280 | 53.272 | 2,828.906 |

SRTM surface obtained from a visual match of lake outlines visible on LANDSAT-7 imagery. The difference reflects the difference in the resolution of the data sets.

We note that Shang et al. [10] report a final lake depth of 54 m. This roughly corresponds to that reported by Zhou et al. [18]. This maximum depth is thus equivalent to a maximum water elevation of 2,264 m a.s.l. and, rounding to 2,265 m a.s.l., we take this as a credible estimate of the maximum water level reached by Yigong Lake on Day 62, the day of the breach, based on reported maximum lake depths.

Using 3D Analyst in GIS we then calculated the area of Yigong Lake and the volume of water impounded at seven lake levels from 2,220 to 2,265 m a.s.l (Table 22.2) using the SRTM topography to approximate the ground surface of the lake basin. We also carried out calculations in GIS for three further hypothetical lake levels of 2,270, 2,275, and 2,280 m a.s.l., elevations higher than our credible maximum (Table 22.2). From the results in Table 22.2 we see that on May 4 (Day 25) the lake had an area of 28.01 $km^2$, a volume of 313 $Mm^3$ at a water level of 2,225 m a.s.l. This area compares to 34.7 $km^2$ reported by Zhou et al. [18] and Shang et al. [10] and 33.659 $km^2$ reported by Wang and Lu [13] on May 4, whereas the volume reported by Wang and Lu [13] was 514 $Mm^3$. On May 20 (Day 41), our analysis shows that the lake had increased to an area of 38.2 $km^2$, a volume of 850.7 $Mm^3$ at a water level of 2,240 m a.s.l. This area compares to 43.121 $km^2$ reported by Wang and Lu [13] on May 20, who also report a volume of 1.23 $Gm^3$. On June 10 (Day 62) on the day of the breach we estimate the lake had an area of 48.9 $km^2$, compared to 52.855 $km^2$ reported by Wang and Lu [13].

In measuring the maximum volume of Yigong Lake, we established a maximum lake level at 2,265 m a. s.l. in the SRTM DEM. At this elevation a lake volume 2.015 $Gm^3$ is calculated (Table 22.2). Given the vertical error in the SRTM data approximated above, volumes were estimated at $\pm 5$ m of the maximum lake elevation. At the limits of this error term (2,260 and 2,270 m a.s.l.) lake volumes of 1.759 and 2.280 $Gm^3$ were calculated from the SRTM data. We suggest that these data represent a minimum and maximum credible maximum volume of Yigong Lake at the time of breaching. We note that the estimate (2.26 $Gm^3$) of Wang and Lu ([13], p. 37) reported without an error term falls within this range.

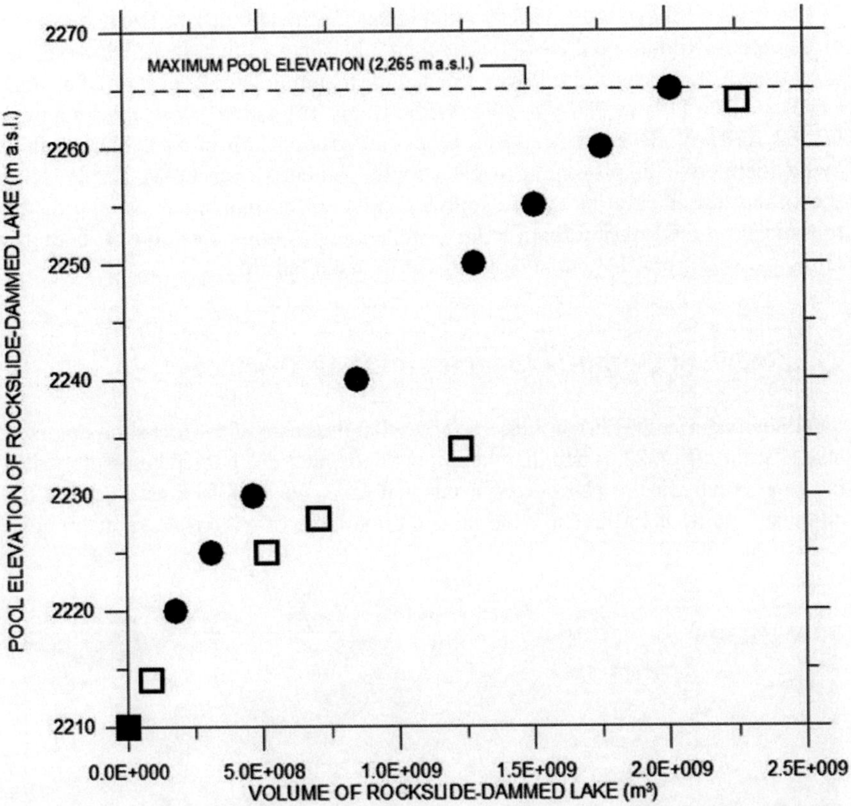


**Fig. 22.5** Comparison in pool elevation – volume relationships for the filling of Yigong rockslide-dammed lake. Data are from Wang and Lu ([13]-*open squares*) and SRTM-3 DEM generated for this study (*solid dots*). Data assumes an original base datum of 2,210 m a.s.l. and a maximum pool elevation of 2,264 m a.s.l., rounded to 2,265 m a.s.l

We compared the pool elevation-volume relationship from the data of Wang and Lu [13] in Table 22.1 to that generated in this study using SRTM-3 data (Fig. 22.5). The difference in the plots reflects the different sources of the topographic data. The estimate of the final volume by Wang and Lu [13] is 12% greater than the volume estimate obtained using SRTM-3 topographic data in this study.

Our best estimate volume of 2.015 $Gm^3$ is equivalent to a minimum average inflow of 376 $m^3/s$ over the 62 days of river damming. For a rockslide-dammed lake of 3 $Gm^3$ (e.g. [10]), an equivalent minimum average inflow of 560 $m^3/s$ is indicated. These inflow data may be compared to the average monthly discharge in the Yigong reach of 88.5 $m^3/s$ (April), 261 $m^3/s$ (May) and 761 $m^3/s$ (June) [10]; averaging these discharges over the length of the impoundment we obtain an average inflow of 283 $m^3/s$, half the average minimum inflow suggested by an impoundment volume of 3 $Gm^3$.

Our analysis shows that it is very unlikely that the area of the lake reached 70 $km^2$ or that the maximum pool elevation reached 2,280 m a.s.l (e.g. [12]). However, as noted above, the excavated spillway reduced the maximum pool elevation of Yigong Lake by 18 m [17] to 2,265 m a.s.l. Without this mitigation effort, the minimum natural spillway elevation would have been at about 2,283 m a.s.l. Thus despite being ineffective in preventing a catastrophic outburst, our DEM (Table 22.2) shows that the excavation of the spillway reduced the maximum volume of the rockslide-dammed lake by about 1 $Gm^3$, a reduction in outburst volume of about one third, and was thus effective in reducing the outburst hazard.

## 7 Estimate of Outburst Volume and Peak Discharge

Satellite imagery obtained on June 16 shows that the basin of the rockslide-dammed lake is empty (Fig. 22.6) and that the impounded water discharged completely during the breach. In the absence of a residual lake, we therefore assume that the outburst volume is 2.015 $Gm^3$, the maximum volume of Yigong Lake determined

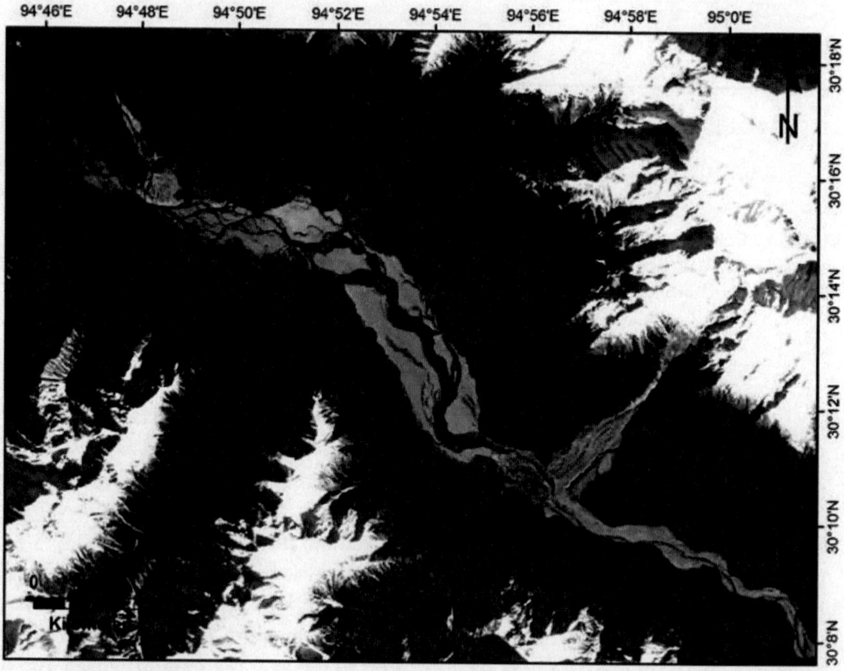


**Fig. 22.6** LANDSAT-7 image obtained on June 16, 2000, 6 days after the breach of the Yigong rockslide dam. Note the rockslide-dammed lake basin is completely empty and extensive erosion has occurred along the Yigong River downstream of the breach

above. If this volume of water drained in 24 h as suggested by Shang et al. [10] the average discharge of the outburst was 23,322 $m^3$/s. This compares to a peak value calculated using the empirical formula of Evans [3] that indicates a maximum breach discharge of 61,461 $m^3$/s.

# 8 Downstream Flood

The outburst flood resulting from the breaching of Yigong Lake has not been documented in detail and only summary details have been published (e.g. [10, 11, 19]). In China, the downstream flood resulted in dramatic erosion immediately downstream of the rockslide dam (Fig. 22.6). A flood discharge of 120,000 $m^3$/s was directly estimated [10] at Tongmai Bridge, located 17 km downstream from the rockslide dam (Fig. 22.1), at 0250 h on June 11, 5 h 50 m after the initiation of the breach. We approximated the cross-sectional area of the Tongmai Bridge reach (ca. 7,000 $m^2$) and for this discharge a flow velocity of 17 m/s is obtained. The flood destroyed 5 highway bridges downstream from Tongmai and channel erosion destroyed a number of sections of the Sichuan-Tibet highway in China alone.

Floodwaters travelled through the mountains of China in the Zangbo River system and into the Dihang River system of India, a tributary of the Brahmaputra (Fig. 22.1). In India, the flood reached the floodplain of the Brahmaputra at a distance of about 500 km. The rising water was measured in a water level gauge at Pasighat (Arunachal Pradesh) [11] at a distance of 462 km from the breached landslide dam (P in Fig. 22.1). The water level began to rise at 12:00 h on June 11, 17 h 50 min after the breach, indicating a downstream flood wave velocity of about 7 m/s. Peak discharge of 44,200 $m^3$/s [11] was measured at Pasighat at 17:00 h, 22 h 50 min after the breach. The maximum rise measured at the Pasighat stage was 4.5 m [11].

# 9 Comparative Events 1841–2010

The initial volume of the 2000 Yigong landslide determined in this paper is ca. 75 $Mm^3$. When compared to other well documented historical rock avalanches that have occurred in the period 1900–2010 the Yigong event is smaller than approximately 14 other rock avalanches that occurred in this period (Evans, unpublished data).

With respect to outburst volumes from landslide-dammed lakes (Table 22.3; Chap. 1 by Evans et al., this volume), the Yigong Lake outburst is exceeded only by the *Great Indus Flood* of 1841 (est. vol. 6.5 $Gm^3$; Chap. 7 by Delaney and Evans, this volume) in the period 1840–2010. The Yigong outburst volume exceeds that of the 1914 outburst from Lago Carri Lauquen, in the Rio Barrancas, Argentina Andes (1.55 G $m^3$ [8]; Chap. 1 by Evans et al., this volume). The volume also exceeds that of the greatest recorded outburst from an artificial dam [4].

**Table 22.3** Comparative data on outburst volumes from rockslide-dammed lakes 1841–2010 (Data from Chap. 1 by Evans et al., this volume)

| Rank | Location | Date of outburst | Impounded volume ($Mm^3$) | Outburst volume ($Mm^3$) |
|---|---|---|---|---|
| 1 | Indus River, Pakistan | 1841 | 6,500 | 6,500 |
| 2 | Yigong Zangbo, Tibet | 2000 | 2,015 | 2,015 |
| 3 | Rio Barrancas, Argentina | 1914 | 1,550 | 1,550 |
| 4 | Hunza River, Pakistan | 1858 | 805 | 805 |
| 5 | Tanggudong, China | 1967 | 680 | 640 |
| 6 | Condor-Sencca, Peru | 1945 | 721 | 664 |
| 7 | Mayunmarca, Peru | 1974 | 352 | 350 |

## 10 Conclusions

This paper identifies important inconsistencies in the published literature on the landslide damming and outburst flood event which occurred on the Yigong Zangbo River (Tibet) in 2000. The inconsistencies included the volume of the damming rock avalanche, the precise evolution of the rockslide-dammed lake, the final volume of the impoundment and its eventual outburst. Using SRTM-3 digital topographic data in conjunction with LANDSAT-7 imagery, we have attempted to resolve these inconsistencies in presenting credible estimates of landslide volume (105 $Mm^3$), landslide-dammed impoundment (2.015 $Gm^3$) and outburst volume (2.015 $Gm^3$) associated with the Yigong event, estimates that are smaller than most published figures. The volume of the landslide was significantly enhanced by massive entrainment (15 $Mm^3$) from Zhamulong Gully.

Our data shows that the whilst the volume of the Yigong landslide was significant it was not exceptional (or "super large") when compared to similar events that have occurred worldwide since 1900. However, the outburst volume in the June 2000 breaching of the Yigong rockslide dam is only exceeded in historical time by that of the 1841 outburst from the Indus rockslide dam in northern Pakistan. With reference to the outburst, we note that although the excavated spillway did not prevent a catastrophic breach it was effective in reducing the maximum possible volume of the rockslide-dammed lake by about 1 $Gm^3$. We conclude that readily available remotely sensed data can be used to calculate reliable estimates for the volumes of large landslides and the volumes of newly-formed landslide-dammed lakes where detailed topographic maps are not available.

## References

1. Becek, K. (2008) Investigating error structure of shuttle radar topography mission elevation data product, *Geophysical Research Letters* **35**, L15403, doi: 10.1029/2008GL034592, 2008.
2. Chen, R.-F., Chang, K.-J., Angelier, J., Chan, Y.-C., Deffontaines, B., Lee, C.-T. and Lin, M.-L. (2006) Topographical changes revealed by high-resolution airborne LiDAR data: The 1999 Tsaoling landslide induced by the Chi-Chi earthquake, *Engineering Geology* **88**, 160–172.

3. Evans, S.G. (1986) The maximum discharge of outburst floods caused by the breaching of man-made and natural dams, *Canadian Geotechnical Journal* **23**, 385–387.
4. Evans, S.G. (2006) The formation and failure of landslide dams: An approach to risk assessment, *Italian Journal of Engineering Geology and Environment* **Special Issue 1**, p. 15–19.
5. Farr, T.G., Rosen, P.A., Caro, E., Crippen, R., Duren, R., Hensley, S., Kobrick, M., Paller, M., Rodriguez, E., Roth, L., Seal, D., Shaffer, S., Shimada, J., Umland, J., Werner, M., Oskin, M., Burbank, D. and Alsdorf, D. (2007) The shuttle radar topography mission, *Reviews of Geophysics* **45**, RG2004, doi: 10.1029/2005RG000183.
6. Han, Z.S. (2003) Large-scale landslide-debris avalanche in Tibet, China (1) April–June 2000 Yigong landslide, Tibet China, *Landslide News* **14–15**, 22–23.
7. Hancox, G.T., McSaveney, M.J., Manville, V.R. and Davies, T.R. (2005) The October 1999 Mt. Adams rock avalanche and subsequent landslide dam-break flood and effects in Poera river, Westland, New Zealand, *New Zealand Journal of Geology and Geophysics* **48**, 683–705.
8. Hermanns, R.L., Naumann, R., Folguera, A. and Pagenkopf, A. (2004) Sedimentologic analyses of deposits of a historic landslide dam failure in Barrancas valley causing the catastrophic 1914 Rio Colorado flood, northern Patagonia, Argentina, *Proceedings of the 9th International Symposium on Landslides*, Rio de Janeiro, Balkema, Rotterdam, Vol. 2, pp. 1439–1445.
9. Ren, J., Shan, X., Shen, J., Deng, G. and Zhang, J. (2001) Geological characteristics and kinematics of the rock-fall landslide in Yi'ong, southeastern Tibet, *Geological Review* **47**, 642–647 (in Chinese with English abstract).
10. Shang, Y., Yang, Z., Li, L., Liu, D., Liao, Q. and Wang, Y. (2003) A super-large landslide in Tibet in 2000: Background, occurrence, disaster, and origin, *Geomorphology* **54**, 225–243.
11. Tewari, P. (2004) A study of soil erosion in Pasighat town (Arunachal Pradesh) India, *Natural Hazards* **32**, 257–275.
12. Wang, Z. (2008) A thunder at the beginning of the 21st century – the giant Yigong landslide, in *Landslides and Engineered Slopes*, Z. Chen, J. Zhang, Z. Li, F. Wu and K. Ho (eds.), *Proceedings, 10th International Symposium on Landslides*, Xian, China, Vol. 2, pp. 2111–2118.
13. Wang, Z. and Lu, J. (2002) Satellite monitoring of the Yigong landslide in Tibet, China, *Proceedings SPIE* **4814**(34), doi:10.11117/12.453739.
14. Wen, B., Wang, S., Wang, E. and Zhang, J. (2004) Characteristics of rapid giant landslides in China, *Landslides* **1**, 247–261.
15. Xu, Q. and Wang, S.T. (2008) Large scale avalanches of Yigong, Tibet (2000), in R.Q. Huang and Q. Xu (eds.), *Catastrophic Landslides in China*. Science Press, Beijing, pp. 406–423.
16. Xu, Q., Wang, S.-T., Chai, H.-J., Zhang, Z.-Y. and Dong, S. (2008) Yigong rock avalanche-flow landslide event, Tibet, China. in *Landslides and Engineered Slopes*, Z. Chen, J. Zhang, Z. Li, F. Wu, K. Ho (eds.), *Proceedings, 10th International Symposium on Landslides*, Xian, China, Vol. 2, pp. 1675–1680.
17. Yang, X., Yang, Z., Cao, S., Gao, X. and Li, S. (2010) Key techniques for the emergency disposal of Quake lakes, *Natural Hazards* **52**, 43–56.
18. Zhou, C.H., Yue, Z.Q., Lee, C.F., Zhu, B.Q. and Wang, Z.H. (2001) Satellite image analysis of a huge landslide at Yi Gong, Tibet, China, *Quarterly Journal of Engineering Geology and Hydrogeology* **34**, 325–332.
19. Zhu, P.Y. and Li, T. (2000/2001) Flash flooding caused by landslide dam failure, *Newsletter of ICIMOD* **38**, 4–5.
20. Zhu, P.Y., Wang, C.H. and Wang, Y.C. (2003) Large-scale landslide-debris avalanche in Tibet, China (2) Formation of an exceptionally serious outburst flood from a landslide dam in Tibet, *Landslide News* **14–15**, 23–25.

# Chapter 23
# The 1999 Tsao-Ling Rockslide: Source Area, Debris, and Life Cycle of Associated Rockslide-Dammed Lake (Central Taiwan)

**Kuo-Jen Chang, Rou-Fei Chen, Hou-Yen Lee, Yu-Chang Chan, and Alfredo Taboada**

## 1 Introduction

Landslides pose significant threats to communities and infrastructure in Taiwan. Among possible triggering factors, heavy precipitation and earthquakes are the most important. Taiwan is situated in a region of active subduction-collision resulting in high seismicity and high relief. The Taiwan orogenic belt results from the rapid convergence between the Luzon volcanic arc (which is part of the Philippine Sea Plate) and the passive Eurasian continental margin. The oblique arc-continent collision began during the late Miocene, creating east-dipping thrusts and folds. Active deformation migrated progressively northward and westward, successively affecting the sediments deposited in foreland basins creating a foreland belt [23].

In addition to its high seismicity, Taiwan is located on the most frequent cyclone path in the northwestern Pacific. Cyclones bring very heavy precipitation that trigger numerous landslides. Each year, an average of about four typhoons either pass over or closely bypass Taiwan. The average precipitation is about 2,515 mm/year while it may reach as much as 3,000–5,000 mm/year in the mountainous areas. Most of the precipitation is concentrated in the summer season. Typhoons are often associated with heavy, intense rainstorms; the maximum precipitation per hour may reach 300 mm and the maximum 24 h precipitation reaches 1,748 mm. The numerous landslides triggered by earthquakes and heavy precipitation, jointly accelerate land erosion and affect the geomorphologic evolution of the Taiwan landscape [11].

The 21st September 1999 Chi–Chi earthquake ($M_L = 7.3$, $M_W = 7.6$) was associated with the reactivation of the Chelungpu thrust fault (central Taiwan), the kinematics of which is coherent with oblique plate convergence [1, 22]. Based on the analysis of SPOT imagery, the total number of landslides triggered by the Chi–Chi earthquake is about 9,272; most of the landslides are located in the hanging wall of the Chelungpu fault [26]. Among the numerous earthquake-triggered landslides,

K.-J. Chang (✉)
Department of Civil Engineering, National Taipei University of Technology, Taipei, Taiwan, Republic of China
e-mail: epidote@ntut.edu.tw

S.G. Evans et al. (eds.), *Natural and Artificial Rockslide Dams*, Lecture Notes in Earth Sciences 133, DOI 10.1007/978-3-642-04764-0_23, © Springer-Verlag Berlin Heidelberg 2011 

the Tsao-Ling rock avalanche is the largest. It is located at 23.58°N/120.67°E, about 31 km south–south-east of the earthquake epicenter and resulted from the sliding of a rock mass on a bedding plane in a dip slope.

This paper presents a summary of the geological and morphological features of the 1999 Tsao-Ling landslide. The morphological analysis was performed by interpreting aerial photos taken before and at several times after the Chi–Chi earthquake. The topography of the landslide was studied by studying several 2 m resolution digital elevation models (DEMs) generated from the aerial photos taken at different times. The comparison between the topographies before and after the slide allows precise calculation of the initial volume of rock and soil mobilized in the landslide, as well as the geometry and thickness of the landslide deposit. Field observations at different scales were also carried along the sliding surface and the landslide deposit and gave new insights into the mechanical behavior of the landslide, and the subsequent geomorphologic evolution of the landslide site.

## 2 The Tsao-Ling Rock Avalanches

### 2.1 *Historical Landslide Events at Tsao-Ling*

As noted above, among the enormous number of landslides triggered in the Chi–Chi earthquake, the Tsao-Ling rock avalanche is the largest [2–4, 6, 7, 10]. At Tsao-Ling, at least five historical catastrophic rockslides, caused either by earthquakes or heavy rainfall, have occurred since 1862 [7, 16]. Events occurred in 1862, 1941, 1942, 1979 and 1999. The Tsao-Ling dip slope has thus been particularly prone to large-scale rock mass failure. Earthquake and heavy rainfall are the two major triggers; the landslides in 1862, 1941 and 1999 were caused by earthquakes, whilst the events of 1942 and 1979 were generated by heavy precipitation.

The debris of the landslides repeatedly dammed the Chingshui River valley. Breaching of the landslide dams formed by historical events occurred in 1898, 1951, and 1979. All the breaching events were caused by overtopping overflow due to heavy rainfall and insufficient reservoir storage. The major historical landslide events at Tsao-Ling area are summarized in Table 23.1.

### 2.2 *Geology of the Tsao-Ling Slope*

The geological structure of the Tsao-Ling slope is particularly favourable to recurrent landslides. The mountain slope relief ranges from 500 to 1,200 m. Bedrock consists of mid-Neogene to Quaternary sedimentary rocks that dip 14° in the same direction as the slope, towards SW forming a dip slope on the east side of the Chingshui River valley [13, 19]. Figure 23.1 shows the general geology in the vicinity of Tsao-Ling.

The Chingshui River cuts through the base of a dip slope and causes the bedding planes to outcrop in the valley sides (i.e., the bedding planes daylight). The incision

**Table 23.1** Historical events in the Tsao-Ling landslide area

| Time | Trigger | Landslide process: Slide volume ($Mm^3$) | Landslide process: Dam height (m) | Effects | References |
|---|---|---|---|---|---|
| 1862/06/06 | Earthquake ($M_L = 6$–$7$) | | | Formation of landslide dam | [8, 14, 17, 24] |
| 1898/?/? | Rainfall | | | Breakage of the 1862 landslide dam | [8, 14, 17, 24] |
| 1941/12/17 | Earthquake ($M_L = 7.1$) | ~10 | 70–140 | 36 persons killed and 59 houses damaged; landslide dam formed | [5, 8, 12, 14, 16, 17, 19] |
| 1942/08/10 | Rainfall (776 m) | ~15 | 140–170 | 1 person killed and 1 house damaged; the dam enlarged | [5, 8, 12, 14, 16, 17, 19] |
| 1951/05/18 | Rainfall (770 m) | 120 | 140–200 | 137 persons killed and 1,200 houses damaged, dam broken (74 die? By [17]) | [5, 8, 12, 14, 16, 19] |
| 1979/08/15 | Rainfall (327 m) | 26 | 90 | Formation of landslide dam | [8, 10, 14, 16, 19] |
| | Rainfall (624 m)/08/24 | 40 | | Breakage of landslide dam | [8, 10, 14, 16, 19] |
| 1999/09/21 | Earthquake (ML = 7.3) | 125 | 45 | 39 persons killed | [8, 10, 16, 20] |

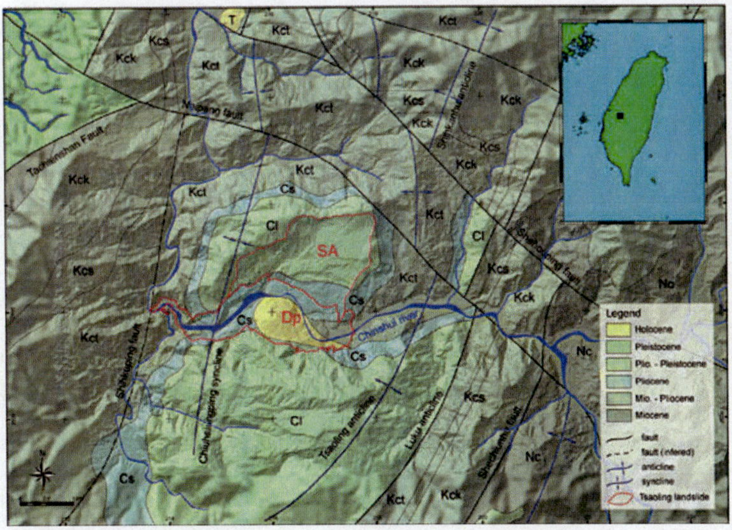


**Fig. 23.1** Regional geological map of the area around the Tsao-Ling area

of the Chingshui River thus facilitates the sliding events; the recurring gigantic landslides reveal the instability of slope conditioned by the geological structure.

From bottom to top, the geological formations exposed around the slide area are Shihliufeng Shale, Tawo Sandstone, Chingshui Shale and Cholan Formation (Fig. 23.2). The Miocene Shihliufeng Shale consists of fine layered green gray to dark gray shale and siltstone. The Late Miocene to Early Pliocene Tawo Sandstone is composed of white to light gray, coarse to fine grain sand, thick to massive sandstone bed. The overall thickness of the stratum is about 200 m. The superimposed Pliocene Chingshui Shale consists of dark gray fine-grained massive mudstone and shale bed, intercalcated occasionally with fine sandy to silty thin bed. The overall thickness of the shaly strata is about 110 m thick (Fig. 23.2). The upper most strata Cholan Formation rests conformably on the Chingshui shale. The stratum has an overall thickness of over 1,000 m and is composed of light gray massive sandstone

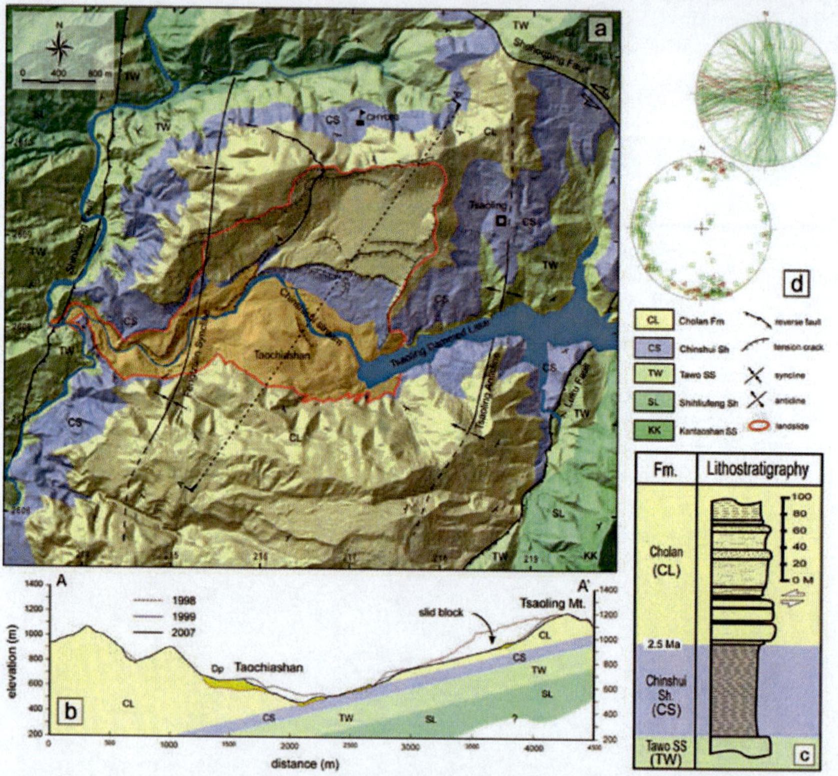


**Fig. 23.2** Geological map, profile and stratigraphic column around the rockslide area. (**a**) Geological map of the Tsao-Ling rock avalanche. (**b**) The geological profile along the *centre* of the rockslide. The profile is parallel to sliding direction, indicated on Fig. 23.2a. (**c**) Stratigraphic column of the studied area, where the major sliding surface is indicated by *double arrows*. (**d**) Summary of structure analysis on the sliding area. Poles of open tension fractures (*squares*), poles to reverse and strike-slip faults (*triangles*), fold axes added as *stars*

beds and metric to sub-metric dark interbeds. Recent terrace deposits, landslide debris and colluvium of various thicknesses overly bedrock and found occasionally in the Chingshui River valley.

## 2.3 Sliding Surfaces

The nature and location of the sliding surfaces of the historical events remains a somewhat debated issue. In most of the events, it is believed that the sliding surface is situated at the boundary between the Chishui Shale and Cholan Formation. The Chihshui Shale consists of a friable silty to muddy shale, that degrades easily and deteriorates rapidly by weathering processes. In the field, a fresh outcrop of the Chihshui Shale can be quickly decomposed in a very short period; this rock propriety is confirmed by a series of laboratory tests [25].

In the 1999 rockslide the sliding surface of the landslide could be divided into more than three surfaces. All of the sliding surfaces are parallel to the stratigraphic bedding planes and movement occurred as large-scale dip-slope sliding. Within the slide mass, most of the detached volume is situated above the highest layer. The major sliding surface is characterized by a smooth planar surface within a fine-layered sub-metric interbed within the Cholan formation, about 50 m above the stratigraphic boundary, as illustrated by double arrows in Fig. 23.2c. The major sliding surface, situated between two massive sandstones by a metric friable fine layer, is composed of fine layered shale, silty mudstone and interlayered beds. The superposed massive sandstone with occasionally intercalated interbeds is over 60 m thick. Figure 23.2c demonstrates the detailed local stratigraphy around the landslide area, which has been reconstructed from the strata exposed on the scarp of the right flank.

The structures observed on the sliding surfaces of the 1999 Tsao-Ling landslide include open fractures, faults, bedding planes and some sliding-related phenomena. The stereonets in Fig. 23.2d illustrates the overall distribution of the discontinuities. Most of the fractures, especially open fissures, are approximately perpendicular to the bedding planes that dip an average of 14° to the southwest. All sites show open fractures that trend approximately E–W and N–S. The nearly E–W fractures generally indicate tension close to the slip direction of the landslide, and display clear correlation with the major detachment scarps of the Tsao-Ling landslide. As illustrated by the comparison between the geological map of Fig. 23.2a and the stereonets of Fig. 23.2d, these fractures and the major scarps left by the landslide are nearly parallel to each other.

Many fractures that trend approximately N–S are also observed. This second family also indicates extension, but in a direction oblique to the slip direction of the landslide. Furthermore, some minor strike-slip and reverse faults occur, which are generally consistent in orientation with local compression induced by the landslide. Most of these faults strike approximately E–W with steep dips. At the same time, a small but steady surface seepage and runoff is observed on several of the open joints situated on the sliding surface, suggesting the presence of a high water table and the fact that the slid rock mass may have been saturated.

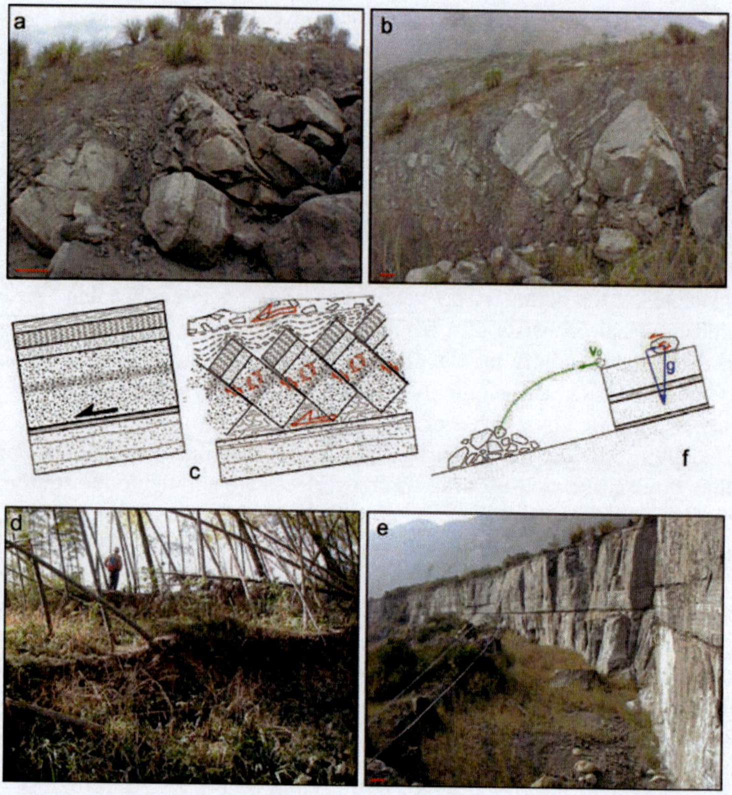


**Fig. 23.3** The possible sliding mechanism in the Tsao-Ling rockslide. (**a**) and (**b**) field photos show the different-scaled domino or bookshelf structures of the slid material, (**c**) schematic diagram demonstrates the mechanism of the domino or bookshelf structures, (**d**) cracks by amphitheatre-shaped extensional zone situated on the southern wall of the graben structure, (**e**) the morphologic characteristics of the toe of the minor scarp, (**f**) schematic diagram illustrates the displacement process and the position of the detached blocks. *Small red bar* denotes 1 m in scale

Behind the major scarp, some large open cracks, over 50 m in length, occur parallel to the crown. The extensional cracks occupy the hill top and exhibit a series of graben-and-horst structures. Figure 23.3d features some of the cracks in the amphitheatre-shaped extensional zone situated on the southern wall of the graben structure. On the sliding surface, some toppled (rotated) blocks reveal the shearing deformation by the superposed slid mass. Figure 23.3a–c illustrate the domino or bookshelf structures in a metric to deca-metric scale. These structures are found especially near the rear of the minor scarps, and facilitate and contribute to block sliding.

Close to the minor scarp, about 20 m height, there is an elongated hump area composed of debris and rock fragments. Between the hump and the minor scarp, a depression is filled with cobbles and sands, as illustrated in Fig. 23.3e. This structure suggests the dropping and sliding process of the metric to deca-metric boulders. The width of the depression zone is larger than 10 m, implying a high initial velocity when the block departs from the slope. Figure 23.3f demonstrates the displacement process of the detached blocks schematically.

To summarize, analysis revealed that the majority of fractures observed on the sliding surfaces (extensional and compressional) were generated by the landslide process itself, with structural grains that were principally slope-parallel and slope-perpendicular. However, some features were also inherited from earlier tectonic deformation, prior to the landslide events, and these features were commonly reactivated as open cracks or minor thrust faults during sliding.

### *2.4 Landslide Debris*

The sliding distance is about 2,250 m on the average, calculated by the boundary of the slid mass and debris deposits, or about 2,000 m if calculated by the horizontal displacement of the centre of mass. The major part of the debris accumulated on the other side (west side) of the Chingshui River valley; about 80% of the debris was transported across the river valley, and rests around the Taochiashan Hill situated on the west side of the river. Taochiashan Hill, composed of rockslide debris, was generated by previous landslides. It is worth noting that the hill was about 140 m higher than the riverbed before the Chi–Chi earthquake. The topography of the debris indicates that the slid materials accelerated to achieve high velocity. The morphology of the deposited debris illustrates that the slid mass surged over Taochiashan Hill, moved further up-hill, and then retreated back to around the Taochiashan Hill. Kuo et al. [18] proposed the scenario and analysed the dynamic of the sliding material via numerical simulation, the results of which confirmed the very mobile phenomena. On the other hand, the trees situated beyond the debris margin were blown down and re-arranged radially outward from the center of mass by air-blast [15].

The yearly monsoon in Central Taiwan extends from May to October, bringing in heavy rains; precipitation peaks are generally linked with destructive typhoons. The Chi–Chi earthquake occurred near the end of the rainy season in 1999 thus the ground water table in the Tsao-Ling slopes was probably at its highest level. Water seepage and small springs were observed on the sliding surface a few days after the landslide indicating the existence of saturated rocks in the overlying layers and reduced effective stress at the time of the triggering earthquake.

## 3 Analysis of Digital Terrain Data

### *3.1 The 1999 Rockslide*

The dimensions of a landslide are a key concern for landslide hazard assessment. The length, width, and sometimes the depth are generally easier to define; estimates can be made from the orthorectified aerial photos or satellite imagery. Accordingly, the volume of a landslide can be easily estimated. In other cases, more sophisticated methods such as digital elevation models (DEMs) at different periods may be useful to measure the volume more precisely, especially for the temporal variation.

The measurement of the volume of the landslide is generally based on the analysis of DEMs before and after sliding process, which can calculate the volume of either the slid materials or the accumulated deposits. In order to estimate the sliding volume more precisely, we generated several different DEMs in different periods which can be used to calculate the different volumes in this study. The 1989 DEM data was the single source of the official quantitative information available and has been used to depict the morphologic condition prior to the 1999 Tsao-Ling landslide [9]. However, the data was obtained in 40 m grid resolution, insufficient for detailed quantitative analysis. Further, the topographic change from 1989 to 1999 remained an unknown factor. To eliminate the ambiguity, a 2 m resolution pre-landslide DEM was generated from 1998 aerial photo stereopairs. After the landslide, aerial photographs were taken regularly to monitor possible subsequent hazards. Based on these images 2 m grid post-landslide DEMs have been generated for the periods of 1999, 2001, 2003, 2004 and 2007. In addition, an airborne LIDAR DEM obtained in 2002 was also incorporated in this study. This integrated dataset allows the detailed analysis of post-seismic morphologic changes in the landslide area as well as in the upstream area of the landslide dammed lake. Figure 23.4 illustrates the six DEMs generated in this study.

To perform calculations with DEMs in the different periods with a single homogenous baseline, we selected several sub-areas situated outside of the sliding area in order to adjust the average altitudes in the DEM. The sub-areas were defined according to the presence of their clear topographic features and the absence of significant topographic changes during the periods analyzed. Since determining volumes is very sensitive to uniform offsets between the elevations models, all homologous points within the overlapping areas should be correlated with satisfactory accuracy, and should be examined carefully. At the end of this adjustment process, all homologous points in the overlapping region could be positioned. The grids were then adjusted and interpolated, based on the difference when compared with the 2003 DEM and verified with the airborne Light Detection and Ranging (LIDAR) data, over all points in the overlapping area of stereoscopic images. Subsequently, all the DEMs were registered and re-sampled within a single grid with 5 m accuracy to allow comparison between all the spatial data.

To illustrate the morphologic evolution of the landscape associated with the earthquake, we compared the relief before and after the Chi–Chi earthquake. Figure 23.5 shows an isopach map illustrating the difference in heights between the DEMs of 1998 and 1999. The isopach lines are shown at 10 m intervals, and the negative and positive values indicate net accumulation and depletion of the rock material, respectively. The major slid mass, situated on the upper slope and adjacent to the main scarp, is an isolated hill composed of successive massive sandstone layers and occasional interbed; the most important feature here is that bedding planes in the rock mass daylighted. The average thickness of the slid mass

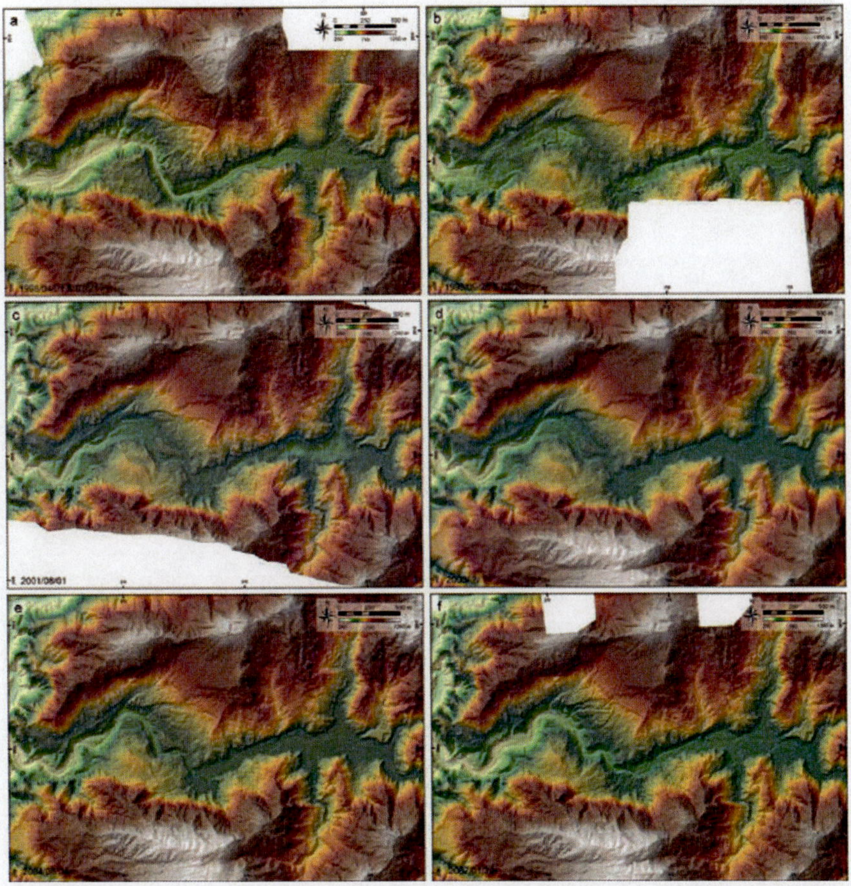

**Fig. 23.4** The 6 DEMs used in this study; a-1998, b-1999, c-2001, d-2003, e-2004, f-2007

is about 150–170 m, whereas the maximum thickness is about 195 m, situated near the southeastern lateral margin (Fig. 23.5)

In the deposit area, the morphology of the deposited debris immediately after the Chi–Chi earthquake indicates that the slid mass crossed over the Chingshui River channel, surged over the Taochiashan Hill and was deposited in the valley. The isopach map reveals that the maximum depth (140 m) of the deposits is coincident with the Chingshui River channel; between 30 and 90 m debris covered the Taochiashan Hill (Fig. 23.5). The debris dammed the valley for a length of about 4.7 km along the original river centerline.

The estimated volume of rock-and-soil material that slid during the main rock avalanche is about 125.5 $Mm^3$ and the volume of the deposit is about 138 $Mm^3$.

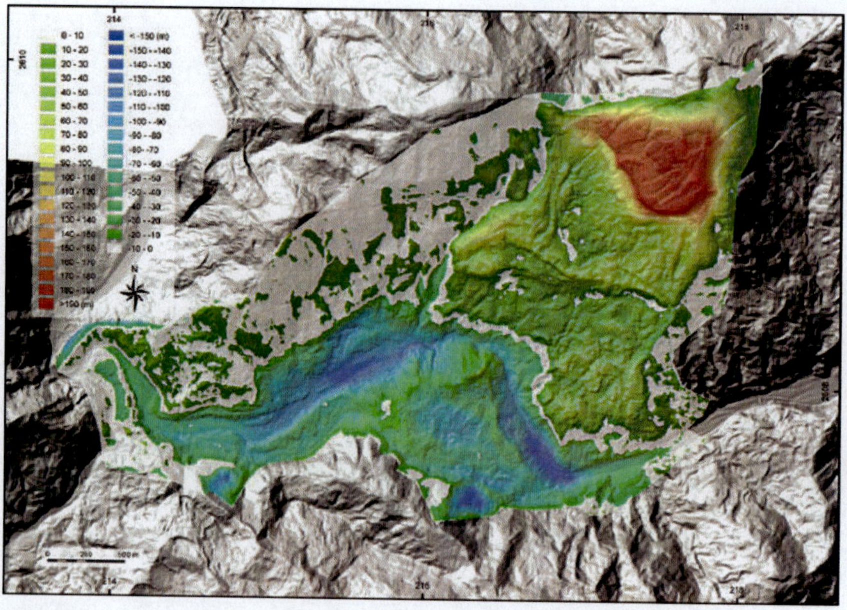


**Fig. 23.5** Isopach map of the landslide illustrates the difference in surface elevation before and after the Chi–Chi earthquake. The isopach lines are shown at 10 m intervals: both negative and positive values indicate net accumulation or depletion of rock material, respectively (explanation in the text)

The difference between the two volumes indicates an increase in volume due to fragmentation. However, several smaller landslides followed the main Tsao-Ling landslide. Thus the calculated volume varies and depends on how the total area is defined. Furthermore, the contribution of the small landslides nearby is not easily separated from the major landslide. Overall, the total slid volume around the Tsao-Ling area is about 142 $Mm^3$, and the total debris deposits is about 160 $Mm^3$.

In the slid area, the overall volume of the major slid mass, situated on the upslope right adjacent to the main scarp, is about 84.6 $Mm^3$ (Fig. 23.5). Some minor slid mass situated on the middle slope is about 27.7 $Mm^3$, where the mean thickness of the slid material is about 70 m. The volume of the debris deposits on the river channel is about 73.6 $Mm^3$. The volume of the accumulation area on Taochiashan Hill is about 26.4 $Mm^3$.

In comparing the accumulated and the depleted volume, a difference of about 17.5 $Mm^3$ is found, suggesting a volume increase of 14% during the landslide event. The volumetric increase of the landslide materials is caused by the fragmentation and dilation during the coseismic landslide event. This ratio is smaller than previously reported. The difference may be explained by contrasting technical approaches and the unknown relationship between pre-landslide and post-landslide erosion. Further, all the reported data here is calculated based on 2 m grid 1998

DEM, instead of the 40 m data generated on 1986. Thus, it is believed that the values estimated in this study may be more precise than the previously reported data.

### 3.2 The 1999 Landslide Dammed Lake

The estimation of the volume of the landslide-dammed lake used the same methodology as above and is derived from the DEMs before and after landslide. The maximum volume of the water body is about 45 $Mm^3$ for an area of 1.8 $km^2$. The maximum depth of the landslide-dammed water body is about 58 m. The backwater length is estimated at about 5.1 km for a rockslide-dammed lake elevation of 540 m a.s.l. After considering the downstream risk posed by a catastrophic outburst flood, the government of Taiwan proposed a spillway to be excavated across the surface of the dam [27].

## 4 Post-seismic Morphologic Evolution 1999–2007

Due to the high annual rainfall and frequent upstream landslides the subsequent topographic evolution of the landslide site is quite dramatic. During the period 1999–2007, several typhoons hit the Tsao-Ling area. Among these were two major typhoons; Toraji on July 30–August 1 in 2001 and Mindulli on July 2–4 in 2004.

Lin et al. [21] document that, judging from SPOT imagery, the subsequent typhoons triggered more landslides than the earthquake itself. Furthermore, in the Chingshui upstream area, numerous landslides have been initiated and identified. Figure 23.6 illustrates the morphologic evolution along three profiles situated mainly in the deposit area of the Tsao-Ling rockslide.

### 4.1 Sliding Area

The surface evolution of the sliding area is estimated by DEMs for different periods and field observations. A shallow slope wash is found on the sliding surface itself. Nevertheless, above the original crown, a significant reteat of the main scarp is observed, with an average rate of over 3 m/year. On the middle slope of the sliding surface where several minor scarps or strata steps have been observed, the retreat of the scarps is 35 m from the period of 1999–2003 and 120 m from 2003–2007 corresponding to an average rate of retreat of the minor scarps of 17.2 m/year.

During the period of 1999–2003, except for the retreat of the minor scarps, no major surface erosion is observed on the sliding surface by analysis of DEMs or aerial photos. Figure 23.7 illustrates the morphologic changes in different periods. However, near the bottom of the sliding surface, where the friable Chingshui shale outcrops, significant material decomposition is noted. The degradation of the muddy material generates small rock blocks and totally broken down rock corresponding

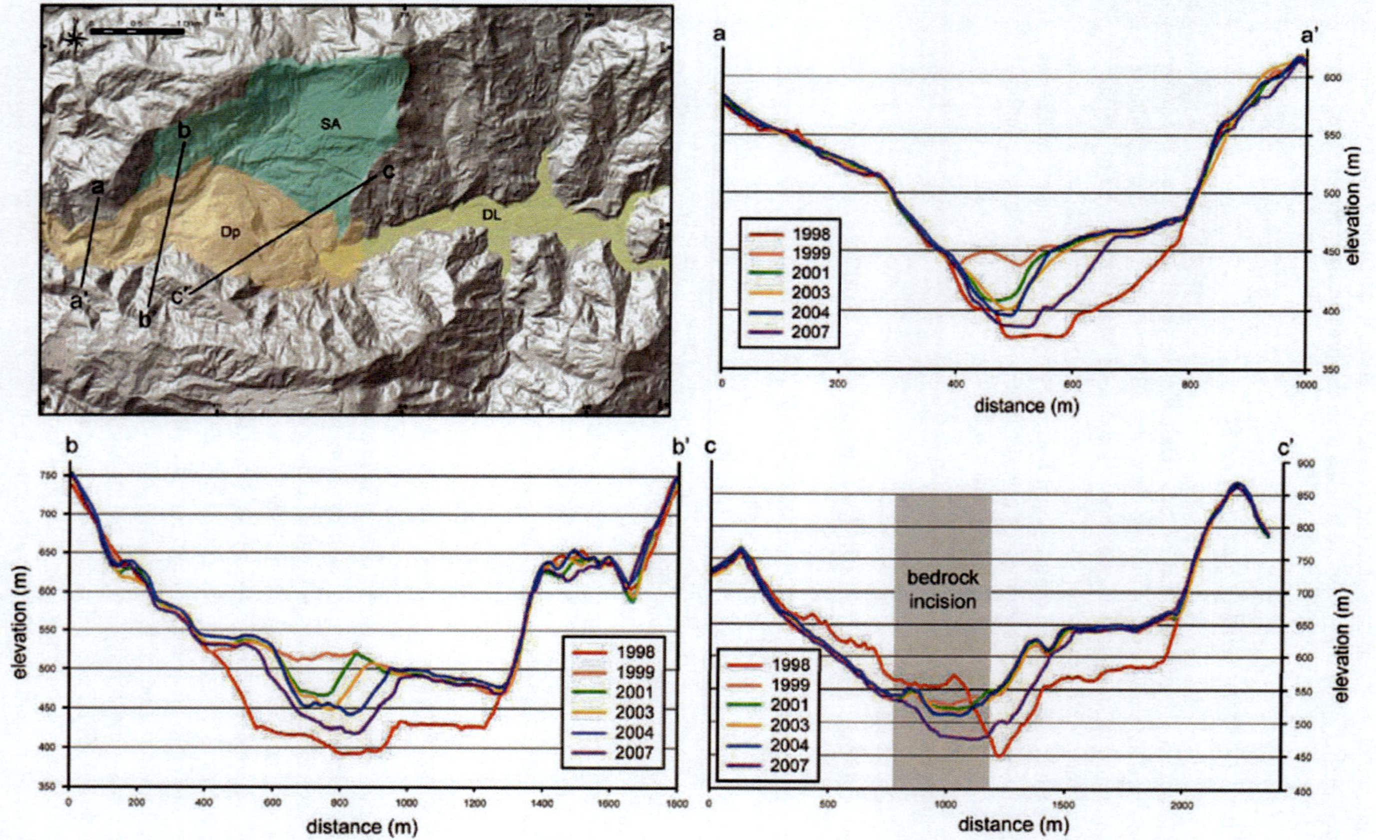


**Fig. 23.6** Morphologic evolution of three profiles situated mainly on the deposit area. The location of the profiles is marked on the *top-left* figure. The b–b' profile illustrates more clearly the incision and lateral erosion of the Chingshui River. The c–c' profile demonstrates bedrock incision and river migration

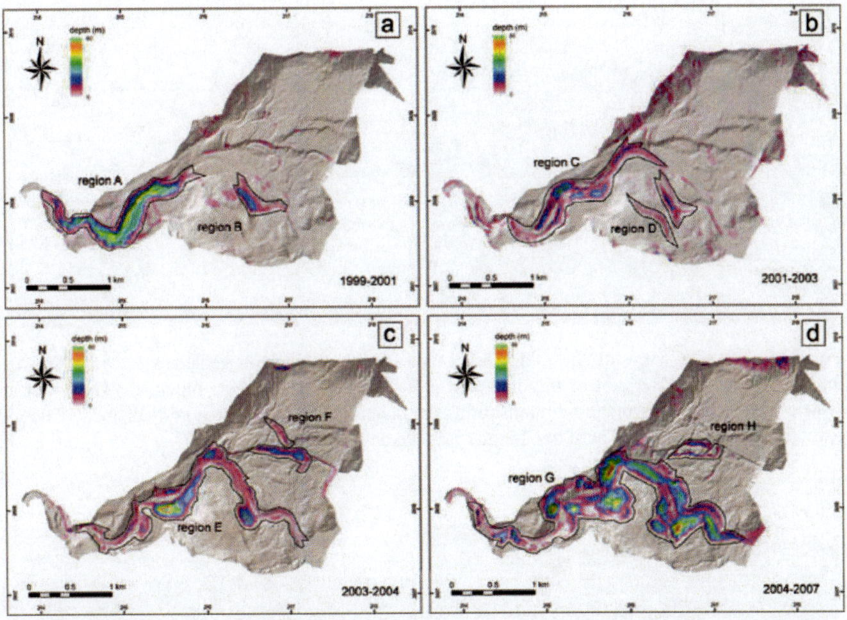


**Fig. 23.7** Isopach map of the landslide illustrates the post-seismic surface processes, determined by DEMs in different periods. The isopach scale is shown by continuous *color bar*. Because few areas of accumulation were found in this study, only depleted areas (areas of negative change) are shown in the figures. Several regions of change have been identified and are listed in Table 23.2

to soils, which can be easily transported by the surficial run-off (Fig. 23.8a). In addition, a deca-metric scale extensional graben-and-host structure is found in the Chingshui shale strata which suggests a slope creeping process by translational block slide (Fig. 23.8b). It is also observed on the Chingshui shale that numerous small gullies formed immediately after the landslide have been beheaded by the strata disruption. Figure 23.8a demonstrates the disrupted gully on the slid and isolated block in a graben. Unfortunately, DEMs can prove only limited indications of sub-metric scale landform changes. Overall, the field investigation deciphers the minor and relatively slower morphologic evolution in the rockslide area during the period of 1999–2003.

When compared to the period 1999–2003, the erosion and landform changes in the period 2003–2007 are more severe, as shown in Fig. 23.7c,d. The major change near main scarp is the transportation of the debris downslope and then being deposited on the main sliding surface. The retreat of the minor scarp line is more significant, up to 120 m. The retreat of the minor scarp line generates postseismic land erosion about 2.7 $Mm^3$ materials in an area of 15.4 ha with a maximum thickness of 33 m. Figure 23.7 demonstrates the location of the eroded area and the enlargement of the landslide scar in different periods. The retreat of the scarp line is caused mainly by rainfall associated with typhoons. Field investigations indicate severe

**Fig. 23.8** Detailed view of the slid block in the friable Chingshui shale. (**a**) Degradation of the muddy material and part of the disrupted gully situated on the glide block. (**b**) Deca-metric scale graben and horst structure demonstrating degradation process of the muddy dislocated block (explanations in the text). These two photos were taken on 2003/01/20

**Fig. 23.9** Detailed view of the outcrops illustrating post-seismic surface processes. (**a**) the incised gully located on the *middle* to upslope sliding surface. (**b**) views of the filled rockslide-dammed lake and river bed incision. The depth of the incised river bed is deeper than 50 m and has occurred since 2004. These two photos were taken on 2009/05/19

gully incisions on the middle to up slope of the sliding surface, with a depth greater than 1.6 m deep. Figure 23.9a illustrates the ditch incision of the middle slope.

## *4.2 Deposit Area*

Compared to the depletion area, the post-earthquake geomorphic evolution of the accumulation area is more significant. Because of material fragmentation, the debris deposit is more easily weathered and mobilized by surface processes. Furthermore, the river incises and transports the debris, accelerating the surface morphologic changes. Figures 23.6 and 23.7 depict the landform evolution around the deposit area, especially on two sides of the river channel.

In the Chingshui stream valley, an elongate area of marked change is observed in the downstream area, where maximum incision is about 58 m over a width of over 300 m, during the period of 1999–2001. In the middle of the deposit area, the maximum incision is about 37 m in depth. The total volume of the river erosion is about 12.78 $Mm^3$ in a relatively narrow and limited area. During this period 1999–2001, typhoon Toraji impacted the Tsao-Ling area. The artificial spillway has been destroyed and new river channels have been rearranged naturally by the erosion associated with the typhoon. In addition to the large volume of debris material produced from upstream landslides, the results demonstrate the power of river erosion and the effect of typhoons on a rockslide dam deposit.

During the period of 2001–2003, debris with an overall volume of 7.5 $Mm^3$ was flushed out of the deposit area, especially from the sides of the river. The curved crowns observed in the downstream area indicate that small landslides developed on the banks of the river, which facilitate lateral river erosion. The orthrectified aerial photos confirm the observation by the presence of unvegetated scars. The major erosion area is about 6.35 $Mm^3$ in volume over an area $45.2\times10^4 m^2$, 43 m depth of about 320 m in width. This is equivalent to an average erosion rate of 7 m/year in the river valley, or a maximum incision rate of 21.5 m/year in a small area of the river bed.

During the period of 2003–2004, immediately after typhoon Mindulli, major river incision took place in the toe of the deposit area and along the river channel. The major incision could be found on the downstream side, where the maximum depth was as great as 83 m deep, as shown in Figs. 23.6 and 23.7c and Table 23.2. The volume of the erosion was about 7.7 $Mm^3$ in an area of $54.2\times10^4 m^2$ along the river valley. The erosion is an important factor in river incision, and is an even more important factor in the lateral erosion of the water course which enlarges the river channel. The lateral enlargement is accelerated by the circular small landslides on the side of the river bank composed of the debris.

In the period of 2004–2007, most of the deposit area underwent extreme erosion, and the land loss was severe. The major incision could be found on the river channel, where the maximum depth was as great as 74 m, as shown in Fig. 23.7d and Table 23.2. The process has been promoted by the lateral erosion of the water

**Table 23.2** Post-seismic erosion in the landslide area

| Region[a] | Surface (ha) | Volume ($Mm^3$) | Maximum depth (m) |
|---|---|---|---|
| A | 42 | 11.76 | 58 |
| B | 6.7 | 1.02 | 37 |
| C | 45.2 | 6.35 | 43 |
| D | 10.8 | 1.03 | 26 |
| E | 54.2 | 7.7 | 83 |
| F | 7.2 | 1.08 | 33 |
| G | 140.3 | 43.23 | 74 |
| H | 8.2 | 1.62 | 30 |

[a]The corresponding regions are denoted on Fig. 23.7.

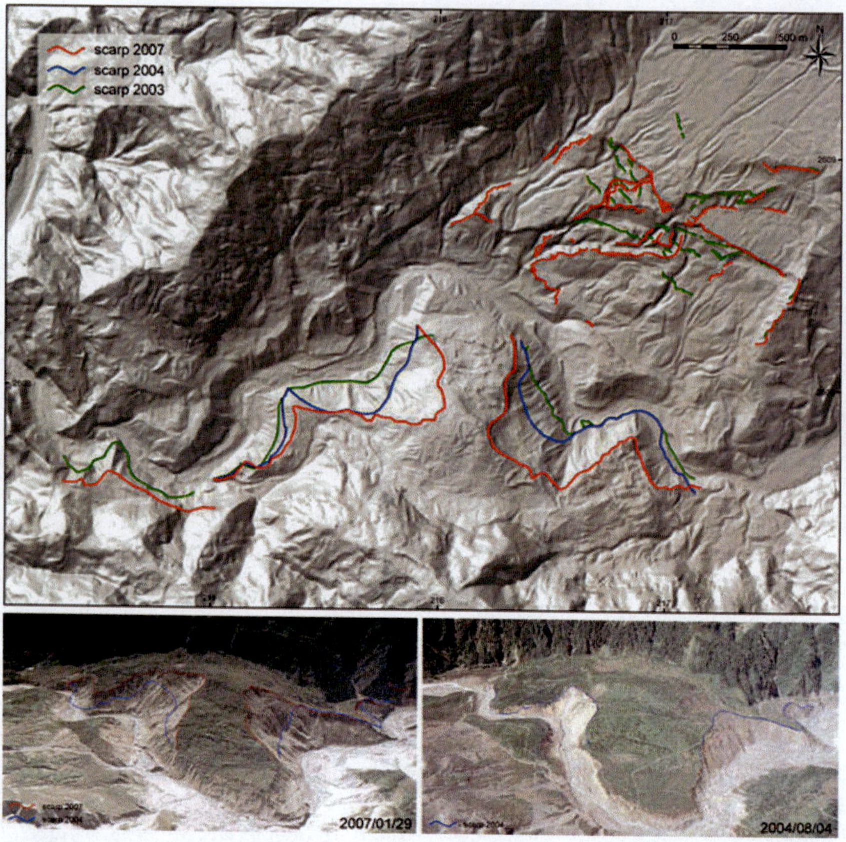


**Fig. 23.10** The retreat of the scarp *lines* on the Chingshui stream river bank on the side of the deposit debris since 2003. The maximum retreat rate could be as high as 93 m/year. However, most of the retreat is generated by small landslides caused by the side erosion and triggered during typhoons or heavy precipitation. The lower images are generated by orthorectified aerial photos draped onto the corresponding DEMs

course which enlarges the river bed. The lateral erosion of the river bed triggers circular landslides on the side of the river bank. Fig. 23.10 demonstrates the retreat of the scarp line along the Chingshui River. On the downstream side around the deposit area, the retreat of the scarp line is as much as 240 m during the period of 2004–2007, equal to 80 m/year rate of retreat. On the upstream side, the erosion is greater, up to 280 m, finally reaching the rate of the scarp retreat as much as 93 m/year. The total volume of erosion is calculated at 43.23 $Mm^3$ in an area of 1.403 $km^2$ along the river valley. The maximum river incision is about 74 m in depth. Thus these estimates of erosion indicate the rapid removal of debris from the landslide dam and dramatic changes in its morphology.

# 5 Life Cycle of Landslide-Dammed Lake

The volume estimation of the dammed lake, as noted above, suggests the maximum volume of about 45 $Mm^3$. To alleviate the possibility of an outburst flood, a spillway had been excavated to assist in draining the rockslide-dammed lake. During the July 2001 Typhoon Toraji, the water of the lake overflowed. Fortunately, in spite of serious economic loss, no major human casualties were caused by the overflow, and the dam survived. During the Toraji overflow the elevation of the outlet was reduced 20 m by erosion. In other words, the estimated volume of the rockslide-dammed water body was reduced to about 25.5 $Mm^3$, in an area of 1.63 $km^2$, whereas the maximum depth of the water body was about 36 m and the length of the backwater was 3.7 km.

The Chi–Chi earthquake had triggered many smaller landslides in Chingshui basin, upstream of the Tsao-Ling rock avalanche. The debris of these upstream landslides was transported in typhoons into the rockslide-dammed lake. During the threat of the Typhoon Mindulli on July 2–4, 2004, the rockslide-dammed lake was filled completely, as shown in Fig. 23.11b. The filled material was composed of fine debris, sands and vegetation (e.g., tree trunks). Figure 23.11a,b were taken before and after the 2004 Mindulli events. Soon after the lake was filled, small river channels developed on the surface of the sediment and became incised. Figure 23.9b demonstrates the power of river incision which affected the filled lake, with a river channel deeper than 50 m. It is worth noting that a flat river terrace was generated by the incised river course. The formation of the river terrace as part of the evolution of a landslide dam is thus different from the general concept related to tectonic uplift.

**Fig. 23.11** The rockslide-dammed dammed lake before and after Typhoon Mindulli on July 2–4, 2004. The dates of the photos are marked on the *lower-left* of each image

# 6 Conclusion

We have used high resolution DEMs to calculate geomorphological changes in the area of Tsao-Ling landlside triggered by the 1999 Chi Chi Earthquake. Taiwan is famous for its high erosion rate on an active orogenic belt. One of the estimate methods is based on the calculation of the river suspended load. By estimating the total volume of sediments transported downslope, it is found that the volume of the Tsao-Ling landslide is comparable with the annual volume of sediments transported by Taiwan's fluvial network to the ocean. Furthermore, the debris delivered during typhoons in each year modifies landforms around the deposit area; the volume of the debris lost from the deposit area is found to be very large when compared with the average river suspended sediment calculated at the downstream station. These results suggest that large landslides have a major impact on the calculation of past erosion rates in the Taiwan orogenic belt. The debris forming the Tsao-Ling landslide dam was heavily modified by erosion processes associated with typhoon-related rainfall in the period 1999–2007. The landslide-dammed lake formed by the 1999 blockage was short-lived and had filled with sediment during Typhoon Mindulli in 2004.

**Acknowledgements** We would like to thank Professor Stephen Evans, Professor Ming-Lan Lin and Professor Chyh-Ti Lee and Professor Jia-Jing Dong for their encouragement. We would also like to thank Professor Fu-Shu Jeng and Professor Chia-Yu Lu who have improved the scientific soundness of this paper. This work is partly supported by the National Science Council of Taiwan (NSC 98-2116-M027-003).

## References

1. Angelier, J., Lee, J.-C., Chu, H.-T., Hu, J.-C., Lu, C.-Y., Chan, Y.-C., Lin, T.-J., Font, Y., Deffontaines, B., and Tsai, Y.-B., (2001). Le Séisme de Chi–Chi (1999) et sa place dans l'orogène de Taiwan, *Comptes Rendus Geosciences* **333**, 5–21.
2. Chang, K.J., Taboada, A., and Chan, Y.C., (2005a). Geological and morphological study of the Jiufengershan landslide triggered by the Chi–Chi Taiwan earthquake, *Geomorphology* **71**, 293–309.
3. Chang, K.J., Taboada, A., Lin, M.L., and Chen, R.F., (2005b). Landsliding by earthquake shaking using a block-on-slope thermo-mechanical model: Example of Jiufengershan landslide, central Taiwan, *Engineering Geology* **80**, 151–163.
4. Chang, K.J., Taboada, A., Chan, Y.C., and Dominguez, S., (2006). Post-seismic surface processes in the Jiufengershan landslide area, 1999 Chi–Chi earthquake epicentral zone, Taiwan, *Engineering Geology* **86**, 102–117.
5. Chang, L.-S., (1951). Topographic features and geology in the vicinity of the nature reservoir near Tsao-Ling, *Taiwan Reconstruction Monthly* **1**, 22–27 (in Chinese).
6. Chen, R.F., Chan, Y.C., Angelier, J., Hu, J.C., Huang, C., Chang, K.J., and Shih, T.Y., (2005). Large earthquake-triggered landslides and mountain belt erosion: The Tsao-Ling case, Taiwan, *Comptes Rendus Geoscience*, **337**, 1164–1172.
7. Chen, R.F., Chang, K.J., Angelier, J., Deffontaines, B., Lee, C.T., Lin, M.L., and Yan, I.H., (2006). Topographical changes revealed by high-resolution airborne LiDAR images: The 1999 Tsao-Ling landslide induced by chichi earthquake, *Engineering Geology*, **88**, 160–172.
8. Chen, T.-C., Lin, M.-L., and Hung, J.-J., 2003. Pseudostatic analysis of Tsao-Ling rockslide caused by Chi–Chi earthquake, *Engineering Geology* **71**, 31–47.
9. Central for Space and Remote Sensing Research (CSRSR). (1989). *1/5000 Digital Terrain Model*. National Central University, Taiwan.
10. Chigira, M., Wang, W.-N., Furuya, T., and Kamai, T., (2003). Geological causes and geomorphological precursors of the Tsao-Ling landslide triggered by the 1999 Chi–Chi earthquake, Taiwan, *Engineering Geology* **68**, 259–273.

11. Dadson, S.J., Hovius, N., Chen, H., Dade, W.B., Shieh, M.L., Willett, S.D., Hu, J.C., Horng, M.J., Chen, M.C., Stark, C.P., Lague, D., and Lin, J.C., (2003). Links between erosion, runoff variability and seismicity in the Taiwan orogen, *Nature* **426**, 648–651.
12. Hsu, T. L., and Leung, H.-P., (1977). Mass movements in the Tsao-Ling area, Yunlin-Hsien, Taiwan, *Proceedings Geological Society of China* **20**, 114–118.
13. Huang, C.S., Ho, H.-C., and Liu, H.-C., (1983). The geology and landslide of Tsao-Ling area. Yunlin Hsien, Taiwan, *Bulletin Central Geological Survey* **2**, 95–112 (in Chinese).
14. Hung, J.J., (1980). A study on Tsao-Ling rockslides, Taiwan, *Journal of Engineering Environment* **1**, 29–39.
15. Hung, J.J., (2000). Chi–Chi earthquake induced landslides in Taiwan, *Earthquake Engineering and Engineering Seismology* **2**, 25–32.
16. Hung, J.-J., Lee C.-T., and Lin M.-L., (2002). Tsao-Ling rockslides, Taiwan, in S.G. Evans, and J.V. DeGraff, (eds.) *Catastrophic Landslides: Effects, Occurrence, and Mechanisms*. Geological Society of America Reviews in Engineering Geology Vol. 15, pp. 91–115.
17. Kawada, S., (1942). Untersuchung des neuen Sees gebildet infolge des Erdbebens vom Jahre 1941 in Taiwan (Formosa), *Bulletin Earthquake Research Institute, University of Tokyo* **21**, 317–325.
18. Kuo, C.Y., Tai, Y.C., Bouchut, F., Mangeney, A., Pelanti, M., Chen, R.F., and Chang, K.J., (2009). Simulation of Tsao-Ling landslide, Taiwan, based on Saint Venant equations over general topography, *Engineering Geology* **104**, 181–189.
19. Lee, C.-T., Hung, J.-J., Lin, M.-L., and Tsai, L.-Y., (1993). *Engineering geology investigations and stability assessments on Tsao-Ling landslide area*. A Special Report prepared for Sinotech engineering consultants. 224 pp (in Chinese).
20. Liao, H.-W., (2000). *Landslides triggered by Chi–Chi earthquake*. Master Thesis, Institute of Geophysics, National Central University, Chung-Li, 90 pp (in Chinese).
21. Lin, C.W., Liu, S.H., Lee, S.Y., and Liu, C.C., (2006). Impacts of the Chi–Chi earthquake on subsequent rainfall-induced landslides in central Taiwan, *Engineering Geology* **86**, 87–101.
22. Malavieille, J., Lallemand, S.E., Dominguez, S., Deschamps, A., Lu, C.-Y., Liu, C.-S., Schnurle, P., and the ACT Scientific Crew. (2002). *Arc-continent collision in Taiwan*. New marine observations and tectonic evolution. in: TB., Byrne, and C.-S., Liu, (eds.), Geology and Geophysics of an Arc-Continent collision, Taiwan, Republic of China. Boulder, Co, Geological Society of America Special Paper. 358, pp. 187–211.
23. Suppe, J., (1981). Mechanics of mountain building in Taiwan, *Memoir Geological Society of China*, **4**, 67–89.
24. Tai-Pei Observatory. (1942). *Report on Chia-Yi Earthquake on 17th Dec. 1941*. Taiwan Governor's Office. 227 pp (in Japanese).
25. Towhata, I., Yamazaki, H., Kanatani, M., Lin C.E., and Oyama, T., (2001). Laboratory Shear Tests of Rock Specimens Collected From Site of Tsao-Ling Earthquake-induced Landslide, *Tamkang Journal of Science and Engineering* **4**, 209–219.
26. Wang, W.N., Chigira M., and Furuya T., (2003). Geological and geomorphological precursors of the Chiu-feng-erh-shan landslide triggered by the Chi–chi earthquake in central Taiwan, *Engineering Geology* **69**, 1–13.
27. Water Conservancy Agency (WCA), Ministry of Economic Affaire. (1999). *The final report of current response measures for Tsao-ling landslide formed in 921 earthquake, Taiwan. R.O.C* (in Chinese).

# Chapter 24
# The Classification of Rockslide Dams

**Reginald L. Hermanns, Kenneth Hewitt, Alexander Strom, Stephen G. Evans, Stuart A. Dunning, and Gabriele Scarascia-Mugnozza**

## 1 Introduction

Landslide damming is a frequent phenomenon in all high mountain settings [6, 14, 45] but has only been systematically studied in few regions of the world [1, 3, 5, 15, 18, 30, 34, 35, 51, 52, 54, 55] (Chap. 5 by Hermanns et al., this volume). In addition, landslide dams with a large regional imprint involve not only the initial landslide impact itself but also upstream inundation due to reservoir impoundment and downstream sediment starving and potential subsequent outburst floods. There are many examples of rock-avalanche dams that survived intact for decades or centuries, and affected upstream conditions for millennia. Some of the largest and most destructive floods known on mountain rivers derive from catastrophic breaching of landslide dams [13, 46]. However, despite a long history of classification systems for landslides themselves e.g. [8, 12, 31, 32, 44, 53], the classification of landslide dams has remained unchanged for a long time. It hardly takes account of the recent, rapid increase in discoveries of rockslide-rock avalanche dams and their complexities. Swanson et al. [51] proposed a classification system based upon the relation of the landslide deposit to valley morphology. The most widely used classification in most regions of the world was proposed by Costa and Schuster [6] in 1988 following the first international meeting focussing on this phenomenon [18, 19, 36, 39, 41]. Their system also emphasises the relation of the landslide deposit to valley and impoundment morphology. It proposed 6 different types of dams, based mainly on two dimensional landslide deposit distribution in a single valley (type I, II, III, V) and the type of dam formation (type IV, VI). It concerned all types of landslide dams, not only those due to massive rock slope failures.

In recent years, studies in a variety of mountain settings have revealed some limitations of this classification system. Other geomorphic attributes of the landslide deposit and dam site need to be taken into account to better describe the landslide dam, its stability, the possibilities of dam failure, and long term geomorphological effects on valley evolution. Of particular concern is that landslide dams are not two dimensional features. Their three dimensional distribution within the

R.L. Hermanns (✉)
International Centre for Geohazards, Geological Survey of Norway, Trondheim, Norway
e-mail: reginald.hermanns@ngu.no

S.G. Evans et al. (eds.), *Natural and Artificial Rockslide Dams*, Lecture Notes in Earth Sciences 133, DOI 10.1007/978-3-642-04764-0_24, © Springer-Verlag Berlin Heidelberg 2011

valley [10, 48, 50] in relation to valley morphology affects the formation and long term stability of impoundments [10] (Chap. 5 by Hermanns et al., this volume). There is also the quite common phenomenon of landslide dam formation at confluences of two or more river valleys that may give rise to more than one lake [29], and the rarer but special case of landslide dams affecting drainage divides [25]. In this paper we propose a classification system, specifically for rockslide-rock avalanche dams. It remains a system focused on the morphological relations of the barrier, but takes into account three dimensional relations between landslide deposits, valley morphology and, where it occurs, landslide emplacement in two or more valleys.

## 2 Rockslide Dam Classification

The proposed system is a three-step classification that considers:

(A) the two dimensional plan view distribution of the landslide deposit within the valley and its relation to impounded water bodies, which extends the Costa and Schuster [6] classification
(B) the cross valley profile of the landslide deposit and its relation to the underlying, buried valley morphology, and
(C) the along-valley profile of the landslide debris and the underlying substrate following the thalweg of the valley.

### *2.1 Two-Dimensional (Plan-View) Distribution of Rockslide Deposit in the Valley*

We recognise five (I–V) main types of two dimensional distributions of rockslide debris on the valley floor (Fig. 24.1):

(I) Ponds on rockslide deposits are frequent on larger rockslide deposits. They can occur with different sizes and in different positions on top of the deposit. These ponds are not connected to the river channel of the dammed river and can occur on all types of rockslides and in combination with other dam types defined here. In general most rockslide dams deposited in wide valleys, where the deposit does not stretch the entire width of the valleys are related to these types of ponds.
(II) Single rockslide barriers can have different forms (Fig. 24.2) related to the rockslide types and rockslide volume in relation to the width of the valley. We distinguish between a rockslide crossing the valley partially or entirely, a rockslide involving the valley floor, and rockslides which after impacting the valley bottom, spread in an up- and downvalley direction.
(III) This type is characterized by two or multiple dams in a line within the same valley. These can form when the rockslide splits in the path of movement into various lobes (Fig. 24.3), or several basins can form in line on top of the deposit if the rockslide fills the valley floor over a larger distance.

(IV) Rockslide dams may form contiguous dams within two or more valleys if either the size is large enough to affect a confluence of valleys or the source location of the rockslide is favorable (IV a). Similarly rockslides with long run-outs can dam tributaries of the main valley or the main valley in the case that the rockslide originated in a tributary valley (IV b). Rockslides with high energy can run over opposing ridges depositing debris within the neighbouring valley (IV c) (Fig. 24.4). Depending on the geometry of confluence and rockslide volume, many more complex confluences can be dammed (IV d). A further sub-type occurs when natural dams fail and the large volume of debris transported during these events can dam tributary valleys in a chain of dams (IV e).

(V) Rockslide dams affecting the drainage divide are rare; however these dams impact the geomorphic and sedimentologic evolution of both valleys actively as they cause river capture resulting in enhanced erosion in one of the valleys and reduced erosion in the other. There are two sub types (Fig. 24.5); type Va is characterized by rockslides deposited directly in the drainage divide; type Vb is characterized by the height of the rockslide dam being higher than the lowest drainage divide. In this last case downcutting would occur in the drainage divide as soon as the lake filled up to the elevation of the divide.

**Fig. 24.1** Five main types (I–V) of plan view distribution of rockslide dams with letter representing sub-morphologies as described in the text and Figs. 24.2, 24.3, 24.4 and 24.5 (In this schematic figure as well as in the schematic Figs. 24.2, 24.3, 24.4, 24.5, 24.6 and 24.7 *dark gray* represents rockslide deposit, *medium gray* valley slopes, *light gray* lakes and *dotted* pattern the valley fill deposits)

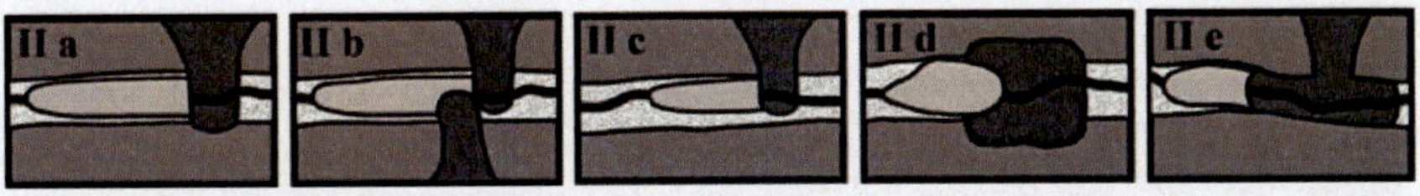


**Fig. 24.2** Five sub-types (IIa–e) of plan view distribution of single rockslide dams in a single valley

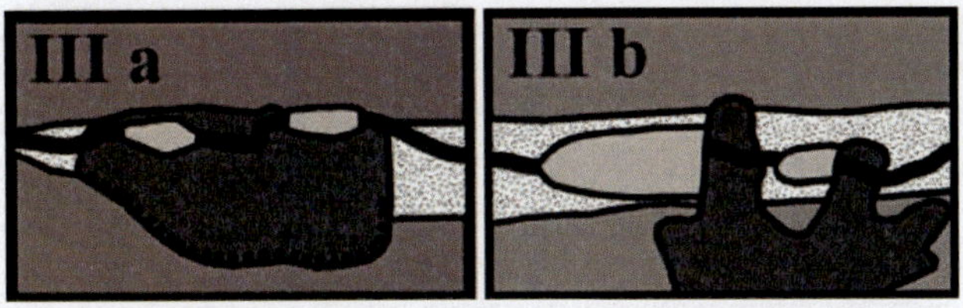


**Fig. 24.3** Two sub-types (IIIa–b) of plan view distribution of multiple rockslide dams in a single valley in a line formed by a single rockslide

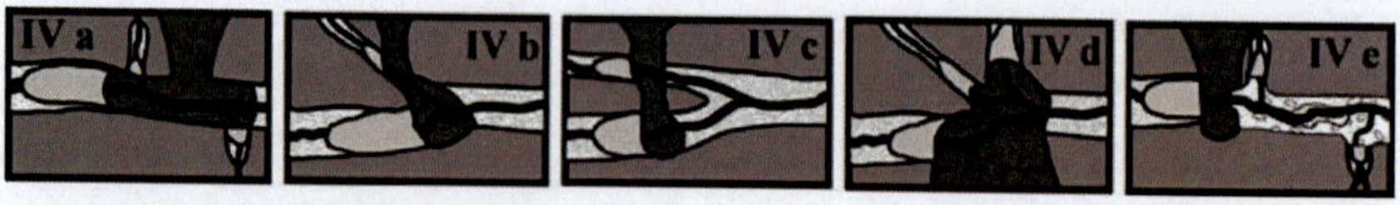


**Fig. 24.4** Five sub-types (IVa–e) of plan view distribution of single rockslide dams forming two or more dams in a single valley in a line

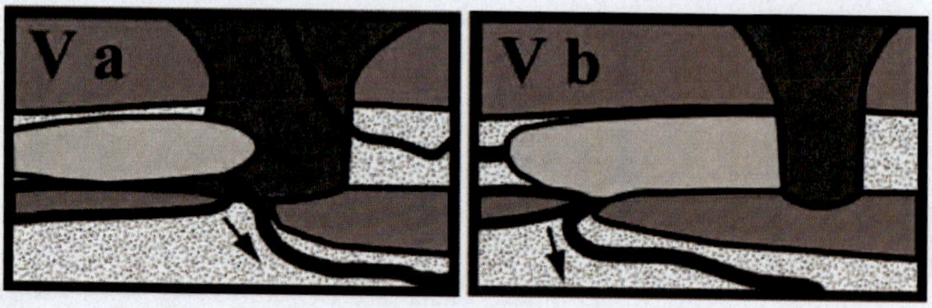


**Fig. 24.5** Two sub-types (Va–b) of plan view distribution of a rockslide dam affecting water divides; (**Va**) with the rockslide deposit depositing directly in the drainage divide, (**Vb**) with the dam higher than the lowest drainage divide up-river of the dam

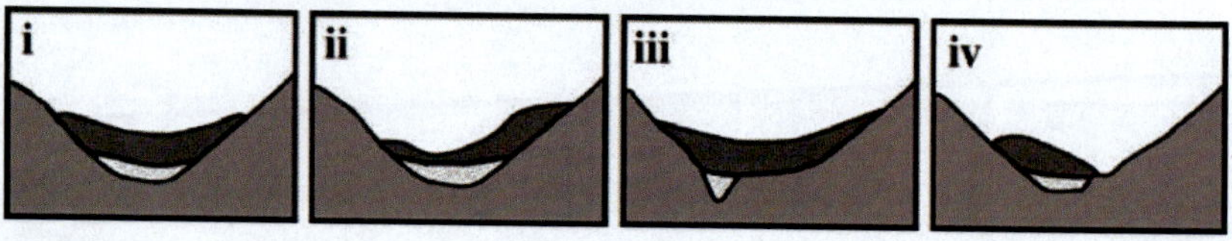


**Fig. 24.6** Four types (i–iv) of cross valley profiles of rockslide deposits and underlying valley morphologies

## 2.2 *The Cross-Valley Profile of the Rockslide Deposit and the Underlying Valley Morphology*

There are four different types (i–iv) of relations of rockslide deposits and valley floor morphologies (Fig. 24.6) in the cross-valley profile.

The first type (i) involves a symmetrical relation between deposit profile and valley profile in such a way that the lowest line (potential spillway) of the dam approximately coincides with the valley floor before landsliding. Therefore this type represents all types of valleys where spillway and former river bed coincides. For example an asymmetric dam in an asymmetric valley would belong to this type if potential spillway and former river bed coincides approximately. In all these cases the dam potentially can breach down to the thalweg of the impounded valley. In cases of high mobility, the landslide may deposit the main mass against the opposing slope (ii). However, in the case of slower moving landslides or one that decelerates rapidly on reaching the valley floor, the main body of the deposit may be against the proximal slope. A similar cross valley profile would be caused by a high-mobility rockslide in which the landslide climbs up the opposing slope but then falls back

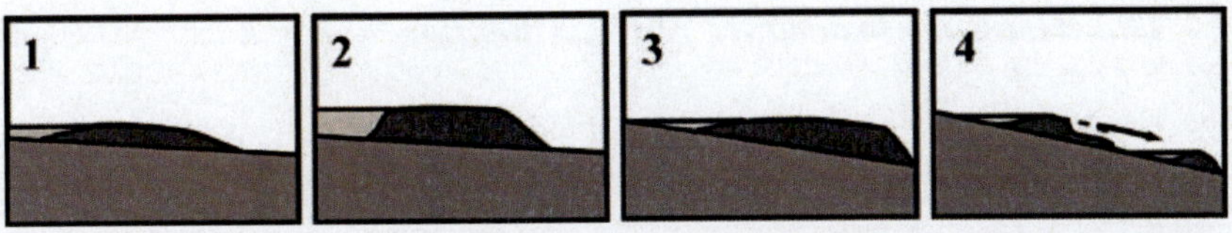


**Fig. 24.7** Four types (1–4) of along-valley profile of rockslide dams

and may, as in the case of Val Pola in Italy [7], recross the valley. In any case the landslide profile would be asymmetric to either side of the valley (ii). As Heim [21] pointed out, the low point and, hence, likely overflow are on opposite sides of the valley and may result in incision of an overflow channel in a thin cover of deposits and underlying bedrock of the distal or proximal slope respectively. The third type represents dams in asymmetric valleys (iii) where the lowest part of the dam does not correspond to the deepest part of the valley. Importantly the potential spillway does not coincide with the valley floor; although the cross valley dam profile might be symmetric or asymmetric. The fourth type Fig. 24.6 (iv) identifies rockslide dams with the deposit only crossing the valley bottom but not the entire width of the valley and hence divert the river to flow around the distal rim of the deposit of what was earlier part of the valley slope. In these cases drainage of the lake will not establish over the dam.

## *2.3 The Along-Valley Profile of the Rockslide Deposit*

The along-valley profile of a dam is influenced by the mobility of the rockslide and geometry of its entrance into the valley in question, as well as the valley thalweg slope, valley width, roughness and the presence of mobile substrates. Again we distinguish four types (1–4 in Fig. 24.7). Rockslides with high mobility and no constraining topography may spread over large areas to produce a relatively gentle deposit of low thickness (1), hence form a shallow lake. Rockslides in a confined setting with a low valley slope or high roughness may stall as deposits many 10s or 100s of meters high (2) and, therefore, can form large lakes. In steeper valleys although strongly confined the rockslide mass continues spreading along the valley for larger distances in the thalweg direction [42, 50] hence resulting in a thin often asymmetric deposit (3), with the main mass in the frontal part of the deposit. Therefore although the dam is relatively high at the distal end, due to the steepness of the valley floor the depth of the impounded lake is reduced which results in only smaller basins. In some cases the rockslide mass splits in two main bodies [50]. This can occur when a rockslide mass impacts a narrow valley perpendicular to the thalweg. While part of the mass runs up the opposite slope and falls back, another part spreads perpendicular to the first movement in direction down valley (4). The same effect is possible in funnel shaped valleys where spreading is restricted by the narrowing valley walls, causing one part of the rockslide to spread ahead. Such slides often produce various dams in one valley or when spreading over river confluences in various valleys.

### *2.4 The Resulting Procedure of Classification*

The classification suggests a three step procedure, depending upon the three-dimensional geometry of the rockslide deposit and the pre-existing valley floor morphology:

(A) two dimensional distribution of rockslide debris,
(B) cross-valley profile relating the emplacement of rockslide deposits to valley floor morphology, and
(C) along-valley profile again relating emplacement patterns of rockslide debris to valley form.

As an example (Fig. 24.8) the rockslide dam forming the Laguna Dial in southern Chile, this dam would be classified as a type IIa/ii/2 rockslide dam.

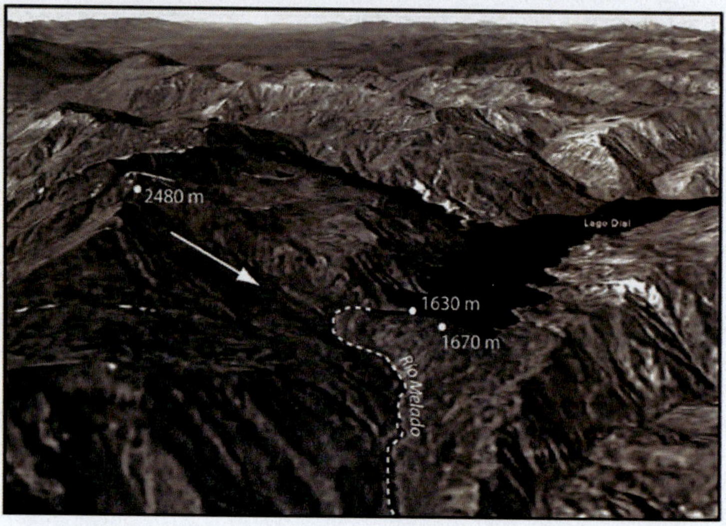


**Fig. 24.8** Oblique view of the rock-avalanche deposit impounding Laguna Dial, Chile (36°25′14″S/70°55′28″W; Google Earth). This rock avalanche forms a single dam in a stretched valley (type II). The cross valley profile has asymmetric distribution of the landslide deposit overlying an apparent symmetric valley floor (type ii). This is a narrow mountain valley and direction of landsliding is normal to the thalweg, the dip of the thalweg is minor, therefore the run out is restricted and the landslide forms a massive thick deposit (type 2)

## 3 Implication of Rockslide Dam Type on Valley Evolution and Dam Stability

As shown in the following the three dimensional emplacement morphology of rockslide debris has a major impact on the size of the impounded basin, the stability of the landslide dam, and its role in subsequent valley evolution, and needs to be considered in any assessment of landslide risk and implications. In the following sections, we elaborate on the ways in which plan view, cross-valley and along-valley distribution of rockslide deposit and valley morphology may each influence the dam problem.

## 3.1 *Impact of the Two Dimensional (Plan View) Distribution of the Rockslide Deposit*

Ponds (I) on rockslide deposits are frequently reported e.g. [16, 19] although rarely considered a significant hazard because of their small size. Deposits from these internal basins may be useful for dating prehistoric rockslides if they contain organic material or tephra [22, 24, 26]. In few occasions water bodies on top of large rockslides are large enough to be considered as a hazard when drained rapidly (Chap. 5 by Hermanns et al., this volume).

Single rockslide barriers (II) are the most frequent types of rockslide dams [6, 38]. The different subforms (a–e) (Fig. 24.2) based upon their formation influence the stability of the dam and the size of the lake (c). While subtype (a), and (c) dams do not contain enough material to spread along the valley they are lower and can therefore easier be eroded. In contrast subtype (d) and (e) dam effect larger portions of the valley and the mass and length of the dam make it much more resistant to erosion.

In most cases sediment flux within the valley and the outburst hazard depend primarily on the stability of single dams. However, two or more dams on the same river (III) involve more complex assessment because the survival of each of the dams may depend upon that of the other(s). This involves various scenarios of dam failure. In one, the upper dam fails first causing an outburst flood into the lower basin. If the lower dam, or basin formed by it, is smaller, it is more likely to fail, and less so if it is larger when it can resist and even contain a catastrophic outburst from the upper smaller basin. In a second scenario the lower dam fails first, initiating or increasing toe erosion of the upper dam which may fail soon after the first and drain catastrophically. These possibilities are important because many rockslide dams do not fail immediately, or catastrophically, but are gradually degraded after overtopping.

Where a rockslide forms parallel dams in two or more valleys (IV) the stability of one portion of the dam may not have any influence on the stability of another portion and their stability must be assessed independently. The total number of sub-dams formed by a rockslide event depends on the type of confluences in the valley and of how far the rockslide spreads along the valley [30]. Multiple dams may also be formed by outburst floods from an existing landslide dam. Due to the large amount of material re-deposited in such an event tributary valleys may be dammed for several km to tens of km [23, 25, 41].

Rockslides impacting a drainage divide rarely form dams, however cases of reorganization of drainage systems including dam formation or even dams in the drainage divide have been reported [16] (Chap. 5 by Hermanns et al., this volume). These dams have an important impact in the morphological evolution of the valleys as they are often related to river capture causing enhanced erosion in one of the valleys and reduced erosion in the other. In addition, dams affecting a drainage divide may potentially fail towards one or the other valley. In contrast dams causing the drainage to be diverted over the drainage divide are extremely stable as dam erosion by overtopping is excluded (Chap. 5 by Hermanns et al., this volume).

### 3.2 Impact of the Cross-Valley Distribution of the Rockslide Deposit

The cross-valley profile of the dam has an important influence on the size of the impounded lake, the stability of the dam, and the size of a potential outburst flood event. For example, in the cross-valley profiles (i) and (iii) the relative difference between the lowest elevation in the profile and the mean elevation is similar, causing the dam to be relatively high and to form a deep lake. In contrast, in the cross-valley profiles (ii) and (iv) the difference of lowest point and highest point is large and the dam therefore forms relatively minor basins.

During failure symmetric profiles of rockslide deposits in a symmetric valley (i) can be eroded down to the valley floor. Hence worst case scenario calculations for outburst floods suggest themselves which should take into account that the entire volume of the stored water would drain [20, 24]. This is different with asymmetric dams in symmetric valleys (ii); during failure erosion is most likely stopped when the breach cuts into the bedrock causing a relatively stable relict lake which may be of considerable size and duration [9, 27, 29] (Fig. 24.9; Chap. 5 by Hermanns et al., this volume). These types of dams are long time traps for sediment flux within a valley. Symmetric dams in asymmetric valleys (iii) behave similar to asymmetric dams in symmetric valleys. Erosion during failure cuts down to the basement, which is not the deepest part of the valley, hence a large relict lake with a stable dam results (Chap. 5 by Hermanns et al., this volume; Fig. 24.9) (iv) Rockslide dams with the deposit higher than the opposite valley wall form lakes and divert the river to flow over the bedrock. Dam failure due to overtopping is impossible making these dams very stable morphological features which can last for several tens ka [17, 38] (Chap. 5 by Hermanns et al., this volume).

### 3.3 Impact of the Along-Valley Distribution of the Rockslide Deposit

Along-valley spreading of rockslide debris may have a variable effect on dam stability and further valley evolution. On one hand reduced dam height due to unconfined spreading within the valley (1) will form shallower lakes, impounding smaller water bodies while at the other hand these smaller basins are likely to fill up faster and dam erosion starts shortly after dam formation. In contrast, rockslide dams which form dams several 10s to 100s of m high (2) can form larger basins which need centuries

**Fig. 24.9** Photo of the base of the breach of the dam of Onilsavatnet [27] at the southern tip of Tafjord, Norway, showing that the breach cut down to the bedrock (*centre* and *left*) leaving the rock avalanche deposits only at the margin of the breach (*right*)

to millennia to fill up [24, 33]. Due to the presence of the highly permeable carapace facies on top of rockslide dams [11, 28, 47], several extremely large dams such as the Usoi dam [2] exist for a very long time, since filtering discharge is nearly equal to the lake inflow. If the dam would be lower it would be breached. On the other hand, the higher the dam is, a more catastrophic outburst flood may occur due to the larger amount of water stored [2, 14, 24, 30, 33, 37, 45]. Moreover, rockslide deposits with less movement in up- and down valley direction (2) form shorter barriers, which could breach faster [43] and in a more catastrophic way than that of widely spread "pancake-like" blockage. Type 3 dams occur only in tributaries in high mountain settings e.g. [16] and basins formed are small. They fill up rapidly but, due to the small volume of the basin, related floods are spatially restricted. The most specific case is type 4 caused by partially spread rockslide debris, some portion of which forms compact dam while other portions spread as long run-out rock avalanche [48–49]. In this case dam heights can be similar to type 2 blockages, but its thickness and, consequently, safety would decrease significantly. It should be noted that such type of blockages can be formed both in large valleys with a relatively gentle thalweg and in short steep gullies.

### *3.4 Rockslide Dam Type As a Matrix of Three Factors*

Our classification system is not meant as three independent classifications considering the plan view, the cross valley and along-valley profile. These morphologic characters are intrinsically linked to each other (Fig. 24.8) and depend on the relative relations of rockslide volume, valley size and morphology and rockslide mobility. For example, a relatively small rockslide impacting at a right angle to the river in a narrow valley, would have restricted mobility and produce a dam with a part of the mass spreading up- and a larger portion spreading downvalley (type II e in plan view). The cross valley profile of such rockslide with larger movement along the valley than across the valley would be rather symmetric (i) and the along-valley profile a massive deposit several 10s to 100s of meters high (2). The same rockslide in a broad valley would have due to lower geomorphic restrictions [42] a higher mobility and would form as a lobate deposit only a few m thick (type II c in plan view) which likely only spans over part of the valley floor (iv in plan view) and form a thin layer in the along-valley profile (1). This influences strongly the amount of water which can be stored behind the dam, the position of an overflow channel and the longevity of the dam. Hence the rockslide deposit distributions in three dimensions of this classification are not factors independent of each other but strongly related to each other.

## 3.5 Overall Impact of Rockslide Dam Type on Hazard Assessment

Our rockslide classification focuses on the morphology of the deposit and its relation to the valley form. These observations make it feasible to consider whether a single outburst flood should occur or multiple outbursts have to be taken into account. In addition, it directs attention to which portion of the dam is likely to fail and, in the case of breaching to less than full valley depth, how much of the stored water volume will be released in a catastrophic event and how much remains. We are well aware that these morphologic relations between the deposit and the valley floor are not the only influences on long term stability of rockslide dams. Other site factors such as type of landslide (Chap. 5 by Hermanns et al., this volume), lithology and grain size distribution of the rockslide mass [4, 10] have to be taken into account. Since our observations are based mainly on rock avalanches, it is important that they are characterized by a boulder deposit at the surface (carapace), while the largest part of the rockslide body is composed of a strongly fractured and relatively fine deposit [11, 28]. Matrix materials, including predominantly fines, comprise the main body of these deposits even where the source rock is not strongly fractured, foliated and tectonized [55]. As few data exists on grain size distributions of rockslide dams to date, this is an important field of future research.

Studies of other geomorphic parameters describing the drainage basin or relations of geomorphic parameters to the dams and impounded water bodies suggest they also have an influence [14, 36]. Comparisons of world wide data seem to indicate different thresholds for different mountain regions [14], which are likely related to climatic settings. However, there is still not enough published evidence to establish how climatic parameters influence rock avalanche dam stability. Meanwhile, large, long-lived rock avalanche dams are known for virtually every mountain region and climatic zone.

**Acknowledgement** This paper was partly written at the Geological Survey of Canada and partly at the Norges geologiske undersøkelse. The work described in this paper is therefore partially supported by the Research Council of Norway through the International Centre for Geohazards (ICG). Their support is gratefully acknowledged. This is ICG contribution No. 283.

## References

1. Adams, J. (1981) Earthquake-dammed lakes in New Zealand, *Geology* **9**, 215–219.
2. Alford, D. and Schuster, R.L. (2000) *Usoi Landslide Dam and Lake Sarez*. ISDR Prevention Series. United Nations Publications, New York, NY and Geneva, p. 113.
3. Casagli, N. and Ermini, L. (1999) Geomorphic analysis of landslide dams in the northern Apenine, *Transaction of the Japanese Geomorphologic Union* **20**, 219–249.
4. Casagli, N., Ermini, L. and Rosati, G. (2003) Determining grain size distribution of the material composing landslide dams in the Northern Apennines: Sampling and processing methods, *Engineering Geology* **69**, 83–97.
5. Clague, J.J. and Evans, S.G. (1994) Formation and failure of natural dams in the Canadian Cordillera, *Geological Survey of Canada, Bulletin* **464**, 1–35.
6. Costa, J.E. and Schuster, R.L. (1988) The formation and failure of natural dams, *Geological Society of America Bulletin* **100**, 1054–1068.
7. Crosta, G., Frattini, P. and Sosio, R. (2006) Formation, characterization and modelling of the 1987 Val Pola rock-avalanche dam (Italy), *Italian Journal of Engineering and Environment*, Special Issue, 1, 145–150.
8. Cruden, D. and Varnes, D. (1996) Landslide types and processes, in K. Turner and R.L. Schuster (eds.), *Landslides. Investigation and Mitigation*. National Academy Press, Washington, pp. 36–75.
9. Duman, T.Y. (2009) The largest landslide dam in Turkey: Tortum landslide, *Engineering Geology* **104**, 66–79.
10. Dunning, S., Petley, D., Rosser, N. and Strom, A. (2005), The morphology and sedimentology of valley confined rock-avalanche deposits and their effect on potential dam hazard, in *Landslide risk management, Vancouver*, Balkema, Amsterdam pp. 691–701.
11. Dunning, S. (2006) The grain size distribution of rock-avalanche deposits in valley confined settings, *Italian Journal of Engineering Geology and Environment*, Special Issue, 1, 117–121.
12. Emilianova, E. (1972) *Main Regularities of Landslide Processes (in Russian)*. Nedra Publishing House, Moscow, 308 p.
13. Ermini, L. and Casagli, N. (2002), Criteria for a preliminary assessment of landslide dam evolution, in *1. European Conference on Landslides, Prague*, 157–162.
14. Ermini, L. and Casagli, N. (2003) Prediction of the behaviour of landslide dams using a geomorphological dimensionless index, *Earth Surface Processes and Landforms* **28**, 31–47.
15. Evans, S.G. (1986) Landslide damming in the Cordillera of western Canada, in R.L. Schuster (ed.), *Landslide Dams: Processes, Risk and Mitigation*. American Society of Civil Engineers, New York, NY, pp. 111–130.
16. Fauqué, L., Cortés, J., Folguera, A. and Etcheverría, M. (2000) Avalanchas de roca asociadas a neotectónica en el valle del río Mendoza, al sur de Uspallata, *Revista de la Asociación Geológica Argentina* **55**, 419–423.
17. Fauqué, L. and Tchilinguirian, P. (2002) Villavil rockslides, Catamarca Province, Argentina, in S.G. Evans and J.V. DeGraff (eds.), *Catastrophic Landslides: Effects, Occurrence, and Mechanism*, Reviews in Engineering Geology. GSA, Boulder, CO, pp. 303–324.
18. Ferrer, C. (1999) Represamientos y rupturas de embalses naturales (lagunas de obstrución) como efectos cosísmicos: Algunos ejemplos en los Andes venezolanos, *Revista Geográfica Venezolana* **40**, 109–121.

19. González Díaz, E., Fauqué, L., Giaccardi, A. and Costa, C. (2000) Las lagunas de Varvar Co Campos y Varvar Co Tapia (N del Neuquén, Argentina): su relación con avalanchas de rocas, *Revista de la Asociacion Geológica Argentina* **55**, 147–164.
20. Groeber, P. (1916) Informe sobre las causas que han producido las crecientes del río Colorado (Territorios del Neuquén y La Pampa) en 1914, Dirección General de Minas, Geología e Hidrogeología, 11 (Serie B, Geología), 1–29.
21. Heim, A. (1932) Bergsturz und Menschenleben, Beiblatt zur Vierteljahresschrift der Naturforschenden Gesellschaft in Zürich: Zürich, 1–217 p.
22. Hermanns, R.L., Trauth, M.H., Niedermann, S., McWilliams, M. and Strecker, M.R. (2000) Tephrochronologic constraints on temporal distribution of large landslides in northwest Argentina, *Journal of Geology* **108**, 35–52.
23. Hermanns, R.L., Naumann, R., Folguera, A. and Pagenkopf, A. (2004) Sedimentologic analyses of deposits of a historic landslide dam failure in Barancas valley causing the 1914 Rio Colorado flood, northern Patagonia, Argentina, in W.A. Lacerda, M. Ehrlich, S.A.B. Fontoura and A.S.F. Sayão (eds.), *Landslides, Evaluation and Stabilization*. Balkema, Leiden, pp. 1439–1445.
24. Hermanns, R.L., Niedermann, S., Ivy-Ochs, S. and Kubik, P.W. (2004) Rock avalanching into a landslide-dammed lake causing multiple dam failure in Las Conchas valley (NW Argentina) – evidence from surface exposure dating and stratigraphic analyses, *Landslides* **1**, 113–122.
25. Hermanns, R.L., Folguera, A., González Díaz, E.F. and Fauque, L. (2006) Landslide dams in the Central Andes of Argentina – showing the need of revising the established landslide dam classification, *Italian Journal of Engineering Geology and Environment*, Special Issue, 1, 55–60.
26. Hermanns, R.L., Niedermann, S., Villanueva Garcia, A. and Schellenberger, A. (2006) Rock avalanching in the NW Argentine Andes as a result of complex interactions of lithologic, structural and topographic boundary conditions, climate change and active tectonics, in S.G. Evans, G. Scarascia Mugnozza, A.L. Strom and R.L. Hermanns (eds.), *Landslide from Massive Rock Slope Failures*, NATO Science Series 4, Earth and environmental Sciences. Springer, Berlin, pp. 497–520.
27. Hermanns, R.L., Blikra, L.H. and Longva, O. (2009) Relation between rockslide dam and valley morphology and its impact on rockslide dam longevity and control on potential breach development based on examples from Norway and the Andes, 2nd International Conference: Long Term Behavior of Dams, Graz, 789–794.
28. Hewitt, K. (1999) Quaternary moraines vs catastrophic rock avalanches in the Karakoram Himalaya, northern Pakistan, *Quaternary Research* **51**, 220–237.
29. Hewitt, K. (2002) Styles of rock-avalanche depositional complexes conditioned by very rugged terrain, Karakoram Himalaya, Pakistan, in S.G. Evans and J.V. DeGraff (eds.), *Catastrophic Landslides: Effects, Occurrence and Mechanisms*. Geological Society of America, Boulder, CO, pp. 345–377.
30. Hewitt, K. (2006) Rock avalanche dams on the transHimalayan Upper Indus streams; a survey and assessment of hazard-related characteristics, *Italian Journal of Engineering Geology and Environment*, Special Issue, 1, 61–65.
31. Hungr, O., Evans, S., Bovis, M. and Hutchinson, J. (2001) Review of the classification of landslides of the flow type, *Environmental and Engineering Geology* **7**, 221–238.
32. Hutchinson, J. (1968) Mass movement, in R. Fairbridge (ed.), Encylo*pedia of Geomorphology, Encyclopedia of Earth Science*. Reinhold, New York, NY, Amsterdam, London, pp. 688–695.
33. Ischuk, A.R. (2006) Usoi natural dam: Problem of security (Lake Sarez, Pamir mountains, Tadjikistan), *Italian Journal of Engineering Geology and Environment*, Special Issue, 1, 189–192.
34. Karakhanyan, A. and Baghdassaryan, H. (2006) Natural dams in Armenia: Landslide hazard and risk assessment, *Italian Journal of Engineering Geology and Environment*, Special Issue, 1, 93–94.

35. Korup, O. (2002) Recent research on landslide dams; a literature review with special attention to New Zealand, *Progress in Physical Geography* **26**, 206–235.
36. Korup, O. (2004) Geomorphometric characteristics of New Zealand landslide dams, *Engineering Geology* **73**, 13–35.
37. Korup, O. (2006) Rockslide and rock-avalanche dams in the Southern Alps, New Zealand, *Italian Journal of Engineering Geology and Environment*, Special Issue, 1, 33–43.
38. Korup, O., Strom, A. and Weidinger, J. (2006) Fluvial response to large rock-slope failures: Examples from Himalaya, the Tien Shan, and the Southern Alsp in New Zealand, *Geomorphology* **78**, 3–21.
39. López, M., Barrios, J. and Barriga, J. (1998) Manual de estabilidad de taludes: Bogota, Escuela Colombiana de Ingeneria, 340 p.
40. Mardones Flores, M. (2002) Evolución morfogenética de la hoya del río Laja y su incidencia en la geomorfología de la región del Biobío, Chile, *Revista Geográfica de Chile Terra Australis* **47**, 97–127.
41. Miller, B. and Cruden, D.M. (2002) The Eureka River landslide and dam, Peace river lowland, Alberta, *Canadian Geotechnical Journal* **39**, 863–878.
42. Nicoletti, P.G. and Sorriso-Valvo, M. (1991) Geomorphic controls of the shape and mobility of rock avalanches, *Geological Society of America Bulletin* **103**, 1365–1373.
43. Plaza-Nieto, G. and Zevallo, O. (1994) The 1993 La Josefina rockslide and Rio Paute landslide dam, Ecuador, *Landslide News* **8**, 4–7.
44. Sassa, K. (1989) Geotechnical classification of landslides, *Landslide News* **3**, 21–24.
45. Schuster, R.L. (2000) A worldwide perspective on landslide dams, in D. Alford and R.L. Schuster (eds.), *Usoi Landslide Dam and Lake Sarez*, ISDR Prevention Series. United Nations Publications, New York, NY and Geneva, pp. 19–22.
46. Schuster, R.L. (2006) Impact of landslide dams on mountain valley morphology, in S.G. Evans, G. Scarascia Mugnozza, A.L. Strom and R.L. Hermanns (eds.), *Landslides from Massive Rock Slope Failures*, Nato Science Series IV, Earth and Environmental Sciences. Springer, Amsterdam, pp. 591–616.
47. Strom, A.L. (1994) Mechanism of stratification and abnormal crushing of rockslide deposits, in *Proc. 7th International IAEG Congress* **3**, Rotterdam, Balkema, 1287–1295.
48. Strom, A.L. (1996), Some morphologic types of long run-out rockslides: effect of the relief on their mechanism and on the rockslide deposits distribution, in *Proceedings of the seventh international symposium on landslides, Trondheim, Norway*, 1977–1982.
49. Strom, A.L. (2006) Morphology and internal structure of rockslides and rock avalanches: Grounds and constraints for their modelling, in S.G. Evans, G. Scarascia Mugnozza, A.L. Strom and R.L. Hermanns (eds.), *Landslides from Massive Rock Slope Failures*, NATO Science Series IV: Earth and Environmental Sciences. Springer, Dordrecht, pp. 305–328.
50. Strom, A.L. and Pernik, L. (2006) Utilisation of the data on rockslide dams formation and structure for blast-fill dams design, *Italian Journal of Engineering Geology and Environment*, Special Issue, 1, 133–136.
51. Swanson, F.J., Oyagi, N. and Tominaga, M. (1986) Landslide dams in Japan, in R.L. Schuster (ed.), *Landslide Dams: Processes, Risk, and Mitigation*. American Society of Civil Engineers, New York, NY, pp. 131–145.
52. Tianchi, L., Schuster, R.L., Asce, F. and Jishan, W. (1986) Landslide dams in south-central China, in R.L. Schuster (ed.), *Landslide Dams: Processes, Risk, and Mitigation*. American Society of Civil Engineers, New York, NY, pp. 146–162.
53. Varnes, D. (1978) Slope movements. Types and processes, in E. Eckel (ed.), *Landslides, Analysis and Control*. Transportation Research Board, Washington, DC, pp. 11–33.
54. Vinnichenko, S.M. (2006) Landslide blockage in Tadjikistan mountains, *Italian Journal of Engineering Geology and Environment*, Special Issue, 1, 81–86.
55. Weidinger, J.T. (2006) Landslide dams in the high mountains of India, Nepal and China, *Italian Journal of Engineering Geology and Environment*, Special Issue, 1, 67–80.

# Chapter 25
# Russian Experience with Blast-Fill Dam Construction

**V.V. Adushkin**

## 1 Introduction

Elaboration of the technology of dams and embankments construction based on utilization of powerful explosions started in Russia in the 1930s of the twentieth century. Advantages of this technology lie in the use of rock material available directly at the dam site, in a shorter construction period, in the possibility to construct large structures in remote areas where few qualified workmen are available, and, finally in the project's cost reduction.

The first explosion for the purpose of river damming was conducted in 1935 during construction of the Chirchik HPP in Uzbekistan. Blasting of three charges with total weight of 2,860 kg created a dam with estimated volume of 20,000 $m^3$. Later on, from 1930s to the 1950s of the twentieth century several blast-fill dams were constructed by the explosions of 5–500 tons, basically for irrigation purposes. These dams were from 5,000 to 160,000 $m^3$ in volume, 4–15 m high and formed blockages 50–120 m along and 10–40 m across the river channels.

Before World War II several rather small dams from 5,000 to 200,000 $m^3$ in volume were constructed by blasts that used from 5 to 500 tons of explosives. Later on this technology developed further and during the period 1959–1989 several large dams were constructed by powerful explosions. They were designed as water-retaining dams, for protection from debris flows, and as tailings storage embankments (Table 25.1).

Explosions carried out at the Alamedin and Shamsi Rivers (northern Kyrgyzstan) have tested a capability to construct dams that had to accumulate sediment runoff. Fragments size distribution within blasted rock masses, filtration and stability of blast-fill dams against water loading have been investigated as well.

V.V. Adushkin (✉)
Institute of Geospheres Dynamics, Russian Academy of Sciences, 119334 Moscow, Russia
e-mail: adushkin@idg.chph.ras.ru

S.G. Evans et al. (eds.), *Natural and Artificial Rockslide Dams*, Lecture Notes in Earth Sciences 133, DOI 10.1007/978-3-642-04764-0_25, 

**Table 25.1** Main parameters of blast-fill dams constructed in the former USSR

| No | Location | Year of the explosion | Rock type | Total amount of the explosives (tons) | Dam height (m) | Dam width along the stream (m) | Dam volume ($m^3$) | Unit rate of the explosives ($kg/m^3$) |
|---|---|---|---|---|---|---|---|---|
| 1 | Alamedin river (Kyrgyzstan) | 1959 | Porphyrite | 153 | 21 | 280 | $1.15 \times 10^5$ | 1.33 |
| 2 | Shamsi river (Kyrgyzstan) | 1960 | Granite | 48 | 19 | 250 | $6 \times 10^4$ | 0.8 |
| 3 | Malaya Almaatinka River (Medeo – Kazakhstan) | 1966, 1967 | Granite | 5,923 (first blast) 3,946 (second blast) | 80 (60 after first blast) | 560 | $3 \times 10^6$ ($1.5 \times 10^6$ after first blast) | 3.95 1st blast 2.63 2nd blast |
| 4 | Baipasa HPP on the Vakhsh river (Tajikistan) | 1968 | Limestone | 2,000 | 50 | 550 | $1.5 \times 10^6$ | 1.33 |
| 5 | Ak-Su river (Dagestan) | 1972 | Limestone | 565 | 84 | 325 | $2.5 \times 10^5$ | 2.2 |
| 6 | Burlykia river (Kyrgyzstan) | 1975 | Granite | 703 | 50 | 320 | $5.4 \times 10^5$ | 1.3 |
| 7 | Alindjai river (Nakhichevan) | 1984 | Porphyrite | 689 | 32 | 200 | $6.1 \times 10^5$ | 1.13 |
| 8 | Kvaisa (Georgia) | 1984 | Porphyrite | 437 | 26 | 250 | $1.1 \times 10^5$ | 3.97 |
| 9 | Uch-Terek river (Kyrgyzstan) | 1989 | Sandstone and siltstone | 1,915 | 45 | 295 | $8.7 \times 10^5$ | 2.2 |

## 2 Modeling Facilities

Since 1972 all large industrial explosions, including those designed for blast-fill dams construction have been tested in the special simulator developed in our Institute (Fig. 25.1) [1, 3–4, 5, 7–10]. This equipment allows us to simulate the explosion site morphology at a scale from 1:100 to 1:1,000 and to evaluate optimal detonation scheme, shape and weight of charges, their positions, and time delays, necessary for construction of a desired dam configuration even in complex relief. It is based on the following modelling principles:

(1) Initial, central-symmetric stage of an explosion evolution- when a cavity is created and surrounding medium is fractured it is considered as completed. It is taken into account by choosing initial conditions of the explosion source and of the fractured rock mass strength.
(2) Modeling is carried out by an energy and geometric scaling law:

$$E/[\rho g W + P_0 W^{\alpha}] = \text{const}; \quad E/(\sigma W^{\alpha}) = \text{const}; \quad W/r_0 = \text{const}$$

where E is the total energy of explosion, $\rho$ – the rock medium density, g – the gravity acceleration, W – minimal distance of charge center from the daylight surface, $r_0$ – the cavity radius, $P_0$ – atmospheric pressure, $\sigma$ – the fractured rock cohesion, $\alpha = 3$ for the spherical charge and $\alpha = 2$ for the elongated (cylindrical) charge.
(3) In the model, explosions are simulated by compressed air placed into an elastic shell of spherical or cylindrical shape.
(4) Fractured rock is modeled by mixture of quartz sand damped by a small amount of glycerin from 0.1 to 0.5% that has low cohesion.

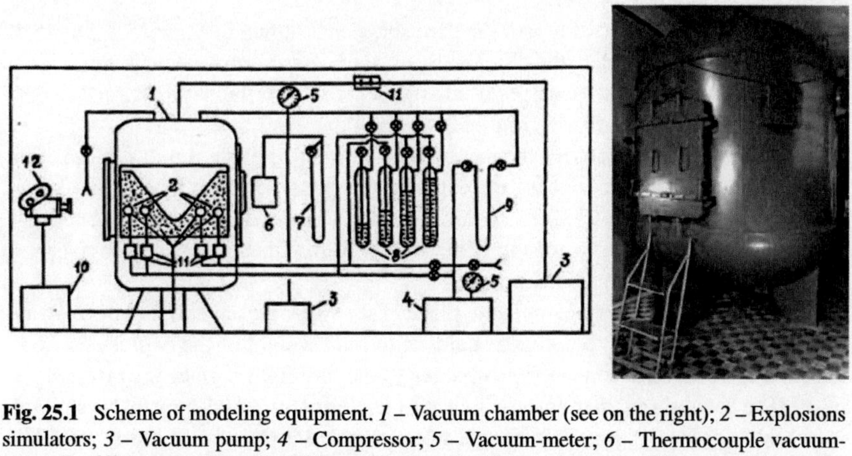

**Fig. 25.1** Scheme of modeling equipment. *1* – Vacuum chamber (see on the right); *2* – Explosions simulators; *3* – Vacuum pump; *4* – Compressor; *5* – Vacuum-meter; *6* – Thermocouple vacuum-meter; *7* – Oil vacuum-meter; *8* – Differential valve; *9* – Mercury gage; *10* – Control and detonation device; *11* – Electromagnetic valve; *12* – High-speed camera

(5) Modelling is conducted in a vacuum chamber with corresponding value of $P_0 \approx (0.001 \div 1)\rho g W$ in order to preserve the natural ratio of rock weight and atmospheric pressure.

The principal element of the equipment is a steel cylindrical vacuum chamber 2.3 m in radius, 3 m high and 12 $m^3$ in volume (see Fig. 25.1). A model of the corresponding rock mass is created in the chamber and relevant explosion simulators are placed inside. The equipment is supplied with detonating instruments, experiment control devices and recording tools. Up to 10 such simulators can be placed inside the model and detonated with desired time delays. Modelling of known industrial explosions has shown that the discrepancy between actual results and modelling does not exceed 5–10% [4–5].

## 3 Case Studies

### *3.1 The Medeo Dam*

Among the blast-fill dams listed in Table 25.1, a dam constructed in 1966–1967 at Malaya Almaatinka River in order to protect Almaty City from debris flows is the largest and best known. This technology was chosen, since it was the only way to construct the dam in a short period between annual floods without making a bypass tunnel and coffer-dams. It was created by two pin-point explosions at both sides of the river valley with nearly symmetric slopes 300 to 500 m high with slope angles from 40° to 50°. The canyon had a 50–70 m wide riverbed and was 250–300 m wide at the design dam crest level. Charges of the first explosion, with total weight of 5,293 tons, were detonated at the right side of the valley on October 21, 1966. The second explosion, with total weight of 3,946 tons, was conducted at the left side of the valley on April 14, 1967.

Scheme of charge location and the dam shape are shown in Figs. 25.2 and 25.3. Such a scheme supported the directness of rock mass emplacement. Individual charge weights, minimal distances of charge centers from the daylight surface (W) and time delays of their detonation are listed in Table 25.2.

To optimise the directivity of blasted rock motion towards the design position and to minimise seismic effect of the explosions separate charges have been blasted with some delay (Table 25.2). It should be noted that no damage to any buildings were caused by these explosions in Almaty City. Before the explosions the river had been diverted into the left-bank tunnel.

Charges of the first explosion were placed by a two-row scheme: the first row up to 200 m long included four coyote-holes with total weight of the explosives 1,689 tons and the shortest distance from charges to the day-light surface (W) from 48 to 58 m. The second row was formed by a single charge of 3,604 tons that was 93 m long and had W=107 m. This charge was detonated 3.6 s after the first one. The first explosion (Fig. 25.4) created the 60-m high dam 560 m along the river with a volume of 1,500,000 $m^3$.

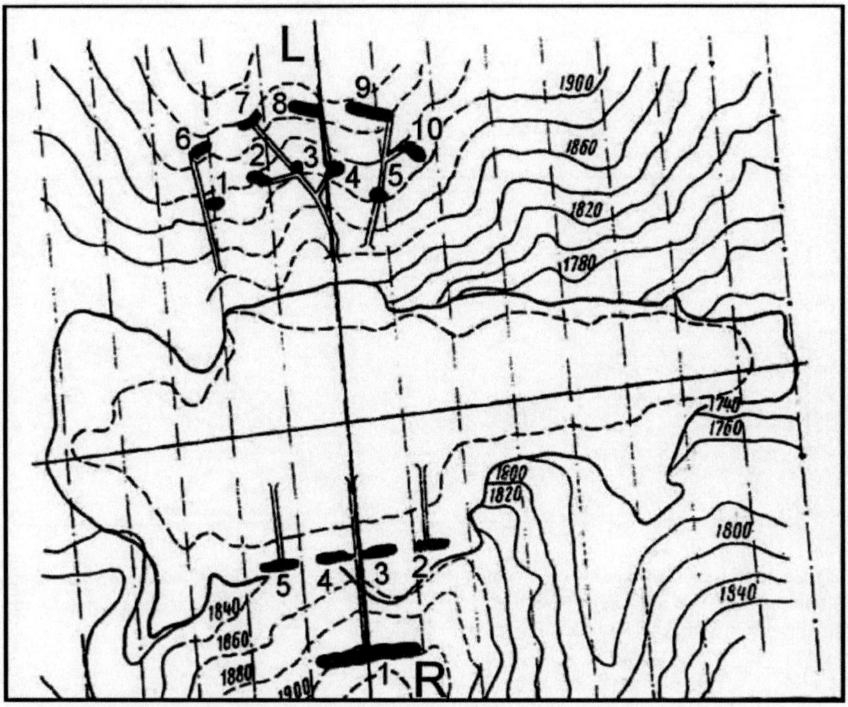


**Fig. 25.2** Scheme of the right and left bank Medeo explosions. *R, L* – Right and left banks respectively. Numbers are number of charges (see Table 25.2). *Dashed line* – contour of the dam created by 1st explosion, *solid line* – that of the second one. Downstream is to the *right*

The second (left bank) explosion was also conducted by the two-row scheme. The first row contained 5 charges with total weight of 1,121 tons and W = 36–46 m. The second row of 5 charges with total weight of 2,852 tons was detonated 2 s after the first one. Total volume of the dam was enlarged up to 3,000,000 $m^3$ and its height up to 80 m (see Fig. 25.3). The average specific consumption of explosives was 3.1 kg/$m^3$, according to the total weight of explosives used for dam creation (9,239 tons).

By use of this technology a large dam (Fig. 25.5) was constructed in a very short time. Later on the dam was enlarged by rockfill. Several years later after its construction, on July 15, 1973, the dam stopped a powerful debris flow and saved Almaty City from a catastrophic disaster. It has stopped several other debris flows since then as well (Chap. 1 by Evans et al., this volume).

### 3.2 The Baipaza Dam

On March 29, 1968 the explosion technique was used to construct the water-retaining dam of the Baipaza HPP on the Vakhsh River in Tajikistan (Figs. 25.6

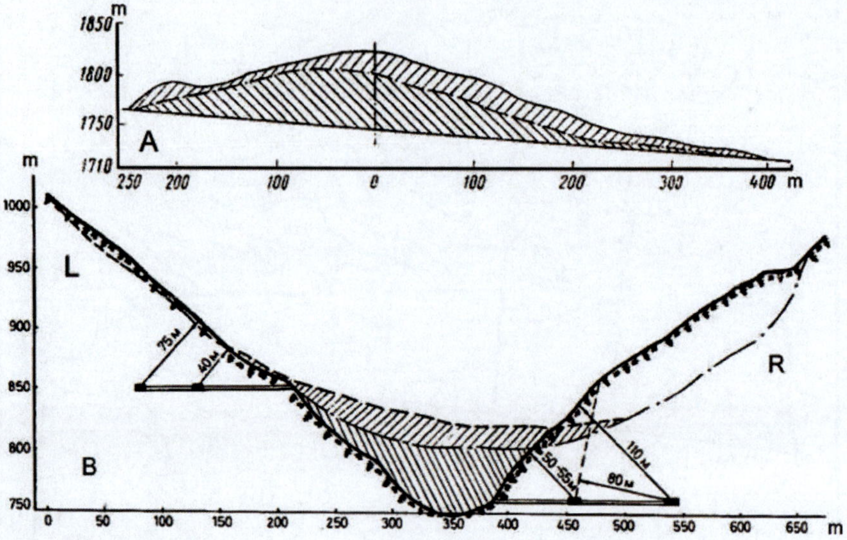


**Fig. 25.3** Longitudinal (**a**) and cross-valley (**b**) profiles of the Medeo dam. *Arrows* with numbers – minimal distance of charges from the daylight surface or from the free surface created by the previous explosion. L = left bank, R = right bank

**Table 25.2** Principal parameters of the Medeo explosions

| Right-bank explosion, October 1966 | | | | | |
|---|---|---|---|---|---|
| Charge No | 1 | 2 | 3 | 4 | 5 |
| Mass of explosives (tons) | 3,604 | 439 | 480 | 275 | 495 |
| W (m) | 85 | 48 | 53 | 48 | 58 |
| Time delay (s) | 3.6 | 0 | 0 | 0 | 0.2 |

| Left-bank explosion, April 1967 | | | | | | | | | | |
|---|---|---|---|---|---|---|---|---|---|---|
| Charge No | 1 | 2 | 3 | 4 | 5 | 6 | 7 | 8 | 9 | 10 |
| Mass of explosives (tons) | 175 | 371 | 205 | 208 | 162 | 440 | 445 | 745 | 845 | 350 |
| W (m) | 36 | 46 | 40.5 | 38 | 33.5 | 75 | 75 | 75 | 75 | 75 |
| Time delay (s) | 0 | 0 | 0 | 0 | 0 | 2.0 | 2.0 | 2.0 | 2.0 | 2.0 |

and 25.7). The specific character of this blast was determined by the presence of the spillway, the intake of the Vaksh-Yavan tunnel and the powerhouse headrace channel constructed beforehand that had to be protected from the seismic effect of the powerful explosion. All material had to be thrown from one bank, in contrast to the Medeo case.

To decrease the seismic effect, and to direct the motion of the blasted rock mass, the total amount (2,000 tons) of explosives was divided into 12 charges located in separate drifts, which were exploded one by one with 0.5–5.5 s delays (Table 25.3).

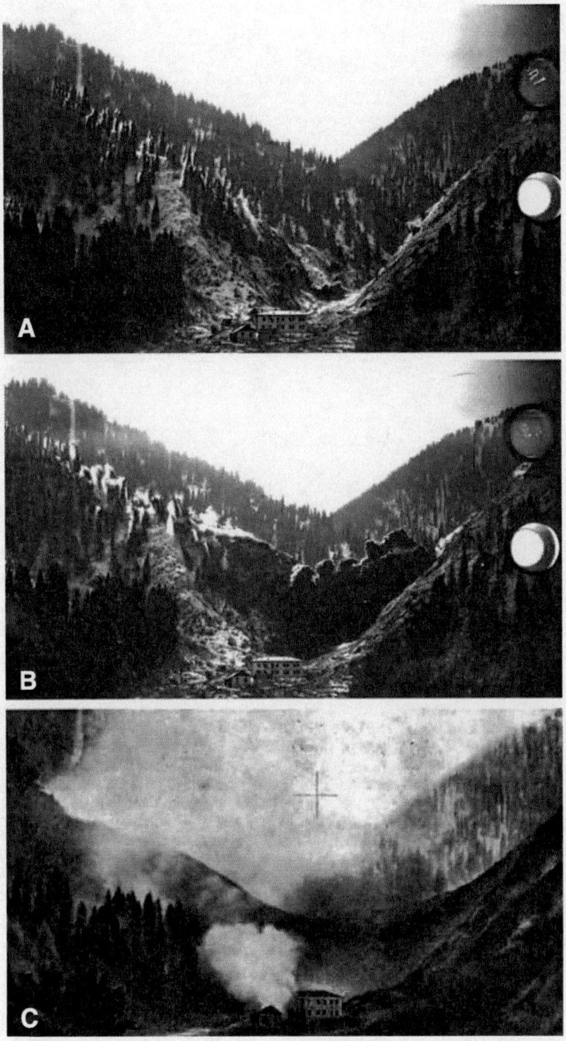


**Fig. 25.4** The first explosion at the Medeo Dam site in 1966. (**a**) – First appearance of the explosive gases at the slopes' foot; (**b**) – Slope starts collapsing; (**c**) – The dam immediately after the explosion (note that a building in front of the dam remained intact)

Charge location is shown on Fig. 25.6. Additionally a special dike had been constructed beforehand to protect the spillway and the headrace channel from boulders that could impact these surface structures.

The blasted slope was 300–350-m high, composed of hard limestone, and was very steep (60–70° approximately). Due to high steepness (see Fig. 25.6a), only part of the blastfill dam body ($8 \times 10^5$ m$^3$) was composed of the material that was exploded and emplaced in the proper position (Fig. 25.7b), while almost the same amount ($7 \times 10^5$m$^3$) just slid down after being unbuttressed under gravity force. The blasted rock mass that forms the lower part of the dam was significantly crushed by the explosion – maximal fragments did not exceed several tens of centimetres. On

**Fig. 25.5** (**a**) – Upstream view of the Medeo dam after its enlargement by the earth fill; (**b**) – Dam retaining debris flow; (**c**) – Downstream view from the high-level Medeo skating rink

the contrary, the rock mass from the upper part of the slope that slid down on top of the blasted material was composed mainly of large angular boulders up to several meters in size.

Slope failure thus increased the total volume of rock fill that formed a dam up to $1.5\times10^6 m^3$ in volume and the unit rate of the explosives was rather low – about 1.33 $kg/m^3$. The minimal dam height at the opposite bank of the river was about 50 m (see Fig. 25.6b). Water seepage through the dam after reservoir filling was very low, and did not cause any superficial or internal erosion. No damage was caused either by seismic effects of the explosion or by boulders ejected from the blasted rock mass.

## 3.3 The Akh-Su Dam

The construction of the Akh-Su dam in Dagestan (northern Caucasus, Russia) at a narrow canyon with very steep slopes (from 60 to 80°) eroded in hard, bedded limestone, was the first project tested using the 1:500 scale model in the vacuum chamber

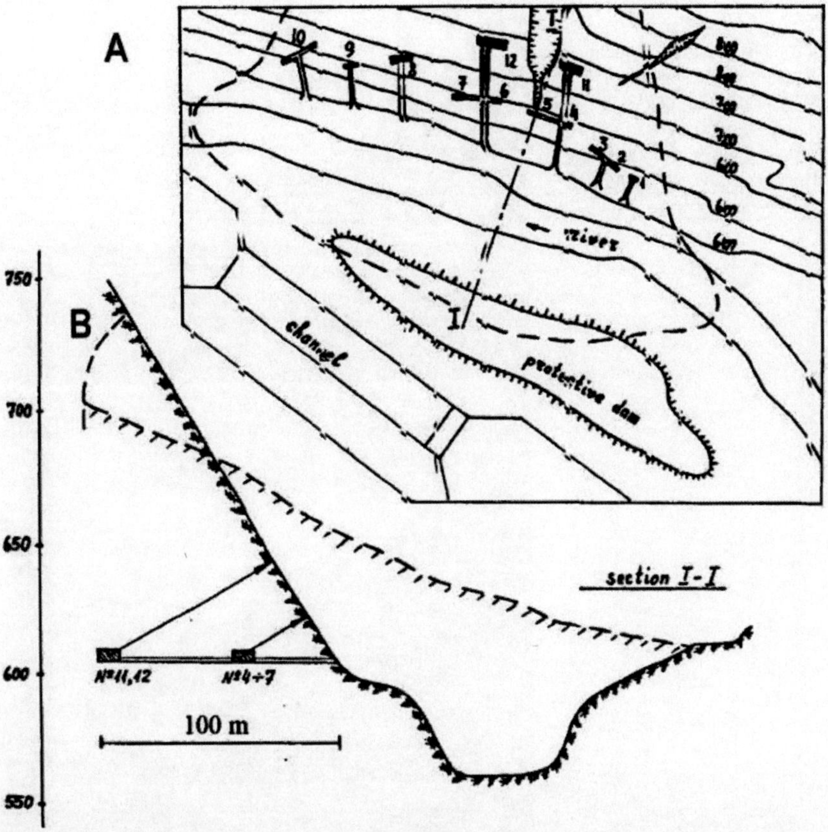


Fig. 25.6 (a) – Scheme of the Baipaza blast-fill dam site and charges location; (b) – Cross-section along the dam axis. Arrows = W – minimal distance of charges from the daylight surface

described above (Fig. 25.1). Modeling had been performed when charge chambers (Figs. 25.8 and 25.9) had been already driven. Different variants of charge masses, of the detonation order and of delay time had been tested to find the optimal one.

After modeling it was recommended to change the weight of some charges and time delays [6]. The dam (Fig. 25.10) was created by blasting of 565 tons of explosives divided in 9 charges (Table 25.4). Basic parameters of the modelled dam created according to the actual explosion scheme (height – 80 m, width – 290 m and volume – $2.7\times10^5$ m$^3$) appeared to be almost the same as of the real structure: height – 84 m, width – 325 m and volume – $2.5\times10^5$ m$^3$.

### 3.4 The Alinjachai Dam

This blast-fill dam was constructed in Azerbaijan in a river canyon at an elevation of 1,300 m a.s.l [2]. The blasted rock mass was composed of diorite porphyry with

**Fig. 25.7** The Baipasa explosion. **A** – The morphology and geological structure of the site; **B** – The dome of blasted rock mass visible above the erupting explosive gases; **C** – Blast-fill dam and spillway that did not undergo any damage

$\rho = 2.4$ g/cm$^3$. Only the right bank of the canyon 130 m high with steepness of about 70°–80° had to be exploded. Water discharge in the river at the period of dam construction was 0.35 m$^3$/s. The 1:750 scale model simulating actual topography had been prepared to test this explosion in the vacuum chamber.

The amount of explosives that could be used in this case was limited in order to reduce the seismic effect at the township located nearby, just 600 m distant from the

**Table 25.3** Principal parameters of the Baipaza explosion

| Charge No | 1 | 2 | 3 | 4 | 5 | 6 | 7 | 8 | 9 | 10 | 11 | 12 |
|---|---|---|---|---|---|---|---|---|---|---|---|---|
| Mass of explosives (tons) | 5.4 | 8.6 | 25 | 57.9 | 62.7 | 61.7 | 31.4 | 289 | 63 | 65 | 542 | 642 |
| W (m) | 19 | 17.5 | 24 | 34 | 33.5 | 35 | 230 | 52 | 32.5 | 32 | 60 | 60 |
| Time delay (s.) | 0 | 0 | 0 | 0 | 0 | 0 | 0 | 1.94 | 4.0 | 5.49 | 0.25 | 0.38 |

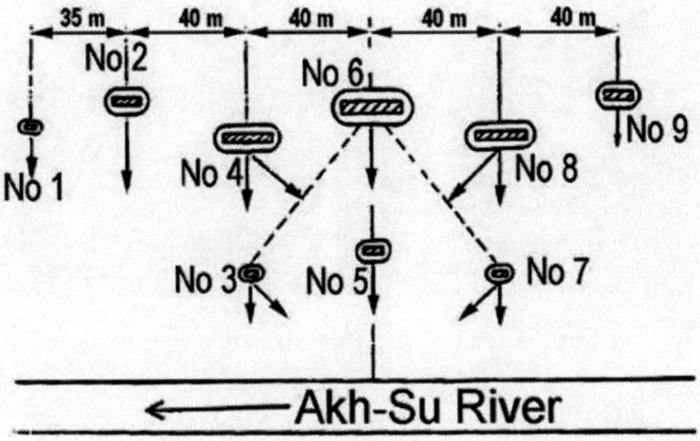


**Fig. 25.8** Scheme of the Akh-Su explosion. Charge numbers correspond to Table 25.4. Arrows show the directivity of each charge explosion with due regard to their detonation consequence

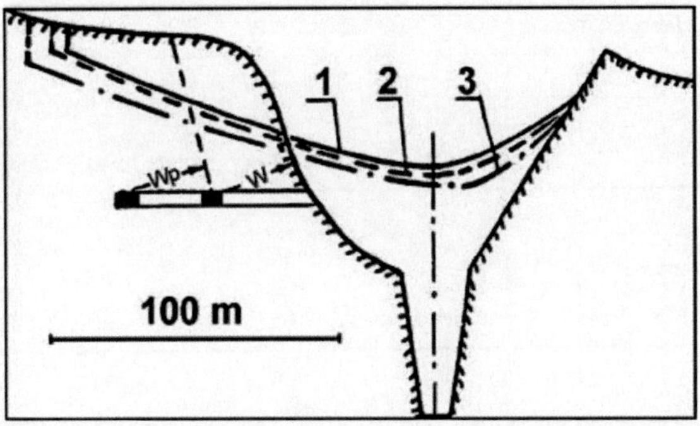


**Fig. 25.9** Axial cross-section of the Akh-Su dam. *1* – actual dams' profile, *2* – profile according to the modeling, *3* – design profile, W – minimal distance of the charge No 5 from the daylight surface, Wp – minimal distance of charge No 6 from the free surface created by the previous explosion

site. For this reason the designed explosion scheme envisaged undercutting of the steep slope and its failure due to gravitational collapse. Based on modeling results the explosion scheme had been calculated. It included three linear charges with an entire amount of explosives of 689 tons (Fig. 25.11). Charge parameters are listed in Table 25.5.

The explosion was detonated on September 4, 1984. It formed a compact 32-m high and 195 m wide (along the riverbed) dam that involved $6.1\times10^5$ $m^3$ of blasted rocks (Fig. 25.12). The specific explosive consumption was the lowest for this type of blockage – only 1.13 $kg/m^3$ [2].

**Fig. 25.10** The Ak-Su River Canyon before (**a**) and after (**b**) the explosion

**Table 25.4** Principal parameters of the Akh-Su explosion

| Charge No | 1 | 2 | 3 | 4 | 5 | 6 | 7 | 8 | 9 |
|---|---|---|---|---|---|---|---|---|---|
| Mass of explosives (tons) | 25.2 | 34.2 | 18.8 | 76 | 40.5 | 163.1 | 13 | 101 | 93.2 |
| W (m) | 20 | 24 | 20 | 30 | 25.6 | 35 | 17 | 33 | 33 |
| Time delay (s) | 1.2 | 1.2 | 0.2 | 0.45 | 0 | 0.2 | 0.2 | 0.45 | 1.2 |

**Table 25.5** Principal parameters of the Alindjachai explosion

| Charge No | Mass of explosives (tons) | Charge length (m) | W (m) | Mass of explosives per charge length unit (t/m) | Time delay (s) |
|---|---|---|---|---|---|
| 1 | 371 | 50 | 60 | 7.4 | 0 |
| 2 | 123 | 23 | 58 | 5.4 | 0 |
| 3 | 195 | 30 | 55 | 6.5 | 0.75 |

## *3.5 The Quaisa Dam*

In the same year (1984) the waste disposal (tailings) dam of the Quisa lead-zinc plant in South Osetia was constructed by the blasting of 437 tons of explosives placed as two linear charges in the left bank of the Bubikamidon River valley (Fig. 25.13). The first charge of 84 tons was 36 m long and the second one of 353 tons was 93 m long. Second charge was detonated 0.25 s after the first one. The charges were used to accumulate blasted rock mass in the central part of the

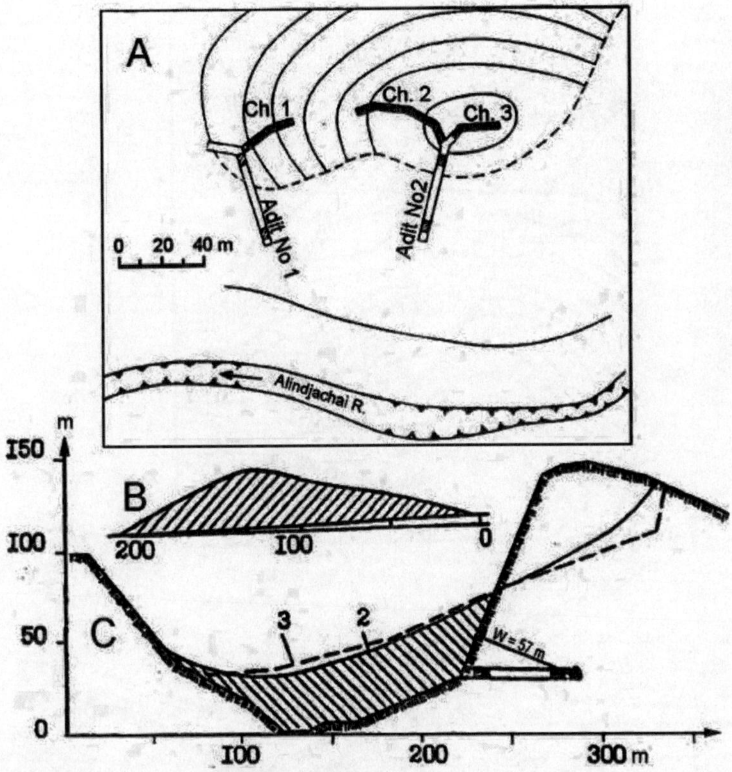


Fig. 25.11 Scheme of the Alinjachai explosion. (a) – Charge location at the right bank of the Alinjachai river valley; (b) – Along-river and (c) – Axial cross-sections. Hatched area – actual blockage, *dashed line* – dams' profile according to modeling result

Fig. 25.12 Alinjachai blast-fill dam site before (a) and after (b) the explosion

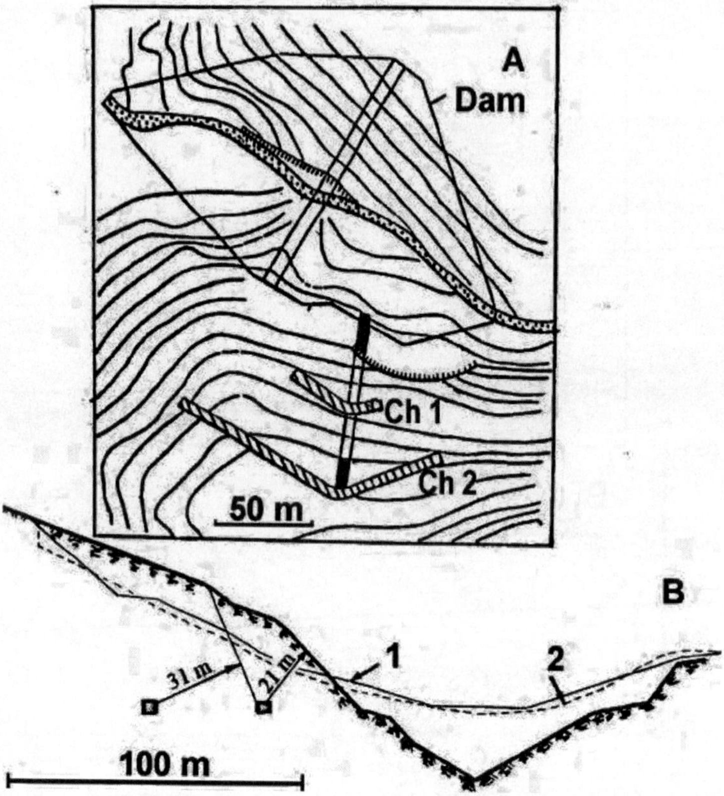


**Fig. 25.13** (**a**) – Scheme of the Quaisa explosion. Ch1 & Ch2 – First and second charges. (**b**) – An axial cross-section. *1* – Actual dam profile, *2* – Predicted by the model. *Arrows* with numbers show the directivity of the exploding charges

dam. This scheme of detonation had been developed by modeling and the dam's post-explosion parameters (26 m high, 250 m along the riverbed and $2 \times 10^5$ $m^3$ in volume) corresponded to the predicted values.

### *3.6 The Kambarata Dam Project and Experimental Burlykia and Uch-Terek Blast-Fill Dams*

In the 1970s and 1980s of the past century several hydraulic projects with blast-fill dams were designed in the middle part of the Naryn River basin in Kyrgyzstan upstream from the Toktogul reservoir. Special emphasis had been devoted to the Kambarata-1 project that included the 270 m high blast-fill dam of 200 millions $m^3$ in volume, that was to be constructed by the blasting of 200–250 thousand tons of explosives. No similar project had been proposed worldwide. Two experimental

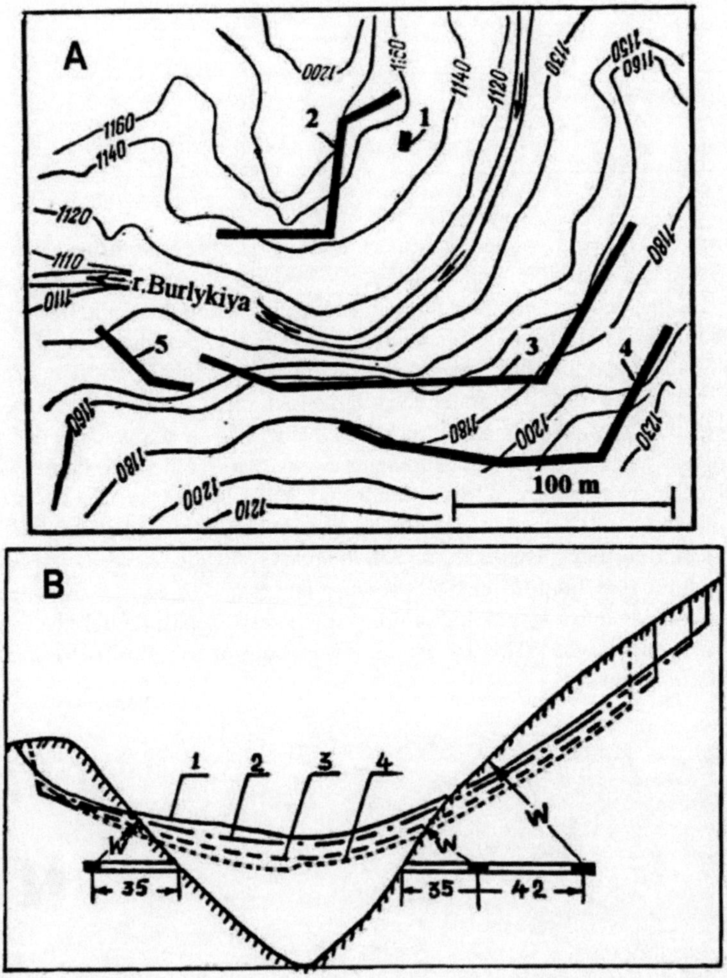


**Fig. 25.14** (**a**) – Scheme of the Burlykia explosion. 1–5 – Charges numbers. (**b**) – An axial cross-section. *1* – Actual dam profile, *2* – Predicted by the model that consider gorge bend, *3* – Predicted by simplified model where gorge bend was not considered, *4* – Predicted for the 500-ton explosion. *Arrows* with "W" show minimal distance from charges to the daylight surface that corresponds to the explosions directivity

blasts were carried out at the Burlykia River on February 8, 1975 and at the Uch-Terek River on June 11, 1989.

Charges of the Burlykia blast had been placed on both sides of the river canyon (Fig. 25.14) to investigate not only the dam formation process, but also the seismic effect of powerful explosion on the nearby underground structures.

Charge parameters are listed in Table 25.6. 703 tons of grained granulite used for this blast corresponds to the 752 tons of trotyl (it has larger specific energy

**Table 25.6** Principal parameters of the Burlykia experimental explosion

| Charge no | 1 | 2 | 3 | 4 | 5 |
|---|---|---|---|---|---|
| Mass of explosives (tons) | 2.3 | 88 | 125 | 470 | 18 |
| W (m) | 10 | 17÷22 | 10÷25 | 40÷55 | 7÷15 |
| Time delay (ms) | 0 | 35 | 162 | 190 | 280 |

than trotyl). Thorough modeling at a scale 1:500 had been performed before this explosion and also after its implementation to evaluate the role of complex topography and actual detonation time delays on dam construction. It should be noted that, though this blast had been designed as an excavation explosion, the upper part of the left-bank slope just slid down. Cracks had formed about 70 m higher than the upper limit of the explosive crater and a large rock mass, about 120-m wide and 7 to 12-m thick, formed the induced rockslide that slid down at a velocity of 3–5 m/s (Figs. 25.15 and 25.16). Smaller sliding also took place at the right bank. Volumes of slid rock masses estimated by surveying and rapid filming was $1.5\times10^5$ m$^3$ at the left bank and $5\times10^4$ m$^3$ at the right bank. Totally they provided about 35% of the entire dam body that was 50 m high, 320 m wide (along the riverbed) and $5.4\times10^5$ m$^3$ in volume (see Table 25.1). The river canyon has a 120° bend at the site of detonation and an almost symmetric profile with 50–60° slopes, though the left-bank slope was much higher. This morphology was more or less similar to that of the Kambarata-1 dam site.

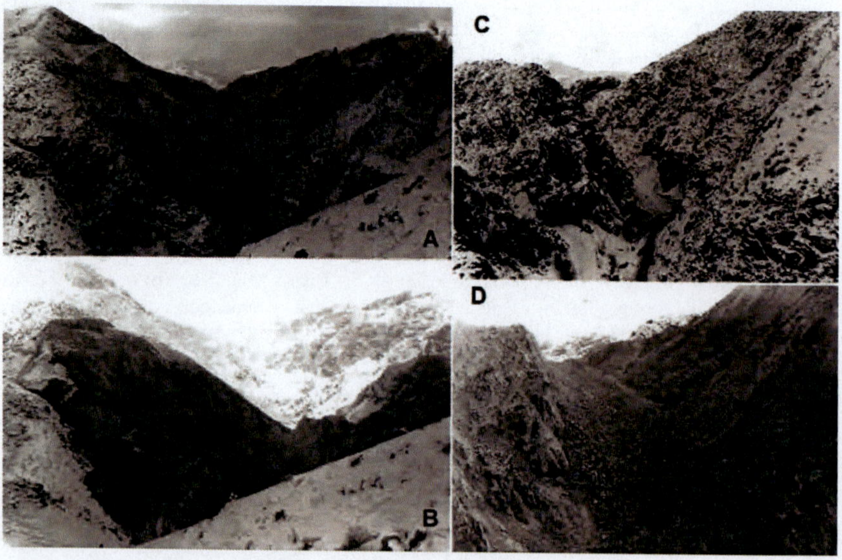


**Fig. 25.15** The Burlykia experimental blast-fill dam site. *Left* – the upstream views before (**a**) and after (**b**) the explosion. *Right* – the downstream views before (**c**) and (**d**) the explosion

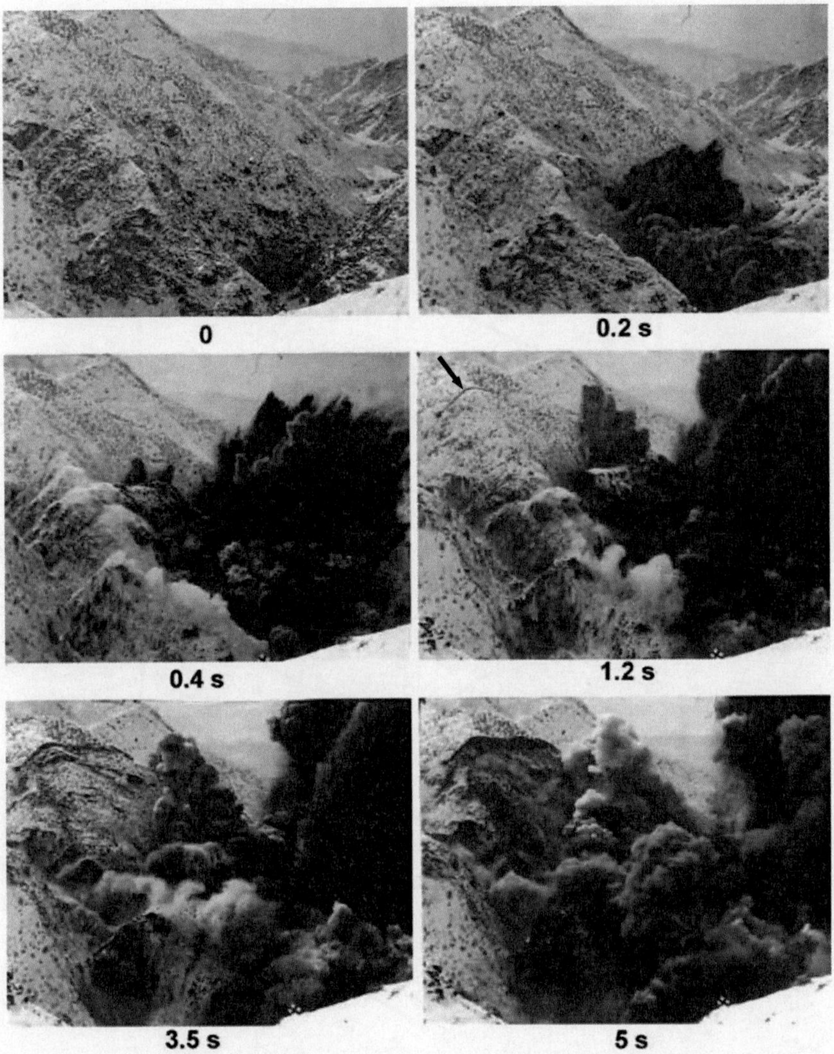


**Fig. 25.16** Development of the Burlykia explosion (view from upstream). Numbers indicate time after start of detonation in seconds. Note first appearance of the gravitational crack 1.2 s after first blast (marked by *arrow*) and further grouth of the collapse

It should be pointed out that consideration of actual conditions of the explosion resulted in good coincidence of the predicted and real dam profiles. The above mentioned gorge curvature also supported concentration of the blasted rock mass and formation of a high blockage.

A large amount of post-explosion field investigation of the dam body was carried out. Main results of these studies are described by Korchevskyi et al. (Chap. 26, this volume).

Though all above mentioned dams had been constructed by pin-point blasting, in many cases there was a significant role of induced gravitational slope failure (i.e. the

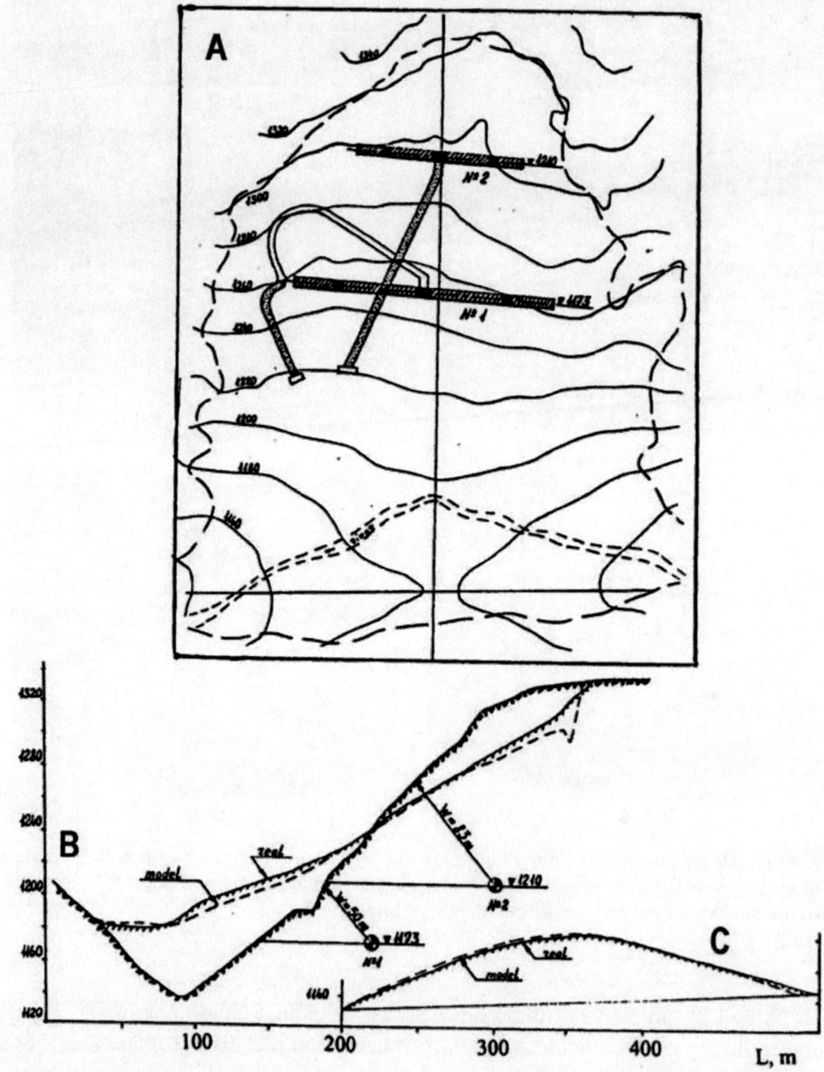


**Fig. 25.17** Scheme of the Uch-Terek explosion. **(a)** – Charge location at the right bank of the Uc-Terek River canyon; **(b)** – Axial and **c** along-river – cross-sections. *Solid line* – actual blockage profile, *dashed line* – dams' profile according to modeling result

Baipaza and Burlykia dams). This mechanism was considered as the main one for the construction of the Kambarata-1 dam [7], the volume of which had to be more than 2 orders of magnitude bigger than of any blast-fill dam constructed up to now. The gigantic explosion had been designed and modeled to form a sliding surface and to cause an artificial rockslide [10] from the left bank of the almost 1-km deep gorge of the Naryn River.

To test the possibility of the realisation of such a mechanism, the explosion of about 2,000 tons of the crude dispersive igdanite (94.5% saltpeter with 5.5% of diesel oil) was performed on June 11, 1989 not far from the Kambarata dam site in the narrow gorge of the Uch-Terek River (Figs. 25.17 and 25.18). Explosives had been placed in two linear charges, the parameters of which were determined by laboratory modeling as listed in Table 25.7. The rock mass is composed of thin-bedded sandstone and siltstone steeply dipping towards the stream. The 200-m deep river canyon had steep (50°) slopes (see Fig. 25.18) and a narrow riverbed 3–6 m wide.

Detonation of the lower charge threw blasted rock into the canyon and formed the cavity through which the rock mass fractured by the detonation of the upper charge could slide down and accelerate. Such a detonation scheme, when the initial pin-point blast undercut the required rock mass and the second one disintegrates it, supports rock-avalanche-like motion and formation of the dam body of the designed size and shape. At the Uch-Terek site, blasting caused a significant slope failure (Fig. 25.19) and formed a dam 45-m high, 295-m wide (along the riverbed) and $8.7\times10^5$ m$^3$ in volume (see Fig. 25.18b).

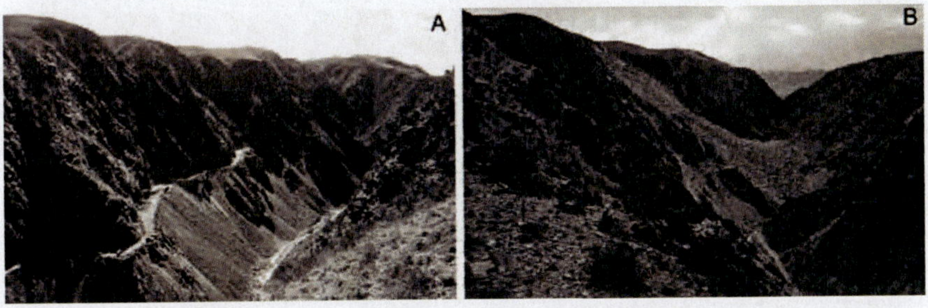


**Fig. 25.18** The Uch-Terek Canyon before (**a**) and after (**b**) the explosion

**Table 25.7** Principal parameters of the Uch-Terek explosion

| Charge No | Mass of explosives (tons) | Charge length (m) | W (m) | Mass of explosives per charge length unit (t/m) |
|---|---|---|---|---|
| 1 | 1,088 | 145 | 50–65 | 7.5 |
| 2 | 827 | 95 | 82–95 | 8.7 |

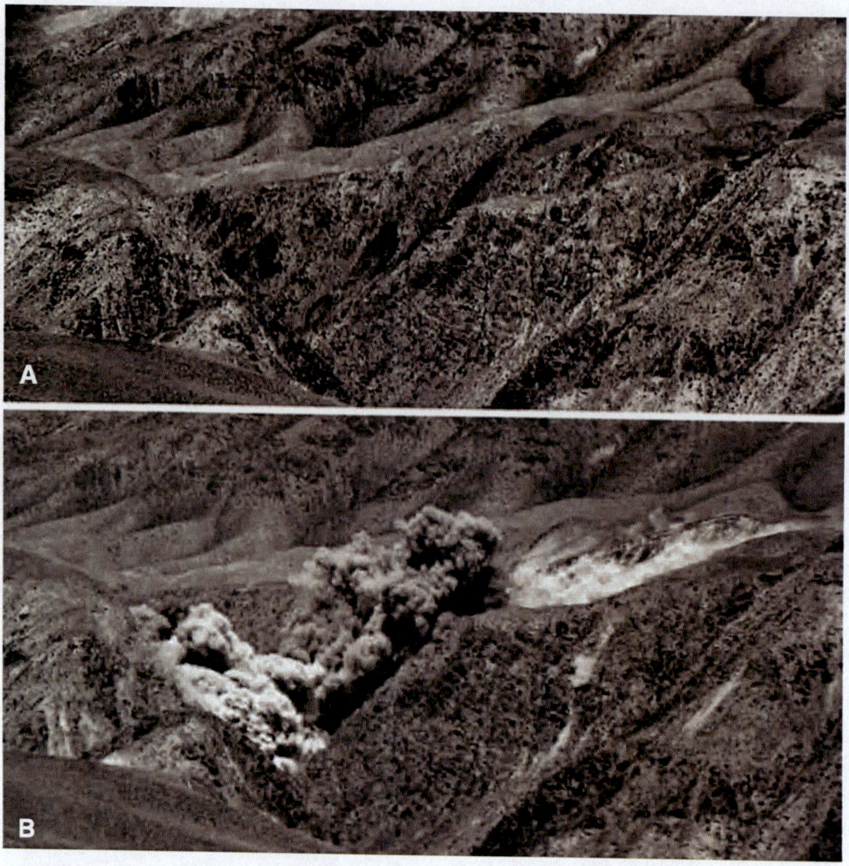


**Fig. 25.19** The Uch-Terek experimental blast. **(a)** – Test site immediately before the explosion; **(b)** – Formation of the artificial rockslide at the upper part of the exploded slope

The specific consumption of explosives was 2.2 kg/m$^3$ which is relatively low. Comparison of the actual and scaled model sites shows good coincidence (within 5% accuracy) and proves the effectiveness of the proposed technology.

The scheme of detonation of three groups of charges at the Kambarata-1 dam site (Fig. 25.20) that should cause slope collapse has been designed on the basis of the Burkykia and Uch-Terek experimental explosions (Fig. 25.21).

The total amount of explosives should be 217.5 kilotons. To prevent spreading of debris along the river valley groups of charges No 1 and 3 should be detonated first and the main group No 2 later. In its turn, to reduce the seismic effects, each charge within these groups should be detonated successively from the lower one to the upper one. The resulting dam is designed to be 270-m high, 2,200-m wide, and about 200 Mm$^3$ in volume. However, after the breakdown of the USSR this project has not been completed as yet. However, construction of Kambarata-2 on the Naryn

**Fig. 25.20** The Kambarata-1 dam site. *Dashed line* corresponds to the designed dam crest; *arrow* shows the direction of rock sliding

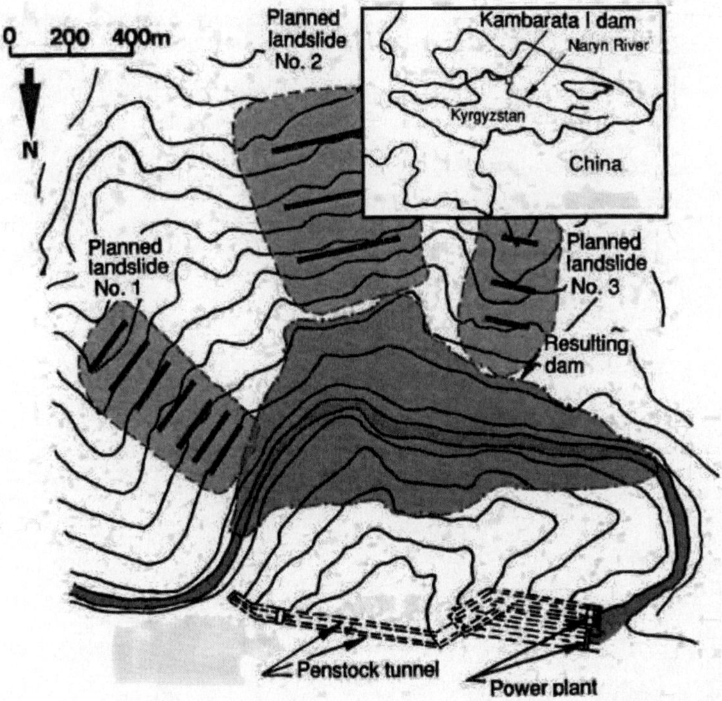


**Fig. 25.21** The scheme of the Kambarata-1 explosion designed according to modeling results

River was initiated in 2009 (see Chap. 1 by Evans et al., this volume) and the first electricity was generated in late 2010.

## References


1. Adushkin, V.V., and Pernik, L.M. (1977) Model study of test dam creation on Burlykiya river, *Hydrotechnical Construction* **5**, 36–38 (in Russian).
2. Adushkin, V.V., Basiev, E.U., Zykov, Yu.N., Pernik, L.M. and Fedorov, A.E. (1986) Construction of rock-thrown dam on River. Alinjachai by collapsing explosion, *Physical and Technical Problems of Deposit Extraction* **2**, 38–44 (in Russian).
3. Adushkin, V.V., Libin, V.Ya. and Pemik, L.M. (1982) Analogue apparatus for study group excavation explosions, *Vzryvnoye delo (Explosion works)* **83/40**, 36–44. Moscow, Nedra (in Russian).
4. Adushkin, V.V., Libin, V.Ya. and Pemik, L.M. (1983) On possibility of large-scale explosion modeling, *Mining Journal* **9**, 38–40 (in Russian).
5. Adushkin, V.V. and Pemik, L.M. (1980) Modeling directed explosions for rock-thrown dams, *Vzryvnoye delo (Explosion works)* **82/39**, 5–17. Moscow, Nedra (in Russian).
6. Adushkin, V.V. and Pemik, L.M. (1973) Prognosis of directed explosion for construction of a dam on the Akh-Su river, *Special works in industrial construction* **2**(80), 7–11. CBNTI Minmontazspecstroi, Moscow (in Russian).
7. Adushkin, V.V., Pemik, L.M. and Zykov, Yu.N. (1993) Modeling of explosion-triggered rock slides for construction of the Kambarata I hydroelectric power dam, Republic of Kyrgyzstan, *Landslide News* **7**, 7–9.
8. Adushkin, V.V., Zykov, Yu.N. and Pemik, L.M. (1991) Specific features of explosion evolution during creation of landslide-type blockage, *Physical and Technical Problems of Deposit Extraction* **2**, 56–59 (in Russian).
9. Rodionov, V.N., Adushkin, V.V., Kostyuchenko, V.N., Nikolaevskyi, V.N., Romashov, A.N. and Tsvetkov, V.M. (1971) *Mechanical Effect of Underground Explosion*. Nedra, Moscow, 224 p (in Russian).
10. Sadovsky, M.A. and Adushkin, V.V. (1985) Dams created by explosion, *Priroda (Nature)* **11**, 3–13 (in Russian).

# Chapter 26
# Utilisation of Data Derived from Large-Scale Experiments and Study of Natural Blockages for Blast-Fill Dam Design

**V.F. Korchevskiy, A.V. Kolichko, A.L. Strom, L.M. Pernik, and K.E. Abdrakhmatov**

## 1 Introduction

Construction of blast-fill dams is a promising though risky technology [5, 21]. This method allows constructing large dams with high safety factor and at acceptable cost at those sites, which geological and seismological conditions complicate or even prevent construction of dams of other types. In remote mountainous areas use of large-scale explosion that combines all operations of constructional material excavation, transportation, and compaction in one process, can decrease the construction time and cost, required efforts, cargo transportation, and also the need for expensive materials and powerful machinery. It is especially efficient with the increase of dam height and volume. At the same time, while the traditional construction methods of both earth-fill and concrete dams, when the structure is being built step by step, allow improving or changing, if necessary, either the material or technology or even the design, the blast-fill dam would be constructed "in one gulp". That is why any possible complexities must be considered beforehand.

Further elaboration of dam construction technology by use of large explosions require extensive field studies and modeling to validate possibility of using a bulk of blasted rock as an engineering structure for which geometric parameters and properties can be predicted beforehand. It relates especially to the technology based on the creation of artificial rock avalanches when explosion is used as a trigger while the main work is done by gravity force. To simulate emplacement of an artificial rockslide, either numerically or physically (in the laboratory), we must know the physical processes that govern its motion, which are not fully understood yet. Modeling in the vacuum chamber described by Adushkin (see Chap. 25, this volume) have been performed to simulate the initial stage of the process, i.e. blasting and initial throw of the exploded rock mass, but not the subsequent runout of a large

A.L. Strom (✉)
Institute of Geospheres Dynamics, Russian Academy of Sciences, Moscow, Russia
e-mail: a.strom@g23.relcom.ru

S.G. Evans et al. (eds.), *Natural and Artificial Rockslide Dams*, Lecture Notes in Earth Sciences 133, DOI 10.1007/978-3-642-04764-0_26, © Springer-Verlag Berlin Heidelberg 2011

amount of debris. Moreover, modeling usually simplifies the real process, while some aspects, that either seem to be negligible or can be hardly simulated, may have a significant effect on the specific conditions of a particular dam site. This is why large-scale field experiments and study of natural rock slope failures of the same size and shape as the designed dam should be considered as the most informative method providing important and reliable information for the blast-fill dam designers.

It should be noted that the pin-point blasting that was used practically for all existing blast-fill dam construction is effective for rather small structures only – generally up to 60–80 m high. For larger dams the amount of explosives increases at a rate of 3.5–4.0 in proportion to the crater's depth that makes such explosion inexpedient both by the explosives cost and by its seismic effect on the structures. Construction of very large blast-fill dams is possible, however, on the basis of the approach proposed by Sadovsky [33], according to which the explosion is considered just as a trigger that releases the potential energy of rocks resting high above a valley bottom. Their further rock avalanche-like emplacement is governed by gravity force only. Such structures could be named "blast-rockslide" dams.

Besides the study of several blast-fill dams and embankments constructed for various purposes (the Medeo debris flow storage dam in Kazakhstan, the Baipaza HPP dam in Tajikistan, the Akh-Su embankment in Dagestan [3]) the most detail investigations in the former USSR were carried out at the Burlykia experimental dam (Fig. 26.1) [3, 25] and, to a lesser extent, at the Uch-Terek experimental dam (Fig. 26.2). Methods of blast-fill dam geotechnical and filtration-piping characteristics prediction have been developed on the basis of these investigations. However, such experiments have some limitations since, due to economical reasons, experimental dams can not be of the same order of size as the largest designed blast-fill dams such as 270-m high dam of the Kambarata-1 Project [2, 24]. This is why

**Fig. 26.1** The Burlykia experimental blast-fill dam. A – adit at the base of the right slope that was made to investigate seismic effect of the explosion on the underground structures. PS – permanent spillway. View from downstream

**Fig. 26.2** The Uch-Terek experimental dam immediately after its creation. View from downstream

the comprehensive study of their natural analogues – blockages formed by rock-slides and rock avalanches has also been carried out. Being of the same order of size as any proposed man-made structure, and even larger, these natural blockages allow a consideration of scale effect, which can not be fully modeled even in large-scale natural experiments.

In this paper we describe the main results of the experimental dams investigation along with several phenomena typical of natural blockages, which should be taken into consideration for large blast-fill dam design.

## 2 Experimental Dams

The experimental Burlykia and Uch-Terek dams were constructed by powerful explosions (Table 26.1) on small tributaries of the Naryn River [3]. They have been studied by repeated geodetic, hydrological and hydrogeological measurements and by excavation of several deep shafts, which were used to analyze grain-size composition, density and filtration of the body of the dams.

### *2.1 Composition and Density of Blast-Fill Dam Material*

It is well known that characteristics of blast-fill dams, such as their filtration properties and deformability, strongly depend on the grain-size composition of the material and on its density [4, 25]. Investigations performed at the Burlykia and Uch-Terek dams showed that grain-size composition of blast-fill dam material results from rock

**Table 26.1** Main Parameters of the Burlykia and Uch-Terek experimental dams

| No | Location | Year of the explosion | Rock type | Total amount of explosives (tons) | Dam height (m) | Dam width along the stream (m) | Dam volume ($m^3$) | Unit rate of the explosives ($kg/m^3$) |
|---|---|---|---|---|---|---|---|---|
| 6 | Burlykia River (Kyrgyzstan) | 1975 | Granite | 703 | 50 | 320 | $5.4 \times 10^5$ | 1.3 |
| 9 | Uch-Terek River (Kyrgyzstan) | 1989 | Sandstone and siltstone | 1,915 | 45 | 295 | $8.7 \times 10^5$ | 2.2 |

crushing both by the explosion and by its compression during the collapse. Initially it depends, mainly, on the natural fracturing of the rock mass (Fig. 26.3). Generally, it can be predicted with high reliability [20]. But resultant grain-size distribution depends to a larger extent on the compression forces, which cause relative motion of rock fragments, their additional crushing and formation of fines that fill the skeleton pores. Role of this factor increases with the increase of dam size that will be exemplified below when we will describe the internal structure of large natural rockslide dams. Large-scale laboratory compression experiments show that a homogeneous rock mass undergoes very intensive crushing, while mixtures that contain enough fines to fill pores practically retain initial grain-size distribution. It was also found that debris crushing depends mainly on the specific compression behavior measured in $N/m^2$ rather than on the momentum (Fig. 26.4) [25]. Thus, processes of debris

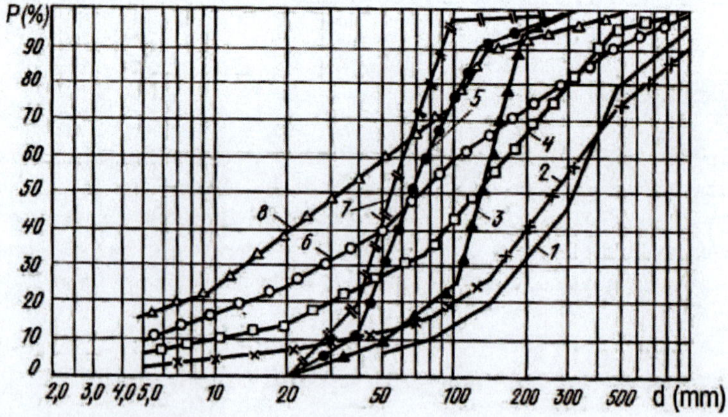


**Fig. 26.3** Grain-size composition of the host rock mass block structure (*1, 3, 5, 7*) and of the corresponding blasted rock masses (*2, 4, 6, 8*). *1, 2* – limestone of the Papan dam site; *3, 4* – porphyry of the Bekabad borrow pit; *5, 6* – granite of the Burlykia dam site; *7, 8* – granite of the Turaigyr area (after [25])

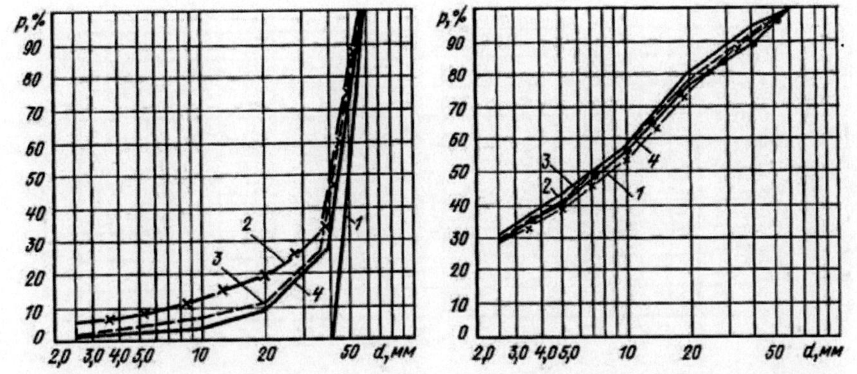

**Fig. 26.4** Dependence of grain-size composition of crushed rock on the specific compression behavior (A – *left*) and on momentum M with constant $A = 1.57 \times 10^6$ N/m$^2$ (*right*). *1* – initial grain-size composition on both plots. *Left*: *2* – $A = 7.6 \times 10^5$ N/m$^2$, *3* – $A = 5 \times 10^5$ N/m$^2$, *4* – $A = 1 \times 10^5$ N/m$^2$; *Right*: *2* – $M = 200$ N × s, *3* – $M = 160$ N × s, *4* – $M = 100$ N × s (after [25])

crushing and compaction seem to be self-regulating; they attenuate as grain-size composition reaches its optimum.

Direct measurements of compressed rock debris density carried out in deep shafts excavated through the Burlykia dam body (Fig. 26.5) show that it exceeds that of traditionally constructed embankment dams, in which compaction has been performed artificially by use of special heavy machines. Though experimental data show rather significant scatter, density gradually increases with depth (Fig. 26.6) assuming that the designed dams, which should be much higher than experimental

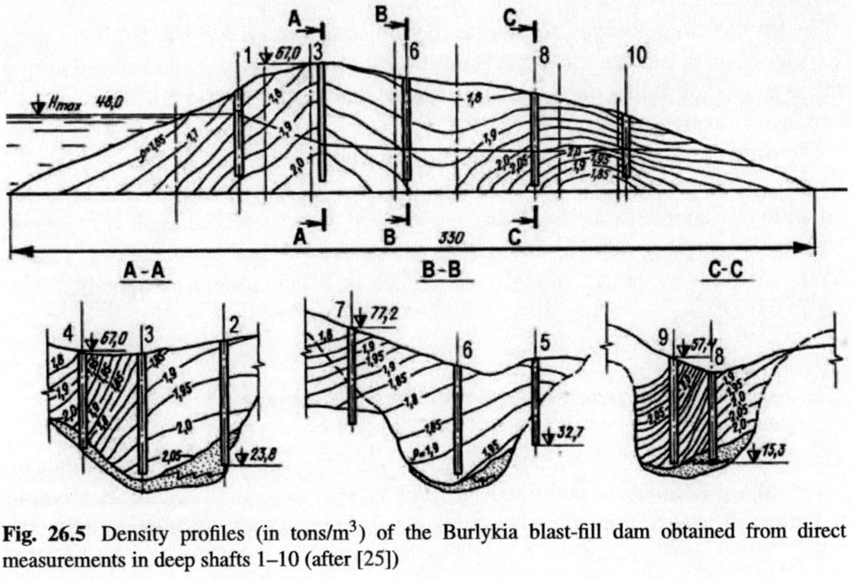

**Fig. 26.5** Density profiles (in tons/m$^3$) of the Burlykia blast-fill dam obtained from direct measurements in deep shafts 1–10 (after [25])

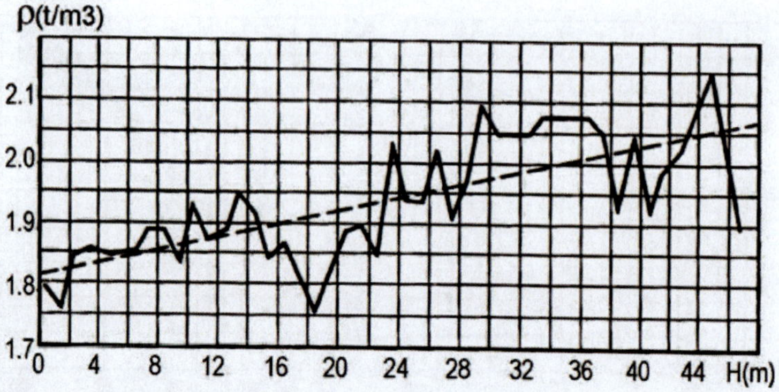


**Fig. 26.6** The Burlykia dam material density ($\rho$) versus depth (H). *Solid line* – data obtained from field measurements, *dashed line* – their linear approximation (after [25], modified)

ones, would have a compact internal "core". It should be noted that the same phenomenon has been observed at large natural blockages deeply incised by river erosion [1, 37] (Chap. 19 by Dunning and Armitage, this volume).

High initial density of the main part of large blast-fill dams causes their relatively low deformability. It can also result from shearing of rock fragments by compression that increases the skeleton strength and, thus, decreases dam deformability. According to on-site observations carried out at the 82-m high Nanshui blast-fill dam in China, dam body density increases with time [29]. Correspondingly its crest settles. During the first year this settling was up to 70–80% of the overall 5-years settlement that was 1.7% of the dam height [5]. Similar processes occur at embankment dams where compaction was performed by use of the "traditional" technologies. Relative settlement of such structures is up to 0.8–1.9% of their height and this process lasts for 9–11 years [8, 28, 42].

A similar effect, though for shorter period, was observed at the Burlykia dam. Repeated measurements show that significant settlement of its crest occurred during the first reservoir filling only, while the second one practically did not affect gradual and insignificant compaction (Fig. 26.7).

Thus monitoring data on dam deformation indicate that this process at blast-fill dams does not exceed that at embankment dams [5]. The peculiar feature of the "traditional" dam deformation is the presence of a comparable lateral component, which is, most likely, due to their rather narrow profile. Since most blast-fill dams have much more "spread" along-river profiles their lateral deformation should be negligible.

## 2.2 Filtration Characteristics of Blast-Fill Dams

The critical parameter of water-retaining dams is their permeability. Lack of data on the ability of fragmented homogeneous rock masses to retain a high water level in the reservoir is one of the main impediments of utilizing this construction method.

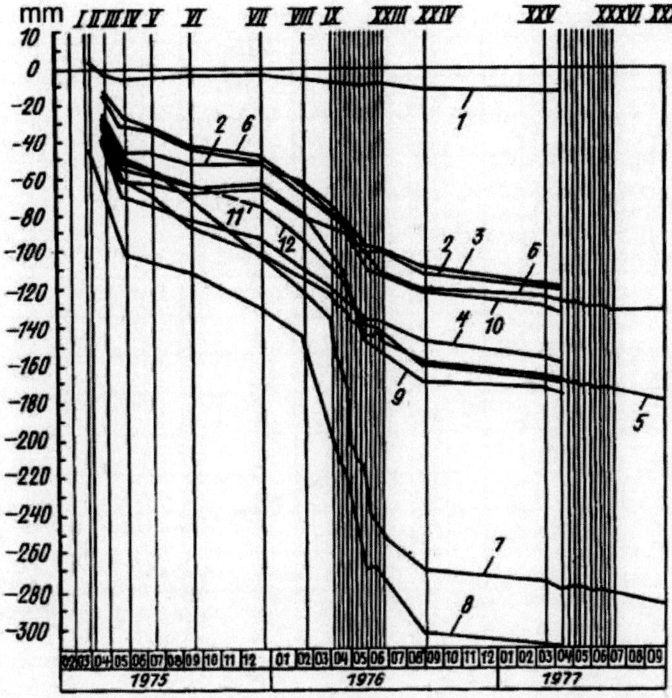


**Fig. 26.7** Set of the benchmarks on the Burlykia dam with time. *1–12* – benchmarks numbers, Roman figures – observational cycles, which were much more frequent during flood periods (after [25])

Crushed rock mass permeability is determined by its density and by content of fines (particles smaller than 5 mm) which, in their turn, determine pores' dimensions. Filtration measurements performed during initial impoundment of the Burlykia reservoir show that some transfer of fines occurred within the dam that led to the formation of reverse filters and to decrease of seepage (Fig. 26.8).

The fundamentally new concept in blast-fill dam construction is the rejection of the effort to ensure "absolute" impermeability of the dam body, which is traditional in hydraulic engineering. This requirement is not important for blast-fill dams constructed for debris flow protection or for irrigation (providing filtration does not exceed water use).

But for hydraulic projects designed for electricity production, filtration discharge through the dam leads to decrease in productivity. Therefore their construction is expedient on the high discharge rivers where relative loss of energy production due to filtration is small and does not exceed several percents, while more than 90% of the river power budget can be utilized for minimal cost. In such cases the decrease of energy loss by the increase in dam impermeability, by its pore space clogging for example, should be economically effective.

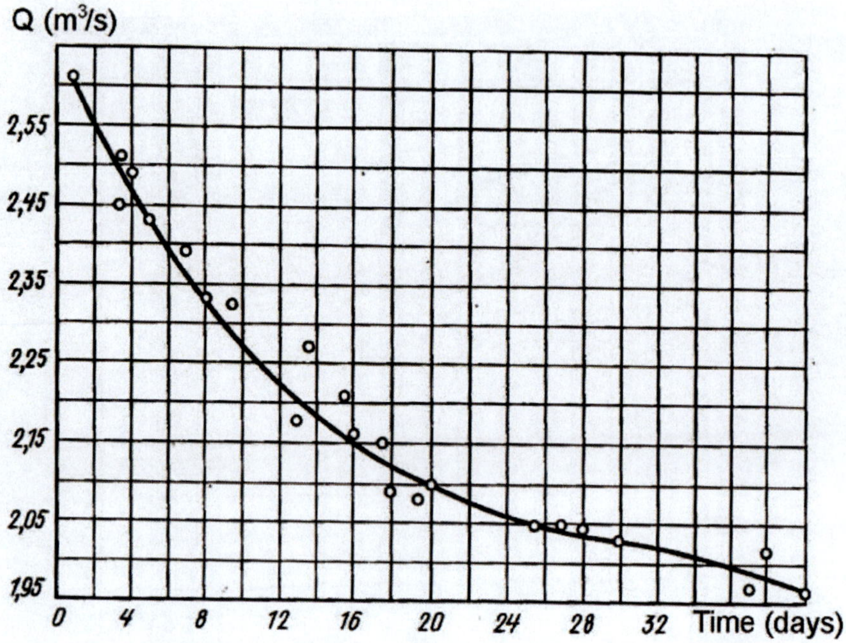


**Fig. 26.8** Change of seepage through the Burlykia dam during the period of constant reservoir level +45 m (after [25])

Permeability of existing blast-fill dams vary from $6 \times 10^{-1}$ to $6 \times 10^{-2}$ cm/s depending on their composition [5]. Ability of such structures to retain high reservoir level is proved by the presence of numerous natural rockslide dammed lakes that, in favorable conditions, exist for hundreds and even thousands years, until they are completely infilled by sediments [26]. Depth of such lakes can exceed several hundreds meters as at the Sarez Lake formed in 1911 in the Pamirs [4, 13, 19]. Absence of piping at these dams testifies to their high suffusion resistance. It should be noted, however, that in existing blast-fill dams the upper parts, up to 8–10 m thick, are quite permeable and dam fill density increases with depth in the upper one third of such structures [5, 25]. Thus, the height of such structures should exceed the designed reservoir level by not less than 12–15 m, a required freeboard much more than for dams constructed by "traditional" technologies.

## 2.3 *Topographic and Geological Conditions Preferable for Blast-Fill Dam Construction*

The technology of dam construction by triggering large-scale artificial rock slope failure (blast-rockslide dams) includes the consecutive (upward) exploding of linear charges located along the designed sliding surface. The amount of explosives is calculated to provide rock mass detachment and its initial push. To prevent debris spreading along the river valley after rock avalanche collision with an opposite slope, small "cofferdams" at the designed dam's up- and downstream limits could be build beforehand by pin-point blasting.

Such technology decreases the total amount of the required explosives more than two times in comparison with the pin-point blasting only, which, along with the delay of blasting of different charges, excludes hazardous seismic effect on the HPP structures and surrounding infrastructure.

The most favorable topographic conditions for blast-rockslide dam construction are at a curved gorge with mean slope steepness of not less than 40–45°. Its plan curvature allows to concentrate the collapsing rock debris from the exploded concave slope and to decrease length of the spillway and of headrace and tailrace tunnels that are constructed in the opposite side of the valley concurrently with blast preparation.

A section of river valley with such morphology was selected for the Burlykia dam site, while Uch-Terek experimental dam was constructed at a more straight section of the river gorge [3].

To assure blast-fill dam operating safety, the properties of the materials that will be used for the dam body and for its foundation should be evaluated.

The first requirement concerns rocks or soils that are used as construction material. They must not be weathering-prone, i.e. the initial properties of the exploded rock mass must not change significantly during dam lifetime. Such rocks as siltstone, mudstone, sandstone and conglomerate with clayey and carbonate-clayey cement and some varieties of shale and gneiss undergo rapid and intensive weathering. Within several months or some years they would weather into sand or rock debris. Intensive weathering is also typical of rocks from fault zones. Weathering leads to significant changes of the density and strength of the dam body and of its filtration properties. Weathering can be non-uniform in time and space. It facilitates formation of density heterogeneity, which, in its turn, creates high filtration gradients that can cause piping and internal erosion. Thus, sites composed of such types of rocks are unfavorable for blast-fill dam construction.

To ensure the blast-fill dam safety its foundation should be specially prepared. It is important since filtration through such dams could be concentrated at their bases composed of coarse talus. Infiltration coincides with the most permeable part of a dam foundation or with that part of the dam-basement boundary, which is located far from charges. If there are piping-prone deposits at the dam foundation, concentrated water flow may cause internal erosion. Presence of buried trees at the dam's foundation can also be hazardous, since wood putrefaction will increase the porosity of the dam-foundation boundary zone and, thus, its permeability. Bedrock, even those types subjected to various weathering processes do not pose any threat since their physical and filtration properties are better than those of the blasted dam fill. Therefore, foundation preparation can be limited to the removal of the piping-prone units (e.g. coarse lowermost parts of the talus fans), characterized by a higher permeability than the dam body and also to the removal of trees if they are widely distributed over the dam site.

For the largest blast-rockslide dams, however, requirements for the preparation of their foundations can be simplified, since a rapidly descending rock mass consisting of many millions cubic meters in volume can "bulldozer" loose material from valley bottom up to the bedrock.

One more problem associated with blast-fill dam design and construction is ensuring slope stability against intensive seismic shaking that can be caused not only by earthquakes that may occur during a project's lifetime, but will definitely be caused by the powerful construction explosion. It is especially important since, in contrast to hydraulic projects with "traditional" types of dams, those that include blast-fill dams, involve construction of the spillway and powerhouse in advance, before dam creation. Explosions used for such dam construction impose a significant seismic effect on these structures and on the host rock mass. It can cause either their destruction or failures of natural and artificial slopes at the dam site and in its vicinity. The latter is especially important since blast-fill dams are constructed, as a rule, in narrow gorges where slope stability could be at a marginal state. Seismic shaking can lead to their failure or to growth of weaknesses that can decrease their stability.

Thus, slope stability assessment should be performed at the initial stages of the design since its results allow suggesting measures that ensure safe and secure work in the hydraulic complex. For example, slope stability studies performed for the design of the Kambarata-1 & 2 HPP on the Naryn River in Kyrgyzstan showed that several tens of thousands cubic meters of rock can collapse from the slope above the Kambarata-2 powerhouse intake due to seismic shaking. To prevent such unfavorable effect the hazardous slope was flattened. It was also found that several millions cubic meters of rocks could slide from the left bank of the river downstream from the Kambarata-1 dam site. Such a rockslide can block the river and lead to powerhouse inundation. Since it is practically impossible to stabilize this slope that will undergo intensive shaking caused by blasting of nearby charges, an artificial channel was designed to guarantee free water flow downstream of the dam site.

To ensure the stability of the host rock mass where the powerhouse and spillway is placed they are usually located within the slope opposite to that which should be exploded at a safe distance from the explosive charges.

## 3 Natural Blockages as Analogues of Large Blast-Fill Dams

The above-mentioned examples summarize experience gained at the experimental dams (about 50 m high) and during construction and operation of blast-fill dams (up to 100–120 m high). Thus, data available can strictly be utilized for such dams only.

Design of larger blast-fill dams, such as that of the Kambarata-1 Project (~275 m high [2]) require experimental data on blockages of the same order of size as the proposed structures. Hereafter we will focus on some processes that can complicate construction of dams formed by artificially-triggered rock avalanches. They are similar to those that take place during natural large-scale rock slope failures.

The important parameters of any blast-fill dam that must be achieved are:

– Height of the dam and of the lowest part of its crest (the effective dam height), considering the above mentioned excess of dam crest (excess freeboard) above the reservoir level;
– Up- and downstream limits of the blockage, which should not extend beyond the designed limits in order neither to block inlets and outlets of the headrace and tailrace tunnels, nor affect powerhouse and other critical structures;
– Homogeneity and grain-size composition of the dam body that must ensure the designed permeability value and large safety factor in the case of a strong earthquake that may occur during the dam lifetime.

Factors governing the achievement (or otherwise) of the above parameters with due regard to the scaling effect have been investigated at numerous natural blockages resulting from massive rock slope failure.

## *3.1 Factors Affecting Dam Size and Shape*

The designed shape and geometrical parameters of the emplaced blast-fill material can be achieved by its optimal across- and along-valley distribution that requires understanding of the processes that govern such distribution. Otherwise it will be necessary to explode a larger rock mass with corresponding growth in the capacity of explosion, which is undesirable for economical, environmental and technical reasons.

### 3.1.1 Across-Valley Rock Avalanche Debris Distribution

The across-valley debris distribution ideal for dam construction corresponds to the case when its crest is flat without prominent rise or lowering (Fig. 26.9A). However this is a rather rare case for natural blockages. We could find few examples of such natural dam morphology including the Ak-Kiol (Fig. 26.10) and the Southern Karakungey rockslides (Fig. 26.11) in the Kokomeren river basin (Central Tien Shan) [39], and the Rio Toro rockslide in Costa Rica [30].

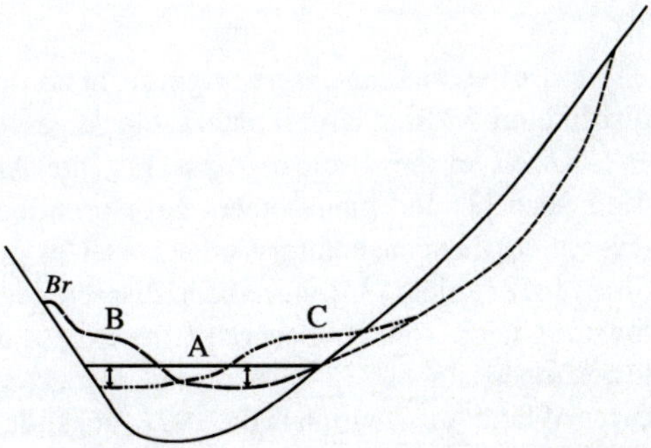


**Fig. 26.9** Across-valley rock avalanche debris distributions. *A* – even crest level optimal for a blast-fill dam; *B* – distally raised profile (*Br* – *brandung*); *C* – profile with proximal accumulation and distal lowering

**Fig. 26.10** The Aik-kiol rockslide that had formed natural dam with a nearly even crest

**Fig. 26.11** The Southern Karakungey rockslide scar (marked by *small arrows*) and a 120-m high dam with nearly horizontal crest and expressive *brandung* (*Br*). Concave shape of its downstream slope reflects presence of a secondary scar above long runout rock avalanche that had moved in direction marked by *white arrow*

More often one can observe significant non-uniformity of the across-valley debris distribution. Most of largest natural blockages, such as the Usoy in the Pamirs [4, 13, 19], Köfels in the Tirolean Alps [11], the Aksu and Djashilkul in the Northern Tien Shan [1] and many others have prominent lowering at their proximal parts (Fig. 26.9B). On the contrary, others, such as giant Kokomeren rockslide in the Tien Shan, for example [35], have been dissected at their fronts due to prominent distal lowering (Fig. 26.9C). Generally, the excess of the highest part of dams over the lowest ones is about 25–30% of the average dam height. One of the most extreme cases of latter distribution is the 1991 rockslide at Randa (Switzerland) where about

$30 \times 10^6$ $m^3$ of rock formed, practically, a steep cone with a lowermost distal part, quite atypical for rockslides of such order of magnitude [14].

If an artificial rock avalanche is characterized by the across-valley distribution as illustrated in Fig. 26.9B or C, the designed dam height may be not achieved at the lowermost part of its crest that would require additional efforts to raise crest level. From the construction point of view lowering in the proximal part (Fig. 26.9B) seems to be worst since it may be dangerous to work just below the rockslide scar where subsequent failure may take place.

The above unfavorable across-valley debris distribution can be amplified by the *brandung* phenomenon first described by A. Heim [15] and very well exemplified in numerous cases in the Karakoram [16, 17] and other regions. Enormous tongues of debris that rise above the average rockslide level for dozens and even hundreds of meters (*Br* on Figs. 26.9 and 26.11) incorporate rather large parts of their entire volume that should be excluded from the designed balance between the exploded and emplaced rock mass. A large *brandung*, about $10^5$ $m^3$ in volume, was found at the Gol-Ghoro rock avalanche in the Indus River [17]. Possibly the most expressive case, that can be classified as a *super-brandung* is the Avalanche Lake rock avalanche [12] in the Mackenzie Mountains where a large portion of debris ($5 \times 10^6$ $m^3$) was thrown out from the gorge onto a shelf about 500 m above the dam surface (the dam itself was only about 100–150 m).

Factors governing across-valley debris distribution (as well as the along-valley distribution that will be described below) are not fully understood. Generally, rockslide dams with proximal lowering correspond to the "primary" type of rock avalanches, while those with distal lowering – to the "secondary" or "spread" types [34, 38]. It is assumed that formation of blockages of these types could be governed by the morphology of the slope foot where initial accelerating motion of the rockslide mass converts into motion according to gained momentum [34, 38].

Further investigations are necessary for a better understanding of the causes of various across-valley debris distributions.

### 3.1.2 Along-Valley Debris Distribution

The most discussed process that can take place during blast-fill dam formation is debris spreading along the river valley [22, 25]. It can have a triple negative effect – (1) lowering of the dam crest, (2) affecting various structures up- and downstream from the dam site, blocking of the inlets and outlets of the HPP headrace and tailrace tunnels in particular, and (3) thinning of the upper part of the blockage that may increase its permeability.

It should be pointed out that even in those cases where a descending rock mass hits the opposite slope almost at a right angle, either compact blockages with steep up- and downstream slopes form (Fig. 26.12), or km-long spreading of debris incorporating significant part of the entire rockslide may occur (Figs. 26.13, 26.14).

Compact blockages are characterized by rather wide crests and steep up- and downstream slopes dipping 30–35° that corresponds to the angle of repose of rockslide debris. They can be exemplified by the Aksu and Djashilkul (Fig. 26.12)

**Fig. 26.12** Up- and downstream views of the Djashilkul rockslide about $2 \times 10^8$ $m^3$ in volume that had formed 200-m high compact dam on the Chon-Kemin River with steep slopes

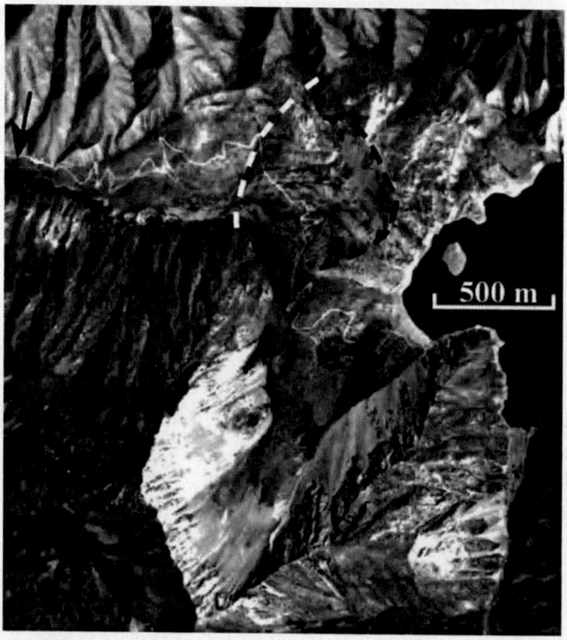


**Fig. 26.13** The Karasu rockslide dam (after [37], modified). *Black dashed line* with bars – secondary scar, *white dashed line* – assumed downstream limit of the initial blockage, *black arrow* – end of the secondary rock avalanche

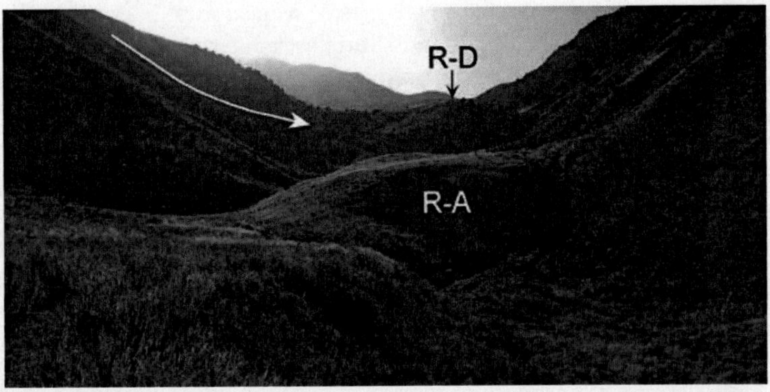


**Fig. 26.14** Northern Karakungey rockslide dam (R-D) about 100-m high accompanied by the 600-m long downstream rock avalanche (R-A). *White arrow* shows the direction of the initial rockslide motion

rockslides in the Northern Tien Shan [1, 39], the Usoi rockslide in the Pamirs [19], and the Köfels rockslide in the Tyrol Alps [11]. It is interesting that all above mentioned rockslides collided with opposite slopes of the dammed river valleys just at the mouths of small tributary valleys. Rockslide fronts had entered these valleys and it can be assumed that increasing friction led to momentum dissipation preventing transverse spreading of debris along the main river valley.

One of the main mechanisms of along-valley spreading, at least in the narrow river gorges where blast-fill dams can be constructed, is the formation of secondary rock avalanches. This phenomenon can be defined as an ejection of debris from the compact rockslide body that accumulates immediately at the foot of the source slope, most likely due to abrupt momentum transfer when the entire sliding mass collides with an obstacle [34, 38], in this case the opposite slope.

In many cases the amount of debris involved in the secondary avalanche motion is more than a half of the total rockslide volume and much bigger than visible dimensions of the secondary scar. It supports the assumption that the initial failure and secondary avalanche formation are consecutive stages of an indivisible process, which can be explained as follows: after collision, debris motion halts abruptly except for some part of it, which retains the possibility of further displacement. Momentum of the entire rock mass transfers to this portion that acquires additional acceleration and forms a long runout avalanche. This process is somewhat similar to that which the armor-piercing shell pierces the target – soft shell deforms during the impact and transfers momentum to its solid heart. The secondary scar just marks a milestone in a continuous process: initial motion → collision → momentum transfer → ejection of secondary rock avalanche.

It can be assumed that the origin of the *brandung* features, noted above can be caused by similar momentum transfer as in the secondary avalanche case.

In some cases, the secondary avalanche moved a long distance along the riverbed practically at a right angle to the direction of the initial failure (see Fig. 26.13). Cases of natural blockages accompanied by secondary rock avalanches are found in the Tien Shan, e.g. South-Karakungey, Kiol-say, Karasu rockslides [34, 38] and, in the Mackenzie Mountains – Nozzle and U-turn rockslides [10].

However, secondary avalanche formation is not the only mechanism that can lead to significant along-valley spreading of rockslide debris. If the base of the rockslide detachment zone is much above the slope foot so that the descending rockslide mass is air launched, it would fall on the valley bottom almost at a right angle – like concrete unloaded from a truck. Due to the large linear dimensions of the descending rock mass, its frontal part reaches the valley bottom first being compressed by following material. It leads to frontal portions of debris being fluidized and its extrusion from under the accumulating blockage. Such a mechanism results in a spread rock avalanche [34, 38]. It can be exemplified by the Northern Karakungey rockslide (Fig. 26.14) that descended from a 500-m high slope, and formed a blockage about 100 m high with steep convex slopes a and 600-m long downstream rock avalanche.

Another example is the Kashkasu blockage in the same name tributary of the Kokomeren River that descended from a steep 600-m high slope while the base of the detachment zone was about 150–200 m above the valley bottom (Fig. 26.15). Its lower part, about 70 m thick (A on Fig. 26.15), forms 300-m long tongue downstream from the main compact body (B on Fig. 26.15) that rises about 200 m above the former surface.

To prevent debris spreading along the river valley due to any of the above mechanisms, special measures should be contemplated. In particular, anticipatory construction of small "cofferdams" by pin-point blasting at the up- and down-stream limits of the designed blast-rockslide dam is proposed as the most effective measure [23, 24].

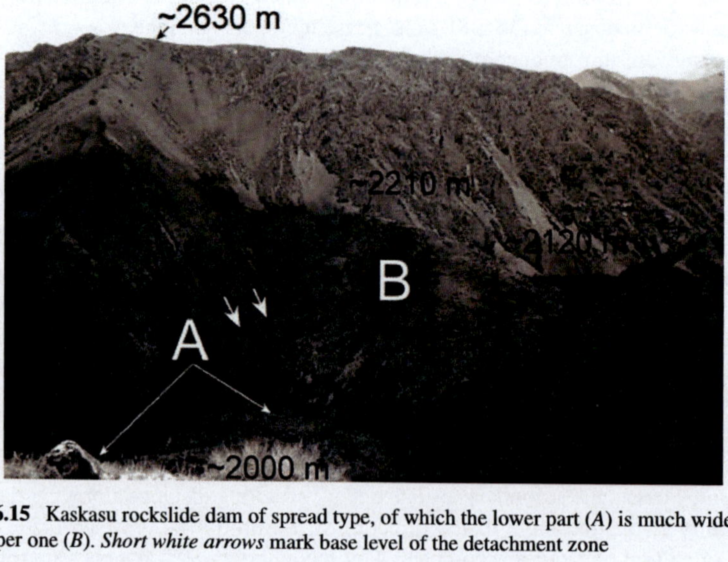


**Fig. 26.15** Kaskasu rockslide dam of spread type, of which the lower part (*A*) is much wider than the upper one (*B*). *Short white arrows* mark base level of the detachment zone

### *3.2 Factors Affecting Dam Permeability and Erosion Resistance*

One of the typical features of large rockslide dams is their distinct dual-zone internal structure. Almost in all cases, when such dams have been dissected by erosion and their composition can be observed in outcrops, one can see intensively comminuted debris that forms a huge internal "core" that is overlaid by a coarse carapace composed of large angular boulders and/or huge blocks. This structure has been described in the Alps at Flims [32] (Chap. 19 by Dunning and Armitage, this volume) and at Köfels [11], in the Karakoram at the Gol-Ghone, Chalt and Haldi rockslides [18], at the numerous cases in the Tien Shan [1, 36, Gaspirovich, personal communication], in the Northern Caucasus [37], and in New Zealand (Chap. 19 by Dunning and Armitage, this volume), irrespective of rock types involved in rock slope failure.

Such composition governs to a large extent both dam permeability and its erosion resistance in the case of overtopping [9] (Chap. 19 by Dunning and Armitage, this volume). Extremely low permeability of the main parts of high natural blockages is well exemplified by the famous Usoy dam where almost the entire seepage takes place through the uppermost blocky unit about 100-m thick [19], while the lower 400 m are practically impermeable (see Chap. 7 by Delaney and Evans, this volume). The same peculiarity characterizes the blockage of Great Almaty Lake in Northern Tien Shan in Kazakhstan [6].

Large blocks and boulders of the upper carapace facies (Chap. 19 by Dunning and Armitage, this volume) are much more permeable. In case of large blockages this facies forms an infiltration zone wide enough to pass the entire discharge and thus prevents dam overtopping. It can also protect the dam against rapid erosion if overtopping occurs. However, the abrupt difference between the permeability of the upper and lower parts may lead to internal erosion if high discharge through the void spaces in the carapace directly affects the underlying comminuted debris. Thus, for water retaining blast-fill dams it would be preferable to keep reservoir level at, or lower than the topmost part of the comminuted impermeable internal core.

The distinct difference between the upper/external and lower/internal parts of rockslide dams [9, 33, 37] (Chap. 19 by Dunning and Armitage, this volume) leads to analyzing their grain-size composition separately. Study of material sampled from the lower comminuted parts of deeply incised natural dams shows that it fits well to the Rosin-Rammler (Weibull) distribution:

$$F(x) = 1 - \exp[-(x/x_0)^n] \qquad (1)$$

Where $x_0$ and $n$ are parameters of distribution [41].

Parameter $n$ characterizes "range" of the grain-size composition: the smaller this value is, the more variable are the size fragments of debris. This ensures dense compaction of rockslide material. In the studied cases coefficient $n$ varies from 0.805 to 0.376, which corresponds to a very "wide" grain size distribution, even wider than that for blasted rock (about 0.8, on an average). It explains the low permeability of the internal parts of large rockslide dams.

It should be mentioned, however, that, according to Chap. 19 by Dunning and Armitage, this volume, grain-size distribution better corresponds to a fractal distribution, rather than to a Weibull one.

### 3.3 Seismic Stability of Natural Blockages

Strong earthquakes are one of the potentially most hazardous natural phenomena that can affect and even destroy high dams. It can be assumed that blast-rockslide dam with an elongated profile and a massive internal "core" composed of highly comminuted debris could be considered as one of the most earthquake-proof type of water retaining structures that can sustain strong motion caused by severe earthquake. Such assumption is proved by numerous examples of natural blockages that withstood extremely strong shaking being located within the epicentral zones of several M 7.4–8.3 earthquakes in the Central Asia region.

The above mentioned dam of the Great Almaty Lake in the Kazakh part of Northern Tien Shan is more than 100 m high [6]. It remained intact after the 1887 M 7.3 Verniy [31] and 1911 M 8.2 Kemin [7] earthquakes, in which intensities at the site were 9–10 points of MSK (MM) scale. Besides it was affected by 7–8 point shaking during another strong earthquake – the 1889 8.3 Chilik earthquake that affected areas east of Almaty City [40].

Several rockslide dams (about 100 m high) in the Kolsai River, a right tributary of the Chilik River, fell within the latter's epicentral zone affected by seismic shaking of not less than 9–10 points of MSK (MM) scale. The next gigantic earthquake in this region, the 1911 M 8.2 Kemin earthquake, had affected these sites with strong motion of nearly the same intensity. Nevertheless no rockslide dam breach was reported.

In 1946 the M 7.6 Chatkal earthquake [27] occurred in Kyrgyzstan just near the 300-m high Sarychelek rockslide dam of more than $2 \times 10^9$ $m^3$ in volume [38]. The earthquake, the intensity of which at the site was not less than 9 points of MSK (MM) scale, caused some deformation of the dam's body, but did not affect its general stability.

In 1949, during the M 7.4 Khait earthquake in Tajikistan, rather small (several dozens meters high) rockslide dam sustained severe shaking that caused catastrophic failure of about $200 \times 10^6$ $m^3$ of bedrock at the adjacent slope and catastrophic rock avalanche [27].

We should point out that no documented failure of a pre-existing rockslide dam in this region, or elsewhere, has been caused by strong earthquake shaking in itself. The most hazardous consequence of strong earthquake shaking that may cause blockage breach is the occurrence of seismically-triggered large-scale slope failure on the lake shoreline and the creation of a tsunami-like wave, as described in the Las Conchas valley in Argentina (Chap. 5 by Hermanns et al., this volume). The same phenomenon poses a threat to the Usoi blockage in Pamirs [4] (see Chap. 7 by Delaney and Evans, this volume).

Since blast-fill and, especially, blast-rockslide dams and their natural analogues (rockslide blockages) do not contain localized impermeable elements, there is no high head pressure within their bodies. This may explain why they can sustain not only intensive seismic shaking, but even surface faulting in the dam foundation, a phenomenon that complicates, if not prevents, construction of any "traditional" type of water retaining structure at a fault-prone site.

As regards the seismic effect of a rockslide-triggering explosion on structures constructed beforehand, the explosion should be designed not to exceed the background seismicity typical of the region and not to cause any damage of these structures [25].

## 4 Conclusions

Large-scale field experiments of blast-fill dams and the study of natural dams of similar size as the designed artificial structures provide unique information about the processes that take place during emplacement of large amount of blasted rock.

Based on detailed study of experimental dams constructed within the framework of the Kambarata Dam Project in Kyrgyzstan, it was found that blast-fill density exceeds that of the traditionally constructed embankment dams. It results in relatively small deformability of such structures during their operating period and in filtration characteristics which are favourable for water-retaining dams. Gradual decrease of dam material density with depth requires 12 to 15-m excess of their height above the designed reservoir level, i.e. excess freeboard.

Since experimental dams cannot be of the same order of size as the largest proposed blast-fill dams hundreds of millions cubic meters in volume, the study of same-size natural blockages is necessary to obtain reliable information on the processes that govern the morphology and internal structure of such structures with due regard to scaling effects.

Large rockslides produce natural blockages characterized by various types of across- and along-valley debris distribution, which, commonly have uneven profiles and include long-runout downstream spreading of debris. Thus, preliminary construction of up- and downstream cofferdams is necessary to ensure formation of an artificial blockage with optimal geometrical parameters. The presence of shattered dense debris forms massive internal cores of large natural blockages that are protected by an external coarse carapace allow us to assert that geotechnical properties of blast-rockslide dam bodies will improve with the increase in dam dimensions.

High seismic stability of natural blockages, exemplified by numerous case studies, shows that large blast-rockslide dams could be considered as one of the most earthquake-proof type of water retaining structures.

## References

1. Abdrakhmatov, K. and Strom, A.L. (2006) Dissected rockslide and rock avalanche deposits; Tien Shan, Kyrgyzstan, in S.G. Evans, G. Scarascia Mugnozza, A. Strom, and R.L. Hermanns (eds.), *Landslides from Massive Rock Slope Failure*, NATO Science Series IV: Earth and Environmental Sciences, Vol. 49. Springer, Dordrecht, pp. 577–599.

2. Adushkin, V.V., Pernik, L.M. and Zykov, Y.N. (1993) Modeling of explosion-triggered rock slides for construction of the Kambarata 1 hydroelectric power dam, Republic of Kyrgyzstan, *Landslide News* **7**, 7–9.
3. Adushkin, V.V. (2004) Russian experience of blast-fill dams' construction, in K. Abdrakhmatov, S.G. Evans, R.L. Hermanns, G. Scarascia Mugnozza, and A. Strom (eds.), *Security of Natural and Artificial Rockslide Dams. NATO ARW, Bishkek, Kyrgyzstan, June 8–13, 2004, Extended Abstracts*, Bishkek, pp. 4–8.
4. Alford, D., and Schuster, R. (eds.) (2000) *Usoi Landslide Dam and Lake Sarez. An assessment of Hazard and Risk in the Pamir Mountains, Tajikistan*, UN Publication. Sales No. E.00.III.M.1.
5. *Blast Fill Dams.* (2000) Bulletin E 01, ICOLD-CIGB.
6. Bochkarev, V.P. (2004) Natural and Artificial Rockslide Dams in Mountain Geosystems of Kazakhstan Orogenic Belt, in K. Abdrakhmatov, S.G. Evans, R.L. Hermanns, G. Scarascia Mugnozza, and A. Strom (eds.), *Security of Natural and Artificial Rockslide Dams. NATO ARW, Bishkek, Kyrgyzstan, June 8–13, 2004, Extended Abstracts*, Bishkek, pp. 9–12.
7. Bogdanovich, M.C., Kark, J., Korolkov, B. and Muchketov, D. (1914) *Earthquake of the 4th January 1911 in the northern districts of the Tien Shan,* Transactions Geol. Com., New Series **89** (in Russian).
8. *Comparative analysis of the state-of-the-art of the Hoabin dam and of the analogues dams world-wide* (2001) Report of Hydroproject Institute, Moscow (in Russian).
9. Dunning, S.A., Petley, D.N. and Strom, A.L. (2005) The morphologies and sedimentology of valley confined rock-avalanche deposits and their effect in potential dam hazard, in O. Hungr, R. Fell., R. Couture, and E. Eberhardt (eds.), *Landslide Risk Management*. A.A. Balkema, Amsterdam, pp. 691–704.
10. Eisbacher, G.H. (1979) Cliff collapse and rock avalanches (sturzstroms) in the Mackenzie Mountains, northwestern Canada, *Canadian Geotechnical Journal* **16**, 309–334.
11. Erismann, T.H. (1979) Mechanisms of large landslides, *Rock Mechanics* **12**, 15–46.
12. Evans, S.G., Hungr, O. and Enegren, E.G. (1994) The Avalanche Lake rock avalanche, Mackenzie Mountains, Northwest Territories, Canada: Description, dating and dynamics, *Canadian Geotechnical Journal* **31**, 749–768.
13. Gaziev, E. (1984) Study of the Usoi Landslide in Pamir, in *Proc. 4th Int. Symp. on Landslides, Toronto*, **1**, pp. 511–514.
14. Götz, A. and Zimmermann, M. (1993) The 1991 rock slides in Randa: Causes and consequences, *Landslide News* **7**, 22–25.
15. Heim, A. (1932) *Bergsturz und Menschenleben*. Fretz and Wasmuth, Zurich.
16. Hewitt, K. (1999) Quaternary moraines vs catastrophic rock avalanches in the Karakoram Himalaya, Northern Pakistan, *Quaternary Research* **51**, 220–237.
17. Hewitt, K. (2002) Styles of rock avalanche depositional complex in very rugged terrain, Karakoram Himalaya, Pakistan, in S.G. Evans and J.V. DeGraff (eds.), *Catastrophic Landslides: Effects, occurrence and Mechanisms, Reviews in Engineering Geology*. Geological Society of America, Boulder, CO, pp. 345–378.
18. Hewitt, K. (2006) Rock avalanches with complex run out and emplacement: Karakoram Himalaya, Inner Asia, in S.G. Evans, G. Scarascia Mugnozza, A. Strom and R.L. Hermanns (eds.), *Landslides from Massive Rock Slope Failure*, NATO Science Series IV: Earth and Environmental Sciences, Vol. 49. Springer, Dordrecht, pp. 545–575.
19. Ischuk, A.R. (2004) Usoy Natural Dam: Problem of Security (Lake Sarez, Pamirs Mountains, Tajikistan), in K. Abdrakhmatov, S.G. Evans, R.L. Hermanns, G. Scarascia Mugnozza, and A. Strom (eds.), *Security of Natural and Artificial Rockslide Dams. NATO ARW, Bishkek, Kyrgyzstan, June 8–13, 2004, Extended Abstracts*, Bishkek, pp. 97–102.
20. Kolichko, A.V. (1988) Forecast of rock mass grain-size composition, *Transactions of Hydroproject Institute* **128**, 19–25 (in Russian).
21. Kolichko, A.V. (2004) Construction and operating safety of the blast-fill dams, in K. Abdrakhmatov, S.G. Evans, R.L. Hermanns, G. Scarascia Mugnozza, and A. Strom (eds.), *Security of Natural and Artificial Rockslide Dams. NATO ARW, Bishkek, Kyrgyzstan, June 8–13, 2004, Extended Abstracts*, Bishkek, pp. 112–114.

22. Korchevskiy, V.F. (2004) Blast-fill dams constructed by pin-point blasting and by artificial rock avalanches, in K. Abdrakhmatov, S.G. Evans, R.L. Hermanns, G. Scarascia Mugnozza, and A. Strom (eds.), *Security of Natural and Artificial Rockslide Dams. NATO ARW, Bishkek, Kyrgyzstan, June 8–13, 2004, Extended Abstracts*, Bishkek, pp. 115–117.
23. Korchevskiy, V.F., Mogilnikov, L.P. and Petrov, G.N. (1989) Formation of blast-fill dams profile, *Energy Construction* **2**, 48–50 (in Russian).
24. Korchevskiy, V.F. and Muratova, M.H. (1991) Construction of dams by collapse, *Hydrotechnical Construction* **3**, 6–11 (in Russian).
25. Korchevskiy, V.F. and Petrov, G.N. (1989) *Blast-fill Dams Design and Investigations*. Moscow (in Russian).
26. Korup, O., Strom, A.L. and Weidinger, J.T. (2006) Fluvial response to large rock-slope failures: Examples from the Himalayas, the Tien Shan, and the Southern Alps in New Zealand, *Geomorphology* **78**, 3–21.
27. Leonov, N.N. (1960) Khait Earthquake of 1949 and the geological conditions of its occurrence, *Geology and Geophysics, Geophysical Series* **3**, 48–56 (in Russian).
28. Leonov, N.N. (1965) Role of geological structure in formation of the residual soil deformations during earthquakes, *Seismic Microzonation, Transactions of the Institute of Physics of the Earth, Russian Academy of Sciences* **36**(203), 132–136.
29. Macedo-Gomes, G., Castro-Abonce, I. and Montanez-Cartaxo, L. (2000) Behavior of Aqampila Dam. *Y. Barry Cooke Volume*, CFRD, Beijing.
30. Mora, S., Madrigal, C., Estrada, J. and Schuster, R.L. (1993) The 1992 Rio Toro Landslide Dam, Costa Rica, *Landslide News* **7**, 19–22.
31. Mushketov, I.V. (1890) Verniy earthquake on 28/05 (9/06) 1887, *Proceedings of the Geological Committee* **10**(1), 1–154 (in Russian).
32. Poschinger, A., Wassmer, P. and Maisch, M. (2006) The Flims rockslide: History of interpretation and new data, in S.G. Evans, G. Scarascia Mugnozza, A. Strom and R.L. Hermanns (eds.), *Landslides from Massive Rock Slope Failure*, NATO Science Series IV: Earth and Environmental Sciences, Vol. 49. Springer, Dordrecht, pp. 347–375.
33. Sadovskiy, M.A. (1999) *Geophysics and Physics of Explosion. Selected Works*. Nauka, Moscow.
34. Strom, A.L. (1994) Mechanism of stratification and abnormal crushing of rockslide deposits, in *Proc. 7th International IAEG Congress* **3**, Balkema, Rotterdam, pp. 1287–1295.
35. Strom, A.L. (1996) Some morphological types of long-runout rockslides: Effect of the relief on their mechanism and on the rockslide deposits distribution, in K. Senneset (ed.), *Landslides. Proceedings of the Seventh International Symposium on Landslides, 1996, Trondheim, Norway*. Balkema, Rotterdam, pp. 1977–1982.
36. Strom, A.L. (1998) Giant ancient rockslides and rock avalanches in the Tien Shan Mountains, Kyrgyzstan, *Landslide News* **11**, 20–23.
37. Strom, A.L. (2005) Rock avalanches at the southern feet of the Northern Caucasus Rocky Range, Ardon River valley, *Landslides* **1**, 237–241.
38. Strom, A.L. (2006) Morphology and internal structure of rockslides and rock avalanches: Grounds and constraints for their modelling, in S.G. Evans, G. Scarascia Mugnozza, A. Strom, and R.L. Hermanns (eds.), *Landslides from Massive Rock Slope Failure*, NATO Science Series IV: Earth and Environmental Sciences, Vol. 49. Springer, Dordrecht, pp. 321–346.
39. Strom, A.L. and Korup, O. (2006) Extremely large rockslides and rock avalanches in the Tien Shan Mountains, Kyrgyzstan, *Landslides* **3**, 125–136.
40. Voznesenskiy, A.V. (1891) The Chilik earthquake on June 30 (July 12), 1889, in I.V. Mushketov (ed.), *Materials for Study of Earthquakes in Russia*, Issue 1, St. Petersburg, pp. 11–46.
41. Weibull, W. (1951) A statistical distribution function of wide applicability, *Journal of Applied Mechanics* **18**(293), 837–843.
42. Zhengguang, P. (2000) Analysis of Deformation of Xibeikou CFRD in Eight Years of Operation. *Proceedings of the 20th Congress ICOLD*, September 18, 2000.

# Index



S.G. Evans et al. (eds.), *Natural and Artificial Rockslide Dams*, Lecture Notes in Earth Sciences 133, DOI 10.1007/978-3-642-04764-0, © Springer-Verlag Berlin Heidelberg 2011